Centaurs

Online at: https://doi.org/10.1088/2514-3433/ada267

AMERICAN ASTRONOMICAL SOCIETY IOP | ebooks™

AAS Editor in Chief

About the program:

AAS-IOP Astronomy ebooks is the official book program of the American Astronomical Society (AAS) and aims to share in depth the most fascinating areas of astronomy, astrophysics, solar physics and planetary science. The program includes publications in the following topics:

- GALAXIES AND COSMOLOGY
- INTERSTELLAR MATTER AND THE LOCAL UNIVERSE
- STARS AND STELLAR PHYSICS
- EDUCATION, OUTREACH, AND HERITAGE
- HIGH-ENERGY PHENOMENA AND FUNDAMENTAL PHYSICS
- THE SUN AND THE HELIOSPHERE
- THE SOLAR SYSTEM, EXOPLANETS, AND ASTROBIOLOGY
- LABORATORY ASTROPHYSICS, INSTRUMENTATION, SOFTWARE, AND DATA

Books in the program range in level from short introductory texts on fast-moving areas, graduate and upper-level undergraduate textbooks, research monographs, and practical handbooks.

For a complete list of published and forthcoming titles, please visit iopscience.org/books/aas.

About the American Astronomical Society

The American Astronomical Society (aas.org), established 1899, is the major organization of professional astronomers in North America. The membership (~7,000) also includes physicists, mathematicians, geologists, engineers, and others whose research interests lie within the broad spectrum of subjects now comprising the contemporary astronomical sciences. The mission of the Society is to enhance and share humanity's scientific understanding of the universe.

Editorial Advisory Board

Centaurs

Edited by
Kathryn Volk
The Planetary Science Institute, Tucson, AZ, USA

Maria Womack
Department of Physics, University of Central Florida, Orlando, FL, USA

Jordan Steckloff
The Planetary Science Institute, Tucson, AZ, USA

IOP Publishing, Bristol, UK

ISBN 978-0-7503-5588-9 (ebook)
ISBN 978-0-7503-5586-5 (print)
ISBN 978-0-7503-5589-6 (myPrint)
ISBN 978-0-7503-5587-2 (mobi)

DOI 10.1088/2514-3433/ada267

Version: 20250501

AAS–IOP Astronomy
ISSN 2514-3433 (online)
ISSN 2515-141X (print)

British Library Cataloguing-in-Publication Data: A catalogue record for this book is available from the British Library.

Published by IOP Publishing, wholly owned by The Institute of Physics, London

IOP Publishing, No.2 The Distillery, Glassfields, Avon Street, Bristol, BS2 0GR, UK

US Office: IOP Publishing, Inc., 190 North Independence Mall West, Suite 601, Philadelphia, PA 19106, USA

Contents

Preface

We gratefully acknowledge the following individuals, as well as several anonymous reviewers, who provided detailed reviews of the individual chapters that resulted in a much improved compendium. We are indebted to Theodore Kareta who facilitated anonymous reviews for our own Introductory chapter. We also thank Jules Pinnick for copyediting.

Sarah Anderson
Erik Asphaug
Dennis Bodewits
Felipe Braga-Ribas
Bobby Bus
Mohamed El-Maarry
Yanga Fernández
Alan Fitzsimmons
Wesley Fraser
Sarah Greenstreet
Bryanna Henderson
Henry Hsieh
Zi Liang Jin
Theodore Kareta
Michael S.P. Kelley
Prajtke Mane
Stephanie Milam
Rosemary Pike
Noemi Pinilla-Alonso
Sean Raymond
Philippe Rousselot
Micah Schaible
Colin Snodgrass
Audrey Thirouin
Yun Zhang

Editor biographies

The Editors from left to right: Maria Womack, Jordan Steckloff, and Kat Volk. Photo taken by Laura Woodney at the 2024 DPS meeting.

Kathryn Volk

Kathryn Volk recieved a PhD in planetary science from the University of Arizona. Her interest in Centaurs began in graduate school with her PhD thesis on the dynamics of Kuiper Belt objects and Centaurs. Much of her research has focused on connecting observational and dynamical studies in the outer solar system. She is an active member of the Vera C. Rubin Observatory's Legacy Survey of Space and Time (LSST) Solar System Science Collaboration and serves on the LSST Survey Cadence Optimization Committee. Dr. Volk was awarded the American Astronomical Society's Division of Dynamical Astronomy's Vera Rubin Early Career Prize for her work on both the dynamics of small bodies beyond Neptune and the long-term dynamics and stability of tightly packed exoplanetary systems. Asteroid 10273 Katvolk was named in her honor. She is currently a Senior Scientist at the Planetary Science Institute.

Maria Womack

Maria Womack received a PhD in physics from Arizona State University. She has studied Centaurs for more than 30 years, primarily using spectroscopy at optical, infrared, and mm-wavelengths. She is an affiliate faculty member at the University of Central Florida and University of Victoria and serves on the American Astronomical Society's Editorial Board. Previously, Dr. Womack worked as a tenured professor for more than 20 years teaching physics and astronomy and supervising over 70 undergraduate and graduate students in research. She received an NSF CAREER award, is a Fellow of the Royal Astronomical Society, and asteroids 229614 Womack and 41030 Mariawomack were named in her honor. She also currently serves at the U.S. National Science Foundation where she oversees the National Center for Atmospheric Research (NCAR) and led the NSF-NASA team that created the NN-EXPLORE exoplanet partnership.

Jordan Steckloff

Jordan Steckloff received a PhD in physics from Purdue University; his interest in the ecliptic comets (TNOs-Centaurs-JFCs) began with his dissertation "On the Interaction of Sublimating Gas with Cometary Bodies", which explored how sublimative back reaction forces can press upon and spin-up comet nuclei to the point of localized mass wasting and/or disruption. Dr. Steckloff is currently a Research Scientist at the Planetary Science Institute in Tucson, Arizona, where he continues investigating the effects of sublimative processes and mass wasting on comet nuclei. Asteroid 37019 Jordansteckloff was named in his honor.

List of contributors

This book would not have been possible without the hard work and dedication from the co-authors who, in many cases, provided the first review on a Centaurs subtopic and ultimately made this book possible.

Alvaro Alvarez-Candal
Instituto de Astrofísica de Andalucía, Granada, Spain

Michele T Bannister
School of Physical and Chemical Sciences–Te Kura Matū, University of Canterbury, Christchurch, New Zealand

James Bauer
Department of Astronomy, University of Maryland College Park, MD, USA

Jürgen Blum
Institut für Geophysik und Extraterrestrische Physik, Technische Universität Braunschweig, Braunschweig, Germany

Julie Brisset
Florida Space Institute, University of Central Florida, Orlando, FL, USA

Rosario Brunetto
Institut d'Astrophysique Spatiale, University Paris-Saclay, Orsay, France

Marc W Buie
Southwest Research Institute, Boulder, CO, USA

Mark Burchell
Centre for Astrophysics and Planetary Science, University of Kent, Canterbury, Kent, UK

Romina P Di Sisto
Facultad de Ciencias Astronómicas y Geofísicas, UNLP and Instituto de Astrofísica de La Plata, CCT La Plata-CONICET-UNLP, Argentina

Luke Dones
Southwest Research Institute, Boulder, CO, USA

Sara Faggi
Department of Physics, American University, Washington, DC, USA
and
NASA Goddard Space Flight Center Planetary System Lab, Greenbelt, MD, USA

Yanga R Fernández
Department of Physics, University of Central Florida, Orlando, FL, USA

Estela Fernandez-Valenzuela
Florida Space Institute, University of Central Florida, Orlando, FL, USA

Tabare Gallardo
Facultad de Ciencias, Udelar, Uruguay

Silvia Maria Giuliatti Winter
Grupo de Dinãmica Orbital & Planetologia, São Paulo State University, Guaratingueta, Brazil

Will Grundy
Lowell Observatory, Flagstaff, AZ, USA

Murthy S Gudipati
Science Division, Jet Propulsion Laboratory, California Institute of Technology, Pasadena, CA, USA

Aurélie Guilbert-Lepoutre
Laboratoire de Géologie de Lyon: Terre, Planètes, Environnement, UMR 5276 CNRS, UCBL, ENSL, Villeurbanne, France

Olga Harrington Pinto
Department of Physics, Auburn University, Auburn, AL, USA

Walter Harris
Lunar & Planetary Laboratory, University of Arizona, Tucson, AZ, USA

Elsa Hénault
Institut d'Astrophysique Spatiale, University Paris-Saclay, Orsay, France

Masatoshi Hirabayashi
Daniel Guggenheim School of Aerospace Engineering and School of Earth & Atmospheric Sciences, Georgia Institute of Technology, Atlanta, GA, USA

Bryan J Holler
Space Telescope Science Institute, Baltimore, MD, USA

Oleksandra Ivanova
Astronomical Institute of the Slovak Academy of Sciences, Tatranská Lomnica, Slovak Republic
and
Main Astronomical Observatory of the National Academy of Sciences of Ukraine, Kyiv, Ukraine
and
Taras Shevchenko National University of Kyiv, Astronomical Observatory, Kyiv, Ukraine

Seth Jacobson
Department of Earth and Environmental Sciences, Michigan State University, East Lansing, MI, USA

Anders Johansen
Globe Institute, University of Copenhagen, Copenhagen, Denmark

Theodore Kareta
Lowell Observatory, Flagstaff, AZ, USA

Matthew M Knight
Physics Department, United States Naval Academy, Annapolis, MD, USA

Rosita Kokotanekova
Sun and Solar System Department, Institute of Astronomy and National Astronomical Observatory, Bulgarian Academy of Sciences, Sofia, Bulgaria

Pedro Lacerda
Instituto de Astrofísica e Ciências do Espaço, Universidade de Coimbra, Coimbra, Portugal
and
Instituto Pedro Nunes, Coimbra, Portugal

Rodrigo Leiva
Instituto de Astrofísica de Andalucía, – Consejo Superior de Investigaciones Científicas (IAA-CSIC), Glorieta de la Astronomía, Granada, Spain

Javier Licandro
Instituto de Astrofísica de Canarias (IAC), La Laguna, Tenerife, Spain
and
Departamento de Astrofísica, Universidad de La Laguna, La Laguna, Tenerife, Spain

Eva Lilly
Planetary Science Institute, Tucson, AZ, USA

Kathleen Mandt
NASA Goddard Space Flight Center, Greenbelt, MD, USA

Raphael Marschall
Centre national de la recherches scientifique, Observatoire de la Cote d'Azur, Nice, France

Zita Martins
Centro de Química Estrutural, Institute of Molecular Sciences and Department of Chemical Engineering, Instituto Superior Técnico, Universidade de Lisboa, Lisboa, Portugal

Adam McKay
Department of Physics and Astronomy, Appalachian State University, Boone, NC, USA

John Noonan
Department of Physics, Auburn University, Auburn, AL, USA

Cathy B Olkin
Muon Space, Mountain View, CA, USA

Nuno Peixinho
Instituto de Astrofísica e Ciências do Espaço, Universidade de Coimbra, Coimbra, Portugal

Darin Ragozzine
Department of Physics & Astronomy, Brigham Young University Provo, UT, USA

Flavia L Rommel
Florida Space Institute, University of Central Florida, Orlando, FL, USA

Nathan X Roth
NASA Goddard Space Flight Center, Greenbelt, MD, USA
and
Department of Astronomy and Carl Sagan Institute, Cornell University, Ithaca, NY, USA

Paul Sanchez
Colorado Center for Astrodynamics Research, University of Colorado Boulder, Boulder, CO, USA

Gal Sarid
Science Systems and Applications, Inc (SSAI), Lanham, MD, USA
and
SETI Institute affiliate, Rockville, MD, USA

Charles Schambeau
Department of Physics and Florida Space Institute, University of Central Florida, Orlando, FL, USA

Megan E Schwamb
Astrophysics Research Centre, School of Mathematics and Physics, Queen's University Belfast, Belfast, UK

Tom Seccull
Independent Researcher, Belfast, Northern Ireland, UK

Darryl Z Seligman
Department of Physics and Astronomy, Michigan State University, East Lansing, MI, USA
and
Department of Astronomy and Carl Sagan Institute, Cornell University, Ithaca, NY, USA

Amanda A Sickafoose
Planetary Science Institute, Tucson, AZ, USA

Kelsi Singer
Southwest Research Institute, Boulder, CO, USA

Ana Carolina Souza-Feliciano
Florida Space Institute, University of Central Florida, Orlanda, Florida, USA

Jordan Steckloff
Planetary Science Institute, Tucson, Arizona, USA
and
Department of Aerospace Engineering and Engineering Mechanics, University of Texas at Austin, Austin, TX, USA

S Alan Stern
Southwest Research Institute, Boulder, CO, USA

Myriam Telus
Earth and Planetary Sciences, University of California Santa Cruz, Santa Cruz, CA, USA

Geronimo L Villanueva
NASA Goddard Space Flight Center, Greenbelt, MD, USA

Jean-Baptiste Vincent
Institute of Planetary Research, German Aerospace Center (DLR), Berlin, Germany

Kathryn Volk
Planetary Science Institute, Tucson, AZ, USA

Kacper Wierzchos
Lunar and Planetary Laboratory, University of Arizona, Tucson, AZ, USA

Maria Womack
Department of Physics, University of Central Florida, Orlando, Florida, USA
and
The National Science Foundation, Alexandria, Virginia, USA
and
Department of Physics and Astronomy, University of Victoria, Victoria, British Columbia, Canada

Laura M Woodney
Department of Physics and Astronomy, California State University San Bernardino, San Bernardino, CA, USA

Chapter 1

Introduction

Kathryn Volk, Maria Womack and Jordan Steckloff

Minor bodies in the solar system, such as comets, asteroids, Centaurs, and trans-Neptunian objects, are the remnants of the primordial planetesimals and thus they offer clues to the chemical composition and conditions of the solar system approximately 4.6 billion years ago. By investigating the physical and chemical properties of these objects, we can test and improve models of the solar system's earliest epochs of planetesimal formation and giant planet migration as well as models of the present-day thermal and physical evolution that drives cometary activity. The global scientific community has invested significant effort in spacecraft missions, ground- and space-based observations, laboratory and theoretical/modeling analysis to support these goals for small bodies in general. However, the Centaurs are a key subset of minor bodies that remains significantly understudied. This serves as the main motivation for this book: *Centaurs*—the first authoritative reference.

1.1 The Motivation for This Book

Centaurs have a unique role in the solar system as objects with transitional orbits between the ancient, minimally processed trans-Neptunian objects (TNOs) and the highly evolved Jupiter-family comets (JFCs) in the inner solar system. They are small, dark bodies comprised of ice, dust, and rock on dynamically short-lived orbits in the giant planet region, just beginning the physical transformation by solar-driven activity that dominates the evolution of comets. New Centaurs are experiencing significant heating from the Sun for the first time after a lifetime in the outer solar system. Most Centaurs are inactive, but a significant fraction (as high as $\sim$20 %) have been observed to develop a cometary-like dust coma at some point. Many important questions remain about the mechanisms and conditions of Centaur activity, including outbursts, and this activity has important differences from what is seen in most comets. Very little is known about the surfaces of inactive Centaurs and even their size distribution is not fully constrained. This is due to their large distances, small sizes, and low albedos, which make them too faint for follow-up

doi:10.1088/2514-3433/ada267ch1

characterization by most observatories, a challenge shared by many outer solar system populations. Some Centaurs have been discovered to have rings and other debris or large fragments near their nucleus, which is another mystery.

The list of Centaurs is expanding steadily, and it is predicted to increase significantly when the NSF-DOE Vera C. Rubin Observatory's Legacy Survey of Space and Time (LSST) begins. Likewise, our compositional understanding of Centaurs, and their transitional neighbors, is expanding now that NASA's JWST is operational and is providing unprecedented high quality spectral imaging data for dominant cometary volatile, ice, and dust features. Laboratory experiments, theoretical models, and computational tools are well-poised to make substantial progress with this plethora of new data in the next decade. Thus, the field of Centaur research is on the cusp of a transformative period, which is the motivation for writing this book *now*.

In Section 1.2, we describe our present-day picture of how the Centaur population fits with the solar system's other minor body populations and a historical overview of how we arrived at that understanding. In Section 1.3, we review the many ways people have divided the continuum from TNO to Centaur to JFC using different definitions for the Centaur population. This section also discusses the definitions of other related populations. We then describe the current census of Centaurs and highlight some of the major mysteries in Section 1.4. Finally, Section 1.5 describes how the Centaur population is critical to improving models of the history and evolution of the solar system. Throughout these sections, we refer the reader to specific chapters that cover various topics in more detail.

1.2 Historical Evolution of the Idea of Centaurs

To provide historical context for how the Centaurs came to be understood as a distinct population of solar system small bodies, we must first very briefly describe our modern picture of the links between TNOs, Centaurs, and JFCs. Today, the outer solar system hosts two major cometary reservoirs: the Oort Cloud (small bodies only very weakly bound to the Sun that reside on nearly isotropically distributed orbits with semimajor axes $a \gtrsim 2000$ au) and the trans-Neptunian population (small bodies on more tightly bound, mostly prograde orbits beyond Neptune; also sometimes referred to as the Kuiper Belt). The trans-Neptunian population can be further divided, with the broadest separation being into the so-called dynamically "cold" TNOs and the dynamically "hot" or excited TNOs. The dynamically cold TNOs have low-eccentricity, low-inclination orbits and are confined to a narrow range of heliocentric distances (these objects are also sometimes called members of the cold classical Kuiper Belt). The cold TNOs are understood to have formed on nearly their current orbits in the 42–46 au region; they are the only known *in situ* remnant of the outer solar system's planetesimal disk. In contrast, the dynamically excited TNOs have a wide range of semimajor axes, eccentricities, and inclinations reflecting their complex dynamical past; these include scattering, resonant, and detached TNOs (these subpopulations and their relationship to the Centaurs are further discussed in Chapter 3). Objects in the Oort Cloud and the dynamically excited TNO populations formed in today's giant planet region and were scattered outward and dynamically evolved into their metastable present-day reservoirs during the epoch of giant planet formation. For more thorough discussions see, e.g., Gladman & Volk (2021)'s recent review of the TNO populations and our understanding of their dynamical origins and Kaib & Volk (2024)'s recent review of cometary reservoirs discussing the dynamical formation and evolution of the Oort Cloud. Chapter 2 discusses planetesimal formation in more detail, focusing on the TNO populations that are linked to the Centaur population.

Objects in the very distant Oort Cloud are subject to perturbations from Galactic tides and passing stars, and their perihelion distances sometimes dip low enough to interact with the solar system's planets which changes their orbits. Some of these objects will enter into the detected Oort Cloud comet population, characterized by their very long orbital periods and nearly isotropic inclinations (hence Oort Cloud comets are also known as long-period comets); see, e.g., Dones et al. (2015) and Kaib & Volk (2024) for reviews of Oort cloud dynamics.

The TNOs are subject to perturbations from the giant planets, which over long timescales cause diffusion in their orbits and supplies a steady influx of new TNOs onto the Neptune-encountering orbits of the scattering TNO population. Most of these newly scattering objects originate from more stable dynamically excited TNO populations such as the resonant TNOs or higher perihelion temporarily detached TNO populations, though objects from the dynamically cold TNOs can also be perturbed onto high-eccentricity planet crossing orbits. Once a TNO has been perturbed onto an orbit that allows close encounters with Neptune, it experiences much larger orbital changes over timescales much shorter than the age of the solar system. Chapter 3 reviews the dynamical evolution of Centaurs and other giant-planet-crossing populations and their links to specific TNO populations in detail. Within typically a few million years of entering the giant planet region, Centaurs are either ejected back into the outer solar system or they evolve into the inner solar system to become part of the JFC population. The JFCs are characterized by their relatively short orbital periods and much flatter inclination distribution compared to the Oort cloud comets (hence they are sometimes referred to as short period comets or ecliptic comets). As with Centaurs, the exact definition for JFCs can vary (see Section 1.3), but they have very short dynamical lifetimes (a few hundred thousand years; Levison & Duncan 1994) due to their strong interactions with Jupiter and even shorter lifetimes as visibly outgassing comets (see, e.g., a recent discussion of inactive comets in Jewitt & Hsieh 2024). Figure 1.1 provides a schematic overview of the solar system's cometary reservoirs and the Centaur population's role as an intermediary between the TNOs and JFCs.

Because of their brightness, detailed observations of comets in the inner solar system go back many centuries (see, e.g., an historical review by Festou et al. 2004). The population statistics of these comets revealed that the orbits of the short-period JFCs are prograde and typically have inclinations below $\sim 30^\circ$, distinguishing them from the isotropically distributed long-period Oort cloud comets by more than just their shorter orbital periods. This difference and the discovery of Pluto in 1930 prompted early ideas about a possible disk of cometary objects in the outer solar

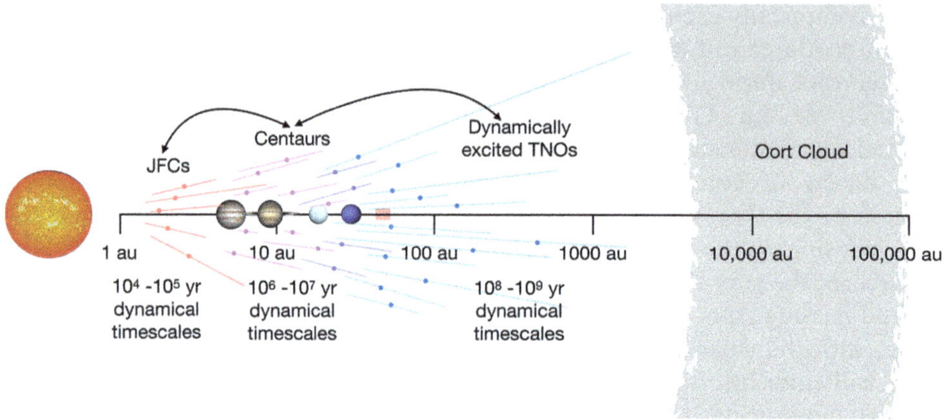

Figure 1.1. Representation of the outer solar system comet reservoirs and their connection to the Centaurs and JFCs. The most distant reservoir is the nearly isotropic Oort Cloud (shaded gray region) whose members are sometimes perturbed by galactic tides and passing stars onto planet-crossing orbits to become Oort Cloud comets (also known as long-period comets) in the inner solar system (not pictured). The dynamically cold (low-eccentricity, low-inclination) TNOs are indicated with a red box in the plane of the planets just beyond Neptune. They are thought to be an *in situ* remnant of the original planetesimal disk; they can be perturbed onto planet-crossing orbits, but are likely a minor contributor to the Centaur and JFC populations. The dynamically excited TNOs (blue points with typical perihelion to aphelion distance ranges shown with lines) reside on eccentric orbits with semimajor axes $a\sim$ 30–1000 au that can have significant inclinations but are nearly all prograde. These TNOs formed in the current giant planet region and were transported to their current orbits during the epoch of giant planet migration. Gravitational interactions with the present-day giant planets (mostly Neptune) can nudge these TNOs into orbits in the giant planet region where they become Centaurs (purple points and lines). The Centaurs evolve among the giant planets and can be ejected back out into the scattering TNO population or onto closer-in JFC orbits (red points and lines). Typical dynamical timescales for each population are indicated. See Chapter 3 for an in-depth discussion of the dynamics of these populations and their relationships to each other.

system beyond Neptune that could be connected to inner solar system comets; see Davies et al. (2008) and Fernández (2020) for detailed historical reviews of ideas on the trans-Neptunian region. In the 1970s, studies continued to examine whether there was a viable dynamical pathway between the distant Oort cloud and the observed prograde, small semimajor axis JFCs or whether the proposed but still unobserved (aside from Pluto) population just beyond Neptune was a better explanation for the JFCs (Everhart 1972; Delsemme 1973; Joss 1973). Then, in 1977, a representative of a new outer solar system small body population was discovered: (2060) Chiron (Kowal & Gehrels 1977), originally classified like an asteroid. Chiron currently resides between the orbits of Uranus and Saturn with a small inclination of $\sim 7°$. Early numerical integrations of Chiron's orbit revealed that it would either be ejected outward within $\sim 10^6$ years or passed inward to Jupiter within $\sim 10^5$ years (Oikawa & Everhart 1979). Other studies confirmed the generally short dynamical lifetimes (10^6–10^7 years) of orbits in the giant planet region (Gladman & Duncan 1990; Wisdom & Holman 1991), implying that similar to comets, Chiron-like objects are on chaotic and unstable rather than primordial orbits and must be supplied from a more long-term stable reservoir. As numerical orbital integration tools and computational resources improved, there were improved models of the supply of new planet-crossing objects from a low-inclination comet belt beyond Neptune (Fernandez 1980; Duncan et al. 1988; Torbett 1989; Quinn et al. 1990; Torbett & Smoluchowski 1990). Then in 1992, the first object residing in such a belt entirely exterior to Neptune's orbit was discovered (Jewitt & Luu 1993), followed quickly by a handful of additional discoveries (Jewitt & Luu 1995). This finally provided direct evidence for the proposed source of the JFCs in what is today called the TNO population.

Around the same time as these early TNO discoveries, new observations of the Centaur population were also unfolding. A second Chiron-like object, (5145) Pholus, was discovered in 1992 with a perihelion near Saturn's orbit and aphelion near Neptune (Scotti et al. 1992). Photometric characterization of Pholus revealed an extremely red surface compared to objects in the main asteroid belt (Mueller et al. 1992). Its reflectance spectra revealed a very steep red slope with no absorption features, further distinguishing it from typical asteroidal spectra (Fink et al. 1992) and contributing to the growing evidence that Chiron and Pholus represented a new class of small bodies (e.g., Buie & Bus 1992). Additional Centaur discoveries and characterization efforts followed shortly, improving and solidifying our understanding of them as the transition population between the TNOs and the JFCs (see, e.g., discussion in an early review by Stern & Campins 1996; see also Section 11.2.1 in Chapter 11 for a discussion of the origin of the name "Centaurs" for this population).

As the census of Centaurs expanded, the orbits of some previously discovered objects could be placed in context as part of this newly recognized population. The earliest example is the object 29P/Schwassmann-Wachmann 1 (29P/SW1), which was discovered in 1927 and classified as a comet due to its activity. Early fits to its observed positions revealed a nearly circular orbit with an unusually large perihelion distance beyond Jupiter, yielding the suggestion of it being an "asteroidal" comet (e.g., Berman & Whipple 1928). We can now recognize 29P/SW1 as the first representative of the Centaur population to have been discovered, though 29P/SW1 is itself in the transition zone to the JFCs (e.g., Sarid et al. 2019) and thus could be classified as either depending on the choice of exact definition (see Section 1.3); 29P/SW1 is one of several notable objects covered in more detail in Chapter 10. Comet 39P/Oterma can also be classified as part of the Centaur population based on its current orbit, and it has an illuminating orbital history. When 39P/Oterma was discovered in 1943, it had a perihelion distance inside Jupiter, placing it firmly among the JFCs. But a subsequent close encounter with Jupiter in 1963 transferred it back onto a higher perihelion orbit in the Centaur region. This provided an early, real-time demonstration of the connections between orbits in the giant planet region and the JFCs (discussed at the time in, e.g., Marsden 1962, 1970 and Kresák 1972) though the full context of this orbital evolution only became apparent after the Centaur population as a whole was better established (see a recent discussion of 39P/Oterma's dynamical pathway in Fernández et al. 2018).

Dynamical studies continued to advance alongside the growing list of observational discoveries (see Figure 1.2). Levison & Duncan (1994) published improved dynamical integrations of JFC orbits, providing more accurate estimates of their

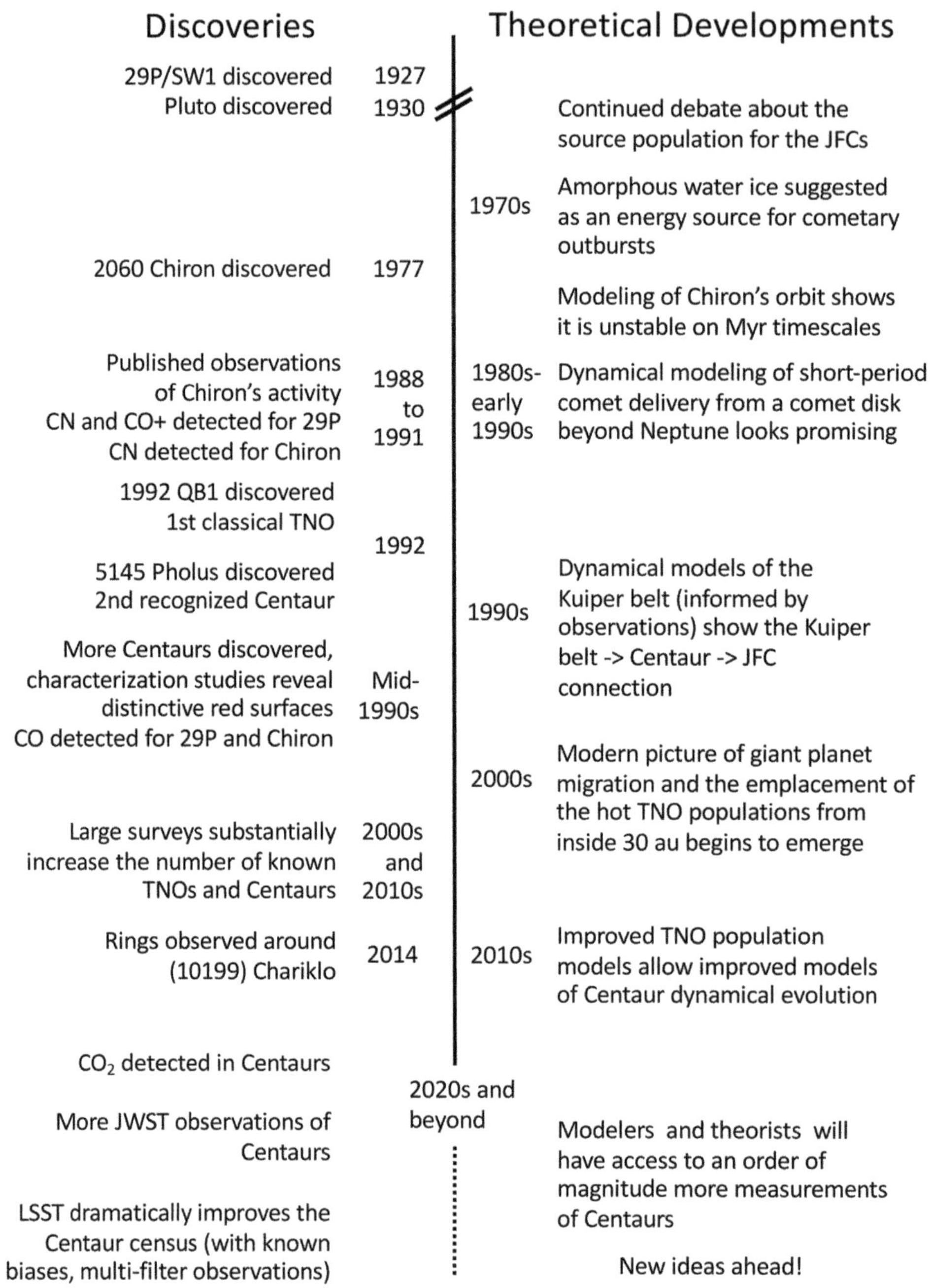

Figure 1.2. An abbreviated timeline of major observational (left) and theoretical (right) developments regarding Centaurs.

short lifetimes in the inner solar system and thus their required replacement rate to maintain a steady-state population. The steady stream of early TNO discoveries provided better constraints on the orbital distribution of TNOs, including the existence of the hot and cold TNO populations; this allowed for improved numerical models of the full dynamical path from the TNO populations, through the giant planet region, and into the JFC population (Levison & Duncan 1997; Duncan et al. 1995). Since those early models, our understanding of the formation and detailed orbital distribution of TNOs has dramatically improved (see, e.g., Gladman & Volk 2021 as well as Chapters 2 and 3), but this basic dynamical connection placing Centaurs as the evolutionary link between the primordial/nearly-primordial TNOs and the active, highly evolved JFCs remains.

The sometimes-cometary nature of Centaurs was revealed shortly after Chiron's discovery when a small dust coma was measured (Meech & Belton 1990). Lightcurve measurements revealed short-term (hours) and long-term (years) variations in brightness, which were used to constrain its rotational period as well as changes in activity levels (e.g., Tholen et al. 1988; Bus et al. 1989; Meech & Belton 1990; Luu & Jewitt 1990). Follow-up observations of Chiron revealed a dust coma and CN, CO^+, and CO gaseous emission, confirming the presence of cometary activity despite its large distance from the Sun (Bus et al. 1989; Cochran & Cochran 1990; Womack & Stern 1999; Luu et al. 2000). Many other active Centaurs have since been observed (see, e.g., Jewitt 2009), with activity levels ranging from relatively constant outgassing to sometimes very bright outburst events. Inner solar system cometary activity (at distances $\leqslant$3 au) is largely driven by water ice sublimation, but other ices, such as carbon dioxide and carbon monoxide, likely contribute to activity at the colder temperature of the Centaur region (e.g., Womack et al. 2017); Chapter 6 provides a detailed overview of volatiles in the Centaur region. Understanding the drivers of cometary activity in Centaurs is a very active area of current research discussed in many chapters throughout this book, with Chapter 8 providing an overview of activity.

1.3 Definition of a Centaur (and Other Notes on Terminology)

Given that Centaurs represent a middle state in the continuum of dynamical evolution between the outer solar system's TNO populations and the inner solar system's JFC population, it is perhaps unsurprising that there is not one single agreed-upon definition for the population. Often, the lines drawn between TNO, Centaur, and JFC depend on whether one is studying these populations from a dynamical perspective, a thermal perspective, an activity perspective, etc. Thus, definitions may differ slightly across different chapters even within this book! Here we will review some commonly used definitions for Centaurs, highlighting our preferred dividing lines. Ultimately, all of these discrepant definitions attempt to capture the Centaurs as a transitory population between their initial state in the TNO reservoir and their final evolved state in the JFC population. **We prefer to define Centaurs as bodies with perihelion distances q exterior to Jupiter and aphelion distances Q interior to Neptune** ($a_J < q$ and $Q < a_N$). Details of other definitions for the Centaur population and the motivation for our preferred definition are given below.

Inner boundary of the Centaur population:
All definitions of the Centaur population start with a requirement that the object's semimajor axis be outside Jupiter's orbit ($a > a_J$ where $a_J \approx 5.2$ au); as far as we can tell, this is the only universally agreed-upon portion of the Centaur definition. The vast majority of definitions also require the perihelion distance to lie outside Jupiter's orbit ($q > a_J$); though JPL's Small Body Database Query system does not have this constraint on the group of objects it labels Centaurs, relying solely on a semimajor axis constraint. We prefer this simple perihelion distance cut for the inner boundary of the Centaur population.

Rather than, or in addition to, a q boundary, some definitions of Centaurs use the Tisserand parameter with respect to Jupiter (T_J) to try to separate Centaurs from the JFC population they feed into; this is defined as

$$T_J = \frac{a_J}{a} + 2\sqrt{\frac{a}{a_J}(1 - e^2)}\cos i, \tag{1.1}$$

where a_J is Jupiter's semimajor axis and a, e, and i are the semimajor axis, eccentricity, and inclination of the small body, respectively. For objects that approach Jupiter's orbit, T_J provides an estimate of the extent to which a small body is dynamically coupled to Jupiter, with the JFC population often defined by having $2 < T_J < 3$ (see, e.g., Carusi et al. 1987; Levison & Duncan 1994). The definition of the Centaur population given by Gladman et al. (2008) requires giant planet region objects with $q < 7.35$ au to have $T_J > 3.05$ to be considered Centaurs; near-Jupiter objects with smaller T_J are labeled "Jupiter-coupled" in that nomenclature scheme. From a dynamical perspective, this separates out the shortest-lived giant planet region orbits from the longer-lived ones, because objects that interact strongly with Jupiter have the most rapid orbital evolution (see Chapter 3).

Figure 1.3 shows the population of observed objects in the giant planet region alongside the closely related scattering TNO and JFC populations. We highlight in this plot 45 objects with semimajor axes between Jupiter and Neptune and $q > a_J$ that are known to be active[1]. In contrast, using the T_J cut from the Gladman et al. (2008) scheme, 32 of the 45 active small bodies with $q > a_J$ would be labeled as active Jupiter-coupled objects rather than active Centaurs. Using the $2 < T_J < 3$ criteria for JFCs, 22 of the 45 could be considered JFCs; this includes the most famous active Centaur, 29P/SW1. Another three of these active objects could be labeled as Halley-type comets (HTCs) because they have $T_J < 2$. However, classifying these objects as comets rather than Centaurs masks the fact that their activity has different drivers than more typical inner solar system cometary activity; activity in objects beyond Jupiter's orbit is not driven by simple water–ice sublimation, which happens inside ~3 au. Like their dynamical evolution, the

[1] The list of active objects was taken in February 2024 from https://physics.ucf.edu/~yfernandez/cometlist.html

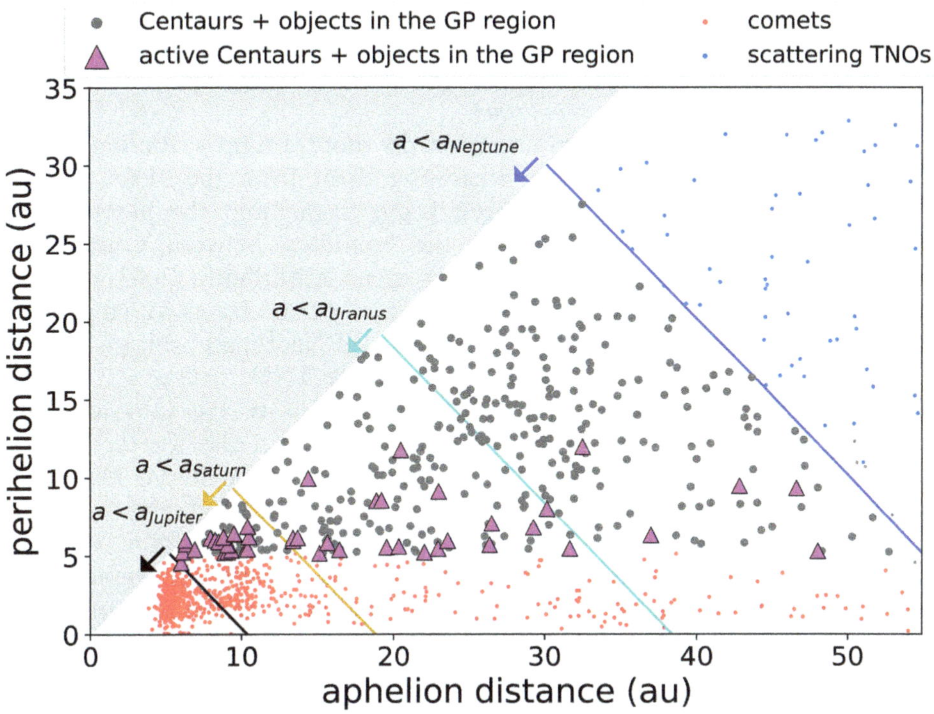

Figure 1.3. Perihelion distance (q) vs. aphelion distance (Q) for Centaurs and other observed objects in the giant planet (GP) region and closely related dynamical populations. The colored lines indicate lines of constant semimajor axis (a) at the semimajor axis of each giant planet; the gray shaded region represents unphysical orbits. The small red points show the orbits of Jupiter-family comets and Halley-type comets, potential end states of Centaur evolution. The small blue points show scattering trans-Neptunian objects (TNOs), the immediate predecessors of Centaurs. Objects with $q > a_{Jupiter}$ and $a < a_{Neptune}$ that are known to display cometary activity are shown as purple triangles (with one notable transition object slightly below that q threshold included in the lower left, 2019 LD2); objects within these orbital constraints that are not known to be active are shown with large gray circles. Exactly which objects within this plot are labeled "Centaurs" can vary (see Section 1.3).

thermal evolution of Centaurs is a continuum and there are no entirely "clean" dividing lines. Water-driven activity can and does occur further out, but it dominates within 2–3 au (Crovisier 1989; Prialnik et al. 2008), which is observationally demonstrated in a recent review paper (e.g., Harrington Pinto et al. 2022). This switch to water-ice-dominated thermal evolution is important to separating the relatively pristine nature of the Centaurs from the more processed JFCs. An inner boundary condition of $q > a_J$ for the Centaurs provides a reasonable overlap between what we understand about the dynamical transition between the giant planet region and the inner solar system comets (e.g., the "Dynamical Gateway" described in Sarid et al. 2019) and the transition to activity driven by water-ice sublimation; the most rapid combined physical and orbital evolution occurs once an object's perihelion distance is inside of Jupiter's orbit. We thus prefer using $a > a_J$ and $q > a_J$ to define the inner boundary of the Centaur population, a boundary that has been used elsewhere in the literature (e.g., Jewitt 2009; Di Sisto & Rossignoli 2020).

Outer boundary of the Centaur population:
The boundary separating the Centaurs from the more distant scattering TNOs is fuzzy just like the inner boundary separating them from the JFCs. Different approaches to this separation have been taken throughout the literature. The Minor Planet Center avoids specifying the boundary between Centaurs and scattering TNOs by listing them together[2]. Tiscareno & Malhotra (2003) performed a dynamical study of observed Centaurs and suggested that objects with a perihelion distance between Jupiter and a few Hill radii beyond Neptune's orbit ($a_J < q < 33$ au) have dynamical lifetimes distinct from scattering TNOs with $q > 33$ au; they would thus include objects with large semimajor axes in the same dynamical category as Centaurs. As with the inner boundary, one can also use the Tisserand parameter (this time, with respect to Neptune) to draw boundaries between the TNOs and giant planet crossing populations (see, e.g., the detailed scheme presented in Horner et al. 2003). However, most conventional Centaur definitions do draw a semimajor axis boundary for the Centaur population, requiring a semimajor axis interior to Neptune's orbit ($a < a_N$) to distinguish them from the scattering TNOs (e.g., Jewitt 2009; Di Sisto & Rossignoli 2020). We prefer to add an additional criteria that objects have aphelion distances inside Neptune ($Q < a_n$), i.e., orbits entirely enclosed in the giant planet region, as suggested by Sarid et al. (2019). Dynamically, this boundary excludes objects on often long-lived Neptune-crossing orbits and highlights the subset of giant planet region objects that have already been transferred sufficiently far inwards that they are affected by the other giant planets; this definition of Centaurs thus yields slightly shorter dynamical lifetimes than the $a < a_N$ boundary. Thermally, this is a region of space in which surface temperature ranges correspond to numerous phase changes/sublimation radii, including CO_2 sublimation and amorphous-to-crystalline ice and cubic-hexagonal ice transitions. These "mesovolatile" phase changes are thermo-physically characteristic of Centaur activity. This is not to suggest that Centaurs cannot produce hypervolatiles or water, which they are indeed observed to produce. The mechanisms driving this activity are currently poorly understood (Chapter 8). The outer edge of this region overlaps with the thermal environment of the TNOs, in which objects become severely depleted in the hypervolatile ices (e.g., CO, N_2) after formation (Lisse et al. 2021; Steckloff et al. 2021; Lisse et al. 2022b); thus the Centaur region defines where these objects would experience their first significant thermally-driven activity since solar system formation. Combined with the inner boundary, which excludes the inner solar system where water sublimation-driven activity rapidly evolves the JFCs, the Centaurs therefore represent the thermochemical state of the nucleus *prior* to any vigorous water-driven activity.

[2] https://minorplanetcenter.net/iau/lists/t_centaurs.html

Before moving on to the current observational constraints on the Centaur population (Section 1.4), we define a few other terms that will come up throughout this book. For the wider population of small bodies beyond Neptune, this book generally uses the terminology of TNOs[3]. On occasion, the terminology Kuiper Belt Object (KBO) might be used in the context of historical discussions or discussions of the classical belt region, especially because the dynamically cold TNOs are often referred to as cold classical KBOs. Likewise, the dynamically hot scattering TNOs are sometimes referred to as scattered disk objects.

The inner solar system comet populations similarly have multiple naming schemes in the literature. Comets are sometimes divided along orbital periods into short period comets and long period comets. We prefer the term Oort cloud comet to long-period comet to emphasize the origin of this population. Similarly, instead of short-period comets, we refer to the JFCs, which are those comets that originate from the TNO populations via the Centaur population and are so-named because they are dynamically controlled by Jupiter; these comets are also sometimes referred to as ecliptic comets in the historical literature to distinguish them from the isotropic Oort cloud comets by their orbital inclinations. Other comet groups with short to moderate orbital periods include HTCs and Encke-type comets; the connections between these comets and the TNOs/Centaurs are less clear and are not a focus of this book (see, e.g., discussion in Kaib & Volk 2024).

1.4 Current Observational Census and Challenges

Depending on which definition you prefer for the Centaur population, the current number of observed objects ranges from ∼250–400[4]. These objects are shown in Figures 1.3 and 1.4; the former highlights the active subset of these objects while the latter shows the correlations between the observed orbital and H-magnitude distributions. Chapter 4 will discuss in detail what we know about the size distribution (and other nucleus properties) of the Centaur population, but Figure 1.4 highlights the strong observational biases present in our Centaur census, which we briefly describe below. The typical H-magnitudes of observed Centaurs range from ∼14 at the inner edge of the population to ∼8–10 at the outer edge; at distances between Uranus and Neptune, only the largest Centaurs are bright enough to have been discovered.

The very wide range of heliocentric distances covered by the Centaur population (both as individual objects over their entire orbital paths and by the population as a whole) and their generally low on-sky density has made them a difficult population to survey efficiently. Their rate of motion on the sky is slower than that of asteroids,

[3] This terminology encompasses both the originally-envisioned low-*e*/low-*i* population as well as the wider TNO populations that became apparent as observational surveys improved. Using TNO instead of "Kuiper Belt object" (KBO) also avoids assigning "discovery" credit for the population as a whole (see reviews by, e.g., Davies et al. 2008; Fernández 2020 for historical reviews of literature on the TNO populations).

[4] Based on queries to JPL's small body database in February 2024 https://ssd.jpl.nasa.gov/tools/sbdb_query.html

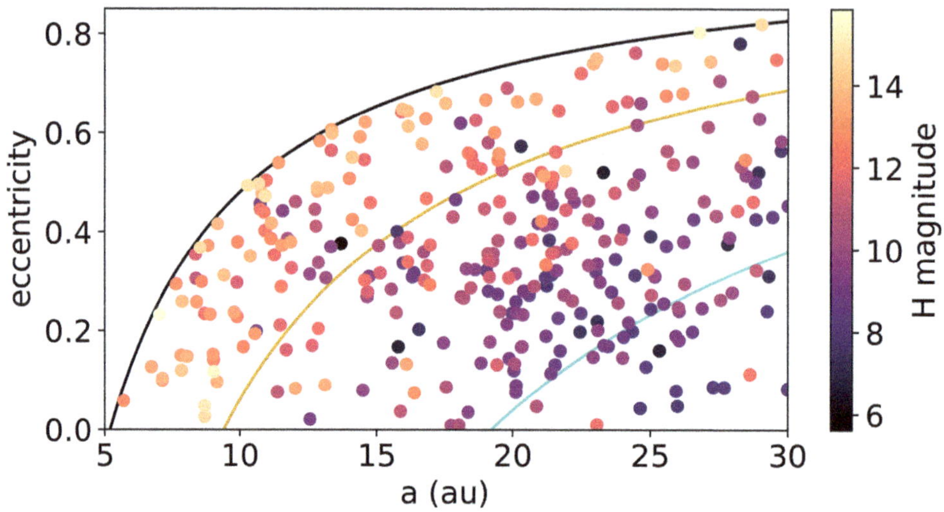

Figure 1.4. Eccentricity vs. semimajor axis for every object in JPL's small body database with a, $q > a_J$ and $a < a_N$ that has a reported H magnitude (color scale). The black, gold, and turquoise curves show orbits with perihelion distances at Jupiter, Saturn, and Uranus, respectively. Note the strong trend in the H-magnitude distribution reflecting the strong observational biases against detecting small Centaurs are larger heliocentric distances. This observational bias is common to all small body populations but is particularly notable for Centaurs because the population spans such a wide range of distances. Our observations of the Centaur population include some very small objects, but only at the innermost edge; we do not have representative observations of small Centaurs in the outer giant planet region when they are just beginning their evolution.

so surveys targeting those closer-in populations are less efficient at detecting Centaurs; similarly, Centaurs move more quickly than TNOs, so TNO surveys are not optimized to discover the closer-in Centaurs (see, e.g., discussions in Bannister et al. 2018; Peixinho et al. 2020). Even moderately deep all-sky surveys such as Pan-STARRS only detected relatively modest numbers of Centaurs due to their limiting magnitude (~150; Weryk et al. 2016). One needs wide sky coverage to discover significant numbers of Centaurs, but because many of them are distant and therefore quite faint, one also needs to reach deep limiting magnitudes. Thus, the majority of Centaur detections have been serendipitous, resulting from searches through data sets not optimized for their detection or tracking.

The lack of optimized tracking introduces unknown biases into the Centaur population beyond the better-understood brightness biases. For example, deep TNO surveys typically cover small sky areas (trading off area for depth to discover their distant targets). Centaurs discovered in these small fields will move out of range of the recovery/tracking fields much faster than the targeted TNOs. If they move out of field before enough observations have been obtained to provide well-determined orbits, these Centaurs will be lost, and the subset of not-lost Centaurs will have some orbit-dependent biases. Overall, we have a very incomplete observational snapshot of the Centaur population. Thankfully this will change over the coming decade as LSST will provide the much needed deep, all-sky survey with extensive follow-up required to detect significant numbers of new Centaurs and track them over long time periods, likely leading to significant insights (see Chapter 14 for more details).

Observational characterization of Centaurs is similarly challenging. As noted above, the faintness of most known Centaurs when compared to inner solar system populations makes follow-up observations more difficult. This is further complicated by the fact that most known Centaurs have elliptical orbits and have peak observability for only a few years out of their many decades-long orbital periods. For example, Chiron was at its 8.5 au perihelion in 1996, passed through aphelion of ~19 au in 2021 and is currently heading in for its next perihelion in 2046. A remarkable contrast to most Centaurs is 29P/SW1, which is observable nearly all the time (when viewing geometry from the Earth allows), due to its nearly circular orbit. In general, measuring Centaur surface properties requires larger telescopes, even for obtaining photometric colors; and relatively few Centaurs are bright enough for spectroscopic or photometric characterization of their surfaces. Chapter 5 discusses what is known about, and the observational constraints on, Centaur surface properties in detail.

Characterizing the volatiles emitted from Centaurs and using these measurements to constrain models driving activity is very challenging with only a few secured volatile detections for a few Centaurs. Table 1.1 contains a list of molecules measured in Centaurs, TNOs, and comets. As of January 2024 the list is brief but complete for Centaurs and TNOs. This briefness largely reflects their faintness from being so far from the Sun and, of course, the lack of spacecraft missions. The list in Table 1.1 for comets is not complete but contains the most abundant molecules. Thus far, there is a great deal of similarity in the dominant volatile species identified in Centaurs, TNOs, and comets: H_2O, CO_2, CO, CH_3OH, and HCN. Detailed comparison beyond that, such as relative abundances or production rates, has been very difficult with current observational limitations (see Fraser et al. 2024). A recent

Table 1.1. Representative Listing of Molecules Detected/Inferred on Surface and in Comae or Atmospheres of TNOs, Centaurs, and Comets (adapted from Fraser et al. 2024. CC BY 4.0)

TNO surface ices	H_2O	CH_3OH	CH_4	N_2	CO
	CO_2	NH_3	C_2H_6		
TNO volatiles	N_2	CO	CH_4		
Centaur surface ices	H_2O	CH_3OH			
Centaur volatiles	CO	H_2O	CO_2	HCN	CN
	CO^+	N_2^+			
Comet surface and coma ices	H_2O	CO_2	COOH-group		
Comet volatiles	H_2O	CO_2	CO	CH_4	HCN
	CH_3OH	H_2CO	NH_3 HNC	C_2H_2	C_2H_6
	HCOOH	$HCOOCH_3$	HNCO	H_2S	OCS
	HC_3N	SO	SO_2	CS	CH_3CO
	S_2	CN	C_2	NH_2	C_3
	CO^+	N_2^+	H_2O^+	CO_2^+	N_2

analysis comparing all the published and simultaneous CO, CO_2 and H_2O production rates in comets and Centaurs may indicate the importance of solar processing to the top layers of Centaur nuclei (Harrington Pinto et al. 2022, 2023), which is discussed in more detail in Chapter 6. Insolation, space weathering, and other surface processing techniques are also highly relevant and are addressed in Chapter 7. Methanol and water absorption have been detected on the surfaces of a small number of inactive Centaurs and TNOs (Chapter 5). Comparison of absorption features on Centaurs with comets is complicated by the fact that it is very difficult to obtain such measurements on inactive comet nuclei but this will also become easier as telescope capabilities improve. The number and amounts of molecules for Centaurs (and TNOs and faint, distant comets) is expected to grow significantly in the coming years based on JWST observations, which have already started coming in; we also anticipate much more information about Centaur outbursts and other unusual variations in lightcurves from LSST (see Chapter 14).

In addition to measuring the chemical compositions of comae and the nucleus surfaces, many observations have also uncovered unusual activity patterns in Centaurs, including developing a significant coma when very far from the Sun (Chiron, 29P/SW1), long term nearly continuous coma (29P/SW1), and dramatic outbursts (Echeclus, Chiron, 29P/SW1). High-resolution spectra reveal the presence of jets on the surface of 29P/SW1 (Bockelée-Morvan et al. 2022). Breakthroughs in Centaur activity are covered in Chapter 8 and this topic would also substantially benefit from increased observing power on all fronts. Occultation data provide accurate information about Centaur sizes and have revealed the surprising presence of debris, satellites and even rings (Chapter 9) and this also requires significant observational resources. Finally, although Arrokoth is the only primordial TNO to be graced with a spacecraft visit (by *New Horizons* in 2019), so far no Centaur has ever had a flyby or orbiter mission, despite their key role in solar system research, which is addressed by Chapter 16.

Only a small subset of Centaurs have benefited from substantial observational resources dedicated to them, and thus our knowledge of Centaurs is heavily weighted on these results (see Chapter 10). Ongoing and new dedicated and coordinated world-wide observational campaigns (see Chapter 11) are required to make additional progress characterizing the known Centaurs. The Vera Rubin Observatory's LSST data will aid these efforts by providing sparse lightcurve and color measurements for some of its newly detected Centaurs, but targeted follow up will still be required. NASA's JWST will also be invaluable for providing new observations of the population (see Chapter 14).

1.5 Overview of Scientific Importance

As the dynamical transitional population between the TNOs and JFCs, Centaur studies are crucial to understanding how TNOs evolve *prior* to entering the JFC population. This evolution is not just dynamical, but also physical, as up to $\sim$20 % of Centaurs are known to be active, ejecting and depositing material, which can lead to significant changes to their nuclei. These processes that occur in the giant planet region could affect the appearance, coma, surface, near surface, as well as deep interiors of Centaurs. They also have important implications for interpreting satellites, debris, and rings in the near-Centaur environment. An improved understanding of Centaur evolution is crucial in understanding the starting states of JFCs.

In the 1950s, Fred Whipple proposed the "Dirty Snowball" model of comet activity, which described comets as small, coherent admixtures of volatile ices and refractory dust (Whipple 1950, 1951). This model explained comet activity as the natural result of frozen water incorporated into the nucleus and sublimating in response to solar heating at the surface; it also countered the "Sandbank" model, in which comets were coherent clouds of grains orbiting the Sun (Lyttleton 1948, 1951). This model is not directly transferable to Centaurs, however, because unlike most comets, Centaur activity occurs at relatively large heliocentric distances that would result in surface temperatures being much too low for water ice to sublimate vigorously. Furthermore, steady sublimation of volatiles typically has a strong heliocentric dependence for comets closer to the Sun. This seemed incapable of explaining the observed short-duration outbursts and transient brightening events that are caused by the sudden injection of large quantities of dust into the coma, especially as many such events occur far from a Centaur's perihelion.

The activity in some Centaurs has led to numerous proposed processes, similar to those proposed for distantly active comets. The sudden outbursts of dust in some distant comets, such as 29P/SW1, led Donn & Urey (1956) to propose that chemical reactions within the nucleus produce compounds such as ammonium azide, which explode violently; further refinements allowed for a wider array of explosive chemical reactions (e.g., Donn 1963). Later, Patashnick (1974) proposed that the crystallization of amorphous water ice, an exothermic reaction, could provide the energy needed to drive activity. Furthermore, laboratory studies indicate that amorphous water ice can trap common cometary volatile species within its amorphous structure, which can be released upon crystallization (Bar-Nun et al. 1985). Although amorphous water ice is unstable at any temperature, the reaction rate is temperature dependent and becomes vigorous starting at 150 K (Kouchi et al. 1994). As a result, the crystallization of amorphous water ice has become a dominant proposed driving mechanism of Centaur activity (e.g., Prialnik 1993; Capria et al. 2000; Sarid et al. 2005; Jewitt 2009; Lisse et al. 2022a). Nevertheless, there are still significant issues with the amorphous ice crystallization model, such as that the ejection of such molecules from the ice in significant quantities causes the crystallization reaction to become *endothermic* (Kouchi & Sirono 2001). Some other exothermic mechanisms have been proposed, including the polymerization of HCN (e.g., Rettig et al. 1992; Matthews 1995) and cryovolcanism (Miles 2016), however, these processes have not been fully modeled (see Womack et al. 2017). Possible energy sources are explored in more detail in Chapter 12 on Centaur interiors. The status of laboratory studies relevant to understanding Centaurs and their activity are discussed in Chapter 13.

The volatiles CO and CO_2 are capable of driving activity at large heliocentric distances but have been detected in only three Centaurs—29P/SW1, Chiron, and 39P/Oterma. This is largely due to the limitations of ground-based telescopes, which are unable to detect CO_2 emission (due to telluric contamination of infrared lines and lack of rotational dipole moment for mm-wavelength transitions) and the faintness of CO lines at mm-wavelengths in smaller Centaurs at large heliocentric (and geocentric) distances. It is still unknown whether one of these mechanisms, several of these mechanisms, or another mechanism altogether is responsible for driving Centaur activity; although significant amounts of CO and CO_2 emission have been detected from two Centaurs (29P/SW1 and 39P/Oterma) implying gas amounts high enough to also release significant dust comae (Senay & Jewitt 1994; Harrington Pinto et al. 2023; Faggi et al. 2024), and two studies show that CO and dust production correlate at least some of the time in the Centaur 29P/SW1 (but not always; Wierzchos & Womack 2020; Bockelée-Morvan et al. 2022). See further discussion of past and future observations of volatiles in Chapters 6 and 14.

In addition to abundant molecules that may drive Centaur activity, the isotopic composition of dust and organic material has important clues to solar system formation models, particularly for studying preserved solar nebular material and possible delivery to primordial Earth. The lack of detailed knowledge about ice-rich planetesimals is an outstanding issue in cosmochemistry, a field which includes the study of mineralogy, elemental abundances, isotope compositions, and radiometric ages of solar system material. Formation models could be significantly constrained by information about how C, H, O, and N isotope abundances varied with heliocentric distance as well as whether a Centaur's soluble organic matter is enriched in D, ^{13}C, or ^{15}N. Comparisons with detailed measurements made from various comet missions would help us understand whether chrondritic material (from primordial granular stony meteorites) that has been found in JFCs is also present in Centaurs, and ultimately establish the geochemical links between JFCs and TNOs. Cosmochemical and astrobiological implications of Centaur observations are discussed in Chapter 15.

Beyond our growing understanding that cometary activity is important in the Centaur population, another startling relatively recent discovery is the presence of rings of material around at least one Centaur. In 2014 Braga-Ribas et al. announced the detection of a ring system around Centaur (10199) Chariklo using stellar occultation observations (Braga-Ribas et al. 2014). A ring-like system or evolving material surrounding Chiron was also reported using other occulation data (Sickafoose et al. 2020, 2023). See Chapter 9 for a detailed discussion of Centaur rings, debris, and moons, Chapter 10 for an in depth look at five particularly interesting Centaurs (95P/Chiron, 29P/SW1, 174P/Echeclus, Chariklo, P/2019 LD2), and Chapter 11 for descriptions of the observational campaigns Centaurs have inspired.

Ultimately, characterizing the processes driving Centaurs' physical evolution are critical to understanding the physical connection between the TNO and JFC populations. The processing that objects experience in the Centaur region plays a crucial role in turning TNOs into JFCs. Thus, evolutionary processes (Chapter 7) are critical to constraining models of both TNOs and Centaurs from data, especially data from missions to the JFCs (e.g., *Deep Space 1, Stardust, Deep Impact, Stardust-NExT, EPOXI,* and *Rosetta*). Although dedicated spacecraft missions have been proposed to study the Centaurs *in situ* (see Chapter 16 on spacecraft exploration), consistent with the recommendations of the 2013 and 2023 National Academy Planetary Science Decadal Reports (National Research Council 2011; National Academies of Sciences, Engineering, and Medicine 2023), no such missions have yet been selected. Future Centaur missions will play an important role in advancing small body science.

In this chapter we have provided only a very brief overview of how the Centaurs emerged as an important piece of the puzzle in understanding the evolution of icy bodies in the solar system. We have summarized some of the key discoveries and theoretical developments in a timeline (Figure 1.2), which also highlights the ongoing nature of new fundamental observations and developments in our models of Centaurs as a population. Observations with JWST and the Vera C. Rubin Observatory (covered in Chapter 14) will dramatically improve our observational characterization of the Centaur population and inspire many new theoretical developments. The rest of this book details the current state of Centaur studies, highlighting open questions and ongoing debates. We will return in Chapter 17 to summarize the exciting developments expected in this field over the coming years.

Acknowledgements

We thank two anonymous referees whose helpful reviews improved our manuscript. This material is based in part on work done by M.W. while serving at the National Science Foundation.

References

Bannister, M. T., Gladman, B. J., Kavelaars, J. J., et al. 2018, ApJS, 236, 18
Bar-Nun, A., Herman, G., Laufer, D., & Rappaport, M. L. 1985, Icar, 63, 317
Berman, L., & Whipple, F. L. 1928, PASP, 40, 34
Bockelée-Morvan, D., Biver, N., Schambeau, C. A., et al. 2022, A&A, 664, A95
Braga-Ribas, F., Sicardy, B., Ortiz, J. L., et al. 2014, Natur, 508, 72
Buie, M. W., & Bus, S. J. 1992, Icar, 100, 288
Bus, S. J., Bowell, E., Harris, A. W., & Hewitt, A. V. 1989, Icar, 77, 223
Capria, M. T., Coradini, A., De Sanctis, M. C., & Orosei, R. 2000, AJ, 119, 3112
Carusi, A., Kresak, L., Perozzi, E., & Valsecchi, G. B. 1987, A&A, 187, 899
Cochran, W., & Cochran, A. L. 1990, IAU Circ., 5144, 1
Crovisier, J. 1989, A&A, 213, 459
Davies, J. K., McFarland, J., Bailey, M. E., Marsden, B. G., & Ip, W. H. 2008, in The Solar System Beyond Neptune, ed. M. A. Barucci, et al. (Tucson, AZ: Univ. Arizona Press) 11
Delsemme, A. H. 1973, A&A, 29, 377
Di Sisto, R. P., & Rossignoli, N. L. 2020, CeMDA, 132, 36
Dones, L., Brasser, R., Kaib, N., & Rickman, H. 2015, SSRv, 197, 191
Donn, B. 1963, Icar, 2, 396
Donn, B., & Urey, H. C. 1956, ApJ, 123, 339
Duncan, M., Quinn, T., & Tremaine, S. 1988, ApJl, 328, L69
Duncan, M. J., Levison, H. F., & Budd, S. M. 1995, AJ, 110, 3073
Everhart, E. 1972, ApL, 10, 131

Faggi, S., Villanueva, G. L., McKay, A., et al. 2024, NatAs, 8, 1237
Fernández, J. 2020, in The Trans-Neptunian Solar System, ed. D. Prialnik, M. A. Barucci, & L. Young (Amsterdam: Elsevier) 1
Fernandez, J. A. 1980, MNRAS, 192, 481
Fernández, J. A., Helal, M., & Gallardo, T. 2018, PSS, 158, 6
Festou, M. C., Keller, H. U., & Weaver, H. A. 2004, in Comets II, ed. M. C. Festou, H. U. Keller, & H. A. Weaver (Tucson, AZ: Univ. Arizona Press) 3
Fink, U., Hoffmann, M., Grundy, W., Hicks, M., & Sears, W. 1992, Icar, 97, 145
Fraser, W. C., Dones, L., Volk, K., Womack, M., & Nesvorný, D. 2024, in Comets III, ed. K. J. Meech, et al. (Tucson, AZ: Univ. of Arizona. Press) 121
Gladman, B., & Duncan, M. 1990, AJ, 100, 1680
Gladman, B., Marsden, B. G., & Vanlaerhoven, C. 2008, in The Solar System Beyond Neptune, ed. M. A. Barucci, et al. (Tucson, AZ: Univ. Arizona Press) 43
Gladman, B., & Volk, K. 2021, ARAA, 59, 203
Harrington Pinto, O., Womack, M., Fernandez, Y., & Bauer, J. 2022, PSJ, 3, 247
Harrington Pinto, O., Kelley, M. S. P., Villanueva, G. L., et al. 2023, PSJ, 4, 208
Horner, J., Evans, N. W., Bailey, M. E., & Asher, D. J. 2003, MNRAS, 343, 1057
Jewitt, D. 2009, AJ, 137, 4296
Jewitt, D., & Hsieh, H. H. 2024, in Comets III, ed. K. J. Meech, et al. (Tucson, AZ: Univ. of Arizona. Press) 767
Jewitt, D., & Luu, J. 1993, Natur, 362, 730
Jewitt, D. C., & Luu, J. X. 1995, AJ, 109, 1867
Joss, P. C. 1973, A&A, 25, 271
Kaib, N. A., & Volk, K. 2024, in Comets III, ed. K. J. Meech, et al. (Tucson, AZ: Univ. of Arizona. Press) 97
Kouchi, A., & Sirono, S.-i. 2001, GRL, 28, 827
Kouchi, A., Yamamoto, T., Kozasa, T., Kuroda, T., & Greenberg, J. M. 1994, A&A, 290, 1009
Kowal, C. T., & Gehrels, T. 1977, IAU Circ., 3129, 1
Kresák, L. 1972, IAU Symp., Vol. 45, The Motion, Evolution of Orbits, ed. G. A. Chebotarev, E. I. Kazimirchak-Polonskaia, & B. G. Marsden (Dordrecht: Reidel)
Levison, H. F., & Duncan, M. J. 1994, Icar, 108, 18
Levison, H. F., & Duncan, M. J. 1997, Icar, 127, 13
Lisse, C. M., Young, L. A., Cruikshank, D. P., et al. 2021, Icar, 356, 114072
Lisse, C. M., Steckloff, J. K., Prialnik, D., et al. 2022a, PSJ, 3, 251
Lisse, C. M., Gladstone, G. R., Young, L. A., et al. 2022b, PSJ, 3, 112
Luu, J. X., & Jewitt, D. C. 1990, AJ, 100, 913
Luu, J. X., Jewitt, D. C., & Trujillo, C. 2000, ApJl, 531, L151
Lyttleton, R. A. 1948, MNRAS, 108, 465
Lyttleton, R. A. 1951, MNRAS, 111, 268
Marsden, B. G. 1962, ASPL, 8, 375
Marsden, B. G. 1970, AJ, 75, 206
Matthews, C. N. 1995, PSS, 43, 1365
Meech, K. J., & Belton, M. J. S. 1990, AJ, 100, 1323
Miles, R. 2016, Icar, 272, 387
Mueller, B. E. A., Tholen, D. J., Hartmann, W. K., & Cruikshank, D. P. 1992, Icar, 97, 150

National Academies of Sciences, Engineering, and Medicine 2023, Origins, Worlds, and Life: A Decadal Strategy for Planetary Science and Astrobiology 2023-2032 (Washington, DC: The National Academies Press)
National Research Council 2011, Vision and Voyages for Planetary Science in the Decade 2013-2022 (Washington, DC: The National Academies Press)
Oikawa, S., & Everhart, E. 1979, AJ, 84, 134
Patashnick, H. 1974, Natur, 250, 313
Peixinho, N., Thirouin, A., Tegler, S. C., et al. 2020, in The Trans-Neptunian Solar System, ed. D. Prialnik, M. A. Barucci, & L. Young (Amsterdam: Elsevier) 307
Prialnik, D. 1993, ApJl, 418, L49
Prialnik, D., Sarid, G., Rosenberg, E. D., & Merk, R. 2008, SSRv, 138, 147
Quinn, T., Tremaine, S., & Duncan, M. 1990, ApJ, 355, 667
Rettig, T. W., Tegler, S. C., Pasto, D. J., & Mumma, M. J. 1992, ApJ, 398, 293
Sarid, G., Prialnik, D., Meech, K. J., Pittichová, J., & Farnham, T. L. 2005, PASP, 117, 796
Sarid, G., Volk, K., Steckloff, J. K., et al. 2019, ApJl, 883, L25
Scotti, J. V., Rabinowitz, D. L., Shoemaker, C. S., et al. 1992, IAU Circ., 5434, 1
Senay, M. C., & Jewitt, D. 1994, Natur, 371, 229
Sickafoose, A. A., Levine, S. E., Bosh, A. S., et al. 2023, PSJ, 4, 221
Sickafoose, A. A., Bosh, A. S., Emery, J. P., et al. 2020, MNRAS, 491, 3643
Steckloff, J. K., Lisse, C. M., Safrit, T. K., et al. 2021, Icar, 356, 113998
Stern, A., & Campins, H. 1996, Natur, 382, 507
Tholen, D. J., Hartmann, W. K., Cruikshank, D. P., et al. 1988, IAU Circ., 4554, 2
Tiscareno, M. S., & Malhotra, R. 2003, AJ, 126, 3122
Torbett, M. V. 1989, AJ, 98, 1477
Torbett, M. V., & Smoluchowski, R. 1990, Natur, 345, 49
Weryk, R. J., Lilly, E., Chastel, S., et al. 2016, arXiv e-prints, arXiv:1607.04895
Whipple, F. L. 1950, ApJ, 111, 375
Whipple, F. L. 1951, ApJ, 113, 464
Wierzchos, K., & Womack, M. 2020, AJ, 159, 136
Wisdom, J., & Holman, M. 1991, AJ, 102, 1528
Womack, M., Sarid, G., & Wierzchos, K. 2017, PASP, 129, 031001
Womack, M., & Stern, S. A. 1999, SoSyR, 33, 187

Chapter 2

Formation of Planetesimals in the Outer Solar System

Anders Johansen, Michele T Bannister, Luke Dones, Seth Jacobson, Kelsi Singer, Kathryn Volk and Maria Womack

The solar system hosts the most studied and best understood major and minor planetary bodies—and the only extraterrestrial bodies to have been visited by spacecraft. The solar system therefore provides important constraints on both the initial stages of planetary growth, communicated to us by its surviving planetesimal populations, and for the final result of the planet formation process represented by the architecture of the system and properties of the individual planets. We review here models of planetesimal formation in the outer solar system as well as the wealth of recent observational constraints that has been used to formulate and refine modern planetesimal formation theory.

2.1 A Brief History of Planetesimal Formation Models

Observations and models of protoplanetary disks have given us a comprehensive insight into the growth of dust to pebbles and the radial drift of the pebbles. The planetesimal formation stage, in contrast, is hard to observe around other stars. The characteristic planetesimal size and birth-size distributions can in principle be inferred from dust production in debris disks, but the conclusions are highly dependent on the collisional cascade model and still give disparate answers (Krivov et al. 2018; Krivov & Wyatt 2021).

The formation of planetesimals is a key step in the planet formation process. Historically, planetesimal formation theory was dominated by two major views, namely that planetesimals in the solar system formed either by a gravitational instability of the sedimented mid-plane layer of grown dust aggregates (Safronov 1969; Goldreich & Ward 1973) or by gradual coagulation from dust to kilometer-sized bodies (Weidenschilling 1980). The gravitational instability model requires unrealistically high densities to be reached by sedimentation to the mid-plane, while

doi:10.1088/2514-3433/ada267ch2

the coagulation model is hampered by the poor sticking of macroscopic dust aggregates (Dullemond & Dominik 2005). We therefore focus this chapter on reviewing newer work on how the streaming instability drives planetesimal formation by giving rise to large fluctuations in the mid-plane dust density that facilitate the gravitational contraction to form planetesimals.

2.2 Overview of the Streaming Instability Model

The grand challenge for planetesimal formation models is to explain the growth from (sub-)micron-sized dust and ice particles to super-km-sized solid bodies (which we refer to as planetesimals). Reaching planetesimal sizes is an important accomplishment in planet formation, since the aerodynamical drag that causes mm-cm-sized pebbles to drift towards the star is strongly diminished when reaching size scales above a few hundred meters. The gas moves at slower than the Keplerian speed around the star, due to the outwards-declining pressure of the gaseous protoplanetary disk, causing dust particles to drift towards the star (Weidenschilling 1977). The drift timescale is proportional to the Stokes number of the particles, defined as $\mathrm{St} = \Omega t_s$ where Ω is the Keplerian frequency at a given distance from the star and t_s is the aerodynamical stopping time of the particle to gas drag (Johansen et al. 2014), when $\mathrm{St} \ll 1$. The Stokes number also determines the collision speed between particles induced by the turbulence (Ormel & Cuzzi 2007) and hence planetesimal formation theories are often formulated in terms of St rather than the particle size.

The protoplanetary disk that orbited our young Sun was endowed with an approximately 1% mass fraction of dust and ice particles inherited from the interstellar medium. Experiments and modeling of collisions between dust aggregates have identified fragmentation as an important barrier to dust coagulation within the protoplanetary disk (Chokshi et al. 1993; Brauer et al. 2008). The collision speed, v_c, between two equal-sized dust aggregates is given approximately by (Ormel & Cuzzi 2007; Birnstiel et al. 2012)

$$v_c = \sqrt{3}\sqrt{\delta}\sqrt{\mathrm{St}}\; c_s, \tag{2.1}$$

where δ is a dimensionless number that quantifies the strength of the turbulence (which we use here as a diffusion analogy to the standard parameter α that describes viscosity in accretion disks), the Stokes number St is (as defined above) a dimensionless number that characterizes the response time of the particles to gas motion, and $c_s = \sqrt{k_B T/\mu}$ is the sound speed of the gas (here k_B is the Boltzmann constant, T is the temperature, and μ is the molecular mass of the gas particles). For both silicate particles and ice particles, the fragmentation threshold speed, v_f, is likely approximately 1 m s^{-1} (Güttler et al. 2010; Musiolik & Wurm 2019). This yields a maximum Stokes number of:

$$\mathrm{St} = \frac{1}{3}\delta^{-1}\left(\frac{v_f}{c_s}\right)^2 = 0.01\left(\frac{\delta}{10^{-4}}\right)^{-1}\left(\frac{T}{100\ \mathrm{K}}\right)^{-1}\left(\frac{v_f}{1\ \mathrm{m\ s^{-1}}}\right)^2. \tag{2.2}$$

This value of St corresponds to dust aggregates between 0.1 and 1 mm in size in the trans-Neptunian region of the solar protoplanetary disk (Johansen et al. 2014). The fragmentation barrier may be circumvented if the dust monomers, the individual building blocks of dust aggregates, are small (sub-micron) and dominated by sticky ice (Okuzumi et al. 2012), but this contrasts with the dominantly micron-sized matrix particles found in meteorites (van Kooten et al. 2019; if the components found within meteorites are indeed representative of the outer solar system) as well as experiments showing that ice particles at low temperatures stick no better than silicates (Musiolik & Wurm 2019). More volatile ices irradiated by UV in the outer disk may nevertheless have higher sticking thresholds (Musiolik 2021), so we emphasize that the exact value of the critical fragmentation speed of dust/ice aggregates in protoplanetary disks is far from settled.

The limited Stokes number reached at the fragmentation limit (Equation 2.2) implies that planetesimal formation must most likely proceed beyond pebble sizes with some aid of gravity. For gravity to become important, dense clumps of pebbles must first emerge within the gas. Many mechanisms have been proposed for such pebble concentration (Johansen et al. 2014). Particles can concentrate passively within gaseous vortices and in pressure bumps (Barge & Sommeria 1995; Klahr & Bodenheimer 2003; Johansen et al. 2009b; Riols et al. 2020). It has also been shown experimentally and numerically that turbulent eddies at the smallest scales of a turbulent flow can concentrate dust particles between the eddies (Cuzzi et al. 2008), but this mechanism does not appear to be efficient at concentrating pebbles in large enough amounts to drive gravitational collapse (Pan et al. 2011). This small-scale particle concentration mechanism may also effectively become similar to trapping in vortices and pressure bumps when applied to larger scales of the turbulent flow (Hopkins 2016; Hartlep & Cuzzi 2020).

The streaming instability is a physical mechanism that is driven by the radial drift of the particles toward the star. This drift is, as discussed above, driven by the headwind from the gas, which moves slightly slower than the Keplerian speed due to its small radial pressure support. The particles concentrate actively in the gas flow by exerting a frictional force onto the gas, which accelerates the gas towards the Keplerian speed and thereby reduces the headwind on the particles (Youdin & Goodman 2005; Johansen & Youdin 2007; Bai & Stone 2010a). This causes pebbles to spontaneously pile up into dense filaments where the headwind speed of the gas is low. This emergence of dense filaments in computer simulations of the streaming instability is illustrated in Figure 2.1. The streaming instability in principle has positive growth rates for any combination of local dust-to-gas ratio ϵ and particle size parameterized through the Stokes number St. The regime of $\epsilon \ll 1$ is nevertheless strongly affected by the size distribution of the particles, with analytical growth rates reduced by orders of magnitude when considering a range of particle sizes rather than a single species (Krapp et al. 2019; Paardekooper et al. 2020, 2021). In contrast, the growth rate and non-linear evolution for $\epsilon \gtrsim 1$ appear relatively unaffected by the size distribution (Zhu & Yang 2021; Yang & Zhu 2021; Schaffer et al. 2021). This implies that significant dust sedimentation is required before the streaming instability will be able to operate to form dense pebble filaments.

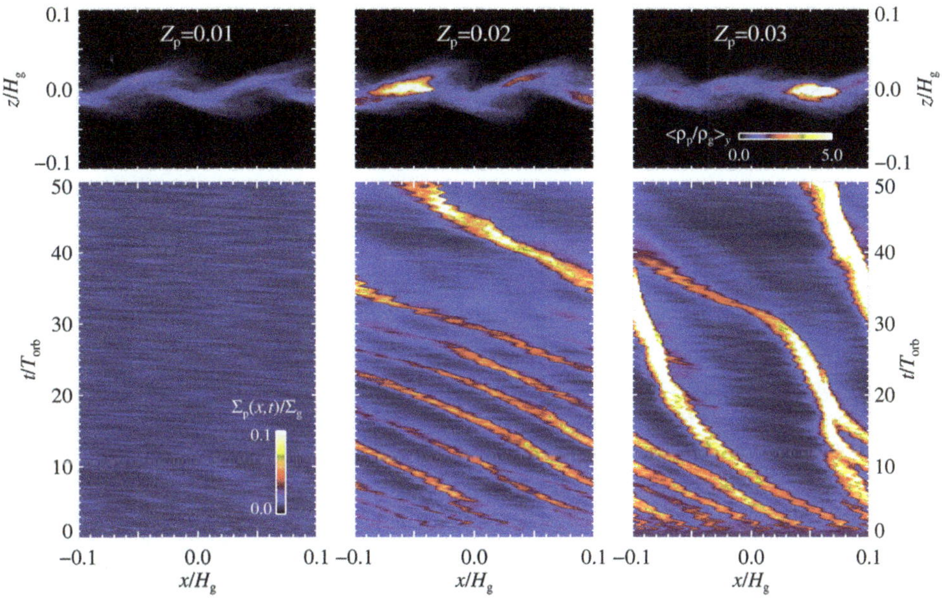

Figure 2.1. Illustration of the formation of dense pebble filaments by the streaming instability. Reproduced from Johansen et al. (2009b) © 2009. The American Astronomical Society. All rights reserved. The top panels show the density of pebbles in the mid-plane at the end of computer simulations that span a timescale of 50 orbits (x represents the radial direction in the disk and z the vertical direction; H_g is the scale height of the gas), while the bottom panels show spacetime plots of the pebble surface density. The three columns display results for three values of the pebbles-to-gas surface density ratio Z_p. The emergence of pebble-trapping filaments is clear for the two high-metallicity cases. These filaments move more slowly towards the star and thereby grow in mass by capturing free-drifting pebbles.

The strength of the turbulent diffusion in the protoplanetary disk is an important quantity that determines the efficiency of the streaming instability, since turbulence sets the scale height of the dust mid-plane layer (Johansen et al. 2014). Background turbulence is also important because it can diminish the growth rate of the streaming instability or even entirely eliminate the growth of the instability (Umurhan et al. 2020; Chen & Lin 2020; Estrada et al. 2022). The dust scale height H_p has been observed to be 10% of the gas scale height H_g in the outer regions of the well-resolved protoplanetary disk HL Tau (Pinte et al. 2016). These observations imply a small dimensionless turbulent diffusion coefficient in the range of $\delta \sim 10^{-4}$ to 10^{-3}. Recently, edge-on observations of another protoplanetary disk, Oph 163131, allowed estimates of a turbulent diffusion coefficient as low as $\delta \sim 10^{-5}$ (Villenave 2022). The turbulent diffusion coefficient can also be calculated from the radial width of pebble rings in protoplanetary disks (Andrews 2020). Here δ is inferred to be somewhat higher, $\delta \sim 10^{-3}$ (Dullemond et al. 2018), but this is still within the range of weak turbulence. The gas turbulence exterior of planetary gaps can nevertheless be enhanced as a consequence of hydrodynamical instabilities that drive extra turbulence at a planetary gap edge (Lyra et al. 2009) and hence may not be representative of the bulk disk properties.

Full hydrodynamical modeling is needed to assess the co-evolution of the streaming instability together with other sources of turbulence (e.g., Lesur et al. 2023). The streaming instability has been studied together with both weak and strong turbulence driven by the magnetorotational instability. Magnetorotational instability can operate where the magnetic field pressure is not too high compared to the thermal pressure (i.e., the magnetic field, likely anchored in the giant molecular cloud, must be weak enough as to not repress the growth of the instability) and is sufficiently ionized such that the magnetic field couples to the gas (Balbus 2003). In the case of weak turbulence ($\delta \sim 10^{-3}$), the streaming instability grows from initial particle concentrations in low-amplitude pressure bumps that form in the turbulence by an inverse cascade of kinetic and magnetic energy (Johansen et al. 2007, 2009a); the case of strong turbulence ($\delta \sim 10^{-2}$) showed no evidence for feedback-regulated particle concentration (Johansen et al. 2011). The magnetorotational instability is nevertheless likely to have been suppressed by the high resistivity to current in most parts of the protoplanetary disk (Turner et al. 2014; Desch & Turner 2015). If the magnetorotational instability operates only in the upper layers of the disk, where the resistivity is lower, then density waves penetrating through the mid-plane stir the pebble layer to large scale heights (Fromang & Papaloizou 2006). Weak pressure bumps forming in the gas were nevertheless again observed to act as seeds for feedback-regulated particle concentration in such setups (Yang et al. 2018; Xu & Bai 2022). The vertical shear instability is a prime candidate for driving turbulence in protoplanetary disks that does not depend on the resistivity level (Nelson et al. 2013). The vertical shear instability evolved without particle feedback on the gas stirs up the particle mid-plane layer to high values (Flock et al. 2020; Schäfer et al. 2020; Schäfer & Johansen 2022). Including the particle feedback on the gas, the pebbles nevertheless sediment to a thin mid-plane, where the friction exerted on the gas suppresses the vertical shear instability (Lin 2019). All in all, the streaming instability appears remarkably resilient to other sources of turbulence when evolved together in hydrodynamical simulations, unless the turbulence reaches very high strengths of $\delta \sim 10^{-2}$.

The growth rate of the streaming instability is normally analyzed in a 2-D (radial-vertical) configuration and for a fixed mean dust-to-gas ratio (Youdin & Goodman 2005; Umurhan et al. 2020), which ignores sedimentation of pebbles towards the mid-plane. Including this sedimentation in computer simulations has nevertheless demonstrated that the ratio of the pebble surface density to the gas surface density plays an important role in determining the non-linear outcome of the instability. Dense filaments only appear above a threshold metallicity (Johansen et al. 2009b; Bai & Stone 2010a). This threshold depends strongly on the Stokes number St (Carrera et al. 2015; Yang et al. 2017; Li & Youdin 2021) and on the strength of the background turbulence (Schäfer & Johansen 2022; Lim et al. 2024). For a realistic Stokes number of 0.01, translating in the outer solar system to a pebble size of 0.1–1 mm, the threshold is approximately 2% metallicity, almost twice solar. For lower values of the metallicity, the streaming instability still develops turbulence that stirs up the mid-plane layer, but dense filaments do not form under such conditions. Increasing the local metallicity to trigger the conditions for planetesimal formation by the streaming instability may be possible by ice condensation near the water ice line (Ros & Johansen 2013; Schoonenberg & Ormel 2017; Drążkowska & Alibert 2017; Ros et al. 2019; Estrada & Umurhan 2023; Ros & Johansen 2024) or by selective photoevaporation of the gas from the disk (Carrera et al. 2017; Ercolano et al. 2017).

2.3 Initial Mass Function of Planetesimals from Streaming Instability Models

The formation of planetesimals with a characteristic size of a few hundred kilometers provided an early success of the streaming instability (Johansen et al. 2007), given its correspondence to the typical size of asteroids in the main belt (Morbidelli et al. 2009). Since then, much effort has gone into quantifying the initial mass function of planetesimals (which is related to, but not necessarily the same as, today's observed size distribution) formed by the streaming instability (Johansen et al. 2015; Simon et al. 2016; Liu et al. 2020; Schäfer et al. 2017; Abod et al. 2019; Li et al. 2019). Resolving the mass distribution below the characteristic mass requires computer simulations of very high resolution because of the multiscale nature of the pebble density field within the filaments formed by the streaming instability (Johansen et al. 2012). The initial mass function appears to be a power law with the differential planetesimal number $\mathrm{d}N/\mathrm{d}M \propto M^{-1.6}$ (Johansen et al. 2015; Simon et al. 2016). The numerical value of the power law exponent is relatively independent of the physical parameters of the simulation (Simon et al. 2017). Interpreting the mass function of planetesimals forming near the simulation resolution limit is nevertheless not straightforward; a resulting lack of small planetesimals could either be due to insufficient resolution (so these small planetesimals would therefore appear once higher resolution simulations are performed) or due to an actual physical turnover of the initial mass function for small masses dictated by the gravitational dynamics of pebbles moving within the turbulent flow (Li et al. 2019; Klahr & Schreiber 2020).

Above the characteristic mass, the mass distribution in simulations appears to transition from a power law to an exponential tapering, a transition that has also recently been observed in the size distribution of cold classical trans-Neptunian objects (TNOs; Kavelaars et al. 2021). This transition makes physical sense, because the power-law exponent of -1.6 would otherwise integrate to infinite total mass. The largest planetesimal masses reach up to 100–1000 times the characteristic mass (Schäfer et al. 2017; Liu et al. 2019), i.e., corresponding to masses like those of Charon and Pluto, respectively, for typical simulation parameters. Simulations of the streaming instability on global disk scales will nevertheless be needed in the future to pin down the value of the most massive planetesimals, because the most massive planetesimals may form from rare, high-density clumps that are statistically unlikely to form in small-domain boxes. The most massive (and very rare) planetesimals residing at the end of the exponential tapering are also the most prone to continue to grow by pebble accretion (Liu et al. 2019; Lyra et al. 2023), with some contribution of planetesimal accretion only in the earliest growth decades (Lorek &

Johansen 2022), particularly in the inner regions of the protoplanetary disk. The mutual stirring between the growing protoplanets additionally helps to limit the number of protoplanets that grow to full planet mass (Levison et al. 2015).

The characteristic mass of planetesimals formed by the streaming instability has been proposed by Liu et al. (2020) to follow the simple scaling:

$$M = 5 \times 10^{-5} M_{\mathrm{E}} \left(\frac{\Gamma}{\pi^{-1}}\right)^{a+1} \left(\frac{H_{\mathrm{g}}/r}{0.05}\right)^{3+b}. \tag{2.3}$$

Here M_{E} is the mass of the Earth, $\Gamma = 4\pi G\rho_{\mathrm{g}}/\Omega^2$ (normalized here to $\Gamma = 1/\pi$) is a parameter that measures the strength of the gas and particle self-gravity (G is the gravitational constant, ρ_{g} is the mid-plane gas density, and Ω is the Keplerian orbital frequency), and H_{g}/r (normalized here to $H_{\mathrm{g}}/r = 0.05$) is the ratio of the gas disk's scale height to the distance from the Sun. We ignored here the weak dependence on the global pressure gradient and on the local metallicity and take $a = 0.5$ and $b = 0$, following Liu et al. (2020). We refer to Abod et al. (2019) for further discussion of these parameters. This expression is a simple consequence of a non-dimensionalization of the hydrodynamical equations that govern local computer simulations performed in the shearing box approximation. Other authors have proposed a characteristic mass that is partially derived from stability analysis of a self-gravitating protoplanetary disk (Li et al. 2019); this yields a similar scaling with H_{g}/r as in Equation (2.3) but with a much stronger dependence on Γ. Equation (2.3) can be transformed to give a characteristic diameter of

$$d = 1000\ \mathrm{km} \left(\frac{\rho_\bullet}{500\ \mathrm{kg\ m^{-3}}}\right)^{1/3} \left(\frac{\Gamma}{0.3}\right)^{1/2} \left(\frac{H_{\mathrm{g}}/r}{0.05}\right), \tag{2.4}$$

where $\rho_\bullet$ is the internal density of a planetesimal. A "pristine" protoplanetary disk with $\Sigma_{\mathrm{g}} \sim 10^3\ \mathrm{kg\ m^{-2}}$ in the trans-Neptunian region (so $\sim 10^4\ \mathrm{kg\ m^{-2}}$ at 1 au) has $\Gamma \sim 1$. This value for the pristine surface density corresponds approximately to the transition to gravitational instability by the Toomre criterion; hence planetesimal formation at larger values of Σ_{g} would require consideration of strong background turbulence caused by the gravitational instability (Gibbons et al. 2015).

In Figure 2.2 we show the characteristic planetesimal diameter from the streaming instability scaling as a function of distance from the star for three different evolution stages of the protoplanetary disk, assuming a low interior density of $500\ \mathrm{kg\ m^{-3}}$. We also indicate the largest object in the cold classical TNO population as well as the characteristic size where the power-law size distribution for cold classical TNOs appears to transition to an exponential tapering (Kavelaars et al. 2021); to transform from the observationally measured absolute magnitude H to an estimated size, we assumed an albedo of $p = 0.06$. The cold classical population fits well with a formation in a protoplanetary disk that is depleted by a factor 10–100 relative to a pristine protoplanetary disk (which we consider here for simplicity to have approximately $\Sigma_{\mathrm{g}} \sim 10^4\ \mathrm{kg\ m^{-2}}$ at 1 au, with $1/r$ scaling with distance). Formation at such a late evolutionary stage of the protoplanetary disk is

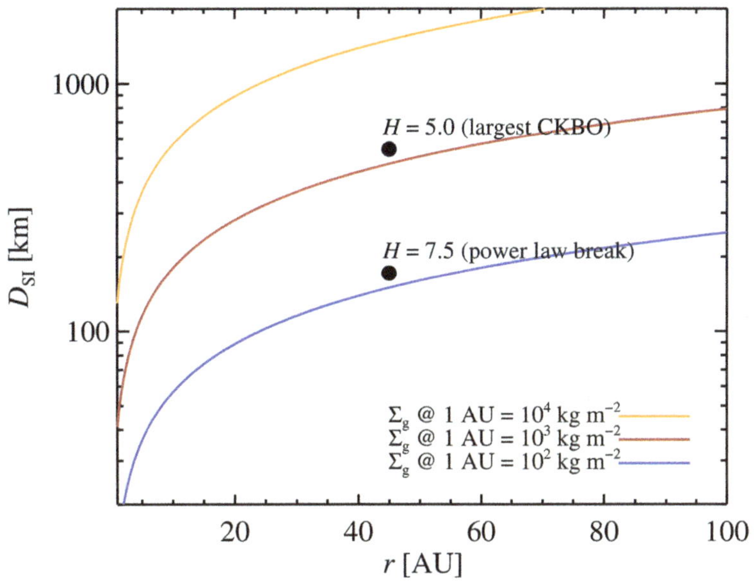

Figure 2.2. The characteristic planetesimal diameter formed by the streaming instability, plotted as a function of the distance from the star for three values of the gas column density at 1 au (Liu et al. 2020). We also show the diameter conversion from the absolute magnitude H of the largest objects among the very low inclination cold classical TNO population ($H = 5$) and the fitted power law break ($H = 7.5$; Kavelaars et al. 2021). These correspond well to the planetesimals formed by the streaming instability at the late stages of protoplanetary disk evolution, where the characteristic column density at 1 au has fallen from 10^4 kg m^{-2} to between 10^2 kg m^{-2} and 10^3 kg m^{-2}.

nevertheless still not consistent with the total mass of the cold classicals of only a few times 10^{-3} Earth masses (Napier et al. 2024); even a narrow planetesimal formation region of 1 au width yields 10 times more mass than that for a depletion factor of 10. However, the scaling law for the planetesimal mass has not been tested for really low values of Γ (e.g., 0.01, corresponding to 100 times depletion from a pristine protoplanetary disk) and it may well be that for very low Γ, the efficiency of turning pebbles into planetesimals is also low. Also, a steepening pressure gradient toward the outer regions of the solar protoplanetary disk would naturally lead to less efficient planetesimal formation (Bai & Stone 2010b; Abod et al. 2019; Baronett et al. 2024). Alternatively, the outward migration of Neptune by planetesimal scattering could have depleted the cold classical population by a factor 10 or more (e.g., Gomes et al. 2018).

We therefore envision a formation of the classical Kuiper Belt late in the evolution of the protoplanetary disk. Perhaps the outer disk lost gas mass by far-ultraviolet photoevaporation at this stage and this put the metallicity above the threshold (Carrera et al. 2017). This late formation could also explain how the densities remained low (Brown 2013), by avoiding heating by ^{26}Al that melted the ice and rock of asteroids. Cañas et al. (2024) explain the trend of the largest objects having higher densities ($\approx$ 2000 kg m^{-3}) with a model in which the first planetesimals are primarily icy, but accrete silicate pebbles after ^{26}Al has decayed.

While the observed properties of the cold classical TNOs discussed above seem consistent with streaming instability formation models, the largest contributors to the Centaur population are the dynamically "hot" TNOs, particularly the scattering TNOs (Levison & Duncan 1997; Duncan & Levison 1997; Di Sisto & Brunini 2007; Volk & Malhotra 2008; Nesvorný et al. 2017, 2019b; Di Sisto & Rossignoli 2020; see also Chapter 3). The dynamically hot TNOs are generally assumed to have formed at smaller heliocentric distances ($r \lesssim 30$ au) and been transported outwards to their present-day orbits during the epoch of giant planet migration (see, e.g., Gladman & Volk 2021 for a recent review). The streaming instability scaling model presented in Figure 2.2 predicts that planetesimal birth diameters are significantly smaller in the 10–30 au region compared to the classical Kuiper Belt objects. The large sizes present among the "hot" TNOs are consistent with formation closer to the Sun at earlier epochs of the protoplanetary disk where the gas and pebble column densities were larger.

2.4 Observational Constraints for Likely Formation Regions of Modern Centaurs

Clues to the histories of Centaurs may be found from both telescopic surveys and from impact crater populations on outer solar system bodies. Other useful observations for testing solar system formation models include individual Centaur sizes, shapes and densities, colors, volatile and dust output, and levels of activity (Fraser et al. 2024; Knight et al. 2024). Here we highlight some key data sets and discuss some implications.

2.4.1 Size Distributions of TNOs and Centaurs as Related to Formation Scenarios

Constraints on the size distributions of the Centaurs and their source TNO populations come from a variety of observations. Optical telescopic surveys provide constraints for objects above ~100 km in diameter, with sizes inferred from the measurement of an object's absolute magnitude (H). As discussed above, the cold classical TNO population is consistent with a power law size distribution that transitions to an exponential cutoff with no objects larger than ~400 km (Kavelaars et al. 2021). The hot population that feeds the Centaurs has recently been found to share a very similar size distribution for the ~100–400 km diameter size range (Petit et al. 2023), though it does not share the ~400 km upper size limit with the cold population; all of the largest known TNOs, including the dwarf planets, belong to the hot population (e.g., Schwamb et al. 2014). Constraints on the smaller-end size distribution from TNO surveys are currently much weaker due to how faint smaller TNOs are at their large heliocentric distances.

Because the Centaur population is closer to the Sun than the TNOs, their size distribution can be measured to larger (i.e., fainter) H magnitudes. Their closer-in orbits also make them more amenable to more direct size measurements through a variety of techniques including thermal infrared measurements and stellar occultations. As of March 2024, the sizes of only about five of the largest Centaurs have been measured by occultations with diameters ranging from ~60 to 250 km (see

Chapter 9), providing some limited size overlap with the well-measured TNO size distribution. Thermal infrared measurements have also been used to infer the diameters of dozens of Centaurs ranging down to $d \approx 20$ km (Bauer et al. 2013). For most Centaurs, however, only photometry at visual wavelengths is available. Assuming a geometric albedo of 0.08 (Bauer et al. 2013), the Centaur size distribution can be inferred for bodies as small as $d \approx 3$–5 km, which is still larger than the well-measured portion of the Jupiter-family comet size distribution. See Chapter 4 for a full discussion of constraints from the intermediate-sized and small Centaurs.

Another source of information about Centaurs and their source populations for objects smaller than $\sim$100 km in diameter comes from the cratering records produced by their impacts on Pluto and Charon, the cold classical TNO Arrokoth, and the satellites of the giant planets. The giant planet satellites in theory should record the Centaur population and the (primarily) scattering TNOs from which they derive more directly than the Pluto system and Arrokoth, which interact with the full range of TNO populations. However, the giant planet satellite surfaces may also record planetocentric debris in some cases, and each satellite has its own unique geologic history which can complicate interpretations of the cratering record. Thus, a comparison to the Pluto system and Arrokoth is still useful. Table 2.1 gives rough size ranges for the smallest and largest craters sampled in each giant planet system and for the worlds visited by New Horizons—Pluto/Charon and Arrokoth.

The smallest crater in each case is limited by image resolution and/or lack of image coverage and is not indicative of the actual smallest crater/impactor on the surface of each of these bodies. We use here the smallest crater sizes that are believed to be from primary impacts (i.e., not likely to be a secondary crater generated from ejecta of a primary impact), and that are well sampled (i.e., we did not consider them if there was just one very high-resolution image with one or a few craters in it). The representative largest craters noted here are features that are clearly visible today, and useful for understanding the shape of the size-frequency distribution up to a given diameter. Again, these are not necessarily the largest craters that have ever formed on these bodies, as those ancient impacts have been erased in many cases. Neptune's large moon Triton has a young and enigmatic surface (Schenk & Zahnle 2007; Mah & Brasser 2019), so we do not include the Neptune system in this analysis. The scaling to impactor sizes from crater sizes depends on the impactor and target properties, such as impact speed, body gravity, and composition, and is different for each body (see, e.g., Holsapple 1993; Housen & Holsapple 2011; see also a discussion of impact cratering in the context of laboratory experiments in Chapter 13). Here we used the same methods as described in the supplement of Singer et al. (2019) for two different endmember materials, a "solid" ice surface and icy regolith, and reran the derivation for each satellite's average impact velocity (Zahnle et al. 2003; Nesvorný et al. 2023) and surface gravity. We give the approximate size ranges of impactors recorded in each satellite system in Table 2.1. The very smallest impactors are tens of meters in diameter, forming the smaller craters on the smaller satellites. More typically, however, the impactors are hundreds of meters in diameter, up to tens of km for the largest impactors.

Table 2.1. Size Ranges For Craters And Impactors For Outer Planet Systems

Location	Smallest Crater Diameter Well Sampled	Representative Largest Crater Diameters	Impactor Diameter Range	Image Sources
Jupiter	~1 km[1]	~150 km (several on Ganymede and Callisto)[2] and ~570 km (Gilgamesh on Ganymede)[3]	~30 m to 10s of km (~50 km for Gilgamesh)	Galileo and Voyager 1 and 2 missions
Saturn (Mid-sized Satellites)	~1 km[4]	~300–600 km (~12 across the satellites with the largest being Turgis basin on Iapetus)[5]	~10–70 m for the smaller craters and 10s of km for the larger basins (up to ~70 km for Turgis)[6]	Cassini and Voyager 1 and 2 missions
Uranus	~1.5 km[7]	~100–340 km[7] (~20 across the satellites, with the largest Gertrude basin on Titania)	10s of m to 10s of km (up to ~40 km for Gertrude on Titania)	Voyager 2 mission
Pluto and Charon	~1.5 km[8]	~250 km (Burney basin)[8] and ~1000 km (Sputnik basin)[9]	~150 m to 40 km[8] (for Burney Basin)	New Horizons mission
Arrokoth	~0.35 km[10]	~7 km (Sky crater)[10]	10s of m to ~0.7 km[11]	New Horizons mission

[1] The smallest and least eroded are on Europa: e.g., Bierhaus et al. (2009)
[2] e.g., Schenk (2002)
[3] Estimated equivalent diameter, Schenk et al. (2004)
[4] This value is for some areas across most of the satellites, e.g., Kirchoff & Schenk (2009, 2010); Bierhaus et al. (2012)
[5] Kirchoff & Schenk (2008); Kirchoff et al. (2018)
[6] Range for this size of craters on various satellites and a range of target materials (from solid ice to icy regolith)
[7] Kirchoff et al. (2022)
[8] e.g., Singer et al. (2019)
[9] e.g., McKinnon et al. (2017)
[10] e.g., Spencer et al. (2020)
[11] McKinnon et al. (2022)

At the smallest sizes, the cratering record yields information on objects at least an order of magnitude smaller (and often two orders of magnitude smaller) than telescopic surveys of the TNO or Centaur populations. There is some overlap between the largest Centaur impactors recorded by craters and the smallest Centaurs and TNOs observed telescopically in the tens of km range. The largest craters are few in number, and the smallest Centaurs are also difficult to observe, so there can be large uncertainties in both types of observations where the overlap occurs. For ease of general comparison, here we summarize the crater size-distribution shapes in terms of one basic parameter: the average differential power-law slope over a given crater size range. A differential slope of −3 (−2 cumulative) is often seen and thus used as a reference as the "typical" or "standard" slope (see Figure 2.3) that all

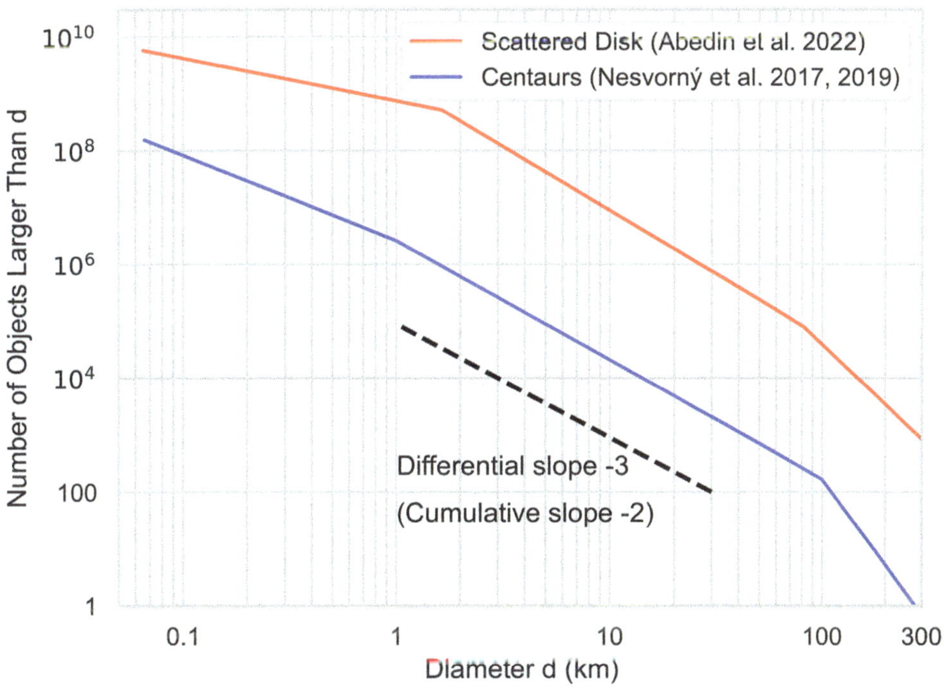

Figure 2.3. Schematic size distributions for scattering TNOs (Abedin et al. 2022) and Centaurs (Nesvorný et al. 2017, 2019b). The size distributions are based on the cratering records of the giant planet satellites for diameters d from ~10 m up to ~70 km, from telescopic observations of Centaurs for $d \gtrsim$ 3–5 km, and from telescopic surveys of TNOs for $d \gtrsim$ 20–50 km (see Table 2.1). Abedin et al. (2022) assume there are 78,000 ± 17,000 "scattering" objects with absolute magnitude $H_r < 8.5$ and slopes α of the magnitude distribution of 0.7, 0.45, and 0.15 between H_r values of 5.5–8.5, 8.5–17, and 17–24, respectively. These values correspond to cumulative power-law size distributions with exponents $\gamma = 5\alpha$ equal to 3.5, 2.25, and 0.75 for diameter ranges of 328–82 km, 82–1.6 km, and 1.6–0.065 km, respectively, assuming a geometric albedo of 0.08. Nesvorný et al. (2017) assume a cumulative size distribution with γ equal to 5, 2.1, and 1.5 for diameter ranges of 300–100 km, 100–1 km, and less than 1 km, respectively, and an estimated population of 21,000 ± 8,000 Centaurs with diameters larger than 10 km, assuming a geometric albedo of 0.06. The population of Centaurs at a given size is ≈ 100–1000 times smaller than the number of scattered disk objects, due to the Centaurs' short dynamical lifetimes (Irwin et al. 1995; Jewitt et al. 1996).

others are either steeper than (e.g., −4) or shallower than (e.g., −2). Additionally, most scaling theories would introduce a slight change in the size distribution slopes in the conversion from crater to impactors (e.g., as described in more detail in Singer et al. 2021 for Pluto impactors). The impactor slope is slightly shallower than that of the craters. For comparison with the IMF obtained from streaming instabilities simulations discussed in Section 2.3, we note that a cumulative power law in diameter $N_{>} \propto d^q$ corresponds to a differential power law in mass $dN/dM \propto M^{q'}$, where $q' = q/3 - 1$, so that $q = -2$ yields $q' = -5/3$ in terms of mass.

The cratering record at Pluto and Charon primarily records impacts from the hot classical TNO population and the rest of the Plutino population rather than from the scattering population (Greenstreet et al. 2015, 2016, 2023). While this means it is not directly probing the dominant Centaur reservoir, the hot classical and resonant TNO populations were likely sourced from a similar region of the planetesimal disk as the scattering TNOs. The largest craters on Pluto/Charon (>10 km craters, or >1 km impactors) display a −3 average slope, similar to the slope inferred from optical surveys for TNOs smaller than about 100 km (Bernstein et al. 2004; Fraser et al. 2014). There is a distinct break to a shallower slope (approximately −1.7) for smaller craters (<10 km craters, or <1 km impactors) on a wide range of geologic terrains (Singer et al. 2019, 2021). The impacting population at Arrokoth is dominated by the moderately dynamically excited "stirred" classical TNOs (see full discussion in Greenstreet et al. 2019). All of the measured craters at Arrokoth are made by impactors less than ≈ 1 km in diameter and display a shallow slope consistent with the Pluto/Charon record (Spencer et al. 2020; McKinnon et al. 2022).

The main impacting small bodies on the giant planet satellites are Jupiter-family comets, Centaurs, and, in the case of Neptune's satellites, the scattering TNOs themselves (Zahnle et al. 2003; Nesvorný et al. 2023). In the Jupiter system, the surfaces where smaller primary craters can be observed are generally the young surfaces on Europa and Ganymede. These terrains show a similar trend to that seen on Pluto and Charon, where the more typical average −3 slope is seen for larger craters, and there is a transition to a similar shallower slope (−2 or shallower) below which craters correspond to an impactor diameter of ≈ 1 km (Schenk et al. 2004; Bierhaus et al. 2009; Singer et al. 2019). The crater distributions at Saturn generally show "typical" or somewhat steeper slopes at large sizes (> a few km) and become shallower at smaller sizes (Kirchoff & Schenk 2009, 2010; Bierhaus et al. 2012; Robbins et al. 2024). These shallow craters occur at impactor sizes of tens to a few hundred meters, which is about an order of magnitude smaller than the ~1 km transition size in the Pluto/Charon system. However, the Saturn system is complex, with different geologic histories for the inner versus outer satellites, the potential for planetocentric crater populations, and ongoing discussions of the origin and overall age of the satellites (Dones et al. 2009; Kirchoff et al. 2018). Thus, it is harder to know what portion of the craters in the Saturn system is representative of the heliocentric Centaur impactor population. Miranda is the Uranian satellite where craters smaller than ≈ 10 km can best be mapped (Strom 1987; Plescia 1988; Kirchoff et al. 2022). Different terrains across Miranda show different size-distribution slopes. The older cratered terrains show a break to a shallow slope similar to that on Pluto/Charon, but at a slightly smaller corresponding impactor size of $\approx$ 300–500 m. However, the craters superposing younger tectonized terrain remain closer to the typical -3 slope even at small crater sizes. The reason for these differences with terrain age are not known, but given the similarity of the younger terrains to Pluto/Charon, they may better represent the heliocentric impactor population (Kirchoff et al. 2022).

Many of these surfaces in the outer solar system are young and show no obvious signs of geologic processes preferentially erasing small craters (e.g., Singer et al. 2019; Bottke et al. 2024). This means the craters record the more recent impact flux with relatively fewer small impactors. The shallower slope at small impactor sizes seen on many surfaces across the outer solar system could indicate that smaller objects were not made in large abundances during either their formation or subsequent collisional evolution. The alternative would be that many small Centaurs and scattering TNOs were made throughout solar system history, but somehow the smaller ones were preferentially removed before impacting these surfaces. The latter scenario seems difficult, given there are currently no known processes that would preferentially remove smaller objects, especially across many TNO sub-populations and from $\approx$ 5 to 50 au. Disruption of small comets with perihelion distances $q_p \lesssim 2$ au does proceed rapidly (Jewitt 2021), but this appears to result from spin-up due to outgassing torques. All known active Centaurs have $q_p < 12$ au (Jewitt 2009; Lilly et al. 2024), so it seems unlikely that small objects can be removed all the way out to 50 au.

If the deficit of objects below about 1 km in diameter seen in the crater records indicates that not as many of those objects formed originally, this could be consistent with direct formation of large TNOs via streaming instabilities and gravitational collapse (Singer et al. 2019) rather than by hierarchical growth from small bodies (Schlichting et al. 2013). For the streaming instability IMF to be consistent with the apparent lack of small planetesimals, this seems to also require a lower-mass cutoff to the bodies that can form by the streaming instability, maybe caused by diffusive outwashing of the smallest pebble clumps (Klahr & Schreiber 2020).

2.4.2 Spectra and Colors of Centaurs and Other Relevant Bodies

Observational clues to chemical compositions and physical conditions may come from icy and refractory abundances measured on the surfaces of inactive Centaurs and in the comae of active Centaurs. For example, some interesting differences were recently found using millimeter- and infrared wavelength spectroscopy of CO and CO_2 produced in the coma of a highly processed Centaur (39P/Oterma) and one that has yet to enter the inner (<4 au) solar system (29P/Schwassmann-Wachmann); however, volatile species have been detected in only a few Centaurs, so it is difficult to draw broader conclusions about Centaurs' chemical compositions (Womack et al. 2017; Wierzchos et al. 2017; Bockelée-Morvan et al. 2022; Harrington Pinto et al. 2023). The relative gas abundance ratio of the highly volatile species CO and CO_2 is high in 29P but much lower in 39P and this may be partly attributed to the objects' very different orbital histories (see discussion in Chapter 3) and thus the different amounts of solar radiation processing they experienced (Harrington Pinto et al. 2022, 2023). Measurements of the relative abundances of CO, CO_2, and other key species are needed in many more Centaurs in order to put tight constraints on formation regions as well as physical evolution models and is discussed further in Chapters 6 and 15.

Unfortunately, most Centaurs are too small and too far from the Sun to readily obtain useful spectra for any but the brightest objects. When objects are too faint for spectroscopic techniques, an alternative way to gain insight is by analyzing their "colors," which are calculated from the measured magnitudes of objects as seen through narrowband or broadband filters. Because this technique collects much more light than most spectra, one can obtain data for much fainter objects. Color diagrams can be used to analyze large-scale surface and gaseous properties of larger ensembles. This allows for an improved description of the Centaur population and can serve as the basis for statistically significant tests for correlations with other observed characteristics and can constrain models of Centaur and TNO formation, which we briefly discuss (see Chapter 5 for further discussion).

The first two Centaurs to be identified, (2060) Chiron and (5145) Pholus, span almost the full range of optical colors seen at visual wavelengths for outer solar system bodies. Chiron's nucleus surface is a nearly gray, or neutral, reflector (e.g., $B–R \sim 1$), while Pholus' is very red ($B–R \sim 2$), once again proving that it is impossible to assess a new population with only two members. Initial Centaur surveys indicated that their colors had a bimodal distribution, but it was unclear whether the two color populations were the result of (1) evolutionary processes such as radiation-reddening, collisions, and sublimation or (2) a primordial, temperature-induced, composition gradient (Peixinho et al. 2003; Tegler et al. 2008; Peixinho et al. 2012).

Intercomparing the colors of Centaurs with those of hot TNO populations can provide strong diagnostics for models (Brown et al. 2011). Interestingly, one of the first color–magnitude surveys of TNOs also found evidence for two distinct populations. Doressoundiram & Peixinho (2002), Doressoundiram et al. (2003), and Trujillo & Brown (2002) found that most cold classicals were very red, while "hot" classicals had a wide range of colors. To explain both the dynamically unexcited orbits of the cold classicals as well as their dominantly red surface colors, they proposed, following Levison & Stern (2001), that the cold classicals formed *in situ*, while the other TNO populations had been implanted from a region of the protoplanetary disk interior to the present-day belt during planetary migration (e.g., Malhotra 1993). Outward migration of the hot TNO populations was elaborated on in the "Nice model" (Tsiganis et al. 2005; Morbidelli et al. 2005; Gomes et al. 2005) and subsequent models in which the giant planets underwent an orbital instability (see reviews by Dones et al. 2015 and Nesvorný 2018). The different proposed migration histories of the giant planets have important implications for the original formation locations of the hot TNOs that feed the present-day Centaur population which may be reflected in their surface colors.

One important test will be to see whether the colors of a large sample of Centaurs are consistent with hot TNO populations (Brown et al. 2011). Peixinho et al. (2003) initially argued that Centaurs have a bimodal color distribution and TNOs do not.

Peixinho et al. (2012) found that both Centaurs and *small* TNOs ($H_r \gtrsim 6.8$, corresponding to diameters $\lesssim$165 km for an assumed geometric albedo of 0.09) had bimodal color distributions, i.e., the bimodality depended on size and not dynamical classification. Wong & Brown (2017a) confirmed this result for small TNOs. However, Figure 2.4 shows the distribution of *B–R* colors measured by Tegler et al. (2016) for a larger survey of 61 Centaurs (twice as many as in the Peixinho analysis). The red peak is broader than it was for the smaller samples of objects in earlier studies with smaller sample sizes, and the color distribution is not bimodal at a statistically significant level (see Chapter 5 for further discussion of the color bimodality). Nonetheless, the greater prevalence of very red Centaurs, as compared with main-belt asteroids (Hasegawa et al. 2021), Jupiter Trojans, and Hildas (Wong & Brown 2015, 2017b) still needs an explanation. Interestingly, no Jupiter-family comets are found with ultra-red surfaces that are commonly seen in Centaurs; some have attributed this difference to the cometary activity of JFCs (Jewitt 2002; Grundy 2009). Melita & Licandro (2012) added that many gray Centaurs likely had cometary activity and a more detailed comparison between active Centaurs and Jupiter-family comets will be useful for constraining models (see Chapter 8 for a discussion of activity).

Many recent works argue that planetesimals in the outer solar system formed with a range of compositions, but with little variation with distance from the Sun. Others propose that the range of colors seen today could result from distance-dependent sublimation of ices due to solar heating (Davidsson 2021; Parhi & Prialnik 2023),

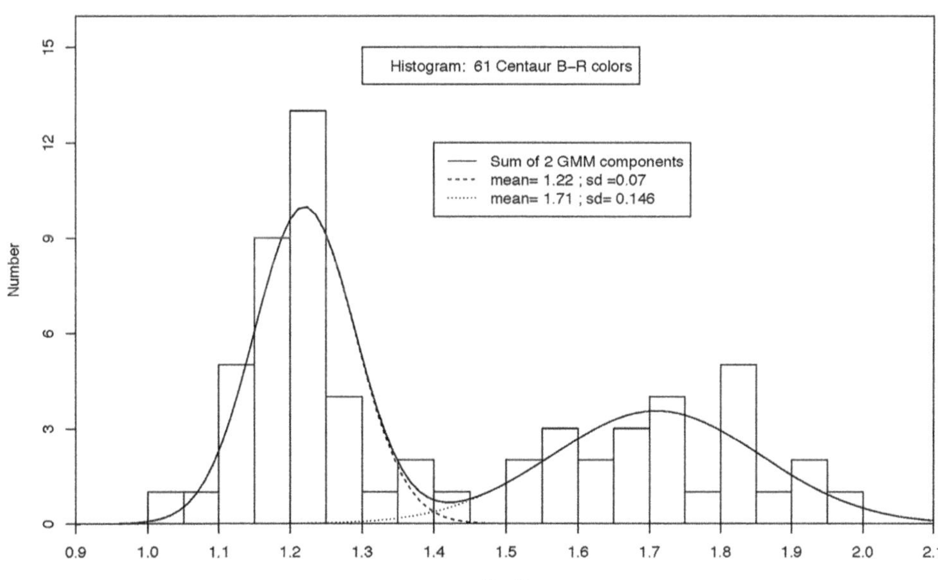

Figure 2.4. The distribution of *B–R* colors measured for 61 Centaurs. The color distribution is not bimodal at a statistically significant level, although a sample size of two or three times larger may be necessary to further test for bimodality. Including this sample, a bimodal color distribution is seen for intrinsically faint (small) and intrinsically bright (large) Centaurs and TNOs, consistent with Peixinho et al. (2012). Reproduced from Tegler et al. (2016) © 2016. The American Astronomical Society. All rights reserved.

followed by irradiation, which we describe below. For example, Brown et al. (2011) proposed that in the innermost primordial Kuiper Belt, water, carbon dioxide, and possibly hydrogen sulfide were the only thermally stable ices. Irradiation would chemically process this mixture into a low-albedo, gray substance. Beyond about 20 au, methanol is also stable, and irradiation would yield red planetesimals with somewhat higher albedos. Finally, beyond about 35 au, ammonia would also be stable, irradiation of which would (hopefully) produce the unique properties of cold classicals. Color measurements, as well as spectroscopic studies, can provide important tests for these models.

Compositional information of Centaur nuclei and comae are expected to be useful to tie back to the formation location in the disk, and thus the conditions for the streaming instability. Additional discussion about Centaur surfaces (e.g., Luu 1993; Campins et al. 1994) and compositions obtained from optical and infrared spectra of Centaur surfaces is provided in more detail in Chapter 5. With the operation of JWST and other next-generation optical telescopes, many Centaur observing programs are underway, which will very likely provide unprecedented spectra for many more Centaurs, as discussed in Chapter 14.

2.4.3 Additional Useful Observational Constraints

Other vital information for constraining solar system formation models may be obtained from observations of Centaurs' shapes, densities, and the prevalence of binaries and/or moons. We briefly consider some of these characteristics here.

We are not aware of specific predictions for Centaurs, although Fraser et al. (2017) suggested that all planetesimals born near the Kuiper Belt formed as binaries, and Nesvorný et al. (2019a) asserted that wide TNO binaries provided evidence for the streaming instability. Thus, the extent to which Centaur binaries may exist is also likely to be an important modeling constraint. Not much is known about how many binary Centaurs exist, although the planet-crossing objects (42355) Typhon/Echidna (semimajor axis $a = 37.5$ au, perihelion distance $q_p = 17.5$ au) and (65489) Ceto/Phorcys ($a = 99.2$ au, $q_p = 17.7$ au) are considered to be Centaur binaries for some extended definitions of the Centaur population (Araujo et al. 2018; Grundy et al. 2007); these two binaries are discussed in detail in Chapter 9. Santos-Sanz et al. (2012) infer low densities for these systems, 360^{+80}_{-70} and 640^{+160}_{-130} kg m^{-3}, respectively. The masses of these systems are known to better than 10%, but the sizes of the objects are less certain. For comparison, Berthier et al. (2020) infer a density of 810 ± 160 kg m^{-3} for the Jovian Trojan binary (617) Patroclus/Menoetius, whose mass is comparable to the masses of the Typhon/Echidna and Ceto/Phorcys systems, while Carry et al. (2023) find a density of 830 ± 50 kg m^{-3} for the less massive bilobate Trojan (17365) Thymbraeus; see Mottola et al. (2024) for a review.

Centaur shapes are still poorly known. Interestingly, many TNOs and JFCs with well-determined shapes from spacecraft encounters and radar have two lobes (see, e.g., Chapter 12). It is unknown whether these bilobate shapes are primordial or due to physical evolution (Hirabayashi et al. 2016; Steckloff et al. 2016; Schwartz et al. 2018; Vavilov et al. 2019; Safrit et al. 2021; Lorek & Johansen 2024). However, the two-lobed structure of the cold classical TNO Arrokoth is believed to be from its initial formation (McKinnon et al. 2020; Spencer et al. 2020; Lyra et al. 2021) and not later breakup and re-accretion as many asteroids binaries are thought to be. Given that Centaurs represent the transitional state between TNOs and JFCs, it would not be surprising to find some bilobate Centaurs. None have been definitively detected thus far, though very recent results by Faggi et al. (2024) suggest that Centaur 29P/SW1 is bilobate. Observations like these and many others may also help us determine which size bodies are primordial, rather than collisional fragments (Morbidelli & Rickman 2015; Benavidez et al. 2022; Jutzi et al. 2017; Davidsson et al. 2016). For collisionally evolved objects, the survival of hypervolatiles in catastrophic collisions in the early solar system has been addressed (e.g., Davidsson 2023) and has implications for Centaurs.

2.5 Future Priorities for Understanding Planetesimal Formation in the Solar System

Planetesimal formation theory has benefited enormously from observations of asteroids, Centaurs, and TNOs, both from the ground and from space-based facilities and missions. In particular, the size distributions and binary and shape properties of planetesimals in the solar system provide excellent measures to test against computer simulations of planetesimal formation. Many aspects of planetesimal formation by the streaming instability are now relatively well-understood (such as the binary properties and the characteristic masses), while others will need significant additional work to fully map (such as the dependence on the background turbulence and the timing of planetesimal formation in the protoplanetary disk). Probing the initial mass function of planetesimals down to much smaller diameters (10 km or even 1 km) will require computer simulations that resolve all the relevant scales of particle concentration and the gravitational collapse. Adaptive mesh refinement methods, such as used in Schäfer et al. (2020), may be a key technology to exploit to reach effective higher resolutions without the necessity of excessive computational resources.

Similarly, measuring the Centaur and TNO source population size distributions down to smaller sizes with telescopic surveys may provide critical constraints on planetesimal formation. Recent and future Earth- and space-based telescopes such as the NASA James Webb Space Telescope, the NSF-DOE Vera Rubin Observatory, and the Nancy Grace Roman Space Telescope will greatly assist in understanding the size distribution of these objects below 50 km, and the latter can even push well into the sub-km regime. Greater sampling of the spectra and color of Centaurs and TNOs will also greatly enhance our statistical understanding of their compositions and also help us understand how much post-formation processing these objects have experienced.

Acknowledgements

We thank two anonymous referees for comments which improved this manuscript. This material is based in part on work done by M.W. while serving at the National Science Foundation. K.V. acknowledges support from NASA (grants 80NSSC21K0376, 80NSSC23K0680, 80NSSC23K1169, and 80NSSC23K0886).

References

Abedin, A. Y., Kavelaars, J. J., Petit, J.-M., et al. 2022, AJ, 164, 261

Abod, C. P., Simon, J. B., Li, R., et al. 2019, ApJ, 883, 192

Andrews, S. M. 2020, ARA&A, 58, 483

Araujo, R. A. N., Galiazzo, M. A., Winter, O. C., & Sfair, R. 2018, MNRAS, 476, 5323

Bai, X.-N., & Stone, J. M. 2010a, ApJ, 722, 1437

Bai, X.-N., & Stone, J. M. 2010b, ApL, 722, L220

Balbus, S. A. 2003, ARA&A, 41, 555

Barge, P., & Sommeria, J. 1995, A&A, 295, L1

Baronett, S. A., Yang, C.-C., & Zhu, Z. 2024, MNRAS, 529, 275

Bauer, J. M., Grav, T., Blauvelt, E., et al. 2013, ApJ, 773, 22

Benavidez, P. G., Campo Bagatin, A., Curry, J., Álvarez-Candal, Á, & Vincent, J.-B. 2022, MNRAS, 514, 4876

Bernstein, G. M., Trilling, D. E., Allen, R. L., et al. 2004, AJ, 128, 1364

Berthier, J., Descamps, P., Vachier, F., et al. 2020, Icar, 352, 113990

Bierhaus, E. B., Dones, L., Alvarellos, J. L., & Zahnle, K. 2012, Icar, 218, 602

Bierhaus, E. B., Zahnle, K., & Chapman, C. R. 2009, in Europa, ed. R. T. Pappalardo, W. B. McKinnon, & K. K. Khurana (Tucson, AZ: Univ. Arizona Press) pp. 161–180

Birnstiel, T., Klahr, H., & Ercolano, B. 2012, A&A, 539, A148

Bockelée-Morvan, D., Biver, N., Schambeau, C. A., et al. 2022, A&A, 664, A95

Bottke, W. F., Vokrouhlický, D., Nesvorný, D., et al. 2024, PSJ, 5, 88

Brauer, F., Dullemond, C. P., & Henning, T. 2008, A&A, 480, 859

Brown, M. E. 2013, ApL, 778, L34

Brown, M. E., Schaller, E. L., & Fraser, W. C. 2011, ApL, 739, L60

Cañas, M. H., Lyra, W., Carrera, D., et al. 2024, PSJ, 5, 55

Campins, H., Telesco, C. M., Osip, D. J., et al. 1994, AJ, 108, 2318

Carrera, D., Gorti, U., Johansen, A., & Davies, M. B. 2017, ApJ, 839, 16

Carrera, D., Johansen, A., & Davies, M. B. 2015, A&A, 579, A43

Carry, B., Descamps, P., Ferrais, M., et al. 2023, A&A, 680, A21

Chen, K., & Lin, M.-K. 2020, ApJ, 891, 132

Chokshi, A., Tielens, A. G. G. M., & Hollenbach, D. 1993, ApJ, 407, 806

Cuzzi, J. N., Hogan, R. C., & Shariff, K. 2008, ApJ, 687, 1432

Davidsson, B. J. R. 2021, MNRAS, 505, 5654

Davidsson, B. J. R. 2023, MNRAS, 521, 2484

Davidsson, B. J. R., Sierks, H., Güttler, C., et al. 2016, A&A, 592, A63

Desch, S. J., & Turner, N. J. 2015, ApJ, 811, 156

Di Sisto, R. P., & Brunini, A. 2007, Icar, 190, 224

Di Sisto, R. P., & Rossignoli, N. L. 2020, CM&DA, 132, 36

Dones, L., Brasser, R., Kaib, N., & Rickman, H. 2015, SSRv, 197, 191

Dones, L., Chapman, C. R., McKinnon, W. B., et al. 2009, in Saturn from Cassini-Huygens, ed. M. K. Dougherty, L. W. Eposito, & S. M. Krimigis (Berlin: Springer) 613

Doressoundiram, A., Peixinho, N., de Bergh, C., et al. 2002, AJ, 124, 2279

Doressoundiram, A., Peixinho, N., & de Bergh, C. 2003, AJ, 125, 1629

Drążkowska, J., & Alibert, Y. 2017, A&A, 608, A92

Dullemond, C. P., & Dominik, C. 2005, A&A, 434, 971

Dullemond, C. P., Birnstiel, T., Huang, J., et al. 2018, ApL, 869, L46

Duncan, M. J., & Levison, H. F. 1997, Sci, 276, 1670
Ercolano, B., Jennings, J., Rosotti, G., & Birnstiel, T. 2017, MNRAS, 472, 4117
Estrada, P. R., Cuzzi, J. N., & Umurhan, O. M. 2022, ApJ, 936, 42
Estrada, P. R., & Umurhan, O. M. 2023, ApJ, 946, 15
Faggi, S., Villanueva, G. L., McKay, A., et al. 2024, NatAs, 8, 1237
Flock, M., Turner, N. J., Nelson, R. P., et al. 2020, ApJ, 897, 155
Fraser, W. C., Brown, M. E., Morbidelli, A., Parker, A., & Batygin, K. 2014, ApJ, 782, 100
Fraser, W. C., Dones, L., Volk, K., Womack, M., & Nesvorný, D. 2024, in Comets III, ed. K. J. Meech, et al. (Tucson, AZ: Univ. of Arizona. Press) 121
Fraser, W. C., Bannister, M. T., Pike, R. E., et al. 2017, NatAs, 1, 0088
Fromang, S., & Papaloizou, J. 2006, A&A, 452, 751
Gibbons, P. G., Mamatsashvili, G. R., & Rice, W. K. M. 2015, MNRAS, 453, 4232
Gladman, B., & Volk, K. 2021, ARA&A, 59, 203
Goldreich, P., & Ward, W. R. 1973, ApJ, 183, 1051
Gomes, R., Levison, H. F., Tsiganis, K., & Morbidelli, A. 2005, Natur, 435, 466
Gomes, R., Nesvorný, D., Morbidelli, A., Deienno, R., & Nogueira, E. 2018, Icar, 306, 319
Greenstreet, S., Gladman, B., Abedin, A., Petit, J.-M., & Kavelaars, J. 2023, BAAS, 55, 8
Greenstreet, S., Gladman, B., & McKinnon, W. B. 2015, Icar, 258, 267
Greenstreet, S., Gladman, B., & McKinnon, W. B. 2016, Icar, 274, 366
Greenstreet, S., Gladman, B., McKinnon, W. B., Kavelaars, J. J., & Singer, K. N. 2019, ApL, 872, L5
Grundy, W. M. 2009, Icar, 199, 560
Grundy, W. M., Stansberry, J. A., Noll, K. S., et al. 2007, Icar, 191, 286
Güttler, C., Blum, J., Zsom, A., Ormel, C. W., & Dullemond, C. P. 2010, A&A, 513, A56
Harrington Pinto, O., Womack, M., Fernandez, Y., & Bauer, J. 2022, PSJ, 3, 247
Harrington Pinto, O., Kelley, M. S. P., Villanueva, G. L., et al. 2023, PSJ, 4, 208
Hartlep, T., & Cuzzi, J. N. 2020, ApJ, 892, 120
Hasegawa, S., Marsset, M., DeMeo, F. E., et al. 2021, ApJL, 916, L6
Hirabayashi, M., Scheeres, D. J., Chesley, S. R., et al. 2016, Natur, 534, 352
Holsapple, K. A. 1993, ARE&PS, 21, 333
Hopkins, P. F. 2016, MNRAS, 456, 2383
Housen, K. R., & Holsapple, K. A. 2011, Icar, 211, 856
Irwin, M., Tremaine, S., & Zytkow, A. N. 1995, AJ, 110, 3082
Jewitt, D. 2009, AJ, 137, 4296
Jewitt, D. 2021, AJ, 161, 261
Jewitt, D., Luu, J., & Chen, J. 1996, AJ, 112, 1225
Jewitt, D. C. 2002, AJ, 123, 1039
Johansen, A., Blum, J., Tanaka, H., et al. 2014, in Protostars and Planets VI, ed. H. Beuther, et al. (Tucson, AZ: Univ. Arizona Press) 547
Johansen, A., Klahr, H., & Henning, T. 2011, A&A, 529, A62
Johansen, A., Mac Low, M.-M., Lacerda, P., & Bizzarro, M. 2015, SciA, 1, 1500109
Johansen, A., Oishi, J. S., & Mac Low, M. M. 2007, Natur, 448, 1022
Johansen, A., & Youdin, A. 2007, ApJ, 662, 627
Johansen, A., Youdin, A., & Klahr, H. 2009a, ApJ, 697, 1269
Johansen, A., Youdin, A., & Mac Low, M.-M. 2009b, ApL, 704, L75
Johansen, A., Youdin, A. N., & Lithwick, Y. 2012, A&A, 537, A125

Jutzi, M., Benz, W., Toliou, A., Morbidelli, A., & Brasser, R. 2017, A&A, 597, A61
Kavelaars, J. J., Petit, J.-M., Gladman, B., et al. 2021, ApL, 920, L28
Kirchoff, M. R., Bierhaus, E. B., Dones, L., et al. 2018, in Enceladus and the Icy Moons of Saturn, ed. P. M. Schenk, et al. (Tucson, AZ: Univ. Arizona Press) 267
Kirchoff, M. R., Dones, L., Singer, K., & Schenk, P. 2022, PSJ, 3, 42
Kirchoff, M. R., & Schenk, P. 2009, Icar, 202, 656
Kirchoff, M. R., & Schenk, P. 2010, Icar, 206, 485
Kirchoff, M. R., & Schenk, P. M. 2008, LPI Contributions, Workshop on the Early Solar System Impact Bombardment Vol 1439, (Houston, TX: LPI) https://ui.adsabs.harvard.edu/abs/2008LPICo1439...37K/abstract
Klahr, H., & Schreiber, A. 2020, ApJ, 901, 54
Klahr, H. H., & Bodenheimer, P. 2003, ApJ, 582, 869
Knight, M. M., Kokotanekova, R., & Samarasinha, N. H. 2024, in Comets III, ed. K. J. Meech, et al. (Tucson, AZ: Univ. of Arizona. Press) 361
Krapp, L., Benítez-Llambay, P., Gressel, O., & Pessah, M. E. 2019, ApL, 878, L30
Krivov, A. V., Ide, A., Löhne, T., Johansen, A., & Blum, J. 2018, MNRAS, 474, 2564
Krivov, A. V., & Wyatt, M. C. 2021, MNRAS, 500, 718
Lesur, G., Flock, M., Ercolano, B., et al. 2023, ASP Conf. Series Vol 534, ed. S. Inutsuka, Y. Aikawa, T. Muto, K. Tomida, & M. Tamura (San Francisco, CA: ASP) 465
Levison, H. F., & Duncan, M. J. 1997, Icar, 127, 13
Levison, H. F., Kretke, K. A., & Duncan, M. J. 2015, Natur, 524, 322
Levison, H. F., & Stern, S. A. 2001, AJ, 121, 1730
Li, R., & Youdin, A. N. 2021, ApJ, 919, 107
Li, R., Youdin, A. N., & Simon, J. B. 2019, ApJ, 885, 69
Lilly, E., Jevčák, P., Schambeau, C., et al. 2024, ApL, 960, L8
Lim, J., Simon, J. B., Li, R., et al. 2024, ApJ, 969, 130
Lin, M.-K. 2019, MNRAS, 485, 5221
Liu, B., Lambrechts, M., Johansen, A., Pascucci, I., & Henning, T. 2020, A&A, 638, A88
Liu, B., Ormel, C. W., & Johansen, A. 2019, A&A, 624, A114
Lorek, S., & Johansen, A. 2022, A&A, 666, A108
Lorek, S., & Johansen, A. 2024, A&A, 683, A38
Luu, J. X. 1993, Icar, 104, 138
Lyra, W., Johansen, A., & Cañas, M. H. 2023, ApJ, 946, 60
Lyra, W., Johansen, A., Klahr, H., & Piskunov, N. 2009, A&A, 493, 1125
Lyra, W., Youdin, A. N., & Johansen, A. 2021, Icar, 356, 113831
Mah, J., & Brasser, R. 2019, MNRAS, 486, 836
Malhotra, R. 1993, Natur, 365, 819
McKinnon, W. B., Stern, S. A., Weaver, H. A., et al. 2017, Icar, 287, 2
McKinnon, W. B., Richardson, D. C., Marohnic, J. C., et al. 2020, Sci, 367, aay6620
McKinnon, W. B., Mao, X., Schenk, P. M., et al. 2022, GeoRL, 49, e98406
Melita, M. D., & Licandro, J. 2012, A&A, 539, A144
Morbidelli, A., Bottke, W. F., Nesvorný, D., & Levison, H. F. 2009, Icar, 204, 558
Morbidelli, A., Levison, H. F., Tsiganis, K., & Gomes, R. 2005, Natur, 435, 462
Morbidelli, A., & Rickman, H. 2015, A&A, 583, A43
Mottola, S., Britt, D. T., Brown, M. E., et al. 2024, SSRv, 220, 17
Musiolik, G. 2021, MNRAS, 506, 5153

Musiolik, G., & Wurm, G. 2019, ApJ, 873, 58
Napier, K. J., Lin, H. W., Gerdes, D. W., et al. 2024, PSJ, 5, 50
Nelson, R. P., Gressel, O., & Umurhan, O. M. 2013, MNRAS, 435, 2610
Nesvorný, D. 2018, ARA&A, 56, 137
Nesvorný, D., Dones, L., De Prá, M., Womack, M., & Zahnle, K. J. 2023, PSJ, 4, 139
Nesvorný, D., Li, R., Youdin, A. N., Simon, J. B., & Grundy, W. M. 2019a, NatAs, 3, 808
Nesvorný, D., Vokrouhlický, D., Dones, L., et al. 2017, ApJ, 845, 27
Nesvorný, D., Vokrouhlický, D., Stern, A. S., et al. 2019b, AJ, 158, 132
Okuzumi, S., Tanaka, H., Kobayashi, H., & Wada, K. 2012, ApJ, 752, 106
Ormel, C. W., & Cuzzi, J. N. 2007, A&A, 466, 413
Paardekooper, S.-J., McNally, C. P., & Lovascio, F. 2020, MNRAS, 499, 4223
Paardekooper, S.-J., McNally, C. P., & Lovascio, F. 2021, MNRAS, 502, 1579
Pan, L., Padoan, P., Scalo, J., Kritsuk, A. G., & Norman, M. L. 2011, ApJ, 740, 6
Parhi, A., & Prialnik, D. 2023, MNRAS, 522, 2081
Peixinho, N., Delsanti, A., Guilbert-Lepoutre, A., Gafeira, R., & Lacerda, P. 2012, A&A, 546, A86
Peixinho, N., Doressoundiram, A., Delsanti, A., et al. 2003, A&A, 410, L29
Petit, J.-M., Gladman, B., Kavelaars, J. J., et al. 2023, ApL, 947, L4
Pinte, C., Dent, W. R. F., Ménard, F., et al. 2016, ApJ, 816, 25
Plescia, J. B. 1988, Icar, 73, 442
Riols, A., Lesur, G., & Menard, F. 2020, A&A, 639, A95
Robbins, S. J., Bierhaus, E. B., & Dones, L. 2024, JGRE, 129, e2023JE007941
Ros, K., & Johansen, A. 2013, A&A, 552, A137
Ros, K., & Johansen, A. 2024, A&A, 686, A237
Ros, K., Johansen, A., Riipinen, I., & Schlesinger, D. 2019, A&A, 629, A65
Safrit, T. K., Steckloff, J. K., Bosh, A. S., et al. 2021, PSJ, 2, 14
Safronov, V. S. 1969, Evoliutsiia Doplanetnogo Oblaka (Nauka: Moskva) https://ui.adsabs.harvard.edu/abs/1969edo..book.....S/abstract
Santos-Sanz, P., Lellouch, E., Fornasier, S., et al. 2012, A&A, 541, A92
Schäfer, U., & Johansen, A. 2022, A&A, 666, A98
Schäfer, U., Johansen, A., & Banerjee, R. 2020, A&A, 635, A190
Schäfer, U., Yang, C.-C., & Johansen, A. 2017, A&A, 597, A69
Schaffer, N., Johansen, A., & Lambrechts, M. 2021, A&A, 653, A14
Schenk, P. M. 2002, Natur, 417, 419
Schenk, P. M., Chapman, C. R., Zahnle, K., & Moore, J. M. 2004, in Jupiter. The Planet, Satellites and Magnetosphere, Vol 1, ed. F. Bagenal, T. E. Dowling, & W. B. McKinnon (Cambridge: Cambridge Univ. Press) 427
Schenk, P. M., & Zahnle, K. 2007, Icar, 192, 135
Schlichting, H. E., Fuentes, C. I., & Trilling, D. E. 2013, AJ, 146, 36
Schoonenberg, D., & Ormel, C. W. 2017, A&A, 602, A21
Schwamb, M. E., Brown, M. E., & Fraser, W. C. 2014, AJ, 147, 2
Schwartz, S. R., Michel, P., Jutzi, M., et al. 2018, NatAs, 2, 379
Simon, J. B., Armitage, P. J., Li, R., & Youdin, A. N. 2016, ApJ, 822, 55
Simon, J. B., Armitage, P. J., Youdin, A. N., & Li, R. 2017, ApL, 847, L12
Singer, K. N., Greenstreet, S., Schenk, P. M., Robbins, S. J., & Bray, V. J. 2021, in The Pluto System After New Horizons, ed. S. A. Stern, et al. (Tucson, AZ: Univ. Arizona Press) 121

Singer, K. N., McKinnon, W. B., Gladman, B., et al. 2019, Sci, 363, 955
Spencer, J. R., Stern, S. A., Moore, J. M., et al. 2020, Sci, 367, aay3999
Steckloff, J. K., Graves, K., Hirabayashi, M., Melosh, H. J., & Richardson, J. E. 2016, Icar, 272, 60
Strom, R. G. 1987, Icar, 70, 517
Tegler, S. C., Bauer, J. M., Romanishin, W., & Peixinho, N. 2008, in The Solar System Beyond Neptune, ed. M. A. Barucci, et al. (Tucson, AZ: Univ. Arizona Press) 105
Tegler, S. C., Romanishin, W., & Consolmagno, G.J.S. 2016, AJ, 152, 210
Trujillo, C. A., & Brown, M. E. 2002, ApJL, 566, L125
Tsiganis, K., Gomes, R., Morbidelli, A., & Levison, H. F. 2005, Natur, 435, 459
Turner, N. J., Fromang, S., Gammie, C., et al. 2014, in Protostars and Planets VI, ed. H. Beuther, et al. (Tucson, AZ: Univ. Arizona Press) 411
Umurhan, O. M., Estrada, P. R., & Cuzzi, J. N. 2020, ApJ, 895, 4
van Kooten, E. M. M. E., Moynier, F., & Agranier, A. 2019, PNAS, 116, 18860
Vavilov, D. E., Eggl, S., Medvedev, Y. D., & Zatitskiy, P. B. 2019, A&A, 622, L5
Villenave, M. 2022, ApJ, 930, 11
Volk, K., & Malhotra, R. 2008, ApJ, 687, 714
Weidenschilling, S. J. 1977, MNRAS, 180, 57
Weidenschilling, S. J. 1980, Icar, 44, 172
Wierzchos, K., Womack, M., & Sarid, G. 2017, AJ, 153, 230
Womack, M., Sarid, G., & Wierzchos, K. 2017, PASP, 129, 031001
Wong, I., & Brown, M. E. 2015, AJ, 150, 174
Wong, I., & Brown, M. E. 2017a, AJ, 153, 145
Wong, I., & Brown, M. E. 2017b, AJ, 153, 69
Xu, Z., & Bai, X.-N. 2022, ApJ, 924, 3
Yang, C.-C., Johansen, A., & Carrera, D. 2017, A&A, 606, A80
Yang, C.-C., Mac Low, M.-M., & Johansen, A. 2018, ApJ, 868, 27
Yang, C.-C., & Zhu, Z. 2021, MNRAS, 508, 5538
Youdin, A. N., & Goodman, J. 2005, ApJ, 620, 459
Zahnle, K., Schenk, P., Levison, H., & Dones, L. 2003, Icar, 163, 263
Zhu, Z., & Yang, C.-C. 2021, MNRAS, 501, 467

Chapter 3

Dynamics of Centaurs and Links to Scattering and Comet Populations

Romina P Di Sisto, Tabare Gallardo and Luke Dones

The region of the giant planets serves as a transition zone for objects traveling from the trans-Neptunian region toward the inner solar system. It also acts as a pathway for small bodies on their way to being ejected from the solar system. Within this zone, these objects, known as Centaurs, are characterized by their relatively rapid dynamical evolution. In this chapter, we will extensively explore our current understanding of Centaur dynamics and the dynamics governing their transition from the trans-Neptunian region through the giant planetary region, eventually leading them into the comet zone. The giant planet region stands out for its substantial diversity, encompassing objects with varying dynamics and physical properties. Therefore, we will not only focus on primary sources of Centaurs but also consider secondary sources that could play a significant role in explaining the observed characteristics of this unique population.

3.1 Introduction

The dynamical evolution of Centaurs has been mainly investigated in relation to their "parental ties", i.e., as progeny of trans-Neptunian objects (TNOs) and as progenitors of Jupiter-family comets (JFCs). In fact, the existence of TNOs was first proposed in order to explain the distinctive dynamical characteristics of the JFC population, particularly their small inclinations to the ecliptic. Therefore, Centaurs emerged as the necessary intermediate population linking JFCs to their source.

The existence of a region beyond Neptune as a source of JFCs was first suggested by Edgeworth (1938) and Kuiper (1951), but it was not until the work by Fernández (1980) that this problem was theoretically analyzed. Fernández proposed a trans-Neptunian belt between 35 and 50 au as a more efficient source for JFCs than the Oort Cloud. Later, Duncan et al. (1988) addressed this problem through numerical simulations. The next two baseline papers that analyzed the TNOs as a source of

doi:10.1088/2514-3433/ada267ch3

JFCs were done by Levison & Duncan (1997) and Duncan & Levison (1997). In Levison & Duncan (1997), the authors studied the evolution of 2200 clones[1] of 20 particles (obtained from a previous simulation) initially on orbits like those of (the now called) classical Kuiper Belt objects (CKBOs) that had evolved onto Neptune-approaching orbits in Duncan et al. (1995). Levison & Duncan (1997) inferred the existence of an excited population in the trans-Neptunian (TN) region, which they called the scattered disk (SD), that could be an order of magnitude larger than the classical Kuiper Belt (CKB). In Duncan & Levison (1997), the authors suggested that the SD should produce more ecliptic comets (ECs; a group that includes JFCs as well as Centaurs) than the CKB.

Because they are transitional objects, the Centaurs' dynamical behavior is conditioned by the intrinsic dynamical features of TNOs. So, individual Centaurs and their dynamical behavior contain physical and dynamical traces of the gestation, formation, and evolution processes in the whole solar system. In the transition from the TN zone, Neptune would be the "nexus" between TNOs and Centaurs, since the eventual gravitational interactions of some TNOs with this planet would transfer them to the planetary zone, then becoming Centaurs. As the structure of the TN region emerged thanks to the discovery of thousands of bodies in observational surveys, the dynamical evolution of Centaurs from their sources in the TN region and the contribution from each subpopulation was investigated by Di Sisto & Brunini (2007), Volk & Malhotra (2008), and others.

In this chapter we review the present-day dynamical transfer of objects from the TNO populations through the giant planet region and into the JFC population. We also delve into the intrinsic dynamics of Centaurs in the giant planet region and make comparisons with the observed population.

3.2 Sources of Centaurs

3.2.1 Population Boundaries

There are 4657 TNOs and Centaurs listed in the Minor Planet Center (MPC) database up to 2023 September 12[2]. The orbital elements of these observed objects are plotted in Figure 3.1. We also include 31 Neptune Trojans and one Uranus Trojan listed by the MPC although, by the definition we use, Trojans are not Centaurs. It can be seen that the semimajor axes extend to ~2000 au and that the inclinations, though mainly less than ~40°, reach retrograde values ($i > 90°$) in some cases.

A rich structure can also be seen in the trans-Neptunian region, with populations with distinctive dynamical characteristics. We can distinguish four subpopulations (also see Lykawka & Mukai 2007; Gladman et al. 2008; Saillenfest 2020; Smullen & Volk 2020; Gladman & Volk 2021; Fraser et al. 2024; Volk & Van Laerhoven 2024):

[1] A clone is a fictitious particle with orbital elements within the uncertainty region of a "real" particle, created to statistically study the orbital evolution. Using clones is necessary because the orbital evolution of bodies that encounter planets is generally chaotic.

[2] https://minorplanetcenter.net/iau/lists/MPLists.html

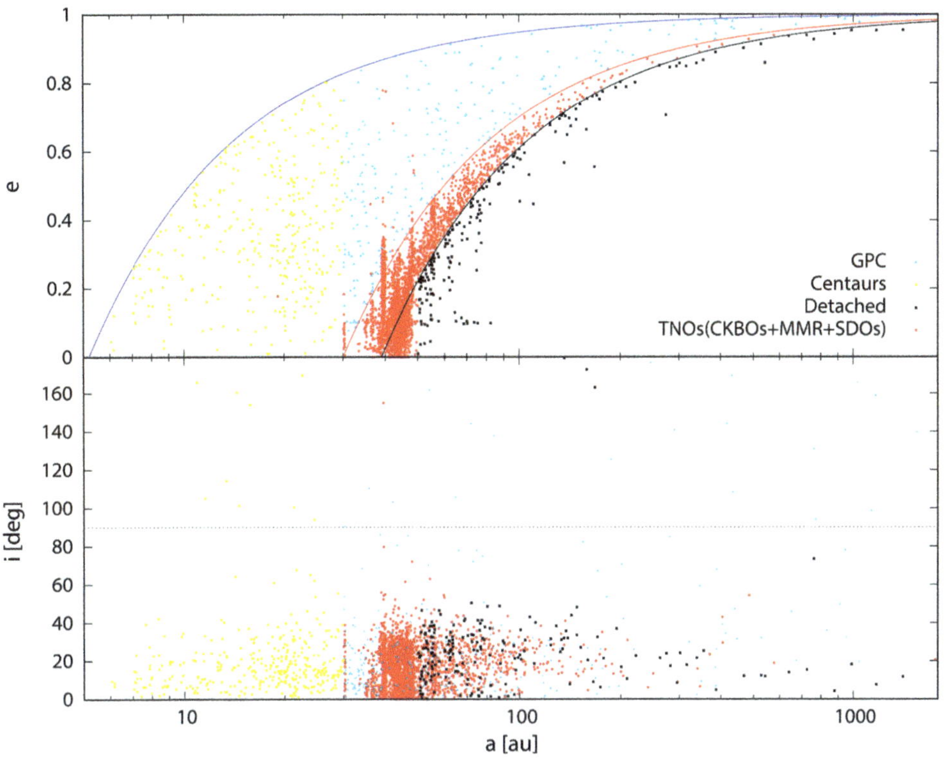

Figure 3.1. Eccentricity *e* and inclination *i* vs. semimajor axis *a* of all observed TNOs and Centaurs as of 2023 September 12. Curves of constant perihelion distance $q = 5.2$ au (blue), $q = 30$ au (red), and $q = 39$ au (black) are shown to display the boundaries between subpopulations. The different colored dots display the different subpopulations. TNOs (red) are the trans-Neptunian objects and include the classical objects (CTNOs), the objects in mean motion resonances with Neptune (MMR) and the scattered disk objects (SDOs). The detached objects (black), Centaur objects (yellow) and giant planet crossers (GPCs; light blue) are also shown. See text for detailed definitions. The dashed line in the lower panel at $i = 90°$ marks the boundary between prograde ($i < 90°$) and retrograde ($i > 90°$) orbits.

- Classical objects (CTNOs or CKBOs): bodies with semimajor axes 40 au $\lesssim a \lesssim$ 50 au with low eccentricities and low-to-moderate inclinations. Near the inner boundary, in the region interior to 42 au, overlapping secular resonances induce instabilities (Duncan et al. 1995; Morbidelli et al. 1995; Huang et al. 2022), while the outer boundary seems to be close to the location of the 1:2 mean-motion resonance (MMR) with Neptune near 47.8 au[3]. That is, the population declines strongly at this distance, but it is not clear why or whether there are many more low-inclination objects well beyond the 1:2

[3] In this paper, an m: n MMR means that a small body completes approximately m orbits for each n orbits of the planet with which it is in resonance. The literature is inconsistent, so what we call the 1:2 resonance is sometimes called the 2:1 MMR, the 2:3 is called the 3:2, and so forth.

MMR (e.g., Gladman & Volk 2021, de la Fuente Marcos & de la Fuente Marcos 2024). This population is usually divided into cold and hot classical objects according to relatively low and high-inclination values (e.g., Levison & Stern 2001; Brown 2001; Petit et al. 2017).

- Resonant objects: bodies in MMRs with Neptune. The resonance with the most *known* TNOs is the 2:3 ($a = 39.4$ au) where Pluto and the plutinos are, although there are other MMRs further from the Sun that may have larger populations of as yet unknown objects (e.g., Gladman et al. 2012; Volk et al. 2016; Crompvoets et al. 2022).
- Scattered Disk Objects (SDOs): bodies with large eccentricities (usually greater than $\sim$0.2), large semimajor axes (the majority with $a \gtrsim 50$ au), and perihelion distances 30 au $< q <$ 39 au (e.g., Hahn & Malhotra 2005; Gomes et al. 2008; Nesvorný et al. 2023a). In this perihelion distance range, Neptune's perturbations are important and capable of scattering the objects.
- Detached objects: bodies with $q > 39$ au and $a > 50$ au. Their perihelion distances are far enough outside of Neptune's orbit that they can be considered detached from planetary perturbations (e.g., Gladman et al. 2002; Morbidelli & Levison 2004; Gomes et al. 2005; Beaudoin et al. 2023). (For objects with $a \gg 50$au, the boundary between SDOs and detached objects is at a somewhat larger perihelion distance; Hadden & Tremaine 2024.)

The boundary between the TN region and the giant planetary region is not well constrained in the literature and the nomenclature is diverse. Since in this chapter we are interested in describing the dynamical transfer from the TN zone to the giant planetary zone and then to the JFC zone, we define these categories, following Di Sisto & Rossignoli (2020):

- Giant planet crossers (GPCs): Objects with 5.2 au $< q <$ 30 au.
- Centaurs: objects with $q > 5.2$ au and $a < 30$ au. They are a subset of GPCs.

GPCs cross the orbits of one or more of the giant planets, and therefore their dynamical evolution is dominated by the gravitational action of these planets. We define Centaurs as those GPCs having q greater than the semimajor axis of Jupiter ($a_J = 5.2$ au) to distinguish them from JFCs and having $a < 30$ au, i.e., the semimajor axis of Neptune, to separate them from the TN zone. Considering this definition, we study the resonant dynamics in the region of the giant planets, i.e., 5.2 au $< a <$ 30 au, as we will see in Section 3.3, in contrast with the MMRs located beyond Neptune ($a > 30$ au). However, it must be taken into account that from a dynamical point of view, these limits are not strict, but rather there is a broader transition region. In any case, it is only a division to facilitate the description of the processes that occur from the trans-Neptunian region to the planetary region and that involve the passage of the same object through the different zones.

Since we are interested in describing the dynamics of both inactive and active bodies in the GPC and Centaur zones, we added the comets in those regions to the list of GPCs and Centaurs from the MPC. Therefore, we downloaded the objects listed as comets, including JFCs, Halley-type comets (HTCs), Chiron-type comets (i.e., active Centaurs), and other comets, from JPL[4] within the ranges of orbital elements described above and included them in the GPC and Centaurs list[5]. Thus, we have 645 GPCs and 344 Centaurs as of September 12, 2023.

3.2.2 The Trans-Neptunian Zone as a Source of GPCs and Centaurs

GPCs and Centaurs are not isolated objects but come from the TN region and are tightly connected to their specific home regions beyond Neptune. Their dynamical evolution is dominated by the giant planets, but also depends on the nature of their orbits when they enter the planetary region. In this section we address the studies on the contribution and dynamics of the entry of TNOs to the region of the giant planets.

Duncan et al. (1995) performed a dynamical evolution of thousands of low-inclination test particles in the classical KB for the age of the solar system. They discovered that there is a dynamical structure in the TN region where gradual instabilities are able to supply a flux of objects to explain the number of observed short-period comets. Levison & Duncan (1997) took particles that left the KB after being stable for $\geqslant 1$ Gyr from the Duncan et al. (1995) simulation as initial conditions for a new numerical simulation. These particles were considered to be representative of SDOs currently leaving the TN region. They numerically integrated those particles plus clones under the action of the Sun and the perturbations of the four giant planets for 1 Gyr or until they were ejected from the solar system or collided with the Sun or a planet. They found the characteristic routes of the transition to GPCs and eventually JFCs. The general path is a "handoff" from the gravitational control of one planet to another, beginning with Neptune. That is, an SDO begins interacting with Neptune and is scattered by this planet. If its pericenter distance (q) decreases and it crosses the orbit of Uranus, this planet now dominates its evolution and can further reduce its q again, until the SDO can cross the orbit of Saturn. The perturbations from Saturn now dominate, until q is reduced up to enter to the Jupiter's domain, becoming then a JFC. This handoff mechanism and the connection of SDOs with JFCs can be better understood considering the Tisserand parameter (T; Levison & Duncan 1997). T is an approximation of the Jacobi constant of the circular, restricted, three body problem (CR3BP), expressed in terms of the orbital elements. It is given by

$$T = \frac{a_p}{a} + 2\sqrt{\frac{a}{a_p}\left(1 - e^2\right)} \cos i, \tag{3.1}$$

[4] https://ssd.jpl.nasa.gov/tools/sbdb_query.html

[5] We use a taxonomy like that of Levison (1996) in which JFCs are periodic comets with Tisserand parameters T_J with respect to Jupiter between 2 and 3 and HTCs are periodic comets with $T_J < 2$. In general, JFCs have smaller orbits and inclinations than HTCs. See the discussion of Equation (3.1) below and in Fraser et al. (2024).

where a, e, and i are the semimajor axis, eccentricity, and inclination of the object, and a_p is the semimajor axis of a planet. This equation is derived from the Jacobi constant in the CR3BP. It is calculated for an object (assumed to be massless) moving under the gravitational action of the Sun and a planet. Since the giant planets have very low eccentricities, the dynamics of a Centaur on a planet-crossing orbit can be approximated by the CR3BP and T can be calculated with respect to that planet. If an object crosses the orbit of a planet, the Tisserand parameter with respect to that planet is always less than, but close to, 3[6]. An SDO with a small inclination that begins crossing the orbit of Neptune will have T with respect to Neptune near 3. When the object is delivered to Uranus and crosses its orbit, T with respect to Uranus will be near 3, and so on down to Jupiter. Therefore, the handoff explains the very narrow range in T_J that JFCs occupy (JFCs have Tisserand parameters with respect to Jupiter, T_J, with $2 < T_J < 3$, but the median value of T_J is 2.8) and the low inclinations of JFCs. That is to say, given the relationship between the inclination and $q = a(1 - e)$, implicit in Equation (3.1) so that T with respect to each planet is close to 3 as it is a crosser of that planet, a smaller value of q implies a small inclination (i.e., $\cos i \approx 1$).

The papers by Duncan et al. (1995) and Levison & Duncan (1997) were followed by other works aiming to explain the population, the orbital element and size distribution, and the origin of the SD population as a source of JFCs. With the development of the models of the formation of the small-body reservoirs based on migration theories (Tsiganis et al. 2005; Walsh et al. 2011), the origin of the SD was investigated (Levison et al. 2008). The migration concept was introduced by Fernandez & Ip (1984), who found that when giant planets (particularly Uranus and Neptune) form by accreting planetesimals from a disk, they move radially inward or outward because of the exchange of angular momentum with planetesimals, and then they migrate from their original locations in the initial planetary disk. The following papers by Malhotra (1993, 1995) first considered the effects of the outward migration of Neptune to explain the capture of Pluto into the 2:3 MMR with Neptune, resulting in its high eccentricity and inclination.

Subsequent models of planetary migration focused on explaining the current orbital architecture of the solar system and its consequences for the populations of small bodies (e.g., Tsiganis et al. 2005; Levison et al. 2008; Walsh et al. 2011; Nesvorný et al. 2013; Nesvorný 2018). It was found that during the migration of the giant planets, planetesimal scattering occurs until Neptune reaches the outer limit of the disk. At that point, the planetary migration slows down, leaving a population of objects scattered by Neptune which constitute the current SD and are now the main source of the current flux of Centaurs and JFCs. For a detailed review of the formation processes in the early solar system, we refer the reader to Chapter 2.

Di Sisto & Brunini (2007) and Di Sisto & Rossignoli (2020) investigated the current contribution of the SD to GPCs and Centaurs by building an intrinsic model of the SD and calculating its dynamical evolution under the action of the Sun and

[6] This can be calculated by the Jacobi constant in terms of the Cartesian coordinates in the rotating frame and considering that the particle is crossing the orbit of the planet.

the four giant planets. They used observed SDOs from the MPC database with 30 au$< q <$ 39 au, $a >$ 40 au, and $e >$ 0.2, and added a set of clones such that the semimajor axis distribution of the particles followed $N(a) \propto a^{-2}$ (Fernández et al. 2004) and the inclination distribution was given by $N(i) \propto \sin(i)\exp(-i^2/2\sigma_i^2)$ (Brown 2001). The numerical simulation was run for the age of the solar system or until the particles were ejected, collided with a planet or entered the region inside Jupiter's orbit ($r <$ 5.2 au). The difference between the models used in 2007 (Di Sisto & Brunini 2007) and in 2020 (Di Sisto & Rossignoli 2020) is the increased number of observed objects in the later study. Di Sisto & Rossignoli (2020) show that the current observations seem to follow the same trend as the debiased semimajor axis and inclination distributions, and therefore the model and results are improved.

Di Sisto & Brunini (2007) and Di Sisto & Rossignoli (2020) found that SDOs evolved in the giant planetary zone by the "handoff" mechanism first found by Levison & Duncan (1997) and explained before. The dynamical evolution through the giant planetary zone is dominated by transfers between MMRs, resonance sticking, chaotic dynamics, and secular evolution in some specific cases. Di Sisto & Brunini (2007) and Di Sisto & Rossignoli (2020) found that the rate of injection of GPCs from the SD is $4.025 \pm 0.008 \times 10^{-10} N_{\rm SDO}$/year and the rate of injection of Centaurs from the SD is $1.796 \pm 0.005 \times 10^{-10} N_{\rm SDO}$/year, where $N_{\rm SDO}$ is the current number of objects in the scattered disk. Note that, because GPCs include Centaurs, the Centaur injection rate involves objects most likely first passing through the GPC population.

The mean lifetime as a GPC is 68 Myr, with the most likely value being between 10 and 100 Myr, and the mean lifetime as a Centaur is 7.2 Myr. There is a correlation between the perihelion distance q of Centaurs and their mean lifetime $\langle t \rangle$: the larger the q, the longer $\langle t \rangle$ is. Figure 3.2 shows the dynamical evolution of a particle initially in the SD when it goes through the Centaur zone. The particle first enters the Centaur zone at t = 1.697 Gyr. It then quickly falls into the von Zeipel-Lidov-Kozai (ZLK) mechanism or resonance (e.g., Ito & Ohtsuka 2019). This secular resonance occurs when the variation of the argument of perihelion $\dot{\omega} = 0$ and ω oscillates around 0° or 180° (or 90° or 270°). The eccentricity and inclination of the object are coupled such that e is a maximum when i is a minimum and vice versa. The particle in Figure 3.2 remains in the ZLK resonance with ω oscillations around 0° until t = 1.706 Gyr. Throughout its subsequent evolution, the particle spends time in the following MMRs with Neptune, Uranus, and Saturn: the 1:1N at a = 30.1 au (it starts stuck to the left side of the 1:1N resonance at a = 29.8 au), 15:13N at a = 27.3 au, 7:8U at a = 21.0 au, 12:13U at a = 20.3 au, 1:1U at a = 19.2 au, 9:8U at a = 17.8 au and 1:2S at a = 15.2 au. The particle's evolution as a Centaur then ends at t = 1.715 Gyr when it enters the JFC zone ($r <$ 5.2 au).

Given that the Di Sisto & Rossignoli (2020) numerical simulation represents the dynamical evolution of a model of an intrinsic current SD, the Centaurs obtained in this simulation will represent a model of an intrinsic Centaur population. In this sense, such a model can be obtained by calculating a normalized time-weighted distribution of Centaurs in the orbital element space. Such maps were presented in Di Sisto & Brunini (2007) and Di Sisto & Rossignoli (2020) for the GPC zone.

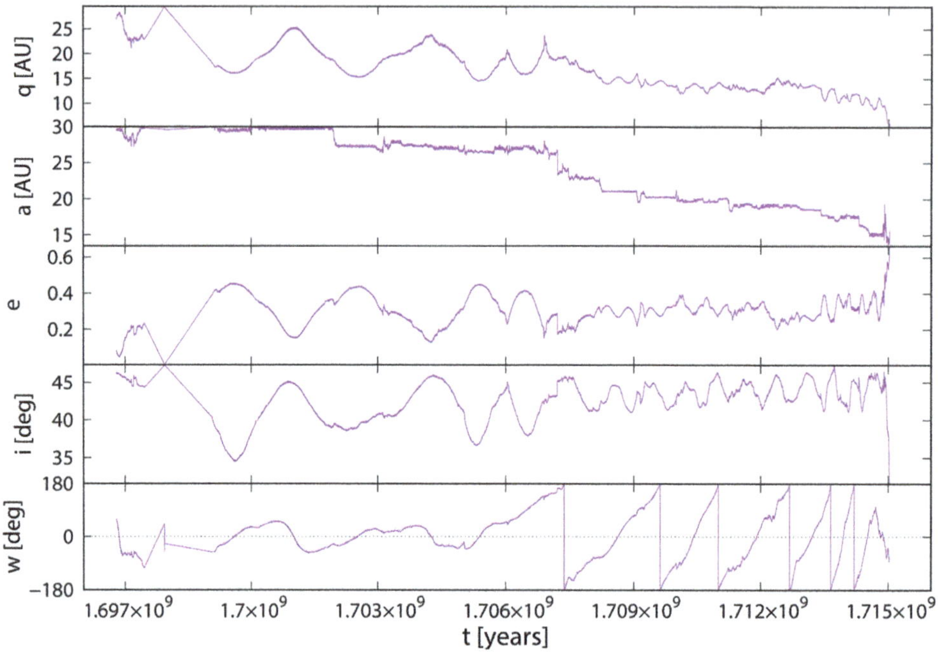

Figure 3.2. Temporal evolution of perihelion distance q, semimajor axis a, eccentricity e, inclination i, and argument of perihelion ω of a particle initially in the SD and shown here when it reaches the Centaur zone (taken from the Di Sisto & Rossignoli 2020 simulation). The particle first enters the Centaur zone at $t = 1.697$ Gyr. It then quickly falls into the ZLK resonance with ω oscillations around 0° and eccentricity and inclination coupled until $t = 1.706$ Gyr. Throughout its following evolution, the particle spends time in MMRs with Neptune, Uranus, and Saturn (see main text for details). Finally, at $t = 1.715$ Gyr, the particle's evolution as a Centaur ends as it enters the JFC zone ($r < 5.2$ au).

However, the particles in those simulations were removed if the heliocentric distance reached $r < 5.2$ au. In that zone, it is necessary to consider the terrestrial planet perturbations and also physical effects such as mass loss and disintegration (Di Sisto et al. 2009) to get a realistic behavior. Now, in order to have a complete model of the Centaur population, we continued the integration of the particles removed at the distance of Jupiter as was done by Di Sisto et al. (2009), but only considering dynamical effects. There were 884 particles that reached $r < 5.2$ au in Di Sisto & Rossignoli (2020) whose position and velocity were stored in a file together with those of the giant planets. Therefore, each particle was integrated separately from this time under the gravitational action of the Sun, the four giant planets, and the terrestrial planets Mars, Earth, and Venus (the mass of Mercury was added to that of the Sun) with the hybrid integrator EVORB (Fernández et al. 2002), using an integration step of 0.01 yr. The orbital elements of the terrestrial planets were the current ones, but their mean anomalies were taken at random. The simulation is followed for 10^8 yr or until a particle collides with the Sun or a planet, or it is ejected from the solar system. The results of this simulation allow us to complete the orbital element distribution in the Centaur zone, since now we have also the objects that evolved as comets and return to Centaurs. We refer to the numerical simulation by Di Sisto & Rossignoli (2020) with the continuation of the evolution of the objects that entered the Jupiter region and returned to Centaurs as the "complete simulation".

From this simulation, we build maps of residence time in the Centaur zone. That is, we divide the (a, e, i) space into bins of size $\delta a = 0.02$ au, $\delta e = 0.005$ and $\delta i = 0.3°$ and calculate the normalized time fraction spent by particles from the SD (simulation by Di Sisto & Rossignoli 2020) and those that return from the JFC zone to the Centaur zone ($q < 5.2$ au and $a < 30$ au). This is represented in Figure 3.3, which shows the probability of finding Centaurs in different regions. Blue shows the most visited regions, while red shows the least visited. The black dots are the observed Centaurs. We have tested that the residence time reached a steady state, i.e., we have time-invariability, since the residence time shows no significant differences as time passes and the same results are reached toward the end of the simulation.

The orbital element region with the greatest probability of occupation is the one near Neptune. As we move away from Neptune and toward Jupiter, the intrinsic

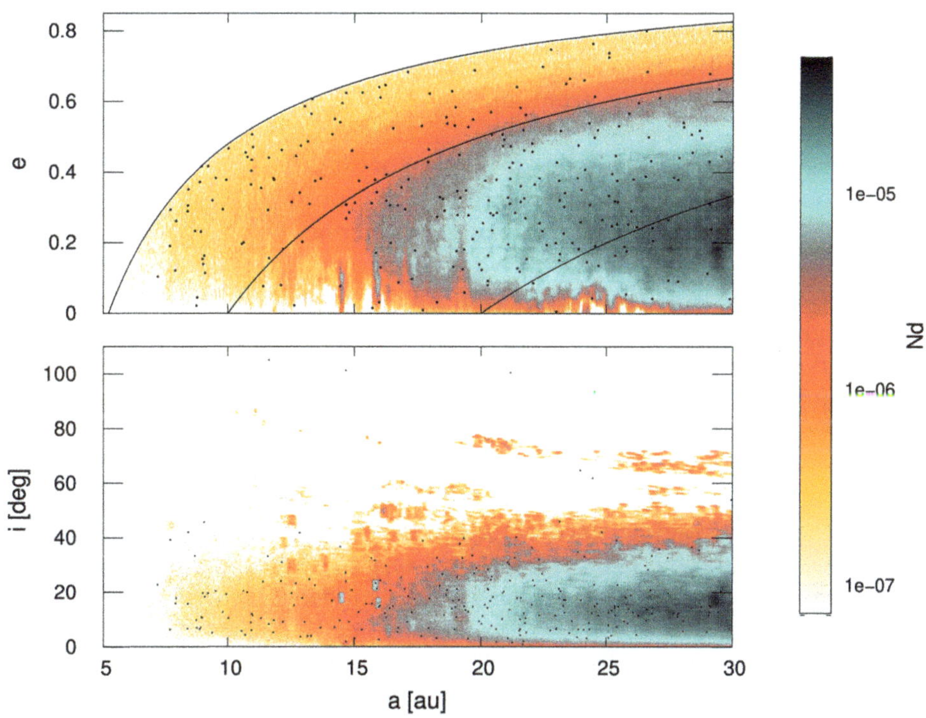

Figure 3.3. Normalized time-weighted distribution (Nd) of the (a, e) and (a, i) distributions of our intrinsic model of Centaurs. The color zones are regions with different degrees of probability where Centaurs can be found (black and turquoise for most visited regions, orange for least visited). The black curves correspond to constant perihelion values equal to the locations of the giant planets. The black dots are observed Centaurs.

density of objects decreases, as indicated by the color scale. Between Jupiter and Saturn, the evolution is very fast due to the stronger gravitational action of the more massive planets Jupiter and Saturn and deeper encounters with these planets. Therefore, the removal is significant, causing there to be fewer objects than in more distant regions. It can be seen that the observed Centaurs cover the entire region and there is no clear trend of more observed Centaurs at larger semimajor axes, like the intrinsic trend in our model. However, discovery of Centaurs is strongly biased toward objects with smaller perihelion distances (Nesvorný et al. 2019), and so it is expected that there will indeed be many more Centaurs as we move away from Jupiter and closer to Neptune when we obtain a more complete census.

Our Centaur model mostly covers the region with inclinations less than 40°, and there are only islands of regions with higher inclinations. We find almost no retrograde particles. Volk & Malhotra (2013) explored the evolution of inclinations in the Centaur region of particles initially in the classical trans-Neptunian and SD regions. They found that there is a very low probability for a high inclination or retrograde Centaur to originate in the TN region. In fact, some authors (e.g., Brasser et al. 2012, de la Fuente Marcos & de la Fuente Marcos 2014b) showed that high inclination and retrograde Centaurs originate mainly in the Oort Cloud. For the transfer to the Centaur region, the perihelion distances of such objects decrease on Gyr timescales until they have encounters with Neptune, which reduces their semimajor axes, introducing them into the GPC zone and eventually to the Centaur zone.

Other mechanisms have been invoked for high-inclination or retrograde Centaurs, but the issue remains under debate. For instance, Greenstreet et al. (2020) propose that the depopulation of huge numbers of primordial near-Earth asteroids might be able to leave a surviving tail of objects which could be the source of the high-i Centaurs. If the hypothetical Planet 9 exists, its perturbations would generate high-inclination Centaur-type objects and GPCs on long timescales, as various works have shown (see, for example, Batygin et al. 2019).

As mentioned, we didn't consider physical effects in the evolution of objects that enter the JFC region within Jupiter's orbit. Therefore, our model for the Centaur population presented in Figure 3.3 could include some objects that would have disintegrated by sublimation or splitting before again entering the Centaur zone (Di Sisto et al. 2009). However, since objects that return to the Centaur region after having passed through the JFC region evolve very quickly compared to objects that enter the region from the SD, we believe that our model is quite realistic, at least for objects with diameters larger than 5 km (Di Sisto et al. 2009).

The dynamical evolution from the SD to the Centaur zone is managed by the giant planet perturbations and encounters in the "handoff" path. The study of encounters with planets is interesting in particular for calculations of cratering on satellites of the giant planets (e.g., Di Sisto & Zanardi 2013; Rossignoli et al. 2019). It is also important for the stability and even survival of Centaur moons and rings to close encounters (see Sicardy et al. 2020; Chapter 9). From our complete simulation, we can calculate the statistics of encounters with the planets. In particular, we calculate the total number of encounters of particles with the giant planets within 1 Hill radius of the planet. The Hill radius of a planet, a measure of the planet's region of gravitational influence, is defined as $a_p(M_p/3(M_\odot + M_p)^{1/3})$, where M_p is the mass of the planet and $M_\odot$ is the mass of the Sun; measured in planetary radii, they range between ~720 for Jupiter and ~4700 for Neptune. From the total number of encounters, 71.4% are with Neptune, 17.7% with Uranus, 5.7% with Saturn and 5.2% with Jupiter. This means that the ratio of the number of encounters with the giant planets is approximately 12.5: 3.1: 1: 0.9. Very similar results were found by Nesvorný et al. (2023b) in their study of the steady-state impact flux of cometary/Centaur impactors on the giant planets and their satellites. They found that the ratio of the number of encounters with Uranus, Saturn, and Jupiter within 0.5 Hill radius is 3.5: 1: 0.9.

Those results are consistent with the increasing number of Centaurs with distance from the Sun as is shown in Figure 3.3. While it crosses the orbit of a planet, we found that on average a particle has an encounter with Jupiter every ~670 years, with Saturn every ~10,100 years, with Uranus every ~11,300 years and with Neptune every ~96,000 years. As can be seen, the frequency of encounters with Jupiter is greater and decreases toward Neptune, which is consistent with the more rapid evolution of the Centaurs in the region of Jupiter and Saturn for a given object. The rates calculated are intrinsic per object rates. The whole population rates of encounters should be scaled by the number of Centaurs at a given size. For example, Di Sisto & Rossignoli (2020) calculated the cumulative size frequency distribution of Centaurs for the whole size range from their dynamical model. Nesvorný et al. (2019) estimated, from OSSOS (Bannister et al. 2018) detections and calibration to Jupiter Trojans, that the number of Centaurs greater than a diameter D ($N(>D)$) is 21,000 ± 8000 for $D > 10$ km, 650 ± 250 for $D > 50$ km, and 150 ± 60 for $D > 100$ km. Lawler et al. (2018) estimated cumulative numbers (considering albedos $p_v = 0.06$) of $N(>D) = 3700^{+2300}_{-2000}$ for $D > 20$ km and $N(>D) = 130^{+80}_{-70}$ for $D > 100$ km. An analysis and discussion on the number and size distribution of known Centaurs can be found in Chapter 4.

It is thought that a Centaur with moons and/or rings could not survive a close encounter with a planet within the Roche limit (Silvia Giuliatti Winter, personal communication). From our complete simulation, we find that only 2% of Centaurs have an encounter with a planet in the age of the solar system within the Roche limit. (The number of passages within the Roche limit is comparable to the number of impacts with the planets; Kary & Dones 1996; Raymond et al. 2020). So, "on average," one should expect that most Centaurs with rings or moons could survive their dynamical evolution. But this could be very dependent on the particular system, e.g., Araujo et al. (2016, 2018), so more work is needed to have a general overview (see Chapter 9).

Volk (2013) and Volk & Malhotra (2008, 2013) studied the dynamical stability of trans-Neptunian subpopulations and their relation to Centaurs and JFCs. For the SD they built an intrinsic model and found that the intrinsic escape rate of SDOs (an object escapes the population if it enters the Hill sphere of a planet) with a 5% error is $(1 - 2) \times 10^{-10}$ yr^{-1}, where the former is the average over the last 1 Gyr and the latter is the average over the first 1 Gyr of the simulation. For the classical belt, they used as initial conditions the observed CTNOs plus clones and obtained an intrinsic escape rate of 0.55×10^{-10} yr^{-1}. Volk (2013) also studied the resonant populations using observed objects as initial conditions. Volk (2013) found an intrinsic escape rate of $(3 - 7) \times 10^{-11}$ yr^{-1} for the 2:3 MMR and $(1 - 2) \times 10^{-10}$ yr^{-1} for the 1:2 MMR.

The dynamical evolution of escaped plutinos was also analyzed by Morbidelli (1997) and Di Sisto et al. (2010). Morbidelli (1997) studied the dynamical structure of the 2:3 MMR and found a slow chaotic diffusion zone related to long-term escaping plutinos that should be a current source of Centaurs and JFCs. They found that only 10% of the Plutinos in this weakly chaotic zone have been delivered to Neptune in the past Gyr. So, the current delivery rate from the 2:3 MMR is $\approx 1 \times 10^{-10}$ yr^{-1}. Di Sisto et al. (2010) performed a numerical study of plutinos to detect long-term escapees following the work of Morbidelli (1997). By selecting those long-term escapees from their simulations, they calculate a current escape rate from the 2:3 MMR of 1.62×10^{-10} yr^{-1} and a rate of injection to the Centaur zone (i.e., the first time plutinos reach $a < 30$ au) of 1.32×10^{-10} yr^{-1} (Di Sisto & Rossignoli 2020).

Muñoz-Gutiérrez et al. (2019) also studied the transfer from the TN region to JFCs. They used the L7 synthetic model based on the CFEPS survey (Petit et al. 2011), which represents the unbiased orbital distribution of TNOs down to an absolute magnitude of $H_g \leqslant 8.5$ (diameter $\gtrsim$140 km for an assumed 5% albedo). The model comprises classical, resonant and scattering TNOs. Muñoz-Gutiérrez et al. (2019) performed a set of numerical simulations to follow the dynamical evolution of those populations for 1 Gyr under the perturbations of the four giant planets and the 34 largest known TNOs (roughly speaking, the known dwarf planets). They studied the evolution to the GPC zone up to the JFC zone ($2 < T_J < 3$ and $q < 2.5$ au). They found that the dwarf planets globally increase the number of JFCs by 12.6%, compared with the number produced by the giant planets alone. The authors also found that of all the MMRs in the TN region, plutinos in the 2:3 MMR contribute the most to the JFC population ($\sim$45 % of all the originally resonant TNOs).

We summarize all the results regarding the intrinsic rate of injection to the GPC, Centaur, and JFC zones in Table 3.1. As we have mentioned, there is no consensus on the definitions of the different populations of the TN region and the region of the giant planets. In Table 3.1 we have considered the injection rates of the listed works, although there are slight differences in terms of the definitions of both the source regions and the sink regions. In any case, it is possible to see that, in general, the greatest intrinsic injection rates to GPCs, Centaurs, and JFCs are those from the SD; though there is still work to be done to evaluate the real contribution of all the MMRs of the TN region. This contribution could be important; in fact, the rate calculated by Muñoz-Gutiérrez et al. (2019) of the contribution of 11 resonances to the JFCs is the highest of all populations.

Table 3.1. Intrinsic injection rates to the GPC, Centaur and JFC.

Source population	$\dot{C}_{\mathrm{GPCs}}$	$\dot{C}_{\mathrm{Centaurs}}$	$\dot{C}_{\mathrm{JFCs}}$
SDOs, Di Sisto & Rossignoli (2020)	4.025	1.796	0.77
SDOs, Brasser & Morbidelli (2013)	1.0–2.3	-	0.10–0.44
SDOs, Volk & Malhotra (2008, 2013)	-	1.05–1.2	0.24–0.54
SDOs, Duncan & Levison (1997)	2.7	-	-
SDOs, Muñoz-Gutiérrez et al. (2019)	-	-	1.5
CTNOs, Duncan et al. (1995)	-	-	0.5
CTNOs, Volk (2013)	-	0.55	0.13–0.26
CTNOs, Muñoz-Gutiérrez et al. (2019)	-	-	0.306
Plutinos, Morbidelli (1997)	1	-	-
Plutinos, Di Sisto et al. (2010)	1.62	1.3	-
11 MMRs, mainly 2:5 and 2:3, Muñoz-Gutiérrez et al. (2019)	-	-	1.07
2:3 MMR, Volk (2013)	0.3–0.7	-	-
1:2 MMR, Volk (2013)	1–2	-	-
Jupiter Trojans (L4–L5), Di Sisto et al. (2019)	0.704–0.756	-	-
Jupiter Trojans, Levison et al. (1997)	0.29	-	-

Note: Intrinsic injection rates to the GPC ($\dot{C}_{\mathrm{GPCs}}$), Centaur ($\dot{C}_{\mathrm{Centaurs}}$), and JFC ($\dot{C}_{\mathrm{JFCs}}$) zones from other populations given in units of 10^{-10} yr^{-1}. Due to lack of consensus on the definitions of the different populations of the TN region and the region of the giant planets, there are slight differences in terms of the definitions of both the source regions and the sink regions in the injection rates of the listed works.

3.2.3 Secondary Sources

The GPC and Centaur regions have a transitional nature for incoming TNOs, but also for other small body populations. Objects like Jupiter Trojans, Hilda asteroids, and even certain main-belt asteroids make their first escapes toward this region. Even objects in the Oort Cloud that reach Neptune for the first time are transferred there (Brasser et al. 2012). The action of the giant planets in the region controls the passage through it and the transfer to other regions through their perturbations and/or encounters. A nice schema of the dynamical evolution of Centaurs and the minor body populations related to them is shown in Figure 7.1 in Chapter 7. In this diagram, the dynamical links between sources and sinks of Centaurs are shown.

Jupiter Trojans are a well-known population occupying relatively stable orbits around the L4 and L5 Lagrangian points of Jupiter. However, there are regions near the nominal resonance with stability timescales less than the age of the solar system, allowing some Trojans to escape from the resonance (Levison et al. 1997; Di Sisto et al. 2014, 2019; Holt et al. 2020). Levison et al. (1997) were the first to investigate the long-term dynamical evolution of Jupiter Trojans through numerical simulations. Their research revealed stable regions within the Trojan population and identified areas from which Trojans slowly disperse and transition onto orbits resembling those of JFCs. It was found that the probability of a Trojan leaving the swarm is approximately 2.9×10^{-11} yr^{-1}.

The primary destinations for escaped Trojans are the adjacent Centaur and JFC zones. In fact, Di Sisto et al. (2019) found that approximately 90% of escaped Trojans pass through the GPC and Centaur regions, thereby populating them with a distribution similar to that of observed Centaurs. Their research also determined a constant escape rate from both Lagrangian L4 and L5 points of 7.04–7.56 $\times 10^{-11}$ yr^{-1}, which is significantly lower than the escape rates from the SD or MMRs (see Table 3.1) and the same order as that found by Levison et al. (1997). Given that the population of Jupiter Trojans is only $\approx 10^{-3}$ that of SDOs (Nesvorný et al. 2017; Uehata et al. 2022), and also that their intrinsic injection rate is very low, their contribution to the GPC and Centaur zones is minimal.

Holt et al. (2020) conducted research on the escape dynamics of Jupiter Trojan swarms, focusing on identified Trojan collisional families through numerical simulations. Their findings revealed that members of known collisional families generally have lower escape rates compared to the overall Trojan swarm, primarily because they tend to be located in the most stable regions. In particular, the (9799) Thronium (1996 RJ), (20961) Arkesilaos (1973 SS_1) and (2247341) Shaulladany (2001 UV_{209}) families are highly dynamically stable over the age of the solar system. The L4 (624) Hektor and L5 (4709) Ennomos families show some escapes. The (3548) Eurybates family, which has the largest population of the known Trojan families, also has a smaller escape rate than the overall population. Eurybates and its small moon Queta are the targets of the Lucy mission's first flyby of a Trojan in 2027 (Noll et al. 2020).

Hildas are another compact group of asteroids, orbiting at ~4 au from the Sun in a 3:2 MMR with Jupiter, characterized by a very stable region in the central zone of the resonance (Ferraz-Mello et al. 1998). However, the stable zone is surrounded by a strongly unstable region, from which an asteroid can escape in a short time. In fact, there is a well-known zone close to the Hilda region where quasi-Hilda comets are found (Kresak 1979). In this region, comets that come from the Centaur zone (and so from the TN region; Fernández 1980; Levison et al. 1997) and escaped Hildas (Di Sisto et al. 2005) coexist. This is an excellent example of the exchange that exists between populations of small bodies in our solar system, including asteroids and comets. In fact, it has been reported that 11 (Gil-Hutton & García-Migani 2016) and 47 (Correa-Otto et al. 2024) quasi-Hilda comet candidates have recently arrived from the Centaur zone. As part of a citizen science project, Oldroyd et al. (2023a) discovered that the quasi-Hilda 2009 DQ_{118} had been active four months before its perihelion passage in 2016. The object was again active near its perihelion passage in 2023, suggesting that its activity is driven by sublimation (Oldroyd et al. 2023b).

Collisions with other asteroids are the most efficient mechanism that injects Hilda fragments into the unstable regions. Di Sisto et al. (2005) found that asteroids leave the resonance, increasing their eccentricity and semimajor axis, which leads them to traverse the GPC zone, where they experience successive encounters with the giant planets, primarily Jupiter. Jupiter dominates the dynamics of escaping Hildas, which usually pass through an MMR with Jupiter.

Kazantsev & Kazantseva (2021) analyzed the possibility that asteroids from the main belt are transferred to the Centaur region. They found that some asteroids initially in the 2:1 MMR near 3.3 au become Centaurs due to encounters with Jupiter. Although the number of transitions is low, some such asteroids may contribute to the Centaurs. Also, several main-belt comets, including the prototype, 133P/Elst–Pizarro, have orbits within or near the Themis family, which is close to the 2:1 MMR (Snodgrass et al. 2017; Hsieh et al. 2020). Liberato et al. (2023) found that 10% of Near-Earth Asteroids reach the Centaur region after spending some time in the JFC zone and have a median residence time inside this region of ~100,000 years.

Neptune Trojans are another important resonant population that could contribute to GPCs and Centaurs. Although the observed sample consists of only 31 objects[7] at present (Markwardt et al. 2023), it is thought to host a large population, perhaps even more numerous than the Jupiter Trojans (Chiang & Lithwick 2005; Sheppard & Trujillo 2010; Alexandersen et al. 2016; Lin et al. 2021). In addition, temporary Trojans of both Neptune and Uranus have been found; these objects are probably recently captured Centaurs or scattering objects (Guan et al. 2012; Horner & Lykawka 2012; Alexandersen et al. 2013). So, this population and the surrounding zone is another excellent example of the exchange and coexistence of small bodies. Horner & Lykawka (2010) found that escaped 1-km Neptune Trojans enter the GPC region every ~60–2000 years. However, the smallest Neptune Trojan known has a diameter of ~50 km (assuming a 5% albedo), so this estimate involves a large extrapolation. Therefore, more work must be done to be able to precisely estimate the number of Neptune Trojans and their contribution to Centaurs.

Secondary sources of Centaurs, while small in number, play a significant role in enriching our understanding of this population, especially considering the physical and dynamical diversity within it. For instance, the presence of retrograde objects, which cannot be readily explained as originating from the TN region, suggests alternative source regions like the Oort Cloud (e.g., Brasser et al. 2012). Additionally, although most Centaurs do not display cometary activity, some, with perihelion distances as large as 12 au, do (Jewitt 2009; Wierzchos et al. 2017; Womack et al. 2017; Pinto et al. 2023; Fraser et al. 2024). The distinction between active and inactive Centaurs hints at potential differences in their origins and evolutionary paths. These secondary sources contribute valuable pieces to the broader puzzle of Centaur dynamics and origins.

3.3 Dynamics in the GPC and Centaur Zones

3.3.1 Atlas of Resonances. Temporary Stability versus Chaos

For GPCs, in the domain of orbital elements there is little room for stable dynamical evolution. Close encounters and random planetary perturbations are the main fuel for the orbital evolution. Close encounters drive the handoff mechanism between the giant planets, which is graphically shown in Figure 3.4. In particular, for objects with perihelion distances within Saturn's orbit ($q \lesssim 10$ au), the evolution is very fast due to typical changes in semimajor axis Δa one or two orders of magnitude greater

[7] https://minorplanetcenter.net//iau/lists/NeptuneTrojans.html

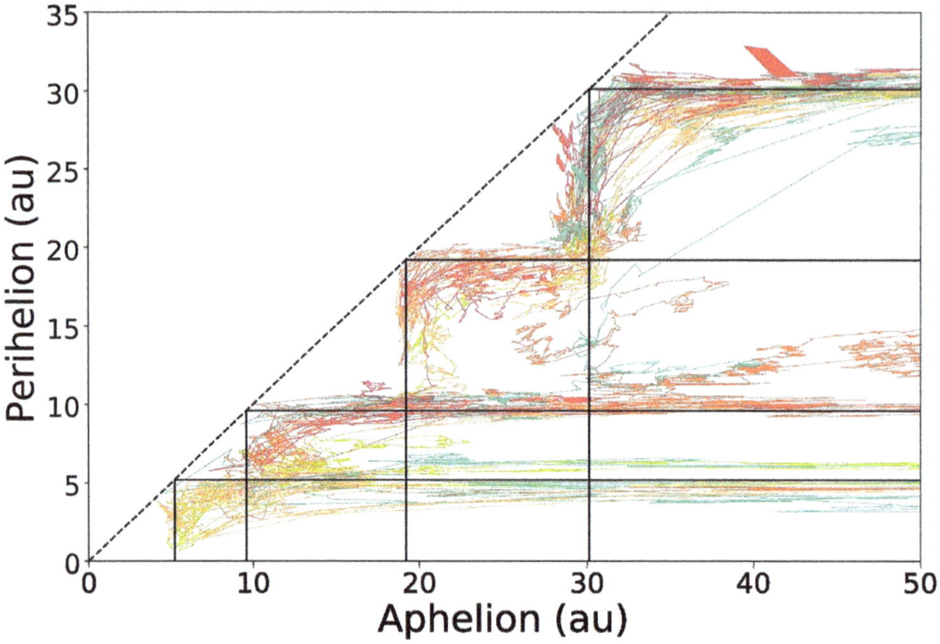

Figure 3.4. Fictitious particles evolving between the giant planets by the handoff mechanism. Different colors indicate different particles. Reprinted from Roberts & Muñoz-Gutiérrez (2021). Copyright 2021, with permission from Elsevier.

than for the rest of the population. This is a known fact for the actual Centaur population (Tiscareno & Malhotra 2003; Di Sisto & Brunini 2007), but we have verified with numerical integrations of fictitious particles that orbits with arbitrary inclinations show the same behavior. So, it is clear that encounters with Saturn strongly accelerate the dynamical evolution of objects with arbitrary inclinations.

The handoff mechanism is not the only possible way to diminish q from the trans-Neptunian region to the Centaur zone. Although it is rarer and valid only for medium and high inclinations, the secular evolution by the ZLK mechanism can reduce the perihelion from Neptune to the interior region, as shown by Gallardo et al. (2012). Figure 3.5, taken from that work, shows a possible path through secular evolution that takes the perihelion towards the region of Saturn without close encounters with the other giant planets.

The only mechanism that gives some protection against jumps in semimajor axis is capture inside strong MMRs with the giant planets. For example, when integrating test particles Gabryszewski & Włodarczyk (2003) found temporary islands of stability associated with MMRs, where particles were trapped for timescales of 10^4–10^5 years. Some years later, by calculating Hurst exponents, Bailey & Malhotra (2009) classified Centaurs into two classes: one is characterized by diffusive evolution of semimajor axis and the other, more long-lived group, is dominated by resonance hopping (Wood et al. 2018). Now, using the semi-analytical model by Gallardo (2020), which is valid for resonant objects with arbitrary orbital

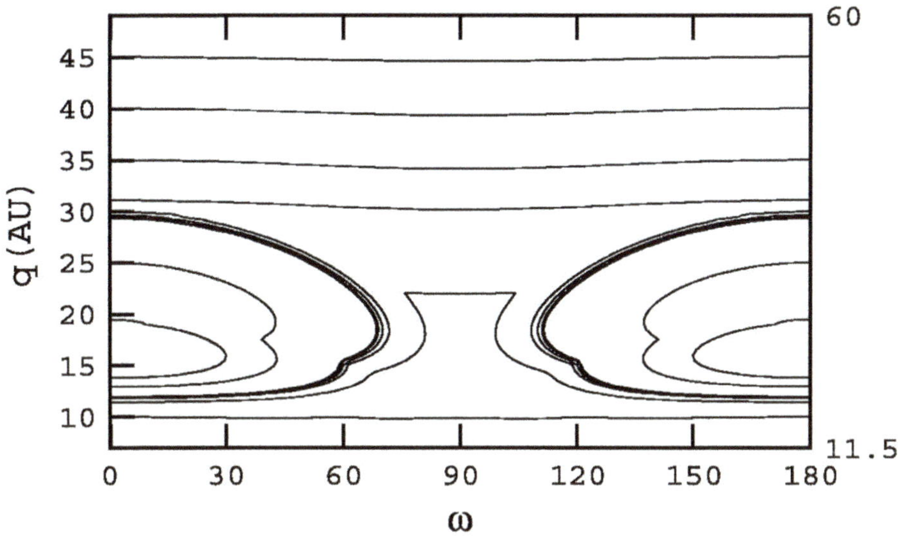

Figure 3.5. Possible dynamical paths in the plane (ω, q) for particles with $a = 50$ au by the ZLK mechanism generated by the giant planets. The right-hand axis represents the orbital inclination in degrees, which has direct correspondence with q. Reprinted from Gallardo et al. (2012). Copyright 2012, with permission from Elsevier.

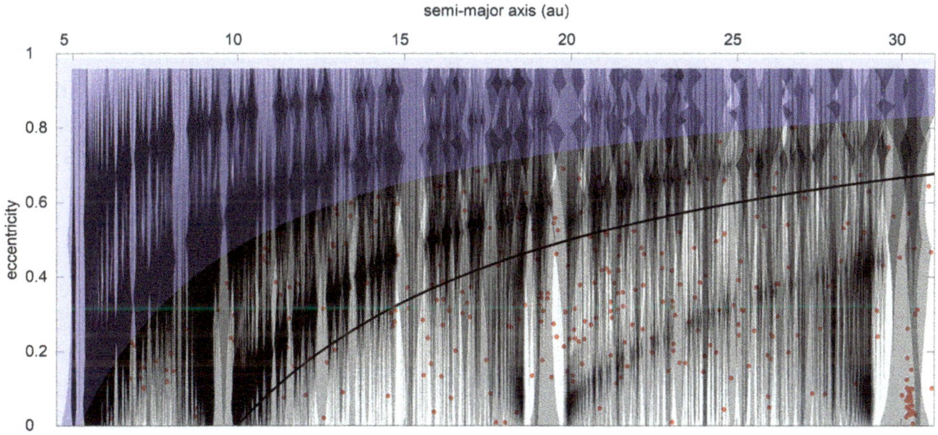

Figure 3.6. Maximum widths of resonances with the giant planets for a test particle with $i = 30°$, $\omega = 60°$. The black curve corresponds to $q = 10$ au. The blue region corresponds to $q < 5.2$ au, which we consider to be the comets' domain. Known direct Centaurs plus Neptune's Trojans are plotted with red points.

elements, we calculate two atlases of resonances, one representative for direct orbits (Figure 3.6) and the other one for retrograde ones (Figure 3.7). These atlases show the total width of each of ~600 MMRs with the giant planets as a function of the orbital eccentricity, assuming fixed values for the other orbital elements. Each resonance domain is plotted in grey, so dark regions correspond to resonance overlap and that means chaos.

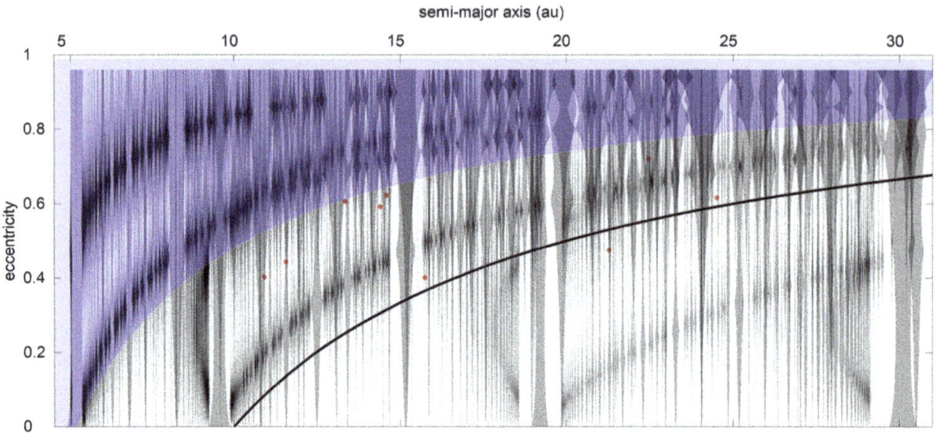

Figure 3.7. Same as Figure 3.6 for orbits with $i = 150°$, $\omega = 60°$. Known retrograde Centaurs are plotted with red points.

Figure 3.6 shows the resonance domains calculated for orbits with $i = 30°$ and $\omega = 60°$. In general, there is an evident overlap of resonances with an extremely chaotic region for $q \lesssim 10$ au; for such perihelion distances, temporary captures only can happen at very strong (or wide) and isolated resonances like the exterior 1:n resonances with Jupiter or Saturn. Figure 3.7 shows resonance domains calculated for orbits with $i = 150°$ and $\omega = 60°$. The known Centaur population is plotted in both figures. It is possible to make some guesses about objects captured in MMRs, but we will study this point below using more detailed figures given by dynamical maps. For retrograde orbits, resonance overlap takes place for very low q, less than 5 au, in the cometary region. More frequent and stable captures in MMRs can be observed for retrograde objects (Gallardo 2019) and the absence of overlap of resonances results in more regular dynamics. Note that the 1:1 resonances of Uranus and Neptune do not overlap with other MMRs, but in the case of Saturn, the direct orbits in the 1:1 resonance are affected by other resonances in its neighborhood, with presumably some effect on its stability. In particular, Zhou et al. (2020) and Wood (2023) showed that only low (e, i) test particles can survive on timescales of 0.1 to 1 Gyr as Uranus Trojans. While for Jupiter and Neptune there are cosmogonic reasons for their respective populations of Trojans (Morbidelli et al. 2005; Nesvorný & Vokrouhlický 2009; Nesvorný et al. 2013), Saturn and Uranus could instead capture coorbitals in the process of handoff from the TN region and also from other minor body populations (see Section 3.2.3). Alexandersen et al. (2021) showed that all the giant planets should have temporary coorbitals of TNO origin and that the durations of the coorbital captures are significantly shorter for Saturn and Jupiter than for Uranus and Neptune.

3.3.2 Dynamical Maps. Centaurs and Resonant Structure

While the atlas gives us a general panorama of all possible resonances, details of the dynamics of particular MMRs and the eventual chaos can be revealed by dynamical maps. They are constructed from numerical integrations of the exact equations of motion of the outer solar system plus thousands of test particles with the same initial orbital elements, except for (a, e), which we vary in a grid covering the region $0 < e < 0.4$ and 5 au $< a <$ 31 au. The numerical integrations are typically carried on for 100 orbital revolutions of the test particle. Then in that interval we calculate changes in the orbital elements that the particle experiences. Representing $\log(\Delta a)$ with a color code scale, we can have an idea of the dynamics of the region. Cold colors (black, blue) represent very small variations in a, which are typical of a particle evolving with stable secular evolution or evolving close to the center of a resonance (with a very small libration amplitude). Hot colors (red, yellow, white) represent larger variations in a, mainly due to particles inside MMRs (far from the center), including chaotic zones due to MMRs overlapping and also large Δa due to close encounters with planets, generally in the color white. Maps vary according to the chosen particle's initial orbital elements, especially the white bands due to close encounters with planets. Moreover, Δa values due to MMRs are sensitive to the chosen initial mean anomaly of the test particle, and sometimes it is not the best value to reproduce the maximum domain of the resonance. For this reason, in general the dynamical maps here tend to show the resonant domains as being somewhat smaller than the maximum domain, which is better defined by the atlas shown in Figure 3.6.

The dynamical maps we show in Figures 3.8–3.10 were computed specially for this work, taking particles' orbits with $i = 30°$, $\Omega = 0°$ and $\omega = 60°$, so they are more or less representative for Centaurs with direct orbits. These maps show that the region is dominated by several MMRs with the giant planets, in agreement with Figure 3.6. Overlap of resonances is very common and only in some low-eccentricity regions is there a place for stable secular evolution. The dynamics inside MMRs

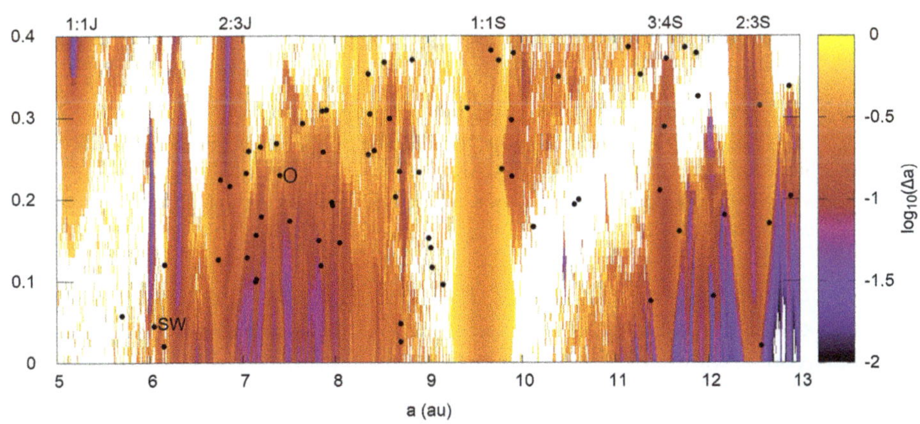

Figure 3.8. Dynamical map for particles with initial $i = 30°$, $\Omega = 0°$, $\omega = 60°$ for the region between 5 au and 13 au with the known Centaur population plotted as black points. Schwassmann–Wachmann 1 and Oterma are labeled. The color scale represents $\log(\Delta a)$. Some relevant resonances are indicated with labels. The 1:1J and 1:1S resonances are not well-mapped due to the chosen initial conditions. See also Table 3.2 for the identification of some resonant Centaurs.

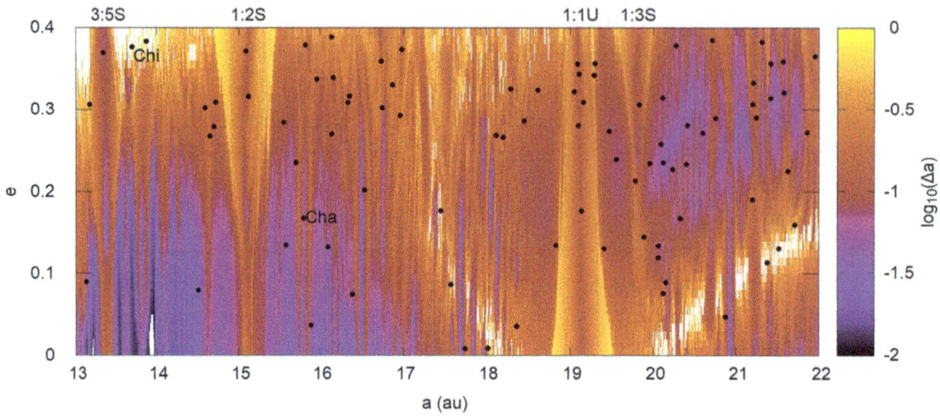

Figure 3.9. Same as Figure 3.8 for the region between 13 and 22 au. Chiron and Chariklo are labeled. The 4:3U resonance is at $a \sim 15.8$ au, the 5:4U is at $a \sim 16.5$ au, and other resonances are labeled. The wide region of coorbitals with Uranus is evident at $a \sim 19.2$ au. The 1:3S resonance shows at high eccentricity for $a \sim 19.8$ au. See also Table 3.2.

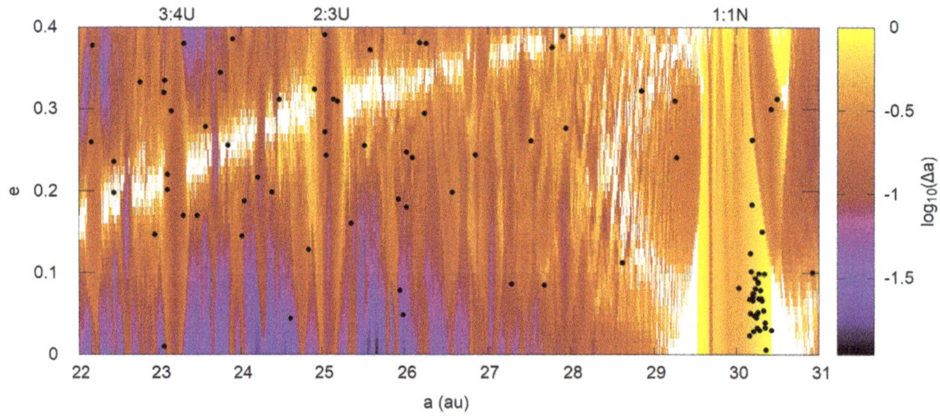

Figure 3.10. Same as Figure 3.8 for the region between 22 and 31 au. Several resonances with Uranus and Neptune dominate here. At $a \sim 24$ au the resonances 5:7U and 7:5N overlap. At $a \sim 25.1$ au several resonances overlap, but 2:3U dominates. The right part of the figure is dominated by the structure of the 1:1N resonance. See also Table 3.2.

deserves more study, but one can guess that Centaurs inside the wide strong resonances could evolve trapped in resonance, while when inside weak resonances, jumps between resonances will dominate and chaotic dynamics will be the rule (Bailey & Malhotra 2009). In Table 3.2 we show some observed Centaurs known to be evolving in MMRs, some of them identified for this work. In Figure 3.11 we show the evolution of the fictitious particle shown in Figure 3.2, but now in the plane (a, e) with the atlas of resonances as background. It begins not inside, but stuck to the left side of 1:1N, then it is temporarily captured in several MMRs, and evolves quickly when $q \lesssim 10$ au.

Table 3.2. Resonant Centaurs

Designation	a (au)	e	i(°)	Code	Res	σ(°)	Reference
698876 (2019 AJ16)	11.48	0.21	10.6	2	3:4S	180	This work
523791 (2015 HT171)	11.52	0.28	33.2	2	3:4S	180	This work
2010 LN68	12.58	0.02	15.0	9	2:3S	180	This work
773892 (2022 SJ21)	12.65	0.16	18.9	1	2:3S	180	This work
472265 (2014 SR303)	15.13	0.31	3.0	2	1:2S	HS	This work
309139 (2006 XQ51)	15.83	0.37	31.6	2	4:3U	180	This work
2001 XZ255	15.88	0.04	2.6	4	4:3U	0	1
472651 (2015 DB216)	19.03	0.32	37.8	2	1:1U	temp HS	2
2002 VG131	19.10	0.34	21.2	9	1:1U	temp QS	3
636872 (2014 YX49)	19.11	0.27	25.5	1	1:1U	Trojan	4
687170 (2011 QF99)	19.16	0.17	10.7	2	1:1U	Trojan	5
83982 Crantor (2002 GO9)	19.47	0.27	17.8	1	1:1U	HS	6
2010 EU65	19.87	0.14	14.6	3	1:1U	HS	7
2003 QD112	19.09	0.58	14.5	4	1:1U	QS	8
2012 DS85	18.84	0.13	16.8	5	1:1U	HS	8
2022 OH10	19.53	0.24	22.1	3	1:1U	HS	8
2005 TH173	19.85	0.30	13.4	6	1:3S	270	This work
55576 Amycus (2002 GB10)	25.00	0.39	13.3	1	2:3U	180	6
2015 DQ249	25.03	0.27	41.9	9	2:3U	HS	This work
533396 (2014 GQ53)	25.16	0.30	22.7	2	2:3U	HS	This work

Note. Some Centaurs identified in mean motion resonances on timescales of 10^4 to 10^6 yr. Code refers to the orbital condition code given by JPL (smaller is better), σ is the characteristic critical angle and its libration center is indicated. Quasi satellites are noted as QS and horseshoes as HS; "temp" means "temporary"and this implies in resonance for less than 10^6 yr. References:

1: Masaki & Kinoshita (2003);

2: de la Fuente Marcos & de la Fuente Marcos (2015);

3: de la Fuente Marcos & de la Fuente Marcos (2014a);

4: de la Fuente Marcos & de la Fuente Marcos (2017);

5: Alexandersen et al. (2013);

6: Gallardo (2006);

7: de la Fuente Marcos & de la Fuente Marcos (2013);

8: Pan & Gallardo (2025)

The region corresponding to $e > 0.4$ for direct orbits not shown in the maps is strongly chaotic. Retrograde orbits exhibit a similar pattern of MMRs, but they are notably weaker as Figure 3.7 shows, so overlap is not very common and the evolution inside MMRs is more stable as long as close encounters with planets are avoided (see Li et al. 2019). Numerical simulations of test particles and clones of real objects show frequent captures of retrograde objects in MMRs with the giant planets, mainly Jupiter, but in general that happens for $q < 5.2$ au, corresponding to the category of comets (Gallardo 2019; Li et al. 2019). However, it is important to note that evolving within an MMR with a giant planet provides some protection against close encounters, but only with that planet. As can be seen in the example in

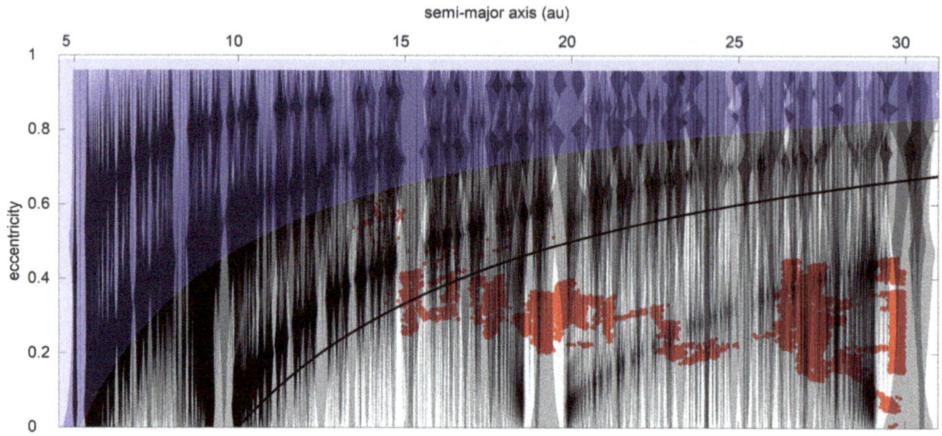

Figure 3.11. Same as Figure 3.6, but with the dynamical evolution over ~16 Myr of the particle shown in Figure 3.2 represented with red points. It begins not inside but stuck to the left side of the 1:1N, then it is temporarily captured in several MMRs, and evolves quickly when $q < 10$ au.

Figure 3.2, the ZLK mechanism within the MMR can change the perihelion distance, enabling close encounters with other planets and disrupting the resonant state.

3.4 Centaurs as a Source of Comets

The need to account for the distinctive characteristics of the JFCs initially drove the proposal of the trans-Neptunian region as a source for these comets (Fernández 1980; Duncan et al. 1988). Subsequently, numerous studies have concentrated on elucidating this connection and quantifying the number of JFCs originating from these populations. This is evident in Table 3.1, which showcases research that directly computes the injection rate of objects into JFCs. However, as we have discussed, the dynamics in the GPC region is fundamental in comprehending how objects become JFCs and manifest their distinct characteristics.

Early dynamical studies of Centaurs integrated the orbits of actual objects and "clones" with slightly different initial conditions to account for their chaotic orbital evolution (Tiscareno & Malhotra 2003; Horner et al. 2004a, 2004b). This work showed that not all Centaurs end up becoming JFCs. For example, Tiscareno & Malhotra (2003) found that most of them escape from the solar system and approximately one third of Centaurs are injected into the JFC zone. Several more recent investigations treat the complete orbital histories of realistic Centaur source populations in the trans-Neptunian region (Di Sisto & Brunini 2007; Nesvorný et al. 2017; Di Sisto & Rossignoli 2020). Di Sisto & Rossignoli (2020) found that 70% of SDOs are delivered to the GPC zone and of those, 22% enter the JFC zone, defined as having perihelion distance $q < 5.2$ au.

Fernández et al. (2018) separately studied the dynamics of active and inactive Centaurs and found different dynamical paths and end states. For objects that evolve to perihelion distances less than 1.3 au, the active ones evolve with the

Tisserand parameter with respect to Jupiter T_J between 2 and 3, while the inactive ones can have arbitrary T_J in their dynamical evolutions. They found that active Centaurs are strongly linked to JFCs and almost always evolve into that class of comets. On the other hand, inactive ones tend to evolve into Halley-type comets (HTCs), and sungrazing is a frequent final state of this class of Centaurs. Indeed, Di Sisto et al. (2009) found that JFCs have significantly shorter mean lifetimes due to erosion processes compared to their dynamical lifetimes. By considering a model that incorporates sublimation and splitting, it was revealed that, for instance, most 1-km comets ultimately disintegrate completely. However, without accounting for these physical processes, they might continue to undergo dynamical evolution, and some could even return to the Centaur region. In fact, larger comets can return to the Centaur zone, but they experience a very quick dynamical evolution among the giant planets until ejection.

So, evolution in the Jupiter-Saturn region can be dramatic for an object, as we commented in Section 3.3. Encounters with those planets accelerate their evolution from and into Centaurs and their final ejection. But also, close encounters promote tidal effects on the objects and its eventual collision with the planet, as was the case of Shoemaker-Levy 9 with Jupiter in 1992–1994.

Sarid et al. (2019) found a region of low-eccentricity orbits with $q > 5.4$ au and aphelion distance $Q < 7.8$ au, which they aptly termed a "Gateway" for transitioning between Centaurs and JFCs. The most notable object in this Gateway region is the well-known active Centaur 29P/Schwassmann–Wachmann 1 (SW1; see Chapter 10 for a detailed discussion of this object). Sarid et al. (2019) performed numerical simulations to integrate test particles representing TNO populations for 500 Myr or until they reached heliocentric distances either greater than 2000 au or less than 4 au. They recorded orbital elements of the particles that entered the Centaur population (defined as $q > 5.2$ au and $Q < 30.1$ au), which allowed them to produce high-resolution residence maps in this zone (Figure 2 of their paper, which we reproduce as Figure 3.12 here). They found that 21% of Centaurs enter the JFC Gateway and spend a very short time there, on the order of thousands of years (see their Table 1 for details). From here, 77% of the objects will become JFCs. Sarid et al. (2019) also found that the transition through this Gateway has important thermophysical consequences on the objects that can experience processing of volatile material on the surface and in the interior. However, Guilbert-Lepoutre et al. (2023) found that the majority of Centaurs, approximately 80%, make their first transition to the JFC population from outside the Gateway. Only about 20% of them pass through the Gateway region. Additionally, more than half of their JFC particles enter the Gateway for the first time after already being JFCs. This suggests that objects within the Gateway may have a higher likelihood of being thermally evolved rather than pristine, as they would have already traversed the JFC region before entering the Gateway. This highlights the complexity of dynamical and thermal evolution in these regions (see Chapter 7 for a review of Centaurs' evolution processes). In fact, Gkotsinas et al. (2022) performed a coupled thermal and dynamical evolution of a sample of model JFCs implanted from the KB and SD. The JFCs of the sample were active and observable, i.e., they reached perihelion distances less than 2.5 au at some

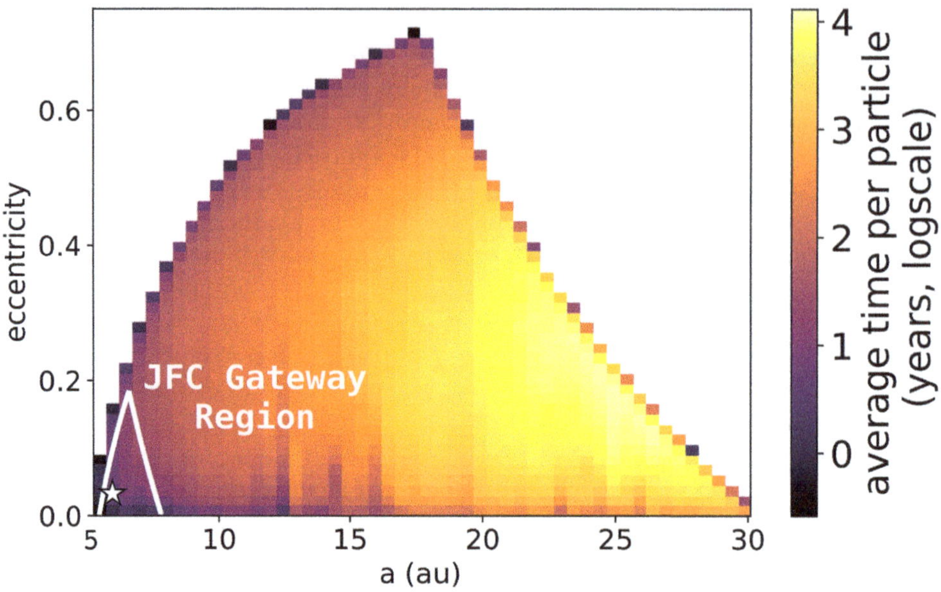

Figure 3.12. Average time spent (log scale; in color) per Centaur particle. The "Gateway" region is indicated by white lines, while comet 29P/Schwassmann–Wachmann 1 is shown as a small star. Reproduced from Sarid et al. (2019).

point. They obtained the internal temperature distribution and evolution of the JFCs, which allowed them to assess the degree of thermal processing. They found that due to the chaotic nature of the orbital evolution of a comet nucleus, which usually enters and leaves regions of significant heating in the inner solar system, multiple heating episodes occur during a JFC's life. This gives rise to thermally induced alteration processes to occur in deep layers below the JFC's surface that can extend as deep as ~4100 m on average. So, they infer that the JFC activity comes from deeper layers within the comet that have been altered by thermal processing, and so the primordial condensed hypervolatiles in those layers should be lost.

Steckloff et al. (2020) investigated the dynamical evolution of the comet P/2019 LD2 through numerical simulations and found that it is currently passing through the dynamical Gateway. In 2063, it will have a close encounter with Jupiter that will transfer it into the JFC region. On the basis of this object, they calculate a median frequency of transition from the Gateway to the JFC region of once every ~3, 70, 340 and 2700 yr for objects of ~1 km, 3 km, 5 km and 10 km, respectively.

Seligman et al. (2021) investigate the dynamical transfer of Centaurs into the inner solar system by performing N-body simulations of a steady-state population of Centaurs from Di Sisto & Brunini (2007). In particular, they investigate the effect of first- and second-order MMRs with Jupiter and Saturn on objects in this region to analyze the mechanism that generates objects like P/2019 LD2 in short timescales. They found that the overlap of MMRs close to Jupiter could be responsible for the injection of bodies and that there may be additional objects transitioning into the inner solar system in the near future. Figure 3.13 shows orbital evolutions for four

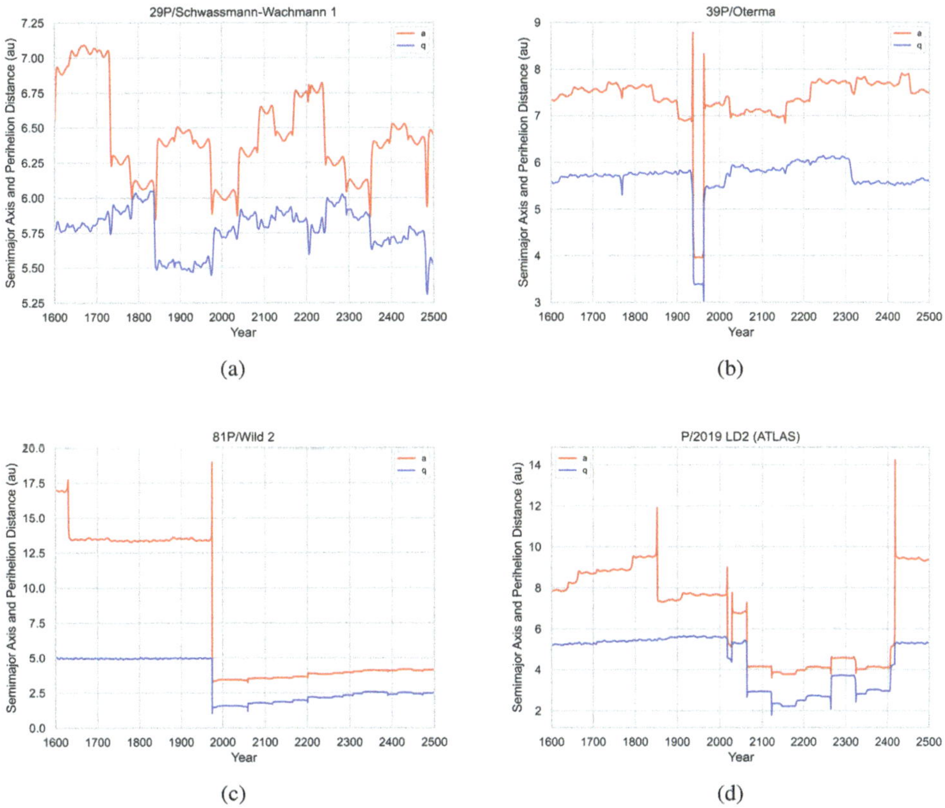

Figure 3.13. Nominal orbital evolution of four Centaurs for 900 years from 1600 to 2500, using orbital elements from JPL. Each panel shows the object's semimajor axis (a) in red and its perihelion distance (q) in blue. (a) The prototype Gateway comet, 29P/Schwassmann–Wachmann 1 (SW1); a varies between 5.8 and 7.1 au, while q varies between 5.3 and 6.1 au. When q is near its maximum value, the orbit is nearly circular, with the eccentricity sometimes less than 0.01. (b) 39P/Oterma was a Centaur that orbited near the 2:3 resonance with Jupiter ($a \approx$ 6.8 au). After an encounter within Jupiter's Hill sphere in 1937, it became a JFC that orbited near the 3:2 resonance ($a \approx$ 3.98 au). Another close encounter with Jupiter in 1963 put 39P back onto a Centaur orbit near the 2:3 resonance. Such "resonance hopping" is common for Centaurs and JFCs (Section 1.3; Belbruno & Marsden 1997; Koon et al. 2001). (c) 81P/Wild 2, the target of the Stardust mission, underwent a very close approach to Jupiter (within the orbit of Ganymede) in 1974 that changed it from a Centaur with perihelion near Jupiter's orbit to a JFC with aphelion near Jupiter's orbit and $q \approx 1.5$ au (Królikowska & Szutowicz 2006). (d) P/2019 LD2 (ATLAS) is a Gateway comet that had a close approach with Jupiter in 2017 and will have another in 2028. After a third close approach in 2063, P/2019 LD2 is expected to become a JFC (Sarid et al. 2019; Steckloff et al. 2020).

Centaurs—two in the Gateway (29P, panel (a); P/2019 LD2, panel (d)), one that became a JFC, then returned to a Centaur orbit less than 30 years later (39P, panel (b)), and one (81P, panel (c)) that became a JFC five decades ago after an exceptionally close approach to Jupiter.

Some secondary sources of Centaurs are also sources of JFCs. The existence of quasi-Hilda comets in the borders of the Hilda stable region show a link between

Hildas and JFCs. In fact, Di Sisto et al. (2005) found that the behavior of 99% of the escaped Hildas resembles that of JFCs, contributing to some extent to this population. In particular, the pre-capture orbital elements of comet Shoemaker-Levy 9 (SL9) overlap the orbital elements of escaped Hildas. The contribution of escaped Trojan asteroids to JFCs is minor from Di Sisto et al. (2019), but it can explain some NEO-JFC orbits, Encke-type comets and the SL9 impacts on Jupiter.

3.5 Conclusions

The Centaur population is probably the most change-prone small body population in a dynamical sense, but also a physical one. A Centaur's journey begins with a crossing of Neptune's path, and as it ventures toward the inner solar system, it undergoes a series of dynamical transformations. These include encounters with giant planets, capture into MMRs, chaos, the ZLK mechanism, etc., all of which significantly alter the characteristics of their orbits. Furthermore, Centaurs experience changes in their observed physical properties, which can occur due to close encounters with planets and the tidal forces they induce, collisions, as well as the heating they undergo as they approach the Sun. This dual dynamical and physical adaptability makes Centaurs a key population in our solar system.

As it resides in the region of the giant planets, the Centaur population serves as a transit zone for various small body populations, extending from those beyond Neptune to objects located interior to Jupiter's orbit. The research discussed in this chapter demonstrates that the most significant contribution to the Centaur population originates from the scattered disk (SD) in the trans-Neptunian region. However, within the SD region, objects in outer MMRs with Neptune also coexist. Identifying objects in distant MMRs is challenging from both observational and dynamical perspectives due to the uncertainties in their orbits. Hence, it's not possible to discount the potential importance of the resonant population as a contributor to the Centaur population (Muñoz-Gutiérrez et al. 2019; Pike et al. 2015; Volk et al. 2016; Crompvoets et al. 2022), even that of the Neptune Trojans, whose number is still uncertain. This complexity underscores the dynamic interplay in this region and the various potential sources of Centaurs.

Secondary sources of Centaurs include CTNOs, main-belt asteroids, Jupiter Trojans, Hildas, and even the Oort Cloud for retrograde Centaurs (Brasser et al. 2012). The physical and dynamic variety of these sources is important to explain particular cases within the Centaur population. The existence of active Centaurs at the same distances from the Sun as inactive Centaurs indicates that there are other factors that control the onset of activity (see Chapter 8 for a discussion of activity). This could be related to different origins or evolutions. For example, Fernández et al. (2018) demonstrate that in the Saturn-Jupiter region, active Centaurs evolve into JFCs, while inactive Centaurs are more likely than active Centaurs to evolve into HTCs and also sungrazers.

When the Centaurs enter the region within Jupiter's orbit, physical effects such as sublimation and splitting are accentuated, leading to disintegration or total loss of volatiles from small comets (e.g., Di Sisto et al. 2009). The transfer between Centaurs and JFCs was studied by Sarid et al. (2019), Steckloff et al. (2020), and Seligman et al. (2021), who found that there is a Gateway region through which the Centaurs that pass evolve into JFCs very quickly. However, Guilbert-Lepoutre et al. (2023) found that this region is also visited by objects that have already evolved in the JFC region from Centaurs, and so it would not be a pristine source of JFCs.

The dynamical evolution of Centaurs is strongly chaotic, with some showing intervals of relative stability due to temporary captures in MMRs. Various works have shown that several Centaurs in the known population experience captures in MMRs for periods of 10^4 to 10^6 years (Table 3.2). The region in space (a, e, i) in which Centaurs evolve is saturated with resonances with the giant planets, many of them overlapping, especially for direct orbits and $q \lesssim 10$ au. In the space of retrograde orbits, the overlap of resonances is not so important (Figure 3.7). Approximately 3% of known Centaurs have retrograde orbits but also high eccentricities ($e > 0.4$), so they are strongly perturbed by close encounters and resonances.

In this chapter, we have provided an overview of the extensive research conducted on the dynamical evolution of Centaurs, the mechanisms through which they transition from their source regions, and their connection with comets. While the dynamics between the giant planets is a well-explored field, and we have an understanding of the transfer of objects from the trans-Neptunian region to the cometary region, Centaurs stand out due to their diversity and evolving properties. Therefore, several questions within this domain still warrant further investigation. One critical aspect is determining the current, precise number of this population, which is crucial for quantifying the flux to comets. This number relies on the count of objects within their source regions, a count that is intricately tied to the limitations of astronomical observations. While there have been studies addressing this matter (e.g., Petit et al. 2017; Bannister et al. 2018; Bernardinelli et al. 2022), there remains a gap in our knowledge. Certainly, a comprehensive understanding of the physical properties of Centaurs, particularly for the smaller objects within this population and their relation to dynamical properties, is a vital area of study (see Chapter 5). In this regard, the upcoming Vera C. Rubin Observatory Legacy Survey of Space and Time (LSST) promises to play a significant role in advancing our understanding of the number of Centaurs and their physical properties (see Chapter 14). LSST will also be crucial in enhancing our understanding of objects in distant MMRs, which can give insights into their contributions to the Centaur population and whether these objects exhibit different dynamical characteristics than other Centaurs.

Acknowledgements

We would like to thank three anonymous reviewers for their detailed comments and suggestions which helped us to significantly improve this article, and also Silvia Giuliatti Winter for valuable discussion about close encounters and the relation to moons and rings.

Symbol and Acronym List

a	semimajor axis
a_p	semimajor axis of a planet
e	excentricity
i	inclination
M_p	mass of planet
$M_\odot$	mass of Sun
ω	argument of perihelion
Ω	longitude of the ascending node
q	perihelion distance
Q	aphelion distance
σ_i	width parameter for Gaussian inclination distribution
σ	resonant argument
au	astronomical unit
CFEPS	Canada-France Ecliptic Plane Survey
CKB	Classical Kuiper Belt
CKBO	Classical Kuiper Belt Object
CR3BP	Circular Restricted Three-Body Problem
CTNO	Classical Trans-Neptunian Object
D	Diameter
EC	Ecliptic Comet
GPC	Giant Planet Crosser
Gyr	Billion years
HS	horseshoes
HTC	Halley-Type Comet
JFC	Jupiter Family Comet
JPL	Jet Propulsion Laboratory
KB	Kuiper Belt
LSST	Legacy Survey of Space and Time
MB	Main Belt
MMR	Mean-Motion Resonance
MPC	Minor Planet Center
Myr	Million years
$N(>D)$	Cumulative diameter distribution
NEO	Near Earth Object
QS	Quasi satellites
SD	Scattered Disk
SDO	Scattered Disk Object
SL9	Shoemaker-Levy 9
SW1	Schwassmann–Wachmann 1
T	Tisserand parameter
T_J	Tisserand parameter with respect to Jupiter
TN	Trans-Neptunian
TNO	Trans-Neptunian Object
ZLK	von Zeipel-Lidov-Kozai

References

Alexandersen, M., Gladman, B., Greenstreet, S., et al. 2013, Sci, 341, 994
Alexandersen, M., Gladman, B., Kavelaars, J. J., et al. 2016, AJ, 152, 111

Alexandersen, M., Greenstreet, S., Gladman, B. J., et al. 2021, PSJ, 2, 212
Araujo, R. A. N., Galiazzo, M. A., Winter, O. C., & Sfair, R. 2018, MNRAS, 476, 5323
Araujo, R. A. N., Sfair, R., & Winter, O. C. 2016, ApJ, 824, 80
Bailey, B. L., & Malhotra, R. 2009, Icarus, 203, 155
Bannister, M. T., Gladman, B. J., Kavelaars, J. J., et al. 2018, ApJS, 236, 18
Batygin, K., Adams, F. C., Brown, M. E., & Becker, J. C. 2019, PhR, 805, 1
Beaudoin, M., Gladman, B., Huang, Y., et al. 2023, PSJ, 4, 145
Belbruno, E., & Marsden, B. G. 1997, AJ, 113, 1433
Bernardinelli, P. H., Bernstein, G. M., Sako, M., et al. 2022, ApJS, 258, 41
Brasser, R., & Morbidelli, A. 2013, Icar, 225, 40
Brasser, R., Schwamb, M. E., Lykawka, P. S., & Gomes, R. S. 2012, MNRAS, 420, 3396
Brown, M. E. 2001, AJ, 121, 2804
Chiang, E. I., & Lithwick, Y. 2005, ApJ, 628, 520
Correa-Otto, J., García-Migani, E., & Gil-Hutton, R. 2024, MNRAS, 527, 876
Crompvoets, B. L., Lawler, S. M., Volk, K., et al. 2022, PSJ, 3, 113
de la Fuente Marcos, C., & de la Fuente Marcos, R. 2013, A&A, 551, A114
de la Fuente Marcos, C., & de la Fuente Marcos, R. 2014a, MNRAS, 441, 2280
de la Fuente Marcos, C., & de la Fuente Marcos, R. 2014b, Ap&SS, 352, 409
de la Fuente Marcos, C., & de la Fuente Marcos, R. 2015, MNRAS, 453, 1288
de la Fuente Marcos, C., & de la Fuente Marcos, R. 2017, MNRAS, 467, 1561
de la Fuente Marcos, C., & de la Fuente Marcos, R. 2024, MNRAS, 527, L110
Di Sisto, R. P., & Brunini, A. 2007, Icar, 190, 224
Di Sisto, R. P., Brunini, A., & de Elía, G. C. 2010, A&A, 519, A112
Di Sisto, R. P., Brunini, A., Dirani, L. D., & Orellana, R. B. 2005, Icar, 174, 81
Di Sisto, R. P., Fernández, J. A., & Brunini, A. 2009, Icar, 203, 140
Di Sisto, R. P., Ramos, X. S., & Beaugé, C. 2014, Icar, 243, 287
Di Sisto, R. P., Ramos, X. S., & Gallardo, T. 2019, Icar, 319, 828
Di Sisto, R. P., & Rossignoli, N. L. 2020, CeMDA, 132, 36
Di Sisto, R. P., & Zanardi, M. 2013, A&A, 553, A79
Duncan, M., Quinn, T., & Tremaine, S. 1988, ApJL, 328, L69
Duncan, M. J., & Levison, H. F. 1997, Sci, 276, 1670
Duncan, M. J., Levison, H. F., & Budd, S. M. 1995, AJ, 110, 3073
Edgeworth, K. E. 1938, Unpublished manuscript, Trustees of the National Library of Ireland, Dublin. Manuscript Nos. 16 869/47 and /48,
Fernández, J. A. 1980, MNRAS, 192, 481
Fernández, J. A., Gallardo, T., & Brunini, A. 2002, Icar, 159, 358
Fernández, J. A., Gallardo, T., & Brunini, A. 2004, Icar, 172, 372
Fernández, J. A., Helal, M., & Gallardo, T. 2018, P&SS, 158, 6
Fernandez, J. A., & Ip, W. H. 1984, Icar, 58, 109
Ferraz-Mello, S., Nesvorny, D., & Michtchenko, T. A. 1998, ASP Conf. Ser., Solar System Formation and Evolution, (Vol 149, ed. D. Lazzaro, et al. San Francisco, CA: ASP) 65
Fraser, W. C., Dones, L., Volk, K., Womack, M., & Nesvorný, D. 2024, in Comets III, ed. K. J. Meech, et al. (Tucson, AZ: Univ. of Arizona. Press) 121
Gabryszewski, R., & Włodarczyk, I. 2003, A&A, 405, 1145
Gallardo, T. 2006, Icar, 184, 29
Gallardo, T. 2019, MNRAS, 487, 1709

Gallardo, T. 2020, CeMDA, 132, 9
Gallardo, T., Hugo, G., & Pais, P. 2012, Icar, 220, 392
Gil-Hutton, R., & García-Migani, E. 2016, A&A, 590, A111
Gkotsinas, A., Guilbert-Lepoutre, A., Raymond, S. N., & Nesvorny, D. 2022, ApJ, 928, 43
Gladman, B., Holman, M., Grav, T., et al. 2002, Icarus, 157, 269
Gladman, B., Marsden, B. G., & Vanlaerhoven, C. 2008, in The Solar System Beyond Neptune, ed. M. A. Barucci, et al. (Tucson, AZ: Univ. Arizona Press) 43
Gladman, B., & Volk, K. 2021, ARA&A, 59, 203
Gladman, B., Lawler, S. M., Petit, J. M., et al. 2012, AJ, 144, 23
Gomes, R. S., Fernández, J. A., Gallardo, T., & Brunini, A. 2008, in The Solar System Beyond Neptune, ed. M. A. Barucci, et al. (Tucson, AZ: Univ. Arizona Press) 259
Gomes, R. S., Gallardo, T., Fernández, J. A., & Brunini, A. 2005, CeMDA, 91, 109
Greenstreet, S., Gladman, B., & Ngo, H. 2020, AJ, 160, 144
Guan, P., Zhou, L.-Y., & Li, J. 2012, RAA, 12, 1549
Guilbert-Lepoutre, A., Gkotsinas, A., Raymond, S. N., & Nesvorny, D. 2023, ApJ, 942, 92
Hadden, S., & Tremaine, S. 2024, MNRAS, 527, 3054
Hahn, J. M., & Malhotra, R. 2005, AJ, 130, 2392
Holt, T. R., Nesvorný, D., Horner, J., et al. 2020, MNRAS, 495, 4085
Horner, J., Evans, N. W., & Bailey, M. E. 2004a, MNRAS, 354, 798
Horner, J., Evans, N. W., & Bailey, M. E. 2004b, MNRAS, 355, 321
Horner, J., & Lykawka, P. S. 2010, MNRAS, 402, 13
Horner, J., & Lykawka, P. S. 2012, MNRAS, 426, 159
Hsieh, H. H., Novaković, B., Walsh, K. J., & Schörghofer, N. 2020, AJ, 159, 179
Huang, Y., Gladman, B., & Volk, K. 2022, ApJS, 259, 54
Ito, T., & Ohtsuka, K. 2019, MEEP, 7, 1
Jewitt, D. 2009, AJ, 137, 4296
Kary, D. M., & Dones, L. 1996, Icar, 121, 207
Kazantsev, A., & Kazantseva, L. 2021, MNRAS, 505, 408
Koon, W. S., Lo, M. W., Marsden, J. E., & Ross, S. D. 2001, CeMDA, 81, 27
Kresak, L. 1979, in Asteroids, ed. T. Gehrels, & M. S. Matthews (Tucson, AZ: Univ. Arizona Press) 289
Królikowska, M., & Szutowicz, S. 2006, A&A, 448, 401
Kuiper, G. P. 1951, PNAS, 37, 1
Lawler, S. M., Shankman, C., Kavelaars, J. J., et al. 2018, AJ, 155, 197
Levison, H. F. 1996, ASP Conf. Ser., Completing the Inventory of the Solar System, Vol 107, ed. T. Rettig, & J. M. Hahn (San Francisco, CA: ASP) 173
Levison, H. F., & Duncan, M. J. 1997, Icar, 127, 13
Levison, H. F., Morbidelli, A., Van Laerhoven, C., Gomes, R., & Tsiganis, K. 2008, Icar, 196, 258
Levison, H. F., Shoemaker, E. M., & Shoemaker, C. S. 1997, Natur, 385, 42
Levison, H. F., & Stern, S. A. 2001, AJ, 121, 1730
Li, M., Huang, Y., & Gong, S. 2019, A&A, 630, A60
Liberato, L., Araújo, R., & Winter, O. 2023, EPJST, 232, 3007
Lin, H. W., Chen, Y.-T., Volk, K., et al. 2021, Icar, 361, 114391
Lykawka, P. S., & Mukai, T. 2007, Icar, 189, 213
Malhotra, R. 1993, Natur, 365, 819

Malhotra, R. 1995, AJ, 110, 420
Markwardt, L., Wen Lin, H., Gerdes, D., & Adams, F. C. 2023, PSJ, 4, 135
Masaki, Y., & Kinoshita, H. 2003, Proc. of the 35th Symp. on Celestial Mechanics ed. E. Kokubo, H. Arakida, & T. Yamamoto (Hayama: The Graduate University for Advanced Studies) 255
Morbidelli, A. 1997, Icar, 127, 1
Morbidelli, A., & Levison, H. F. 2004, AJ, 128, 2564
Morbidelli, A., Levison, H. F., Tsiganis, K., & Gomes, R. 2005, Natur, 435, 462
Morbidelli, A., Thomas, F., & Moons, M. 1995, Icar, 118, 322
Muñoz-Gutiérrez, M. A., Peimbert, A., Pichardo, B., Lehner, M. J., & Wang, S. Y. 2019, AJ, 158, 184
Nesvorný, D. 2018, ARA&A, 56, 137
Nesvorný, D., Bernardinelli, P., Vokrouhlický, D., & Batygin, K. 2023a, Icar, 406, 115738
Nesvorný, D., Dones, L., De Prá, M., Womack, M., & Zahnle, K. J. 2023b, PSJ, 4, 139
Nesvorný, D., & Vokrouhlický, D. 2009, AJ, 137, 5003
Nesvorný, D., Vokrouhlický, D., Dones, L., et al. 2017, ApJ, 845, 27
Nesvorný, D., Vokrouhlický, D., & Morbidelli, A. 2013, ApJ, 768, 45
Nesvorný, D., Vokrouhlický, D., Stern, A. S., et al. 2019, AJ, 158, 132
Noll, K. S., Brown, M. E., Weaver, H. A., et al. 2020, PSJ, 1, 44
Oldroyd, W. J., Chandler, C. O., Trujillo, C. A., et al. 2023a, RNAAS, 7, 42
Oldroyd, W. J., Chandler, C. O., Trujillo, C. A., et al. 2023b, ApJL, 957, L1
Pan, N., & Gallardo, T. 2025, Celest. Mech. Dyn. Astron., 137, 2
Petit, J.-M., Kavelaars, J. J., Gladman, B. J., et al. 2011, AJ, 142, 131
Petit, J. M., Kavelaars, J. J., Gladman, B. J., et al. 2017, AJ, 153, 236
Pike, R. E., Kavelaars, J. J., Petit, J. M., et al. 2015, AJ, 149, 202
Pinto, O. H., Kelley, M. S. P., Villanueva, G. L., et al. 2023, PSJ, 4, 208
Raymond, S. N., Kaib, N. A., Armitage, P. J., & Fortney, J. J. 2020, ApJL, 904, L4
Roberts, A. C., & Muñoz-Gutiérrez, M. 2021, Icar, 358, 114201
Rossignoli, N. L., Di Sisto, R. P., Zanardi, M., & Dugaro, A. 2019, A&A, 627, A12
Saillenfest, M. 2020, CeMDA, 132, 12
Sarid, G., Volk, K., Steckloff, J. K., et al. 2019, ApJL, 883, L25
Seligman, D. Z., Kratter, K. M., Levine, W. G., & Jedicke, R. 2021, PSJ, 2, 234
Sheppard, S. S., & Trujillo, C. A. 2010, ApJL, 723, L233
Sicardy, B., Renner, S., Leiva, R., et al. 2020, in The Trans-Neptunian Solar System, ed. D. Prialnik, M. A. Barucci, & L. Young (Amsterdam: Elsevier) 249
Smullen, R. A., & Volk, K. 2020, MNRAS, 497, 1391
Snodgrass, C., Agarwal, J., Combi, M., et al. 2017, A&ARv, 25, 5
Steckloff, J. K., Sarid, G., Volk, K., et al. 2020, ApJL, 904, L20
Tiscareno, M. S., & Malhotra, R. 2003, AJ, 126, 3122
Tsiganis, K., Gomes, R., Morbidelli, A., & Levison, H. F. 2005, Natur, 435, 459
Uehata, K., Terai, T., Ohtsuki, K., & Yoshida, F. 2022, AJ, 163, 213
Volk, K., & Malhotra, R. 2008, ApJ, 687, 714
Volk, K., & Malhotra, R. 2013, Icar, 224, 66
Volk, K., & Van Laerhoven, C. 2024, RNAAS, 8, 36
Volk, K., Murray-Clay, R., Gladman, B., et al. 2016, AJ, 152, 23
Volk, K. M. 2013, PhD thesis, Univ. Arizona

Walsh, K. J., Morbidelli, A., Raymond, S. N., O'Brien, D. P., & Mandell, A. M. 2011, Natur, 475, 206
Wierzchos, K., Womack, M., & Sarid, G. 2017, AJ, 153, 230
Womack, M., Sarid, G., & Wierzchos, K. 2017, PASP, 129, 031001
Wood, J. 2023, MNRAS, 519, 812
Wood, J., Horner, J., Hinse, T. C., & Marsden, S. C. 2018, MNRAS, 480, 4183
Zhou, L., Zhou, L.-Y., Dvorak, R., & Li, J. 2020, A&A, 633, A153

Chapter 4

Centaur Nuclei: Sizes, Shapes, Spins, and Structure

Y R Fernández, M W Buie, P Lacerda and R Marschall

We present a wide-ranging but in-depth analysis of Centaurs, focusing on their physical and structural aspects. Centaurs, originating from the Scattered Disk and Kuiper Belt, play a crucial role in our understanding of solar system evolution. We first examine how biases in discovery and measurement affect our understanding of the Centaur size distribution. In particular we address the strong dependence of the census on perihelion distance and the broad distribution of Centaur geometric albedos. We explore the rotational characteristics derived from lightcurves, revealing a diverse range of spin rates and photometric variabilities, with most Centaurs showing low amplitude lightcurves, suggesting near-spherical shapes. Additionally, we investigate the relationships between Centaur orbital parameters, surface colors, and physical properties, noting a lack of correlation between rotational dynamics and orbital evolution. We also address the influence of sublimation-driven activity on Centaur spin states, and the rarity of contact binaries. We then discuss some observational and modeling limitations from using common observations (e.g., visible or infrared photometry) to determine diameters and shapes. Following that, we give some points on understanding how Centaur diameters and shapes can reveal the 'primitive' nature of the bodies, emphasizing the important role occultation observations play. We also then assess how the Centaur size distribution we see today has been influenced by the collisions in both the primordial Kuiper Belt and in the subsequent Scattered Disk. Finally, we end the chapter with a short narrative of future prospects for overcoming our current limitations in understanding Centaur origins and evolution.

4.1 Introduction

As described by other works in this volume (e.g., Chapters 1 and 7), the current sizes, shapes, spin states, and structures of Centaurs are manifestations of the evolutionary

doi:10.1088/2514-3433/ada267ch4

processes that they have suffered since the original icy planetesimals were first accreted (see, e.g., Chapter 2). In principle, if we were to know these physical properties of the entire Centaur population, and of the small-body groups that the Centaurs are related to, then we would have useful constraints that would have to be matched by any model purporting to explain the evolution.

Centaurs play a unique role in our quest to understand solar system evolution, since they come from the Scattered Disk (i.e., as Scattered Disk Objects, SDOs) and the Kuiper Belt, and since many of them end up as Jupiter-family comets (JFCs); see, e.g., Duncan & Levison (1997), Duncan et al. (2004), Volk & Malhotra (2008), and Chapter 3. A broad objective with studying Centaur physical properties is thus to contextualize what we see among the SDOs and JFCs. Part of this objective is an understanding of if and how the Centaurs change before and during the dynamical cascade from the Scattered Disk, and what happens to the Centaurs in the (typically) few 10^6 yr of their residence in the giant-planet region.

While there is much ongoing work to understand the physical properties of Centaurs, a murky fog hinders us in several ways. First, as with many groups of small bodies in the solar system, the characterization of Centaurs lags somewhat behind their discovery. There are many Centaurs with only minimal characterization. Second, the discovery of Centaurs itself has biases, so even if we were characterizing all known Centaurs well, there would still be gaps in our knowledge because we are simply ignorant of many of the faintest and smallest objects in the population. Third, the techniques that we use to measure Centaur brightnesses and thereby extract physical properties have their own set of limitations. All of these problems conspire to bias our picture of what we know about Centaur ensemble properties, and eventually they must be all well understood if we are to properly interpret our data.

In this chapter we focus on two themes that touch on these issues. (1) What are the current constraints on Centaur physical properties, and how might limitations in our measurement techniques be biasing those results? (2) What is the context of Centaurs in terms of primitive-body evolution in the solar system?

4.2 Size Distribution

Much of the physics of this topic is discussed in Chapter 7, so we describe a few statistical points here. A very basic look at the current absolute magnitude (H) distribution of the known Centaurs is shown in Figure 4.1. We extracted all asteroid entries in the JPL Horizons database (as of 2023 November 21) that had $q > 5.203$ au (i.e., Jupiter's orbital semimajor axis a_J) and $a < 30.1$ au (i.e., Neptune's orbital semimajor axis a_N), but excluding objects that are (or could be) from different source regions: Jovian and Neptunian Trojans, parabolic and hyperbolic objects, and nine additional objects with inclinations above 80°. This left 308 Centaurs. We note that this a_N cutoff in the Centaur definition does introduce some bias in the distant Centaurs since there is no real physical/dynamical break here between the Scattered Disk and the Centaur region; e.g. (523746) 2014 UT_{114} with $a = 30.09$ au makes our cut, but (309239) 2007 RW_{10} with $a = 30.16$ au

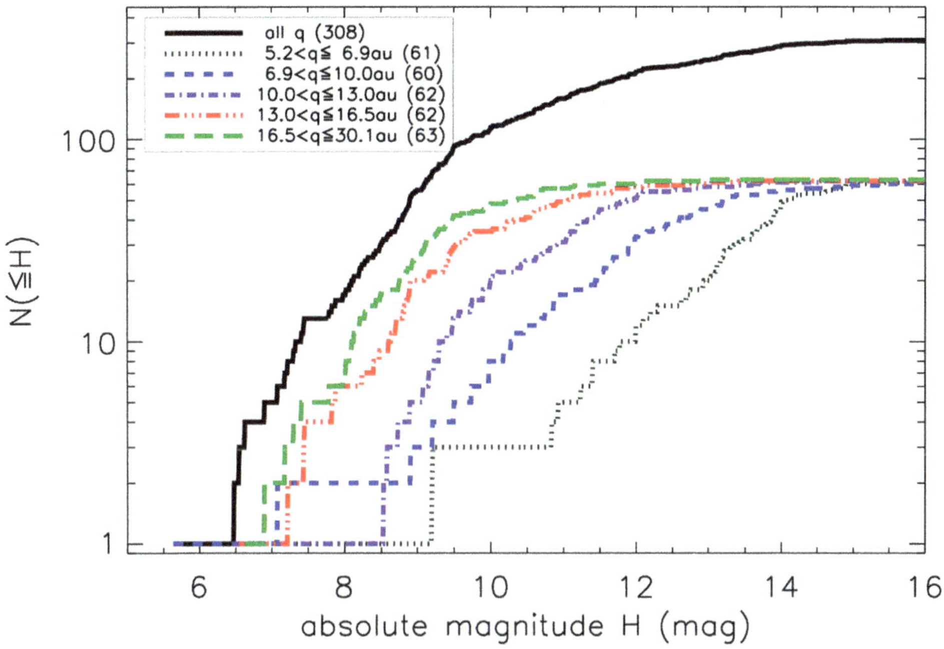

Figure 4.1. Cumulative magnitude distribution (CMD) of known asteroidal Centaurs. All objects (as described in the text) are in the thick black line. The other five thick linestyles with color show CMDs for various perihelion (q) bins, tuned so that each bin has approximately the same number of Centaurs. There are very strong differences in the five curves indicating that the completion of the Centaur census does not extend to very faint magnitudes yet.

does not. We also note that even if the high-inclination/retrograde Centaurs are indeed from the Scattered Disk anyway (see, e.g., Namouni & Morais 2020), this is a small number and does not affect the point we are making here.

The magnitude distribution of those 308 Centaurs are represented by the thick solid line in Figure 4.1. The curves with other linestyles/colors show the distribution broken up into perihelion bins, as listed in the figure's legend. We decided to create bins where approximately equal number of Centaurs would be in each one. The idea is to show that the discovery rates very strongly depend on how close the Centaurs come to our telescopes here in the inner solar system. For the smallest perihelion bin (dark green dotted line), the magnitude distribution seems to be a single power law that turns over due to lack of detected objects around $H = 14$. In contrast, the other curves are less straight and of course also turn over at brighter values of H. The most distant bin (light green dashed line) has a turnover about 5 full magnitudes brighter than the closest bin. One could argue from this that we are only really confident in our census of the Centaurs down to about $H = 9$, and we certainly have not yet discovered enough Centaurs in the $H = 9$ to 14 range to permit characterizing the population in this size range without large corrections due to discovery biases.

Another demonstration of the aforementioned perihelion effect is shown in Figure 4.2, which plots the discovery time versus a Centaur's H. The color coding here is the same as in the previous figure. For the Centaurs that are presumably easier to discover, i.e., those in the closest perihelion bin (dark green squares), there is some indication that we have discovered all or nearly all of the relatively large Centaurs ($H \approx 13$ and brighter). In the last few years, the only such Centaurs discovered in that bin are around $H = 13$ to 15. This is true even though clearly the sky surveys are still capable of finding brighter Centaurs, since those are in fact being found in the more distant perihelion bins. Interestingly, no Centaur brighter than about $H = 8.7$ has been discovered since late 2016 (7 years ago at time of writing).

The fundamental property that we wish to understand is the size distribution, not the magnitude distribution, and so the geometric albedo is required to make the conversion. (See, e.g., Pravec & Harris 2007 for a definition of this albedo.) With JFCs, the spread of albedos seems to be fairly limited (see, e.g., Lamy et al. 2004), but this is not the case for Centaurs. Using a recent compilation of outer solar system albedos by Müller et al. (2020), we make use of 42 Centaur albedos in their Table 7.1. (The table itself lists 55 Centaurs, but here we exclude 3 active comets for which the nucleus albedo may be difficult to get (29P, 167P, and C/2011 KP36) as well as 10 other objects that Müller et al. (2020) included as Centaurs even though their perihelia are smaller than a_J.) A scatter plot of those 42 albedos is shown in the top of Figure 4.3. The color coding for perihelion bins is again the same as the previous figures. There is no trend of albedo with diameter, and no significant

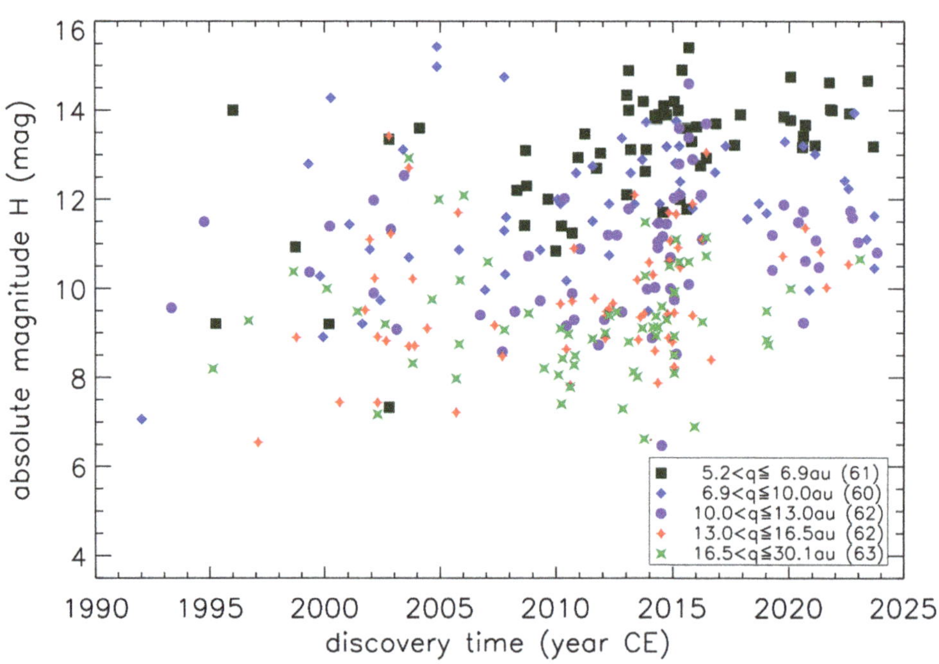

Figure 4.2. Scatter plot of when asteroidal Centaurs of various q and H were discovered. Color coding and perihelion binning match that in Figure 4.1.

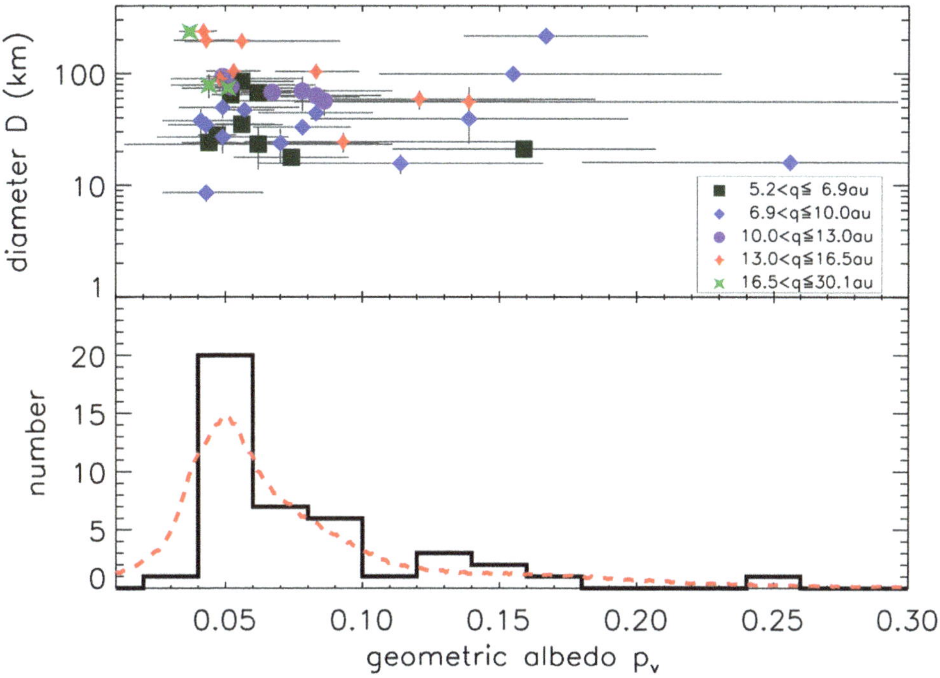

Figure 4.3. Top: Scatter plot of 42 Centaur diameters D and geometric albedos p_v as summarized by Müller et al. (2020). Color coding matches that in Figures 4.1 and 4.2. Bottom: Histogram of those 42 albedos. The red dashed curve is our estimate of the albedo probability density function (PDF) based on the given albedos. We estimated the PDF by generating simulated albedos based on the measured values and their uncertainties.

difference between the albedos in any perihelion bin (via a check of the Kolmogorov–Smirnov (KS) test). At the moment, if one were to assume that albedo is controlled by the surface evolution of the Centaur (see Chapter 5; Chapter 6; Chapter 7; Chapter 8), there is apparently no indication that surfaces of Centaurs with smaller q have evolved differently than those of the Centaurs farther out. This might be simply a manifestation of the fact that Centaurs do jostle toward and away from the Sun in the Centaur region, even if there is an overall trend of the Centaurs (at least the ones that survive to become JFCs) slowly moving inward (Duncan et al. 2004; Fraser et al. 2024).

Given the absence of a strong trend of albedo with perihelion or semimajor axis, we will take the current albedo distribution for those 42 Centaurs as being representative of the population as a whole, and use it as a probability density function (PDF) in order to explore what the size distribution might be as derived from the H distribution. In other words, since we have albedos for less than 1/7 of the Centaurs (42 out of 308), the conversion requires assuming an albedo distribution. The histogram of those 42 albedos is shown in the bottom panel of Figure 4.3. We estimated a PDF from those 42 measurements and their error-bars by simulating 10,000 albedo distributions Monte Carlo-style. The result is the red dashed curve in the bottom plot of Figure 4.3. Note that this is not a fit to the albedo

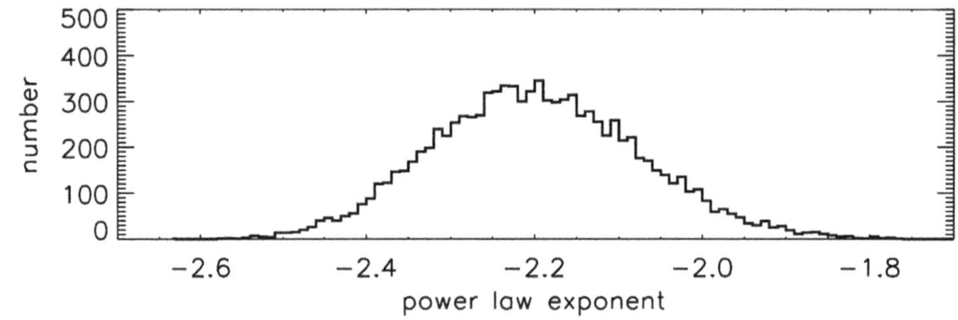

Figure 4.4. Distribution of power-law slopes from our simulation of 10,000 cumulative size distributions created as described in the text. While the most likely result of our particular fit scheme is around −2.2, there is a wide range of other possible exponents, and this must be taken into account when interpreting the Centaur size distribution. Note that we are not claiming a value for the actual Centaur size distribution, we are merely demonstrating that whatever power-law is produced has appreciable uncertainty.

distribution, simply an empirical assessment of what the true underlying distribution might be.

Using this PDF, we then simply ran 10,000 simulations that each created a size distribution from the magnitude distribution by randomly extracting an albedo for each object according to the probability density. We then fit a power law to each cumulative size distribution with diameters between 100 and 200 km. This range was chosen as one for which the distribution seems to follow a single power-law; i.e., it made the fitting seem sensible. The distribution of those 10,000 power-law exponents is shown in Figure 4.4. While there is a most-likely result near a power-law exponent of −2.2, there is a reasonable probability of getting an exponent that is well off from that answer. We emphasize that the main point here is not the actual value of the exponent, but rather that until a higher fraction of the Centaur diameters and albedos (not just their H magnitudes) have actually been measured, there is intrinsic uncertainty in any characterization of the Centaur size distribution. This has ramifications for any attempt to use a physical model to explain such a size distribution or to explain how Centaurs fit into the SDO and JFC size distributions. This is above and beyond other problems that could make such hypothesizing difficult—e.g., the fact that there is almost no overlap between the known JFC diameters and the known Centaur diameters.

It can be argued that the size distribution is not actually the most fundamental property we might use to gauge formation and evolutionary processes. Rather in some contexts the mass distribution might be more important. This is currently even more fraught than the size assessment is since to go from size to mass requires a density, and we have precious little info on the densities, porosities, and overall structure of the Centaurs (see Section 4.3.6 below, and Chapters 9 and 12).

4.3 Shape and Spin Distributions

Determining the shapes and spins of Centaurs involves studying their lightcurves, which record periodic oscillations in brightness over time. Well-sampled lightcurves are processed to extract rotational properties: the period, P, and the full range of photometric variation, Δm. The former corresponds to the spin period, while the latter constrains the shape of individual objects. If measured at different observing geometries, time series data can be inverted to extract more precise shapes and spin pole orientations. In some cases, under favorable observing geometries, lightcurves can also reveal whether the object has multiple components (bilobed shape as the nucleus 67P, or two separate components as KBO Arrokoth). The distributions of rotational properties are useful for comparing different populations when trying to decode evolutionary links between them.

Other techniques exist to decipher the shapes of Centaurs. For bright objects, adaptive optics systems on ground-based telescopes can also be used to estimate their shapes and dimensions. Stellar occultations are a powerful technique to constrain shapes, capable to achieving extraordinary resolution on chords across the sky-projected cross-section of individual objects. These techniques are often complementary.

The Asteroid Lightcurve Database (Warner et al. 2009) currently includes lightcurves properties, spin period and photometric range, for 16 Centaurs. Of those, 7 have been observed at more than one observing geometry, revealing changing variability. Table 4.1 summarizes the known Centaur rotational properties, listed together with other orbital, physical and surface parameters. Figures 4.5–4.11 highlight a few aspects of data discussed in this chapter.

4.3.1 Period and Δm Distributions

Figures 4.5 and 4.6 show histograms and cumulative distributions of spin frequency and lightcurve variability. Even though the period is often easier to grasp, frequency is the more relevant physical quantity; we will refer to both interchangeably. Six of the 16 Centaurs cluster between 2.5 and 3 rotations per day ($P \sim 9$ hr) near the center of the distribution. A long tail extends to much slower rotations, with 3 objects spinning slower than $P = 24$ hr and Centaur 2013 XZ8 spinning with $P \sim 88$ hr. The two fastest rotators are Chiron and 2002 GZ32 with periods near 6 hr.

The Δm distribution is skewed toward low values, with roughly half the Centaurs within $\Delta m < 0.2$ mag. At the opposite end of the distribution, Bienor is the most variable object at $\Delta m = 0.75$ mag. As of yet no Centaurs display the extreme variability ($\Delta m > 0.9$ mag) characteristic of extreme bilobed or compact binary shapes (although it is possible that there is a bias against discovering such objects, since perhaps the large lightcurve amplitude means it is less likely for such objects to be seen at follow-up observations right after discovery). We discuss binarity in more detail below. Another bias against discovery highly variable objects is that larger objects, which are easier to detect, will tend to be more spherical (although high-Δm Bienor is the fourth largest object in Table 4.1).

Figure 4.7 compares the spin rate distribution of Centaurs with different dynamical families of trans-Neptunian objects. We use the two-sample KS test[1] to

[1] We used the implementation provided in Python module `scipy.stats.ks_2samp`.

Table 4.1. Known Centaur Rotational Properties.

Centaur	Lightcurve Period[a] [hr]	Min. Δm [mag]	Max. Δm [mag]	q [au]	e	i [deg]	Abs. Mag. H [mag]	$B-R$ Color [mag]	Surface Geometric Albedo	Diameter D [km]	Notes[b]
2060 Chiron (1977 UB)	5.92	0.04	0.09	8.55	0.376	6.9	6.56	1.01	0.16 ± 0.03	218 ± 20	U3, C, R?
5145 Pholus (1992 AD)	9.98	0.15	0.6	8.67	0.573	24.7	7.64	1.97	0.16 ± 0.08	99 ± 15	U3
8405 Asbolus (1995 GO)	8.94	0.32	0.55	6.88	0.619	17.6	9.19	1.228	0.06 ± 0.02	85 ± 9	U3
10199 Chariklo (1997 CU26)	7.00		0.11	13.14	0.168	23.4	6.65	1.299	0.04 ± 0.01	248 ± 18	U2, R
31824 Elatus (1999 UG5)	26.82		0.1	7.22	0.386	5.3	10.42	1.672	0.05 ± 0.03	50 ± 10	U2
32532 Thereus (2001 PT13)	8.34	0.16	0.38	8.50	0.199	20.4	9.29	1.19	0.07 ± 0.02	74 ± 17	U3
52872 Okyrhoe (1998 SG35)	9.72	0.07	0.4	5.82	0.305	15.7	11.23	1.237	0.06 ± 0.02	36 ± 1	12.17,U2
54598 Bienor (2000 QC243)	9.14	0.08	0.75	13.20	0.203	20.7	7.69	1.158	0.04 ± 0.02	198 ± 7	U3
60558 Echeclus (2000 EC98)	26.80		0.24	5.84	0.457	4.3	9.55	1.376	0.05 ± 0.02	65 ± 2	U2, C
83982 Crantor (2002 GO9)	13.94		0.14	14.1	0.274	12.8	9.17	1.864	0.12 ± 0.06	59 ± 12	19.34,U1
95626 (2002 GZ32)	5.8		0.08	18.01	0.219	15.0	7.39	1.199	0.04 ± 0.01	237 ± 8	U3
136204 (2003 WL7)	8.24		0.05	14.92	0.258	11.2	8.73	1.23	0.05 ± 0.01	105 ± 7	S,U2-
145486 (2005 UJ438)	8.32		0.13	8.22	0.532	3.8	10.89	1.64	0.24 ± 0.12	16 ± 2	A,U2-
250112 (2002 KY14)	8.50	0.09	0.13	8.62	0.314	19.5	9.74	1.75	0.12 ± 0.09	43 ± 6	U2
459865 (2013 XZ8)	87.74		0.26	8.42	0.369	22.6	9.53	1.17		69 ± 35[c]	U2
471931 (2013 PH44)	22.16		0.15	15.57	0.213	33.5	9.38			74 ± 37[c]	A,U1

Note: Lightcurve properties taken from Lightcurve Database (Warner et al. 2009), updated October 2023. Diameters and albedos come from Herschel (Fornasier et al. 2013, Duffard et al. 2014), or NEOWISE (Mainzer et al. 2019) surveys; the least uncertain of the two measurements is taken when they agree within 1σ, otherwise the mean is shown. Colors ($B-R$) taken from Peixinho et al. (2012) and Tegler et al. (2016).

[a] Lightcurve period, where all are double peaked except where indicated in notes.

[b] "Ux(-)" is the quality of the lightcurve from $x = 1$ (worse) to $x = 3$ (best) and "-" is a half-step, "S" indicates a single-peaked period is given, "A" indicates ambiguity between double-peaked and single-peaked period, and a number indicates another possible period. "C" indicates cometary activity, "R" means rings detected (followed by "?"" if unconfirmed).

[c] An albedo of 0.057 is assumed to calculate diameter from H.

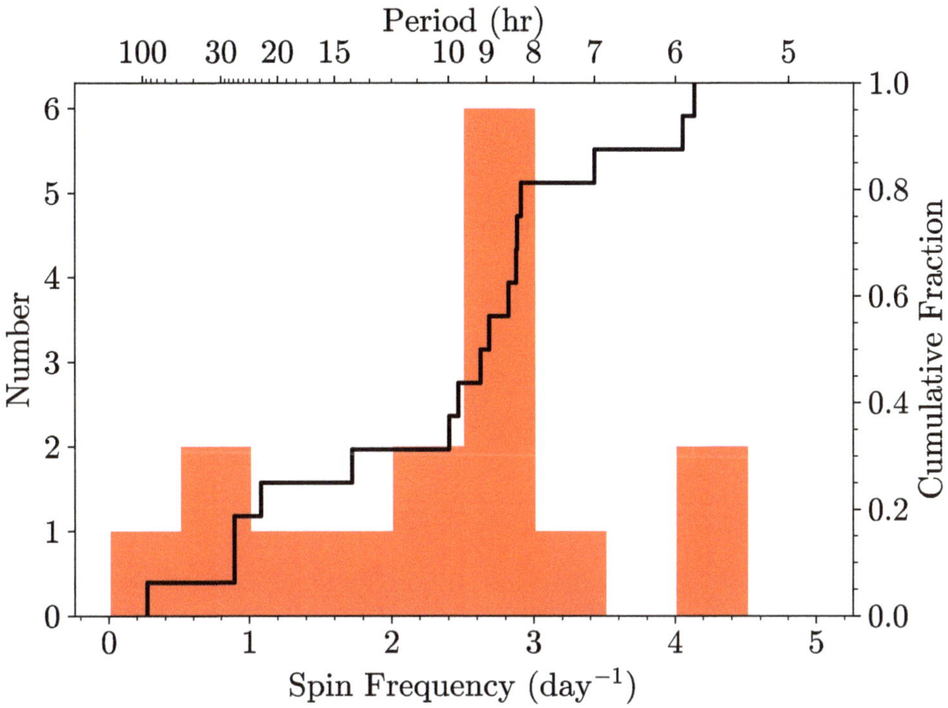

Figure 4.5. Centaur spin rate distribution. Cumulative distribution (solid black line) is plotted on the right vertical axis.

compare the Centaurs with the other classes of objects. Cold Classicals have the least compatible spin distribution with the Centaurs (p_{KS}-value of 0.05). The KS-test does not rule that Centaur spins come from the same distribution as Plutinos ($p_{KS} = 0.15$), SDOs ($p_{KS} = 0.33$) or Hot Classicals ($p_{KS} = 0.44$). However, the sample sizes are small, so more data is needed before strong conclusions can be made. A similar exercise comparing the max Δm distributions of Centaurs and the same trans-Neptunian dynamical families shows no discernible differences, with $p_{KS} > 0.66$ in all cases.

4.3.2 Relation to Orbits

Centaurs display no obvious relation between orbital parameters (perihelion distance, orbital eccentricity and inclination) and lightcurve properties. To zeroth order, this suggests that the processes responsible for the orbital evolution and the spin evolution are unrelated. One possible exception is that higher perihelion objects have lower Δm lightcurves (although again there may be some bias against recovering such objects in the first place). Since Centaur activity correlates inversely with perihelion distance, this suggests a possible link between shape and activity. We discuss this possibility below. However, it would be interesting to study whether close approaches to giant planets can significantly modulate Centaur spins and shapes.

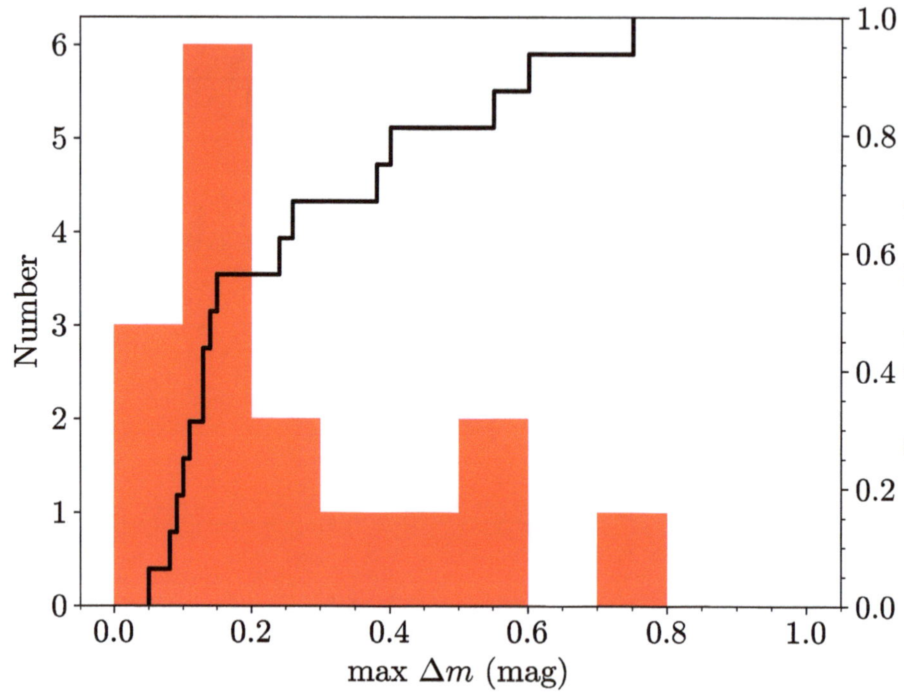

Figure 4.6. Distribution of maximum lightcurve variability for Centaurs. Cumulative distribution (solid black line) is plotted on the right vertical axis.

4.3.3 Relation to Surface Properties

The surface colors of Centaurs and small Kuiper Belt objects span a broad range from a solar $B - R \approx 1$ mag to a very red $B - R \approx 2$ mag. Furthermore, the color distribution appears bimodal, with a gap at $B - R \sim 1.5$ mag (Peixinho et al. 2012; Peixinho et al. 2020). The reason for this surface color bimodality is unclear (Peixinho et al. 2020); see also discussion in Chapter 5. An initial explanation based on collisional resurfacing of radiation reddened surface with freshly excavated, neutrally colored material (Luu & Jewitt 1996) was ruled out (Luu & Jewitt 1998; Thébault 2003). Cometary activity may also cause the resurfacing, affecting the colors of Centaurs and causing the bimodality (Delsanti et al. 2004). Indeed, all active Centaurs but one[2] occupy the bluer of the two color clumps. If low level, undetectable activity is present, this can also be due to intrinsically bluer coma dust diluting the redder color of the nucleus, or simply to nongeometric scattering in optically small particles dominating the coma and making it appear bluer (Jewitt 2009). Another possibility is that the bimodality stems from a composition gradient with heliocentric distance in the planetesimal disk from which the Centaurs originate (e.g., Wong & Brown 2017; Liu & Ip 2019). Basically, depending on where a

[2] The exception is 166P ($q = 8.6$ au, $Q = 19.2$ au, unknown rotational properties) which falls in neither the blue nor the red clump.

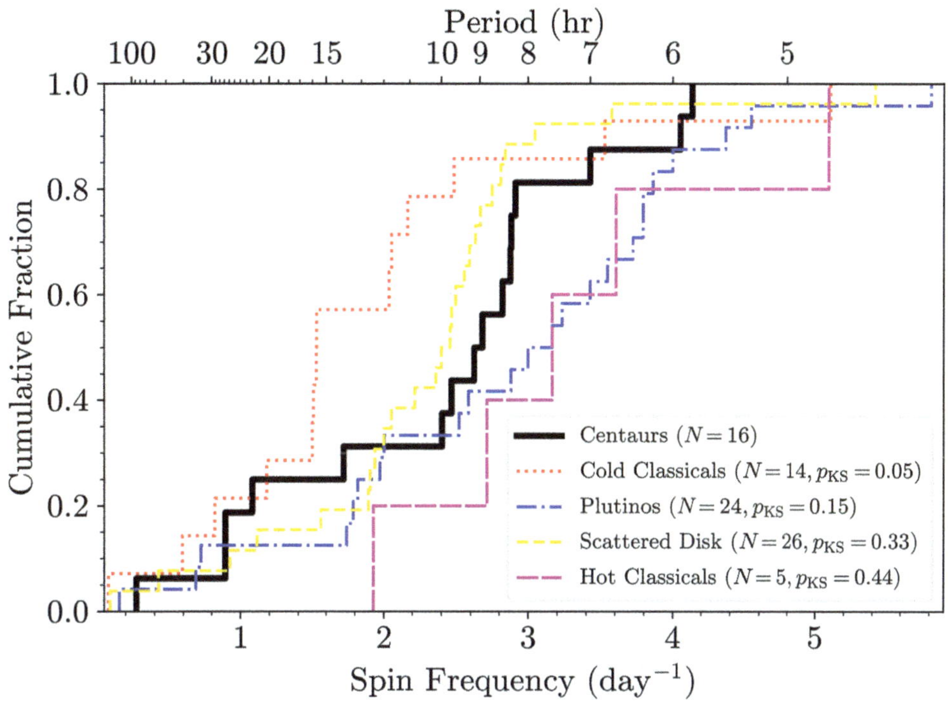

Figure 4.7. Cumulative spin rate distributions of Centaurs and other trans-Neptunian objects. Legend indicates, for each dynamical group, sample size and the KS test p-value that the sample and the Centaurs have the same spin rate distribution.

Centaur progenitor actually formed with respect to the condensation lines of various red species, a surface could evolve differently due to the subsequent radiation and thermal processing, perhaps leading to the discontinuity of colors that we see today.

In any case these processes do not appear to correlate with the rotational properties of Centaurs, as shown in Figure 4.8. The different types of surfaces scatter across the entire range of observed spin frequencies. The red clump of Centaurs concentrates at low Δm, with the exception of Pholus, but this may be a small number fluctuation. An interesting observation not related to rotational properties is that blue Centaurs are darker. This is also seen in other outer solar system populations (Lacerda et al. 2014b).

4.3.4 Relation to Centaur Size

Figure 4.9 (left panel) plots Centaur size against spin frequency and period. Smaller Centaurs ($D < 100$ km) span a wide range of observed spin frequencies, from 0.5 to 3 rotations per day, with a median spin period of ~14 hr. Larger Centaurs spin faster than 2.5 rotations per day, with a median period of ~7 hr. The three largest Centaurs are also the fastest rotators, with $P < 7$ hr.

If a large proportion of the Centaurs are collisional fragments (see Section 4.5.2), then collisional evolution prior to leaving the source region must have played a role

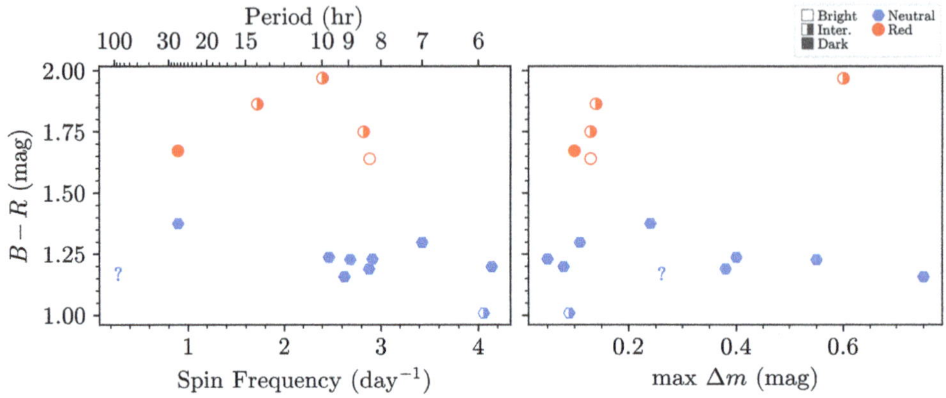

Figure 4.8. $B - R$ color vs. spin frequency (left panel) and maximum Δm (right panel). Point color and shape highlight the bimodality and the fill pattern indicates surface albedo (filled symbols for "Dark" albedos from 0.04 to 0.07; half-filled symbols for "Intermediate" albedos from 0.12 to 0.16; open symbols for "Bright" albedo 0.24; question mark indicates unknown albedo).

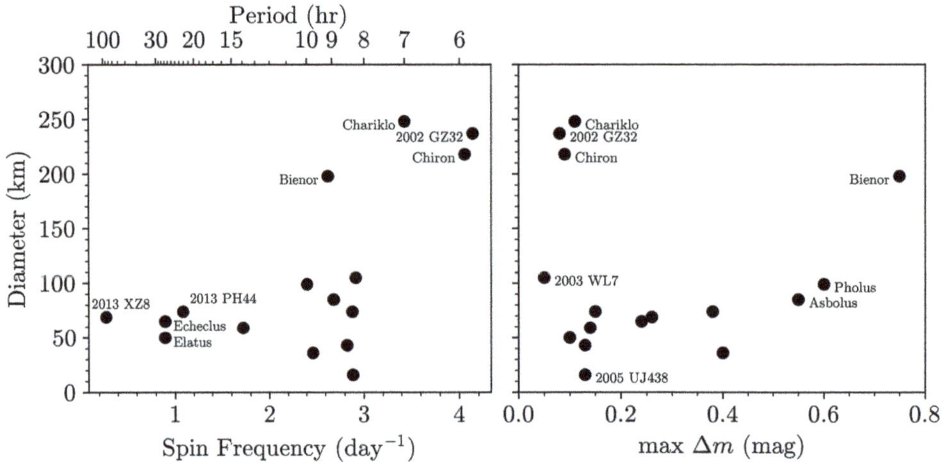

Figure 4.9. Centaur diameter vs. spin rate (left panel) and maximum lightcurve range (right panel).

in setting rotations (see Section 4.3.9). However, energy equipartition would predict smaller Centaurs to spin faster, which argues against the observed distribution of spin frequencies being driven by collisional evolution. It may be that the smaller Centaurs hit a spin barrier at three rotations per day, which would suggest that they are gravitationally reaccumulated outcomes of collisions. If so, then the larger Centaurs were never disrupted by collisions, and retain their original spin rates.

Another possible explanation for smaller Centaurs displaying slower spins has to do with torques due to activity. If spin-up and spin-down are equally likely, this tends to increase the median spin period as spun-up Centaurs may be rotationally disrupted leaving a population of survivors biased toward slower rotators. This effect, investigated in Jewitt (2021) for JFCs, is a function of size, with small objects more affected than larger ones: a factor of 10 increase in size corresponds to a factor 100 increase in spin-up/down timescale (see also Samarasinha & Mueller 2013; Steckloff & Samarasinha 2018; Safrit et al. 2021).

It is unlikely that activity can explain the presence of slower rotators and absence of faster rotators among smaller Centaurs relative to the spin rates of larger Centaurs. According to Figure 1 in Jewitt (2021), notional Centaur-sized nuclei ($D \sim 100$ km) should have spin-up timescales $\sim 10^5$ yrs, comparable to the dynamical lifetimes of Centaurs. However, it is important to note that the figure applies to JFCs with $q \sim 1$ to 2 au and hence larger mass-loss rates. An interesting complication is that some of the Centaurs may have dipped into temporary low-q orbits where activity could have played a more important role (Chapter 7; Chapter 3). More work is needed to understand the effect of orbital evolution and activity on Centaurs spins.

4.3.5 Shapes

A simple way to translate lightcurve variability into a shape model is to assume that it is caused by a triaxial ellipsoid object rotating around its shortest principal axis. The caveat here is that in reality many objects are not ellipsoidal but have more complex shapes (e.g., Stern et al. 2019; Buie et al. 2018) making lightcurve interpretation more difficult. Multiple lightcurves obtained at several orbital longitudes can help with this (Kaasalainen & Torppa 2001), but that can be a lengthy proposition for a study of Centaurs. Nonetheless, we can continue with the simplifying assumption, and the ellipsoid minimum and maximum sky-projected cross-section areas are given by

$$C_{\min} = \pi b(a^2 \cos^2 \theta + c^2 \sin^2 \theta)^{1/2} \tag{4.1}$$

and

$$C_{\max} = \pi a(b^2 \cos^2 \theta + c^2 \sin^2 \theta)^{1/2}, \tag{4.2}$$

where $a \geqslant b \geqslant c$ are the semi-axes of the ellipsoid, and $0 \leqslant \theta \leqslant \pi/2$ is the aspect angle, measured between the line of sight and the spin pole direction.

The photometric range, Δm, is related to ratio of those areas and given by

$$\Delta m = -2.5 \log \frac{C_{\min}}{C_{\max}} = -1.25 \log \left(\frac{\cos^2 \theta + (c/a)^2 \sin^2 \theta}{\cos^2 \theta + (c/b)^2 \sin^2 \theta} \right), \tag{4.3}$$

which depends only on the ellipsoid axis ratios and the aspect angle. If the ellipsoid is observed equator-on ($\theta = \pi/2$), the axis ratio can be obtained directly from:

$$b/a = 10^{-0.4\Delta m}. \tag{4.4}$$

The pole orientation is generally unknown, so an observed Δm sets an upper limit on b/a.

Figure 4.9 (right panel) shows maximum lightcurve variability, $\Delta m_{\max}$, plotted against Centaur size. The larger Centaurs ($D > 150$ km) display very "flat" lightcurves ($\Delta m \lesssim 0.1$ mag), indicative of more spherical shapes. Bienor is an interesting outlier: with $D \approx 200$ km it has the largest observed variability $\Delta m \sim 0.75$ mag, which implies an axis ratio ~1:2.

The icy moons of the giant planets larger than $D = 400$ km tend to be spherical (Lineweaver & Norman 2010), suggesting a transition to a regime where gravitational pull takes over material strength and causes a body to become spherical. The data on Centaurs are sparse, but indicate a transition at smaller sizes, around $D = 100$ km. If confirmed, this difference in size at which the gravity becomes dominant should imply different bulk composition and/or internal structures between the two populations.

4.3.6 Constraints on Density

A useful family of triaxial shapes to consider are the Jacobi ellipsoids. These are the hydrostatic equilibrium shapes of self-gravitating, fluid bodies of uniform density spinning with constant angular frequency (Chandrasekhar 1969). The shape (axis ratios) is set by each combination of spin frequency and density, so the latter can be estimated from the lightcurve information, subject to the hydrostatic fluid behavior assumption.

Figure 4.10 plots lines of Δm against spin frequency for Jacobi ellipsoids of different densities seen equator-on. Centaur lightcurve properties are overplotted and shaded according to each body's size. If Centaurs are in hydrostatic equilibrium, none spins fast enough to require bulk densities much larger than 1000 kg m^{-3}. Larger Centaurs ($D \gtrsim 200$ km) concentrate near the density upper limit, mainly due to their fast spins. Smaller Centaurs spin slowly and only require densities of a few hundred kg m^{-3} to sustain their shapes. We note, however, that occultation measurements of Chariklo and its rings have revealed a shape that is not well matched by a figure of hydrostatic equilibrium (Morgado et al. 2021). If we simply balance centrifugal and gravitational accelerations at the tip of prolate ellipsoids to obtain a minimum bulk density, then all Centaurs spin below the critical rate for a bulk density of 500 kg m^{-3} (see black, slanted lines in Figure 4.10), which is suggestive of a spin barrier (see Section 4.3.9).

4.3.7 Binaries, Contact Binaries and Rings

Two known binaries, (42355) Typhon and Echidna and (65489) Ceto and Phorcys, traverse the Centaur region. They were resolved in HST observations (Noll et al. 2006; Grundy et al. 2007) with apparent separations 0.1 to 0.2 arcsec (1330 ± 130 and 1840 ± 48 km, respectively). However, both have orbits extending far beyond Neptune[3] and are thus excluded from this review. None of the Centaurs in Table 4.1 is a resolved, wide binary (Chapter 9).

[3] $Q_{T\&E} = 57.6$au, $Q_{C\&P} = 181.1$ au.

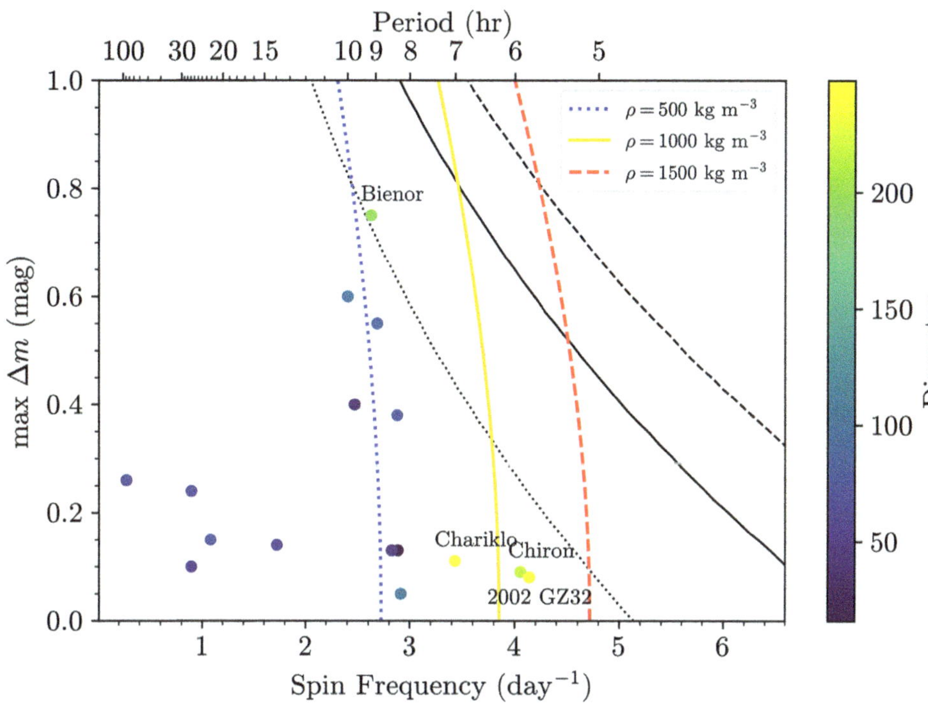

Figure 4.10. Lightcurve variability vs. spin rate. Points are shaded according to their diameter in km. Three colored lines correspond to Jacobi triaxial ellipsoids with densities 500 (dotted), 1000 (solid) and 1500 kg m^{-3} (dashed), viewed equator-on. Three (more slanted and thinner) black lines indicate the minimum density that balances centripetal and gravitational acceleration at the tips of prolate ellipsoids (dotted, solid and dashed correspond to the same densities as for the colored lines.)

Neither do the Centaur rotational properties suggest strong contact binary candidates, which are usually inferred from large lightcurve variability, $\Delta m > 0.9$ mag (Sheppard & Jewitt 2004; Gnat & Sari 2010; Lacerda et al. 2014a; Thirouin & Sheppard 2018). Bilobed shapes can produce similarly large variation (Descamps 2016). Contact binaries seem abundant in many small body populations (Sheppard & Jewitt 2004; Mann et al. 2007; Lacerda 2011; McNeill et al. 2018; Thirouin & Sheppard 2018; Showalter et al. 2021). Even though none of the Centaurs in Table 4.1 display such extreme variability, Bienor's $\Delta m = 0.75$ mag can in principle be caused by an eclipsing binary of two equal spheres. Recent work suggests the contact binary/bilobed shapes observed in JFCs visited by spacecraft can originate in the Centaur region via spin-up due to sublimation torques (Safrit et al. 2021). KBO Arrokoth is evidence that such a shape can already exist in the Kuiper Belt. We note that unambiguous shape determination for these objects requires complementary observations. From the ground, dense, multi-chord occultation data may help solve the ambiguity, but only spacecraft flybys can provide ground truth.

A number of factors could contribute to attenuating a lightcurve's range. For instance, the presence of an unknown secondary in unresolved observations from the ground can complicate the interpretation of the lightcurve. The additional cross section, if present, dampens the lightcurve variation giving the impression of a less elongated primary object. Rings such as Chariklo's (Braga-Ribas et al. 2014), and possibly Chiron's (Ortiz et al. 2015; Ruprecht et al. 2015; Chapter 9), and debris or dust in the coma of active Centaurs also contribute additional cross section in unresolved lightcurve observation (Chapter 9). The result is a lower, contaminated lightcurve Δm_c, given by

$$\Delta m_c = -2.5 \log\left(\frac{C_{\rm min} + C_{\rm coma}}{C_{\rm max} + C_{\rm coma}}\right) = -2.5 \log\left(\frac{k + (b/a)}{k + 1}\right) \tag{4.5}$$

where $k = C_{\rm coma}/C_{\rm max}$ is the rings/coma cross section relative to the maximum nucleus cross section (Luu & Jewitt 1990). Observations of active Centaurs (Luu & Jewitt 1990; Jewitt 2009) and Chariklo's ring (Braga-Ribas et al. 2014) show that $k < 0.5$ for the objects in Table 4.1. We note that the level of activity (and k) varies with time, so they are relevant if occurring at the time the lightcurve is measured. For example, if Chiron's cross-section would include a contemporaneous $k = 0.5$ contribution from activity, debris, rings, or a small unresolved companion, then its undiluted lightcurve would vary by $\Delta m = 0.14$ mag, compared to the apparent $\Delta m = 0.09$ mag. For Bienor, the most photometrically variable Centaur, a relative contribution of $k = 0.25$ would imply an undiluted $\Delta m = 1.1$ mag. Attempts to detect activity for this Centaur have only rendered an upper limit $k < 0.001$ (Jewitt 2009; Dobson et al. 2023), but rings or an unresolved companion cannot be ruled out. Finally, significant obliquity of the spin pole, combined with an observing geometry away from equinox (equator-on), would also "hide" an elongated shape (see Equation 4.3).

4.3.8 Spin Vectors and Obliquities

Lightcurve measurements at different aspect angles (the angle between the line of sight and the spin pole) are needed to constrain the spin pole orientation. Such studies of the changing lightcurves of Pholus and Bienor over 13 and 16 years, respectively, have yielded spin poles directed toward ecliptic latitude $\beta = +30 \pm 5°$ and longitude $\lambda = 145 \pm 5°$ for Pholus, and $\beta = 50 \pm 3°$ and longitude $\lambda = +35 \pm 8°$ for Bienor (Tegler et al. 2005; Fernández-Valenzuela et al. 2017). Reliable pole solutions are challenging (reliable lightcurves are needed spanning a wide range of ecliptic longitudes) and hence still rare: Pholus and Bienor are the only cases. For distant objects of moderate inclination, two measurements of Δm taken sufficiently far apart along the orbit of an object can set limits on its obliquity, ε, which is the angle between the pole and the normal to the orbit plane (Lacerda 2011). At $\varepsilon = 90°$ there is a big variation in Δm (from a maximum value when seen equator-on down to 0 at pole-on geometry), while for $\varepsilon = 0$ there is none.

Figure 4.11 plots lines of minimum versus maximum Δm for Jacobi ellipsoids of different obliquities. A given shape (*b*/*a*) implies a maximum $\Delta m = -2.5 \log(b/a)$ and the obliquity defines what the minimum Δm will be according to Equation (4.3), where the minimum possible aspect angle is $\theta = 90 - \varepsilon$ in degrees. Overplotted are the Centaurs with measured minimum and maximum lightcurve ranges, which are,

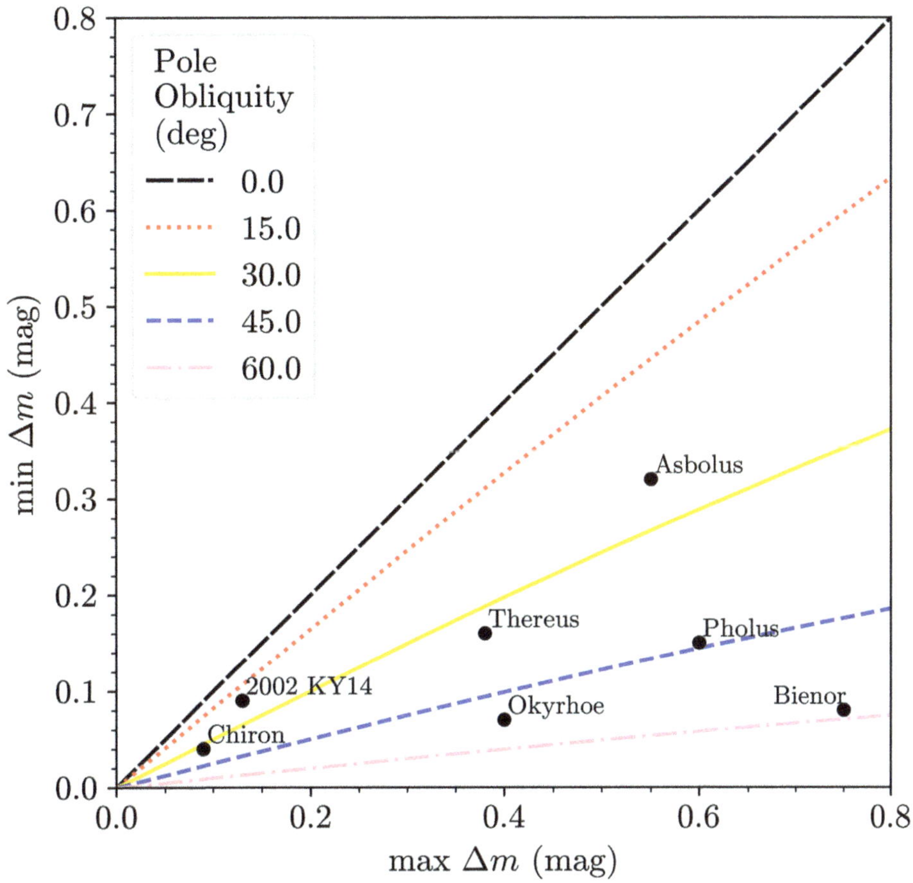

Figure 4.11. Minimum vs. maximum lightcurve variability of Centaurs. Lines correspond to Jacobi ellipsoids spinning around the minor principal axis seen equator-on (max Δm) and at minimum aspect angle along the orbit (min Δm).

respectively, upper and lower limits to the real min Δm and max Δm. Under the figure's idealized shape assumptions, the lines indicate minimum obliquity for each Centaur. Pholus's pole obliquity, calculated between the pole solution and the normal to its orbit plane (inclination $i = 24.7°$ and longitude of ascending node $\Omega = 119.4°$) is $\varepsilon \approx 73°$. For Bienor ($i = 20.7°$ and $\Omega = 337.7°$) the pole solution implies an obliquity $\varepsilon \approx 58°$. These do not contradict Figure 4.11's minimum obliquities of $\varepsilon_{\min} \sim 45°$ for Pholus and $\varepsilon_{\min} \sim 60°$ for Bienor. The figure suggests that Centaurs have considerable obliquities. Except for Asbolus and 2002 KY14, the remaining five Centaurs have obliquities in excess of 30°, and three have obliquities larger than 45°, if their axes ratios are approximately Jacobi-ellipsoidal. Significant obliquity can cause interesting surface illumination and heating patterns, which are important for understanding Centaur activity (see, e.g., Chapters 7 and 8).

4.3.9 Processes Affecting Rotation

As mentioned above, a number of processes can affect the rotational properties of Centaurs. Collisions, discussed in detail in Section 4.5.2, will most likely lead to spin-up, which for the smaller objects can lead to rotational breakup (Lacerda 2005). Collisions are unlikely in the Centaur phase, so spins imposed by this process should have been set before the objects left their reservoir in the trans-Neptunian disk.

Torques from cometary activity can cause both spin up or spin down, with equal probability. Such sublimation torques are likely negligible at Centaur distances and certainly for the larger objects, but a scenario in need of more careful study are those objects that have dipped closer to the Sun during their dynamical evolution (see, e.g., 39P/Oterma and P/2019 LD2 discussed in Chapter 3). Sublimative torques may play a role in those cases.

Processes that lead to the formation of binaries or rings, and their dynamical evolution, may also set or affect rotation. This remains largely unexplored for Centaurs, mostly because of the lack of data to constrain any proposed models.

These processes would manifest in the observations in different ways. Collisions cause net spin-up and eventually lead to rotational breakup, setting a spin-barrier and a lower limit to the size of the survivors. Figure 4.10 hints at a rotational limit even though the data are still sparse. A simple collisional evolution model in an assumed massive, primordial Kuiper Belt would result in a minimum size of survivors of a few tens of km in radius (Lacerda 2005). Centaurs and descendant JFCs an order of magnitude smaller are observed, but the model is too rudimentary to allow conclusions to be drawn. Sublimative torques lead to a random walk in spin rate, causing the survivor spin rate distribution to be skewed toward slow rotators while also imposing a maximum spin barrier.

4.4 Limitations of Measurement Methods

Many observational techniques to derive physical properties suffer from limitations. This does not mean that the techniques are not useful, only that one must interpret results carefully. We describe some of the particulars here. Many of these issues are of course common to all small bodies, not just to Centaurs.

As was mentioned earlier, visible-wavelength photometry of a bare (inactive) Centaur can eventually yield an absolute magnitude H but requires an albedo in order to find the diameter. More specifically, the well-known relationship (see Pravec & Harris 2007) is

$$D_{km} = \frac{2a_{km}10^{0.2m_{\odot\lambda}}}{\sqrt{p_\lambda}}10^{-0.2H_\lambda}, \tag{4.6}$$

where D_{km} is the diameter in km, a_{km} is a constant, the number of km in 1 au, subscript λ indicates a quantity at a specific wavelength, $m_\odot$ is the monochromatic magnitude of the Sun and p is the geometric albedo. Thus we see that this conversion in fact requires additional information such as the geometric albedo (p), and the phase darkening law and rotational context (for a better *H*). Extensive photometry on rotational and seasonal timescales can eventually resolve some of this problem.

But even overcoming all of these issues, there is still a problem of what does it mean to have a diameter of a nonspherical object (see Section 4.3). With visible photometry, this is often effectively the geometric-mean of the cross-section axes, but this isn't necessarily the best answer. If the ultimate goal is, say, to understand the mass distribution of Centaurs, then one could argue that the "'best" diameter to use is one that is the geometric mean of the three ellipsoid axes, $\sqrt[3]{abc}$. In a typical scenario, if one measures the object's light curve, assumes an equator-on view, and assumes that $b = c$, then one's calculation of the volume-derived radius—i.e., $a(b/a)^{2/3}$—is off by a factor of $\sqrt[3]{c/b}$ from the true answer. If every nucleus had the same c/b axial ratio, then this would just mean that the size distribution would as a whole be shifted sideways. Of course it is likely the case that c/b is not constant across objects, and so there could be a slope bias in the size distribution. This would be especially deleterious if the largest objects in the size distribution have c/b values significantly below unity; usually the largest objects in a size distribution are few in number, and if they are widely incorrect, it can significantly affect the resulting power-law exponent.

To return to the problem of determining the size, certainly infrared techniques, using measurements of the thermal radiation, have given us great improvements. Since the square of the diameter is, broadly speaking, proportional to the thermal infrared flux density divided by $(1 - pq_{ph})$, where q_{ph} is the phase integral, and since pq_{ph} is generally small, the systematic uncertainty is in principle lower. Nonetheless, the rotational context can still be a problem, and the phase darkening of thermal emission is not well understood. The temperature map of a Centaur's surface may not be well known, either because the object's shape is poorly constrained or because the fundamental thermal quantities (thermal inertia, thermal diffusivity) are not well constrained, or both. Detailed work on the thermal properties of near-Earth asteroids (e.g., Wright 2007; Wright et al. 2018; Howell et al. 2018; Mommert et al. 2018; Myers et al. 2023) reveals the simplistic thermal models to interpret single-snapshot thermal-IR photometry of highly nonspherical objects can give diameters that are significantly off. Such a problem can be even more severe for distant objects like Centaurs where photometry in the vicinity of 10 μm lies on the Wien-law side of a thermal continuum. This is demonstrated for a particular case in Figure 4.12, where there is just one band of mid-IR photometry and so NEATM and a beaming parameter range are assumed. For example, suppose we have thermal-IR photometry at a wavelength of 10 microns that has yielded a $S/N = 10$ measurement of an object that is 10 au from the Sun. Assuming that the beaming parameter is between 0.8 and 1.3, we find from the plot that a wide range of radii fit the data point, with the maximum possible radius being $\mathcal{R} = 2.3$ times larger than the minimum. This corresponds to a fractional uncertainty in the radius of $(\mathcal{R} - 1)/(\mathcal{R} + 1) = 39\%$, quite significant. The figure shows that the result has some dependence on *S*/*N*, since less accurate photometry will let more possible values of NEATM parameters work, but this problem persists even for high quality photometry ($S/N = 100$ in the plot). In

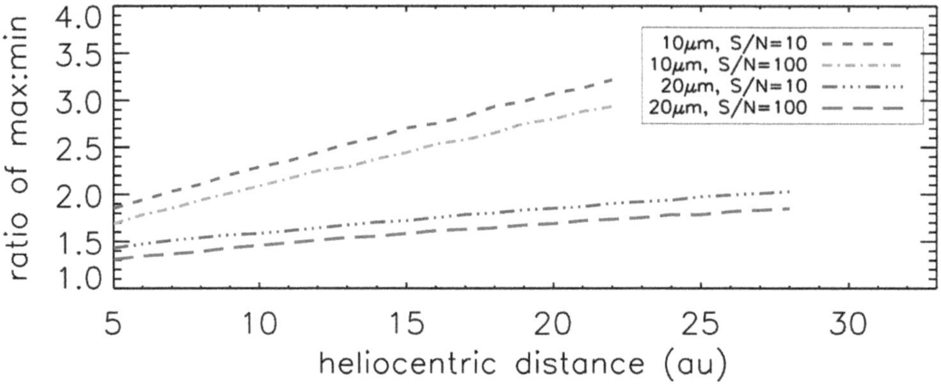

Figure 4.12. Example of the problem with single-band photometry of the thermal-IR emission from a Centaur. The plot shows how one-band mid-IR photometry of a Centaur at a given heliocentric distance results in a large uncertainty in the diameter. The *y*-axis shows the ratio of the maximum and minimum possible diameters. In this case, we assumed a geometric albedo of 0.08, emissivity of 0.95, and a phase angle of 2°. We used NEATM (Harris 1998), and assumed a beaming parameter range of 0.8 to 1.3. We tried two wavelengths (10 and 20 μm) and two photometric *S*/*N* values (100 and 10). The 10 μm lines stop at 22 au since past that, that wavelength is dominated by reflected sunlight. The 20 μm lines stop at 28 au since a phase angle of 2 degrees is impossible past that. For more distant (and thus cooler) Centaurs, the range of possible diameters spans a large factor. This uncertainty ignores the applicability of NEATM in the first place, since more distant objects are more likely to be fast rotators.

any case, this demonstrates that multiband photometry improves the ability to apply more sophisticated thermal models.

As discussed in Equation (4.5), a Centaur's coma will complicate finding a Centaur's physical properties. In principle an extraction technique can be done (Lamy & Toth 1995; Lisse et al. 1999; Hui & Li 2018) but the method loses reliability the coarser the spatial resolution of the imaging and the more dominant the coma's flux is against the nucleus's. While traditionally one of the criteria for determining if an object has cometary activity is to look for extended emission beyond the PSF (see, e.g., Jewitt 1991), it is possible for comets to be active and yet hide coma within the PSF, as is the case apparently for ecliptic-comet 2P/Encke (Fernández et al. 2005). For more distant objects such as Centaurs, it would be even easier to be fooled into thinking that an ostensibly PSF-looking Centaur is bare when it in fact has some cometary activity.

We mention one further limitation here: the spatial resolution limitation for identifying a Centaur as a binary (see Section 4.3.7). HST has been used to show that at least about one-fifth of cold classical TNOs are binaries (Noll et al. 2008; Porter et al. 2024), with the limitation there for detection being that the secondary needs to be bright enough and at least several hundred kilometers from the primary. Searches that use HST to look for binarity among Centaurs include, e.g., that of Li et al. (2020), who looked at 23 objects that would be included in our Centaur definition here (among many other objects) but found no binaries, even though their linear spatial resolution was roughly about twice as good as it is for cold-classical TNOs. In any case, if there is a population of "tight' Centaur binaries, it may be necessary to conduct a search where spatial resolutions better than $\sim$100 km are possible. This is 14 milliarcsec for an object 10 au away, i.e., just under 0.5 JWST NIRCam short-wavelength pixels. It would certainly be interesting to determine definitively whether the binary fraction of the Centaurs were much lower than that of the TNO population.

4.5 Comparative Planetology of Outer Solar System Bodies

The term "comparative planetology" evokes an approach of taking a set of diverse measurements and properties of individual objects in search of a deeper understanding of the set. Much of what we call comparative planetology has been driven by comparing and contrasting terrestrial planets, giant planets, dwarf planets where individual properties matter a great deal. When applying this notion to small bodies, a different framework and approach becomes important.

For small bodies, individual properties are still important but the distribution of those properties across the full population provide far more insight. A fantastic example of this approach is the data from the Sloan Digital Sky Survey (SDSS; Ivezić et al. 2001, 2002). A prominent example of this work is showing the compositional ties within and between asteroid collisional families. While we do not yet have a data set comparable to SDSS that covers the entire solar system, we do expect such fundamental data to come from future surveys. In the meantime, we make do with such data as we have. The critical point here is to focus on the properties of populations of small bodies and look for insight in comparing and contrasting populations. For this section, comparative planetology will refer to the comparison of population properties and what it can tell us.

4.5.1 Centaur Shapes

The concept of a "primordial" object is unfortunately an overused and poorly codified term used widely in planetary science. At one time or another the term primordial has probably been applied to every small body in the solar system.

With the successful flyby of Arrokoth by New Horizons, we finally have an object truly worthy of the term. As discussed by McKinnon et al. (2020), Arrokoth is now helping to more clearly define the concept of what a primordial object is. Here we have an example of an object that has remained essentially untouched since the end of the accretion phase of building our solar system (or at least since the time of thermal equilibration once the disk cleared). As such, it directly informs and constrains theories of how the accretion process works. As a cold-classical Kuiper Belt object (CCKBO), we can look upon these objects as being truly primordial. It would appear that every other population of small bodies, while made up of primordial material, has been more affected by post-accretional processes than what we understand for CCKBOs.

An important question to address is the extent and nature of any processing that has been at work on Centaurs. Quantifying the amount of processing will go far in placing them in context within the other populations of small bodies and help identify the source population and pathway by which an object becomes a Centaur.

Our best prospect, short of spacecraft missions, for assessing the primordial nature of Centaurs lies with detailed stellar occultation observations. Observations of Jupiter Trojans conducted in support of the Lucy mission show that all the targets have significantly more complicated shapes than exhibited by Arrokoth (e.g., Buie et al. 2021; Mottola et al. 2023, 2024). In this case, this constraint is not about the contact binary shape of Arrokoth. None of the Lucy targets is an obvious contact binary though we cannot exclude some being badly distorted contact binaries. The more interesting constraint is that the Lucy targets do not have smooth (i.e., elliptical) shapes and all show significant modification with respect to Arrokoth. This result runs over the range of 30–120 km for the Lucy targets. It remains to be seen if there are any remaining signatures of the primordial shape for these objects and the definitive answers will come with the culmination of that mission.

There are three key observables in the shape of a small body that are likely to be important clues tracing back to primordial shape and thus the degree of processing that Centaurs have undergone: (1) a contact binary shape that is consistent with a gentle merger (though see Safrit et al. 2021), (2) a highly oblate shape, and (3) a smooth shape. There may be other signatures we do not yet know enough about, such as the frequency of tight binaries, since we do not have good knowledge of this for any other distant population.

4.5.2 Collisions

Collisions can significantly alter small body populations. This is particularly evident in the main asteroid belt. Its size distribution shows clear signs of being sculpted by collisions (Bottke et al. 2005, 2020). Furthermore, the main asteroid belt contains many collisional families, which can be traced back to catastrophic disruptions of larger bodies (e.g., Vokrouhlickýct al. 2010; Licandro et al. 2012; Spoto et al. 2015; Bolin et al. 2017). In contrast, the Jupiter Trojans have only one major collisional family, the one of Eurybates (Brož & Rozehnal 2011; Marschall et al. 2022).

The amount of collisional evolution is driven by the dynamical evolution of a small body population. There is now a general consensus that planetesimals that end up as Centaurs and JFCs formed in an original reservoir, the primordial Kuiper Belt (PKB), located between 20 and 40 au from the Sun (e.g., Nesvorný 2018). From there, they were scattered into the current trans-Neptunian region. From that region, the scattered disk in particular acts as the source reservoir for Centaurs and JFCs (e.g., Duncan & Levison 1997). Crucially, the scattering phase of planetesimals must occur as the gas disk dissipates to prevent the damping of the scattered orbits. This is well described by the final phase of planetesimal-driven migration, particularly Neptune's migration (e.g., Nesvorný et al. 2018; Nesvorný 2021).

The collisional environment of a Centaur can thus be divided into three phases: (1) the period in the PKB and subsequent scattering into the current trans-Neptunian region; (2) the period in the scattered disk; and (3) the Centaur phase (Bottke et al. 2023). Phase 1 in the PKB may only last a few tens of millions of years. Nevertheless, it is the phase with the largest number of impacts because of the large mass of the PKB (several tens of Earth masses) compared to the scattered disk

(1/400 of the initial PKB; Bottke et al. 2020; Nesvorný et al. 2020). Phase 2 lasts for most of the lifetime of the solar system (4.5 Gyr) and may still contribute moderately to the collisional evolution. The final phase, when objects are scattered into the giant planet region, lasts only a few million years and will not contribute in any notable way to the collisional evolution of Centaurs. In this phase, sublimation-driven activity will dominate any collisional evolution (see, e.g., Chapters 5, 7, and 8).

Another crucial ingredient to understanding collisional evolution is the initial size distribution. Current planetesimal formation models suggest that planetesimals form via the streaming instability (see Chapter 2). This disk instability accumulates dust particles in clouds massive enough for self-gravity to cause the cloud to collapse and directly form large planetesimals. This mechanism predicts an initial Gaussian size distribution centered around $D = 100$ km (e.g., Simon et al. 2016; Schäfer 2017; Klahr & Schreiber 2020). While these are models, the current Kuiper Belt has a distinct bump around 100 km. This is in line with a leftover streaming instability size distribution.

Bottke et al. (2023) showed that an initial streaming instability size distribution can indeed evolve into the current scattered disk size distribution. They showed that this evolution is consistent with (i) crater SFDs on icy satellites and KBOs (e.g., Singer et al. 2019; Zahnle et al. 2003; Schenk et al. 2004; Kirchoff & Schenk 2010), and (ii) observed SFDs of populations derived from the PKB (e.g., Jupiter's Trojans Wong & Brown 2015; Yoshida & Terai 2017). The craters on icy satellites were used to infer the size distribution of the PKB population scattered onto planet-crossing orbits and those that went to the scattered disk. Figure 4.13 shows the initial and evolved Kuiper Belt size distribution. It exhibits a wavy shape characteristic of collisional evolution.

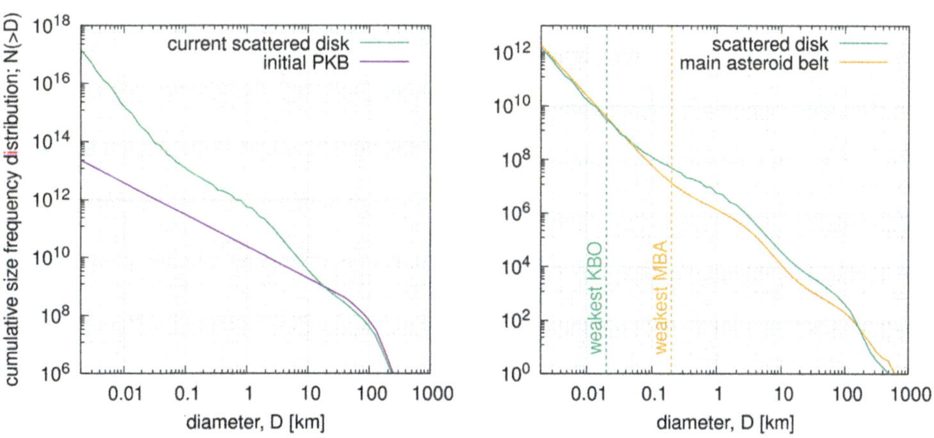

Figure 4.13. The left panel shows the cumulative size frequency distribution (CSFD) of the initial primordial Kuiper-belt (PKB) and the collisionally evolved CSFD of the current day scattered disk according to Bottke et al. (2023). The dynamical depletion of the current scattered disk is not shown. The right panel shows the CSFD of the current day scattered disk (Bottke et al. 2023) and the main asteroid belt (Bottke et al. 2020). The dashed lines show the approximate location of the weakest Kuiper-belt object (KBO) and main belt asteroid (MBA). The scattered disk SFD has been scaled to be on the same order of magnitude of the main belt.

The Kuiper Belt thus appears to be similarly collisionally evolved as the main asteroid belt with one significant difference (Figure 4.13). While the weakest body in the asteroid belt has a diameter of roughly 200 m, it is only 20 m in the scattered disk (Figure 4.13). This shift in the weakest bodies shifts the location of the "collisional wave" of the size distribution of asteroids to KBOs to smaller sizes.

The fact that the scattered disk size distribution can be explained by collisionally evolving a streaming instability distribution indicates that most smaller bodies in the scattered disk, and by extension the Centaur population, are collisional fragments. Marschall et al. (2023) argue that 50% of SDOs/comets/centaurs larger than 10 km should be collisional fragments. For objects larger than 1 km, at least 95% of objects are fragments. In contrast, large Centaurs (100 km and larger) are, to a large extent, untouched by disruption and thus the most "pristine" in this respect.

4.6 Future Prospects

As with many things in astronomy, one way to address some of the issues we have described above is with large amounts of data, and especially data that sample new regimes of faintness. Number statistics are crucial. New data sets from Vera C. Rubin Observatory and NEO Surveyor promise to improve the discovery rate and characterization of many Centaurs (Mainzer et al. 2023; Schwamb et al. 2023; Chapter 14). Further data from existing surveys like Pan-STARRS will continue to help us with photometric coverage. This is especially useful for objects like Centaurs that tend to have multidecade orbital periods, so it takes a while for our vantage point (i.e., sub-Earth latitude) to change significantly. Disentangling the photometric manifestations of a Centaur's shape, rotation period, spin axis direction, color/albedo variations, and phase darkening can be challenging, just as it is with any asteroid. Add onto that the possibility that some Centaurs could have hiccups or outbursts of activity, that may or not be resolved, and it is clear that denser photometric coverage, at a range of cadences, can be necessary.

Insights into cometary activity (see Chapter 8), especially such activity at distances from the Sun where water ice sublimation is *not* the main driver, will be useful for further studies of how Centaur surfaces change. This can include more detailed assessment of Rosetta's observations of 67P when that comet was far from the Sun. But it may also require sending a spacecraft to study a Centaur up close (see Chapter 16). Once we can treat a Centaur as a geologic object it will surely improve our understanding of the observations of a Centaur as an astronomical object.

Acknowledgements

We thank two anonymous referees whose helpful reviews improved our manuscript.

References

Bolin, B. T., Delbo, M., Morbidelli, A., & Walsh, K. J. 2017, Icar, 282, 290

Bottke, W. F., Durda, D. D., Nesvorný, D., et al. 2005, Icar, 175, 111

Bottke, W. F., Vokrouhlický, D., Ballouz, R. L., et al. 2020, AJ, 160, 14

Bottke, W. F., Vokrouhlický, D., Marshall, R., et al. 2023, PSJ, 4, 168

Braga-Ribas, F., Sicardy, B., Ortiz, J. L., et al. 2014, Natur, 508, 72
Brož, M., & Rozehnal, J. 2011, MNRAS, 414, 565
Buie, M. W., Zangari, A. M., Marchi, S., Levison, H. F., & Mottola, S. 2018, AJ, 155, 245
Buie, M. W., Keeney, B. A., Strauss, R. H., et al. 2021, PSJ, 2, 202
Chandrasekhar, S. 1969, Ellipsoidal figures of Equilibrium (New Haven, CT: Yale University Press) 1
Delsanti, A., Hainaut, O., Jourdeuil, E., et al. 2004, AA, 417, 1145
Descamps, P. 2016, Icar, 265, 29
Dobson, M. M., Schwamb, M. E., Benecchi, S. D., et al. 2023, PSJ, 4, 75
Duffard, R., Pinilla-Alonso, N., Santos-Sanz, P., et al. 2014, AA, 564, A92
Duncan, M., Levison, H., & Dones, L. 2004, in Comets II, ed. M. C. Festou, H. U. Keller, & H. A. Weaver (Tucson, AZ: Univ. Arizona Press) 193
Duncan, M. J., & Levison, H. F. 1997, Sci, 276, 1670
Fernández, Y. R., Lowry, S. C., Weissman, P. R., et al. 2005, Icar, 175, 194
Fernández-Valenzuela, E., Ortiz, J. L., Duffard, R., Morales, N., & Santos-Sanz, P. 2017, MNRAS, 466, 4147
Fornasier, S., Lellouch, E., Müller, T., et al. 2013, AA, 555, A15
Fraser, W. C., Dones, L., Volk, K., Womack, M., & Nesvorný, D. 2004, in Comets III, ed. K. J. Meech, et al. (Tucson, AZ: Univ. of Arizona. Press) 121
Gnat, O., & Sari, R. 2010, ApJ, 719, 1602
Grundy, W. M., Stansberry, J. A., Noll, K. S., et al. 2007, Icar, 191, 286
Harris, A. W. 1998, Icar, 131, 291
Howell, E. S., Magri, C., Vervack, R. J., et al. 2018, Icar, 303, 220
Hui, M.-T., & Li, J.-Y. 2018, PASP, 130, 104501
Ivezić, Ž., Tabachnik, S., Rafikov, R., et al. 2001, AJ, 122, 2749
Ivezić, Ž., Lupton, R. H., Jurić, M., et al. 2002, AJ, 124, 2943
Jewitt, D. 1991, IAU Colloq. 116: Comets in the post-Halley era, ed. J. Newburn, R. L. M. Neugebauer, & J. Rahe (Cambridge: Cambridge Univ. Press)
Jewitt, D. 2009, AJ, 137, 4296
Jewitt, D. 2021, AJ, 161, 261
Kaasalainen, M., & Torppa, J. 2001, Icar, 153, 24
Kirchoff, M. R., & Schenk, P. 2010, Icar, 206, 485
Klahr, H., & Schreiber, A. 2020, ApJ, 901, 54
Lacerda, P. 2005, PhD thesis, Leiden Observatory https://ui.adsabs.harvard.edu/abs/2005PhDT........21L/abstract
Lacerda, P. 2011, AJ, 142, 90
Lacerda, P., McNeill, A., & Peixinho, N. 2014a, MNRAS, 437, 3824
Lacerda, P., Fornasier, S., Lellouch, E., et al. 2014b, ApJL, 793, L2
Lamy, P. L., & Toth, I. 1995, A&A, 293, L43
Lamy, P. L., Toth, I., Fernandez, Y. R., & Weaver, H. A. 2004, in Comets II, ed. M. O. Festou, H. U. Keller, & H. A. Weaver (Tucson, AZ: Univ. Arizona Press) 223
Li, J., Jewitt, D., Mutchler, M., Agarwal, J., & Weaver, H. 2020, AJ, 159, 209
Licandro, J., Hargrove, K., Kelley, M., et al. 2012, AA, 537, A73
Lineweaver, C. H., & Norman, M. 2010, in Proc. of the 9th Australian Space Science Conf., 67 arXiv e-prints, arXiv:1004.1091
Lisse, C. M., Fernández, Y. R., Kundu, A., et al. 1999, Icar, 140, 189

Liu, P.-Y., & Ip, W.-H. 2019, ApJ, 880, 71
Luu, J., & Jewitt, D. 1996, AJ, 112, 2310
Luu, J. X., & Jewitt, D. C. 1990, AJ, 100, 913
Luu, J. X., & Jewitt, D. C. 1998, ApJL, 494, L117
Mainzer, A. K., Bauer, J. M., Cutri, R. M., et al. 2019, NEOWISE Diameters and Albedos V2.0, NASA Planetary Data System https://ui.adsabs.harvard.edu/abs/2019PDSS..251.....M/abstract
Mainzer, A. K., Masiero, J. R., Abell, P. A., et al. 2023, PSJ, 4, 224
Mann, R. K., Jewitt, D., & Lacerda, P. 2007, AJ, 134, 1133
Marschall, R., Morbidelli, A., Bottke, W. F., et al. 2023, Asteroids, Comets, Meteors Conference 2023 (Houston, TX: LPI) 2470
Marschall, R., Nesvorný, D., Deienno, R., et al. 2022, AJ, 164, 167
McKinnon, W. B., Richardson, D. C., Marohnic, J. C., et al. 2020, Sci, 367, aay6620
McNeill, A., Fitzsimmons, A., Jedicke, R., et al. 2018, AJ, 156, 282
Mommert, M., Jedicke, R., & Trilling, D. E. 2018, AJ, 155, 74
Morgado, B. E., Sicardy, B., Braga-Ribas, F., et al. 2021, AA, 652, A141
Mottola, S., Britt, D. T., Brown, M. E., et al. 2024, SSRv, 220, 17
Mottola, S., Hellmich, S., Buie, M. W., et al. 2023, PSJ, 4, 18
Müller, T., Lellouch, E., & Fornasier, S. 2020, in The Trans-Neptunian Solar System, ed. D. Prialnik, M. A. Barucci, & L. Young (Amsterdam: Elsevier) 153
Myers, S. A., Howell, E. S., Magri, C., et al. 2023, PSJ, 4, 5
Namouni, F., & Morais, M. H. M. 2020, MNRAS, 494, 2191
Nesvorný, D. 2018, ARAA, 56, 137
Nesvorný, D. 2021, ApJL, 908, L47
Nesvorný, D., Vokrouhlický, D., & Bottke, W. F. 2018, NatAs, 2, 878
Nesvorný, D., Vokrouhlický, D., Alexandersen, M., et al. 2020, AJ, 160, 46
Noll, K. S., Grundy, W. M., Stephens, D. C., Levison, H. F., & Kern, S. D. 2008, Icar, 194, 758
Noll, K. S., Levison, H. F., Grundy, W. M., & Stephens, D. C. 2006, Icar, 184, 611
Ortiz, J. L., Duffard, R., & Pinilla-Alonso, N. 2015, AA, 576, A18
Peixinho, N., Delsanti, A., Guilbert-Lepoutre, A., Gafeira, R., & Lacerda, P. 2012, AA, 546, A86
Peixinho, N., Thirouin, A., Tegler, S. C., et al. 2020, in The Trans-Neptunian Solar System, ed. D. Prialnik, M. A. Barucci, & L. Young (Amsterdam: Elsevier) 307
Porter, S. B., Benecchi, S. D., Verbiscer, A. J., et al. 2024, PSJ, 5, 143
Pravec, P., & Harris, A. W. 2007, Icar, 190, 250
Ruprecht, J. D., Bosh, A. S., Person, M. J., et al. 2015, Icar, 252, 271
Safrit, T. K., Steckloff, J. K., Bosh, A. S., et al. 2021, PSJ, 2, 14
Samarasinha, N. H., & Mueller, B. E. A. 2013, ApJL, 775, L10
Schäfer, U., Yang, C.-C., & Johansen, A. 2017, AA, 597, A69
Schenk, P. M., Chapman, C. R., Zahnle, K., & Moore, J. M. 2004, in Jupiter: The Planet, Satellites and Magnetosphere, ed. F. Bagenal, T. E. Dowling, & W. B. McKinnon (Cambridge: Cambridge Univ. Press) 427
Schwamb, M. E., Jones, R. L., Yoachim, P., et al. 2023, ApJS, 266, 22
Sheppard, S. S., & Jewitt, D. 2004, AJ, 127, 3023
Showalter, M. R., Benecchi, S. D., Buie, M. W., et al. 2021, Icar, 356, 114098
Simon, J. B., Armitage, P. J., Li, R., & Youdin, A. N. 2016, ApJ, 822, 55
Singer, K. N., McKinnon, W. B., Gladman, B., et al. 2019, Science, 363, 955

Spoto, F., Milani, A., & Knežević, Z. 2015, Icar, 257, 275
Steckloff, J. K., & Samarasinha, N. H. 2018, Icar, 312, 172
Stern, S. A., Weaver, H. A., Spencer, J. R., et al. 2019, Sci, 364, aaw9771
Tegler, S. C., Romanishin, W., & Consolmagno, G. J. J.S. 2016, AJ, 152, 210
Tegler, S. C., Romanishin, W., Consolmagno, G. J., et al. 2005, Icar, 175, 390
Thébault, P. 2003, EM&P, 92, 233
Thirouin, A., & Sheppard, S. S. 2018, AJ, 155, 248
Vokrouhlický, D., Nesvorný, D., Bottke, W. F., & Morbidelli, A. 2010, AJ, 139, 2148
Volk, K., & Malhotra, R. 2008, ApJ, 687, 714
Warner, B. D., Harris, A. W., & Pravec, P. 2009, Icar, 202, 134
Wong, I., & Brown, M. E. 2015, AJ, 150, 174
Wong, I., & Brown, M. E. 2017, AJ, 153, 145
Wright, E., Mainzer, A., Masiero, J., et al. 2018, arXiv e-prints, arXiv:1811.01454
Wright, E. L. 2007, arXiv e-prints, arXiv:astro-ph/0703085
Yoshida, F., & Terai, T. 2017, AJ, 154, 71
Zahnle, K., Schenk, P., Levison, H., & Dones, L. 2003, Icar, 163, 263

Chapter 5

Surface Properties and Composition

Nuno Peixinho, Javier Licandro, Eva Lilly, Alvaro Alvarez-Candal, A C Souza-Feliciano and Tom Seccull

This chapter reviews the current knowledge of Centaur surface properties and composition derived from photometry, polarimetry, and spectroscopy. Photometry has been critical in providing initial insights into their surface characteristics and enabling comparisons with other minor body populations, despite its rather limited depth. Centaurs do reveal themselves to be somewhat distinct from their trans-Neptunian progenitors, but some early-found trends are still subject to debate. Polarimetric studies offer additional information, though they are challenging due to the faintness of Centaurs and remain too scarce to show their full potential. Spectroscopy remains the method of choice for studying surface composition and it has been used on the brighter Centaurs, allowing for the first detection of ices—water ice, mostly. While constrained by the limits of ground-based telescopes and transparency of our atmosphere, the recent advances in near-infrared observations with JWST allow the detection of a whole new range of chemical species and foreshadow a new era of discovery.

5.1 Introduction

When (2060) Chiron was discovered in 1977 and announced as Slow-Moving Object Kowal (getting in that same month the provisional designation of 1977 UB; Kowal & Gehrels 1977; Kowal et al. 1977), its colors were not measured. The initial debate of its asteroidal or cometary nature led to highly uncertain diameter estimates between 180 and 790 km (Wallentineen 1978). The first visible and near-IR measured colors revealed it as a dark-surfaced object with a rather neutral spectral behavior (e.g., $B - R = 1.06 \pm 0.03$; Hainaut et al. 2012) not compatible with the presence of clean ice. Instead it displayed a drop in the UV, appearing to be similar to a C-type asteroid, with an albedo of about 0.1. Time-series observations revealed a linear phase coefficient β of 0.05 mag/°, a rotational period of about 6 hours, and a faint cometary behavior (Hartmann et al. 1981; Lebofsky et al. 1984; Bus et al. 1989;

doi:10.1088/2514-3433/ada267ch5

Meech & Belton 1989, 1990). The second Centaur to be discovered fifteen years later, (5145) Pholus (Scotti et al. 1992), was quite distinct, exhibiting an ultra-red surface in the visible (e.g., $B - R = 2.05 \pm 0.13$; Hainaut et al. 2012) and never showing signs of cometary behavior. We would soon find out that these first two objects were members of two distinct Centaur surface groups seen later on in visible photometric studies (see reviews by Tegler et al. 2008; Peixinho et al. 2020).

Characterizing the physical and chemical properties of the surfaces of Centaurs is critical to obtaining a broader understanding of the processes involved in the formation and evolution of planetary systems. Studies of the dynamical evolution of Centaurs show that they most likely originate among populations of icy minor bodies residing beyond Neptune, namely the Kuiper or trans-Neptunian belt (e.g., Levison & Duncan 1997; Di Sisto & Brunini 2007; Volk & Malhotra 2008; Di Sisto & Rossignoli 2020), and perhaps in some cases the Oort Cloud (Brasser et al. 2012; Volk & Malhotra 2013). Their giant planet crossing orbits are short-lived and chaotic, leading to them being either re-ejected into the trans-Neptunian region or further injected into the inner solar system, ultimately becoming Jupiter-family comets (JFCs; e.g., Duncan et al. 2004; Sarid et al. 2019); see Chapter 3 for a full discussion of Centaur dynamics.

The dynamical lifetimes of Centaurs are short in comparison to those of minor body populations on stable orbits (only 10^6–10^7 yrs; Levison & Duncan 1997; Dones et al. 1999; Tiscareno & Malhotra 2003; Horner et al. 2004; Di Sisto & Brunini 2007). However, this brief period of transformation from trans-Neptunian object (TNO)—a.k.a. Kuiper belt object (KBO)—to JFC opens multiple observational windows that are key to understanding the evolution of the surface of an icy minor body as it experiences greater thermal, photolytic, and radiolytic processing by the Sun. How the onset of cometary activity as a result of heating affects the surface color, the surface composition, and other physical properties of Centaur surfaces, is an active subject of research (Jewitt 2009, 2015; Mazzotta Epifani et al. 2018; Seccull et al. 2019; Wong et al. 2019; Zubko et al. 2020; Licandro et al. 2024). One needs to unravel how the surfaces of Centaurs change on their journey from the outer solar system into the inner solar system. Understanding that is also critical to understanding similar evolutionary processes involving icy planetesimals that most likely have occurred in the early history of our solar system. The ultimate accretion or dispersal of most of the dust and gas in the Sun's protoplanetary disk, which had plausibly shielded icy planetesimals in a cold core, would have resulted in increased exposure of the surfaces of those planetesimals to solar heat, light, and particle radiation (e.g., Öberg & Bergin 2021). As a result, the surfaces of those primitive objects may have been changed in similar ways to those observed in today's Centaur population.

It is also noteworthy that, because of their smaller geocentric distances and smaller average intrinsic size (e.g., Lawler et al. 2018), the Centaurs accessible to observation with current facilities are much smaller than than the average observationally accessible TNO. This means that Centaurs provide a convenient way to study the surface properties of small objects from the trans-Neptunian populations so they can be compared to those measured for the surfaces of larger TNOs still beyond Neptune (with due caveats because of their distinct thermal, orbital, and collisional evolution environments and to consequences not yet sufficiently understood). The closer proximity of Centaurs additionally permits their observation over a wider range of phase angles ($\alpha \lesssim 8°$) than what is possible for TNOs ($\alpha \lesssim 2°$), increasing the parameter space reachable for Earth-based observational measurement of phase curves (e.g., Alvarez-Candal et al. 2016; Ayala-Loera et al. 2018; Dobson et al. 2023) and polarimetry (e.g., Belskaya et al. 2008).

5.2 Properties from Photometry

Broadband color photometry can give us a low-resolution surface reflectivity spectrum of any minor body of the solar system. When photometric measurements are taken nearly simultaneously across different wavelength regions, to avoid possible rotational brightness variations, or when rotational variation is negligible or can be averaged out, the magnitude differences between different filters (a.k.a., bands) determine the object's color. Most color photometry of Centaurs has been carried out using some of the $U\ B\ V\ R\ I\ J\ H\ K$ filters (extensions to the Johnson system) and more recently using the SDSS $u\ g\ r\ i\ z$ filter system. Table 5.1 shows a compilation of the quasi-simultaneous $B\ V\ R$ colors (those with smallest errors) of both active and inactive Centaurs available in the literature alongside with those objects' orbital parameters as well as any detected ices. Colors, as low-resolution surface reflectivity, have shown a large variety of surface properties among Centaurs.

Looking at the most frequently employed color parameters, $B-V$, $V-R$, and $B-R$, Centaurs exhibit a large diversity, ranging from neutral (solar-like) colors $B-V \approx 0.67$, $V-R \approx 0.36$, or $B-R \approx 1.03$ to exceptionally red colors $B-V = 1.30$, $V-R = 0.79$, or $B-R = 2.09$. The literature can be confusing when referring to the Centaurs with neutral colors, i.e., surfaces that, regardless of the albedo, reflect both filter frequencies with the same intensity. Neutral colors are sometimes called gray colors. However, these neutral colors have been frequently called "blue", because they were the least red colors observed and seemed to cluster in a group contrasting with another existing cluster of very red colors. Since truly blue colors among Centaurs have not been detected, it became a habit to call those ranging from neutral to slightly red "blue" Centaurs to differentiate them from the red group. Only those ranging from red to very red have been called the red Centaurs.

The intensity of redness in the visible colors of these icy objects, translating to a higher spectral slope, suggests a larger concentration of complex organics such as tholins because they tend to efficiently absorb shorter optical wavelengths. Lower spectral slopes (more neutral colors) are related to highly processed surfaces covered by dark carbonaceous materials (Andronico et al. 1987), although other compounds can also exhibit similar behavior. Strongly weathered organics may also exhibit flat/neutral spectra (Thompson et al. 1987; Kaňuchováet al. 2012); see also discussions in Chapter 6 on volatiles and Chapter 13 on laboratory studies. In practice, the low-resolution reflectance spectrum provided by surface colors offers insights into the

Table 5.1. Compilation of BVR colors, confirmed ices, and orbital parameters of Centaurs

Object	$B - V$	$V - R$	$B - R$	$q[au]$	$i[°]$	e	a	Ices*;	References Phot.	References Spec.
Inactive objects										
(10199) Chariklo	0.81 ± 0.05	0.49 ± 0.02	1.30 ± 0.05	13.13	23.40	0.168	15.78	H_2O	a	b, c, d, e, f
(10370) Hylonome	0.77 ± 0.08	0.38 ± 0.06	1.15 ± 0.10	18.90	4.14	0.244	25.00		g, h	
(119315) 2001 SQ_{73}	0.67 ± 0.02	0.46 ± 0.03	1.13 ± 0.04	14.42	17.44	0.175	17.47		a	
(120061) 2003 CO_1	0.74 ± 0.03	0.48 ± 0.05	1.22 ± 0.06	10.95	19.76	0.471	20.71		a	
(121725) Aphidas	1.28 ± 0.11	0.65 ± 0.07	1.93 ± 0.13	9.59	6.77	0.464	17.90		g	
(136204) 2003 WL_7	0.74 ± 0.04	0.49 ± 0.02	1.23 ± 0.04	14.92	11.18	0.259	20.13	H_2O $c + a$	h	i
								CO_2		i
(145486) 2005 UJ_{438}	1.01 ± 0.03	0.63 ± 0.03	1.64 ± 0.04	8.22	3.80	0.532	17.55		j	
(148975) 2001 XA_{255}	0.81 ± 0.05	0.68 ± 0.05	1.49 ± 0.07	9.35	12.64	0.674	28.67		f	
(160427) 2005 RL_{43}	1.12 ± 0.04	0.73 ± 0.06	1.85 ± 0.07	23.58	12.26	0.043	24.65		f	
1994 TA	1.26 ± 0.14	0.67 ± 0.08	1.93 ± 0.16	11.71	5.40	0.301	16.76		a	
2000 FZ_{53}	0.61 ± 0.03	0.60 ± 0.08	1.21 ± 0.09	12.35	34.87	0.479	23.70		a	
2001 XZ_{255}	1.17 ± 0.02	0.75 ± 0.07	1.92 ± 0.07	15.29	2.61	0.037	15.88		a	
(427507) 2002 DH_5	0.66 ± 0.07	0.47 ± 0.05	1.13 ± 0.09	13.91	22.47	0.366	21.94		a	
2002 PQ_{152}	1.13 ± 0.04	0.72 ± 0.05	1.85 ± 0.06	21.03	9.33	0.189	25.93		j	
(523597) 2002 QX_{47}	0.70 ± 0.04	0.38 ± 0.04	1.08 ± 0.06	16.08	7.26	0.371	25.57		j	
(523620) 2007 RH_{283}	0.72 ± 0.03	0.43 ± 0.02	1.15 ± 0.04	10.60	21.34	0.336	15.97		j	
(527328) 2007 TK_{422}	0.71 ± 0.03	0.51 ± 0.02	1.22 ± 0.04	17.21	3.06	0.189	21.21		j	
(527443) 2007 UM_{126}	0.74 ± 0.03	0.39 ± 0.03	1.13 ± 0.04	8.55	41.74	0.337	12.89		j	
2007 VH_{305}	0.69 ± 0.03	0.49 ± 0.02	1.18 ± 0.04	8.15	6.22	0.663	24.18		j	
2010 BL_4	0.86 ± 0.05	0.39 ± 0.03	1.25 ± 0.06	8.54	20.86	0.539	18 53		f	
2010 TH	0.72 ± 0.02	0.46 ± 0.02	1.18 ± 0.03	12.61	26.69	0.323	18 63		j	
(523676) 2013 UL_{10}	0.97 ± 0.02	0.67 ± 0.02	1.64 ± 0.03	6.16	19.19	0.379	9.92		j	
(463663) 2014 HY_{123}	0.67 ± 0.05	0.49 ± 0.02	1.16 ± 0.05	6.98	13.94	0.625	18.62		f	

(523785) 2015 CM_3	1.21 ± 0.06	0.57 ± 0.05	1.78 ± 0.08	6.91	16.77	0.507	14.03			f	
(248835) 2006 SX_{368}	0.74 ± 0.02	0.47 ± 0.02	1.21 ± 0.03	11.99	36.25	0.461	22.23			j	
(250112) 2002 KY_{14}	1.06 ± 0.02	0.69 ± 0.02	1.75 ± 0.03	8.63	19.46	0.314	12.57	H_2O		e	k
								CO_2			i
(281371) 2008 FC_{76}	0.97 ± 0.02	0.63 ± 0.02	1.60 ± 0.03	10.18	27.10	0.309	14.74			j	
(309139) 2006 XQ_{51}	0.74 ± 0.03	0.41 ± 0.03	1.15 ± 0.04	9.85	31.63	0.378	15.84			j	
(309737) 2008 SJ_{236}	1.01 ± 0.02	0.58 ± 0.02	1.59 ± 0.03	6.11	6.06	0.439	10.89			j	
(309741) 2008 UZ_6	0.92 ± 0.04	0.59 ± 0.03	1.51 ± 0.05	10.54	35.88	0.612	27.13			j	
(315898) 2008 QD_4	0.74 ± 0.02	0.46 ± 0.02	1.20 ± 0.03	5.40	42.08	0.354	8.36			j	
(31824) Elatus	1.05 ± 0.02	0.64 ± 0.02	1.69 ± 0.03	7.22	5.25	0.385	11.75	H_2O		g, h	b
(32532) Thereus	0.77 ± 0.03	0.45 ± 0.03	1.22 ± 0.04	8.51	20.38	0.199	10.61	H_2O	$c + a$	k	i, m, n, o, p
								CO_2			i
(341275) 2007 RG_{283}	0.79 ± 0.03	0.47 ± 0.03	1.26 ± 0.04	15.28	28.76	0.236	20.00			j	
(346889) Rhiphonos	0.82 ± 0.02	0.55 ± 0.02	1.37 ± 0.03	5.97	19.95	0.445	10.76			j	
(349933) 2009 YF_7	0.72 ± 0.02	0.46 ± 0.02	1.18 ± 0.03	6.50	31.03	0.460	12.04			j	
(382004) 2010 RM_{64}	1.00 ± 0.02	0.55 ± 0.02	1.55 ± 0.03	6.13	27.00	0.690	19.80			j	
(447178) 2005 RO_{43}	0.77 ± 0.03	0.47 ± 0.03	1.24 ± 0.04	13.88	35.46	0.521	28.96			j	
(449097) 2012 UT_{68}	1.02 ± 0.02	0.66 ± 0.02	1.68 ± 0.03	12.61	15.41	0.379	20.31			j	
(459865) 2013 XZ_8	0.72 ± 0.02	0.45 ± 0.02	1.17 ± 0.03	8.42	22.56	0.369	13.35			j	
(459971) 2014 ON_6	0.97 ± 0.02	0.58 ± 0.02	1.55 ± 0.03	5.84	3.94	0.563	13.36			j	
(463368) 2012 VU_{85}	1.07 ± 0.06	0.63 ± 0.04	1.70 ± 0.07	20.20	15.05	0.311	29.33			j	
(471339) 2011 ON_{45}	1.11 ± 0.03	0.71 ± 0.02	1.82 ± 0.04	9.82	8.19	0.161	11.70			j	
(49036) Pelion	0.74 ± 0.08	0.56 ± 0.09	1.30 ± 0.12	17.40	9.35	0.133	20.07			a	
(5145) Pholus	1.30 ± 0.10	0.79 ± 0.03	2.09 ± 0.10	8.66	24.70	0.573	20.29	H_2O		g, j	m
								CH_3OH			l
(52872) Okyrhoe	0.74 ± 0.07	0.48 ± 0.05	1.22 ± 0.09	5.82	15.65	0.305	8.37	H_2O		a	b
(52975) Cyllarus	1.17 ± 0.05	0.71 ± 0.04	1.88 ± 0.06	16.26	12.63	0.380	26.23			g, j	
(54598) Bienor	0.67 ± 0.03	0.44 ± 0.03	1.11 ± 0.04	13.20	20.74	0.203	16.56	H_2O		g, q	b, f, r
(55576) Amycus	1.11 ± 0.02	0.70 ± 0.02	1.81 ± 0.03	15.22	13.34	0.391	24.99	H_2O		j	k

(*Continued*)

Table 5.1. (*Continued*)

Object	$B-V$	$V-R$	$B-R$	$q[au]$	$i[°]$	e	a	Ices*;	References	
									Phot.	Spec.
(60558) Echeclus	0.86 ± 0.04	0.56 ± 0.03	1.42 ± 0.05	5.85	4.34	0.456	10.75		j	
(63252) 2001 BL_{41}	0.72 ± 0.05	0.50 ± 0.04	1.22 ± 0.06	6.95	12.43	0.297	9.89		a	
(7066) Nessus	1.09 ± 0.04	0.79 ± 0.04	1.88 ± 0.06	11.91	15.62	0.516	24.59		g	
(83982) Crantor	1.10 ± 0.04	0.76 ± 0.04	1.86 ± 0.06	14.12	12.77	0.274	19.44	H_2O	a	b, f, s
(8405) Asbolus	0.75 ± 0.04	0.51 ± 0.07	1.26 ± 0.08	6.89	17.61	0.619	18.07	H_2O	g, h	t
(88269) 2001 KF_{77}	1.08 ± 0.04	0.73 ± 0.01	1.81 ± 0.04	19.81	4.36	0.239	26.03		a	
(95626) 2002 GZ_{32}	0.71 ± 0.13	0.58 ± 0.05	1.29 ± 0.14	18.01	15.03	0.219	23.05	H_2O	a	k
(342842) 2008 YB_3	0.74 ± 0.02	0.51 ± 0.02	1.25 ± 0.03	6.46	105.12	0.442	11.57		l	
166P/NEAT	0.87 ± 0.03	0.73 ± 0.03	1.60 ± 0.04	8.56	15.37	0.383	13.88		u	
(2060) 95P/Chiron	0.60 ± 0.03	0.32 ± 0.03	0.92 ± 0.04	8.55	6.92	0.377	13.71	H_2O	u	v
29P/Schass.-Wachmann 1	0.78 ± 0.03	0.50 ± 0.03	1.28 ± 0.04	5.78	9.36	0.045	6.05		u	
39P/Oterma	0.74 ± 0.07	0.45 ± 0.05	1.19 ± 0.09	5.71	1.48	0.228	7.40		u	
(523676) 2013 UL_{10}	1.13 ± 0.10	0.75 ± 0.12	1.88 ± 0.16	6.16	19.19	0.379	9.92		q	
C/2001 M_{10}	0.66 ± 0.10	0.51 ± 0.03	1.17 ± 0.10	5.30	28.08	0.801	26.66		l	
C/2011 P_2 PANSTARRS	0.81 ± 0.06	0.43 ± 0.05	1.24 ± 0.08	6.15	8.99	0.370	9.76		l	
C/2012 Q_1 Kowalski	0.91 ± 0.02	0.58 ± 0.01	1.49 ± 0.02	9.48	45.18	0.637	26.13		l	
C/2013 C_2	0.87 ± 0.02	0.55 ± 0.02	1.42 ± 0.03	9.13	21.34	0.431	16.06		l	
C/2013 P_4	0.82 ± 0.02	0.49 ± 0.02	1.31 ± 0.03	5.97	4.26	0.596	14.77		l	
P/2010 C_1 Scotti	0.92 ± 0.05	0.49 ± 0.03	1.41 ± 0.06	5.23	9.14	0.259	7.07		w	
P/2011 S_1 Gibbs	0.96 ± 0.11	0.59 ± 0.03	1.55 ± 0.11	6.89	2.68	0.203	8.65		l	
167P/CINEOS	0.76 ± 0.03	0.53 ± 0.02	1.29 ± 0.04	11.78	19.13	0.270	16.14		j	

Notes:* Ices include water ice: H_2O (*c*—crystalline, *a*—amorphous); carbon dioxide ice: CO_2; methanol ice: CH_3OH.
References: a: Peixinho et al. (2015); b: Barkume et al. (2008); c: Brown & Koresko (1998); d: Dotto et al. (2003b); e: Guilbert et al. (2009b): f: Guilbert et al. (2009a); g: Tegler et al. (2003); h: Peixinho et al. (2003) i: Licandro et al. (2024); j: Tegler et al. (2016); k: Barucci et al. (2011); l: Jewitt (2015); m: Cruikshank et al. (1998); n: Barucci et al. (2002); o: Licandro & Pinilla-Alonso (2005); p: Merlin et al. (2005); q: Mazzotta Epifani et al. (2018); r: Dotto et al. (2003a); s: Doressoundiram et al. (2005); t: Kern et al. (2000); u: Jewitt (2009); v: Romon-Martin et al. (2003); w: Mazzotta Epifani et al. (2014).

spectral continuum of any studied object at a relatively low observational time and resource cost. Actually, both Centaurs and TNOs have revealed themselves to be quite featureless in visible wavelengths, so visible colors are indeed a good proxy for their visible spectra. However, the same does not hold over the full set of near-IR (NIR) colors obtained so far (e.g., when using $J\ H\ K$, or $i\ z$ filters).

Many authors convert the colors into reflectivity $S_\lambda = 10^{-0.4\,(c-c_\odot)}$, where c is the object's color, $c_\odot$ is the solar color, and λ is the central wavelength of the filter in question in Å; They then determine a normalized reflectivity reddening gradient between two filters $S' = 100\,(S_{\lambda 2} - S_{\lambda 1})/[(\lambda 2 - \lambda 1)/1000]$, in %/1000 Å. After conversion, a solar-like color translates to $S' = 0\%/1000$ Å (i.e., a flat spectra in the wavelength range used), a color of $B - R = 1.5$ translates to $S' \approx 18\%/1000$ Å, and a color of $B - R = 2$ corresponds to $S' \approx 30\%/1000$ Å. Mathematically, it is equivalent to describe the observations using either colors or the normalized reflectivity gradient, which corresponds to a spectral slope. When using S', a blue surface has a negative value, a neutral surface has a zero value, and a red surface has a positive value; when using colors directly one needs to compare them with the solar value.

5.2.1 The Visible Color Bimodality

Centaurs appear to have a bimodal photometric color distribution in $B - R$, and their surfaces fall into two groups. Part of the population is red with optical colors $1.6 \lesssim (B - R) \lesssim 2$, like the primordial cold classical TNOs. The rest have more neutral surfaces similar to comets with $1 \lesssim (B - R) \lesssim 1.4$. Figure 5.1 shows the two

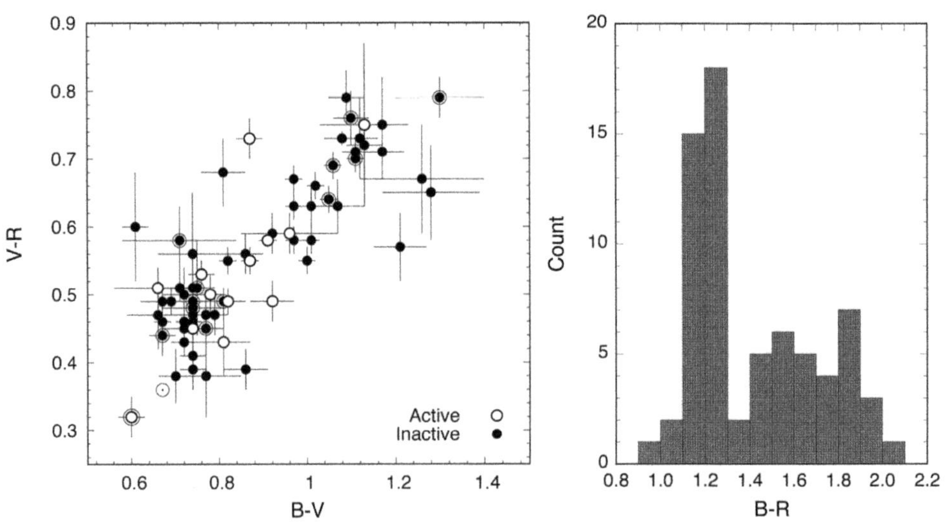

Figure 5.1. Left: $(V - R)$ *vs.* $(B - V)$ color–color diagram, showing both the active (open white dots) and the inactive (solid black dots) Centaurs. Objects with known spectra are indicated with an extra circle. There is a clustering for high $(B - V)$ and $(V - R)$ colors and for low $(B - V)$ and $(V - R)$ colors. Note: solar colors are $(B - V)_\odot = 0.67$ and $(V - R)_\odot = 0.36$ being indicated with the symbol $\odot$. Produced with data from Tegler & Romanishin (2003); Peixinho et al. (2003, 2015); Tegler et al. (2016); Mazzotta Epifani et al. (2018). Right: $B - R$ color histogram showing the two-color clustering, although with some overlap between them.

groups of Centaur surface colors are seen in a 2-D plot of just $B-V$ and $V-R$ colors but become also evident in the $B-R$ color distribution. The Centaur color bimodality was first noted in the early optical surveys of TNOs and Centaurs (Tegler & Romanishin 1998; Peixinho et al. 2003; Tegler et al. 2003; Peixinho et al. 2015). Its existence has been strengthened by numerous observing campaigns since, though, with higher sampling, some overlap is now seen between the color groups (Bauer et al. 2003; DeMeo et al. 2009; Fraser & Brown 2012; Tegler et al. 2016; Mazzotta Epifani et al. 2018; Wong et al. 2019). The difference in colors could be: (i) primordial, in the sense that each group was already intrinsically different when residing in the trans-Neptunian region, having formed in different regions of a planetesimal belt with heliocentric composition gradients; or (ii) the result of some evolutionary mechanism occurring during the Centaurs' orbital evolution with long residency of large incursions at lower heliocentric distances (see Tegler et al. 2008 for a review).

Similarly to what has been done for asteroids since the 1970s (e.g., Chapman et al. 1975), a taxonomy scheme for Centaurs and TNOs using surface properties has been attempted. Using $B-V$, $V-R$, $V-I$, $V-J$ colors and clustering algorithms, a taxonomy consisting of four clustered groups of objects of increasing redness in the visible colors has been proposed. These groups are designated by BB, BR, IR, and RR—"blue", intermediate "blue-red", intermediate red, and red, respectively— and no Centaurs are found in the intermediate IR group (Barucci et al. 2005; Perna et al. 2010). Using the spectral slopes obtained from the surface colors and including albedo, another taxonomy with ten groups (or *taxa*) has been proposed. These are designated by BL ("Blue Light", subdivided into BL1 and BL2), BD ("Blue Dark"), GD ("Grey Dark"), VD ("Very Dark"), RM ("Red Moderate"), VR ("Very Red"), RD ("Red Dark"), RI ("Red Intermediate"), and RL ("Red Light"); no Centaurs are found belonging to the BL1, BL2, BD, and the VR taxa (Dalle Ore et al. 2013).

The visible color bimodality was found to be common to both Centaurs (which are all small-sized, i.e., $H_R \sim 6$ or $D \lesssim 250$ km) and small-sized TNOs; so it has been hypothesized that such bimodality could be a size-related phenomenon (Peixinho et al. 2012). Simultaneously, Fraser & Brown 2012 found that small objects from the low-perihelion scattered disk and resonant populations of TNOs were also bimodal in colors, arguing for the existence of two distinct classes of objects in the protoplanetary disk. That is, it is hypothesized that a less red class (LR) and a very red class (VR) were superimposed at a later stage in the early solar system into the trans-Neptunian region over top of a native and also very red class of objects, usually called cold classical TNOs. More recently, using PCA analysis on a large sample of optical and near-IR colors of TNOs, Fraser et al. 2023 conclude that globally there are only two classes of objects: the "BrightIR" and the "FaintIR" TNOs.

The ultra-red colors of some TNOs have been attributed to irradiated, space-weathered simple organic compounds, including ammonia, that might exist in certain outer solar system regions (e.g., Brown 2014) and which take time on the order of 10^6 yrs to form (Brunetto et al. 2007). The overall color distribution of TNOs has been modeled using mixtures of neutral silicate materials and red water-rich organic materials (Fraser & Brown 2012; Fraser et al. 2023). However, even though links between the colors of Centaurs and their dynamical evolution point to an evolutionary mechanism (Melita & Licandro 2012; see also discussion in Chapter 7), this bimodality is not a closed subject and, as we will see in Section 5.3, disregarding phase angle effects may also cause a falsely detected bimodal behavior.

5.2.2 The Lack of Red Colors among Active Centaurs

Given the fact that the TNO colors span at the whole extent of the $B-V$ *versus* $V-R$ space and inactive Centaurs show both neutral and red to extremely red surfaces similar to cold classical TNOs (Jewitt 2009, 2015; Mazzotta Epifani et al. 2018), it seems peculiar that this bimodality doesn't apply to active members of the Centaur population. Figure 5.2 shows that almost all active Centaur nuclei have neutral (gray) colors, similar to long-period comets and short-period comets—including the JFCs. Because the lack of red surfaces and emergence of activity in the Centaur population appear to occur at similar perihelion distances (~10 au; see Figure 5.2), it has been speculated that the activity itself could play a part in this process. The Centaur color transformation from red to neutral could be the result of a fallout ejecta gradually blanketing the surface with the neutrally colored material characteristic of cometary dust coma (Delsanti et al. 2004; Jewitt 2015). Covering a Centaur's surface with gray material should occur on very short timescales, on the order of months to decades (Jewitt 2015; Mazzotta Epifani et al. 2018); this would erase the red surface in at most 1–2 orbits, which is the blink-of-an-eye compared to their dynamical evolution timescales. The blanketing process appears to be more likely as a culprit in the disappearance of the primordial ultra-red surfaces that other mechanisms; the

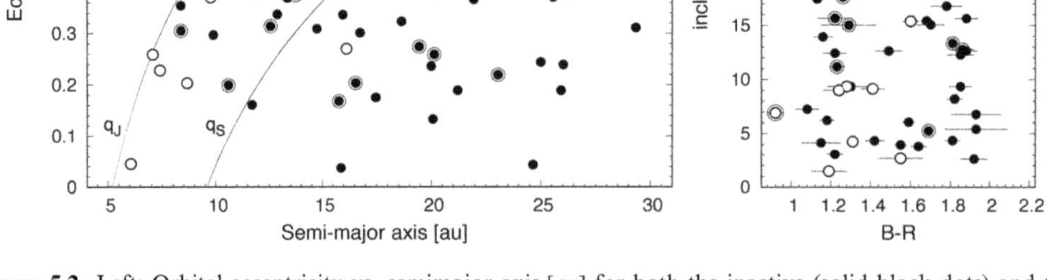

Figure 5.2. Left: Orbital eccentricity vs. semimajor axis [*au*] for both the inactive (solid black dots) and the active (open white dots) Centaurs with measured $(B-R)$, $(V-R)$, and $(B-R)$ colors described in this work. Objects with known spectra are indicated with an extra circle. The Jupiter-crossing and the Saturn-crossing perihelia lines are drawn (q_J and q_S, respectively). All active Centaurs, except one, have their perihelia below ~10 *au*, i.e., roughly below q_S. Right: $(B-R)$ color vs. orbital inclination [°] of the same Centaurs shown in the left panel. There is a lower range of inclinations among the Centaurs of the reddest group.

collisional timescales in the Centaur region are more than an order of magnitude longer than their dynamical lifetime (Durda & Stern 2000), and the irradiated red organics have high molecular weight and therefore are not volatile at the temperatures typical for Centaur region. Interestingly, several active and inactive Centaurs were recently observed to have colors in-between the two distinctive groups. One active Centaur, (523676) 2013 UL_{10}, falls into the red category (Mazzotta Epifani et al. 2018), suggesting that it might be possible to observe the blanketing process in real time and either confirm or reject this hypothesis.

5.2.3 The Color Inclination Correlation

In 2008, it was noticed that the two distinct color groups of Centaurs occupy different inclination ranges (Peixinho et al. 2008). This was confirmed in 2016 with the analysis of a much larger sample (Tegler et al. 2016; see Figure 5.2) and extended to all dynamically hot TNO populations, the commonly accepted precursors of Centaurs within the trans-Neptunian region (e.g., Tiscareno & Malhotra 2003; Volk & Malhotra 2008), in 2019 (Marsset et al. 2019). Because the inclinations of TNOs tend to be preserved even after they enter the Centaur region (Volk & Malhotra 2013), it could be expected that the inclination dichotomy between gray and red Centaurs occurs in the TNOs as well. Showing that this dichotomy is indeed present in all classes of the dynamically hot TNOs would favor the hypothesis of different formation locations for these objects and a compositional gradient in the original protoplanetary disk—stated frequently as "nature"—rather than being explained by different evolutionary pathways—stated frequently as "nurture".

The question of the color grouping and classification, combined with dynamical evolution, is crucial for the understanding of the formation of the trans-Neptunian populations, and their Centaur escapees, and for addressing the major question about the surface evolution of these objects: are the surface properties, or at least the present distinctions among them, mostly primordial ("nature") or mostly evolutionary ("nurture")? Even though, with our current understanding, it seems likely that distinct primordial groups were mixed together and slow and long surface weathering/evolution did not erase their distinct surfaces, we still do not have an answer. See Chapter 2 for further discussion of formation and Chapter 7 for discussions of evolutionary processes.

5.3 Properties from Phase Curves

The phase curve of a small body describes its magnitude change with the viewing phase angle, α (i.e., the Sun-Target-Observer angle; see Figure 5.3). Because the magnitude is strongly dependent on the distance to the Sun and the distance to the observer, it is customary to compute the absolute magnitude, H, that is: the magnitude the object would have if observed at $r = 1$ au from the Sun and $\Delta = 1$ au from the observer and without phase curve effects. It is straightforward to compute the so-called reduced magnitude, denoted by $M(\alpha)$ or $H(\alpha)$, i.e., the magnitude the object would have if $r = 1$ au and $\Delta = 1$ au, from $M(\alpha, r = 1, \Delta = 1) \equiv M(\alpha) \equiv H(\alpha) = m - 5\log(r\Delta)$, where m is the apparent magnitude, r is the heliocentric distance in au, and Δ is its

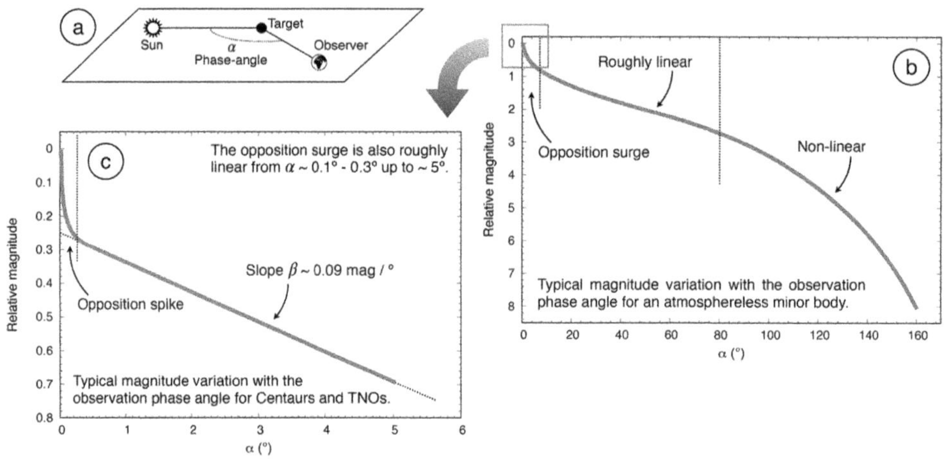

Figure 5.3. (a) Diagram of the Sun-Target-Observer phase-angle α. (b) Typical magnitude/brightness variation with the observation phase angle for an atmosphereless minor body. The effect is strongly non-linear for $\alpha \gtrsim 80°$, approximately linear in the range $7° \lesssim \alpha \lesssim 80°$, and exhibits the so-called opposition surge for $\alpha \lesssim 7°$. (c) All Centaurs, as well as TNOs, are observed within the expected opposition surge which shows a linear behavior from $\alpha \sim 0.1°$–$0.3°$ to $\sim 5°$. An exponential opposition spike has been observed for $\alpha \lesssim 0.2°$, but data is scarce.

distance to the observer in au. However, it is not straightforward to determine the precise correction for the phase angle, α, which is related to the fraction of the illuminated surface as seen by the observer. Such a correction has to be made through a phase function fitted into the phase curve.

The $H - G$ magnitude phase function (commonly used for asteroids; Bowell et al. 1989) necessary to fit the non-linear behavior seen for $\alpha \gtrsim 80°$ (see Figure 5.3) is rather complex. Given the narrow observational phase angle range of Centaurs and TNOs from ground-based observations (going no further than 7°–9°, and 2°, respectively), the phase angle correction is frequently made using a simple linear model. In this case, the absolute magnitude is computed by $H = H(\alpha) - \beta\alpha$, where β is the phase coefficient in magnitude per degree. Surges in magnitude have been measured for TNOs when at opposition, consisting on a quasi-exponential spike for $\alpha \lesssim 0.2°$, usually modeled by $H(\alpha) = H - a/(1 + \alpha) + b\alpha$ (with a and b being empirically determined); however, data on such phase angle values is still scarce for both Centaurs and TNOs (Belskaya & Shevchenko 2000; Belskaya et al. 2003). Using the New Horizons space mission data, Verbiscer et al. (2019) measured phase curves of 6 TNOs at very high phase angles (up to 60°–130°), but no Centaurs were observed.

While the absolute magnitude is related to the size and albedo of the object, β is related in an a priori unknown way to the surface properties. Phase curves of Centaurs can be observed from the ground up to larger phase angles than those of TNOs, reaching $\alpha \simeq 8°$ for a low orbital eccentricity ($e < 0.1$). Several phase curves of Centaurs have already been obtained, showing β values ranging from 0.03 to 0.10

with typical errors of 0.01 (see Figure 5.3; Bauer et al. 2003; Rabinowitz et al. 2007; Alvarez-Candal et al. 2016; Dobson et al. 2023).

The phase-angle correction is necessary for asteroids, which can be easily observed at large phase angles. However, given the range of phase angles covered by Centaurs, and most particularly by TNOs, and with small slope parameters of $\beta \sim 0.09$, it is not uncommon to disregard the phase correction to obtain the absolute magnitude, H. This is because the absolute magnitude is usually used to estimate the diameter of the object assuming it is spherical and it has a geometric albedo of $p = 0.09$, i.e., to produce only a rough estimate. Nonetheless, a more detailed knowledge of the phase function of Centaurs, as well as other icy bodies of the outer solar system is still required (e.g., Verbiscer et al. 2013). Phase curves of a sample of both asteroids and TNOs, restricted to phase angles compatible with those observed for the TNO/Centaur population, have shown that redder objects become more neutral with increasing phase angle, while the opposite happens for the neutral objects (Alvarez-Candal et al. 2016, 2019; Alvarez-Candal et al. 2022; Dobson et al. 2023).

Hence, the phase correction appears to be wavelength/filter-dependent, and colors computed from magnitudes uncorrected for the phase effect (or corrected assuming a β equal for all filters) may have larger errors than what has been published. Additionally, as further indication of how this issue requires further study, an analysis of surface colors computed from magnitudes corrected for phase effects no longer shows the bimodal behavior as described in Section 5.2.1 (Alvarez-Candal et al. 2019). This does not invalidate the links found between the colors of Centaurs and their dynamical evolution (e.g., Melita & Licandro 2012), but it may invalidate the existence of a color bimodality; further investigation is required.

5.4 Properties from Polarimetry

Polarization is a remote-sensing method which can reveal physical properties of airless bodies whose observational characteristics are governed by small scatterers, such as dusty, regolith-covered surfaces. For polarimetry of minor bodies of the solar system, one usually addresses the four parameters of the Stokes vector—intensity, I, the two components of linear polarization, Q and U, and the circular polarization component, V—relative to the Sun-Target-Observer plane, the scattering plane. The scattering angle, θ, defined as the angle between the observer and the direction of the light propagation, is replaced by the phase angle $\alpha = 180° - \theta$. The linear polarization Q—equal to the intensity of light in the perpendicular plane, $I_\perp$, minus the intensity of light in the parallel plane, $I_\parallel$—is reduced to $P_Q = -Q/I = (I_\perp - I_\parallel)/(I_\perp + I_\parallel)$; this is usually given in percentage. The 45° and 135° polarization U is reduced to $P_U = U/I = (I_{\swarrow} - I_{\nearrow})/(I_{\swarrow} + I_{\nearrow})$. In the absence of surface anisotropy, U and P_U are expected to be zero.

The variation of polarization with the phase angle is called the polarization phase curve. Airless solar system bodies show negative polarization values for $\alpha \lesssim 25°$; i.e., the electric vector oscillates predominantly in the Sun-Target-Observer scattering plane, contrarily to what is typical for smooth homogeneous surfaces. The behavior of the negative polarization with phase angle, $P_Q(\alpha)$, depends on the size, composition and packing of the scatterers. Therefore, simultaneous observations of negative polarization and photometry of small solar system bodies allow us to infer properties of the surface texture of these objects if compared to laboratory models of the light scattering by the dust or regolith (Belskaya et al. 2005; Rosenbush et al. 2005). To date, only a handful of the largest Centaurs—(2060) Chiron, (5145) Pholus, and (10199) Chariklo (Bagnulo et al. 2006; Belskaya et al. 2008, 2010)—have been observed with polarimetric methods due to the population's intrinsic faintness.

Fitting a Lumme and Muinonen function $P_Q(\alpha) = b \sin^{c_1}(\alpha) \cos^{c_2}(\alpha/2) \sin(\alpha - \alpha_0)$ to the data (with $c_1 \approx 0.5 - 0.7$, but variable, $c_2 = 0.35$, and both b and α_0 being free parameters) we obtain the polarization minimum, $P_{Q\,min}$, and the inversion angle, $\alpha_{inv} \equiv \alpha_0$, i.e., the phase-angle beyond which the polarization becomes positive (Lumme & Muinonen 1993; Penttiläet al. 2005). The three measured Centaurs reveal negative polarization in the observed α phase-angle range 0.5°–4.4°. Data for (5145) Pholus and (10199) Chariklo is too scarce to find a fit of the aforementioned function, but for (2060) Chiron one obtains a best fit with $c_1 = 0.340$, $b = 54.479$, and $\alpha_0 = 6.475$. Therefore, for (2060) Chiron, $P_{Q\,min} = -1.374\%$ at $\alpha = 1.643°$ and $\alpha_{inv} = 6.475°$.

These Centaurs show a similar polarimetric behavior to the $D \lesssim 1000$ km TNOs for which polarimetry has been obtained, like (28978) Ixion, (38628) Huya, (20000) Varuna, (26375) 1999 DE_9, and (29981) 1999 TD_{10}; they are also similar to the Uranian icy moons Ariel, Umbriel, Oberon, and Titania, but quite different from the large TNOs, and dwarf planets, (136199) Eris, (136472) Makemake, and (134340) Pluto (Bagnulo et al. 2008; Rosenbush et al. 2015; Belskaya & Bagnulo 2015). It was suggested that the presence of a small amount of water frost on a dark surface could explain the lower $P_{Q\,min}$ and α_{min} values of Centaurs, small TNOs, and the Uranian icy moons, but polarimetric observations are still too scarce to draw any robust conclusion (see Rosenbush et al. 2015; Belskaya & Bagnulo 2015; Figure 5.4).

5.5 Properties and Composition from Spectroscopy

It comes as no surprise that the spectral properties of Centaurs cannot be studied and understood disconnected from those of TNOs (see the reviews by Barucci & Merlin 2020; Peixinho et al. 2020). However, the detection of different molecules and silicates on their surfaces has long been limited by the terrestrial atmosphere and the sensitivity of the available instrumentation. In the visible, the spectra of Centaurs are featureless showing a range of spectral slopes from the roughly neutral $S' = -0.5 \pm 0.2\%/1000$ Å to the very red $S' = 47.8 \pm 0.7\%/1000$ Å (e.g., Duffard et al. 2014). Although the spectral slope does not provide the signature of any particular component, as discussed on Section 5.2, it does give some insights on the surface composition of the object.

In the pre-James Webb Space Telescope (JWST) era, NIR detection of signatures from water ice in Centaurs has been the most common finding (e.g., Chariklo, Pholus, and with variability, Chiron; Foster et al. 1999; Romon-Martin et al. 2003; Belskaya et al. 2010), although it is not detected among all Centaurs (Barkume et al. 2008).

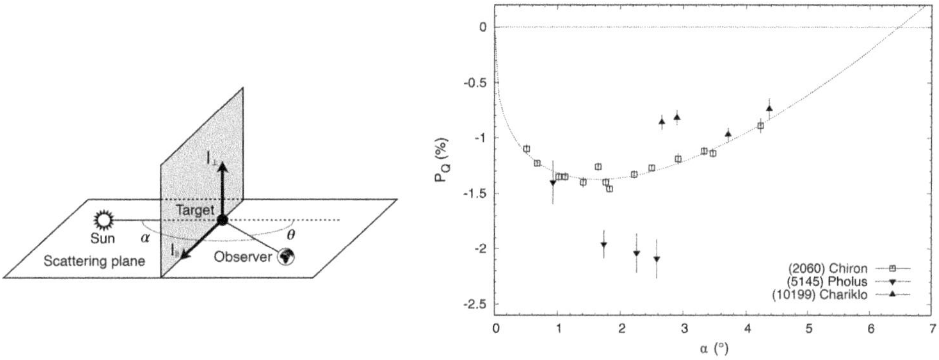

Figure 5.4. Left: Diagram of the Sun-Target-Observer scattering plane and the crossing plane where the intensity of light is measured as perpendicular $I_\perp$ and parallel $I_\parallel$, as normally used for polarimetry of solar system bodies. Given that the scattering angle θ equals 180° minus the phase-angle α, polarimetry results are given relative to α. Right: Linear polarization P_Q(%) vs. phase-angle α(°) for Centaurs (2060) Chiron, (5145) Pholus, and (10199) Chariklo. A Lumme and Muinonen function fit to (2060) Chiron's data with a polarization inversion at $\alpha_{inv} = 6.475°$ is shown. Data for (5145) Pholus and (10199) Chariklo are insufficient for the function fit. Data retrieved from Bagnulo et al. (2006) and Belskaya et al. (2008; 2010).

Methanol ice was identified on (5145) Pholus throughout the NIR feature at 2.27 μm (Cruikshank et al. 1998)—see the review by Barucci & Merlin 2020. No other ices were detected pre-JWST, but the access that JWST provides at longer wavelengths is allowing us to probe for a whole new range of molecules that have their fundamental signatures beyond 2.5 μm. Notably, this spectral range allows us to probe the existence of ices like carbon dioxide, solid complex organic materials, silicates, and stronger spectral features of water and methane; new discoveries are to be expected (see also Chapter 14).

5.5.1 Brief Note on the TNO Progenitors

A recent JWST study by Pinilla-Alonso et al. (2024) (hereafter NPA24) of 54 TNOs on the 0.6–5.3 μm region (Pinilla-Alonso et al. 2021) identified three distinct spectral and compositional groups, characterized by the shape of the spectrum in the 3μm region. These groups are named according to their water ice spectral shapes: Bowl (pictogram: ◡∧, 25% of the sample), Double-Dip (pictogram: ◡υ, 43% of the sample), and Cliff (pictogram: ⁄∕, 32% of the sample)—see Figure 5.5. Additionally, De Prá et al. (2024) has shown that all TNOs in that sample exhibit clear signatures of CO_2 ice, with roughly half of them displaying strong indications of CO ice as well; the Bowl-type show the fundamental CO_2 feature only, whereas Double-Dip and Cliff TNO types display CO_2 combination bands, $^{13}CO_2$ isotopologue, and CO fundamental features. Broad absorptions attributed to O–H, C–H, and N–H bands of complex organics have also been detected around the 1.5 and 2.0 μm regions of the Double-Dip and Cliff TNO types. The new JWST data suggest that only Bowl-type TNOs clearly exhibit the 1.5 and 2.0 μm water ice bands. Figure 5.5 summarizes the aforementioned main

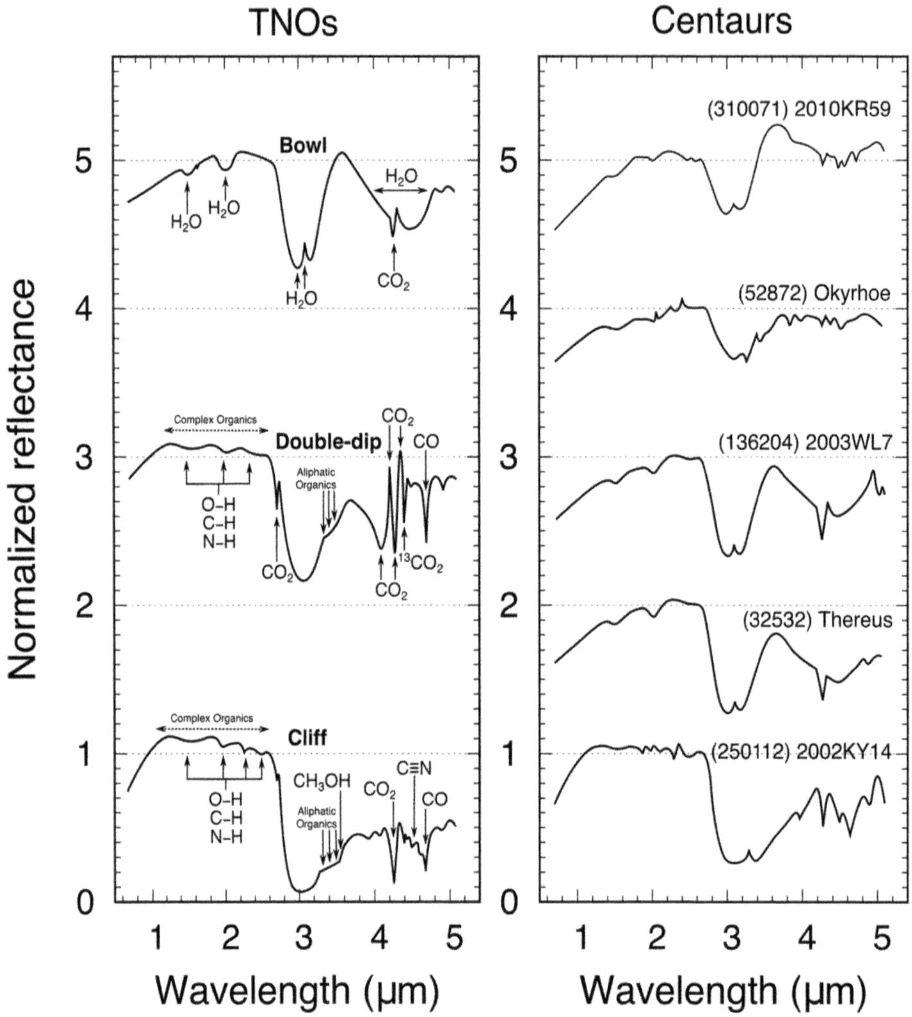

Figure 5.5. Left: Simplified sketch of the reflectance spectra of the three groups of TNOs found by Pinilla-Alonso et al. (2024): Bowl, Double-Dip, and Cliff. The location of the most prominent H_2O, CO_2, $^{13}CO_2$, CO, and CH_3OH ice lines/overtones/bands are indicated. The regions of signatures of complex and aliphatic organics bearing O–H, C–H, N–H, and C ≡ N are also indicated. Spectra are oversmoothed and shifted on the y-axis for clarity. For the detailed spectra see Pinilla-Alonso et al. (2024). Right: Simplified sketch of the reflectance spectra of the five Centaurs discussed by Licandro et al. (2024). (52872) Okyrhoe and (310071) 2010 KR_{59}, do not match with any of the TNOs groups and were designated as "Shallow-type" Centaurs. (32532) Thereus and (136204) 2003 WL_7 are Bowl-type, and (250112) 2002 KY_{14} is Cliff-type. Spectra are oversmoothed and shifted on the y-axis for clarity. For the detailed spectra see Licandro et al. (2024).

features. Given these results, De Prá et al. (2024) propose that the three compositional groups of TNOs give us a picture of ice retention lines that occurred in the outer protoplanetary disk before the onset of planetary migration and strong scattering of planetesimals.

5.5.2 Discussion on the Current Status of Centaurs

Focusing exclusively on Centaurs, Licandro et al. (2024) (hereafter Lic24) studied the 0.6–5.3 μm reflectance spectra of five inactive Centaurs (see Figure 5.5), as part of the DiSCo-TNOs program on JWST (Pinilla-Alonso et al. 2021). Comparing those Centaurs with the TNOs compositional groups found by NPA24, Centaurs (32532) Thereus and (136204) 2003 WL_7 are Bowl-type (pictogram: ), and (250112) 2002 KY_{14} is Cliff-type (pictogram: ). The other two Centaurs, (52872) Okyrhoe and (310071) 2010 KR_{59}, did not align with any of NPA24's groups, exhibiting a much shallower Bowl-like band around 3 μm, faint or absent water-ice bands at 1.5 and 2.0 μm, and no significant CO_2 and/or CO bands. Lic24 names this unique group "'Shallow-type" Centaurs (pictograms:  and 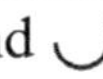).

Of the four Centaurs observed in the framework of the 1st Cycle program of Guaranteed Time of JWST ("Kuiper Belt Science with JWST", Stansberry et al. 2017), (55576) Amycus, and (281371) 2008 FC_{76} have been reported as being Cliff-type, and (10199) Chariklo and (459865) 2013 XZ_8 have been reported as Shallow-type, the group so far seen only among Centaurs (Licandro et al. 2023). The reflectance spectrum of the active Centaur C/2014 OG_{392} (PANSTARRS)[1] and of the active Centaur 39P/Oterma both also seem compatible with this Shallow-type group (see Figure 12 in Harrington Pinto et al. 2023). Even though the sampling is still rather small, it seems that there are significant differences between the distribution of spectral groups for Centaurs and TNOs.

As TNOs leave the Kuiper Belt and cross the giant planet region, becoming shorter-lived Centaurs, they will suffer higher surface temperatures. Those higher temperatures should cause phase changes, such as the sublimation of volatiles like CO_2 and CO; surface spectra do show a lower level of CO_2 and an almost absence of CO among Centaurs (Lic24). The almost absence of CO suggests that its presence on the TNO progenitors results from irradiation processes and is not significantly present in the TNOs' interiors; therefore it should not be a main driver for activity (see De Prá et al. 2024, Lic24, Parhi & Prialnik 2023, and references therein). The activity observed in some Centaurs is more likely due to crystallization of amorphous water ice, which is efficient up to 10–12 au, and releases trapped volatiles (Jewitt 2009; Guilbert-Lepoutre 2012). See Chapter 8 for further discussion of Centaur activity.

The Bowl-type Centaurs (32532) Thereus and (136204) 2003 WL_7 do possess the well-defined 3 μm band mainly produced by water ice, like the Bowl-type TNOs; these Centaurs seem to possess, at most, shallower water-ice absorption bands at 1.5 and 2.0 μm and slightly lower continuums at 3.6 μm than their TNO counterparts (see Figure 5.5). Shallow-type Centaurs (52872) Okyrhoe and (310071) 2010 KR_{59} have no parallel among TNOs and show higher levels of amorphous silicates, indicating a dominance of primitive cometary-like dust (Wooden et al. 2017; Lic24); note that the aforementioned Centaurs C/2014 OG_{392} and 39P/Oterma are also Shallow-type. The transition from ice-rich surfaces to less icy and more refractory

[1] C. Schambeau (2023), personal communication.

ones is very important to understand all other populations of presumably ice-bearing small bodies with thermally evolved surfaces; these bodies presumably evolved under similar processes, such as the Jupiter Trojans, Main Belt Comets—a.k.a. Activated Asteroids—, and D-type Asteroids. The shallower 3 μm bands and lower CO and CO_2 levels of Cliff-type Centaur (250112) 2002 KY_{14} compared to Cliff-type TNOs also suggests that it has suffered thermal resurfacing effects (see Figure 5.5).

Interestingly, while 43% of the TNOs in the NPA24 sample are reported as Double-Dip-type ◡∪, no Centaur has been observed as Double-Dip so far. Lic24 proposes that Double-Dip TNOs may quickly evolve into Shallow-type Centaurs upon entering the giant planets region through the loss of CO and CO_2; this has already been identified as a cometary activity driver at large heliocentric distances (Jewitt et al. 2017; Meech et al. 2017; Kelley et al. 2022). Understanding the reason for the non-existence of Double-Dip-type Centaurs, if confirmed with larger samples, will be a key issue to this field of work.

Another key issue Lic24 finds is that the Cliff-type Centaurs belong to the visible "red" population, all the Bowl-type belong to the visible "neutral/gray" population, and two of the three Shallow-type belong to the "neutral/gray" population (the other one, (310071) 2010 KR_{59}, is "red"). With due caution because of the small sample size, these findings suggest once more that the visible color dichotomy is, in essence, determined by their original composition ("nature"). This is contrary to what was expected in the past, i.e., that surface evolution processes dominate ("nurture"; see NPA24).

5.6 Conclusion and Future

Centaurs have been studied for more than 45 years, yet they remain a puzzling family of minor bodies in the solar system. There remains very little doubt, not to say no doubt, as to their main dynamical origin in the trans-Neptunian populations and as to their role as progenitors of Jupiter-family comets. The particular contribution of each subfamily of TNOs to the Centaur family is less clear, though. The task of correlating the orbital histories of these objects with their spectral/surface characteristics is still tremendously difficult. This complexity stems largely from the challenges in obtaining accurate data on each object's unique dynamical history, which can be tracked with accuracy only for several hundred to thousand years into the past, typically only up to the last close encounter with Jupiter or Saturn, which usually lead to orbital changes, sometimes quite significant.

Although Centaurs are, on average, brighter than TNOs, identifying their spectral signatures and surface composition has been extremely difficult. Nonetheless, using photometry, a visible two-color group (bimodality) has been detected (Peixinho et al. 2003; Tegler et al. 2003), with the redder group showing a smaller inclination range than the more neutral group (Tegler et al. 2016). This has become a key issue to the understanding of these objects.

Orbital changes, due to the interaction with the giant planets, can lead to dramatic changes of the thermal environment on the surface and inside the nucleus of Centaurs, which may trigger cometary activity (e.g., Fernández et al. 2018; Lilly et al. 2024) and subsequent thermal processing and surface alteration. There is, indeed, an absence of ultra-red colors among active Centaurs but it is close to impossible to determine if a particular orbital change was responsible for a certain spectral signature or photometric color change, as Centaurs on average undergo multiple heating cycles during their lifetime (Jewitt 2009; Gkotsinas et al. 2022).

With the beginning of operations of JWST a whole new window for the identification of spectral signatures among the Centaurs has been opened; see Chapter 14 for additional discussion of this. Although TNOs probably preserve most of their original properties from the time when they were formed in the protoplanetary disk, prior to their migration/scattering in to their current orbits, the most recent results suggest that the Centaurs might suffer some significant changes when (re)injected into the Jupiter-Neptune region (Pinilla-Alonso et al. 2024; Licandro et al. 2024). These fossil remnants of the solar system's formation evolve endlessly and, inevitably, erase more and more of the ancient history written in them. Deciphering what is happening requires not only larger but longer photometric, polarimetric, radiometric, and spectroscopic studies and surveys, accompanied by improved modeling. Great progress has been made and many advancements are still to come. The new window opened by JWST will be long lasting. Other long lasting windows will certainly be opened by the future Vera C. Rubin Observatory, the Extremely Large Telescope, and other facilities, including laboratories; a space mission to a Centaur is also envisaged to open new opportunities.

Acknowledgements

NP acknowledges funding from Fundação para a Ciência e a Tecnologia (FCT), Portugal, through the research grants UIDB/04434/2020 and UIDP/04434/2020. JL acknowledges support from the ACIISI, Consejería de Economía, Conocimiento y Empleo del Gobierno de Canarias and the European Regional Development Fund (ERDF) under grant with reference ProID2021010134 and support from the Agencia Estatal de Investigacion del Ministerio de Ciencia e Innovacion (AEI-MCINN) under grant "'Hydrated Minerals and Organic Compounds in Primitive Asteroids" with reference PID2020-120464GB-100. EL acknowledges financial support from the NASA SSO Award 80NSSC23K1169. AAC acknowledges financial support from the Severo Ochoa grant CEX2021-001131-S funded by MCIN/AEI/10.13039/501100011033.

References

Alvarez-Candal, A., Ayala-Loera, C., Gil-Hutton, R., et al. 2019, MNRAS, 488, 3035
Alvarez-Candal, A., Jimenez Corral, S., & Colazo, M. 2022, A&A, 667, A81
Alvarez-Candal, A., Pinilla-Alonso, N., Ortiz, J. L., et al. 2016, A&A, 586, A155
Andronico, G., Baratta, G. A., Spinella, F., & Strazzulla, G. 1987, A&A, 184, 333
Ayala-Loera, C., Alvarez-Candal, A., Ortiz, J. L., et al. 2018, MNRAS, 481, 1848
Bagnulo, S., Belskaya, I., Muinonen, K., et al. 2008, A&A, 491, L33
Bagnulo, S., Boehnhardt, H., Muinonen, K., et al. 2006, A&A, 450, 1239

Barkume, K. M., Brown, M. E., & Schaller, E. L. 2008, AJ, 135, 55
Barucci, M. A., Alvarez-Candal, A., Merlin, F., et al. 2011, Icar, 214, 297
Barucci, M. A., Belskaya, I. N., Fulchignoni, M., & Birlan, M. 2005, AJ, 130, 1291
Barucci, M. A., & Merlin, F. 2020, in The Trans-Neptunian Solar System, ed. D. Prialnik, M. A. Barucci, & L. Young (Amsterdam: Elsevier) 109
Barucci, M. A., Boehnhardt, H., Dotto, E., et al. 2002, A&A, 392, 335
Bauer, J. M., Meech, K. J., Fernández, Y. R., et al. 2003, Icar, 166, 195
Belskaya, I., & Bagnulo, S. 2015, in Polarimetry of Stars and Planetary Systems, ed. L. Kolokolova, J. Hough, & A.-C. Levasseur-Regourd (Cambridge: Cambridge Univ. Press) 405
Belskaya, I. N., Bagnulo, S., Barucci, M. A., et al. 2010, Icar, 210, 472
Belskaya, I. N., Barucci, A. M., & Shkuratov, Y. G. 2003, EM&P, 92, 201
Belskaya, I. N., Levasseur-Regourd, A. C., Shkuratov, Y. G., & Muinonen, K. 2008, in The Solar System Beyond Neptune, ed. M. A. Barucci, et al. (Tucson, AZ: Univ. Arizona Press) 115
Belskaya, I. N., & Shevchenko, V. G. 2000, Icar, 147, 94
Belskaya, I. N., Shkuratov, Y. G., Efimov, Y. S., et al. 2005, Icar, 178, 213
Bowell, E., Hapke, B., Domingue, D., et al. 1989, in Asteroids II, ed. R. P. Binzel, T. Gehrels, & M. S. Matthews (Tucson, AZ: Univ. Arizona Press) 524
Brasser, R., Schwamb, M. E., Lykawka, P. S., & Gomes, R. S. 2012, MNRAS, 420, 3396
Brown, A. J. 2014, Icarus, 239, 85
Brown, M. E., & Koresko, C. D. 1998, ApJL, 505, L65
Brunetto, R., Orofino, V., & Strazzulla, G. 2007, MSAIS, 11, 159
Bus, S. J., Bowell, E., Harris, A. W., & Hewitt, A. V. 1989, Icar, 77, 223
Chapman, C. R., Morrison, D., & Zellner, B. 1975, Icar, 25, 104
Cruikshank, D. P., Roush, T. L., Bartholomew, M. J., et al. 1998, Icar, 135, 389
Dalle Ore, C. M., Dalle Ore, L. V., Roush, T. L., et al. 2013, Icar, 222, 307
De Prá, M. N., Hénault, E., Pinilla-Alonso, N., et al. 2025, NatAs, 9, 252
Delsanti, A., Hainaut, O., Jourdeuil, E., et al. 2004, A&A, 417, 1145
DeMeo, F. E., Fornasier, S., Barucci, M. A., et al. 2009, A&A, 493, 283
Di Sisto, R. P., & Brunini, A. 2007, Icarus, 190, 224
Di Sisto, R. P., & Rossignoli, N. L. 2020, CeMDA, 132, 36
Dobson, M. M., Schwamb, M. E., Benecchi, S. D., et al. 2023, PSJ, 4, 75
Dones, L., Gladman, B., Melosh, H. J., et al. 1999, Icar, 142, 509
Doressoundiram, A., Barucci, M. A., Tozzi, G. P., et al. 2005, P&SS, 53, 1501
Dotto, E., Barucci, M. A., Boehnhardt, H., et al. 2003a, Icar, 162, 408
Dotto, E., Barucci, M. A., Leyrat, C., et al. 2003b, Icar, 164, 122
Duffard, R., Pinilla-Alonso, N., Santos-Sanz, P., et al. 2014, A&A, 564, A92
Duncan, M., Levison, H., & Dones, L. 2004, in Comets II, ed. M. C. Festou, H. U. Keller, & H. A. Weaver (Tucson, AZ: Univ. Arizona Press) 193
Durda, D. D., & Stern, S. A. 2000, Icar, 145, 220
Fernández, J. A., Helal, M., & Gallardo, T. 2018, P&SS, 158, 6
Foster, M. J., Green, S. F., McBride, N., & Davies, J. K. 1999, Icar, 141, 408
Fraser, W. C., & Brown, M. E. 2012, ApJ, 749, 33
Fraser, W. C., Pike, R. E., Marsset, M., et al. 2023, PSJ, 4, 80
Gkotsinas, A., Guilbert-Lepoutre, A., Raymond, S. N., & Nesvorny, D. 2022, ApJ, 928, 43
Guilbert, A., Alvarez-Candal, A., Merlin, F., et al. 2009a, Icar, 201, 272
Guilbert, A., Barucci, M. A., Brunetto, R., et al. 2009b, A&A, 501, 777

Guilbert-Lepoutre, A. 2012, AJ, 144, 97
Hainaut, O. R., Boehnhardt, H., & Protopapa, S. 2012, A&A, 546, A115
Harrington Pinto, O., Kelley, M. S. P., & Villanueva, G. L. 2023, PSJ, 4, 208
Hartmann, W. K., Cruikshank, D. P., Degewij, J., & Capps, R. W. 1981, Icar, 47, 333
Horner, J., Evans, N. W., & Bailey, M. E. 2004, MNRAS, 354, 798
Jewitt, D. 2009, AJ, 137, 4296
Jewitt, D. 2015, AJ, 150, 201
Jewitt, D., Hui, M.-T., Mutchler, M., et al. 2017, ApJL, 847, L19
Kaňuchová, Z., Brunetto, R., Melita, M., & Strazzulla, G. 2012, Icar, 221, 12
Kelley, M. S. P., Kokotanekova, R., Holt, C. E., et al. 2022, ApJL, 933, L44
Kern, S. D., McCarthy, D. W., Buie, M. W., et al. 2000, ApJ, 542, L155
Kowal, C., Dressler, A., Adams, R., et al. 1977, IAU Circ., 3134, 6
Kowal, C. T., & Gehrels, T. 1977, IAU Circ., 3129, 1
Lawler, S. M., Shankman, C., Kavelaars, J. J., et al. 2018, AJ, 155, 197
Lebofsky, L. A., Tholen, D. J., Rieke, G. H., & Lebofsky, M. J. 1984, Icar, 60, 532
Levison, H. F., & Duncan, M. J. 1997, Icar, 127, 13
Licandro, J., & Pinilla-Alonso, N. 2005, ApJL, 630, L93
Licandro, J., Pinilla-Alonso, N., Stansberry, J., et al. 2023, LPI Contributions, Asteroids, Comets, Meteors Conf. Vol 2851, (Houston, TX: LPI) 2255
Licandro, J., Pinilla-Alonso, N., Holler, B., et al. 2025, NatAs, 9, 245
Lilly, E., Jevčák, P., Schambeau, C., et al. 2024, ApJL, 960, L8
Lumme, K., & Muinonen, K. O. 1993, LPI Contributions, Asteroids, Comets, Meteors Vol 810, (Houston, TX: LPI) 194
Marsset, M., Fraser, W. C., Pike, R. E., et al. 2019, AJ, 157, 94
Mazzotta Epifani, E., Dotto, E., Ieva, S., et al. 2018, A&A, 620, A93
Mazzotta Epifani, E., Perna, D., Licandro, J., et al. 2014, A&A, 565, A69
Meech, K. J., & Belton, M. J. S. 1989, IAU Circ., 4770, 1
Meech, K. J., & Belton, M. J. S. 1990, AJ, 100, 1323
Meech, K. J., Kleyna, J. T., Hainaut, O., et al. 2017, ApJL, 849, L8
Melita, M. D., & Licandro, J. 2012, A&A, 539, A144
Merlin, F., Barucci, M. A., Dotto, E., de Bergh, C., & Lo Curto, G. 2005, A&A, 444, 977
Öberg, K. I., & Bergin, E. A. 2021, PhR, 893, 1
Parhi, A., & Prialnik, D. 2023, MNRAS, 522, 2081
Peixinho, N., Delsanti, A., & Doressoundiram, A. 2015, A&A, 577, A35
Peixinho, N., Delsanti, A., Guilbert-Lepoutre, A., Gafeira, R., & Lacerda, P. 2012, A&A, 546, A86
Peixinho, N., Doressoundiram, A., Delsanti, A., et al. 2003, A&A, 410, L29
Peixinho, N., Lacerda, P., & Jewitt, D. 2008, AJ, 136, 1837
Peixinho, N., Thirouin, A., Tegler, S. C., et al. 2020, in The Trans-Neptunian Solar System, ed. D. Prialnik, M. A. Barucci, & L. Young (Amsterdam: Elsevier) 307
Penttilä, A., Lumme, K., Hadamcik, E., & Levasseur-Regourd, A. C. 2005, A&A, 432, 1081
Perna, D., Barucci, M. A., Fornasier, S., et al. 2010, A&A, 510, A53
Pinilla-Alonso, N., Bannister, M., Brunetto, R., et al. 2021, JWST Proposal, Cycle 1, ID. #2418
Pinilla-Alonso, N., Brunetto, R., et al. 2024, NatAs, 9, 230
Rabinowitz, D. L., Schaefer, B. E., & Tourtellotte, S. W. 2007, AJ, 133, 26

Romon-Martin, J., Delahodde, C., Barucci, M. A., de Bergh, C., & Peixinho, N. 2003, A&A, 400, 369

Rosenbush, V., Kiselev, N., & Afanasiev, V. 2015, in Polarimetry of Stars and Planetary Systems, ed. L. Kolokolova, J. Hough, & A.-C. Levasseur-Regourd (Cambridge: Cambridge Univ. Press) 340

Rosenbush, V. K., Kiselev, N. N., Shevchenko, V. G., et al. 2005, Icar, 178, 222

Sarid, G., Volk, K., Steckloff, J. K., et al. 2019, ApJL, 883, L25

Scotti, J. V., Rabinowitz, D. L., Shoemaker, C. S., et al. 1992, IAU Circ., 5434, 1

Seccull, T., Fraser, W. C., Puzia, T. H., Fitzsimmons, A., & Cupani, G. 2019, AJ, 157, 88

Stansberry, J. A., Holler, B. J., Pinilla-Alonso, N., & Rieke, M. J. 2017, JWST Proposal. Cycle 1, ID. #1191

Tegler, S. C., Bauer, J. M., Romanishin, W., & Peixinho, N. 2008, in The Solar System Beyond Neptune, ed. M. A. Barucci, et al. (Tucson, AZ: Univ. Arizona Press) 105

Tegler, S. C., & Romanishin, W. 1998, Natur, 392, 49

Tegler, S. C., & Romanishin, W. 2003, Icar, 161, 181

Tegler, S. C., Romanishin, W., & Consolmagno, G. J. 2003, ApJ, 599, L49

Tegler, S. C., Romanishin, W., & Consolmagno, G. 2016, AJ, 152, 210

Thompson, W. R., Murray, B. G. J. P. T., Khare, B. N., & Sagan, C. 1987, JGR, 92, 14933

Tiscareno, M. S., & Malhotra, R. 2003, AJ, 126, 3122

Verbiscer, A. J., Helfenstein, P., & Buratti, B. J. 2013, in The Science of Solar System Ices, ed. M. S. Gudipati, & J. Castillo-Rogez (New York: Springer Science & Business) 47

Verbiscer, A. J., Porter, S., Benecchi, S. D., et al. 2019, AJ, 158, 123

Volk, K., & Malhotra, R. 2008, ApJ, 687, 714

Volk, K., & Malhotra, R. 2013, Icar, 224, 66

Wallentineen, D. 1978, MPBu, 5, 30

Wong, I., Mishra, A., & Brown, M. E. 2019, AJ, 157, 225

Wooden, D. H., Ishii, H. A., & Zolensky, M. E. 2017, RSPTA, 375, 20160260

Zubko, E., Videen, G., & Kulyk, I. 2020, RNAAS, 4, 75

Chapter 6

Volatiles

Kathleen Mandt, Oleksandra Ivanova, Olga Harrington Pinto, Nathan X Roth and Darryl Z Seligman

Small bodies are the remnant building blocks from the time when the planets formed and migrated to their current positions. Their volatile composition and relative abundances serve as time capsules for the formation conditions in the protosolar nebula. By constraining the volatile composition of Centaurs, we can fill in important gaps in understanding the history of our solar system. We review the state of knowledge for volatiles in small bodies, processes that influence volatile composition and activity in small bodies, and future capabilities that can be leveraged to advance our understanding of volatiles in Centaurs.

6.1 Introduction

Determining how our solar system formed and evolved is critical for answering several of the fundamental questions identified in the most recent planetary science decadal survey, Origins Worlds and Life (OWL; NASEM 2023a). Furthermore, figuring out why our solar system architecture – or the number, size, and locations of the planets and small bodies – appears to be unique compared to exoplanet and planetary disk systems is a top priority for investigation recognized in the most recent astrophysics decadal survey, Pathways to Discovery in Astronomy and Astrophysics for the 2020s (Astro 2020; NASEM 2023b).

Our solar system's architecture is the result of the timing and location of the giant planets' formation and their subsequent migration. We know from exoplanetary systems hosting hot Jupiters that inward migration of Jupiter could have destroyed the terrestrial planets — or at least removed them from their current day orbits (Fogg & Nelson 2007a, 2007b; Walsh et al. 2011; Madhusudhan et al. 2014). Furthermore, it is possible that the migration of the giant planets is responsible for delivering volatiles to the inner planets (Raymond & Izidoro 2017; Kane et al. 2021; see also Chapter 15). Therefore, understanding the formation conditions, dynamical history, current distribution, and volatile composition of small bodies in our solar system is critical for determining how our solar system formed and evolved. There are four primary populations of small bodies within the solar system – asteroids, Centaurs, trans-Neptunian objects (TNOs), and Oort Cloud objects. Comets are a class of small body that originate as both TNOs and Oort Cloud objects. At the present time, the bulk volatile composition of Centaurs as a population is the least understood.

6.1.1 Volatiles and Solar System Formation

Dynamical models for solar system formation have so far provided a range of solutions that are able to reproduce solar system architecture constraints, including the current locations of the planets; the distribution of small bodies throughout the solar system; and the small mass of Mars. However, they provide a wide range of potential formation locations and overall migration scenarios for the giant planets.

Early studies of orbital migration suggested that proto-Uranus and proto-Neptune could smoothly migrate 5–10 au while embedded in the disk (Fernandez & Ip 1984), allowing the giant planets to form closer together with Saturn, Uranus, and Neptune migrating outward while Jupiter migrated inward (Hahn & Malhotra 1999). In the "Nice" model – initially developed in Nice, France – the giant planets formed within a range of ~5.5–17 au with a population of small icy planetesimals distributed between ~17 and ~35 au. Some of these small bodies interacted with Saturn, Uranus, and Neptune and were scattered inward to the inner solar system (Tsiganis et al. 2005) while the planets migrated outward. This outward migration created a dynamical instability that led to the current structure of the Kuiper Belt (Levison et al. 2007). Jupiter's interaction with the small bodies that were scattered inward then caused them to be ejected outward into the Oort Cloud or out of the solar system (Tsiganis et al. 2005; Levison et al. 2007). Neptune is also thought to have scattered bodies outward. Alternatively, the "Grand Tack" model proposes that Jupiter formed at 3.5 au and migrated inward to 1.5 or 2 au (Walsh et al. 2011; Brasser et al. 2016a, 2016b). Saturn formed at a slower rate, then migrated inward at a higher velocity (e.g., Masset & Papaloizou 2003), allowing it to catch up to and capture Jupiter in a 3:1 resonance (Masset & Snellgrove 2001; Morbidelli & Crida 2007; Pierens & Nelson 2008). After this, both giant planets migrated outward to their current locations. In this scenario Jupiter crossed the asteroid belt twice leading to its current mass and distribution (Walsh et al. 2011). More recent studies suggest that a fifth giant planet, potentially another ice giant, formed as well but was ejected from the solar system (e.g., Deienno et al. 2017; Nesvorný 2011). Current observational constraints are not sufficient to determine which dynamical model scenario provides the best solution.

Models for how the giant planets formed have also attempted to constrain their formation locations, with little success. Originally, each planet's core was assumed to have grown hierarchically from small grains to a planet many times the mass of Earth. However, the current locations of Uranus and Neptune did not have sufficient solid material available at that distance. Therefore, they either must have formed closer to the Sun, as in the Nice model, or formed through a different method. The streaming instability offers a promising alternative, where collective aerodynamic interactions in the protosolar nebula (PSN) produced high density filaments and streams that allowed pebbles to form (Youdin & Goodman 2005; Johansen & Youdin 2007; Johansen et al. 2015; Simon et al. 2016) and clouds of pebbles to contract into planetesimals (Nesvorný et al. 2019). This mechanism permits formation at greater distances given sufficient dust to gas ratios. Pebble accretion also works at greater distances, in which case pebbles are slowed by gas drag in the PSN and rapidly accumulate onto existing planetesimals (Lambrechts & Johansen 2014). For a detailed review see Chapter 2. These models, therefore, are not able to limit the formation locations of the giant planets to a specific region of the PSN.

The remaining approach for constraining solar system formation and evolution is to compare models for the volatile composition of the solid building blocks of the planets as a function of time and distance in the PSN with the composition of small body populations and of the giant planet atmospheres. Combining this with dynamical models can help to identify the formation locations and times for the giant planets and for the various small body populations remaining today. Recent modeling evaluated the influence of ice lines of specific species in the PSN and the type of water ice to determine the composition of Jupiter's atmosphere based on when and where it formed (Mousis et al. 2019; Aguichine et al. 2022; Schneeberger et al. 2023). Although these models provide new information for understanding the formation processes of solid building blocks, the observational constraints remain limited. Future observations of the bulk elemental composition of the giant planet atmospheres and small body populations, as well as isotopic ratios of several elements, would help to constrain formation locations.

6.1.2 Role of Centaurs in Understanding Solar System Formation

We know that small icy bodies, including Centaurs, formed in an extensive region of the PSN with temperatures that could have ranged between 200 K at about 5 au to approximately 30 K near 40 au. The models for ice formation in the PSN show that the composition of volatile ices formed within this wide temperature range can vary significantly (Mousis et al. 2019; Aguichine et al. 2022; Schneeberger et al. 2023). Variability in the volatile composition of small body populations could indicate different regions where different populations formed.

Centaurs are proposed to originate as TNOs and to represent a steady-state population that transfers material from the trans-Neptunian region to the inner solar system eventually as Jupiter-family comets (JFCs) (Emel'yanenko 2005; Brasser et al. 2012; Sarid et al. 2019; Chapter 3). For a time, they reside within the orbits of the giant planets and are subject to repeated dynamical interactions with the giant planets. If this is the case, then their formation location was the same as comets and the initial volatile inventory should be similar. Any differences would be due to volatile evolution over time. If this is not the case, then their volatile composition would indicate their formation region. Constraining the dynamical history of Centaurs compared to TNOs and JFCs requires a comparison of the volatile inventories of Centaurs with what is known of the volatile inventories of TNOs and JFCs.

In either case, evolution of volatiles needs to be considered. The primary type of evolution would be preferential removal of volatiles that sublimate at lower temperatures than water when the surface and interior are heated. Although studies have been done on volatile evolution in comets (e.g., Guilbert-Lepoutre et al. 2015) and Centaurs (e.g., Fernández et al. 2018; Chapter 7), more work in the form of observations with new state of the art facilities on Earth and in space, as well as laboratory and theoretical studies is required to fully understand how volatile depletion works.

A second question about Centaurs and their volatile inventories is what impact they may have had on the type of volatiles delivered to the inner planets (see Chapter 15). The inner planets' current atmospheres formed in part from internal degassing, and in part due to the impacts of water-rich small bodies scattered into the inner solar system (Marov & Ipatov 2001; Schönbächler et al. 2010). The relative contribution of volatiles from icy bodies and asteroids may be constrained by their atmospheres' noble gas content and isotopic composition (e.g., Owen & Bar-Nun 2001) given sufficient information on small body composition.

6.2 Volatiles in Small Body Populations

The bulk volatile composition of different small body populations is constrained by observations of volatiles that are on the surface and volatiles released into a coma through heating and other processes. Understanding what the observations mean for the bulk composition requires understanding volatile phases and processes that change the phases of volatiles. We outline in this section the volatile phases relevant to small body populations and describe observations of small body populations including Centaurs.

6.2.1 Volatile Phases

The ice in small bodies can be present in different phases, or forms, depending on the temperature of formation, the thermal processing experienced over the history of the small body, and any coma activity that leads to redeposition of volatiles. Ice phases relevant to Centaurs include amorphous, condensed, crystalline, and clathrates.

6.2.1.1 Amorphous Ice

Amorphous ice is formed when water molecules freeze rapidly under low-temperature conditions (<136 K), preventing the formation of a well-defined crystalline structure found in water ice on Earth (Mastrapa et al. 2013). The low temperature conditions where cometary nuclei formed would allow for condensation as amorphous ice that could still be preserved (Patashnick 1974). However, amorphous ice is unstable thermodynamically and can undergo an exothermic transformation into a crystalline state when exposed to warmer temperatures or additional energy (Bar-Nun et al. 1985). The temperature of the ice influences the rate at which this conversion occurs.

Amorphous ice is ~2 times more dense than crystalline ice and will crystallize when heated to a temperature of ~150 K. Crystallization occurs exothermically and

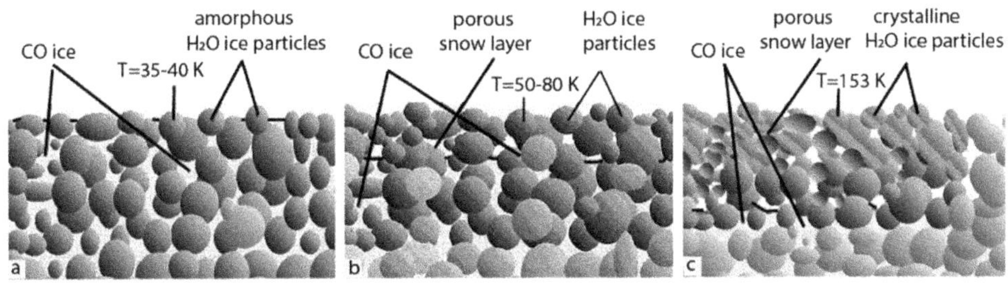

Figure 6.1. Illustration of the process of crystallization of amorphous ice. (a) The initial state of the object's surface with amorphous water ice particles embedded in a CO ice substrate; (b) A surface porous layer forms from microcrystals of amorphous ice at the onset of surface heating; (c) The crystallization front of amorphous ice moves deeper into the nucleus (Adapted from Shulman 2007, CC BY 4.0).

rapidly. The temperature rises to 190 K, releasing 0.02 eV per molecule. This energy is insufficient to cause evaporation of the water ice, as shown in Figure 6.1. However, crystallization models depend on poorly constrained parameters leading to uncertainties regarding processes related to amorphous ice in cometary nuclei (Kouchi & Sirono 2001). It is possible that the transition in natural ice might not be exothermic due to the sublimation of trapped volatiles, but because the trapped volatiles are in buried ice, they cannot escape directly to space and the phase transition energy would remain trapped.

Crystallization of amorphous ice can occur at lower temperatures over longer timescales. However, at temperatures below 77 K, the crystallization time exceeds the age of the solar system (Notesco & Bar-Nun 1996; Notesco et al. 2003). The potential for phase transition in a specific body can be estimated by comparing the crystallization time to the orbital period where amorphous ice is expected to have crystallized on objects with crystallization time shorter than the orbital period (Notesco & Bar-Nun 1996; Notesco et al. 2003).

6.2.1.2 Condensed Ice

Volatiles condense to form ice at temperatures and pressures specific to each molecule. The left panel of Figure 6.2 illustrates the condensation curves for volatiles observed in comets at pressures and temperatures relevant to the PSN. The heavy noble gases, CH_4, CO, and N_2 condense at the coldest temperatures while NH_3, CO_2, and H_2S condense at warmer temperatures. Water condenses at temperatures above the range in this figure.

6.2.1.3 Crystalline Ice and Clathrate Hydrates

Crystalline ice, including clathrate hydrates, is known to form under a range of thermodynamic conditions in small bodies (Kargel & Lunine 1998). Unlike amorphous ice, which only forms at very low temperatures, crystalline ice forms at temperatures >110 K and has a regular lattice-like structure. It is worth noting that crystalline ice can also be converted to amorphous ice via exposure to high energy radiation, or cosmic rays. This high energy radiation or cosmic rays disrupts

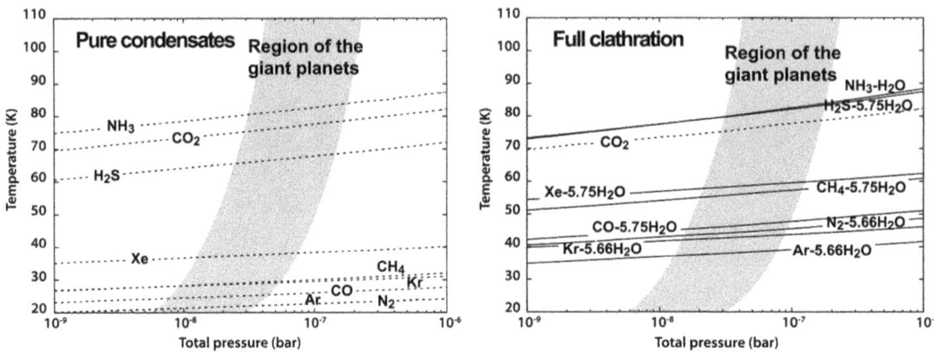

Figure 6.2. Condensation curves serve as a tool for predicting building block composition as a function of conditions in the PSN. Here they are compared to the predicted cooling curve of the PSN where the giant planets may have formed. (Left) Equilibrium curves for pure condensates (dashed lines) where species are in the gas phase above the curves. (Right) Scenario for clathrate formation including a combination of condensate, hydrate (NH_3–H_2O), and clathrate formation (X–5.75H_2O or X–5.67H_2O; solid lines), along with crystallization of pure CO_2 condensate (dotted line). Reproduced from Mousis et al. (2020), with permission from Springer Nature.

the crystalline structure and creates voids and other amorphous properties (Hansen & McCord 2004). Therefore, it is possible that amorphous ice is created and retained in TNOs that are not at sufficient ambient temperatures for the amorphous ice to re-crystallize.

Clathrate hydrates are a unique form of crystalline ice and are compounds where water molecules form a lattice structure with small cages that trap guest molecules, such as methane or carbon dioxide. Two common clathrate structures are structures I and II, characterized by different water cells (Sloan & Koh 2007). Structure I has pentagonal and tetrakaidekahedral cells, while structure II has pentagonal and hexakaidecahedral cells. The structure selection relies on the dimensions and configuration of the guest molecule. For example, CH_4 and CO_2 are characteristic of structure I, while N_2 favors structure II (Sloan & Koh 2007). The right panel of Figure 6.2 illustrates one potential scenario for the formation of crystalline ice and clathrates in small bodies (Mousis et al. 2020).

Early on it was proposed that CH_4, NH_3, and CO_2 in cometary ices exist as clathrate hydrates rather than macroscopic inclusions (Delsemme & Swings 1952) and thermodynamic models suggest that CH_4 should always be trapped in clathrates beyond the water snowline (Mousis et al. 2015; Luspay-Kuti et al. 2016). Furthermore, the condensation process would involve the construction of hydrates instead of direct condensation from the gas phase, which could explain the presence of CH_4 and NH_3 in cometary nuclei because their high vapor pressures would cause immediate evaporation if not trapped in clathrates.

Additionally, thermodynamic calculations suggest that N_2 would have condensed as pure ice at lower temperatures than required for clathration, while CH_4 and CO would have remained stable (Mousis et al. 2015; Luspay-Kuti et al. 2016).

Any radiogenic heating would lead to the devolatilization of N_2 while preserving CH_4 and CO in cometary nuclei.

6.2.1.4 Observations of Volatile Phases

Observations regarding the physical state of ice in small bodies is limited. Because amorphous ice is expected to form at the temperatures where TNOs orbit and crystalline ice can be converted back to amorphous ice over $\sim$10 Myr timescales, it is expected to exist on the surfaces of TNOs. However, the surfaces of some TNOs exhibit reflectance spectra typical of crystalline ice. For example, the crystalline ice on the surface of (50000) Quaoar (Jewitt & Luu 2004) suggests that it was created in the recent past.

Although distinguishing amorphous ice from crystalline ice on the surfaces of active small bodies at large heliocentric distances remains challenging, the presence of clathrates can be inferred. Methane has been observed in several cometary comae while crystalline water ice was detected on the surface of comet 9P/Tempel 1 (Sunshine et al. 2006), both supporting the hypothesis of a methane-rich clathrate layer. Further support has been provided by Rosetta observations of comet 67P/Churyumov-Gerasimenko (67P/C-G) (Mousis et al. 2015; Luspay-Kuti et al. 2016).

While clathrates in comets are actively studied, indirect observational evidence for amorphous ice in Centaurs is found with no evidence for water ice clathrates (Klinger et al. 1996; Devlin 2001). Furthermore, the perihelion distribution and formation conditions of Centaurs suggest that amorphous forms of ice are more likely than clathrates (Jewitt 2009).

Volatile composition is an important tool for evaluating the form of ice that is present in a small body. For example, the presence of highly volatile ices, such as N_2, Ar, CO, and O_2 in these objects seems to suggest that they formed at temperatures within the sublimation range of these species, 22–25 K (Bar-Nun et al. 1987; Meech & Svoren 2004). However, it is also possible that these species were trapped in amorphous or crystalline ice. Models of the relative composition of these species in specific ices formed in conditions within the PSN serve as useful tools for evaluating the form of ice in small bodies (Mousis et al. 2019; Aguichine et al. 2022; Schneeberger et al. 2023).

6.2.2 How Volatiles Are Released

Volatiles are released by several different mechanisms. Work is ongoing to understand what mechanisms may drive activity in Centaurs compared to comets (see, e.g., Chapter 8). We review here three possible mechanisms for volatile release. Depending on the mechanism there may be nongravitational accelerations that influence the dynamics of comets and Centaurs.

6.2.2.1 Sublimation

Sublimation is the primary driver of activity and coma formation in most comets within the solar system. Solar radiation received at the surface of a small body provides energy for a phase transition from ice to gas that produces an outflow of gas. After release, the radiation energy is converted into kinetic energy as the now liberated gas molecules are heated to a thermal speed that is approximately the local sound speed. Dust particles that are trapped within the bulk ice matrix or residing on the surface of the comet are also liberated with the outgassing volatiles, producing a dusty tail.

Sublimation lines, also called snow lines, determine which volatiles are sublimated and are similar to the freeze out lines in protoplanetary disks of various volatiles. Three major snowlines relevant to comet activity are those for H_2O, CO_2 and CO, which are currently located approximately at the orbits of Jupiter, Saturn, and Neptune, respectively. Interior to 4 au, within the vicinity of the water snow line (or sublimation front), activity is driven mostly by direct sublimation of H_2O, and H_2O production rates, Q, increase with decreasing distance (A'Hearn et al. 1995). At further distances, past the nominal H_2O snowline, activity has been attributed to direct sublimation of more tenuous volatile species such as CO_2 and CO.

6.2.2.2 Crystallization of Amorphous Ice

Amorphous ice is highly porous and contains trapped molecules that are released upon crystallization (Bar-Nun et al. 1985), which involves rearranging water molecules into an ordered, crystalline lattice (Mastrapa et al. 2013). The initial step is nucleation i, where a small cluster of water molecules organizes itself in a repeating pattern, initiating the growth of a crystal (Kouchi et al. 2002). The presence of impurities, or specific conditions such as mineral surfaces or radiation exposure, can influence nucleation. Surface conditions also affect the conversion process. Solar radiation, for example, can induce structural changes in amorphous ice, facilitating its conversion into the crystalline form. Heliocentric distance, surface roughness, topography, and volatile compounds can further influence crystallization (Gibb et al. 2000; Protopapa et al. 2018).

Crystallization models have been studied extensively for cometary activity. Because this process involves a release of energy, crystallization is suggested to provide additional energy sources that can explain mass loss from comets at low temperatures (Enzian et al. 1997; Prialnik 1997). Studies of crystallization in relation to cometary activity are also relevant to Centaurs. All active Centaurs have small enough perihelia for the crystallization of amorphous water ice to be a contributing factor to any observed activity, indirectly suggesting the presence of amorphous ice in Centaurs (Notesco & Bar-Nun 1996). However, translating laboratory experiment results to cometary and Centaur timescales poses challenges due to the vast difference in timescales.

6.2.2.3 Clathrate Destabilization

Clathrate hydrates consist of water cages that enclose guest molecules. The characteristics of the guest molecule determine the clathrate structure and impact the stability and subsequent destabilization processes (Sloan & Koh 2007). Explosive outgassing of clathrates as they become destabilized has been proposed as an explanation for the presence of chaotic terrains found on many icy bodies of the outer solar system.

The stability of clathrates is affected by many factors, including temperature, pressure, composition, and impurities (Mousis et al. 2015). Elevating temperatures can destabilize clathrates, resulting in the liberation of trapped gases. Temperature increases may be caused by solar radiation, internal heat sources, or impact events (Mousis et al. 2015). Reductions in pressure caused by sublimation, outgassing, or shock events can disrupt the clathrate lattice, destabilizing the clathrate and facilitating the release of trapped gases. Irradiation, thermal cycling, and mechanical disruption can also destabilize clathrates in small bodies (Davidsson et al. 2016).

Destabilization of clathrates has long-term implications for the evolution of small bodies. This is because the release of volatiles modifies surface properties and can result in the formation of secondary compounds (Kargel & Lunine 1998).

6.2.3 Activity Drivers for Centaurs

The primary driver of activity in Centaurs is thought to be sublimation caused by the Centaur crossing sublimation fronts of volatiles as it approaches perihelion. As noted above, the H_2O sublimation front is located at $\sim$4 au, depending on characteristics of the particular Centaur such as albedo. The sublimation fronts of CO and CO_2 are at greater distances than this, allowing them to begin sublimating at greater distances from the Sun. The Centaur 29P/Schwassmann–Wachmann 1 (hereafter 29P/S–W1) is a prime example of activity initiated by species that are more volatile than water. It has been observed to be active in a state of constant outgassing since its initial discovery. Its semi-major axis of 5.986 au is interior to the CO ice line and it has a dust coma with CO-dominated outgassing (Senay & Jewitt 1994; Crovisier et al. 1995; Gunnarsson et al. 2008; Paganini et al. 2013). Recent observations with the James Webb Space Telescope (JWST) have also detected CO_2, finding distinct jets with heterogeneous composition (Faggi et al. 2024). Similarly, the Centaurs (60558) 174P/Echeclus and (2060) Chiron display similar (hereafter Echeclus and Chiron), but lower levels of production of CO (Wierzchos et al. 2017; Womack & Stern 1999). See Chapter 14 for further discussion of JWST results for Centaurs.

In addition to sublimation, the crystallization of amorphous ice may drive activity in some Centaurs. The subsurface temperatures of Centaurs are consistent with crystallization fronts of amorphous water ice at approximately 10–12 au (Guilbert-Lepoutre 2012). There is indirect evidence of this process occurring in some active Centaurs, as activity is seen at distances that are too great to correspond to sublimation of H_2O. Active Centaurs – defined as having non point-source comae – are a small fraction of all of the known Centaurs and have lower mean perihelia distance (5.9 au) than inactive Centaurs (8.7 au; Jewitt 2009). This is too far from the Sun for sublimation of H_2O ice to drive activity, while sublimation of CO ice would be *too* efficient to explain the mass loss seen. This suggests that the crystallization of amorphous water ice is responsible for the activity (Jewitt 2009).

Further evidence for amorphous conversion driving activity on Centaurs is found in spectroscopic measurements of Centaur (10199) Chariklo at different dates indicating both the presence and absence of H_2O surface ice. This behavior was consistent with crystalline ice in the center of the Centaur with amorphous ice on the surface, the crystallization of which could drive the sporadic activity (Guilbert-Lepoutre 2011). Detailed three-dimensional thermal evolutionary models found that outgassing due to the crystallization of amorphous ice cannot be sustained for greater than $\sim 10^4$ yr, implying that the active Centaurs only recently experienced changes in orbits that triggered activity (Guilbert-Lepoutre 2012).

Impacts of smaller objects with Centaurs could also cause activity by heating volatiles in the region where the impact took place (Wierzchos et al. 2017; Womack et al. 2017). The volatiles released would include those from the impactor and from the Centaur. In this case, the relative abundances of volatiles released from the Centaur would depend on their volatility, for example more CO and N_2 would be released than water because they sublimate at much lower temperatures (see discussion in Mandt et al. 2022).

Additionally, the release of icy dust from the nucleus can contribute volatiles as an extended source. This has been suggested as a source of additional activity for comets C/1995 O1 Hale–Bopp (Dello Russo et al. 2000) and 17P/Holmes (de Almeida et al. 2016). Studies of the target comet for the Rosetta mission, 67P/C-G found through observations in the coma that distributed hydrogen halides, HF, HCl, and HBr – species that are known to freeze out on icy grains in molecular clouds – likely originated from icy grains in the dust coma (De Keyser et al. 2017). Some of the CO observed in the coma of Centaur 29P/S–W1 is proposed to originate from icy dust grains in the coma (Gunnarsson 2003), although observations exist of entirely unrelated dust and CO outbursts in 29P/S–W1 (Wierzchos & Womack 2020). Some water is observed in the coma of 29P/S–W1, with an observed relative production of $Q\text{CO}/Q\text{H}_2\text{O}$ of ~ 4.64 at a heliocentric distance of 6.18 au (Ootsubo et al. 2012), despite being far outside of the H_2O sublimation front. This water is proposed to originate from icy dust grains (Bockelée-Morvan et al. 2022).

6.2.4 Observing Volatiles and Connecting to Bulk Composition

As described in Section 6.1, we need to understand the bulk composition of small bodies to connect them to the processes involved in the formation and early evolution of the solar system. We review the methods currently used to observe volatiles in comets and Centaurs, including observations of surface volatile composition and observations of volatiles in the coma. We then discuss how volatiles observed on the surface and in the coma identify which species are present, but require further analysis to determine the bulk composition of the body. The task of determining the comprehensive composition is particularly challenging for both comets and Centaurs, given the constraints of limited observational data, underscoring the need for ongoing research in this field.

6.2.4.1 Surface Volatiles

A detailed overview of the surface composition of Centaurs is provided in Chapter 5. Volatiles located on the surface can provide insight into the bulk volatile composition of Centaurs. These volatiles are studied through observations of their surface spectra, surface colors, and through polarimetry.

Spectroscopic observations can detect specific ice species, but this can be challenging for distant and faint objects like Centaurs and TNOs (Delsanti et al. 2006; Barucci et al. 2008; Fornasier et al. 2009; Harrington Pinto et al. 2022). Laboratory experiments play a crucial role in interpreting spectra (see Chapter 13 for more discussion). Detailed studies of crystalline water ice, pure CH_4, CH_4 diluted in N_2, N_2 (alpha and beta phases), and CO have been done (see, e.g., Schmitt et al. 2012 and references therein) as well as CO_2 and CH_3OH (Fulchignoni et al. 2008; de Bergh et al. 2013) and irradiation effects on ice and their role in creating organic refractory residues. Figure 6.3 illustrates the species that have been detected in spectral observations of Centaurs and TNOs, including CH_4, N_2, and CO surface ice with percentages representing the percent of the objects that have been observed for which each species has been detected so far (de Bergh et al. 2013; Dalle Ore et al. 2015).

In addition to direct detection of surface volatiles, the surface colors may provide insights into their volatile composition. Surface colors are categorized based on how reddened they are, with an index such as B–R colors or spectral slopes. Figure 6.4 illustrates the B–R color distributions of several small body populations within the solar system, including Centaurs. The majority of small bodies have slightly reddened colors but the cold classical KBOs have an excess of what is commonly referred to as ultra-red material. This ultra-red material presumably results from processing. The colors of Centaurs appear to be bimodal (Jewitt 2009), with a mix of mildly reddened Centaurs and ultra-red Centaurs.

Figure 6.5 illustrates the inclination distributions of red and blue Centaurs, including both active and inactive Centaurs. It would appear feasible that the redder Centaurs could come from reservoirs with more reddened material, specifically the cold classical KBOs, which is the only solar system small body reservoir that also has

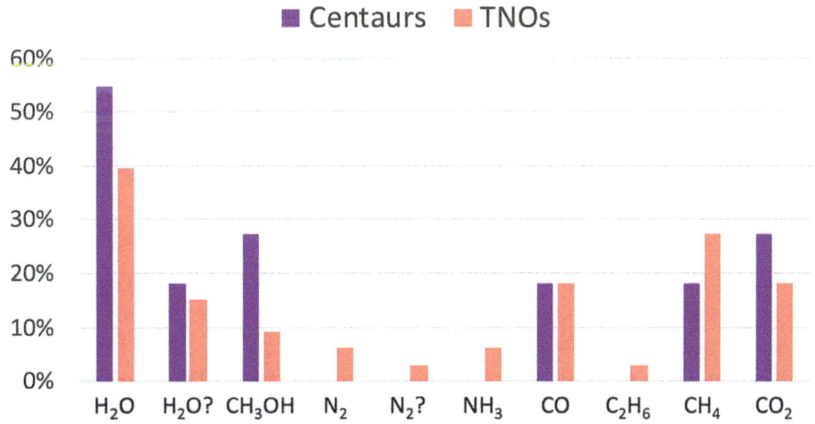

Figure 6.3. Percentage of Centaurs and TNOs on which specific volatile species have been detected (or are suspected to have been detected when "?" is present) in spectral observations of the surface ices. (Data combined from de Bergh et al. 2013 and Dalle Ore et al. 2015).

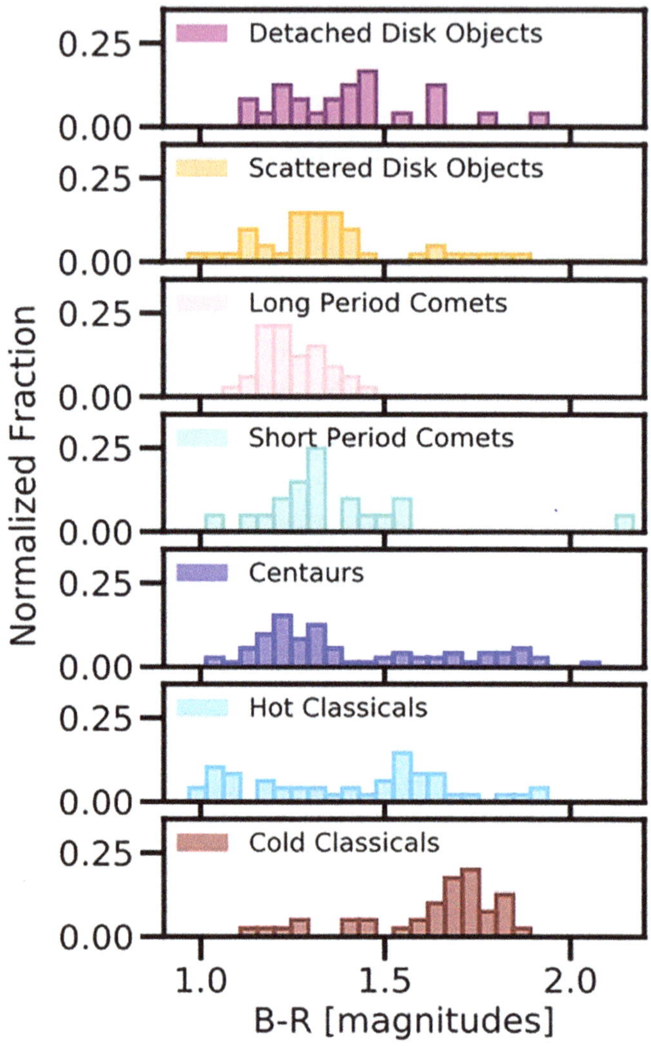

Figure 6.4. Color distributions of various populations of small bodies. Shown on the x-axis for each panel are the B–R measured colors of the surfaces, or the extended comae for active objects. Each panel depicts the color distributions measured in different populations of distant and cometary objects, with Centaurs shown in dark blue (data from Hainaut et al. 2012).

ultra-red material (see Figure 6.4). However, if the redder Centaurs originated as cold classical KBOs, this should be evident in their inclination distributions. Because there is no obvious dependence on the inclination distributions for either the red or the blue Centaurs, and because the cold classical KBOs are suggested by long term dynamical simulations to be dynamically very stable (Nesvorný et al. 2017), the colors of the Centaurs are not likely to correlate with the colors of their formation reservoirs.

As Figure 6.6 illustrates, inactive Centaurs have a bimodal distribution including some ultra-red material and some bluer material, while active Centaurs are blue with

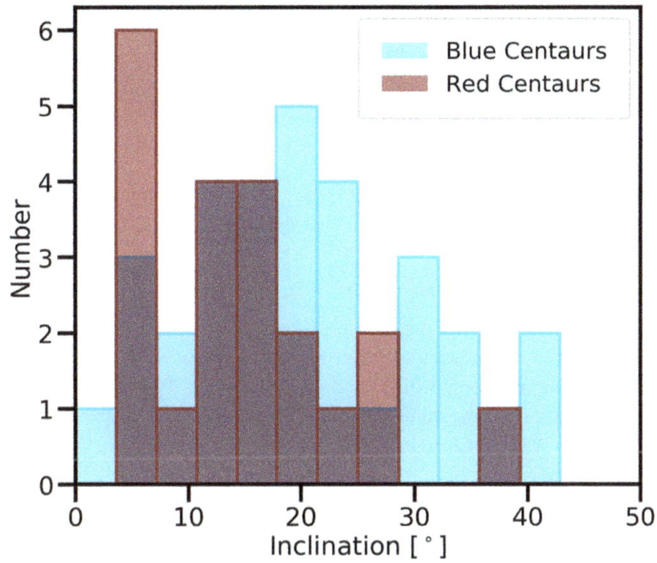

Figure 6.5. Inclination distribution of Centaurs sorted by their colors. The "red" Centaurs have B–R > 1.5 and the "blue" have B–R < 1.5. These include active and inactive Centaurs. The data is drawn from Hainaut et al. (2012) and the orbital elements (inclination in this case) are drawn from the Minor Planet Center database.

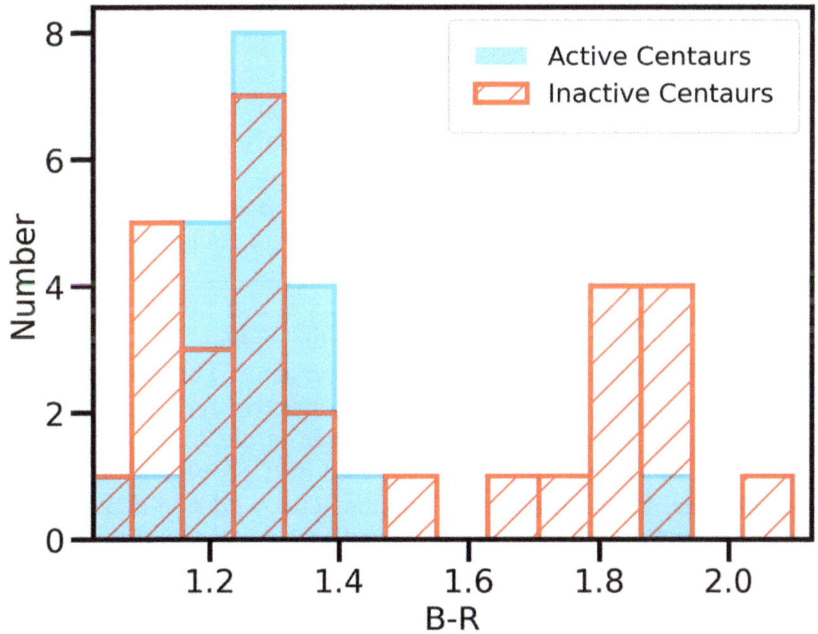

Figure 6.6. Photometric colors of active and inactive Centaurs. Data is taken from Jewitt (2015). The blue histograms show measured B–R colors of active Centaurs, while the red hatched histogram show the B–R colors of inactive Centaurs. The active red Centaur is 523676 (2013 UL_{10}) from Mazzotta Epifani et al. (2018).

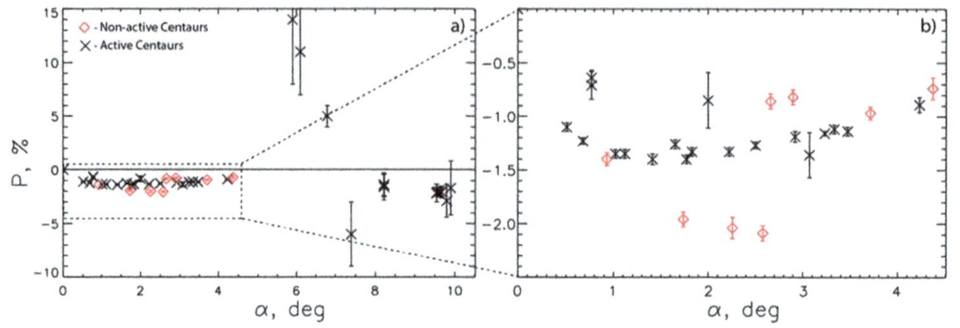

Figure 6.7. The phase-angle (α) dependency of linear polarization degree (P,%) of Centaurs provides useful insights into the surface properties, composition, and dust dynamics. This is an illustration of polarization measurements of Centaur surfaces during non-active time periods measured in the R band (a) Polarization vs. phase angle, excluding 29P/S–W1. (b) Close-in view.

the exception of 523676 (2013 UL_{10}; Mazzotta Epifani et al. 2018). The possible interpretation is that activity removes reddened material from the surface of active Centaurs (Jewitt 2015), although the details are not clear given that there exists observations of one reddened active Centaur to date. Further studies to explain the color distribution and determine any connection to volatile composition are clearly needed.

Polarimetric observations provide useful insights into the surface properties, composition, and dust dynamics of Centaurs. When an active Centaur is producing a coma, the sublimation of volatiles releases dust and gas, and the coma can contain ice particles. This influences the scattering properties of light, leading to different polarization signatures for the coma compared to the surfaces of Centaurs. Additionally, activity may change the surface polarization signatures of active Centaurs when they are observed during inactive time periods.

Observations have been made of the surfaces of most Centaurs during inactive periods, with the exception of 29P/S–W1, which is found to be active in all observations. Deep negative polarization has been observed in smaller-sized Centaurs, and one possible explanation for this is the presence of an optically heterogeneous surface with dark and bright scatter regions, such as a dark surface covered by a very thin layer of submicron water crystals. The observed variation in linear polarization (P, %) with phase angle between active and inactive Centaurs (see Figure 6.7; Kiselev & Chernova 1979; Kochergin et al. 2021) may be related to differences in the abundance of surface ices and dust properties on their surfaces. Results of polarimetric observations in specific active and non-active Centaurs are provided in Section 6.2.5.

6.2.4.2 Coma Morphology

Active Centaurs are not necessarily always active. Currently activity in ~20 Centaurs has been detected, primarily in the form of a dust coma. The composition of their comae can be analyzed through spectroscopic observations, providing insights into the nucleus composition and its evolution.

The morphology of the coma in active Centaurs provides insights into the underlying physical processes and dynamical behavior. We describe here the morphology and evolution of comae in active Centaurs as they relate to volatile composition. Evolutionary processes and Centaur activity are described in greater detail in Chapters 7 and 8.

Morphological structures (tails, jets, fans, shells, arcs, etc.) of active Centaurs can be formed by dust, neutral gas, and ions, and indicate processes that drive activity (Ivanova et al. 2016; Picazzio et al. 2019; Korsun et al. 2008, Miles et al. 2016a, Miles 2016b; Rousselot et al. 2016, 2021; Kulyk et al. 2016). Activity can be short- or long-term, periodic or spontaneous, and features can be symmetric or asymmetric.

The coma morphology of 29P/S–W1 is highly variable. The total coma magnitude brightens for a few days after an outburst, likely due to subsequent sublimation and fragmentation of particles (Trigo-Rodríguez et al. 2008, 2010). Outburst comae generally fade over weeks, but sometimes a new condensed coma appears after another outburst (Miles 2016b). Asymmetric, fan-shaped comae, sometimes described as radial jets, suggest dust grain acceleration by volatile ice sublimation and imply continuous feeding from an active source on the nucleus. Four jet-like structures observed over long time periods in 29P/S–W1 (Shubina et al. 2023) suggest that the coma is formed by active regions located within a narrow belt near the equator. Additionally, CO^+ emission and continuum revealed differing distributions of CO^+ ions and dust in the coma, depending on the comet's activity level (see Figure 6.8 for CO^+; Ivanova et al. 2019a). The CO^+ ions were more concentrated towards the nucleus than the dust continuum. Notably CO^+ production appears to be correlated with solar wind intensity suggesting that the solar wind could be the dominant source for ion production at this distance from the Sun compared to closer in comets where ion production is dominated by photoionization (Ivanova et al. 2019a). Finally, recent JWST observations found separate CO and CO_2 jets that indicate possible differences in nucleus composition (Faggi et al. 2024).

Some Centaurs demonstrate unpredictable outbursts that are not correlated with their perihelion passage. Echeclus is an interesting case. Its primary outburst, which

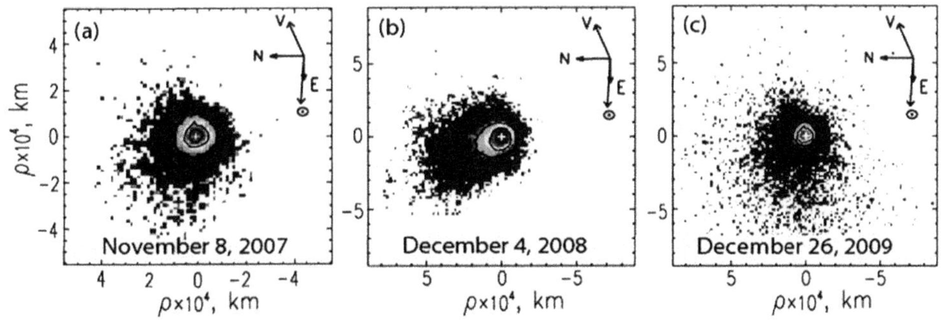

Figure 6.8. Observations of how the morphology of 29P/S–W1 has changed over time based on CO^+ in the coma: (a) November 8, 2007, (b) December 4, 2008 and (c) December 26, 2009. Produced using data from Ivanova et al. (2019a).

lasted several months, was detected during its pre-perihelion phase at a distance of 13 au. An apparent source seemed to move away from its nucleus, initially thought to be caused by impacts and fragmentation, but observations did not support this. Three smaller subsequent outbursts were also observed. Notably, observations revealed an unusual blue dust coma, possibly indicating a carbon-rich dust composition. A faint emission line of CO was detected at a distance of 6 au from the Sun (Wierzchos et al. 2017).

6.2.4.3 Coma Volatile Composition

The coma of an active small body provides direct access to volatiles in the nucleus, whether in ice form in the nucleus or condensed onto dust that is also released into the coma. The volatile ices are thought to primarily consist of H_2O ice with additions of macroscopic inclusions of ices such as CO, CO_2, CH_4, and other ices such as CH_3OH, NH_3, HCN, etc. (Barucci et al. 2008; Wierzchos & Womack 2020; Harrington Pinto et al. 2022). However, connecting observations in the coma to the bulk composition of a comet or Centaur is challenging because the composition of a coma varies as a function of the orbit due to changes in the temperature of the nucleus. The most comprehensive evaluation so far has been done for the comet 67P/C-G using observations from Rosetta (Rubin et al. 2019).

Neutral species observed in a coma fall into three categories: (1) "parent" or "primary" species that sublimed directly from the nucleus and are indicative of native ice composition, (2) "daughter" or "product" species produced by photolysis of gas-phase species in the coma used to determine parent species abundances, and (3) "distributed sources" or species whose distribution is incompatible with gas-phase chemistry and likely associated with photo or thermal degradation of refractories or volatile sublimation from icy grain in the coma. Ions observed in the coma are produced by ionization of neutral species and provide an indication of the parent species present in the nucleus.

Observations of primary species in the coma are straightforward and provide direct evidence that a species is present in the nucleus ice. Primary molecules H_2O, CO, CO_2, and HCN have been observed in multiple comets and the comae of at least three active Centaurs.

Uncertainties exist in determining the relative abundance of primary molecules in a coma based on their product species. The radical CN is a great example of this. During or shortly after an outburst, CN was observed around the Centaur Chiron when it was at a distance of ~11.26 au from the Sun (Bus et al. 1991). A steady-state HCN outgassing rate of 3.7×10^{25} s^{-1} was estimated based on a total CN gas production rate of 5.3×10^{29} s^{-1}. However, other observations of Chiron when it was closer to perihelion did not detect CN (Barucci et al. 1999; Rauer et al. 1997). Although HCN has been detected in comets as far out as 6.2 au (Biver et al. 1998), its relatively high sublimation temperature might prevent sublimation at greater distances. It is therefore likely that the production of CN at large distances could be due to a combination of various volatile parent molecules, including HCN, but also small dust grains (composed of CHON) that can reach temperatures higher than the blackbody equilibrium temperature.

The radical CN has also been detected in the coma of 29P/S–W1 (Cochran & Cochran 1991; Korsun et al. 2008; Ivanova et al. 2016, 2018) with estimated production rates between 8.4×10^{24} s^{-1} and 3.34×10^{25} s^{-1}. In this case it is likely a decay product of HCN, which has been detected at 29P/S–W1 and suggested to be produced by icy-grain sublimation (Bockelée-Morvan et al. 2022). The HCN abundance relative to water was a factor of 10 higher than values found in comets at 1 au the Sun.

Distributed sources create further challenges because they influence the local composition of a coma such that it is different from the bulk coma composition. Extended sources can play an important role in coma composition for some small bodies, referred to as "hyperactive" comets, or comets whose active surface areas cannot account for their overall H_2O production rate. The EPOXI mission to comet 103P/Hartley 2 revealed a complex coma environment. Strong CO_2 outgassing from the nucleus lobe region dragged water-ice coated grains into the coma that then sublimed and added to the overall gas content (A'Hearn et al. 2011) while H_2O production in the nucleus waist region was dominated by direct release. Ground-based radio observations detected additional molecular contributions from distributed sources, including HCN and CH_3OH (Drahus et al. 2012; Boissier et al. 2014). Recent work has demonstrated that icy grains are also a viable extended source for active Centaur comae with neutral gas such as HCN and H_2O (Bockelée-Morvan et al. 2022).

The measurement of the N_2/CO ratio in comets plays a particularly important role in understanding planetesimal formation models and determining the solar nebula's physical properties during their creation. The condensation temperatures of N_2 and CO are much lower than most other species of volatiles in the PSN. Their presence in planetary building blocks provides indications of the formation temperature of these planetesimals and potentially the form of ice. Although laboratory experiments with amorphous ice suggest that CO and N_2 should be released simultaneously in the same proportion as they exist in the ice (Bar-Nun et al. 1988), some predict that dynamically new comets should have N_2/CO ratios in their coma similar to the solar value that decreases as comets are continuously exposed to solar radiation in the inner solar system due to the preferential loss of any N_2 in their outer layers (Owen & Bar-Nun 1995). Observations of this ratio in solar system small bodies are too few to make any firm conclusions and more work is needed (Anderson et al. 2022, 2023).

Because N_2 is difficult to detect through remote sensing, the only direct measurement of N_2 in a comet was made by the ROSINA instrument in comet 67P/C-G (Rubin et al. 2015), which allowed precise determination of N_2/CO of 5.7×10^{-3}. This ratio showed a significant depletion of N_2 compared to the proto-solar value of $\sim$0.148, providing useful constraints on the formation temperature and ice type (Rubin et al. 2015; Mousis et al. 2016).

Ions are a useful tool for determining the abundance of species that are difficult to measure directly, specifically, CO and N_2 (Cochran 2002; Fortenberry et al. 2021). The integrated intensities of the CO^+ and N_2^+ bands are typically used to estimate the ratio. Figure 6.9 illustrates a comparison of the N_2/CO derived from N_2^+/CO^+

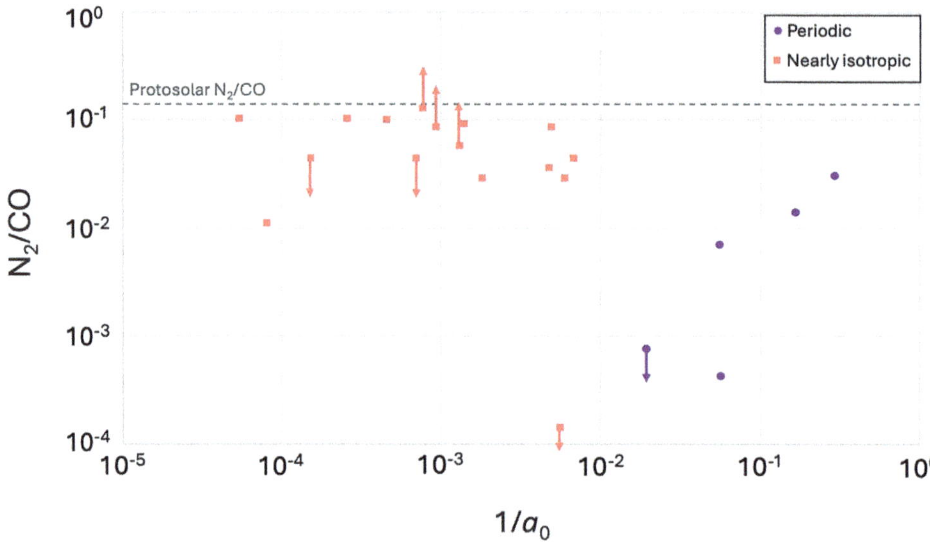

Figure 6.9. The ratio of N_2^+/CO^+ has been used as a proxy for determining the N_2/CO ratio in several comets. Compilation of estimates of N_2/CO for different comets and one Centaur depending on the bodies' semi-major axis show that the ratio may increase with increasing $1/a_0$ for periodic comets. Most of the nearly isotropic comets are closer to the solar value than periodic comets (Anderson et al. 2023). However, more observations are needed to determine if this is an artifact of observational limitations.

for comets and one Centaur as a function of the semimajor axis, a_0. This shows that most nearly isotropic comets have ratios closer to the protosolar value than periodic comets as Owen & Bar-Nun (1995) predicted based on potential evolutionary processes that preferentially deplete the nucleus of N_2. The current data also potentially suggests an increase in N_2/CO with increasing $1/a_0$ for periodic comets. However, Cochran et al. (2000) pointed out that a similar trend may only be partially objective, based on approximate estimates for some objects. Therefore, it is crucial to gather new observations and broaden the range of objects from different dynamical groups to draw reliable conclusions about determining the intrinsic values of N_2/CO for them.

Furthermore, caution is needed in confirming detections of N_2^+. Low-resolution spectral observations have indicated the presence of N_2^+ emissions (see, for example, Wyckoff & Theobald 1989; Womack et al. 1994; Korsun et al. 2006, 2008, 2014; Ivanova et al. 2016, 2018) but high-resolution spectral observations have not confirmed their presence (Cochran et al. 2000; Cochran 2002). This could suggest that N_2 may be rare in comets, although it is important to note that increasing spectral resolution requires a narrow slit that may miss the ion tail. One exception is comet C/2016 R2 (Pan-STARRS), which exhibited strong CO^+ and N_2^+ emissions at a heliocentric distance beyond 3 au (Cochran & McKay 2018; Opitom et al. 2019; Anderson et al. 2022). The only Centaur where spectral emissions of CO^+ and N_2^+ have been detected is 29P/S–W1 (Figure 6.10).

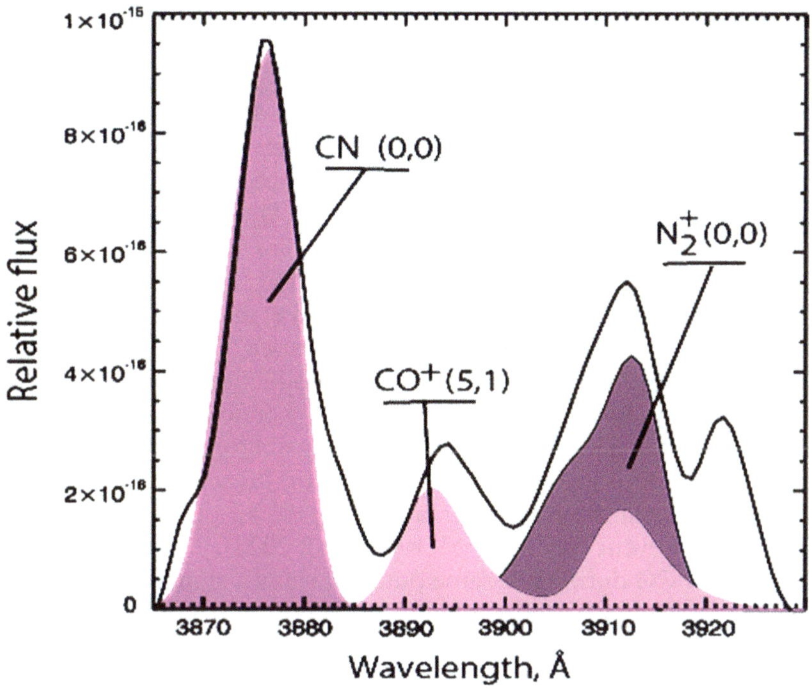

Figure 6.10. N_2^+/CO^+ ion ratios have been used as a proxy for the N_2/CO ratio in 29P/S–W1. Observed (black solid curve) and calculated (filled) spectroscopic profiles of the N_2^+ (dark purple), CN (purple), and CO^+ (5, 1) band (lavender) for Centaur 29P/S–W1. Reprinted from Korsun et al. (2008). Copyright 2008, with permission from Elsevier.

6.2.5 Current Understanding of Centaur Volatiles

We summarize below the state of knowledge for the surface and coma volatiles of seven Centaurs. They include active Centaurs 29P/S–W1, Echeclus, Chiron, (29981) 1999 TD_{10}, and 39P/Oterma along with non-active Centaurs (10199) Chariklo and (5145) Pholus (hereafter TD10, Oterma, Chariklo, and Pholus). Several of these objects are also covered in detail in Chapter 10.

6.2.5.1 29P/Schwassmann–Wachmann 1

Centaur 29P/S–W1 is an active Centaur also sometimes classified as a JFC originating from the Kuiper Belt (Sarid et al. 2019). It has displayed significant activity since its discovery in 1927, remaining active throughout all observed time periods (Whipple 1980; Jewitt 1990; Miles et al. 2016a; Miles 2016b, 2016c). Its orbit extends beyond Jupiter, so its continuous activity is not driven solely by water ice sublimation. However, the possibility of water ice evaporation cannot be entirely ruled out even at its location around 6 au, as it can be stimulated by absorbing contaminants such as carbonaceous materials or silicates with a significant iron content (Hanner et al. 1981; Beer et al. 2006). The estimated radius of the 29P/S–W1 is ~30 km. This is significantly larger than most known comets, which have radii less than 10 km (Schambeau et al. 2015; Sierks et al. 2015; Kelley et al. 2017). The rotational period is still uncertain, ranging from ~12 days based on coma features (Ivanova et al. 2012) to ~57 days based on suggested regularities in the outburst activity (Miles et al. 2016a), but these regularities have not been confirmed. In either case, the rotational period of 29P/S–W1 appears to be slower compared to most other small bodies in the solar system. This is discussed in more detail in Chapter 10.

Because of the high activity levels, there are no surface observations. Observations of H_2O in the coma are of interest. One study of a large outburst found that the estimated temperature at the subsolar point required 440% of the crystalline ice in the nucleus to be exposed and heated to explain the observed H_2O production rate of 4.1×10^{27} s^{-1} (Bockelée-Morvan et al. 2022). This scenario is not physically possible. The small velocity offset observed for the H_2O line indicated that the nucleus contributes minimally to water production, while an analysis of the temperature, velocity, and sublimation lifetime as a function of grain size revealed that long-lived icy grains with sizes exceeding a few micrometers played a significant role in the production of water molecules. Based on the water abundance observed in the coma, a lower limit of approximately 2×10^8 kg of icy grains was released during the outburst event (Bockelée-Morvan et al. 2022).

Observations of CO during this same outburst show a similar mass of this species released, indicating a substantial amount of material. As CO levels decreased, H_2O levels exhibited a corresponding decrease. CO is often detected with production rates varying between 1×10^{28} and 7×10^{28} s^{-1} (Womack et al. 2017; Wierzchos & Womack 2020; Harrington Pinto et al. 2022). Further observations provide upper limits on abundance ratios of CH_4, C_2H_6, CH_3OH, and H_2CO relative to CO that are consistent with the values in C/2016 R2 (PanSTARRS), a CO-rich Oort Cloud Comet (OCC; Roth et al. 2023). Thus, CH_4 may be preferentially stored with (polar) H_2O ice in the Centaur nucleus rather than with CO (Table 6.1).

Table 6.1. Abundance Measurements and Upper Limits for 29P/S-W1 Compared to Comets.

Ratio	29P/S–W1 (%)	C/2016 R2 (%)	Avg. Comets (%)
CH_4/CO	<0.98[a]	0.59 ± 0.09[b]	4.6–164[a]
C_2H_6/CO	<9.1[a]	<0.089[b]	2.3–98[a]
CH_3OH/CO	<21[a]	1.04 ± 0.08[b]	10–500[a]
H_2CO/CO	<1.6[a]	0.043 ± 0.006[b]	1–81[a]
OCS/CO	<22[a]	<0.24[b]	1.5–14[a]
N_2/CO	0.014[c]	0.089[c]	n/a
HCN/CO	0.12 ± 0.03[d]	$(3.8 \pm 1.0) \times 10^{-3}$[b]	5.5 ± 1.4[b]
CO_2/CO	$4–18 \times 10^{-3}$[e]	0.05–0.16[f]	>0.75[f]
Perihelion	5.8 au	2.6 au	n/a

[a] Roth et al. (2023);
[b] McKay et al. (2019);
[c] Anderson et al. (2023);
[d] Bockelée-Morvan et al. (2022);
[e] Faggi et al. (2024);
[f] Harrington Pinto et al. (2022).

As discussed earlier, HCN was observed in the coma of 29P/S–W1 and was found to be 10 times higher than comets at 1 au (Bockelée-Morvan et al. 2022). Finally, as discussed earlier, observations have shown that the coma of 29P/S–W1 is CO^+ and $N2^+$-rich. The abundant presence of these highly volatile species could serve as drivers for its unique outburst behavior. However, there are no other large Centaurs similar to 29P/S–W1 within the orbits of Jupiter and Saturn, making it challenging to generalize its characteristics to other Centaurs.

6.2.5.2 39P/Oterma

39P/Oterma (39P) is an active Centaur, which has undergone major orbital changes in the last 90 yr that had it fluctuating from a Centaur orbit to a closer-in Jupiter-family comet orbit and then back out to Centaur status. In 1943 Liisi Oterma discovered it with a visual magnitude of $m \sim 15$ (Marsden 1962), when it had just passed perihelion ($r_{helio} = 3.4$ au). Since its orbit maintained a heliocentric distance between 3.4 and 4.5 au for several years, it was possible to observe it continuously with visual apparent magnitudes ranging from 15 to 19 from when it was discovered 1943–1961. When recovered in 2001, 39P appeared dimmer until 2023 with visual apparent magnitudes ranging from 22 to 24.

Recent photometry and image analysis from Gemini and the Lowell Discovery Telescope (LDT) indicated no extended emission, suggesting that its bare nucleus was observed in the r′ filter. 39P's continuum shows absorption features at 2.0 and 3.1 μm attributed to water ice similar to what was seen in the JFC 103P (Harrington Pinto et al. 2023). The results are consistent with a water ice grain size that is similar to what Protopapa et al. (2014) measured in the coma of 103P: about 1 μm and are more likely in the coma rather than on the surface. 39P was also observed with JWST and it is the first Centaur to have CO_2 emission detected (Harrington Pinto et al. 2023). JWST observations provide a CO_2 detection with a production rate, Q, of $(5.96 \pm 0.80) \times 10^{23}$ molecules s^{-1} which is the lowest detection of CO_2 yet for any comet or Centaur. While CO and H_2O were not detected, 3-sigma upper limits of $QCO < 12.1 \times 10^{23}$ molecules s^{-1} and $QH_2O < 10.0 \times 10^{23}$ molecules s^{-1} were obtained. The CO_2/CO ratio for Oterma is >0.49 (Harrington Pinto et al. 2023), which is much larger than the ratio observed for 29P/S-W and indicates that the composition of Centaurs could be very diverse. This large difference could also be at least partly due to the different thermal heating that each of these objects have undergone.

6.2.5.3 (60558) 174P/Echeclus

Echeclus is an active Centaur demonstrating outburst activity that does not depend on its location over its orbit and sometimes at distances as great as 13 au. It was initially discovered on March 3, 2000, has a diameter of 64.6±1.6 km, and an albedo of around 5% (Duffard et al. 2014). Spectral absorption features of water ice have not been observed on its surface (Seccull et al. 2019). It has a steeper red color in visible wavelengths compared to near-infrared wavelengths. Preliminary modeling for analyzing the average polarization values indicates a surface mixture of carbon, silicates, and some ice – consistent with Rosetta measurements of 67P/C-G.

However, the presence of ice is suggested to be only a few percent, significantly lower than the few tens of percent observed in distant comets.

CO emission was detected in the coma of Echeclus at ~6 au (Wierzchos et al. 2017) with an estimated production rate of 7.7×10^{26} s^{-1}. No other volatiles were detected in Echeclus to date. Echeclus has been observed to have a blue coma (Seccull et al. 2019), similar to observations of the CO-rich comet C/2016 R2 (Biver et al. 2018).

6.2.5.4 (2060) Chiron

Chiron is an active Centaur that was discovered in 1977, demonstrating the recurrent occurrence of comet-like activity (Duffard et al. 2002). Chiron's mean radius is estimated to be approximately 107.8 ± 4.95 km, with an albedo of 0.08 (Fornasier et al. 2013). Detailed analysis of its light curve has yielded a well-defined rotational period of 5.918 hr with a minor brightness variation ranging from 0.05 to 0.09 magnitudes suggesting a predominantly spheroidal shape (Marcialis & Buratti 1993) and possible surface volatile activity. Furthermore, evidence suggests the possible presence of rings (Ortiz et al. 2015).

Spectral observations have shown water absorption features on the surface, but the data quality is insufficient to definitively determine if the ice is crystalline or amorphous (Foster et al. 1999). The surface reflectivity gradient appears to more closely resemble that of C-type asteroids than that of cometary nuclei. Small variations in the optical reflectivity gradient correlating with episodes of activity may be attributed to fluctuations in dust production. Long-term variations in brightness indicate surface activity and the existence of volatile substances. The polarization phase behavior of Chiron in its non-active period is significantly different from any other Solar system bodies studied so far (Bagnulo et al. 2006).

CO has been observed in the coma at a production rate of 1.3×10^{27} s^{-1} (Womack & Stern 1999). This CO may have resulted from an outburst, although the production rate is also sufficient to drive the observed dust coma activity. As noted earlier, CN was observed in the coma when Chiron was at a distance of ~11.26 au from the Sun (Bus et al. 1991) but not when it was closer to perihelion (Rauer et al. 1997; Barucci et al. 1999) suggesting that HCN may not be the main source of this radical.

6.2.5.5 (29981) 1999 TD_{10}

This active Centaur has a perihelion distance greater than 12 au, a diameter of 103.7 ± 13.5 km (Barkume et al. 2008), and a period of 15.448 ± 0.012 hr (Choi et al. 2003; Mueller et al. 2004; Rousselot et al. 2005). It exhibits weak activity (Choi et al. 2003). The normalized reflectivity gradient on the surface is similar to the median value observed for Centaurs (Mueller et al. 2004), and no water has been detected in surface spectral observations (Barkume et al. 2008).

6.2.5.6 (10199) Chariklo

Chariklo is an inactive Centaur that was discovered in 1997 and has a diameter of approximately 248 ± 18 km (Fornasier et al. 2013), ranking among the largest known Centaurs. Its rotation period is ~7 hr. Photometric analysis of observations has unveiled the presence of two distinct narrow rings (Braga-Ribas et al. 2014). Infrared observations of Chariklo's surface have periodically indicated the presence of water ice (Fornasier et al. 2013 and references therein), which may be located within its rings. It is not clear if the water ice is amorphous or crystalline (Dotto et al. 2003). Polarimetric observations revealed evidence of surface heterogeneity (Belskaya et al. 2010).

6.2.5.7 (5145) Pholus

This inactive Centaur is one of the most primitive objects known within its class (Cruikshank et al. 1998), with a diameter of approximately 107 ± 19 km and an albedo of 0.12 (Duffard et al. 2014). Its rotation period is 9.98 ± 0.02 hr (Buie & Bus 1992; Farnham 2001; Tegler et al. 2005). The surface displays a highly conspicuous reddish hue (Cruikshank et al. 1998; Fornasier et al. 2009) and the reflectance spectrum reveals the presence of two distinct components that are spatially separated. These components include dark amorphous carbon and an intimate mixture of water ice, methanol ice, olivine grains, and complex organic compounds known as tholins. Polarimetric observations have unveiled a significant negative polarization at specific phase angles that is distinctly different from that observed in TNOs (Belskaya et al. 2010).

6.3 Centaurs in Solar System Context

Centaurs play an important role in understanding the formation and evolution of the solar system. Their volatile composition can tell us where they formed and how they relate to other small body populations, providing critical pieces of a dynamical puzzle that is difficult to constrain with the current data available. We review the current state of knowledge for the volatile composition of other small body populations providing a baseline for future comparisons as more data becomes available for Centaurs.

6.3.1 Census of Observed Small Body Volatiles

6.3.1.1 Oort Cloud and Jupiter-Family Comets

In general, comets originating from the Oort Cloud are called OCCs, while ones that originated as TNOs are referred to as JFCs. However, there are exceptions where comets originating from the Oort Cloud are captured into shorter orbits causing them to exhibit similar orbital characteristics to JFCs. Chemical diversity is evident among comets both in parent volatiles and daughter species. A review of the average mixing ratios of eight parent volatiles in 30 comets (CH_3OH, HCN, NH_3, H_2CO, C_2H_2, C_2H_6, CH_4, CO) suggest that OCCs are volatile-rich (Figure 6.11). This review also proposed an overall depletion of these volatiles in JFCs compared to OCCs that was most pronounced for the four species of highest volatility (C_2H_2, C_2H_6, CH_4, and CO), suggesting that thermal evolution may affect the compositional differences between the populations (Dello Russo et al. 2016). There were weak positive correlations between NH_3 and HCN, and between NH_3 and H_2CO in

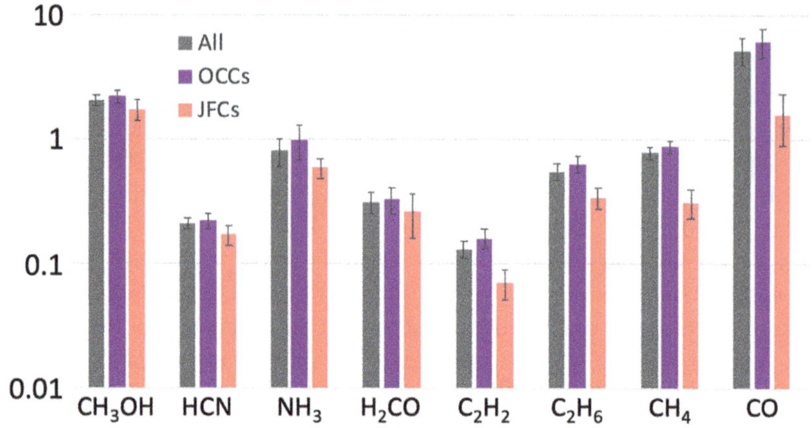

Figure 6.11. The relative abundance of different volatile species in comet families can help to constrain their origins. Average mixing ratios (% with respect to H_2O) for up to 27 comets based on availability of measurements compared to the mixing ratios for the OCCs (up to 19) and JFCs (up to 8) in the sample (adapted from data by Dello Russo et al. 2016).

both OCCs and JFCs and strong correlations between HCN, C_2H_2, and C_2H_6 (Dello Russo et al. 2016). However, it should be noted that more recent work with improved statistics indicates broader compositional diversity in each dynamical class, and found that with the exception of CO (for which JFCs are on average depleted compared to OCCs) there are no significant compositional differences between parent volatile abundances among the dynamical families (Harrington Pinto et al. 2022, Biver et al. 2024). This suggests that many proposed differences may be related to sample size biases but heliocentric effects may still occur (Harrington Pinto et al. 2022).

The presence of hypervolatiles in OCCs can help to constrain models to determine whether Oort Cloud objects were ejected to this region by giant planet migration early (<20 million yr) or later (0.05–2.0 billion yr). Dynamical models suggest that hypervolatile-rich planetesimals in the Oort Cloud may have been placed within approximately 20 million yr of solar system formation and represent the earliest objects in the Oort Cloud providing potential insights into $CO/N_2/CH_4$ ratios in the protoplanetary disk (Davidsson 2021; Steckloff et al. 2021).

Research on radicals (OH, CN, C_2, C_3, NH) in 85 comets indicated that about half of JFCs are classified as carbon-chain depleted, a much larger percent than the observed OCCs (A'Hearn et al. 1995). This is interpreted to be caused by differences in carbon-chain chemistries in the initial conditions where JFCs formed and not caused by their extensive thermal processing due to numerous perihelion passages (A'Hearn et al. 1995; Fink 2009). If this is the case, it could also imply that pre-cometary ices that formed JFCs were subjected to conditions that, on average, selectively depleted the most volatile species.

However, our understanding of comet composition and how it varies from one comet to the next remains limited. In general, we have found that the observed volatiles in typical OCCs and JFCs primarily consist of H_2O ice. CO_2 and CO ice are

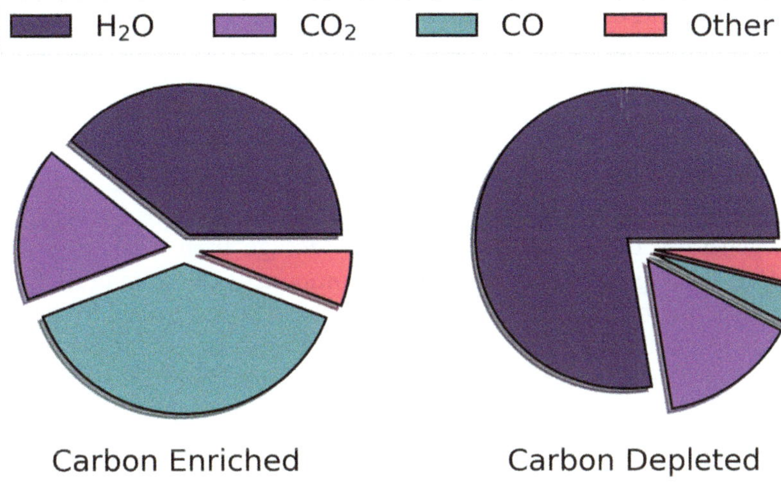

Figure 6.12. Typical compositions of carbon-enriched and -depleted comets illustrate two compositional families of comets that may indicate different formation regions. The carbon-enriched comet is C/2006 W3 Christensen (Ootsubo et al. 2012). The carbon depleted pie chart is representative of most comets for which production rate measurements of CO_2, CO, and H_2O exist. This is a generalized version of an analogous figure in McKay et al. (2019) and Seligman & Moro-Martín (2022), reprinted by permission of the publisher (Taylor & Francis Ltd, http://www.tandfonline.com.)

the two next most abundant volatiles followed by carbon- and nitrogen- bearing species such as CH_4, C_2H_2, C_2H_6, CH_3OH, NH_3 and HCN (Dello Russo et al. 2016; Biver et al. 2024). The discovery of C/2006 W3 Christensen followed by C/2016 R2 (Pan-STARRS) indicated the potential existence of a different class of comets in the solar system based on composition and described as carbon-rich. The composition of C/2006 W3 Christensen compared to the average composition of JFC and OCC comets is illustrated in Figure 6.12. Carbon-enriched comets are proposed to have formed in the region of (Mousis et al. 2021), at (Price et al. 2021), or outside of the CO ice-line in the PSN (Seligman et al. 2022).

Recent work has compared the composition of 29P/S–W1 during an outburst against parent volatiles measured in comets at radio and near-infrared wavelengths (Roth et al. 2023). Stringent upper limits on abundance ratios of hypervolatiles (CH_4/CO and C_2H_6/CO) and hydrocarbons (CH_3OH/CO and H_2CO/CO) were most consistent with values measured in C/2016 R2 (PanSTARRS). However, compositional comparisons between Centaurs and comets are challenging, as the vast majority of parent volatile studies in comets take place in the inner solar system (heliocentric distances $<\sim 2$ au) where H_2O is vigorously sublimating, in contrast to the much larger heliocentric distances of Centaurs.

6.3.1.2 Active TNOs

TNOs are categorized into two main groups: the classical and resonant entities of the Kuiper belt and the scattered disc and detached objects, with the sednoids representing the most distantly located members. Observations of activity for distant

TNOs are limited and direct observation of sublimation of gases is limited to Pluto (Gladstone & Young 2019; Young et al. 2020). Therefore, additional mechanisms are required to explain activity at significant distances from the Sun. What is notable is that many distant comets exhibit long duration activity that cannot be explained by crystallization of amorphous water ice, so further studies are needed.

Dust production in these distant comets is significantly higher than that of short-period comets, similar to new comets entering the inner solar system for the first time. It is generally higher before perihelion passage than after, and appears to contain relatively large particles with an icy component (Korsun et al. 2006, 2010, 2014) with sublimation times lasting several years. The structures in distant comet comae do not resemble the comet tails that form at close heliocentric distances. They typically do not have an internal structure, display an almost constant width along the tail, and are often strongly bent. Some distant comets have asymmetric and elongated comae, without tails.

Polarimetric studies suggest that the composition of particles in distant comets differs from the composition of dust in short-period comets (Dlugach et al. 2018; Ivanova et al. 2015a, 2015b, 2019b, 2021, 2023). For example the coma of C/2014 A4 (SONEAR) is dominated by submicron particles composed of a large amount of ice and tholin-like organic substances (Ivanova et al. 2019b) while the coma of C/2011 KP36 (Spacewatch) is formed by particles of various sizes consisting of water ice, CO_2 ice, and refractory material (Ivanova et al. 2021).

Spectral observations of comet C/2002 VQ94 (LINEAR) when it was at 7.33 au from the Sun revealed the presence of CO^+ and $N2^+$ emissions (Korsun et al. 2006, 2008) interpreted to mean that this comet is enriched in CO and N_2. The emissions disappeared when the comet was at a distance of 9.86 au.

6.3.1.3 Active Asteroids

The main asteroid belt contains rare asteroids that are classified as active asteroids based on the Tisserand parameter, the presence of a coma, and the detection of water and other volatiles (Jewitt et al. 2015). In these small bodies, manifestations of activity can be more diverse than comets. Similar to comets, active asteroids shed a noticeable amount of material, with activity varying from short-lived events (Neslusan et al. 2016; Ivanova et al. 2023), to recurring activities, exemplified by objects like 133P/Elst-Pizarro (Hsieh et al. 2010; Jewitt et al. 2014) and (6478) Gault (Chandler et al. 2019; Ivanova et al. 2020).

Several mechanisms that can trigger such activity have been proposed including ice sublimation with associated dust expulsion, rotational breakup, impact events leading to dust ejection, thermal fractures, and rotational fission of contact binary asteroids. However, it is easier to rule out specific processes than to identify a definitive cause. This is partly due to limitations in data, but it also reflects the intricate nature of the observed activity. For example, in cases where sublimation is believed to be the primary driver of activity, an impact or other disruptive event may be required to expose buried ice and trigger sublimation, with rapid rotation potentially aiding mass loss.

Active asteroids demonstrate remarkable orbital stability over extremely long timescales, often exceeding 100 million yr (Jewitt 2009; Hsieh et al. 2012, Stevenson et al. 2012). However, exceptions like 238P and 259P have been identified as unstable over shorter timescales, around 10 million yr. Several active asteroids have been associated with collisional asteroid families and clusters, which could play a role in preserving ice over long timescales through shielding within a larger parent body until more recent exposure to direct solar heating triggers activity.

Recently, a new class of potentially active asteroids was also discovered (see Figure 6.13), the so-called "dark comets" and include (523599) 2003 RM, 1998 KY_{26}, 2005 VL_1, 2016 NJ_{33}, 2010 VL_{65}, 2006 RH_{120}, and 2010 RF_{12}. These dark comets are photometrically inactive asteroids that exhibit significant nongravitational accelerations that are (1) too strong in magnitude and (2) in the wrong direction to be driven by typical radiation based effects seen on asteroids, such as the Yarkovsky effect or radiation pressure (Seligman & Moro-Martín 2022). Because these objects are photometrically inactive, it is unclear whether the nongravitational accelerations are driven by outgassing. However, to date there is no alternative mechanism known that could explain the observed nongravitational accelerations.

If these objects are actively outgassing, it is possible that they were delivered to their current orbits from the Jupiter-family comets, and presumably from the Centaur region prior to that. Therefore, it is possible that these dark comets represent one possible end state of Centaur evolution into the inner solar system. Future observations will be required to measure the outgassing volatile content of these objects, and comparisons to JFCs and Centaurs will likely be revealing if these objects share similar evolutionary histories. Excitingly, one of the dark comet candidates, 1998 KY_{26}, is already the target for the extended Hayabusa2 mission and exhibits favorable viewing geometry before 2025. The Legacy Survey of Space and Time (LSST) planned to be conducted with the forthcoming Vera Rubin

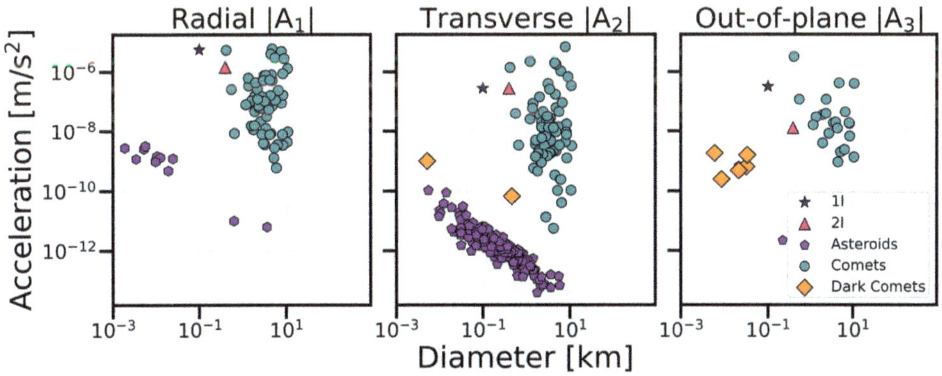

Figure 6.13. A new class of potentially active asteroids, referred to as dark comets, are photometrically inactive but exhibit significant nongravitational accelerations. Measured components of nongravitational accelerations of small bodies with known diameters in the radial A1 direction (left), transverse A2 direction (middle) and out-of-plane A3 direction (right). Figure from Seligman & Moro-Martín (2022), reprinted by permission of the publisher (Taylor & Francis Ltd, http://www.tandfonline.com.)

Observatory is poised to further transform our understanding of these classes of objects.

6.3.1.4 Interstellar Comets

We have limited knowledge of the census of volatiles on interstellar comets, given only two known objects. The first interstellar object 1I/'Oumuamua displayed no detectable cometary activity either in volatiles or dust reflectance but exhibited nongravitational acceleration (Micheli et al. 2018). The mechanisms proposed include outgassing invoked compositions of H_2, N_2, CO or H_2O (Seligman & Laughlin 2020; Desch & Jackson 2021; Jackson & Desch 2021). The interstellar comet 2I/Borisov, which had a closest approach to the Sun of $\sim$2 au, displayed a significant and distinct cometary tail and was enriched in hypervolatiles compared to H_2O (Cordiner et al. 2020; Bodewits et al. 2020).

Thus, it is challenging to draw population level comparisons with well characterized solar system small body populations. However, 2I/Borisov was enriched in hypervolatiles to a higher extent than observed in Centaurs and it could be explained as having formed at further stellocentric distances in its host disk. It is proposed to have formed at or exterior to the CO snowline, possibly in an M-dwarf system where a larger fraction of the total area in the protostellar disk would be exterior to the CO snowline (Cordiner et al. 2020). However, observations of Centaurs are typically taken at much larger heliocentric distances (>3 au) than the observations that were taken of 2I/Borisov (2–2.7 au), so it is likely that some volatile reservoirs (H_2O) in Centaurs that have been observed were not active during the observations.

6.3.2 Comparison to Solar System Models

A central goal in the study of primitive small bodies is using their dynamical history and composition to trace back to the chemistry and physics present in the protoplanetary disk midplane at the time of planet formation. Such an analysis requires a statistically significant sample, which has historically been difficult to obtain for volatile composition studies of Centaurs based on the few active Centaurs available and observational challenges associated with the limitations of ground-based facilities.

Overall the current state of knowledge of primordial solar system volatiles is limited. Figure 6.14 illustrates the current state of knowledge for the bulk composition of volatile elements in the giant planet atmospheres and known small body populations compared to the composition of the PSN (Lodders 2021). From this figure we can see that Jupiter is the only giant planet with a complete set of observations thanks to the *in situ* measurements made by the Galileo Probe Mass Spectrometer (GPMS; Mahaffy et al. 2000; Wong et al. 2004) and the Juno remote observations with the Microwave Radiometer (MWR; Li et al. 2017, 2020).

Bulk composition constraints for small bodies are currently limited to chondrites (Lodders 2021) and cometary ices (Le Roy et al. 2015; Rubin et al. 2019) and show significant fractionation of the elements compared to solar values. The heavy noble gas relative abundances are particularly interesting, as shown in Figure 6.15 where solid materials differ by orders of magnitude compared to the solar values.

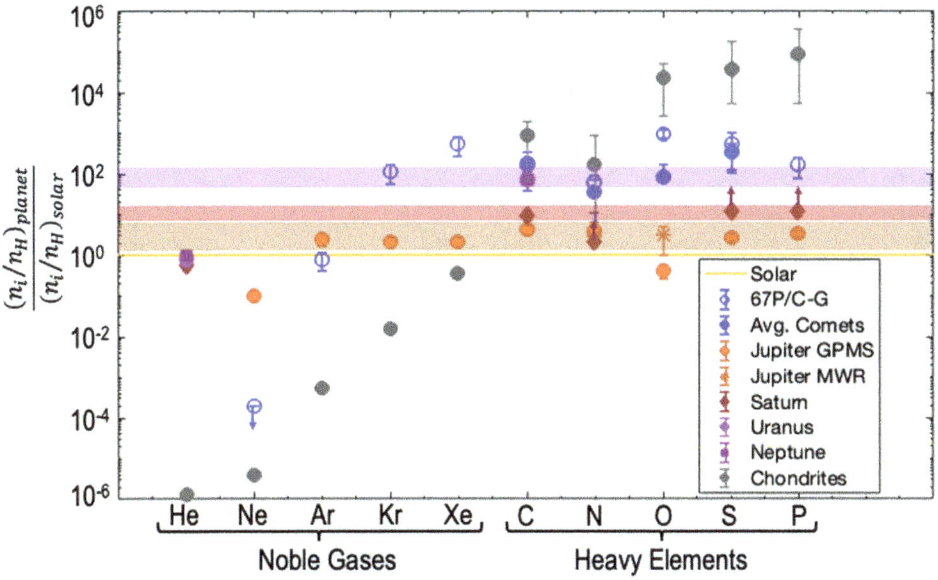

Figure 6.14. Available observations of the heavy element and noble gas abundances in the atmospheres of the four giant planets relative to PSN composition along with analogs for planetesimals from the time of solar system formation can help to constrain the building blocks for the giant planets. The color of the points indicate the origin of the measurements, and references are provided in the main text.

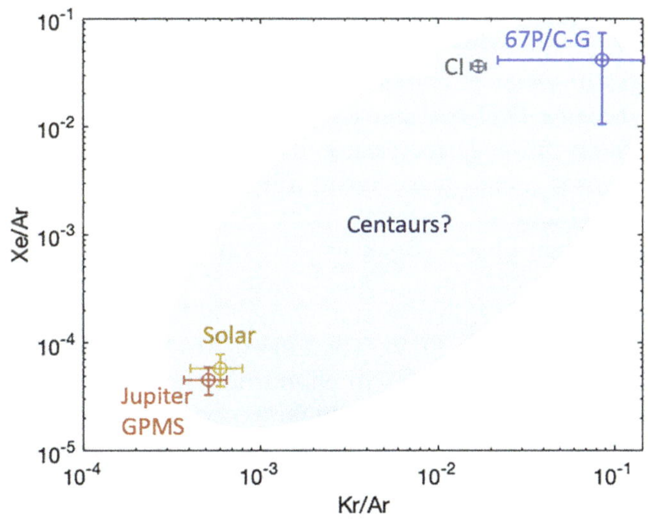

Figure 6.15. Available relative noble gas abundances in Jupiter's atmosphere, the Sun, and the limited knowledge for analogs for solid building block materials. Adding more information on the noble gas abundances can help us to understand when and where each giant planet formed and how they migrated after formation.

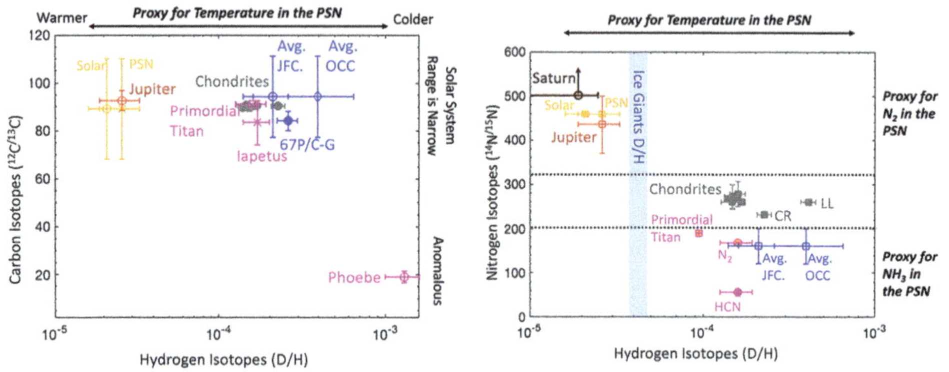

Figure 6.16. Current state of knowledge for isotope ratios that can be used to determine the origin and evolution of materials throughout the solar system (see text for individual measurement references).

Chondrites found in meteorites that impact the Earth provide an analogue for the volatiles contained in asteroids and the refractory material of icy bodies like comets. The comet observations shown in Figure 6.14 are based on icy materials observed in comae that either originate as TNOs (JFCs) or from the Oort Cloud (OCCs). Centaurs are intermediate in distance from the Sun between asteroids and TNOs, residing within the orbits of the giant planets.

Figure 6.16 illustrates the current state of knowledge for three isotope ratios of interest within various solar system bodies. Carbon isotope ratios, $^{12}C/^{13}C$, are limited to a narrow range around a value of ~90 with error bars representing a range between 65 and 110 (e.g., Alexander et al. 2013; Mandt et al. 2009). The one exception is the Saturnian satellite Phoebe, which is highly enriched in the heavy isotope compared to any other solar system body (Clark et al. 2019).

The D/H ratio in water is a useful proxy for the formation location of solid building blocks because D/H in water varies as a function of distance from the Sun as the solar system formed, increasing in value with greater distance as the temperature decreased (Aikawa & Herbst 1999). As Figure 6.16 illustrates water incorporated into chondritic material (Alexander et al. 2012) likely formed in warmer conditions than OCCs (Bockelée-Morvan et al. 1998; Meier et al. 1998; Biver et al. 2006; Hutsemékers et al. 2008; Weaver et al. 2008; Villanueva et al. 2009; Bockelée-Morvan et al. 2012; Brown et al. 2012; Gibb et al. 2012; Biver et al. 2016; Paganini et al. 2017; Lis et al. 2019). The D/H in JFCs suggested that this class of comets formed in similar conditions to chondrites (Hartogh et al. 2011; Lis et al. 2013, 2019). Initial reports of one of the highest reported D/H values in the Rosetta JFC target comet, 67P/C-G appeared to contradict this possibility (Altwegg et al. 2015). However, a recent reanalysis of the Rosetta observations making use of over 4000 data points from the mission discovered that the D/H in coma near the spacecraft location was highly influenced by enriched ice sublimating from dust grains (Mandt et al. 2024). When accounting for the role of dust in the measurements, the D/H was revised to a lower value that is much closer to chondrites (Mandt et al. 2024).

Three moons of Saturn are also illustrated. Titan's observation is in methane, which is different from the D/H in water on Enceladus (not shown; Waite et al. 2009) and Iapetus suggesting that Titan's methane is primordial and not produced from interacting with water through hydrothermal chemistry in Titan's interior (Mousis et al. 2009). As with the carbon isotopes, Saturn's moon Phoebe is enriched in the heavy isotope far more than any other solar system body (Clark et al. 2019).

Finally, the nitrogen isotope ratio, $^{14}N/^{15}N$, is a useful tracer for the source of nitrogen in the PSN. The bulk of nitrogen in the PSN was in the form of N_2, which had a high ratio. Other species, including organics found in chondrites as well as NH_3 and HCN ices, were more enriched in the heavy isotope.

At the present time insufficient information is available about the bulk composition of Centaurs to include them in Figure 6.14 and no noble gas abundances or isotope ratios have been measured in Centaurs. These measurements are needed because as likely dynamical progenitors to JFCs, Centaurs may provide a window into TNO material that has undergone less potential thermal processing than that preserved in short period comets. Although earlier studies proposed that OCCs formed between heliocentric distances $\sim$5 – 30 au and JFCs formed at larger heliocentric distances, the presence of crystalline silicates in some comets require that processed material from the hot inner disk was incorporated into their nuclei along with hypervolatiles from the cold outer disk. This suggests a more "spatially mixed" comet formation region.

Furthermore, the composition of ices preserved in Centaurs can improve comparisons with ice-phase astrochemical models of protoplanetary disk midplanes. Recent work comparing the composition of JFCs and OCCs measured to date found that no single time or location in the midplane could fully account for the molecular diversity measured in comets (Willacy et al. 2022). As volatile inventories for Centaurs continue to improve with the latest ground- and space-based facilities, similar comparisons between the primitive small bodies and midplane ice-phase chemistry will become possible.

6.4 Future Prospects

Volatiles in Centaurs provide important clues to the conditions in which they formed and can help us to better understand the bigger picture of solar system formation. Furthermore, the form of volatiles may play an important role in Centaur activity, more so than regular comets where activity is driven by sublimation. We close this chapter by outlining the most important measurements needed through future exploration of Centaur volatiles and describe the future prospects based on upcoming capabilities.

6.4.1 Open Questions

We need to learn more about the composition and form of ice in Centaurs compared to other small bodies in order to understand their role in the history of the solar system. This information will provide vital clues about the formation conditions of small bodies and how they may have contributed to the formation of the solar system.

The summary of observed volatile elements that can be used to trace formation and evolution processes for the solar system illustrated in Figure 6.14 is clearly incomplete. More definitive constraints on the abundances of carbon-, nitrogen-, and sulfur-bearing volatiles relative to water in Centaurs are needed to place them in the context of comets and asteroids. Additionally, an improved understanding of comet composition is needed. Most comet measurements are from a single point in a comet's orbit, and it is not clear if these observations are reflective of the bulk composition. Measuring the composition of the comae of several comets over a range of distances from the Sun is needed to resolve this question.

Additionally, measurements of noble gases are needed in Centaurs and in more comets. Noble gases are particularly useful for tracing the sources of volatiles because they are nonreactive and not influenced by chemistry. As Figure 6.15 showed, the relative abundances of the heavy noble gases can separate solid materials from the bulk composition of the PSN and we need to determine where Centaurs fit in. It should be noted that because noble gases are non-reactive they cannot be measured remotely and require spacecraft observations *in situ*. Opportunities to determine noble gases should be prioritized in planning future missions to Centaurs.

Finally, measurements of stable isotope ratios in Centaur volatiles are needed. The D/H ratio in water will provide clues to the temperature conditions where the water ice formed. The nitrogen isotopes can provide indications of the source of nitrogen and the temperature conditions for formation while carbon isotopes can provide constraints on volatile evolution.

6.4.2 Recommended Activities

A combination of archival studies to search for activity not previously detected in known Centaurs, Earth-based observations from the ground and space, as well as spacecraft missions either to fly by or orbit Centaurs is needed to obtain the observations described above.

6.4.2.1 Ground-based

Ground-based observations of Centaurs, facilitated by utilizing surveys with large telescopes such as the LSST, can play a pivotal role in obtaining many of the needed observations. Large telescopes are particularly important because of the inherently low brightness of Centaurs. A quasi-simultaneous approach to ground-based observations that combine photometry, spectroscopy, and polarization studies are of high value and offer a comprehensive view of Centaurs. This approach has proven effective for several distant comets (Ivanova et al. 2019b, 2021, 2023).

Furthermore, long-term monitoring of Centaurs is indispensable for uncovering variations with time and heliocentric distances. These variations hold the key to understanding the surface dynamics and composition of both non-active and active Centaurs. Through extended observation campaigns, astronomers can capture the subtle changes that occur on these distant objects over time, shedding light on their non-typical behavior and evolutionary processes.

6.4.2.2 Space-based from Earth

Observations with the recently launched James Webb Space Telescope are already taking place and revolutionizing our understanding of the distant universe and our solar system. The infrared capabilities of JWST allows us to characterize the volatile content of active Centaurs. Prior to JWST, 29P/S–W1 was the only Centaur with measured CO production rates and upper limits on H_2O and CO_2 was (see Section 6.3.1). So far, JWST Cycle 1 observations have included observations of 29P/S–W1, 39P/Oterma, C/2004 A1 (LONEOS), C/2008 CL94 (Lemmon), C/2014 OG392 (PanSTARRS), and C/2019 LDS (ATLAS). Successful detection of CO_2 gas emission by 39P/Oterma and upper limits of CO and H_2O provide limits on the mixing ratios of $CO/CO_2 \leqslant 2.03$ and $CO_2/H_2O \geqslant 0.60$ (Harrington Pinto et al. 2023). Moreover, JWST imaging provides constraints on the radius of the nucleus between 2.21 to 2.49 km. Further observations of 29P/S–W1 revealed detections of CO and CO_2 isotopologues, $^{13}C^{16}O_2$, $^{13}C^{16}O$, $^{12}C^{17}O$, and $^{12}C^{18}O$, marking their first detection in a Centaur (Faggi et al. 2024). While these are some of the first measurements of active Centaurs with JWST, they demonstrate how characterization of volatile content of Centaurs in the near future will greatly advance with the capabilities of JWST (see Chapter 14).

The forthcoming Habitable Worlds Observatory (HWO; NASEM 2020) poses significant prospects to advance our understanding of the formation and evolution of the solar system via Centaur science. HWO is recommended to have ultraviolet spectroscopic capabilities that would enable measurements of the production rates in active Centaurs of nitrogen-bearing species such as N_2, NH_3 and HCN, nitrogen isotopes, sulfur bearing species and isotopes, and oxygen isotopes. It is also possible that this observatory could measure D/H in surface water ice or water released from grains in the coma, which would provide information into the still-open question of volatile delivery to the terrestrial planets.

6.4.2.3 Centaur Spacecraft Missions

Spacecraft missions, whether flying by a Centaur or orbiting one, would provide access to information that is not available from a distance. A flyby of a Centaur would provide an opportunity to image the surface and make spectroscopic observations of surface ices. Thermal imaging of the nucleus could provide important constraints on the conditions for activity if the Centaur is observed to be active. Mass spectrometers capable of measuring the composition of ions and neutral particles could detect or set upper limits for activity during the flyby and measure the composition of volatiles being released if the Centaur is active. Dust detectors could provide composition measurements for any dust being released. An orbital mission would provide even more information through detailed mapping of the surface geology and composition. Furthermore, ongoing monitoring for activity and characterization of any volatiles and dust released would provide a greater understanding of the composition of the nucleus itself. The Centaur 29P/S–W1 is the most compelling target for a spacecraft mission because it is renowned for its considerable outburst activity that is constant and has already provided tantalizing clues to the volatile composition of this Centaur.

The application of advanced observational techniques, including mass spectrometry for isotopic and noble gas analyses, combined with new telescopes like JWST and the Vera Rubin Observatory, promise to revolutionize our understanding of Centaur volatiles. JWST can provide valuable insights into the composition and activity of Centaurs, especially those exhibiting volatile-rich surfaces. Long-term observations of select Centaurs will help us to understand how Centaurs change as they transition between their distant orbits and closer encounters with the Sun. By comparing observations taken at various points in their orbits, we can understand how factors such as solar distance, temperature, and radiation influence their behavior and surface characteristics.

Acknowledgements

K.E.M. acknowledges support from NASA ROSES RDAP grant 80NSSC19K1306. O.I. was supported by the Slovak Academy of Sciences (grant Vega 2/0059/22) and by the Slovak Research and Development Agency under Contract no. APVV-19-0072. N.X.R. acknowledges support by the Planetary Science Division Internal Scientist Funding Program through the Fundamental Laboratory Research (FLaRe) work package. D.Z.S. is supported by an NSF Astronomy and Astrophysics Postdoctoral Fellowship under award AST-2303553. This research award is partially funded by a generous gift of Charles Simonyi to the NSF Division of Astronomical Sciences. The award is made in recognition of significant contributions to Rubin Observatory's Legacy Survey of Space and Time.

References

A'Hearn, M. F., Millis, R. C., Schleicher, D. O., Osip, D. J., & Birch, P. V. 1995, Icar, 118, 223
A'Hearn, M. F., Belton, M. J. S., Delamere, W. A., et al. 2011, Sci, 332, 1396
Aguichine, A., Mousis, O., & Lunine, J. I. 2022, PSJ, 3, 141
Aikawa, Y., & Herbst, E. 1999, ApJ, 526, 314
Alexander, C. M. O., Bowden, R., Fogel, M. L., et al. 2012, Sci, 337, 721
Alexander, C. M. O., Howard, K. T., Bowden, R., & Fogel, M. L. 2013, GeCoA, 123, 244
Altwegg, K., Balsiger, H., Bar-Nun, A., et al. 2015, Sci, 347, 1261952
Anderson, S. E., Rousselot, P., Noyelles, B., Jehin, E., & Mousis, O. 2023, MNRAS, 524, 5182
Anderson, S. E., Rousselot, P., Noyelles, B., et al. 2022, MNRAS, 515, 5869
Bagnulo, S., Boehnhardt, H., Muinonen, K., et al. 2006, A&A, 450, 1239
Bar-Nun, A., Kleinfeld, I., & Kochavi, E. 1988, PhRvB, 38, 7749
Bar-Nun, A., Dror, J., Kochavi, E., & Laufer, D. 1987, PhRvB, 35, 2427
Bar-Nun, A., Herman, G., Laufer, D., & Rappaport, M. L. 1985, Icar, 63, 317
Barkume, K. M., Brown, M. E., & Schaller, E. L. 2008, AJ, 135, 55
Barucci, M. A., Brown, M. E., Emery, J. P., & Merlin, F. 2008, in The Solar System Beyond Neptune, ed. M. A. Barucci, et al. (Tucson, AZ: Univ. Arizona Press) 143
Barucci, M. A., Lazzarin, M., & Tozzi, G. P. 1999, AJ, 117, 1929
Beer, E. H., Podolak, M., & Prialnik, D. 2006, Icar, 180, 473

Belskaya, I. N., Bagnulo, S., Barucci, M. A., et al. 2010, Icar, 210, 472
Biver, N., Bockelée-Morvan, D., Crovisier, J., et al. 2006, A&A, 449, 1255
Biver, N., Winnberg, A., Bockelée-Morvan, D., et al. 1998, BAAS, 30, 1452
Biver, N., Moreno, R., Bockelée-Morvan, D., et al. 2016, A&A, 589, A78
Biver, N., Bockelée-Morvan, D., Paubert, G., et al. 2018, A&A, 619, A127
Biver, N., Dello Russo, N., Opitom, C., & Rubin, M. 2024, in Comets III, ed. K. J. Meech, et al. (Tucson, AZ: Univ. of Arizona. Press) 459
Bockelée-Morvan, D., Gautier, D., Lis, D. C., et al. 1998, Icar, 133, 147
Bockelée-Morvan, D., Biver, N., Schambeau, C. A., et al. 2022, A&A, 664, A95
Bockelée-Morvan, D., Biver, N., Swinyard, B., et al. 2012, A&A, 544, L15
Bodewits, D., Noonan, J. W., Feldman, P. D., et al. 2020, NatAs, 4, 867
Boissier, J., Bockelée-Morvan, D., Biver, N., et al. 2014, Icar, 228, 197
Braga-Ribas, F., Sicardy, B., Ortiz, J. L., et al. 2014, Natur, 508, 72
Brasser, R., Duncan, M. J., Levison, H. F., Schwamb, M. E., & Brown, M. E. 2012, Icar, 217, 1
Brasser, R., Mojzsis, S. J., Werner, S. C., Matsumura, S., & Ida, S. 2016a, E&PSL, 455, 85
Brasser, R., Matsumura, S., Ida, S., Mojzsis, S. J., & Werner, S. C. 2016b, ApJ, 821, 75
Brown, R. H., Lauretta, D. S., Schmidt, B., & Moores, J. 2012, P&SS, 60, 166
Buie, M. W., & Bus, S. J. 1992, Icar, 100, 288
Bus, S. J., A'Hearn, M. F., Schleicher, D. G., & Bowell, E. 1991, Sci, 251, 774
Chandler, C. O., Kueny, J., Gustafsson, A., et al. 2019, ApJL, 877, L12
Choi, Y. J., Brosch, N., & Prialnik, D. 2003, Icar, 165, 101
Clark, R. N., Brown, R. H., Cruikshank, D. P., & Swayze, G. A. 2019, Icar, 321, 791
Cochran, A. L., & Cochran, W. D. 1991, Icar, 90, 172
Cochran, A. L. 2002, ApJL, 576, L165
Cochran, A. L., Cochran, W. D., & Barker, E. S. 2000, Icar, 146, 583
Cochran, A. L., & McKay, A. J. 2018, ApJL, 854, L10
Cordiner, M. A., Milam, S. N., Biver, N., et al. 2020, NatAs, 4, 861
Crovisier, J., Biver, N., Bockelee-Morvan, D., et al. 1995, Icar, 115, 213
Cruikshank, D. P., Roush, T. L., Bartholomew, M. J., et al. 1998, Icar, 135, 389
Dalle Ore, C. M., Barucci, M. A., Emery, J. P., et al. 2015, Icar, 252, 311
Davidsson, B. J. R. 2021, MNRAS, 505, 5654
Davidsson, B. J. R., Sierks, H., Güttler, C., et al. 2016, A&A, 592, A63
de Almeida, A. A., Boice, D. C., Picazzio, E., & Huebner, W. F. 2016, AdSpR, 58, 444
de Bergh, C., Schaller, E. L., Brown, M. E., et al. 2013, ApSSL, 356, 107
De Keyser, J., Dhooghe, F., Altwegg, K., et al. 2017, MNRAS, 469, S695
Deienno, R., Morbidelli, A., Gomes, R. S., & Nesvorný, D. 2017, AJ, 153, 153
Dello Russo, N., Mumma, M. J., DiSanti, M. A., et al. 2000, Icar, 143, 324
Dello Russo, N., Kawakita, H., Vervack, R. J., & Weaver, H. A. 2016, Icar, 278, 301
Delsanti, A., Peixinho, N., Boehnhardt, H., et al. 2006, AJ, 131, 1851
Delsemme, A. H., & Swings, P. 1952, AnAp, 15, 1
Desch, S. J., & Jackson, A. P. 2021, JGR (Planets), 126, e06807
Devlin, J. P. 2001, JGR, 106, 33333
Dlugach, J. M., Ivanova, O. V., Mishchenko, M. I., & Afanasiev, V. L. 2018, JQSRT, 205, 80
Dotto, E., Barucci, M. A., Leyrat, C., et al. 2003, Icar, 164, 122
Drahus, M., Jewitt, D., Guilbert-Lepoutre, A., Waniak, W., & Sievers, A. 2012, ApJ, 756, 80
Duffard, R., Pinilla-Alonso, N., Santos-Sanz, P., et al. 2014, A&A, 564, A92

Duffard, R., Lazzaro, D., Pinto, S., et al. 2002, Icar, 160, 44
Emel'yanenko, V. V. 2005, E&MP, 97, 341
Enzian, A., Cabot, H., & Klinger, J. 1997, A&A, 319, 995
Faggi, S., Villanueva, G. L., McKay, A., et al. 2024, NatAs, 8, 1237
Farnham, T. L. 2001, Icar, 152, 238
Fernandez, J. A., & Ip, W.-H. 1984, Icar, 58, 109
Fernández, J. A., Helal, M., & Gallardo, T. 2018, P&SS, 158, 6
Fink, U. 2009, Icar, 201, 311
Fogg, M. J., & Nelson, R. P. 2007a, A&A, 472, 1003
Fogg, M. J., & Nelson, R. P. 2007b, A&A, 461, 1195
Fornasier, S., Barucci, M. A., de Bergh, C., et al. 2009, A&A, 508, 457
Fornasier, S., Lellouch, E., Müller, T., et al. 2013, A&A, 555, A15
Fortenberry, R. C., Bodewits, D., & Pierce, D. M. 2021, ApJS, 256, 6
Foster, M. J., Green, S. F., McBride, N., & Davies, J. K. 1999, Icar, 141, 408
Fulchignoni, M., Belskaya, I., Barucci, M. A., de Sanctis, M. C., & Doressoundiram, A. 2008, in The Solar System Beyond Neptune, ed. M. A. Barucci, et al. (Tucson, AZ: Univ. Arizona Press) 181
Gibb, E. L., Whittet, D. C. B., Schutte, W. A., et al. 2000, ApJ, 536, 347
Gibb, E. L., Bonev, B. P., Villanueva, G., et al. 2012, ApJ, 750, 102
Gladstone, G. R., & Young, L. A. 2019, AREPS, 47, 119
Guilbert-Lepoutre, A., Besse, S., Mousis, O., et al. 2015, SSRv, 197, 271
Guilbert-Lepoutre, A. 2011, AJ, 141, 103
Guilbert-Lepoutre, A. 2012, AJ, 144, 97
Gunnarsson, M. 2003, A&A, 398, 353
Gunnarsson, M., Bockelée-Morvan, D., Biver, N., Crovisier, J., & Rickman, H. 2008, A&A, 484, 537
Hahn, J. M., & Malhotra, R. 1999, AJ, 117, 3041
Hainaut, O. R., Boehnhardt, H., & Protopapa, S. 2012, A&A, 546, A115
Hanner, M. S., Giese, R. H., Weiss, K., & Zerull, R. 1981, A&A, 104, 42
Hansen, G. B., & McCord, T. B. 2004, JGR (Planets), 109, E01012
Harrington Pinto, O., Kelley, M. S. P., Villanueva, G. L., et al. 2023, PSJ, 4, 208
Harrington Pinto, O., Womack, M., Fernandez, Y., & Bauer, J. 2022, PSJ, 3, 247
Hartogh, P., Lis, D. C., Bockelée-Morvan, D., et al. 2011, Natur, 478, 218
Hsieh, H. H., Jewitt, D., Lacerda, P., Lowry, S. C., & Snodgrass, C. 2010, MNRAS, 403, 363
Hsieh, H. H., Yang, B., Haghighipour, N., et al. 2012, ApJL, 748, L15
Hutsemékers, D., Manfroid, J., Jehin, E., Zucconi, J.-M., & Arpigny, C. 2008, A&A, 490, L31
Ivanova, A. V., Afanasiev, V. L., Korsun, P. P., et al. 2012, SoSyR, 46, 313
Ivanova, O., Skorov, Y., Luk'yanyk, I., et al. 2020, MNRAS, 496, 2636
Ivanova, O., Rosenbush, V., Luk'yanyk, I., et al. 2021, A&A, 651, A29
Ivanova, O., Neslušan, L., Krišandová, Z. S., et al. 2015b, Icar, 258, 28
Ivanova, O., Agapitov, O., Odstrcil, D., et al. 2019a, MNRAS, 486, 5614
Ivanova, O., Luk'yanyk, I., Kolokolova, L., et al. 2019b, A&A, 626, A26
Ivanova, O., Rosenbush, V., Luk'yanyk, I., et al. 2023, A&A, 672, A76
Ivanova, O. V., Dlugach, J. M., Afanasiev, V. L., Reshetnyk, V. M., & Korsun, P. P. 2015a, P&SS, 118, 199
Ivanova, O. V., Luk`yanyk, I. V., Kiselev, N. N., et al. 2016, P&SS, 121, 10

Ivanova, O. V., Picazzio, E., Luk'yanyk, I. V., Cavichia, O., & Andrievsky, S. M. 2018, P&SS, 157, 34
Jackson, A. P., & Desch, S. J. 2021, JGR (Planets), 126, e06706
Jewitt, D. 1990, AJ, 351, 277
Jewitt, D. 2015, AJ, 150, 201
Jewitt, D., Hsieh, H., & Agarwal, J. 2015, in Asteroids IV, ed. P. Michel, F. E. DeMeo, & W. F. Bottke (Tucson, AZ: Univ. Arizona Press) 221
Jewitt, D. 2009, AJ, 137, 4296
Jewitt, D., Ishiguro, M., Weaver, H., et al. 2014, AJ, 147, 117
Jewitt, D. C., & Luu, J. 2004, Natur, 432, 731
Johansen, A., & Youdin, A. 2007, ApJ, 662, 627
Johansen, A., Mac Low, M.-M., Lacerda, P., & Bizzarro, M. 2015, SciA, 1, 1500109
Kane, S. R., Arney, G. N., Byrne, P. K., et al. 2021, JGR (Planets), 126, e06643
Kargel, J. S., & Lunine, J. I. 1998, in Solar System Ices, ed. B. Schmitt, C. de Bergh, & M. Festou (Dordrecht: Kluwer Academic Publishers) Astrophysics and Space Science Library Series 227, 97
Kelley, M. S. P., Woodward, C. E., Gehrz, R. D., Reach, W. T., & Harker, D. E. 2017, Icar, 284, 344
Kiselev, N. N., & Chernova, G. P. 1979, SvAL, 5, 156
Klinger, J., Levasseur-Regourd, A.-C., Bouziani, N., & Enzian, A. 1996, P&SS, 44, 637
Kochergin, A., Zubko, E., Chornaya, E., et al. 2021, Icar, 366, 114536
Korsun, P. P., Ivanova, O. V., & Afanasiev, V. L. 2006, A&A, 459, 977
Korsun, P. P., Kulyk, I. V., Ivanova, O. V., et al. 2010, Icar, 210, 916
Korsun, P. P., Ivanova, O. V., & Afanasiev, V. L. 2008, Icar, 198, 465
Korsun, P. P., Rousselot, P., Kulyk, I. V., Afanasiev, V. L., & Ivanova, O. V. 2014, Icar, 232, 88
Kouchi, A., & Sirono, S. 2001, GeoRL, 28, 827
Kouchi, A., Kudo, T., Nakano, H., et al. 2002, ApJL, 566, L121
Kulyk, I., Korsun, P., Rousselot, P., Afanasiev, V., & Ivanova, O. 2016, Icar, 271, 314
Lambrechts, M., & Johansen, A. 2014, A&A, 572, A107
Le Roy, L., Altwegg, K., Balsiger, H., et al. 2015, A&A, 583, A1
Levison, H. F., Morbidelli, A., Gomes, R., & Backman, D. 2007, in Protostars and Planets V, ed. B. Reipurth, D. Jewitt, & K. Keil (Tucson, AZ: Arizona Press) 669
Li, C., Ingersoll, A., Janssen, M., et al. 2017, GeoRL, 44, 5317
Li, C., Ingersoll, A., Bolton, S., et al. 2020, NatAs, 4, 609
Lis, D. C., Biver, N., Bockelée-Morvan, D., et al. 2013, ApJL, 774, L3
Lis, D. C., Bockelée-Morvan, D., Güsten, R., et al. 2019, A&A, 625, L5
Lodders, K. 2021, SSRv, 217, 44
Luspay-Kuti, A., Mousis, O., Hässig, M., et al. 2016, SciA, 2, 1501781
Madhusudhan, N., Amin, M. A., & Kennedy, G. M. 2014, ApJL, 794, L12
Mahaffy, P. R., Niemann, H. B., Alpert, A., et al. 2000, JGR, 105, 15061
Mandt, K. E., Mousis, O., Hurley, D., et al. 2022, NatCo, 13, 642
Mandt, K. E., Waite, J. H., Lewis, W., et al. 2009, P&SS, 57, 1917
Mandt, K. E., Lustig-Yaeger, J., Luspay-Kuti, A., Wurz, P., Bodewits, D., et al. 2024, SciAd, 10, eadp2191
Marcialis, R. L., & Buratti, B. J. 1993, Icar, 104, 234
Marov, M. Y., & Ipatov, S. I. 2001, ApSSL, 261, 223

Marsden, B. G. 1962, ASPL, 8, 375
Masset, F., & Snellgrove, M. 2001, MNRAS, 320, L55
Masset, F. S., & Papaloizou, J. C. B. 2003, ApJ, 588, 494
Mastrapa, R. M. E., Grundy, W. M., & Gudipati, M. S. 2013, ApSSL, 356, 371
Mazzotta Epifani, E., Dotto, E., Ieva, S., et al. 2018, A&A, 620, A93
McKay, A. J., DiSanti, M. A., Kelley, M. S. P., et al. 2019, AJ, 158, 128
Meech, K. J., & Svoren, J. 2004, in Comets II, ed. M. C. Festou, H. U. Keller, & H. A. Weaver (Tucson, AZ: Univ. Arizona Press) 317
Meier, R., Owen, T. C., Matthews, H. E., et al. 1998, Sci, 279, 842
Micheli, M., Farnocchia, D., Meech, K. J., et al. 2018, Natur, 559, 223
Miles, R., Faillace, G. A., Mottola, S., et al. 2016a, Icar, 272, 327
Miles, R. 2016b, Icar, 272, 387
Miles, R. 2016c, Icar, 272, 356
Morbidelli, A., & Crida, A. 2007, Icar, 191, 158
Mousis, O., Lunine, J. I., Luspay-Kuti, A., et al. 2016, ApJL, 819, L33
Mousis, O., Aguichine, A., Atkinson, D. H., et al. 2020, SSRv, 216, 77
Mousis, O., Aguichine, A., Bouquet, A., et al. 2021, PSJ, 2, 72
Mousis, O., Chassefière, E., Holm, N. G., et al. 2015, AsBio, 15, 308
Mousis, O., Lunine, J. I., Pasek, M., et al. 2009, Icar, 204, 749
Mousis, O., Ronnet, T., & Lunine, J. I. 2019, ApJ, 875, 9
Mueller, B. E. A., Hergenrother, C. W., Samarasinha, N. H., Campins, H., & McCarthy, D. W. 2004, Icar, 171, 506
National Academies of Sciences, Engineering, and Medicine (NASEM) 2023a, Origins, Worlds, and Life: A Decadal Strategy for Planetary Science and Astrobiology 2023-2032 (Washington, DC: The National Academies Press)
National Academies of Sciences, Engineering, and Medicine (NASEM) 2023b, Pathways to Discovery in Astronomy and Astrophysics for the 2020s (Washington, DC: The National Academics Press)
Neslusan, L., Ivanova, O., Husarik, M., Svoren, J., & Krisandova, Z. S. 2016, P&SS, 125, 37
Nesvorný, D. 2011, ApJL, 742, L22
Nesvorný, D., Li, R., Youdin, A. N., Simon, J. B., & Grundy, W. M. 2019, NatAs, 3, 808
Nesvorný, D., Vokrouhlický, D., Dones, L., et al. 2017, ApJ, 845, 27
Notesco, G., Bar-Nun, A., & Owen, T. 2003, Icar, 162, 183
Notesco, G., & Bar-Nun, A. 1996, Icar, 122, 118
Ootsubo, T., Kawakita, H., Hamada, S., et al. 2012, ApJ, 752, 15
Opitom, C., Hutsemékers, D., Jehin, E., et al. 2019, A&A, 624, A64
Ortiz, J. L., Duffard, R., Pinilla-Alonso, N., et al. 2015, A&A, 576, A18
Owen, T., & Bar-Nun, A. 1995, Icar, 116, 215
Owen, T. C., & Bar-Nun, A. 2001, OLEB, 31, 435
Paganini, L., Mumma, M. J., Gibb, E. L., & Villanueva, G. L. 2017, ApJL, 836, L25
Paganini, L., Mumma, M. J., Boehnhardt, H., et al. 2013, ApJ, 766, 100
Patashnick, H. 1974, Natur, 250, 313
Picazzio, E., Luk'yanyk, I. V., Ivanova, O. V., et al. 2019, Icar, 319, 58
Pierens, A., & Nelson, R. P. 2008, A&A, 482, 333
Prialnik, D. 1997, ApJL, 478, L107
Price, E. M., Cleeves, L. I., Bodewits, D., & Öberg, K. I. 2021, ApJ, 913, 9

Protopapa, S., Sunshine, J. M., Feaga, L. M., et al. 2014, Icar, 238, 191
Protopapa, S., Kelley, M. S. P., Yang, B., et al. 2018, ApJL, 862, L16
Rauer, H., Biver, N., Crovisier, J., et al. 1997, P&SS, 45, 799 803
Raymond, S. N., & Izidoro, A. 2017, Icar, 297, 134
Roth, N. X., Milam, S. N., DiSanti, M. A., et al. 2023, PSJ, 4, 172
Rousselot, P., Levasseur-Regourd, A. C., Muinonen, K., & Petit, J.-M. 2005, EM&P, 97, 353
Rousselot, P., Korsun, P. P., Kulyk, I., Guilbert-Lepoutre, A., & Petit, J.-M. 2016, MNRAS, 462, S432
Rousselot, P., Kryszczyńska, A., Bartczak, P., et al. 2021, MNRAS, 507, 3444
Rubin, M., Altwegg, K., Balsiger, H., et al. 2015, Sci, 348, 232
Rubin, M., Altwegg, K., Balsiger, H., et al. 2019, MNRAS, 489, 594
Sarid, G., Volk, K., Steckloff, J. K., et al. 2019, ApJL, 883, L25
Schambeau, C. A., Fernández, Y. R., Lisse, C. M., Samarasinha, N., & Woodney, L. M. 2015, Icar, 260, 60
Schmitt, B., de Bergh, C., & Festou, M. 2012, *Solar System Ices: Based on Reviews Presented at the International Symposium "Solar System Ices"* Vol. 227 (Berlin: Springer Science & Business Media)
Schneeberger, A., Mousis, O., Aguichine, A., & Lunine, J. I. 2023, A&A, 670, A28
Schönbächler, M., Carlson, R. W., Horan, M. F., Mock, T. D., & Hauri, E. H. 2010, Sci, 328, 884
Seccull, T., Fraser, W. C., Puzia, T. H., Fitzsimmons, A., & Cupani, G. 2019, AJ, 157, 88
Seligman, D., & Laughlin, G. 2020, AJ, 896, L8
Seligman, D. Z., & Moro-Martín, A. 2022, ConPh, 63, 200
Seligman, D. Z., Rogers, L. A., Cabot, S. H. C., et al. 2022, PSJ, 3, 150
Senay, M. C., & Jewitt, D. 1994, Natur, 371, 229
Shubina, O., Kleshchonok, V., Ivanova, O., Luk'yanyk, I., & Baransky, A. 2023, Icar, 391, 115340
Shulman, L. 2007, Izvestiya Krymskoi Astrofizicheskoi Observatorii, 103, 209
Sierks, H., Barbieri, C., Lamy, P. L., et al. 2015, Sci, 347, aaa1044
Simon, J. B., Armitage, P. J., Li, R., & Youdin, A. N. 2016, ApJ, 822, 55
Sloan, E. D., & Koh, C. A. 2007, Clathrate Hydrates of Natural Gases (Boca Raton, FL: CRC Press)
Steckloff, J. K., Lisse, C. M., Safrit, T. K., et al. 2021, Icar, 356, 113998
Stevenson, R., Kramer, E. A., Bauer, J. M., Masiero, J. R., & Mainzer, A. K. 2012, ApJ, 759, 142
Sunshine, J. M., A'Hearn, M. F., Groussin, O., et al. 2006, Sci, 311, 1453
Tegler, S. C., Romanishin, W., Consolmagno, G. J., et al. 2005, Icar, 175, 390
Trigo-Rodríguez, J. M., García-Melendo, E., Davidsson, B. J. R., et al. 2008, A&A, 485, 599
Trigo-Rodríguez, J. M., García-Hernández, D. A., Sánchez, A., et al. 2010, MNRAS, 409, 1682
Tsiganis, K., Gomes, R., Morbidelli, A., & Levison, H. F. 2005, Natur, 435, 459
Villanueva, G. L., Mumma, M. J., Bonev, B. P., et al. 2009, ApJL, 690, L5
Waite, J. H., Lewis, W. S., Magee, B. A., et al. 2009, Natur, 460, 487
Walsh, K. J., Morbidelli, A., Raymond, S. N., O'Brien, D. P., & Mandell, A. M. 2011, Natur, 475, 206
Weaver, H. A., A'Hearn, M. F., Arpigny, C., et al. 2008, Asteroids, Comets, Meteors (Houston, TX: LPI) 8216
Whipple, F. L. 1980, AJ, 85, 305
Wierzchos, K., Womack, M., & Sarid, G. 2017, AJ, 153, 230

Wierzchos, K., & Womack, M. 2020, AJ, 159, 136
Willacy, K., Turner, N., Bonev, B., et al. 2022, ApJ, 931, 164
Womack, M., Sarid, G., & Wierzchos, K. 2017, PASP, 129, 031001
Womack, M., Lutz, B. L., & Wagner, R. M. 1994, ApJ, 433, 886
Womack, M., & Stern, S. A. 1999, SoSyR, 33, 187
Wong, M. H., Mahaffy, P. R., Atreya, S. K., Niemann, H. B., & Owen, T. C. 2004, Icar, 171, 153
Wyckoff, S., & Theobald, J. 1989, AdSpR, 9, 157
Youdin, A. N., & Goodman, J. 2005, ApJ, 620, 459
Young, L. A., Braga-Ribas, F., & Johnson, R. E. 2020, TrNSS, 127

Chapter 7

Evolutionary Processes in the Centaur Region

Rosita Kokotanekova, Aurélie Guilbert-Lepoutre, Matthew M Knight and Jean-Baptiste Vincent

Centaurs populate relatively short-lived and rapidly evolving orbits in the giant-planet region, and are believed to be one of the solar system's most complex and diverse populations. Most Centaurs are linked to origins in the dynamically excited component of the trans-Neptunian region, and are often considered an intermediate phase in the evolution of Jupiter-family comets (JFCs). Additionally, the Centaur region hosts objects from various source populations that have different dynamical histories. In this chapter, we focus on the physical processes responsible for the evolution of this heterogeneous population in the giant-planet region. The chapter begins with a brief review on the origin and early evolution of Centaurs that determines their properties prior to entering the giant-planet region. Next, we discuss the thermal, collisional, and tidal processes believed to drive the changes Centaurs experience. We provide a comprehensive review of the evidence for evolutionary changes derived from studies of the activity, physical properties, and surface characteristics of Centaurs and related populations, such as trans-Neptunian objects, JFCs, and Trojans. This chapter reveals a multitude of gaps in the current understanding of the evolution mechanisms acting in the giant-planet region. In light of these open questions, we conclude with an outlook on future telescope and spacecraft observations, detailing how they are expected to elucidate Centaur evolutionary processes.

7.1 Introduction

Centaurs are a population of small solar-system bodies that orbit the Sun in the realm of the giant planets, but outside of 1:1 mean motion resonances with them. These objects populate chaotic orbits with short dynamical lifetimes estimated from a few Myr up to tens of Myr (see Chapter 3 for a review). As was introduced in Chapter 1, the relatively short lifetime of the population requires the Centaur region to be continuously replenished with objects from other reservoirs. The main source

doi:10.1088/2514-3433/ada267ch7

of Centaurs has been identified as the dynamically excited trans-Neptunian population consisting of the dynamically excited or hot classical Kuiper Belt objects (KBOs), the Scattered Disk, and various resonant populations below 50 au (e.g., Nesvorný et al. 2017; Chapter 3). At the other end of Centaurs' dynamical evolution, objects that are not ejected typically evolve toward the inner solar system as Jupiter-family comets (JFCs). This characteristic pathway defines Centaurs as an intermediate evolutionary phase between the trans-Neptunian objects (TNOs) and comets.

Since the discovery of the Kuiper Belt in the 1990s, objects from the TNO-Centaur-JFC continuum have been regarded as valuable probes preserving evidence for the conditions in the early solar system. This promising lead was among the main arguments supporting the spectacular lineup of spacecraft missions to comets (Snodgrass et al. 2024), NASA's New Horizons exploration of the Kuiper Belt (Stern et al. 2015, 2019), as well as the recently proposed Centaur mission concepts described in Chapter 16. Spacecraft missions along with ground-based observational and theoretical advances of the past decades have resulted in a new, more complex view on these populations. It is now widely accepted that JFCs and Centaurs have undergone significant processing throughout their history since formation (e.g., Gkotsinas et al. 2022; Pajola et al. 2024; Filacchione et al. 2024) and should be perceived as complex geological bodies. Naturally, this poses challenges to the notion that we can derive unambiguous constraints on models of the early solar system. Nevertheless, it opens the possibility to acquire a better understanding of the rich variety of evolutionary processes thought to have shaped today's populations.

Centaurs are key for making progress on this task, but to date they remain the least understood population of the TNO-Centaur-JFC continuum. Moreover, the Centaur population also provides a valuable resource for understanding other small-body populations. Centaurs cannot be viewed simply as a phase of JFC evolution. In addition to perturbed TNOs, the Centaur region is thought to host objects originating from the Oort Cloud, as well as the Trojans, Hildas and the outer main asteroid belt (see Section 7.2.1 and Chapter 3). The reader should therefore keep in mind throughout the chapter that the Centaur region can be regarded as an agglomeration of objects with potentially different formation histories and different levels of evolutionary processing.

In this chapter we review the current understanding of the processes governing the evolution experienced by these diverse objects in the giant planet region. The processes described below have been observed or proposed to act on Centaurs to produce physical and chemical changes to the surface layers of individual objects, and as a result to shape the cumulative population properties. In order to focus on discussing processes characteristic of the giant planet region, we will generally stick to the strict Centaur definition requiring orbits enclosed in the giant planet region ($q \geqslant a_{\mathrm{Jupiter}}$, $Q \leqslant a_{\mathrm{Neptune}}$; see Chapter 1).

The current review of Centaur evolution is a natural continuation of an increasing body of work from the past few decades. However, past works predominantly considered Centaurs in the context of their source populations in the trans-Neptunian region (see Davies & Barrera 2004 and Barucci et al. 2008) while active

Centaurs, in particular, were reviewed together with comets (see Festou et al. 2004). Notably, an increasing number of works have focused on discussing Centaurs as part of the TNOs-Centaurs-JFCs continuum (see Levison & Duncan 1997; Jewitt 2015; Peixinho et al. 2020; Jewitt & Hsieh 2024; Fraser et al. 2024). This volume is the inaugural standalone review of Centaurs and this chapter offers the first opportunity to spotlight Centaurs' evolution processes.

To this end, the chapter aims to combine evidence from all aspects of Centaur research in order to review our current understanding of the processes evolving Centaur surfaces and upper layers. This paper is therefore closely related to several other chapters in this issue. In particular, we would like to draw the reader's attention to the chapters that provide comprehensive reviews on the topics that are key for understanding Centaur evolution. We refer the reader to Chapters 2 and 3 for comprehensive reviews on the formation and dynamical evolution of Centaurs. Reviews of the observational evidence discussed in this paper can be found in: Chapter 4, focusing on the physical properties of Centaurs; Chapter 5, surface composition; nearby environment; and Chapter 6, volatile composition. We would like to highlight that this chapter is closely related to Chapter 12. While Chapter 12 focuses on inferring the deep interior properties of Centaurs, the current chapter explores the contribution that evolutionary processes in the giant planet region have on Centaur surface layers and on the overall processing/aging of the population.

The review is structured as follows: in Section 7.2 we provide an overview of the dynamical pathways of Centaurs and their links to other solar-system populations. Section 7.3 reviews the physical processes that are thought to drive the evolution of Centaurs in the giant planet region, followed by Section 7.4 where we present the observational evidence motivating and constraining the physical models of Centaur evolution. We conclude the chapter by discussing the advances of Centaur evolution we foresee for the coming years (Section 7.5) and a chapter conclusion (Section 7.6).

7.2 The Life Cycle of Centaurs

In this chapter, we explore the evolutionary processes that shape the Centaurs in the present epoch. This, however, is a notoriously complex endeavor. The Centaur region is a melting pot of small bodies with different origins, dynamical histories, and past processing. Each object reflects not only the conditions at its formation in the early solar system, but also the subsequent alterations affecting it throughout its lifetime. We begin our effort to disentangle the complexity of this region with a short census of the Centaur region. We discuss the life cycle of Centaurs and identify the other small-body populations likely to provide key evidence for Centaur evolution. In the interest of space, this review omits a number of details, such as specific formation mechanisms and the exact pathways of dynamical evolution, which the interested reader can find in other chapters (see Chapter 3).

7.2.1 Origins of Centaurs

The provenance of Centaurs can be traced back to a formation in the outer planetesimal disk at $\sim$20–30 au from the Sun (Nesvorný 2018). Following the gas disk dissipation (within the first 10 Myr of the solar system formation), this region, also known as the primordial Kuiper Belt (PKB), is thought to have undergone extensive dispersal. Our understanding of this period ~4.2 Gyr ago comes predominantly from the family of planetary instability models Nesvorný (2018) derived from the original Nice model by Tsiganis et al. (2005). According to these dynamical simulations, most objects from the PKB (with the exception of Cold Classical TNOs which have remained on approximately the same orbits since formation) were ejected into interstellar space. A small fraction of the perturbed planetesimals remained in the solar system and populated today's asteroid belt, Jupiter and Neptune Trojans, irregular satellites of the giant planets, dynamically excited TNOs, and the Oort Cloud (Figure 7.1).

At the current epoch, a variety of destabilizing mechanisms can act on objects stored in the Trans-Neptunian region (Nesvorný et al. 2017) to force them back into the inner regions of the solar system. Due to gravitational interactions with the giant planets, they enter a regime of chaotic motion that gradually reduces their perihelion distance until they reach low-inclination Neptune-crossing orbits, followed by a dynamical cascade that supplies the giant planet region with icy objects now called Centaurs (Duncan & Levison 1997). Pathways from the outer to the inner solar system can be very complex (e.g., Fraser et al. 2024), described by two modes of migration through the giant planet region: either dynamical chaos or mean motion

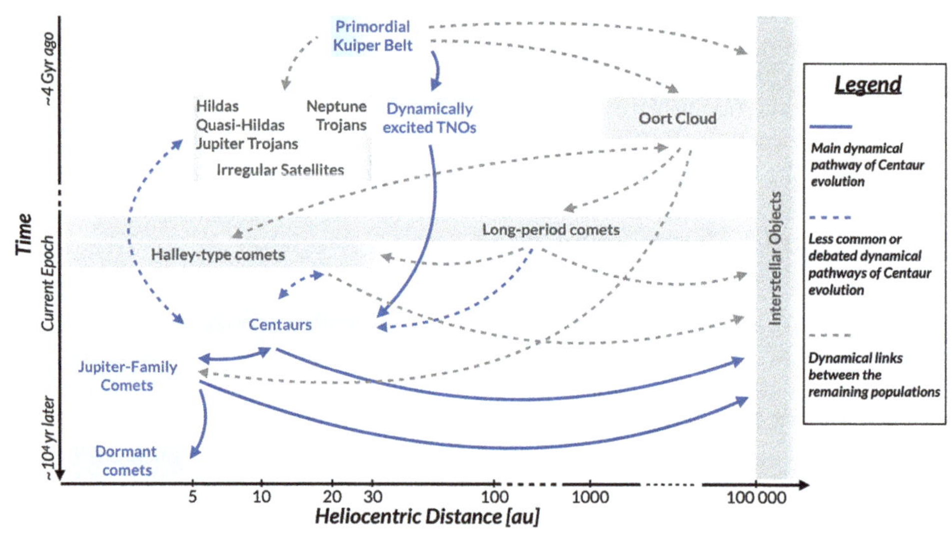

Figure 7.1. Schematic representation of the dynamical evolution of Centaurs and the minor-planet populations related to them. This diagram highlights the main pathway of the TNOs-Centaurs-JFCs continuum (blue arrows) and depicts the complex dynamical links of Centaurs with other small-body populations. The approximate range of typical orbits is shown by the horizontal extent of a given box. The blue boxes indicate the populations representing the main stages of the Centaur life cycle. In the cases when the population density and/or the definition of a given population varies across different works the shading of the boxes indicates the uncertainty in the heliocentric range. For simplicity, Hildas, Quasi-Hildas, Trojans and irregular satellites have all been depicted together within the heliocentric distance range of the giant planet orbits.

resonance hopping (Bailey & Malhotra 2009; Seligman et al. 2021). These pathways can be fast or rather lengthy if the object remains trapped in the Uranus-Neptune region, with characteristic dynamical timescales reaching up to gigayears (Di Sisto & Brunini 2007). On average however, Centaurs remain in the giant planet region for $\sim$10 Myr (Levison & Duncan 1997; Tiscareno & Malhotra 2003; Volk & Malhotra 2008; Bailey & Malhotra 2009). Each Centaur thus follows a unique and distinct dynamical track to reach the orbit where it is currently observed. When the orbit becomes controlled by interactions with Jupiter, multiple passages from the giant planet region to the inner solar system (where Centaurs would typically be called Jupiter-Family Comets) are possible (Sarid et al. 2019; Seligman et al. 2021; Guilbert-Lepoutre et al. 2023b). This constitutes the main dynamical pathway of Centaurs and gives rise to the most prevalent view of Centaurs as a part of the "TNO-Centaur-JFC continuum." Even though the TNO region is considered the main Centaur reservoir (Di Sisto & Rossignoli 2020), the Centaur region is likely to also host objects sourced from other populations. Minor contributions are thought to originate from Jupiter Trojans (see Di Sisto et al. 2019 and references therein), Neptune Trojans (e.g., Horner & Lykawka 2010), Quasi-Hildas (Gil-Hutton & García-Migani 2016), and Haumea family members (Lykawka et al. 2012). Other possible source populations, specifically of high-inclination and retrograde Centaurs (see Di Sisto & Rossignoli 2020; Kaib & Volk 2024) are proposed to be Oort-Cloud comets (Brasser et al. 2012; Fouchard et al. 2014, de la Fuente Marcos & de la Fuente Marcos 2014) and escaped asteroids from the inner solar system (Greenstreet et al. 2020).

Most, if not all, of these Centaur source populations can be traced back to a formation in the PKB (see Nesvorný 2018). It is, therefore, plausible to expect that all Centaurs still share some common properties. However, the different Centaur source populations have evolved in different parts of the solar system and have therefore undergone various levels of collisional and/or thermal processing before entering the giant planet region. Once in the Centaur region, the heterogeneity is expected to increase further due to the differences in the individual objects' dynamical paths. This suggests heterogeneity among Centaurs that goes beyond the established classifications such as active/inactive (see Chapter 8 and red/neutral Centaurs (see Section 7.4.3 and Chapter 5) To understand this complexity, it is therefore important to consider the processes that have determined the properties of incoming Centaurs.

7.2.2 Early Evolution

In this section, we briefly review the early evolution of Centaurs' progenitors within the PKB, during the epoch of planetary migration and subsequently within the Centaur reservoirs. The formation and very early evolution of PKB planetesimals prior to and during planet migration determined the initial chemical, physical, and thermal properties of the Centaur parent populations. Constraining the degree of processing sustained during the earliest stages of a Centaur's evolution is extremely difficult. To provide a quantitative assessment, we would need to have identified their formation mechanism, the time at which these objects formed, and their initial composition. Most of our current understanding thus relies on theoretical expectations derived from models that are based on the properties of the currently observed objects and populations, and are applied to explore the free parameter space corresponding to the PKB and the outer solar system reservoirs. Bearing these caveats in mind, in the following, we outline a few aspects of the early evolution of Centaur progenitors that are most relevant to the discussion of today's evolution in the giant planet region.

7.2.2.1 Early Thermal Evolution

Thermal processing during the early solar-system stages includes internal heating by radioactive elements (e.g., Prialnik et al. 1987; Haruyama et al. 1993). The possible outcomes of this process strongly depend on the sizes of the objects, their thermal conductivities, compositions (in particular their dust-to-ice mass ratios), and the timing of their formation. With low thermal conductivities, heating is more effective and deep interiors are thus more likely to reach temperatures triggering phase transitions (e.g., Prialnik & Podolak 1995). Additional sources of energy can also be considered: for example accretional energy, although most likely insignificant for most small objects, could yield more diversity to the expected outcomes if added to radiogenic heating (e.g., Merk & Prialnik 2006). The aforementioned studies suggest that it is more probable to find pristine material in the outermost layers of icy objects rather than in their deep interiors. By considering the possibility of collisional heating in addition to radiogenic heating, Golabek & Jutzi (2021) estimated that the preservation of highly and moderately volatile species such as CO and CO_2 would only be possible for objects smaller than 20 km that formed relatively late ($>$3.5 Myr).

Finally, insolation remains a prevalent cause for thermal processing. Recently, simulations by Davidsson (2021), Steckloff et al. (2021), and Parhi & Prialnik (2023) suggested that all icy objects would lose their highly volatile content present in the form of pure ice on timescales ranging from a few thousand to 200 Myr. Those species would need to be trapped in a more stable phase (like water ice or CO_2 ice in amorphous or crystalline form) to survive on longer timescales. We expect that chemical differentiation should take place on longer timescales, as equilibrium temperatures met in the trans-Neptunian region ($\sim$30–50 K) exceed the sublimation temperature of pure icy species like CO, N_2, or O_2 (De Sanctis et al. 2001; Choi et al. 2002; Merk & Prialnik 2003; Malamud & Prialnik 2015; Kral et al. 2021; Lisse et al. 2021; Loveless et al. 2022; Malamud et al. 2022; Parhi & Prialnik 2023). As a consequence, Centaurs that originate from the Kuiper Belt and the Scattered Disk are expected to be depleted to some degree of their most volatile free-condensed (e.g., pure ice, not trapped) species, within at least a subsurface layer of a few kilometers, even before they enter a pathway that brings them toward the inner solar system. The fact that such species are currently observed in active Centaurs (see Chapters 6 and 14) indicates that we do not yet have a full understanding of how the most volatile species are retained for the age of the solar system.

7.2.2.2 Early Collisional Evolution

Another highly-debated aspect of the early evolution of Centaur progenitors is their past collisional evolution. On the one hand, the characteristic physical properties of JFCs (e.g., their shapes, high porosities, low tensile strength, high volatile content, etc.) have been interpreted as clear signatures of their primitive nature (Davidsson et al. 2016). On the other hand, there is building evidence that JFCs (and therefore Centaurs as their progenitors) are the outcome of collisional processing in the PKB, during the disk instability phase and while residing in the Scattered Disk (Morbidelli & Rickman 2015; Bottke et al. 2023). Furthermore, in support of this view, it has been shown that properties that were previously thought to exclude comets from being collisionally evolved, such as high porosities and the presence of volatiles, can survive catastrophic disruptions (e.g., Jutzi & Benz 2017; Schwartz et al. 2018; Steckloff et al. 2023). For more details on the subject of early collisional evolution, we refer the reader to Chapter 4.

7.2.2.3 Space Weathering

During their residence in the outer solar system, icy objects are subject to long-term irradiation by solar wind, UV photons, and cosmic rays, as well as by the impacts of micrometeorites. The effects of such continuous bombardment are known to affect the surface composition, color, albedo, and structure of the surface material (e.g., Hapke 2001; Cooper et al. 2003; Hudson et al. 2008; Pieters & Noble 2016). Prolonged ion irradiation can also act to amorphize water ice (e.g., Dartois et al. 2015). The impact of space weathering on surface chemistry strongly depends on the initial composition, since the net effect is to break molecular bonds to produce irradiation products in the form of new surface and subsurface molecules.

Centaurs are subjected to space weathering during their prolonged residences in their reservoirs. Therefore, it could be expected that the surfaces of recently escaped Centaurs should reflect the influence of long-term irradiation, unless they are modified by some other process once they enter the giant planet region (see Section 7.3). However, it is not considered likely that space weathering during the Centaur phase is responsible for noticeable surface evolution. Due to its location in the middle heliosphere, the Centaur region experiences moderate irradiation from solar energetic ions (whose flux decreases with increasing heliocentric distance) and galactic cosmic rays and ion fluxes from the interstellar environment (which increase further away from the Sun). For the radiation dose at $\sim$1 μm depth for objects at 20 au estimated by Hudson et al. (2008), and the saturation doses for volatile ice irradiation found by Brunetto et al. (2006), the irradiation timescale (i.e., the time needed to change the properties of the material down to a given depth) is between 50 Myr and 1 Gyr (Wong & Brown 2016). Similarly long timescales of 10^8 years have been estimated for forming a meter-deep irradiation crust depleted of volatiles (Shul'man 1972; Luu & Jewitt 1996). These timescales are longer than both the typical dynamical lifetime in the Centaur region ranging from a few Myr up to tens of Myr (Peixinho et al. 2020), and the really short blanketing timescales (the time over which a comet's surface is covered by a layer of fallback coma material, i.e., Jewitt 2002, 2015). It is therefore unlikely that space weathering drives the surface evolution during the Centaur phase and it is not discussed further in the following section.

7.3 Processes Driving the Evolution of Centaurs

7.3.1 Thermal Processing

Thermal processing is identified as the main driver of Centaur evolution in the giant planet region. The course of an object's thermal evolution is tightly connected to the long-term orbital evolution, which determines its thermal environment. As insolation increases with decreasing heliocentric distance, Centaurs periodically reach regions where surface and subsurface temperature may favor various phase transitions (described below) that may result in observable cometary activity. As of now, Centaurs with perihelia located beyond Saturn appear to lack detectable activity (Cabral et al. 2019; Li et al. 2020; Lilly et al. 2021), which is generally taken as a sign that sublimation of hypervolatile species such as CO or CH_4 is not driving the activity of Centaurs at these distances in contrast to what has been inferred for Long period comets (e.g., Jewitt et al. 2021). Crystallization of amorphous water ice has instead been proposed as an efficient source of activity (Jewitt 2009; Prialnik & Jewitt 2024), at least in the 5–10 au region and possibly up to 12 au (Guilbert-Lepoutre 2012). Beyond about 10 au, crystallization rates strongly decline and amorphous water ice crystallization is thought to be less efficient than sublimation of CO_2 ice or the sublimation of segregated hypervolatile species from CO_2 ice (Davidsson 2021). The latter process, in which hypervolatiles such as CH_4 (Luna et al. 2008), N_2 (Satorre et al. 2009), or CO (Simon et al. 2019) are trapped in CO_2 ice and released at temperatures above their individual sublimation temperatures, is similar to the trapping and release of molecules during the crystallization of amorphous water ice.

As the heat wave and phase-transition fronts progressively move inward below the surface of Centaurs (see an example in Figure 7.2), the composition of the top layer could change to a degree that prevents any further activity. For example, Guilbert-Lepoutre (2012) suggested that activity driven by the crystallization of amorphous water ice could only be sustained for 10^4 yrs at most, a small fraction of the typical dynamical timescale. This implies that Centaurs that are currently active should have recently undergone a sunward orbital change (e.g., Lilly et al. 2024) allowing for phase transitions to be triggered deeper below the surface and producing some observable cometary activity. Empirical data seem to confirm that active Centaurs have undergone drastic drops in their perihelion distances within the past 10^2 to 10^3 yrs (Fernández et al. 2018; Lilly et al. 2024).

The emerging picture is that the activity of Centaurs should be closely related to their orbital evolution. For example, Cabral et al. (2019) argued that crystallization-driven activity should be observed for objects dynamically new to the Jupiter-Saturn region, as others having previously stayed in this region would have exhausted their amorphous water ice content in the near-surface layers. The extent of the thermal modifications sustained during the dynamical evolution of Centaurs is only starting to be quantified. As a result of the complex and varied dynamical pathways from the outer solar system, it is impossible to constrain the exact orbital evolution that any

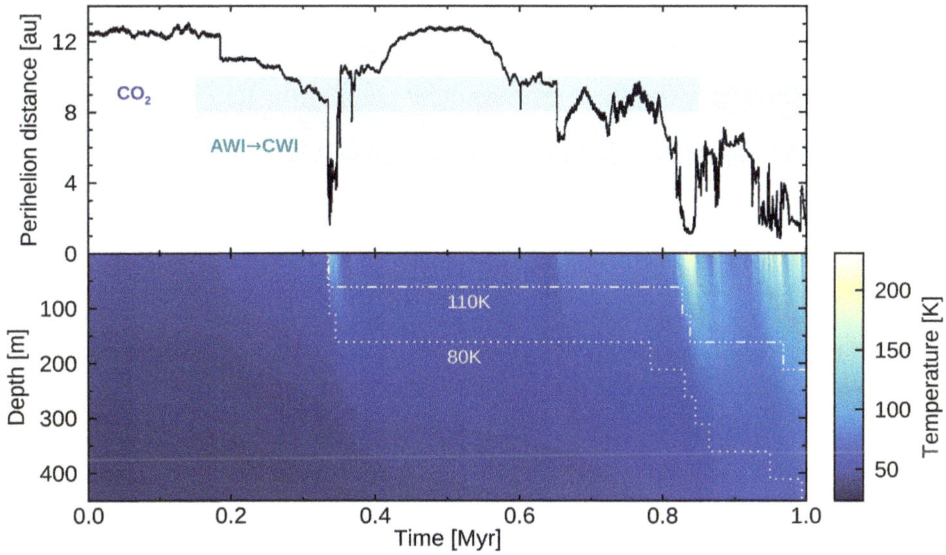

Figure 7.2. Example dynamical and resulting thermal evolution for an object leaving the trans-Neptunian region, reaching the giant planet region, and eventually reaching a JFC orbit (Nesvorný et al. 2017; Gkotsinas et al. 2022). The top panel shows the evolution of the perihelion distance as a function of time for the last Myr of evolution before the object is ejected out of the solar system. Shaded areas illustrate the most likely drivers for Centaurs' activity: CO_2 segregation or sublimation (in blue), amorphous water ice (AWI) to crystalline water ice (CWI) transition (in green). We note that these phase transitions can also occur at smaller heliocentric distances. The bottom panel shows the distribution of the internal temperature as a function of depth and time, resulting from the orbital evolution. The depths of two isotherms are given as guides: a dotted line for 80 K and a dash-dotted line for 110 K.

given Centaur has followed beyond the last interaction with Jupiter, nor is it possible to know whether it is on its first pass to the inner solar system. Constraining the degree of thermal processing sustained by each individual Centaur is, consequently, extremely challenging. We expect that distinct individual orbital pathways should result in significant differences in the internal composition and structures when Centaurs are observed. As water ice does not efficiently sublimate in the giant planet region, erosion is limited and chemical differentiation could occur down to several hundred meters (depending on thermo-physical properties; Guilbert-Lepoutre et al. 2016). An effort to provide more realistic constraints on the degree of thermal processing sustained during the Centaur phase has recently been made, by coupling both the thermal and the orbital evolution of Centaurs and JFCs. Indeed, Gkotsinas et al. (2022) showed that as a result of the stochastic nature of the dynamical trajectories, each Centaur may experience multiple, long-lasting heating episodes, leading to chemical alteration of their upper layers, down to several hundred meters, allowing for a substantial depletion of their super-volatile content (Figure 7.2). The relative lack of compositional signatures in the coma of active Centaurs is impeding further progress, a status that is currently changing with JWST observations of these objects.

Recently, attempts to identify Centaurs on the edge of transitioning to a JFC-like orbit have been made. These would allow for an efficient and detailed investigation of how dynamical and thermal evolution alters comet nuclei before they become JFCs (Sarid et al. 2019; Steckloff et al. 2020; Kareta et al. 2021; Seligman et al. 2021), especially as potential targets for future space missions. However, Guilbert-Lepoutre et al. (2023b) showed that most objects found near the orbit of Jupiter are statistically more altered than the rest of the Centaur population, due to thermal processing sustained in previous stages of dynamical evolution since these objects might have already been JFCs in the past. In addition to long-term effects discussed above, seasonal and diurnal patterns of activity should act on Centaurs as they do for JFCs. For these smaller timescales, the orientation of the spin axis is key (see Chapter 4), as it can cause insolation patterns on the surface that trigger thermal behavior relatively unique to each active object.

7.3.2 Impacts and Cratering

Cratering is a prominent shaping process in the solar system, and every object experiences impacts during its lifetime. Throughout the solar system, craters are used to date the surfaces of objects, to derive the size distribution of the impactors and to constrain the dynamical history of different populations. However, identifying craters on the surfaces of active bodies is complicated by activity-driven erosion. Thus, there are few unambiguous crater detections on the surfaces of comets visited by spacecraft besides the crater produced by Deep Impact on 9P/Tempel 1 (see review by Pajola et al. 2024, for details). In this section, we consider the limited time period during which an object is classified as a Centaur. As the population of Centaurs is not well constrained, it is difficult to establish reliable probabilities of collisions within the population. The best alternative is to consider other, better calibrated sources of impactors.

In the outer solar system, the main reservoir of projectiles is the population of eccentric comets, which are known to collide with giant planets and their moons. Impacts on Jupiter are routinely observed and amount to about one event per year for 10 m diameter projectiles (Hueso et al. 2010). A recent review of cratering records on the moons of giant planets, combined with dynamical models of the solar system, allows an estimate of the collision rate for small objects impacting the moons (Nesvorný et al. 2023). The impact rate and velocity are variable and depend on both heliocentric and planetocentric distances. For projectiles larger than 1 km, the average time between collisions varies from 2.7 Myr to 42 Myr, which is comparable to the dynamical lifetime of Centaurs. At the same heliocentric distance, collisions on objects unbound from planets are less frequent as the impact rate on a moon is magnified by the focusing presence of its host planet. We can therefore conclude that larger impacts, which would reshape or destroy Centaurs, are very unlikely to occur in the dynamical lifetime of these objects. In other words, the current population is not shaped by impacts. Smaller scale collisions are possible, and future missions to Centaurs may observe craters in the surface of these objects, but most impacts will have occurred in earlier dynamical phases (see Section 4.5.2 of Chapter 4).

7.3.3 Close Encounters

Due to gravitational interactions with the giant planets, Centaurs' orbits are chaotic, and an object's orbital elements can vary considerably during its dynamical lifetime. One consequence is that Centaurs may experience close encounters ($<$5 planet radii) with the giant planets. Earlier dynamical calculations for the observed population showed that at least 10% of the known Centaurs will encounter a giant planet within its Roche limit (Tiscareno & Malhotra 2003; Hyodo et al. 2016). However, more recent works suggest that close encounters with the giant planets are very rare (e.g., Araujo et al. 2016; Safrit et al. 2021). The orbital integration performed and discussed in Chapter 3 suggests that only $\sim$2% of the Centaur population have an encounter within the Roche limit of a giant planet in the age of the solar system.

Such a close encounter would induce strong tidal forces in the Centaur, which can have significant effects on its surface morphology and global shape. During a close encounter, the gravitational pull of the planet creates a differential force that may stretch the Centaur along an axis aligned with the planet, or at the very least lower the effective gravity in areas facing toward and away from the planet. As a consequence, the local slope of the terrain, measured as the angle between the surface normal vector and the effective gravity (self gravity + tidal force + centrifugal force), can be modified. Even if the body is not stretched by tides, the direction of the effective gravitational acceleration will change during the encounter, effectively raising or lowering the local slopes by a few degrees. This effect has been studied and modeled for close encounters of Near Earth Objects with our planet, for instance asteroid (99942) Apophis (Kim et al. 2023). While modeling this process for Centaurs remains challenging given our limited knowledge of the mechanical properties of these objects, simulations of tidal effects on Near Earth asteroids show that tidal forces could be contributing to Centaur evolution (Kim et al. 2023). At their most extreme end, tidal forces can lead to a complete shearing of the body and catastrophic disruption. A prominent example of such an event is comet D/1993 F2 (Shoemaker-Levy 9), which was disrupted by a close encounter with Jupiter in July 1992 and eventually collided with the planet two years later (Nakano et al. 1993).

Recent observations have revealed the existence of icy rings around two large Centaurs: (10199) Chariklo (Braga-Ribas et al. 2014) and (2060) 95P/Chiron (Ortiz et al. 2015). Understanding the mechanisms that could lead to the formation of Centaur rings is still in its early phase and a few scenarios that involve a close encounter with a giant planet are under consideration (see Chapter 9 for a review). For instance, one of the proposed hypotheses (Hyodo et al. 2016) is that a close encounter with a giant planet could have stripped the icy mantle of a Centaur. The resulting debris would reassemble within the Roche limit of the remaining core, forming a ring.

While the mechanisms responsible for ring formation remain poorly constrained (see Chapter 9), the first attempts to investigate the possible connection between close encounters and ring survivability provide instructive results. Araujo et al. (2016, 2018) performed numerical simulations to study the encounter probabilities of Chariklo and Chiron with the giant planets. Their work identified that tidal forces are less efficient at disrupting the ring systems of more massive bodies, as well as of objects with high-inclination and low-eccentricity orbits. Recently, the possibility that a close encounter with another small body could disturb a Centaur's ring system was considered by Ikeya & Hirata (2024). They found that while it is possible for an encounter with a similar-sized body to perturb an object's rings, such disruptive close encounters are extremely rare and unlikely to occur in 4 Gyr.

7.4 Evidence for Evolutionary Processes

The overview in Section 7.3 clearly exposes the limitations of our current knowledge regarding the evolutionary processes in the Centaur region. In this section, we focus on key observational and modeling results. Our goal is to review how they have shaped the present understanding of evolution mechanisms and to highlight some intriguing manifestations of Centaur evolution worthy of future investigation.

7.4.1 Activity

A considerable fraction (estimated at 10%–15%) of the known Centaurs display comet-like activity (Jewitt 2009; Bauer et al. 2013), which is driven by the dynamical and consequently thermal evolution of the object (see Section 7.3.1). Hypothetically, Centaur activity opens the possibility to characterize the volatile content of the population. However, we know from a modeling point of view that single measurements of gas production rates in the coma of an active object can be misleading in relation to the abundance of volatile species inside that object's nucleus (e.g., Benkhoff & Huebner 1995; Prialnik 2006). This diagnosis can be extended to Centaurs, which evolve in a region of the solar system where critical phase transitions for a variety of volatile species occur, so that relative abundances in their coma do not necessarily depend only on heliocentric distance, but also on seasonal and possibly diurnal cycles (Guilbert-Lepoutre et al. 2024). Centaur nuclei can furthermore be heterogeneous in composition (as recently evidenced by the JWST observations of 29P/Schwassmann-Wachmann 1 by Faggi et al. 2024), either because they formed heterogeneously, or because the layer of material contributing to the observed activity has inevitably developed a non-uniform structure and composition as a result of prior thermal evolution (see Guilbert-Lepoutre et al. 2024 and Section 7.3.1 above).

Detecting any volatile signature in the comae of active Centaurs has proven elusive up to now because the main process by which gas is observable is through its fluorescence excitation (see reviews by Biver et al. 2024; Bodewits et al. 2024). The combination of large heliocentric and geocentric distances combined with very low levels of activity makes the direct detection of volatiles extremely challenging. For instance, no Centaur has been detected to be active beyond $\sim$12 au (e.g., Li et al. 2020), and no definitive detection of gaseous CO was made in the pre-JWST era, except for 29P (see Chapter 10 for a detailed review of 29P's activity characterization). Marginal detections of CO were reported for Chiron (Womack & Stern 1999; Womack et al. 2017), and (60558) 174P/Echeclus (Wierzchos et al. 2017). These observations are consistent with the activity of Centaurs not being primarily driven by the sublimation of CO (Drahus et al. 2017) although neither Chiron nor

Echeclus exhibited significant activity when the CO observations were obtained. A few fragment (CN) and ionic (CO^+, N_2^+) species have been detected in 29P (Cochran & Cochran 1991; Korsun et al. 2008) but we highlight that this object has exceptionally high activity. The recent detections of gas signatures in the coma of several active Centaurs by JWST are game changers for understanding how the evolutionary processes work in the giant planet region, and to pinpoint the mechanisms of activity. For instance, CO_2 was detected in the coma of active Centaur 39P/Oterma (Harrington Pinto et al. 2023), and both CO and CO_2 are detected in the coma of 29P (Faggi et al. 2024). Emission features from CO_2 and CH_4 are detected in the spectrum of 95P/Chiron, suggesting a substantive level of activity, although the coma itself is not directly detected (Pinilla-Alonso et al. 2024). Only a careful analysis of this object's phase curve provides hints for a new epoch of activity (Dobson et al. 2024). We point to Chapters 6, 8, and 14 for further discussion.

Patterns of activity can yield information about the surficial layer of material, as a number of Centaurs have been reported to be active several times. For instance, the activity of 29P has been documented for several decades, as this Centaur appears to be nearly always active with dramatic outbursts reported on a regular basis (see Chapters 10 and 11 for comprehensive reviews). Chiron has been reported active at least twice since its discovery: a first period of activity lasted $\sim$6 yrs when the object was between 9 and 12 au (see review by Womack et al. 2017), and a more recent epoch of outbursts in 2021 near aphelion at $\sim$18.8 au (Dobson et al. 2021; Ortiz et al. 2023), with the activity from 2021 apparently continuing through at least early 2023 (Dobson et al. 2024). A puzzling aspect of Chiron is that basic characteristics such as its activity, near-infrared spectrum, brightness, and colors vary with no clear correlation with heliocentric distance (Womack et al. 2017, and references therein). The detection of a ring system around Chiron has provided new insights into these unusual changes (Ortiz et al. 2015; Ruprecht et al. 2015), however, these cannot explain the occasional outbursts and development of a coma. Another Centaur, Echeclus is characterized by repeated outbursts of activity, with at least five events reported to date (see Rousselot et al. 2021, and references therein). Each outburst could have its own origin (e.g., fragmentation events, collisions), however outbursts could also reflect internal heterogeneities as described by Rosenberg & Prialnik (2010). Indeed, those authors show that non-uniform structures (both in composition and thermal properties) can cause erratic behaviors, with outbursts of diverse intensities and durations being triggered at any point of the orbit, even at large heliocentric distances, with gas production rates that can even exceed those expected at perihelion. Continuous monitoring of active Centaurs, coupled with gas detections in their comae, are therefore necessary to provide crucial insights into the mechanisms driving the activity evolution of this population.

7.4.2 Physical Properties

7.4.2.1 Size and Shape

The Centaur size distribution and how it compares to other populations is discussed in Chapter 4. That chapter mainly focuses on the extent to which the size distribution of today's Centaurs is informative about the original planetesimal properties. Here, we pose a different question: do the sizes of Centaurs contain any indications of significant evolutionary processing during the Centaur phase? In Section 7.3 we discussed tidal interactions and impacts. In addition, the sizes and shapes of Centaurs can also be modified by outbursts, rotationally-driven instabilities and splitting. While each of these processes can influence the fate of an individual object, it is unclear whether a large fraction of the population is subject to significant size modifications during their lifetime in the Centaur region.

Answering this question is unlikely to be possible with current data given the uncertainties in the size distributions of Centaurs and the other populations from the main evolutionary pathway (excited TNOs and JFCs, see Chapter 4). A major challenge is in detecting similarly sized objects for comparison; most JFCs are much smaller than the known Centaurs, while most TNOs are larger. Future studies, comparing the small members of each population are essential (see Section 7.5).

Considering the global size distribution of potential impactors, we established in Section 7.3.2 that there is no evidence for collisional evolution playing a major role in shaping the current size distributions of objects while in a Centaur phase. Other effects, like sublimation-driven activity, are even less likely to modify this size distribution. We have so far not observed outbursts large enough to indicate a total disruption of a Centaur (see Chapter 8) and observations indicate that a dust-production rate similar to that observed for 29P (e.g., Stansberry et al. 2004) would not be efficient at modifying the size of a Centaur such as 29P within the typical dynamical lifetime in this region. It is therefore likely that the current size distribution is a mix of the size distributions inherited from the parent groups of the observed objects, and does not inform us on processes taking place in the Centaur phase.

To study Centaur evolution, it is perhaps more informative to consider the global shape rather than size, especially when it departs from a simple ellipsoid. Of particular interest are objects like binaries (either in contact or separated), or Centaurs with rings, as these features can be the signatures of recent large scale reshaping events and would then bring insights into the physical and mechanical response of the objects.

Radar and spacecraft observations have revealed that bilobate shapes could be widespread among comet nuclei. Out of the seven comets whose shapes have been determined, five appear bilobate (Keller et al. 2015; Hirabayashi et al. 2016; Safrit et al. 2021). Binaries are also common in the TNO region (Stephens & Noll 2006; Noll et al. 2020), see also the discussion in Bernardinelli et al. (2023). However, the absolute percentage of binaries among TNOs transitioning to JFCs is unknown, as is their formation mechanism, but we know that bilobate small bodies may be formed at large heliocentric distances, as evidenced by the images of Arrokoth returned by NASA's New Horizons mission (Spencer et al. 2020). Two other TNOs crossing the Centaur region, are also known to be binaries: (42355) Typhon and Echidna (Noll et al. 2006) and (65489) Ceto and Phorcys (Grundy et al. 2007). Statistical analysis of the available observations sets a 3σ limit $< 8\,\%$ for the fraction of binaries in the Centaur population (Li et al. 2020).

Although the data are too sparse to conclude that there is an excess of bilobate objects in the JFC population, compared to their progenitor groups, it is worth considering whether there could exist a mechanism to transform singular objects into binaries before they become JFCs. A promising candidate would be rotational instability triggered by sublimation torques (Safrit et al. 2021), but it remains speculative until the detection of a bilobate and/or fast-rotating Centaur. We discuss rotational stability in more details in the next section.

7.4.2.2 Spin State, Density and Strength

As has been discussed in detail for JFCs (e.g., Knight et al. 2024), Long period comets (e.g., Jewitt 2022), and Centaurs (Chapter 4), activity-driven torques can change an objects' spin state significantly and even lead to rotational instability or breakup. The rotational period of small bodies can thus be used to estimate a lower bound for their internal cohesion. Beyond a critical spin rate, the resulting tensile stresses overcome gravity and cohesion, leading to a catastrophic disruption of the object. For bodies with extremely low cohesion (e.g., as observed with asteroid rubble piles), balancing gravitational and centrifugal accelerations on the surface gives a typical critical period of 2.2 h: $P_0 = \sqrt{\frac{3\pi}{G\rho}}$ with ρ the density (typically 2300 kg m^{-3}), and G the gravitational constant. We do not know the density of Centaurs but it is reasonable to consider the range [500–2500] kg m^{-3}, i.e., comet-like to Trojan-like. For the lowest density, the critical period becomes 3.3 h. Any object spinning at a faster rate will start shedding material from its surface, and eventually be destroyed. This is a well accepted hypothesis to explain the dearth of fast rotating asteroids at sizes larger than a couple hundred meters (e.g., Pravec & Harris 2000). We note that this preceding discussion is only for gravity-dominated objects; data are insufficient to speculate on strength-dominated objects.

When considering cohesion and inner structure, calculating the spin limit requires more complex models which properly account for the distribution of stresses (e.g., Scheeres & Sánchez 2018; Li & Scheeres 2021). However, one can approximate the maximum rotation rate within an order of magnitude with $\omega_c = \sqrt{\frac{2Y_t}{\rho d^2}}$ where Y_t is the tensile strength and d is the object diameter (Sánchez & Scheeres 2014; Safrit et al. 2021). The critical period is then simply $P_c = 2\pi/\omega_c$.

Very little is known about the mechanical properties and internal structure of Centaurs, and we only have measurements of size and rotation period for 16 objects (Chapter 4). But we can nevertheless calculate what would be the minimum tensile strength required by Centaurs to withstand their current observed spin rate. Although we do not have measurements of their density, it is reasonable to assume that it lies somewhere in between the typical densities of comets (500 kg m^{-3}) and Trojans (2500 kg m^{-3}). Using this range, we calculated the lower bound for the tensile strength of all objects (Figure 7.3). We find that even the fastest observed Centaurs could sustain their current rotation rates with as little as 1 Pa of cohesion, which is lower than the typical tensile strength of objects of comparable sizes and heliocentric distances, see discussion in Biele et al. (2022). The actual tensile strength

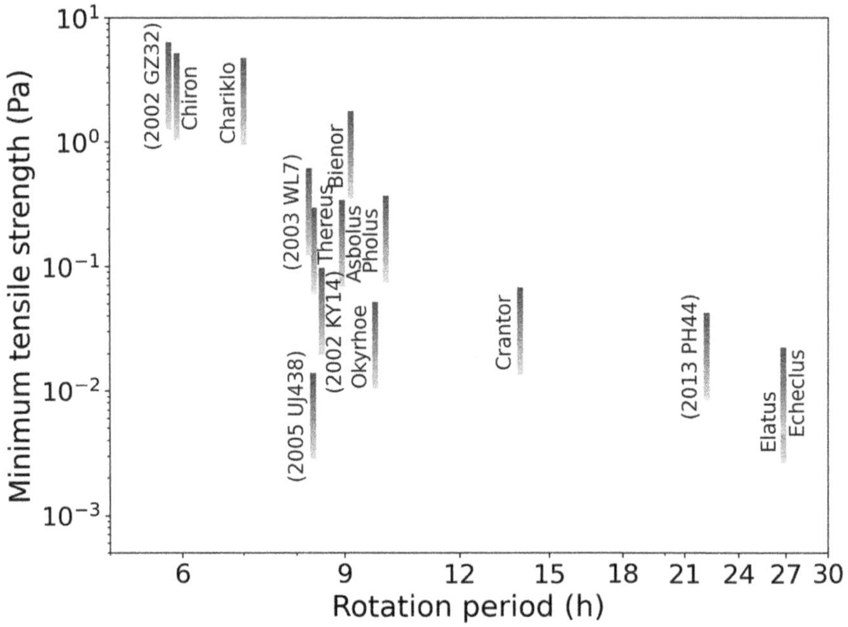

Figure 7.3. Minimum tensile strength required for Centaurs to spin at their observed rotation period without breaking apart. For each object, the line gradient represents a range of possible densities (from light to dark: 500 kg m^{-3} to 2500 kg m^{-3}). To preserve figure readability, Centaur 2013 XZ_8 is not displayed ($P = 88$ h, $Y_t < 10^{-3}$ Pa).

of Centaurs is almost certainly larger than this boundary. At comparable size range and heliocentric distances, comets would be the weakest objects known and they are stronger than this limit (see Figure 8 in Biele et al. 2022).

It is therefore unlikely that any of the current Centaurs with known rotation periods will break up due to their spin. Future surveys, however, may reveal fast rotators in the Centaur population, which would bring significant insight into the mechanical properties of these objects.

7.4.3 Surface Properties

The surface properties of Centaurs, and especially their colors have been used extensively to characterize the population. This book includes chapters on the surface chemical composition (Chapter 5) and on the recent JWST spectra beginning to revolutionize our understanding of Centaur surfaces (Chapter 14). We refer the reader to these papers for details, and here we focus on selected observational evidence that can provide insight into the evolutionary processes introduced in Section 7.3.

7.4.3.1 Color Index

The most widely characterized Centaur property is undoubtedly optical photometric color. Early studies of the visible spectra of TNOs, Centaurs, and JFCs revealed predominantly featureless surface spectra. The reflectance properties of Centaurs can thus be reliably described using parameters such as the normalized reflectivity gradient between two end points, S' (A'Hearn et al. 1984), and color indices, which measure the photometric difference between the object's magnitude in standard broadband filters (e.g., $B-R$, $g' - r'$, etc.). Centaurs and TNOs display a large diversity of colors, spanning from solar-like neutral or grey surfaces with $B - R \sim 1.0$ mag to very red colors, $B - R \sim 2.0$ mag.

A remarkable feature of the color distribution of small bodies in the outer solar system is the existence of very red objects. Most works in the literature use a limit of $B - R > 1.6$ mag (which corresponds to $g' - r' > 0.75$ mag in the SDSS system) or spectral gradient $S' \geqslant 25\%$ per 100 nm to define the so-called "'ultrared" population (e.g., Jewitt 2002; Lacerda et al. 2014; Schwamb et al. 2019). A series of multiband photometric studies have found evidence that the color distribution of Centaurs is bimodal (see Peixinho et al. 2015, Tegler et al. 2016 and references therein). Even though the color bimodality is most pronounced in Centaurs, it has also been found in dynamically excited TNOs (e.g., Peixinho et al. 2015; Tegler et al. 2016; Wong & Brown 2017; Marsset et al. 2019, and references therein). The two surface-color types are also characterized by different geometric albedos. The very red objects have more reflective surfaces with geometric albedos ~ 0.15 while the more neutral objects have small geometric albedos of ~ 0.05 (e.g., Lacerda et al. 2014).

A few hypotheses have been proposed (see Chapter 5 for a review), but the underlying reason for the existence of the two surface types remains uncertain. Likewise, the nature of the ultrared matter remains elusive. The current prevailing hypothesis is that ultrared objects have a thin ($\sim$meter thick) shell of complex organic matter resulting from prolonged space weathering in the outer solar system (e.g., Thompson et al. 1987; Cruikshank et al. 1998; Jewitt 2002; Dalle Ore et al. 2015). The existence of very red objects is thus thought to reflect compositional differences in the PKB.

Interestingly, red Centaurs have a smaller orbital inclination angle distribution than gray Centaurs (Tegler et al. 2016). This color-inclination relationship is considered to be inherited from the color-inclination correlation observed throughout the Kuiper Belt (e.g., Marsset et al. 2019). In the case of Centaurs, however, the color distribution is hypothesized to also be modified by ongoing evolution in the giant planet region. One of the most remarkable observations in favor of this argument is the finding that ultrared objects have not been discovered at heliocentric distance below $\sim$10 au (Jewitt 2009, 2015)[1]. Moreover, this location coincides with the heliocentric distance where Centaur activity begins, which motivated the hypothesis that the onset of activity leads to the disappearance of ultrared matter (Jewitt 2002, 2009, 2015). This hypothesis is supported by further evidence in the

[1] Although there is consensus in the literature that ultrared matter disappears below $\sim$10 au, we note that Lamy & Toth (2009) identified a few JFCs with nucleus color indices $V - R > 0.63$ mag which corresponds to a normalized spectral gradient larger than 20% per 100 nm and could be classified as ultrared. We caution that without more sensitive photometric observations and a wider band coverage, this remains tentative evidence.

orbital and color distributions of Centaurs. Dynamical simulations of Centaur orbits show that neutral Centaurs have longer dynamical lifetimes than red ones (Melita & Licandro 2012) and that inactive Centaurs have longer dynamical lifetimes than active ones (Fernández et al. 2018). These findings can be explained by postulating that a sufficiently long dynamical lifetime orbiting among the giant planets allows activity onset and leads to loss of the very red surface colors.

To explain the disappearance of ultrared matter in the inner solar system, Jewitt (2002) explored mechanisms in which the onset of activity would destroy or occlude the very red colors of Centaur surfaces. Their work, followed by further investigations (see Jewitt 2015) established that blanketing of the surface by fallback of material ejected due to outgassing is a plausible mechanism to explain the rapid disappearance of ultrared colors with the onset of activity (see also the discussion on surface mobilization in Section 7.4.4.2). This fallback process was, in fact, observed on comet 67P/Churyumov–Gerasimenko by Rosetta (Thomas et al. 2015; Keller et al. 2017; Davidsson et al. 2021), and it was associated with color changes on the surface of the comet (see Filacchione et al. 2020). However, because of our limited understanding of Centaur activity, it is unclear whether parallels could be made with processes taking place in the giant planet region. It is also important to note that the global optical color change of 67P's nucleus could not be determined precisely because of uncertainties in the variation of the phase-reddening curve (defined below) and can only be approximated to changing from $V - R \sim 0.52$ mag in August 2015 to ~ 0.48 mag during perihelion (Fornasier, private communication 2024).

Even though the hypothesis of Centaur surface evolution has been widely accepted, it is now being re-examined after the discovery of the first active red Centaur (523676) 2013 UL_{10} (Mazzotta Epifani et al. 2018). To explain the existence of this object, and also invoking the dynamical trend that all neutral Centaurs (inactive and active) have orbits with smaller perihelia than red Centaurs, Wong et al. (2019) proposed that thermal processing can be sufficient for the disappearance of ultrared matter without the need for activity onset. Alternatively, Ortiz et al. (2015) speculated that since the two Centaurs with ring detection (Chariklo and Chiron) are both neutral in color, the optical colors of Centaurs can be related to the presence of rings. Even though the evidence supporting this claim is scant, as more occultation events are observed (potentially detecting or setting constraints on the existence of rings; see Chapter 9), they might shed light onto the possible connections between surface evolution, activity and rings/debris around Centaurs.

Another noteworthy characteristic of the color bimodality of Centaurs is its dependence on size. The color bimodality is evident only among small (diameter <120 km) Centaurs (Duffard et al. 2014). However, the dependence of color on size has not been the focus of Centaur investigations, and it remains unclear whether it is a consequence of Centaur evolution, or whether it can be attributed to the properties of the source populations of Centaurs and/or observational biases. Notably, recent observations have indicated that very red Neptune Trojans also show a possible trend with absolute magnitude (Markwardt et al. 2023b). It will remain difficult to investigate the existence of relationships between Centaurs' size and surface properties unless significant efforts are directed into thermal infrared surveys of outer solar system populations (see Section 7.5).

7.4.3.2 Surface Spectra

Before JWST, near-infrared spectra had been obtained for a small sample of Centaurs as parts of surveys primarily targeting TNOs (Barkume et al. 2008; Guilbert et al. 2009; Barucci et al. 2011; Brown 2012). Recent and future observations with the JWST Near Infrared Spectrograph (NIRSpec) will allow for a more detailed understanding of which surface properties Centaurs inherit from their TNO parents, and how their heightened thermal processing affects their surface properties (see Chapter 14 for further discussion). JWST is also opening the possibility to compare Centaur surfaces with those of more evolved objects, especially JFCs (see Kelley et al. 2016).

Some of the most intriguing JWST results on the links between Centaur and TNO surface properties have so far come from the large JWST program “Discovering the composition of the TNOs, icy embryos for planet formation” (DiSCo-TNOs, Program ID: 2418). Using these data, three spectral types have been identified in the trans-Neptunian region: (1) a first group with spectra dominated by water ice, (2) a second group with spectra dominated by CO_2 and CO ices, and (3) a third group with spectra dominated by carbon-bearing ices and organic species (Pinilla-Alonso et al. 2024, De Prá et al. 2024). The first and third groups (dominated by water ice and carbon-bearing ices and organics, respectively) are also detected in the sample of Centaurs targeted by JWST (see Licandro et al. 2025): some grey Centaurs display water-dominated spectra, while some red Centaurs display C-bearing and organic species. We note that both spectral types, when observed on Centaurs, appear slightly depleted in volatile species compared to TNOs, with shallower absorption bands on average testifying to lower amounts of the corresponding ice at the surface. As of now, the second group (CO_2- and CO-ice dominated) appears absent in the giant planets region, which might suggest that increased surface temperatures lead to the sublimation of these volatile species.

Interestingly, a distinct spectral group is detected in the Centaur sample, which does not correspond to any parent spectral type in the trans-Neptunian region (see “misfits” in Licandro et al. 2025): these objects display both red and grey optical colors. Whether these coincide with a thermally-processed version of one or several of the TNO parent spectral types remains to be resolved.

Additional insights on the compositional differences between objects with different surface types and on the processing responsible for surface evolution at different environments in the solar system can be obtained from comparisons with other populations. The first steps in that direction are already being taken in the analysis of recent JWST spectra of Jupiter Trojans (Wong et al. 2024) and Neptune Trojans (Markwardt et al. 2023a).

7.4.3.3 Other Spectrophotometric Properties

Although the surface colors of a large fraction of the known Centaurs have been measured, the other spectrophotometric properties of the surfaces remain poorly constrained. There are two main limitations in characterizing the reflectance properties of Centaurs from the Earth: the faintness of the targets, and the small phase-angle (the Sun-target-observer angle) range, which is limited to a maximum of $\sim 10^{\circ}$.

Geometric albedo is defined as the ratio of the reflected light to the incident light at zero phase angle. It has been determined for more than 40 Centaurs (Müller et al. 2020). This relatively large sample has been achieved mainly as a result of large thermal-IR observing programs of TNOs with Spitzer (see Lisse et al. 2020) and Herschel (Müller et al. 2009). Deriving the geometric albedos is possible when the optical brightness (absolute magnitude) of the object is compared to the size estimate derived from thermal IR observations. The latter relies on the use of thermal models (e.g., NEATM, Harris 1998) and assumptions about the surface properties of Centaurs (via the emissivity parameter of the models).

Compared to JFCs, whose geometric albedos range from 0.01 to 0.06 (Knight et al. 2024), the albedos of Centaurs span a much larger range from ~ 0.04 to ~ 0.25 (Müller et al. 2020). Notably, Centaur albedos are neither correlated with their orbital parameters, nor with the objects' diameters (see Duffard et al. 2014 and Chapter 4). The only pronounced trend identified in the literature is that red Centaurs have a larger median albedo (Duffard et al. 2014; Lacerda et al. 2014; Tegler et al. 2008, 2016; Romanishin & Tegler 2018). Additionally, the albedos of grey Centaurs are found to be similar to the albedos of JFCs (Duffard et al. 2014; Müller et al. 2020; Knight et al. 2024) and Jupiter Trojans (Romanishin & Tegler 2018). However, these limited clues are insufficient to derive any meaningful constraints on the evolutionary processes proposed for the Centaur region.

As can be seen in Table 7.1 in Müller et al. (2020) and Figure 4.3 in Chapter 4, the uncertainties of the albedo estimates are very large. It is possible to derive more precise estimates of the object size and therefore of the geometric albedo in the cases of occultation observations (see Chapter 9). Additionally, accurate albedo determinations can be improved by more precise estimates of the absolute magnitude (defined as the apparent magnitude the object would have if it were placed at 1 au from the Sun and the Earth and at zero phase angle). Therefore, the calculation of an object's geometric albedo depends also on the phase function assumed.

The phase function describes the variation of the reflectance with phase angle. Typically, the small phase angles at which Centaurs are observable are characterized by an opposition effect or a steep increase of the reflectance (Gehrels 1956; Belskaya & Shevchenko 2000). However, instead of the broad and sharp opposition effect characteristic of asteroids with moderate and large albedos (Belskaya & Shevchenko 2000), Centaurs and TNOs display a shallow linear slope down to $0.1^{\circ} - 0.2^{\circ}$ and a narrow, weakly pronounced opposition effect (Belskaya & Shevchenko 2000; Belskaya et al. 2006). Most Centaur phase functions are observed to be linear and are described by a linear slope, β, known as the phase coefficient (see Belskaya et al. 2008). This behavior is similar to that of comets (see Knight et al. 2024) and Jupiter Trojans (Shevchenko et al. 2012).

Even with the limited observational evidence on Centaur phase functions, it is interesting to make a comparison between Centaurs and JFCs. Kokotanekova et al.

(2018) suggested that the possible correlation of smaller phase coefficient for decreasing albedos may be related to the level of sublimation-driven erosion on their surfaces, which is hypothesized to decrease the phase coeffient and albedo of the comets with time (see also Section 7.4.4). In Figure 7.4 we plot the Centaurs and JFCs with known phase coefficients and albedos. We note that the distributions of Centaurs in this plot is in agreement with the prediction that objects with less evolved surfaces should have larger albedos and phase function slopes. However, this plot is based on very limited data and should not be overinterpreted before further photometric data can be used to derive more reliable phase-function estimates for a larger sample of Centaurs and JFCs.

Finally, a review of the surface properties of Centaurs would not be complete without mentioning polarization. Polarization observations of outer solar system objects are challenging due to the targets' faintness and limited phase angle range (Belskaya & Bagnulo 2015; Belskaya et al. 2019). To date, only the polarization of a handful of the brightest Centaurs has been observed and these observations with their possible implications have already been reviewed by Belskaya & Bagnulo (2015) and Belskaya et al. (2019). Within the narrow phase-angle range

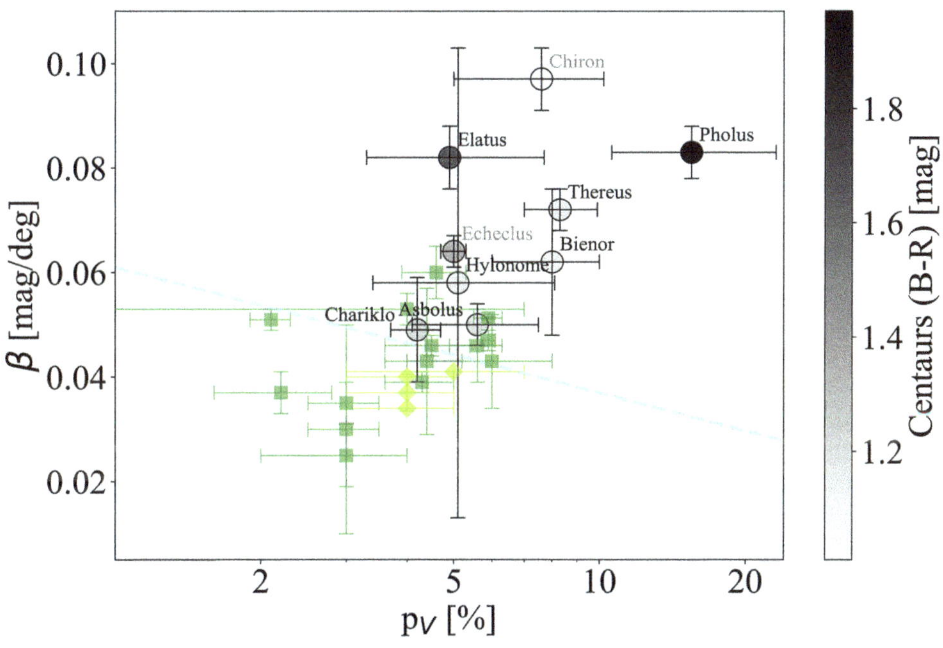

Figure 7.4. Apparent correlation between the phase coefficient (β) and geometric albedo (p_V) of JFCs (green squares) and other comets (light green diamonds) from Knight et al. (2024). Centaurs are overplotted as circles with colors corresponding to their $B-R$ color index. The phase coefficients of Chiron and Echeclus were derived from observations when they were weakly active (Dobson et al. 2023). With the exception of the very-red (Elatus and Pholus) and the active (Chiron and Echeclus) Centaurs, other Centaurs appear to follow the correlation of increasing phase coefficient with geometric albedo identified by Kokotanekova et al. (2018) and do not agree with the phase-coefficient–albedo correlation found for Main-Belt asteroids by Belskaya & Shevchenko (2000), plotted as a dashed line.

(0.5° – 4.4°), Centaurs display a great diversity of their polarization properties, which has been attributed to different abundance of ices on their surface (Belskaya & Bagnulo 2015). Polarization is therefore showing promise to serve as a diagnostic of the surface evolution of Centaurs, but this would only be possible if the number of Centaurs observed increases, polarization is measured in multiple bands, and advances in numerical and laboratory work are made.

7.4.4 Surface Morphology

7.4.4.1 Evolution of Surface Topography

A number of recent studies have established a possible link between features observed at the surface of JFCs (like topography or roughness), and the evolution state of that surface. With the detection of numerous pits at the surface of 9P, Belton et al. (2008, 2013) tried to link the occurrence of mini-outbursts with the origin of such morphologic features. Later, through a statistical analysis of the distribution of large-scale topographic features imaged on 67P and other comet nuclei imaged by spacecraft missions, Vincent et al. (2017) identified that the height of cliffs correlates with the rate of surface erosion. This suggested that topography could be used to trace a surface's erosional history, which is a conclusion independently derived by Steckloff & Samarasinha (2018) using a theoretical approach.

Inspired by these results, Kokotanekova et al. (2018) proposed that the correlation they had previously reported in the phase function of 14 JFCs (Kokotanekova et al. 2017) could be related to surface topography. In this framework, rough and more primitive surfaces, characterized by a stronger phase darkening, would gradually evolve toward smoother terrains with smaller phase coefficients. This is supported by theoretical calculations showing that pure geometric effects stemming from sharp surface topography represents a non-negligible contribution to the phase darkening (Vincent 2019). A physical framework can be provided by establishing quantitive trends from the thermal processing of cometary surfaces. Thermal evolution calculations suggest that cometary activity tends to erase sharp morphological features, which become wider and shallower over time (Benseguane et al. 2022; Guilbert-Lepoutre et al. 2023a). Consequently, pits observed at the surface of JFCs cannot be carved on the orbit where these objects are currently observed. Finally, following the interplay between surface topography and cometary activity observed on 67P (Vincent et al. 2019), Steckloff & Melosh (2016), followed by Kelley et al. (2021), hypothesized that cometary mini-outbursts observed from the ground could be associated with steep terrain features like cliffs and scarps.

All the aforementioned results converge toward a unique evolutionary sequence evidenced from independent measurables to transform "young" cometary surfaces, with sharp surface topography prone to outbursts, into "old" cometary surfaces. Such an evolutionary sequence is of particular interest to the study of Centaurs. Indeed, Guilbert-Lepoutre et al. (2023a) suggest that in order to carve the deepest circular pits, water ice should not sublimate, otherwise these topographic features would concurrently be eroded. The Centaur phase, experienced during the dynamical evolution of all JFCs, thus appears critical to understand surface properties.

The best way to verify the validity of this sequence would be to increase the number of Centaurs and comets for which multiple independent observational techniques are combined (i.e., albedos, phase functions, rotational properties, mini-outburst rates, colors, compositions). For example, a decreasing phase function coefficient would provide a useful observable to characterize the level of erosion of a cometary surface, so would a decreasing rate of observed mini-outbursts. Figure 7.4 indicates that exploring the possible phase-coefficient–albedo correlation for Centaurs is a promising step in this direction. However, more and better observational data are needed to test the evolutionary hypothesis. In Section 7.5 we discuss how the current observing limitations can be overcome in the future using new instruments as well as spacecraft photometric observations covering a larger phase angle range than is accessible from the ground.

7.4.4.2 Surface Mobilization

Surface mobilization describes the ensemble of processes that may transport material on an object, locally or globally. The associated physics is typically a competition between the intrinsic properties of the material, which tend to keep it in place, and other forces acting on the surface. For instance, landslides occur when the gravitational pull overcomes the material cohesion or the surface is oriented at an angle larger than its critical angle of repose. In both cases, the material will fail and physically move toward a lower potential energy, downslope.

On active bodies, such as comets and Centaurs, the destabilizing force may come from the sublimation of subsurface volatiles, which may facilitate motion of the surface by lowering material cohesion, or directly accelerating the material against gravity if the gas pressure is significant enough. Other effects such as saltation, electromagnetic forces, or thermal cracking, can also trigger the motion of particles across the surface.

Once mobilized, the material may be redistributed locally, over large distances, or even escape its parent body, depending on the forces at play. This has been well documented on comets, especially with the Rosetta mission, which revealed extensive material transport from the southern to the northern regions of the comet, driven by sublimation at perihelion (e.g., Thomas et al. 2015; Lai et al. 2016; Davidsson et al. 2021). Such transport may lead to detectable color changes (e.g., Fornasier et al. 2016) and different surface properties between dust sources, which become depleted and more rock-like in one case, and the blanketed terrains as depositional sinks in another (Davidsson et al. 2021). Due to the lack of observational evidence, it is not clear whether similar surface evolution takes place as a result of activity in the Centaur region.

Although there is no direct evidence for surface mobilization on Centaurs, it is very likely that such processes occur, since the same forces are at play as on comets or asteroids. In particular, landslides have often been associated with cometary activity, and may be triggered by outbursts like those occurring on Centaurs. The literature distinguishes two possible cases: (1) large scale outbursts create deep pits whose vertical side walls are inherently unstable. At any time, additional triggers may lead to further collapse of the walls, concurrently expanding the pit and exposing fresh subsurface material that can then sublimate and lead to more mobilization. This has been invoked as one of the main sources of dust activity on comets (Britt et al. 2004; Farnham et al. 2013), and several cliff collapses were directly observed by Rosetta (Vincent et al. 2016; Pajola et al. 2017). (2) Even in the absence of cliffs, dusty deposits can still become unstable if their angle of repose becomes too large. This can happen during material deposition, or from a change of local gravitational acceleration that may be triggered by evolution of the rotation state of the body, or tidal forces during a close encounter with larger objects. The material thus accelerated may deposit further downslope, or be lofted off by surface activity. This process is likely occurring on comet 103P/Hartley 2 (Steckloff et al. 2016). Laboratory experiments (Kossacki et al. 2020, 2022) have shown that sublimation-triggered landslides can occur on comets on slopes with inclination as low as 10°. In these experiments, the effect of gas flow within a layer of dust particles is considered. They show that sublimation facilitates sliding and rolling of grains, effectively lowering the critical angle of repose of the material and triggering the sliding slopes that would otherwise be stable. The sliding threshold is discussed through a mobility coefficient, which encompasses multiple physical parameters such as particle size, pore size, dust/ice ratio, temperature, and cohesion. If Centaurs are comet-like, with global or local deposits of porous aggregates of dust and ice, this process must be taking place in the giant planet region as well.

7.5 Outlook to the Future

As discussed previously in this chapter, our knowledge of Centaurs and, thus, our ability to understand the evolutionary processes at work upon them, is limited by the relative dearth of observations. Owing to their large heliocentric distances, only the largest and/or brightest Centaurs are detected. While most Centaurs are likely rich in volatiles, they reside too far from the Sun for significant water ice sublimation, and many may be depleted in more volatile ices near the surface, limiting or precluding activity. Furthermore, the viewing geometry from Earth changes little, restricting our ability to probe surface properties via changing phase angles.

However, we are on the cusp of dramatic increases in observational capabilities and survey power, potentially augmented by one or more missions through the outer solar system. Collectively, these should revolutionize our understanding of Centaurs. We detail below observational advances we expect to most significantly influence our understanding of Centaur evolution over the next decade or two. Undoubtedly, novel data will drive innovation in other areas, such as thermal modeling, dynamical modeling, and laboratory studies. Aside from obvious increases in computational power, the directions that advances in these areas will take are harder to predict and so are not discussed further.

7.5.1 Current and Very Near-Future Facilities

7.5.1.1 JWST

With the successful start of JWST observations in 2022, we are firmly in a new era for solar system studies. At the time of this writing, the first JWST results are just being published (see Chapter 14). JWST is capable of imaging and spectroscopy from 0.6 to 28.5 μm. Most of this wavelength range is inaccessible from the ground, but is critical for identifying key volatile species expected to be contained in Centaurs, notably H_2O, CO, and CO_2. JWST's unprecedented sensitivity will allow detection of very low levels of activity in the comae of Centaurs and/or detection of absorption features due to ices on the surface. Detections at a single epoch for a number of Centaurs will reveal the diversity of properties of the population and may yield identifiers that can discriminate past evolutionary histories. JWST also has the capability to make coronagraphic images; by occulting a Centaur, deep searches might be conducted for rings or faint binary companions.

The expected lifetime of JWST is 20 years, so Centaurs with semimajor axes of up to 7.4 au could be monitored for an entire orbit. This could be transformative for our understanding of Centaurs' evolution since it would demonstrate how/if the surface properties change with insolation, while a quantification of coma volatile abundances as a function of heliocentric distance would provide unprecedented constraints for numerical models of Centaurs' interiors.

7.5.1.2 LSST

The Vera Rubin Observatory's Legacy Survey of Space and Time (LSST) is scheduled to begin operating in 2025. LSST will survey the southern hemisphere sky every few nights to $r \approx 24.7$ mag, much deeper than existing surveys. Simulations of its detection capabilities (Silsbee & Tremaine 2016; Ivezić et al. 2019; Schwamb et al. 2023) do not specify Centaur discovery rates, but an order of magnitude increase in the number of known Centaurs is consistent with expectations for other small body populations. A debiased analysis of the size-frequency distribution would likely yield significant new insight into how collisionally evolved the population is and to what extent the population is modified as objects transition from TNOs to JFCs. LSST will obtain data in the standard broadband *u, g, r, i, z*, and *y* filters, allowing robust spectrophotometric color measurements to be made of the Centaur population. When compared with other physical properties and dynamical simulations, these colors might yield insight into past dynamical histories and links with other well-characterized populations.

Depending on the brightness around the orbit and the extent to which LSST observes the northern part of the ecliptic (the survey design is not yet finalized; see discussion in Schwamb et al. 2023), LSST will likely detect each Centaur tens to a few hundred times during its planned 10-year operating time, which should allow rotational lightcurves to be measured or constrained. Given how few Centaur rotation periods and axial ratio measurements currently exist, even poorly constrained lightcurves will be informative for understanding the prevalence of highly elongated objects (Donaldson et al. 2023, 2024) and associated implications for Centaurs' internal strength. The regular observations will also yield robust limits on activity and the frequency of outbursts. While 29P is well-known for its frequent outbursts and sustained activity (e.g., Miles et al. 2016; Wierzchos & Womack 2020), it is currently unknown how common such behavior is across the Centaur population. As with other properties just discussed, comparison of activity and outburst rates with other physical and dynamical properties may yield insight into the mechanisms at work in Centaurs.

7.5.1.3 SPHEREx

NASA's SPHEREx (Spectro-Photometer for the History for the Universe, Epoch of Reionization, and Ices Explorer; Doré et al. 2018), is a space-based all-sky spectral survey from 0.75 to 5.0 μm due to launch in 2025. Its core mission will survey the entire sky every six months for two years, collecting data in 96 spectral channels for each pixel to approximately the same limiting magnitude as WISE (Ivezić et al. 2022). Owing to the instrument design, the spectral channels will not be obtained simultaneously, but over several days for a given object. Thus, it will likely provide low resolution ($R = 35$–130), rotationally smeared spectra for several dozen Centaurs (based on WISE discovery rates; Bauer et al. 2013). Since this wavelength region is diagnostic of the presence of various key ices (see Section 7.4.3), SPHEREx should be a valuable complement to targeted observations by JWST (notably the DiSCo-TNOs large program) for investigating compositional differences between the TNO, Centaur, and comet populations. SPHEREx contains no major expendables (Lisse et al. 2024), so it could observe significantly longer than the core mission, potentially allowing shift-and-stack detection of fainter objects, or catching more objects as they brighten near perihelion.

7.5.2 Future Telescopes

Looking beyond JWST, LSST, and SPHEREx several next-generation facilities have sufficiently developed plans to warrant discussion. First, up to three 30-m class ground-based optical telescopes are likely to become operational in the next decade. With mirror diamcters roughly three times as large as today's largest facilities, these will have unrivaled sensitivity and angular resolutions. Most compellingly, the diffraction-limited angular resolutions of these facilities will be 0.005–0.010 arcsec, which will result in some of the larger/closer Centaurs subtending more than one pixel on the detector. Deconvolution techniques (e.g., Marchis et al. 2006) should allow at least hints of the shapes for Centaurs like 29P (diameter of $\sim$ 0.014 arcsec at perihelion) and Chiron (diameter of $\sim$ 0.027 arcsec at perihelion), and might allow the identification of close binaries (see Noll et al. 2008; Agarwal et al. 2017). Thus, it will be possible to measure some Centaur nucleus shapes directly, to make rotationally resolved maps of their surface properties, and to quantify the frequency of close binaries.

Second, NASA's planned NEO Surveyor will be a space-based IR (4 to 10 μm) survey telescope designed to detect the thermal emission from near-Earth objects. Designed as a more powerful successor to WISE/NEOWISE (25% larger mirror), NEO Surveyor will be sensitive to a narrower spectral range that the WISE primary mission, but a wider spectral range than NEOWISE, and is expected to operate for 12 years as compared to WISE's 10 months. WISE constrained sizes and albedos for 41 Centaurs (Bauer et al. 2013); NEO Surveyor should obtain comparable data for smaller, darker, and/or more distant Centaurs. This will allow the size-frequency distribution to be extended closer to the sizes of comet nuclei, thus probing the extent to which the Centaur population has been collisionally modified and thereby constraining models of solar system evolution (e.g., Bottke et al. 2023). Comparison of the albedo with other photometric properties such as color and phase function will yield new insight into the processes affecting Centaur surfaces.

Finally, several space-based facilities that might contribute new knowledge about Centaurs have either recently launched (ESA's Euclid) or are planned to launch by the end of the decade (NASA's Nancy Grace Roman, ESA's PLATO, and China's Xuntian). All will operate at optical or IR wavelengths, where Centaurs can be best studied. However, since none will prioritize solar system science, their major contributions to our understanding of Centaurs is likely to be via serendipitous observations (see Carry 2018). These are likely to include new discoveries, spectro-photometric measurements, and possibly detections of wide binaries. Roman may support a guest investigator program; ideas for Centaur science are discussed in Holler et al. (2018) and include detection of faint activity, measurement of colors, and IFU spectroscopy. Roman will have a coronagraphic imager, potentially enabling searches for rings or faint companions.

7.5.3 Ongoing and Future Space Missions

Launched in 2021, NASA's Lucy mission is on a 12-year journey to visit both of Jupiter's Trojan asteroid populations (Levison et al. 2021). Since the Centaur region is likely populated with some asteroids that leaked from Jupiter Trojan orbits (see Section 7.2), high resolution observations of the seven Trojans that Lucy will visit should provide robust information about this population. This may prove useful for interpreting and contextualizing Centaur observations, e.g., potentially identifying observable properties by which escaped Trojans can be identified within the Centaur population. Furthermore, observations of Centaurs during the nominal mission or during a potential extended mission would be capable of sampling Centaurs at larger phase angles than is possible from Earth. Observations acquired beyond the opposition effect region will be sensitive to the surface topography, as the effects of shadows on the photometry become more pronounced. Such observations would allow comparison with comet nuclei and the TNOs that have been sampled in a similar manner by the New Horizons mission (Porter et al. 2016; Verbiscer et al. 2019, 2022), further illuminating our understanding of how surface roughness evolves.

ESA's Comet Interceptor mission is slated to launch around 2029, and will fly-by a yet to be discovered dynamically new or returning Oort Cloud comet in the early 2030s (Snodgrass & Jones 2019; Jones et al. 2024). All previous missions to comets have visited Jupiter-family or Halley-type comets which have been through the inner solar system many times. Comparison of the surface properties of the Comet Interceptor target with past mission targets will give novel insight into how comet surfaces evolve. As Centaurs are at an intermediate evolutionary step, these results will provide context for interpreting Centaur surface properties, and may yield insight into how to use observable surface properties as diagnostics of Centaurs' evolutionary differences, e.g., identifying ones that have previously been at substantially smaller heliocentric distances.

7.5.4 Potential Future Space Missions

Several space missions that could influence our understanding of Centaur evolution might plausibly be selected in the next decade. The most obviously relevant mission would be a mission to a Centaur. Two such missions were proposed in the 2019 NASA Discovery mission call, Centaurus (Singer et al. 2019) and Chimera (Harris et al. 2019), and the NRC Planetary Science Decadal Survey 2023–2032 ranked a Centaur orbiter and lander among the top seven mission themes for the New Frontiers 6 call. See Chapter 16 for further discussion on the motivation and science goals of the proposed mission concepts.

A Uranus orbiter and probe, to launch in the 2030s, was the highest priority flagship mission by the 2023-2032 Decadal survey (Mandt 2023), while ESA's Voyage 2050 recommended a mission to the moons of giant planets as a top three priority for a future large-class mission. An advanced mission concept for an interstellar probe has been developed for the next U.S. Solar and Space Physics Decadal Survey, and ESA's Voyage 2050 advocated for potential participation. This mission would launch in 2036–2041 and rapidly leave the solar system, traveling at 6–7 au per year. Any of these missions would necessarily traverse the Centaur region and could potentially be designed to fly by one or more Centaurs en route. Even without a fly-by encounter, remote observations of Centaurs during the cruise phase of a mission would sample an even larger range of phase angles than could the Lucy mission (see previous subsection), and might be more sensitive to gas and dust activity than Earth-based observations.

7.6 Concluding Remarks

After the discovery of the Kuiper Belt and the recognition of the TNO-Centaur-JFC continuum, the Centaur population was largely regarded as the transition phase experienced by TNOs on their way to becoming JFCs. Growing interest in the unique properties of Centaurs, however, has led to more thorough studies, which have unveiled them as highly interesting and complex in their own right. First, the dynamical evolution of Centaurs into JFCs is now seen as an intricate dance combining alternating residences in the colder Centaur region with periods as JFCs. Moreover, even though most Centaurs are sourced from the Scattered Disk, other populations throughout the solar system have been identified as potential contributors to the Centaur region. Finally, the time spent on orbits in the giant planet region has been found to lead to significant evolution of these objects. We now think that characterizing the evolution in the Centaur region is crucial for understanding the life cycle of outer-solar-system planetesimals and for drawing accurate conclusions about the early stages of the solar system.

The main goal of this chapter has been to review the current understanding of the processes driving the current-epoch Centaur evolution in the giant planet region. We set out to describe the mechanisms shaping Centaur evolution and the evidence that informs our understanding of these processes. At a quick glance, the rapidly growing body of work we have reviewed paints a relatively straightforward picture: the main cause for the changes experienced by Centaurs is orbital evolution. Their orbital dynamics is what controls the environmental factors that enable thermal, collisional, or tidal-force processing to take place. However, a closer look at individual objects instantly reveals a much greater complexity. The extent to which an object gets altered depends on the intricate interplay of its physical properties, structure and composition. As one of the most diverse populations in the solar system, Centaurs thus pose significant challenges to identifying the unique signatures of evolution. Nevertheless, we are optimistic that these obstacles will be overcome with the spectacular observing prospects presented in Section 7.5. These expected advances make us confident that the coming decades will offer remarkable progress in our understanding of Centaurs and their related populations, which will enable us to acquire a thorough insight into their evolution in the giant planet region.

Acknowledgements

This work was supported by the International Space Science Institute (ISSI) in Bern, through ISSI International Team project 504 "The Life Cycle of Comets". RK would like to acknowledge the support from "L'Oreal UNESCO For Women in Science" National program for Bulgaria. AGL received funding from the European Research Council (ERC) under the European Union's Horizon 2020 research and innovation programme (Grant Agreement No. 802 699).

References

Agarwal, J., Jewitt, D., Mutchler, M., Weaver, H., & Larson, S. 2017, Natur, 549, 357

A'Hearn, M. F., Schleicher, D. G., Millis, R. L., Feldman, P. D., & Thompson, D. T. 1984, AJ, 89, 579

Araujo, R. A. N., Sfair, R., & Winter, O. C. 2016, ApJ, 824, 80

Araujo, R. A. N., Winter, O. C., & Sfair, R. 2018, MNRAS, 479, 4770

Bailey, B. L., & Malhotra, R. 2009, Icar, 203, 155

Barkume, K. M., Brown, M. E., & Schaller, E. L. 2008, AJ, 135, 55

Barucci, M. A., Alvarez-Candal, A., Merlin, F., et al. 2011, Icar, 214, 297

Barucci, M. A., Boehnhardt, H., Cruikshank, D. P., & Morbidelli, A. 2008, in The Solar System Beyond Neptune, ed. M. A. Barucci, H. Boehnhardt, D. P. Cruikshank, A. Morbidelli, & R. Dotson (Tucson, AZ: Univ. Arizona Press) 3

Bauer, J. M., Grav, T., Blauvelt, E., et al. 2013, ApJ, 773, 22

Belskaya, I., & Bagnulo, S. 2015, in Polarimetry of Stars and Planetary Systems (Cambridge: Cambridge Univ. Press) 405

Belskaya, I., Cellino, A., Levasseur-Regourd, A.-C., & Bagnulo, S. 2019, in Astronomical Polarisation from the Infrared to Gamma Rays, Vol. 460, ed. R. Mignani, et al. (Switzerland: Springer Nature) 223

Belskaya, I. N., Levasseur-Regourd, A. C., Shkuratov, Y. G., & Muinonen, K. 2008, in The Solar System Beyond Neptune, ed. M. A. Barucci, et al. (Tucson, AZ: Univ. Arizona Press) 115

Belskaya, I. N., Ortiz, J. L., Rousselot, P., et al. 2006, Icar, 184, 277

Belskaya, I. N., & Shevchenko, V. G. 2000, Icar, 147, 94

Belton, M. J. S., Feldman, P. D., A'Hearn, M. F., & Carcich, B. 2008, Icar, 198, 189
Belton, M. J. S., Thomas, P., Carcich, B., et al. 2013, Icar, 222, 477
Benkhoff, J., & Huebner, W. F. 1995, Icar, 114, 348
Benseguane, S., Guilbert-Lepoutre, A., Lasue, J., et al. 2022, A&A, 668, A132
Bernardinelli, P. H., Bernstein, G. M., Jindal, N., et al. 2023, ApJS, 269, 18
Biele, J., Vincent, J.-B., & Knollenberg, J. 2022, Univ, 8, 487
Biver, N., Dello Russo, N., Opitom, C., & Rubin, M. 2024, in Comets III, ed. K. J. Meech, et al. (Tucson, AZ: Univ. of Arizona. Press) 459
Bodewits, D., Bonev, B. P., Cordiner, M. A., & Villanueva, G. L. 2024, in Comets III, ed. K. J. Meech, et al. (Tucson, AZ: Univ. of Arizona. Press) 407
Bottke, W. F., Vokrouhlický, D., Marshall, R., et al. 2023, PSJ, 4, 168
Braga-Ribas, F., Sicardy, B., Ortiz, J. L., et al. 2014, Natur, 508, 72
Brasser, R., Schwamb, M. E., Lykawka, P. S., & Gomes, R. S. 2012, MNRAS, 420, 3396
Britt, D. T., Boice, D. C., Buratti, B. J., et al. 2004, Icar, 167, 45
Brown, M. E. 2012, AREPS, 40, 467
Brunetto, R., Vernazza, P., Marchi, S., et al. 2006, Icar, 184, 327
Cabral, N., Guilbert-Lepoutre, A., Fraser, W. C., et al. 2019, A&A, 621, A102
Carry, B. 2018, A&A, 609, A113
Choi, Y.-J., Cohen, M., Merk, R., & Prialnik, D. 2002, Icar, 160, 300
Cochran, A. L., & Cochran, W. D. 1991, Icar, 90, 172
Cooper, J. F., Christian, E. R., Richardson, J. D., & Wang, C. 2003, EM&P, 92, 261
Cruikshank, D. P., Roush, T. L., Bartholomew, M. J., et al. 1998, Icar, 135, 389
Dalle Ore, C. M., Barucci, M. A., Emery, J. P., et al. 2015, Icar, 252, 311
Dartois, E., Augé, B., Boduch, P., et al. 2015, A&A, 576, A125
Davidsson, B. J. R. 2021, MNRAS, 505, 5654
Davidsson, B. J. R., Sierks, H., Güttler, C., et al. 2016, A&A, 592, A63
Davidsson, B. J. R., Birch, S., Blake, G. A., et al. 2021, Icar, 354, 114004
Davies, J. K., & Barrera, L. H. 2004, The First Decadal Review of the Edgeworth-Kuiper Belt Vol 92, (Dordrecht: Kluwer Academic)
de la Fuente Marcos, C., & de la Fuente Marcos, R. 2014, Ap&SS, 352, 409
De Prá, M. N., Hénault, E., Pinilla-Alonso, N., et al. 2025, NatAs, 9, 252
De Sanctis, M. C., Capria, M. T., & Coradini, A. 2001, AJ, 121, 2792
Di Sisto, R. P., & Brunini, A. 2007, Icar, 190, 224
Di Sisto, R. P., Ramos, X. S., & Gallardo, T. 2019, Icar, 319, 828
Di Sisto, R. P., & Rossignoli, N. L. 2020, CeMDA, 132, 36
Dobson, M. M., Schwamb, M. E., Fitzsimmons, A., et al. 2021, RNAAS, 5, 211
Dobson, M. M., Schwamb, M. E., Benecchi, S. D., et al. 2023, PSJ, 4, 75
Dobson, M. M., Schwamb, M. E., Fitzsimmons, A., et al. 2024, PSJ, 5, 165
Donaldson, A., Kokotanekova, R., Rożek, A., et al. 2023, MNRAS, 521, 1518
Donaldson, A., Snodgrass, C., Kokotanekova, R., & Rożek, A. 2024, PSJ, 5, 162
Doré, O., Werner, M. W., Ashby, M. L. N., et al. 2018, Science Impacts of the SPHEREx All-Sky Optical to Near-Infrared Spectral Survey II: Report of a Community Workshop on the Scientific Synergies Between the SPHEREx Survey and Other Astronomy Observatories, arXiv e-prints, arXiv:1805.05489
Drahus, M., Yang, B., Lis, D. C., & Jewitt, D. 2017, MNRAS, 468, 2897
Duffard, R., Pinilla-Alonso, N., Santos-Sanz, P., et al. 2014, A&A, 564, A92

Duncan, M. J., & Levison, H. F. 1997, Sci, 276, 1670
Faggi, S., Villanueva, G. L., McKay, A., et al. 2024, NatAs, 8, 1237
Farnham, T. L., Bodewits, D., Li, J. Y., et al. 2013, Icar, 222, 540
Fernández, J. A., Helal, M., & Gallardo, T. 2018, P&SS, 158, 6
Festou, M. C., Keller, H. U., & Weaver, H. A. 2004, in Comets II, ed. M. C. Festou, H. U. Keller, & H. A. Weaver (Tucson, AZ: Univ. Arizona Press) 3
Filacchione, G., Ciarniello, M., Fornasier, S., & Raponi, A. 2024, in Comets III, ed. K. J. Meech, et al. (Tucson, AZ: Univ. of Arizona. Press) 315
Filacchione, G., Capaccioni, F., Ciarniello, M., et al. 2020, Natur, 578, 49
Fornasier, S., Mottola, S., Keller, H. U., et al. 2016, Sci, 354, 1566
Fouchard, M., Rickman, H., Froeschlé, C., & Valsecchi, G. B. 2014, Icar, 231, 110
Fraser, W. C., Dones, L., Volk, K., Womack, M., & Nesvorný, D. 2024, in Comets III, ed. K. J. Meech, et al. (Tucson, AZ: Univ. of Arizona. Press) 121
Gehrels, T. 1956, ApJ, 123, 331
Gil-Hutton, R., & García-Migani, E. 2016, A&A, 590, A111
Gkotsinas, A., Guilbert-Lepoutre, A., Raymond, S. N., & Nesvorny, D. 2022, ApJ, 928, 43
Golabek, G. J., & Jutzi, M. 2021, Icar, 363, 114437
Greenstreet, S., Gladman, B., & Ngo, H. 2020, AJ, 160, 144
Grundy, W. M., Stansberry, J. A., Noll, K. S., et al. 2007, Icar, 191, 286
Guilbert, A., Alvarez-Candal, A., Merlin, F., et al. 2009, Icar, 201, 272
Guilbert-Lepoutre, A. 2012, AJ, 144, 97
Guilbert-Lepoutre, A., Benseguane, S., Martinien, L., et al. 2023a, PSJ, 4, 220
Guilbert-Lepoutre, A., Davidsson, B. J. R., Scheeres, D. J., & Ciarletti, V. 2024, in Comets III, ed. K. J. Meech, et al. (Tucson, AZ: Univ. of Arizona. Press) 249
Guilbert-Lepoutre, A., Gkotsinas, A., Raymond, S. N., & Nesvorny, D. 2023b, ApJ, 942, 92
Guilbert-Lepoutre, A., Rosenberg, E. D., Prialnik, D., & Besse, S. 2016, MNRAS, 462, S146
Hapke, B. 2001, JGR, 106, 10039
Harrington Pinto, O., Kelley, M. S. P., Villanueva, G. L., et al. 2023, PSJ, 4, 208
Harris, A. W. 1998, Icar, 131, 291
Harris, W., Woodney, L., & Villanueva, G. 2019, EPSC-DPS Joint Meeting 2019, Vol. 2019, (Washington, DC: AAS) EPSC–DPS2019–1094
Haruyama, J., Yamamoto, T., Mizutani, H., & Greenberg, J. M. 1993, JGR, 98, 15079
Hirabayashi, M., Scheeres, D. J., Chesley, S. R., et al. 2016, Natur, 534, 352
Holler, B. J., Milam, S. N., Bauer, J. M., et al. 2018, JATIS, 4, 034003
Horner, J., & Lykawka, P. S. 2010, MNRAS, 402, 13
Hudson, R. L., Palumbo, M. E., Strazzulla, G., et al. 2008, in The Solar System Beyond Neptune, ed. M. A. Barucci, et al. (Tucson, AZ: Univ. Arizona Press) 507
Hueso, R., Wesley, A., Go, C., et al. 2010, ApL, 721, L129
Hyodo, R., Charnoz, S., Genda, H., & Ohtsuki, K. 2016, ApJL, 828, L8
Ikeya, R., & Hirata, N. 2024, Icar, 418, 116153
Ivezić, Ž, Kahn, S. M., Tyson, J. A., et al. 2019, ApJ, 873, 111
Ivezić, Ž, Ivezić, V., Moeyens, J., et al. 2022, Icar, 371, 114696
Jewitt, D. 2009, AJ, 137, 4296
Jewitt, D. 2015, AJ, 150, 201
Jewitt, D. 2022, AJ, 164, 158

Jewitt, D., & Hsieh, H. H. 2024, in Comets III, ed. K. J. Meech, et al. (Tucson, AZ: Univ. of Arizona. Press) 767
Jewitt, D., Kim, Y., Mutchler, M., et al. 2021, AJ, 161, 188
Jewitt, D. C. 2002, AJ, 123, 1039
Jones, G. H., Snodgrass, C., Tubiana, C., et al. 2024, SSRv, 220, 9
Jutzi, M., & Benz, W. 2017, A&A, 597, A62
Kaib, N., & Volk, K. 2024, in Comets III, ed. K. J. Meech, et al. (Tucson, AZ: Univ. of Arizona. Press) 97
Kareta, T., Woodney, L. M., Schambeau, C., et al. 2021, PSJ, 2, 48
Keller, H. U., Mottola, S., Davidsson, B., et al. 2015, A&A, 583, A34
Keller, H. U., Mottola, S., Hviid, S. F., et al. 2017, MNRAS, 469, S357
Kelley, M. S. P., Woodward, C. E., Bodewits, D., et al. 2016, PASP, 128, 018009
Kelley, M. S. P., Farnham, T. L., Li, J.-Y., et al. 2021, PSJ, 2, 131
Kim, Y., DeMartini, J. V., Richardson, D. C., & Hirabayashi, M. 2023, MNRAS, 520, 3405
Knight, M. M., Kokotanekova, R., & Samarasinha, N. H. 2024, in Comets III, ed. K. J. Meech, et al. (Tucson, AZ: Univ. of Arizona. Press) 361
Kokotanekova, R., Snodgrass, C., Lacerda, P., et al. 2017, MNRAS, 471, 2974
Kokotanekova, R., Snodgrass, C., Lacerda, P., et al. 2018, MNRAS, 479, 4665
Korsun, P. P., Ivanova, O. V., & Afanasiev, V. L. 2008, Icar, 198, 465
Kossacki, K. J., Skóra, G., & Czechowski, L. 2020, Icar, 348, 113781
Kossacki, K. J., Wesołowski, M., Skóra, G., & Staszkiewicz, K. 2022, Icar, 379, 114946
Kral, Q., Pringle, J. E., Guilbert-Lepoutre, A., et al. 2021, A&A, 653, L11
Lacerda, P., Fornasier, S., Lellouch, E., et al. 2014, ApJL, 793, L2
Lai, I.-L., Ip, W.-H., Su, C.-C., et al. 2016, MNRAS, 462, S533
Lamy, P., & Toth, I. 2009, Icar, 201, 674
Levison, H. F., & Duncan, M. J. 1997, Icar, 127, 13
Levison, H. F., Olkin, C. B., Noll, K. S., et al. 2021, PSJ, 2, 171
Li, J., Jewitt, D., Mutchler, M., Agarwal, J., & Weaver, H. 2020, AJ, 159, 209
Li, X., & Scheeres, D. J. 2021, PSJ, 2, 229
Licandro, J., Pinilla-Alonso, N., & Holler, B. J. 2025, NatAs, 9, 245
Lilly, E., Hsieh, H., Bauer, J., et al. 2021, PSJ, 2, 155
Lilly, E., Jevčák, P., Schambeau, C., et al. 2024, ApL, 960, L8
Lisse, C., Bauer, J., & Kim, Y. 2024, arXiv e-prints, arXiv:2402.08705
Lisse, C., Bauer, J., Cruikshank, D., et al. 2020, NatAs, 4, 930
Lisse, C. M., Young, L. A., Cruikshank, D. P., et al. 2021, Icar, 356, 114072
Loveless, S., Prialnik, D., & Podolak, M. 2022, ApJ, 927, 178
Luna, R., Millán, C., Domingo, M., & Satorre, M. Á. 2008, Ap&SS, 314, 113
Luu, J., & Jewitt, D. 1996, AJ, 112, 2310
Lykawka, P. S., Horner, J., Mukai, T., & Nakamura, A. M. 2012, MNRAS, 421, 1331
Malamud, U., Landeck, W. A., Bischoff, D., et al. 2022, MNRAS, 514, 3366
Malamud, U., & Prialnik, D. 2015, Icar, 246, 21
Mandt, K. E. 2023, Sci, 379, 640
Marchis, F., Kaasalainen, M., Hom, E. F. Y., et al. 2006, Icar, 185, 39
Markwardt, L., Holler, B. J., Lin, H. W., et al. 2023a, arXiv e-prints, arXiv:2310.03998
Markwardt, L., Wen Lin, H., Gerdes, D., & Adams, F. C. 2023b, PSJ, 4, 135

Marsset, M., Fraser, W. C., Pike, R. E., et al. 2019, AJ, 157, 94
Mazzotta Epifani, E., Dotto, E., Ieva, S., et al. 2018, A&A, 620, A93
Melita, M. D., & Licandro, J. 2012, A&A, 539, A144
Merk, R., & Prialnik, D. 2003, EM&P, 92, 359
Merk, R., & Prialnik, D. 2006, Icar, 183, 283
Miles, R., Faillace, G. A., Mottola, S., et al. 2016, Icar, 272, 327
Morbidelli, A., & Rickman, H. 2015, A&A, 583, A43
Müller, T., Lellouch, E., & Fornasier, S. 2020, in The Trans-Neptunian Solar System, ed. D. Prialnik, M. A. Barucci, & L. Young (Amsterdam: Elsevier) 153
Müller, T. G., Lellouch, E., Böhnhardt, H., et al. 2009, EM&P, 105, 209
Nakano, S., Kobayashi, T., Meyer, E., et al. 1993, IAU Circ., 5800, 1
Nesvorný, D. 2018, ARA&A, 56, 137
Nesvorný, D., Dones, L., De Prá, M., Womack, M., & Zahnle, K. J. 2023, PSJ, 4, 139
Nesvorný, D., Vokrouhlický, D., Dones, L., et al. 2017, ApJ, 845, 27
Noll, K., Grundy, W. M., Nesvorný, D., & Thirouin, A. 2020, in The Trans-Neptunian Solar System, ed. D. Prialnik, M. A. Barucci, & L. Young (Amsterdam: Elsevier) 201
Noll, K. S., Grundy, W. M., Chiang, E. I., Margot, J. L., & Kern, S. D. 2008, in The Solar System Beyond Neptune, ed. M. A. Barucci, et al. (Tucson, AZ: Univ. Arizona Press) 345
Noll, K. S., Levison, H. F., Grundy, W. M., & Stephens, D. C. 2006, Icar, 184, 611
Ortiz, J. L., Duffard, R., Pinilla-Alonso, N., et al. 2015, A&A, 576, A18
Ortiz, J. L., Pereira, C. L., Sicardy, B., et al. 2023, A&A, 676, L12
Pajola, M., Vincent, J.-B., El-Maarry, M. R., & Lucchetti, A. 2024, in Comets III, ed. K. J. Meech, et al. (Tucson, AZ: Univ. of Arizona. Press) 289
Pajola, M., Höfner, S., Vincent, J. B., et al. 2017, NatAs, 1, 0092
Parhi, A., & Prialnik, D. 2023, MNRAS, 522, 2081
Peixinho, N., Delsanti, A., & Doressoundiram, A. 2015, A&A, 577, A35
Peixinho, N., Thirouin, A., Tegler, S. C., et al. 2020, in The Trans-Neptunian Solar System, ed. D. Prialnik, M. A. Barucci, & L. Young (Amsterdam: Elsevier) 307
Pieters, C. M., & Noble, S. K. 2016, JGRE, 121, 1865
Pinilla-Alonso, N., Licandro, J., Brunetto, R., et al. 2024, A&A, 692, L11
Porter, S. B., Spencer, J. R., Benecchi, S., et al. 2016, ApJL, 828, L15
Pravec, P., & Harris, A. W. 2000, Icar, 148, 12
Prialnik, D. 2006, Asteroids, Comets, Meteors, Proc. of the 229th Symp. of the IAU (Cambridge: Cambridge University Press) 153–170
Prialnik, D., Bar-Nun, A., & Podolak, M. 1987, ApJ, 319, 993
Prialnik, D., & Jewitt, D. 2024, in Comets III, ed. K. J. Meech, et al. (Tucson, AZ: Univ. of Arizona. Press) 823
Prialnik, D., & Podolak, M. 1995, Icar, 117, 420
Romanishin, W., & Tegler, S. C. 2018, AJ, 156, 19
Rosenberg, E. D., & Prialnik, D. 2010, Icar, 209, 753
Rousselot, P., Kryszczyńska, A., Bartczak, P., et al. 2021, MNRAS, 507, 3444
Ruprecht, J. D., Bosh, A. S., Person, M. J., et al. 2015, Icar, 252, 271
Safrit, T. K., Steckloff, J. K., Bosh, A. S., et al. 2021, PSJ, 2, 14
Sánchez, P., & Scheeres, D. J. 2014, M&PS, 49, 788
Sarid, G., Volk, K., Steckloff, J. K., et al. 2019, ApL, 883, L25

Satorre, M. Á., Luna, R., Millán, C., Santonja, C., & Cantó, J. 2009, P&SS, 57, 250
Scheeres, D. J., & Sánchez, P. 2018, PEPS, 5, 25
Schwamb, M. E., Fraser, W. C., Bannister, M. T., et al. 2019, ApJS, 243, 12
Schwamb, M. E., Jones, R. L., Yoachim, P., et al. 2023, ApJS, 266, 22
Schwartz, S. R., Michel, P., Jutzi, M., et al. 2018, NatAs, 2, 379
Seligman, D. Z., Kratter, K. M., Levine, W. G., & Jedicke, R. 2021, PSJ, 2, 234
Shevchenko, V. G., Belskaya, I. N., Slyusarev, I. G., et al. 2012, Icar, 217, 202
Shul'man, L. M. 1972, in The Motion, Evolution of Orbits, and Origin of Comets,Vol 45, ed. G. A. Chebotarev, E. I. Kazimirchak-Polonskaia, & B. G. Marsden, (Dordrecht: Reidel) 265
Silsbee, K., & Tremaine, S. 2016, AJ, 152, 103
Simon, A., Öberg, K. I., Rajappan, M., & Maksiutenko, P. 2019, ApJ, 883, 21
Singer, K. N., Stern, S. A., Stern, D., Verbiscer, A., & Olkin, C. 2019, EPSC-DPS Joint Meeting 2019, (Vol. 2019; Washington, DC: DAAS) EPSC–DPS2019–2025
Snodgrass, C., Feaga, L., Jones, G. H., Kueppers, M., & Tubiana, C. 2024, in Comets III, ed. K. J. Meech, et al. (Tucson, AZ: Univ. of Arizona. Press) 155
Snodgrass, C., & Jones, G. H. 2019, NatCo, 10, 5418
Spencer, J. R., Stern, S. A., Moore, J. M., et al. 2020, Sci, 367, aay3999
Stansberry, J. A., Van Cleve, J., Reach, W. T., et al. 2004, ApJS, 154, 463
Steckloff, J., & Melosh, H. J. 2016, DPS Meeting, 48, 206.06
Steckloff, J. K., Graves, K., Hirabayashi, M., Melosh, H. J., & Richardson, J. E. 2016, Icar, 272, 60
Steckloff, J. K., Lisse, C. M., Safrit, T. K., et al. 2021, Icar, 356, 113998
Steckloff, J. K., & Samarasinha, N. H. 2018, Icar, 312, 172
Steckloff, J. K., Sarid, G., & Johnson, B. C. 2023, PSJ, 4, 4
Steckloff, J. K., Sarid, G., Volk, K., et al. 2020, ApL, 904, L20
Stephens, D. C., & Noll, K. S. 2006, AJ, 131, 1142
Stern, S. A., Bagenal, F., Ennico, K., et al. 2015, Sci, 350, aad1815
Stern, S. A., Weaver, H. A., Spencer, J. R., et al. 2019, Sci, 364, aaw9771
Tegler, S. C., Bauer, J. M., Romanishin, W., & Peixinho, N. 2008, in The Solar System Beyond Neptune, ed. M. A. Barucci, et al. (Tucson, AZ: Univ. Arizona Press) 105
Tegler, S. C., Romanishin, W., Consolmagno, G., & J, S. 2016, AJ, 152, 210
Thomas, N., Davidsson, B., El-Maarry, M. R., et al. 2015, A&A, 583, A17
Thompson, W. R., Murray, B. G. J. P. T., Khare, B. N., & Sagan, C. 1987, JGR, 92, 14933
Tiscareno, M. S., & Malhotra, R. 2003, AJ, 126, 3122
Tsiganis, K., Gomes, R., Morbidelli, A., & Levison, H. F. 2005, Natur, 435, 459
Verbiscer, A. J., Porter, S., Benecchi, S. D., et al. 2019, AJ, 158, 123
Verbiscer, A. J., Helfenstein, P., Porter, S. B., et al. 2022, PSJ, 3, 95
Vincent, J. B. 2019, A&A, 624, A5
Vincent, J.-B., Farnham, T., Kührt, E., et al. 2019, SSRv, 215, 30
Vincent, J. B., Oklay, N., Pajola, M., et al. 2016, A&A, 587, A14
Vincent, J. B., Hviid, S. F., Mottola, S., et al. 2017, MNRAS, 469, S329
Volk, K., & Malhotra, R. 2008, ApJ, 687, 714
Wierzchos, K., & Womack, M. 2020, AJ, 159, 136
Wierzchos, K., Womack, M., & Sarid, G. 2017, AJ, 153, 230
Womack, M., Sarid, G., & Wierzchos, K. 2017, PASP, 129, 031001

Womack, M., & Stern, S. A. 1999, SoSyR, 33, 187
Wong, I., & Brown, M. E. 2016, AJ, 152, 90
Wong, I., & Brown, M. E. 2017, AJ, 153, 145
Wong, I., Mishra, A., & Brown, M. E. 2019, AJ, 157, 225
Wong, I., Brown, M. E., Emery, J. P., et al. 2024, PSJ, 5, 87

AAS | IOP Astronomy

Centaurs

Kathryn Volk, Maria Womack and Jordan Steckloff

Chapter 8

Activity, Outbursts and Explosions

James Bauer, Oleksandra Ivanova, Adam McKay and Gal Sarid

This chapter provides a discussion and overview of Centaur activity, including outbursts, comae, ejections, and disruptions, among other phenomena relating to morphology. It discusses the proposed drivers and energy sources behind these events, such as volatile sublimation and the crystallization of amorphous water ice. The effects of such activity on Centaurs are analyzed in detail, offering insights into their dynamic behavior and evolutionary processes. Furthermore, the chapter explores how these phenomena are observed and interpreted from Earth, providing a valuable perspective on the observational techniques and data analysis methods used in Centaur research.

8.1 Introduction

Centaurs, the population of small bodies that occupy the region of our solar system between the orbits of giant planets, are inextricably linked with the phenomenon of cometary activity, including as the direct source of short period comets (e.g., Horner et al. 2004; Volk & Malhotra 2013; see Chapter 3 for a review of Centaur dynamics). The first recognized member of the Centaur population discovered was 95P/Chiron (Kowal et al. 1977), and it was active (Tholen et al. 1988) as discerned from the Centaur's brightening after its discovery. This is not surprising in retrospect, as active bodies produce dust comae that present a larger surface area and reflect significantly more light than the nucleus alone, and therefore appear much brighter than when they are inactive. Active near its aphelion distance of $\sim$18.7 au, what was unexpected was that an object so distant from the Sun would exhibit such a high level of activity. As Bus et al. (2001) showed, (2060) 95P/Chiron, while active, exhibited an absolute magnitude that was over a full magnitude brighter than when inactive. However, the physical processes and species that were normally attributed with driving cometary activity, namely the sublimation of water-ice (e.g., Delsemme 1982), was clearly not the driver of Chiron's activity given its large heliocentric distance of $\sim$18 au, a heliocentric distance where water-ice sublimation is very

doi:10.1088/2514-3433/ada267ch8

inefficient. Other species have been identified as drivers of cometary activity, and many associated with Centaurs (Womack & Stern 1999; Bauer et al. 2015; Harrington Pinto et al. 2022), as well as processes apart from direct surface or near-surface sublimation (e.g., Prialnik et al. 1995, 2004; Lisse et al. 2022).

Other bodies were subsequently identified as members of the Centaur population, and several were active. 29P/Schwassmann–Wachmann 1 (29P/SW1) was discovered in 1927 (van Biesbroeck 1927), but later recognized as dynamically qualifying as a member of the Centaur population. The case was further strengthened when Senay & Jewitt (1994) discovered CO emission as the main driver of activity. 39P/Oterma was discovered in 1948, but while in a Jupiter-family comet (JFC) orbit.[1] After a close encounter with Jupiter, the comet was recovered in a Centaur orbit (Fernandez et al. 2001). Comets 165P/LINEAR and 166P/NEAT were discovered in Centaur orbits while active and therefore only received comet designations. However, (60558) 174P/Echeclus exhibited a strong outburst of activity nearly 8 yr after its discovery (Choi et al. 2006), and thus it shares the rare distinction of having comet and asteroid designations. See Chapter 10 for discussions of some of these particularly notable individual Centaurs.

Jewitt (2009) conducted one of the earliest studies to determine the fraction of active Centaurs. From his own observations, he determined a fraction of 13%. Using the same orbital criteria, where a small body is defined as on a Centaur orbit if both its semi-major axis and perihelion are bracketed by the Jovian and Neptunian semi-major-axes, there are about 310 Centaurs currently as derived from the Minor Planet Center's orbit database, with 39 that are active, or, remarkably close to the originally derived ~13%, as listed in Table 8.1. Figure 8.1 reprises the summary figure from Jewitt (2009), updated with the current Centaur and JFC populations, and shows a remarkably similar distribution with respect to semi-major axis and eccentricity. Presumably, a larger fraction of these bodies will undergo activity at some time, provided they evolve into JFC orbits. However, a formal removal of any selection bias from this rate, say for active bodies possibly being more easily discovered owing to their brightening during outbursts, has not really been conducted. Horner et al. (2004) suggests that approximately a third of the present Centaurs will evolve into short-period comet orbits (see also Chapter 3).

The compositional components of the Centaurs and the fraction that have been preserved from the solar system's genesis needs to be further explored. In fact, the Centaurs are related to several of the key solar system science gaps identified in the 2023 decadal survey (*Origins, Worlds, and Life: A Decadal Strategy for Planetary Science and Astrobiology 2023–2032*; NASEM 2023; see Table 17.1 in Chapter 17), such as where and how volatile deposition, sublimation, transport, redeposition, and mass loss take place, now and in the past. Also the length of "interaction timescales" and in particular "how TNOs are modified into the Centaur and Jupiter-family comet

[1] JFCs are defined as comets with orbital periods under 20 years (e.g., Dones et al. 2004), and they tend to have orbital inclinations less than 40°. This sometimes overlaps with the definition of a Centaur, as with the case of 39P/Oterma's orbit after its 2001 recovery. However, when 39P/Oterma was discovered in 1948, its orbital semimajor axis value near 3.4 au was too low to be considered a Centaur.

Table 8.1. Orbital Elements of Active Centaurs

full_name	e	q[au]	i(°)	Asc. Node	Node of Peri.	Q [au]	tp_cal	per_y
29P/Schwassmann–Wachmann 1	0.0448	5.777	9.36	312.39	50.91	6.32	2019-04-19.7	14.9
39P/Oterma	0.2287	5.709	1.48	303.95	89.37	9.09	2023-04-07.5	20.1
165P/LINEAR	0.6216	6.83	15.91	0.64	126.21	29.27	2000-06-15.7	76.7
166P/NEAT	0.3831	8.564	15.37	64.49	321.88	19.2	2002-05-20.8	51.7
167P/CINEOS	0.27	11.783	19.13	295.84	343.65	20.5	2001-04-09.8	64.9
423P/Lemmon	0.1201	5.419	8.35	33.35	80.71	6.9	2021-09-24.0	15.3
450P/LONEOS	0.3129	5.448	10.57	124.94	21.56	10.41	2004-09-12.6	22.3
467P/LINEAR-Grauer	0.0739	5.264	2.55	44.15	262.52	6.1	2008-12-17.8	13.6
C/2001 M10 (NEAT)	0.8011	5.303	28.08	293.92	5.48	48.02	2001-06-21.2	138
P/2005 S2 (Skiff)	0.1967	6.398	3.14	161.27	229.93	9.53	2006-06-30.7	22.5
P/2005 T3 (Read)	0.1738	6.202	6.26	28	8.16	8.81	2006-01-12.4	20.6
C/2007 S2 (Lemmon)	0.5571	5.558	16.86	296.25	210.47	19.54	2008-09-14.9	44.4
P/2010 C1 (Scotti)	0.2591	5.235	9.14	142.03	3.67	8.9	2009-12-01.6	18.8
P/2010 H5 (Scotti)	0.1564	6.026	14.09	24.9	175.13	8.26	2010-04-17.7	19.1
P/2011 C2 (Gibbs)	0.2685	5.389	10.91	12.2	160.56	9.34	2012-01-07.9	20
C/2011 P2 (PANSTARRS)	0.3698	6.148	8.99	204.02	76.38	13.36	2010-09-13.5	30.5
P/2011 S1 (Gibbs)	0.2032	6.894	2.68	218.9	193.37	10.41	2014-08-24.0	25.4
C/2013 C2 (Tenagra)	0.4313	9.131	21.34	247.52	308.76	22.98	2015-08-29.4	64.3
C/2013 P4 (PANSTARRS)	0.5961	5.967	4.26	256.61	113.56	23.58	2014-08-12.3	56.8
C/2014 F3 (Sheppard-Trujillo)	0.6442	5.707	6.55	326.86	2.67	26.37	2021-05-16.0	64.2
C/2014 OG392 (PANSTARRS)	0.1812	9.967	9.04	145.84	254.3	14.38	2021-12-07.7	42.5
P/2015 B1 (PANSTARRS)	0.3824	5.976	18.03	353.36	188.27	13.38	2015-09-19.7	30.1
C/2015 D2 (PANSTARRS)	0.5683	5.606	31.83	162.86	291.92	20.36	2013-09-27.6	46.8
P/2015 M2 (PANSTARRS)	0.1791	5.913	3.97	86.61	224.79	8.49	2015-08-31.1	19.3
C/2015 T5 (Sheppard-Tholen)	0.6661	9.338	11.05	311.19	61.84	46.59	2016-01-24.5	148
C/2015 V4 (PANSTARRS)	0.7056	5.46	60.75	179.9	306.91	31.63	2016-08-27.9	79.9
C/2016 Q4 (Kowalski)	0.5783	7.084	7.26	271.33	99.47	26.51	2018-01-28.1	68.9
P/2017 P1 (PANSTARRS)	0.3085	5.438	7.7	221.39	122.27	10.29	2018-06-18.5	22.1

(*Continued*)

Table 8.1. (*Continued*)

full_name	e	q[au]	i(°)	Asc. Node	Node of Peri.	Q [au]	tp_cal	per_y
C/2018 S2 (TESS)	0.6149	5.471	64.22	85.13	290.68	22.94	2018-11-05.1	53.5
C/2019 A5 (PANSTARRS)	0.7079	6.324	67.54	146.62	356.04	36.97	2019-06-09.4	101
P/2019 LD2 (ATLAS)	0.1266	4.552	11.6	179.55	119.29	5.99	2020-02-28.3	12.1[a]
P/2020 B4 (Sheppard)	0.1925	6.439	11.6	186	342.46	9.51	2021-11-25.3	22.5
P/2020 MK4 (PANSTARRS)	0.0205	6.024	6.69	1.89	169.99	6.28	2015-05-06.4	15.3
C/2020 Q2 (PANSTARRS)	0.5056	5.399	3.31	179.95	118.45	16.44	2020-02-06.4	36.1
P/2020 V3 (PANSTARRS)	0.2549	6.23	23.03	198.29	249.24	10.49	2021-02-07.6	24.2
P/2020 W1 (Rankin)	0.2649	5.288	10.79	124.26	264.77	9.1	2020-04-04.1	19.3
P/2023 B1 (PANSTARRS)	0.129	6.141	14.59	78.59	81.39	7.96	2023-04-29.1	18.7
C/2023 H3 (PANSTARRS)	0.6164	5.233	2.49	55.1	193.34	22.05	2024-02-19.1	50.4
(2060) 95P/Chiron	0.377875	8.462271	6.9423	209.2572	339.58	18.74	2046-08-03.0	50.17
(60558) 174P/ Echeclus	0.45692	5.835077	4.3438	173.2801	163.3	15.65	2023-02-25.0	35.22

[a] The definition of a Centaur is not universally agreed upon, but we use for this table the definition from Jewitt (2009), with the addition of P/2019 LD2.

population – which have different characteristics (e.g., colors) – as they migrate closer to the Sun." Active Centaurs are at the critical intersection point of cometary and primitive bodies. Centaur activity manifests itself through multiple phenomena: the presence of dust or icy grained comas with particulates ranging from submicron to centimeters in diameter (Bockelée-Morvan et al. 2022), volatile ejection of species such as CO and CO_2 (Wierzchos & Womack 2020; Harrington Pinto et al. 2022), often identified through spectroscopic or narrow-band observation of gas emission lines, and sudden brightening caused by outbursts of materials (Miles et al. 2016; Ivanova et al. 2019a), over a wide range of intensities (Rousselot et al. 2021). These phenomena are observed and quantified by increasingly sensitive instrumentation and analysis techniques that may reveal or constrain the objects' composition and structure.

Knowledge of the mechanisms of Centaur activity allows us to better understand the structure and composition of the objects when they formed. Any activity is driven by energy, which may be intrinsic to the body or extrinsic, coming from the surrounding environment. Changes in environment, for example through alteration of the Centaur's orbit, induce evolutionary change in a Centaur. Insolation, or the absorption of solar radiation, by the Centaur's surface or immediate subsurface, is extrinsic, and likely the main energy source that drives activity, as it triggers the sublimation of volatiles from ice into gas states. Another extrinsic source of energy could be from gravitational sources,

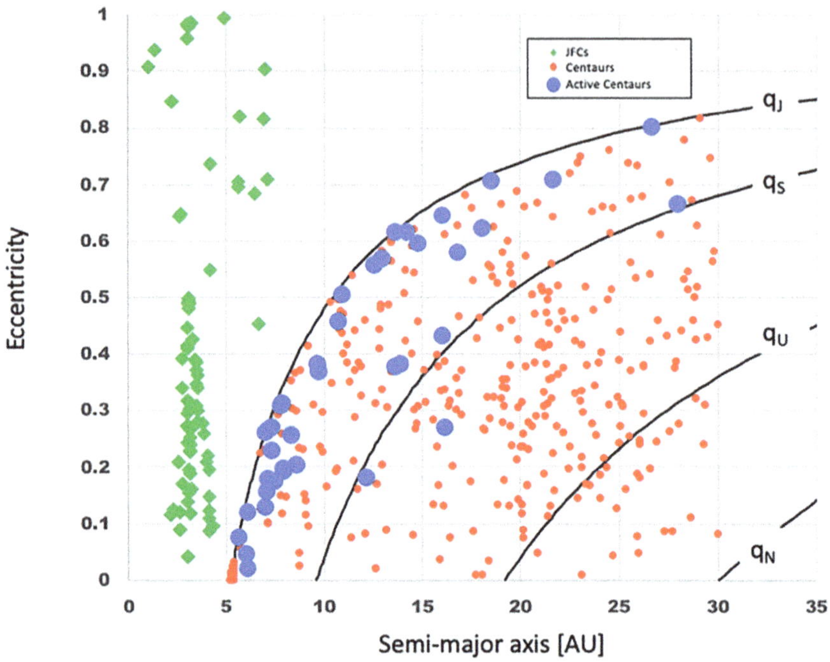

Figure 8.1. Reprise of the summary figure from Jewitt (2009) replotted with updated JFC and Centaur semimajor axes and eccentricities from the Minor Planet Center's orbit database. The distribution of the active Centaurs (blue circles) are similar to the original work's relative to the JFCs (green diamonds) and quiescent Centaur (red circles) population. The active Centaurs are found mainly between Jupiter and Saturn.

such as a close encounter with Jupiter, where tidal forces may be intense enough to disrupt the structural integrity of the comet as well. Radioactive decay, an intrinsic energy source, (Guilbert-Lepoutre 2011; Wood et al. 2017) mainly affects the deep interior of the nucleus since the outer layers are efficiently cooled by the emission of thermal radiation. Secondary sources that may be triggered by the either intrinsic or extrinsic sources include latent heat of phase transition, mostly sublimation, but also possibly heat released in crystallization of amorphous water ice (Prialnik et al. 2004; Prialnik 2006; Guilbert-Lepoutre 2012; Peixinho et al. 2020), and perhaps other exothermic chemical reactions (Neslušan 2014).

Within the protoplanetary disk, the presence of grains resulted in reduced radiative cooling and increased mid-plane kinetic temperature (Jewitt et al. 2007). As solid particles grew and moved within the disk, they influenced the opacity, subsequently altering the position of the snow line. If Centaurs initially formed beyond the snow line, it is highly likely that they still retain ice. It is worth noting that some Centaurs do not exhibit spectral signatures associated with water ice or hydration (e.g., Brown 2000; Barkume et al. 2008), meaning their ice components remained unexposed and never transitioned into a liquid phase, and that the ice may remain frozen within their interiors. Detection of water ice on the surface of

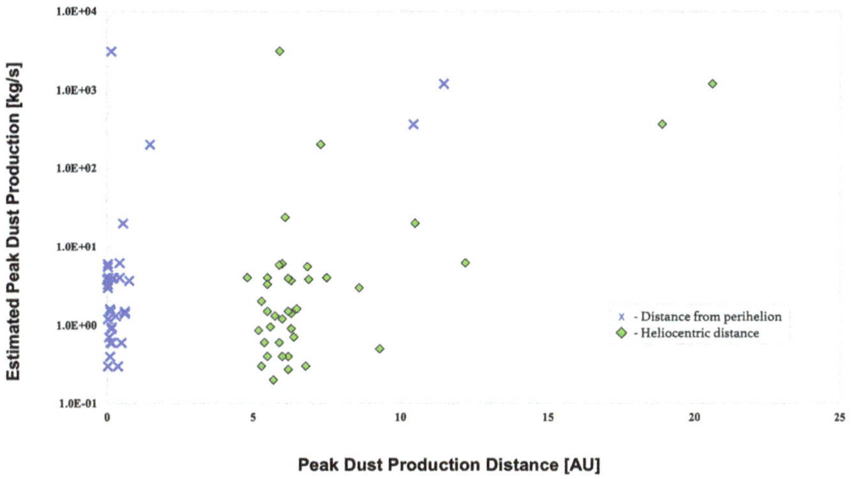

Figure 8.2. The peak dust production rate estimates and distances based on photometry reported to the Minor Planet Center. Distance values from perihelion (blue x) and heliocentric distances at the time of peak dust production (green diamonds) are shown. The production rates assume a 1-micron average dust grain diameter (e.g., Meech et al. 1997), a grain density near that of water-ice (1 g cm^{-3}), and a standard ejection velocity relation of $\sim {r_{helio}}^{-0.5} \times 1$ km s^{-1}, where r_{helio} is in units of au. Almost all the observed Centaurs have peak activity near their perihelion (see text for exceptions). Note that while the average grain diameter may vary between individual comets, the dust production will scale as the cube of the average dust grain diameter.

95P/Chiron, and CO, CO_2, and CH_4 volatile emission lines amongst Centaurs (Luu et al. 2000; Harrington Pinto et al. 2023; Pinilla-Alonso et al. 2024; Faggi et al. 2024) strongly supports the sublimation-driven explanation for the activity of some Centaurs.

As selected active small bodies of the solar system, Centaurs exhibit various activities about the surface's complex processes and evolution. Most active Centaurs showed outbursts (Bus et al. 2001; Rousselot 2008; Trigo-Rodríguez et al. 2008, 2010). From photometry reported to the Minor Planet Center, the vast majority (98%) of these have their peak reported outburst within 1 au of their perihelion, while only 8% have their full orbital heliocentric distance range less than 1 au (see Figure 8.2). The interesting outliers are 95P/Chiron, C/2014 F3 (Sheppard-Trujillo), and 174P/Echeclus. Whether this is an indicator of more exotic composition remains to be seen, but recently reported spectra of Chiron (Pinilla-Alonso et al. 2024) suggest that other species than the more commonly observed three (CO_2, CO, and H_2O; see Section 8.4) may be driving activity.

One-time activity in Centaurs is often associated with sporadic outbursts or events that occur suddenly and are typically short-lived, leading to an extended coma and tail formation. A prime example is Centaur 174P/Echeclus (Choi et al. 2006; Bauer et al. 2008; Rousselot et al. 2021), which exhibits a sudden outburst of activity from time to time. Short-lived activity in Centaurs may include transient phenomena that do not fit the classical definition of a "cometary" coma but involve the temporary release of material or the presence of dust structures. Observations of Centaurs like

(10199) Chariklo (1997 CU_{26}) have revealed short-lived structures, possibly due to outbursts, non-gravitational forces, or the sublimation of volatile ices (Peixinho et al. 2001; Gutiérrez et al. 2001). Active Centaurs exhibit, on average, closer perihelion distances (near 6 au) compared to the median perihelion distance of all known Centaurs (after 12 au). Most probably, active Centaurs are associated with higher surface temperatures, since they are located closer to the Sun. Understanding the triggers and mechanisms behind these sporadic events is essential for understanding the nature of Centaurs.

Some Centaurs exhibit continuous or long-term activity. This activity is similar, for example, to that of comets and persists over extended periods. For instance, we can consider the case of Centaur 29P/SW1 (Trigo-Rodríguez et al. 2008, 2010; Miles et al. 2016; Picazzio et al. 2019), which has displayed prolonged activity over much of its orbit, occasionally punctuated by short-term outbursts. This long-term activity contrasts with the typical one-time outbursts observed in other Centaurs. A preliminary estimate of the total mass loss from Centaur 29P/SW1 over a specific observation period was calculated using the dust production rate from Ivanova et al. (2011) to be from (2.47-21.5) $\times$ 10^5 kg, for several model variants. It was assumed that the sublimation of highly volatile ice is the driving force behind the ejection of dust particles from the nucleus. This calculation enabled a tentative exploration of potential scenarios for explaining the long-term activity of the Centaur. The results indicated that continuous removal of the outer surface layer of the Centaur is required to account for the observed long-term high activity. It is worth noting that the prevailing hypothesis proposed for the activity of distant objects, the crystallization of amorphous water ice, may not fully account for the observed long-term activity at such large heliocentric distances, at least in its current form (Jewitt 2009; Ivanova et al. 2011). Long-lasting activity in Centaurs can also be attributed to the sublimation of volatile ices.

Centaurs may display periodic activity linked to their rotation rates and orientation due to the absorption and re-emission of solar radiation, which can cause variations in the rotation rate of Centaurs, leading to periodic changes in their brightness. For example, the Centaur (5145) Pholus exhibits periodic brightness variations associated with its non-principal-axis rotation (Hoffmann et al. 1993). See Chapters 4 and 7 and for further discussion of Centaur rotation states.

An analysis of cometary activity reveals that morphological features can arise from unrelated physical processes (Belton 2010; Sierks et al. 2015; Vincent et al. 2019); this may also be typical for some active Centaurs. When a coma displays anisotropies and exhibits activity in collimated streams of gas and dust, there is likely an uneven distribution of gas emission sources across the object's surface. This non-uniform distribution could be due to areas with more favorable conditions for material release, such as localized increases in volatile materials, small-size dust more easily lifted, or craters that accumulate heat (De Sanctis et al. 2015). However, the object's topography sometimes plays a more significant role in the outflow of material from the surface than presents an active area on the surface. Variations in topography can lead to focus and channel gas flows, creating regions of higher density where dust particles can be more effectively accelerated. In contrast, the presence of observed ions suggests that some localized regions are richer in volatiles, sometimes with exposed surface ice. Consequently, these regions with a greater abundance of volatile materials can sustain activity for longer durations compared to depleted regions, resulting in more robust gas and dust emissions and the formation of jet-like features.

In summary, Centaurs represent a unique and dynamically evolving category of objects within the solar system, demonstrating a wide range of activity patterns, including isolated events, periodic fluctuations, brief episodes, and enduring processes. Only a small fraction (12.6%) of the known Centaurs demonstrate activity (e.g., Ivanova et al. 2011; Jewitt 2009; Mazzotta Epifani et al. 2017, 2018; Steckloff et al. 2020; de la Fuente Marcos et al. 2021; Rousselot et al. 2021). The available data indicate that temperature-related mechanisms are the primary drivers of this activity, emphasizing the significant role of highly volatile constituents in facilitating mass loss from active Centaurs. Observations have indicated that the sources of highly volatile components are likely located relatively close to the surface of the Centaurs rather than deep within the objects (Jewitt 2002; Barucci et al. 2008; de Bergh et al. 2013). New observations may elucidate the reasons behind the inactivity of some Centaurs, even when their perihelion distances are small. Mass loss from these Centaurs occurs intermittently or in bursts over timescales longer than their orbital periods, possibly due to instabilities associated with the crystallization process. Another hypothesis is that these seemingly quiescent, high-temperature Centaurs may have experienced devolatilization of their near-surface regions as a result of prior activity. In other words, they might have inhabited their current orbits for a more extended duration compared to active Centaurs as a group. Alternatively, some Centaurs might lack an ample supply of volatile materials, particularly the amorphous ice phase, that are crucial energy sources for active Centaurs. These diverse, active manifestations underscore the intricate interplay of physical and chemical factors that dictate the behavior and evolution of Centaurs, rendering them a captivating subject of study.

8.2 Outbursts and Explosions

Comae of active Centaurs demonstrate a range of features, from long-term, stable formations enduring for months, to rapid variations occurring within minutes. These features can be formed by dust, neutral gas, and ions, resulting in diverse forms and structures within comae. The level of Centaur activity covers fluctuations in brightness, differences in color, variations in polarization, and temporal changes. Studying the diverse activity levels (changes in the brightness of objects over time, including outburst) in active Centaurs allows us to understand the nature of objects, their volatile composition, surface characteristics, and the dynamic processes shaping their evolution. Long-lasting observations in a wide range of wavelengths and progress in observational techniques play a significant role in studying the

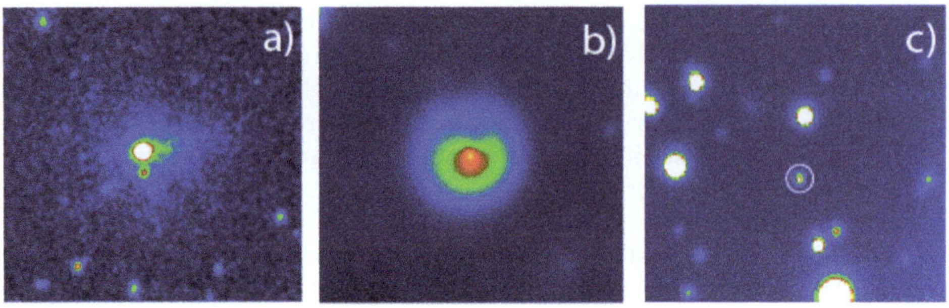

Figure 8.3. Centaur (60558) 174P/Echeclus images in the R-filter at the 6-m telescope Special Astrophysical Observatory (SAO; Minor Planet Center code 115): (a) observations activity in 2016, (b) activity observed in 2017 and (c) no cometary activity was detected in 2021, for the Centaur indicated by the white circles (previously unpublished data).

Centaur activity in the outer solar system (Figure 8.3). Also, observations of stellar occultation by Centaurs (Pereira et al. 2024) allow us to detect activity and obtain detailed information about the ejected material, such as optical depth of the dust, in the coma (see, e.g., Sickafoose et al. 2020, 2023 and Chapter 9).

Low activity: this activity is often challenging to detect and may manifest as subtle variations in brightness (typically less than 1 magnitude) over time. This outgassing is most likely linked to the sublimation of volatile ice from the surface, contributing to a faint coma (for example, Centaur 2020 MK_4; de la Fuente Marcos et al. 2021). Observations in different wavelengths, particularly in the infrared and visible spectrum, help identify highly volatile components in a coma.

Moderate activity: Centaurs with moderate-level activity show more pronounced outbursts or episodic releases of dust or gas material. These events can significantly alter the brightness (from 1 to 3 mag) and appearance of the Centaur. This activity can often be associated with specific events, for example, perihelion passages. For instance, 95P/Chiron exhibits moderate activity levels, with periodic outbursts of material from the surface when it approaches perihelion (Duffard et al. 2002). These events, accompanied by a noticeable change in the light curve, allow us to study surface properties.

High activity: Centaurs demonstrating high-level activity most probably undergo intense and frequently occurring outbursts, leading to dramatic changes in their overall brightness (up to 3 mag) and appearance. These events can indicate significant changes in conditions on the surface of a Centaur. Examples of such active Centaurs are 174P/Echeclus (Kareta et al. 2019; Rousselot et al. 2021) and 29P/SW1 (Lin 2023; Miles et al. 2016; Trigo-Rodríguez et al. 2008, 2010; Voitko et al. 2022).

Centaurs can probably transit between different activity levels throughout their evolution. Most importantly, long-term observations in this situation allow us to discern activity patterns and investigate underlying mechanisms. See Chapter 11 for a discussion of Centaur observational campaigns.

8.3 Effects of Activity

8.3.1 Color

Studying color variations in Centaurs provides valuable insights into their composition, surface properties, and dynamic processes (see Chapter 5). Only a small number of Centaurs are active (Tegler et al. 2008; Jewitt 2009), revealing different colors in their coma and nucleus (Seccull et al. 2019). For active Centaurs, the coma was bluer than the solar radiation in only three active Centaurs, 29P/SW1, 95P/Chiron and 174P/Echeclus. Different factors contribute to the observed color diversity among Centaurs during activity, including the changing composition of the surface (opening the fresh layers of material), sublimation, and outgassing (ejected dust or ice grains). Most importantly for non-active Centaurs, the influence of solar and radiation processing of the surface (chemical reactions and altering the properties of surface materials) and changes in the topography of the surface can lead to differences in color. It is known that the dust color depends on the microphysical properties of particles. Sen et al. (2017) found that dust color is primarily dependent on the particle size distribution, their chemical composition, and independent of porosity.

Rapid variations of the dust color within the formed coma (during the object's activity) with optocentric distance are the evident result of changes in the population of dust grains. Table 8.2 shows some of the reported colors from the literature while the Centaurs were known to be coma-dominated. For reference, B-R colors > 1.4, or V-R equivalent colors in excess of $\sim$0.52, are considered "red" (e.g., Jewitt 2009), like 166P/NEAT. Chiron's coma color would be considered "blue" for comparison. These changes can depend on the terminal velocities of particles when solar radiation pressure sorts grains by their sizes within the coma (Betzler et al. 2017). Also, particles can fragment, and volatile species can sublimate as a particle moves outward (Betzler et al. 2017; Ivanova et al. 2019b). The small number of results (Zubko et al. 2020; Rousselot et al. 2021) can be explained by a relatively poor base

Table 8.2. Select Known Active Centaur Coma Colors

Centaur	Color	Reference
29P/Schwasmann-Wachmann 1	B-R: 1.21 ± 0.09	Jewitt (2009); Betzler & de Sousa (2023).
P/2020 MK4 (PanSTARRs)	$(g'' - r'')$: 0.42 ± 0.04	de la Fuente Marcos et al. (2021)
P/2004 A1 (LONEOS)	V-R: 0.5 ± 0.2	Mazzotta Epifani et al. (2011)
P/2011 S1 (Gibbs)	B-R:1.25 ± 0.05	Lin et al. (2014)
(60558) 174P/Echeclus	B-R: 1.3 ± 0.1	Rousselot et al. (2021); Bauer et al. (2008)
(2060) 95P/Chiron	B-R: 0.92 ± 0.06	Jewitt (2009)
166P/NEAT	B-R: 1.60 ± 0.05	Jewitt (2009)
C/2001 M10 (NEAT)	B-R: 1.16 ± 0.11	Jewitt (2009)

of observations, as such Centaurs (the only exception is 29P/SW1) are observed rather episodically. The Centaur 29P/SW1 monitoring program in 2018 and 2021 (Voitko et al. 2022; Lin 2023) allowed detection of rapid color variations in a coma during the outburst. The observations in 2018 showed the activity variation had provoked the change in the color slope from very red down to blue. This behavior of the dust color may mean that the outburst has thrown into the coma new particles from the nucleus, which differ from the previous population. According to the modeling, the reddest color corresponds to the Fe/Mg silicate and/or organic particles in the coma, and the bluest color can be explained by water ice and/or Mg-rich silicates, but size distribution has most likely remained the same (Voitko et al. 2022).

Another Centaur, 174P/Echeclus, exhibited a blue coma during its active period, while the nucleus displayed a red color (Seccull et al. 2019). Authors suggested that the color indicated two different dust populations and that the deposition of the blue coma particles back onto the nucleus could neutralize its red color. Zubko et al. (2020) showed that a similar phenomenon was found in laboratory optical measurements of single-scattering feldspar particles and a surface comprised of these particles. Thus, the observations do not necessarily suggest different chemical composition and size distribution of dust in the coma and surface of 174P/ Echeclus. Instead, the observed difference could arise from multiple-scattering phenomena that significantly contribute to light scattering by a regolith but do not appear in an optically thin coma. Therefore, questions about color variation and their reasons are still open and need new monitoring observations in a wide range of wavelengths.

8.3.2 Morphology

Various physical characteristics define active Centaurs' morphology and provide valuable insights into their nature and evolution. In most cases, active Centaurs exhibit cometary activity shown by the extended and asymmetrical coma and, in some cases, tails or active structures (for example fans, jets, shells, etc.). The morphology of the coma (in most cases dusty) is different among active Centaurs. Analysis of observations revealed differences in coma size, shape, and heterogeneity, suggesting various processes of activity (Jewitt 2009; Miles et al. 2016; Rousselot et al. 2016; Mazzotta Epifani et al. 2011, 2018; Ivanova et al. 2019a; Shubina et al. 2023). The feature detections in the coma of active Centaurs have become increasingly more common, and spatial structures are seen in a wide range of wavelengths. Similar to comets, a Centaur's surface (irregular shapes, craters, and sublimation-driven erosion) plays an important role in coma morphology. Structures can appear as long-term, steady-state formations that persist for months or vary on timescales as short as minutes (e.g., Miles & Faillace 2011). Observed active structures can be spontaneous or show remarkable regularity. They can exhibit apparent disorder, lack organization, or display balanced and well-defined forms. This fact encompasses changes in brightness, distinctions in color, polarization degree values, and temporal fluctuations. Regardless of the specific details, each structure holds data connected to the comet's nucleus and activity. The main source of information regarding the morphology produced by activity comes through photometric imaging of active Centaurs over a wide range of heliocentric distances. Most active structures observed in active Centaurs can be organized into morphological categories.

Coma shape: Most active Centaurs exhibit a coma dominated by a diffuse nebulosity surrounding the nucleus. The coma may be spherical, elongated, or asymmetric, with the nucleus offset from the center in any direction. The level of condensation also can be variable. Sometimes, variations in brightness are observed.

Jets and fans: The most commonly reported feature types are fans and jets. Fans are broad extensions of the coma, frequently exhibiting a wedge or triangular shape. The orientation of structures can be varied. Jet structures emanate from the nucleus but tend to be narrow and more sharply defined than fans. Both active structures can be stable, long-lived, and highly dynamic, changing their appearance. The orientation of jets can be different. The form can be radial but also shows curvature at increasing distances from the nucleus.

Halos, shells, spirals: These structures primarily take on an azimuthal shape, encircling the nucleus in incomplete arcs or forming spiral patterns. These features may exhibit a diffuse appearance or possess well-defined boundaries.

There are definite relationships between the different types of active structures listed above. For example, jets, fans, or shells may be the same physical entity, taking on different appearances depending on the particular geometry and conditions of observations. Also, active Centaurs demonstrate other features that can all be classified as large-scale structures, including dust or ion tails.

Historical observations (Roemer 1958) and recent studies (Trigo-Rodríguez et al. 2008, 2010; Miles et al. 2016; Shubina et al. 2023) provide the most detailed description of the morphological evolution of the coma for Centaur 29P/SW1 before and after outbursts. The authors highlight the initial stellar appearance during an outburst, followed by the development of an extended coma with varying shapes and characteristics. Typically, an asymmetric coma (which changes over time), fan-shaped comae (consistent with an initial expansion phase), and occasional brightening of the outer border of the coma are detected in the Centaur (Figure 8.4). During outbursts, one-, two- and fourfold symmetries were observed in a coma.

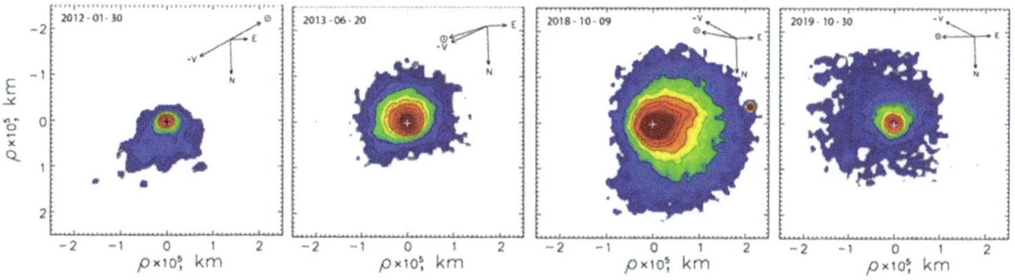

Figure 8.4. Intensity maps of Centaur 29P/Schwassmann–Wachmann 1 in the R filter obtained on 2012-01-30, 2013-06-20, 2018-10-09, and 2019-10-30 (images adapted from Shubina et al. 2023). The directions to the North, the East, the Sun, and the negative heliocentric velocity vector of the Centaur as seen in the observer's plane of the sky are noted for each date. The optocenter is indicated by the cross mark. Reprinted from Shubina et al. (2023). Copyright 2023, with permission from Elsevier

Miles (2016) pointed out that observed symmetry may arise when a crustal plate flexes and lifts, allowing near-simultaneous material to escape from fissures around its perimeter driven by the accumulated pressure in a subsurface cavity. Moreover, Miles et al. (2016) noted recurring patterns of outbursts within the expanding coma during 2010, 2011, and 2012, appear to be separated by intervals of 51–65 days. This may suggest that a specific source on the nucleus undergoes periodic outbursts, maintaining its characteristics afterward. Using a simple geometric model (Shubina et al. 2023) to describe the jet structure for the entire observation period made it possible to determine some parameters of the nucleus. From this scenario, the rotation period of the nucleus equals 57±2 d (similar was obtained from Spitzer data; Stansberry et al. 2004), and the axis of rotation is directed approximately towards the observer with the propagated direction of spin. Since the active areas that form the jets are located in a narrow belt, the jet structures practically lie in the sky plane (for observations of Centaur 29P/SW1 from 2012 to 2019).

Another active Centaur for which model interpretation of the coma morphology was made is 174P/Echeclus (Rousselot et al. 2016). A series of results from existing studies impose constraints on the model's input parameters. 174P/Echeclus was active for 1 yr, from the end of 2005 to the end of 2006, so the active processes on the nucleus's surface had started sometime before this period and ceased in 2006 (Bauer et al. 2008; Rousselot et al. 2016). The primary outburst documented in 2005 was succeeded by three subsequent smaller events: one in May 2011 (Jaeger et al. 2011), exhibiting a 2–3 visual magnitude increase; another in August 2016, showing a 2.5–3 visual magnitude rise; and a third in December 2017, marked by a 4–4.5 visual magnitude escalation (Rousselot et al. 2021). The morphology of the coma images cannot be explained within the framework of isotropic substance outflow. Several individual active zones were likely activated on the nucleus's surface. The most prominent morphological detail in the Centaur's coma cannot be attributed to a fragment of the nucleus that could have been ejected from its surface, as within this detail, a local maximum associated with this hypothetical fragment cannot be identified (Rousselot et al 2016). If we do not insist on the ejection of a fragment from the nucleus, most likely, we are dealing with a process of short-term activity, possibly induced by a collision with an external body (Neslušan 2014). The coma's morphological structure in the Sun's direction formed due to the outflow of substance from the nucleus, which started earlier than the outflow of the substance that formed the bright detail in the tail direction of the object. The lack of a brightly pronounced maximum in the brightness distribution of this detail indicates that the activity of the substance outflow process significantly decreased before the observations.

Moreover, the orientation of this feature may be determined by the tilt of the nucleus's rotation axis. The model calculations clarified the possible scenario of the formation of the brightness distribution inside the coma. The morphology, which was analyzed using images of 174P/Echeclus obtained in March 2006 on the 8.2 m VLT telescope (ESO), is explained by the occurrence of two short-term active processes on the surface of the nucleus in a brief interval around December 7, 2005. Since the processes were short-lived (a few hours), the outflow of material from these areas on the nucleus's surface was no longer occurring during observation. Only a residual dust cloud formed due to these processes was recorded. The non-collimated particle source became active earlier, around November 15, 2005. The activity of the source subsided by March 2006, but it was still active at a significantly lower level compared to the maximum on December 22, 2005.

Modern morphology models, applicable to comets (Vincent et al. 2010; Ivanova et al. 2023; Samarasinha et al. 2023; Kleshchonok & Sierks 2025) and active Centaurs (Moreno 2009), utilize anisotropic emission to explain the presence of features. In these models, isolated active regions on the nucleus release gas and dust, creating spatial variations in the coma. As material from different jets may originate from various sources (Voitko et al. 2022), this mechanism naturally accounts for different properties (density, ejection velocity, particle sizes, etc.). Active regions can represent areas with higher or lower gas and dust production, possibly due to compositional variations in ices-inducing activity, or they may manifest as craters or fractures on the surface.

The collimated jet model explains all active structures. For instance, narrow jets are highly collimated flows that create a high-density region expanding radially from the optocenter. Fans are most probably generated by larger active areas or as the overlay of multiple narrow jets. Depending on specific observational conditions, spiral structures of various forms may form active regions at the equator. Similar to comets, the surface layer of an individual active Centaur plays a crucial role in regulating and sustaining the activity of these celestial objects. Model analyses have revealed that even with homogeneous gas outflow, features in the coma can form due to the effects of large-scale core topography (Rousselot et al. 2016).

Kareta et al. (2021) present visual wavelength images (sloan g', r', and i') of P/2019 LD2 (ATLAS), an object on a very-near Centaur orbit. With its orbital perihelion distance just inside that of Jupiter's, some would categorize it as excluded from the Centaur population. However, as most of its orbit falls within the Giant Planet region and since it has recently evolved inward (e.g., Steckloff et al. 2020, Lilly et al. 2024), it makes sense to consider it as a Centaur. Its morphology appears typically comet-like, as it exhibits a pronounced extended tail (Kareta et al. 2021).

Despite the dominance of collimated jet models in comet morphology analysis, recent models (Crifo et al. 2002; Szegö et al. 2002) have become more complex, incorporating more detailed gas dynamic physics. Comprehensive models must consider multidimensional, time-dependent imaging of gas and dust in various flow regimes (collisional, transitional, and free-streaming), dissociation and photochemistry in the coma, thermal properties of the core, surface activity, and core topography effects. Hydrodynamic and collimated jet models represent different approaches to studying comet and active Centaur coma. Each has advantages and limitations, depending on observational constraints and adopted assumptions.

In general, despite the relatively limited research on the morphology of active Centaurs, the physical and dynamic properties of the object are key parameters determining the form and structure of the coma. Modeling using collimated jets, especially with the Monte Carlo method (Rousselot et al. 2016), is predominantly applied to analyze comets and active Centaurs. Observations of temporal changes in the coma provide information about the rotational and seasonal characteristics of the object, as well as the distribution of active areas on the surface. Simultaneous images at different wavelengths reveal information about the sizes of dust particles (e.g., Bauer et al. 2011), while differences in the color of jets and the overall coma indicate variations in material. Polarization measurements in the coma provide information about dust properties and constrain the models' parameters in combination with color analysis. Even without direct observations of structures in the coma, specific methods, such as analyzing spectral lines and color variations, allow conclusions to be drawn about their presence.

8.3.3 Scattering Properties of Dust in Active Centaurs

Advancements in the technology of developing modern light radiation detectors, coupled with the utilization of large telescopes, enable a significant expansion of the list of Centaurs capable of displaying activity. This also leads to the acquisition of new insights into the physical properties of dust within these objects. The primary source of information regarding the dust component of Centaurs is the radiation scattered by these objects. Among the most extensively developed and widely used remote methods for determining their physical characteristics is polarimetry. Our understanding of regolith particles and dust, including their sizes, structure, and chemo-mineralogical composition, heavily relies on the analysis of phase functions of brightness and polarization, as well as the wavelength dependence of these parameters. These dependencies are determined by the mechanisms of light scattering and are linked to the physical properties of the scattering medium, including the composition, size, shape, structure, and other characteristics of the particles. By comparing measurement data with the results of theoretical modeling and laboratory measurements, it becomes possible to deduce the most likely composition, structure, and certain other characteristics of the substances on the surfaces of quiescent Centaurs and in the atmospheres of active Centaurs. For well-known active Centaurs such as 95P/Chiron, (5145) Pholus, (10199) Chariklo, (29981) 1999 TD_{10}, and 174P/Echeclus, polarization observations have only been conducted during their quiescent phases (Rousselot et al. 2005; Bagnulo et al. 2006; Belskaya et al. 2010; Belskaya 2020; Ivanova et al. 2023). To date, polarimetric research during the active phase has been conducted only for one Centaur, namely, 29P/SW1.

The substantial perihelion distance of the Centaur of nearly 5.7 au indicates a low temperature of the nucleus of 29P/SW1, making it unlikely to experience significant water ice evaporation. This is in contrast to many other Jupiter-family comets, where water ice evaporation is a well-known driver of activity. Instead, it is presumed that the activity of 29P/SW1 is primarily governed by the evaporation of super volatile ices, with their emission lines being identified in spectroscopic observations (e.g., as reported by Korsun et al. 2008; Ivanova et al. 2018). On the other hand, the temporal variations in gaseous emissions within the coma of 29P/SW1 do not seem to closely align with changes in the brightness of the continuum, which corresponds to the elastic scattering of light by Centaur dust (Wierzchos & Womack 2020). This disparity introduces uncertainty regarding the mechanisms responsible for the release of dust from the nucleus of 29P/SW1. Moreover, the complexities surrounding the mechanisms driving the activity of this Centaur are compounded by its regular outburst events. Centaur 29P/SW1 is known to have experienced around a dozen major outbursts, each causing a substantial increase in brightness of approximately 5 magnitudes, within a single terrestrial year. This phenomenon has been well-documented, as reviewed by Miles et al. (2016).

The presence of morphological structures in the coma of active Centaurs and a change in color during the outbreak period should also appear in the form change in the scattering properties of dust in their comas. There are practically no polarimetric observations of Centaurs during their activity. For the first time, the results of polarimetric observations of Centaur 29P/SW1 were reported in the literature by Kiselev & Chernova (1979). They found high positive linear polarization and these values exceed all other measured polarizations observed at similarly small phase angle α (Ivanova et al. 2019b, 2021, 2023). Next polarimetric observations of the Centaur were conducted by Ivanova et al. (2016), including circular polarization (Rosenbush et al. 2014). Negative linear polarization in the coma of the Centaur was detected. The most recent polarimetric observations of 29P/SW1 were obtained during five nights in 2021 (Kochergin et al. 2021). It is the first polarimetric investigation of this Centaur measured during a period of quiescent activity. They found a nearly constant negative polarization during all periods of observations. In Figure 8.5, we plotted all polarimetric observations that were obtained for Centaur 29P/SW1 (data in the green and red filters) and presented in the polarimetric database (Kiselev et al. 2017) and paper (Kochergin et al. 2021).

Modeling the polarization of Centaur 29P/SW1 (Kochergin et al. 2021) suggests that its coma consists of at least two components, Mg-rich silicates (36%) and amorphous carbon (64%). It is worth noting that the presence of Mg-rich silicates in

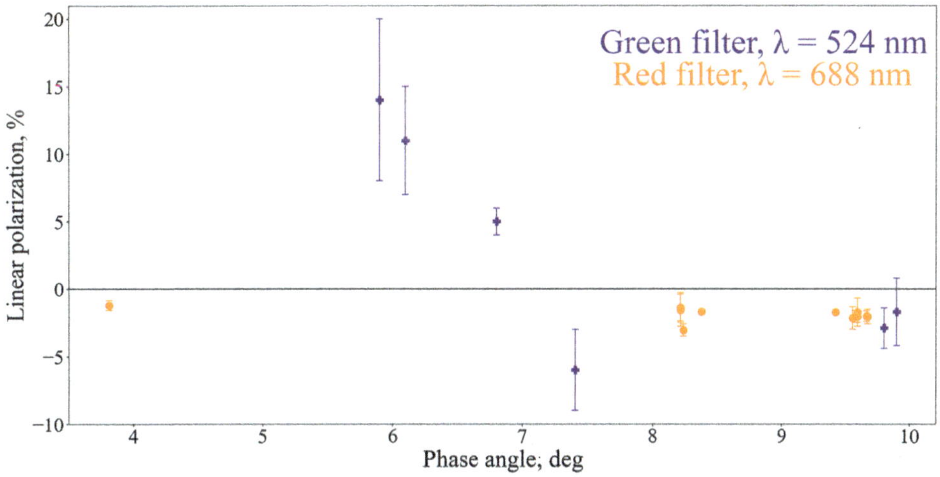

Figure 8.5. The phase-angle dependency of the linear polarization degree for Centaur 29P/Schwassmann–Wachmann 1 in an active period (data from Kiselev et al. 2017; Kochergin et al. 2021).

the 29P/SW1 was also suggested by Voitko et al. (2022) based on color variation in the dust coma of the Centaur. Nevertheless, the chemical composition of these particles remains somewhat uncertain, primarily because of the restricted range of phase angles involved in the observations. To enhance our understanding and gain more confidence in the microphysical properties of the dust particles, additional polarimetric observations are required to further constrain the shape of the negative-polarization branch in Centaur 29P/SW1. A more detailed comparison of polarimetric investigations between active and inactive Centaurs is presented in Section 8.6.

8.4 Gas Production

Despite plentiful evidence for dust comae around many Centaurs, due to their large heliocentric distances detection of the volatiles in the gas phase responsible for the observed activity has proven difficult to achieve for more than a handful of Centaurs. In this section we discuss observation of gas production in Centaurs to date; we also refer the reader to Chapter 6 for further discussion of volatiles in Centaurs.

8.4.1 Early Observations of Optical Emission Lines

The first gas emission detected from a Centaur was CO^+ emission in 29P/SW1, observed at optical wavelengths. Unlike "normal" comets closer to the Sun, the optical spectrum is dominated by CO^+ emission features (with some sign of CN as well), providing the first indications that 29P/SW1's gas production may be dominated by CO (Cochran et al. 1980, 1982; Larson 1980; Cochran & Cochran 1991). Later, ${N_2}^+$ emission was also reported in the spectrum of 29P/SW1 (Korsun et al. 2008), suggesting an N_2/CO ratio of $\sim$1%. The only other reported detection of optical emission lines in an active Centaur is CN emission in 95P/Chiron (Bus et al. 1991).

8.4.2 Sub-mm Emission Lines

Most of the reports of emission from neutral species in active Centaurs come from sub-mm observations of CO. CO emission was reported for 29P/SW1 by Senay & Jewitt (1994), and has been routinely detected in 29P/SW1 ever since (e.g., Gunnarsson et al. 2008; Wierzchos & Womack 2020). Womack & Stern (1999) reported the detection of CO emission in 95P/Chiron. The Centaur with the most recent detection of CO emission is 174P/Echeclus, for which CO was detected during an outburst in 2017 (Wierzchos et al. 2017).

Recent observations of 29P/SW1 have extended the inventory of molecules observed in this object at sub-mm wavelengths to include H_2O and HCN. Bockelée-Morvan et al. (2022) used Herschel observations to provide spectrally resolved observations of H_2O, where the much broader line width compared to CO suggested an extended source for the water production. HCN also showed a similar line width, indicating an extended source for these species as well. NH_3 was searched for but not detected.

8.4.3 IR Observations

While very powerful for searching for a number of species, such as CO, CO_2, H_2O, C_2H_6, and CH_4, IR observations of gas-coma volatiles in Centaurs have been difficult to achieve, and detections are currently limited to 29P/SW1 (Faggi et al. 2024), 39P (Harrington Pinto et al. 2023), and 95P/Chiron (Pinilla-Alonso et al. 2024) though this is changing quickly in the era of JWST (see Section 8.6 and Chapter 14). In all three of these JWST-observed active Centaurs, CO_2 coma emission has been reported with other emission species detected, such as CO in 29P/SW1 (Faggi et al. 2024). Emission of CH_4 is also seen in the JWST observation of 95P/Chiron (Pinilla-Alonso et al. 2024) showing different gas emission morphologies between the CO_2 and CH_4 coma species, which may suggest different source regions for these gases. In any case, it is clear that these species are replete (especially CO_2) amongst the Centaurs, and are drivers of activity in some significant sense. Observations with AKARI revealed H_2O and CO emission, and placed a stringent upper limit on the CO_2/CO ratio of <1%, though much of the H_2O emission observed was likely due to an icy grain source (see Bockelée-Morvan et al. 2022 and Section 4.2). Observations with CRIRES on the VLT detected CO, and provided upper limits on production rates of other species such as C_2H_6 and CH_4 (Paganini et al. 2013). Later observations during a major outburst using IRTF iSHELL (Roth et al. 2021) did not result in any new detections, but improved the upper limits previously established by Paganini et al. (2013) and performed searches for sulfur bearing compounds CS and OCS. NEOWISE observations in 2010 revealed infrared excess in the WISE spacecraft's 29P/SW1 images that could be attributable to CO production at levels of several times 10^{28} molecules per second (Bauer et al. 2015; Fernández et al. 2020). Yet, about 56% of the Centaurs have their complete orbits within 16 au (Table 8.1), where CO should be quickly depleted (Brown 2000) from their surfaces within the $\sim 10^6$–10^7 yr Centaur lifetimes (see Chapter 3 and references therein), so these reservoirs of CO, and possibly other volatiles, are likely subsurface.

Observations of Centaurs in the infrared have derived several size measurements and constraints of the nuclei (Stansberry et al. 2008; Bauer et al. 2013; Schambeau et al. 2019, 2021). For the Centaurs observed in the infrared while active, infrared images taken of 174P/Echeclus in 2006 (Bauer et al. 2008), are morphologically similar to those taken of 29P/SW1 in 2010 by the WISE spacecraft (Fernandez et al. 2020), as demonstrated in Figure 8.6. Both dominated by dust emission, at the longer wavelengths the signals were likely dominated by emission from larger particles. Analysis of the 174P/Echeclus data suggested that the majority of the mass ejected by larger particles was several times greater than residing in the smaller coma particles. If the majority of dust particles are from more explosive outbursts, then more shallow particle size distributions may result, and so be consistent with the 174P/Echeclus analysis.

8.5 Dynamics

The static picture of the orbital element distribution of active Centaurs lends little beyond what has been explained in that the orbital dynamics of the Centaur can in

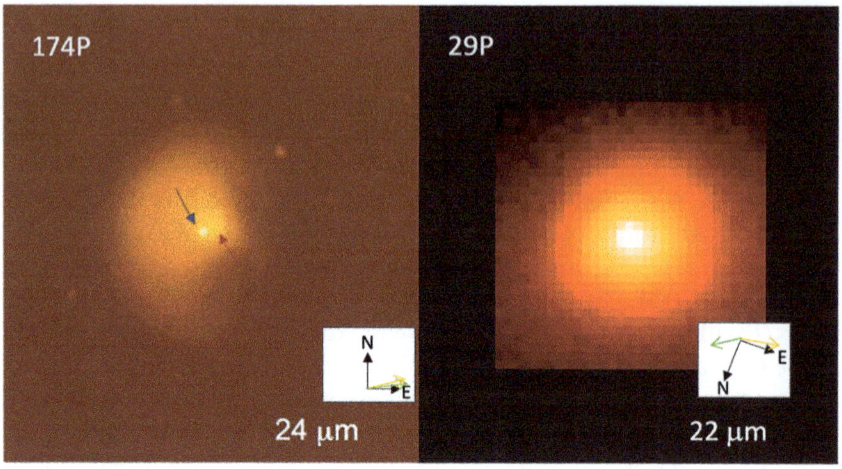

Figure 8.6. Comparison images of 174P/Echeclus (Bauer et al. 2008) and 29P/SW1 (from observations described in Fernandez et al. 2020) from Spitzer's 24 micron channel and WISE/NEOWISE 22 micron channel, respectively. The left panel shows the image, approximately 5 arcmin on a side, taken on 2006 February 24, while 174P/Echeclus was still active. The red and blue arrows indicate the coma peak and the nucleus, respectively. The right panel was from an image stack of 29P/SW1 with mid-exposure date on 2010 May 3, at the same angular scale. The North (N) and East (E) directions on each image are shown in the white box insets, as are the Sun-comet vector (yellow arrow) and the comet's anti-velocity vector. Left panel reproduced from Bauer et al. (2008)

part determine the level of activity (see Chapter 3 for a complete discussion of Centaur dynamics). Table 8.1 lists the osculating elements of the active Centaur population. Figure 8.1 shows that most of the active Centaurs are on orbits that bring them close to Jupiter's, i.e., are part of the Centaur population that experience the most insolation. The notable exceptions were few, namely C/2014 F3, 95P/Chiron, and 174P/Echeclus most prominently. Comets tend to be discovered when they are most active, i.e., near their perihelion, as with the active Centaurs (Figure 8.7), neglecting any potential discovery biases. Other comparisons of osculating orbital elements in total reveal no significant trends with active Centaurs that set them apart from the rest of the Centaur population; the orbital distribution is largely driven by giant planet perturbations. Studies of orbital integrations of Centaurs over time indicate that a Centaur's inclination is quickly evolved to the mean distribution by close planetary encounters (e.g., Volk & Malhotra 2013). Recent work by Lilly et al. (2024) reveals that all active Centaurs recently underwent large changes in semi-major axis on timescales of less than decades. Such "a-jumps" may trigger activity through mechanisms like the amorphous-to-crystalline ice transition (e.g., Prialnik et al. 2004; Lisse et al. 2022) that may release interstitially trapped volatiles like CO, or alternatively expose subsurface volatile reservoirs to inward propagation of heat from insolation, driving the sublimation of the volatiles.

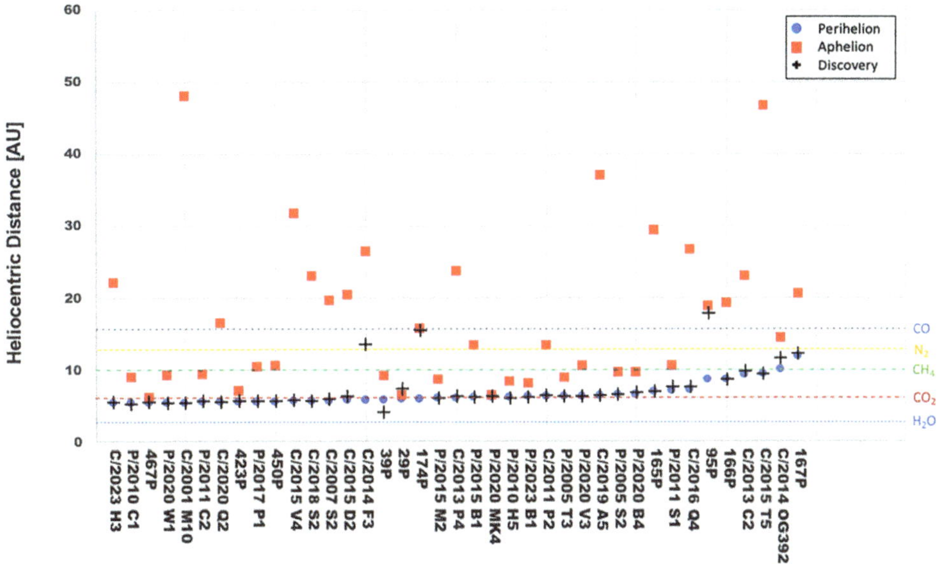

Figure 8.7. The discovery distance (black +) relative to the comet's perihelion (blue circles) and aphelion distances (red squares) at the current epoch. Note that some orbital parameters have changed since the object's discovery. Distances for when surface production rates reach $\sim 10^{21}$ molecules m^{-2} s^{-1} for the common volatile species CO, CO_2, CH_4, and N_2 (Meech & Svoren 2004; Cowan & A'Hearn 1979; Brown 2000). With the notable exception of 95P/Chiron, 174P/Echeclus, and C/2014 F3 (Sheppard-Trujillo), most Centaurs are discovered near perihelion.

8.6 Future Observations

8.6.1 Remote Observations

8.6.1.1 JWST

As discussed in Section 8.3, direct detections of the gas phase in Centaurs has been rare, with 29P/SW1 being by far the best characterized object. However, this is changing in the era of JWST, which is making observations of volatiles like H_2O, CO_2, and CO much more practical for a sample of objects. Harrington Pinto et al. (2023) reported the first detection of CO_2 in a Centaur, 39P/Oterma, based on analysis of JWST NIRSpec observations. Five other Centaurs including 29P/SW1 were also observed with JWST NIRSpec, with early results for 29P/SW1 presented in Harrington Pinto et al. (2023), McKay et al. (2023) and Faggi et al. (2024). Preliminary analysis of the other observed objects in the sample shows a diversity of CO_2/CO ratios (Harrington Pinto et al. 2023). See Chapter 14 for further discussion of Centaur observations in the JWST era.

8.6.1.2 HST

The Hubble Space Telescope (HST) will continue to view Centaurs at high resolution in the optical and UV wavelengths. Fainter fine-grained coma can be detected using HST imaging. Many species that were used as standard comparisons for activity like OH (e.g., A'Hearn et al. 1995), can be observed more readily with this space-based platform. Furthermore, though not yet seen in any Centaur, OH may be produced as a water-dissociation product following sublimation of icy grains. Species like CN have been detected in Centaurs (Bus et al. 1991), and can also be measured at lower activity with HST observations. Polarization and the quantity of reflected light reflected from smaller grains can complement observations from JWST. The recently-implemented single-gyro mode will limit HST's ability to track objects within the orbit of Mars, but will allow for tracking objects at Centaur distances.[2]

8.6.1.3 LSST (and Other Ground Based Surveys)

The next generation of future NEO surveys from ground based observation platforms will be able to provide constant coverage of the available night sky. The Vera C. Rubin Observatory's Legacy Survey of Space and Time (LSST) will provide the most coverage "at depth", i.e., to the faintest levels afforded by the observatory's 8-meter telescopic collection aperture. In combination with other sky surveys, which often employ more solar system discovery optimized exposure cadences, monitoring of Centaur activity will likely be routine enough to identify most outburst events. While the LSST's exposure cadence is a compromise to support solar system moving object and inertial frame object programs of study, the survey will have the distinct advantage of reaching single-exposure visual magnitudes as faint as 24.5 (Juric et al. 2023) and will be able to detect extended PSFs. These observations will not only provide context for targeted observations, they can serve to identify events of emerging activity. LSST will facilitate community-based analysis through its data-broker systems, and can enable target-of-opportunity observations on other high-demand facilities (like HST and JWST); see Chapter 14 for further discussion of LSST. Larger (25–30 meter class) ground-based observation platforms can still add to the study of active Centaurs. Larger-aperture telescopes may increase sensitivity to weakly active Centaurs and the study of their coma morphology, expanding the number of those with manifest activity, even beyond Saturn's orbit.

8.6.1.4 NEO Surveyor

The Near Earth Object Surveyor mission (NEO Surveyor) will be amongst the set of next-generation surveys that will contribute significantly to our understanding of cometary behavior. For Centaurs, this survey will be less sensitive than LSST, but, as the WISE survey demonstrated, will promote thermal emission from large-grain dust from active Centaurs in the vicinity spanned by the Jovian and Saturnian orbits. As Bauer et al. (2013) notes, coma dust thermal signal for the Centaur 29P/SW1 was present in the WISE 11.56 micron channel, for single exposures of approximately 9 s durations. NEO Surveyor will provide exposures of 145 s (Mainzer et al. 2023) per visit, or about 580 s per 6–9 hr period. Hence, the dust-signal will likely be apparent

[2] E.g., https://science.nasa.gov/mission/hubble/observatory/design/hubble-one-gyro-mode/

at sensitivity thresholds several times lower than NEOWISE. The NEO Surveyor will also be sensitive to CO and CO_2 emission, and will serve to identify those objects that produce large quantities of these species, i.e., those active Centaurs whose activity is driven by CO or CO_2. Because NEO Surveyor has a regular survey cadence, these observations will be untargeted, but will sample the activity at regular intervals, so that the Centaur's nucleus size, the large grain dust production, and the $CO+CO_2$ production will be constrained or measured throughout a Centaur's orbit.

8.6.2 Future Missions

The NASA Discovery 2019 proposal round had two missions proposed that would be investigations of active Centaurs. The proposed *Chimera* mission (Harris et al. 2019) would have investigated 29P/SW1, using its unique proximity to Jupiter to put the spacecraft in orbit around 29P/SW1 and monitor the body through much of its orbit. Another mission, Centaurus (Singer et al. 2019), proposed flybys of 29P/SW1 and 95P/Chiron to obtain first-image snapshots of two active Centaur bodies. These missions have laid the groundwork for future mission concepts, like the CORAL mission study for the future New Frontiers mission call.

8.6.2.1 New Frontiers Missions – CORAL

CORAL[3] is a mission concept to study a Centaur from orbit and *in situ*, exploring one of a population of dynamically evolved but compositionally primitive small icy bodies from the Kuiper belt that currently reside between Jupiter and Neptune. The proximity of Centaurs provides an opportunity to conduct a comprehensive study of the geochemical and physical properties of primordial ice-rich planetesimals, which trace the composition of nebular volatiles such as H_2O, CO_2, CO, and NH_3, revealing the nature of early solar system compositional reservoirs. The mission concept would map the surface and measure the ices and organics *in situ* with the following science objectives:

- Determine the chemical and physical properties of a Centaur to understand the nature of primitive planetesimals.
- Perform *in situ* elemental, isotopic, and organic analyses of a Centaur to develop a comprehensive understanding of the composition and initial conditions of the protoplanetary disk.
- Determine the shape, topography, geological landforms, and density of a Centaur to understand the evolutionary history of this population of objects.
- Determine degree of aqueous alteration on a Centaur to investigate the biologic potential of icy planetesimals and potential brine reservoirs.

See Chapter 16 for an in-depth discussion of Centaur missions.

[3] As detailed in NASEM (2023), Appendix C: Technical Risk and Cost Evaluation of Priority Missions and NASA (2021), CORAL: Centaur Orbiter and Lander. Mission Concept Study Report for the Planetary Science and Astrobiology Decadal Survey 2023–2032. Greenbelt, MD: NASA Goddard Space Flight Center. https://tinyurl.com/2p88fx4f.

8.7 Conclusions

Centaur activity spans the gaps between primitive bodies in the outer solar system and their evolution into the routinely active Jupiter-family comet population. Though they represent a fraction of the total known Centaur populations (~13%), recent studies suggest:

1. CO and CO_2 are ubiquitous and play an important role in driving the activity of Centaurs at Jovian distances and beyond. Other species, like CH_4, also may drive activity at further heliocentric distances.
2. Dynamically, rapid evolution, in the form of changes in semi-major axis over decade timescales, are commonly linked with activity.

Further study of dust and gas emissions of active Centaurs remains the most likely way of filling the knowledge gaps of volatile emplacement and transport throughout the early and present solar system.

References

A'Hearn, M. F., Millis, R. C., Schleicher, D. O., Osip, D. J., & Birch, P. V. 1995, Icar, 118, 223
Bagnulo, S., Boehnhardt, H., Muinonen, K., et al. 2006, A&A, 450, 1239
Barkume, K. M., Brown, M. E., & Schaller, E. L. 2008, AJ, 135, 55
Barucci, M. A., Brown, M. E., Emery, J. P., & Merlin, F. 2008, in The Solar System Beyond Neptune, ed. M. A. Barucci, et al. (Tucson, AZ: Univ. Arizona Press) 143
Bauer, J. M., Walker, R. G., Mainzer, A. K., et al. 2011, ApJ, 738, 171
Bauer, J. M., Choi, Y.-J., Weissman, P. R., et al. 2008, PASP, 120, 393
Bauer, J. M., Grav, T., Blauvelt, E., et al. 2013, ApJ, 773, 22
Bauer, J. M., Stevenson, R., Kramer, E., et al. 2015, ApJ, 814, 85
Belskaya, I. N., Bagnulo, S., Barucci, M. A., et al. 2010, Icar, 210, 472
Belskaya, I. N. 2020, Polarimetry of Transneptunian Objects and Centaurs V1.0. urn:nasa:pds:compil.tno-centaur.polarimetry::1.0. NASA Planetary Data System, http://doi.org/10.26033/8mqj-m072
Belton, M. J. S. 2010, Icar, 210, 881
Betzler, A. S., Almeida, R. S., Cerqueira, W. J., et al. 2017, AdSpR, 60, 612
Betzler, A. S., & de Sousa, O. F. 2023, ATel, 16137, 1
Bockelée-Morvan, D., Biver, N., Schambeau, C. A., et al. 2022, A&A, 664, A95
Brown, M. E. 2000, AJ, 119, 977
Bus, S. J., A'Hearn, M. F., Schleicher, D. G., & Bowell, E. 1991, Sci, 251, 774
Bus, S. J., A'Hearn, M. F., Bowell, E., & Stern, S. A. 2001, Icar, 150, 94
Choi, Y.-J., Weissman, P., Chesley, S., et al. 2006, CBET, 563, 1
Cochran, A., Barker, E. S., & Cochran, W. 1980, AJ, 85, 474
Cochran, A. L., Cochran, W. D., & Barker, E. S. 1982, ApJ, 254, 816
Cochran, A. L., & Cochran, W. D. 1991, Icar, 90, 172
Cowan, J. J., & A'Hearn, M. F. 1979, M&P, 21, 155
Crifo, J.-F., Rodionov, A. V., Szegö, K., & Fulle, M. 2002, EM&P, 90, 227
de Bergh, C., Schaller, E. L., Brown, M. E., et al. 2013, ApSSL, 356, 107
de la Fuente Marcos, C., de la Fuente Marcos, R., Licandro, J., et al. 2021, A&A, 649, A85
De Sanctis, M. C., Capaccioni, F., Ciarniello, M., et al. 2015, Natur, 525, 500

Delsemme, A. H. 1982, IAU Colloq., 61, 85
Dones, L., Weissman, P. R., Levison, H. F., & Duncan, M. J. 2004, in Comets II, ed. M. C. Festou, H. U. Keller, & H. A. Weaver (Tucson, AZ: Univ. Arizona Press) 153
Duffard, R., Lazzaro, D., Pinto, S., et al. 2002, Icar, 160, 44
Faggi, S., Villanueva, G. L., McKay, A., et al. 2024, NatAs, 8, 1237
Fernandez, Y. R., Meech, K. J., Pittichova, J., et al. 2001, IAU Colloq., 7689, 3
Fernandez, Y. R., Bauer, J. M., Kramer, E. A., et al. 2020, 51st Annual Lunar and Planetary Science Conf. (Houston, TX: LPI) 1802
Guilbert-Lepoutre, A. 2011, AJ, 141, 103
Guilbert-Lepoutre, A. 2012, AJ, 144, 97
Gunnarsson, M., Bockelée-Morvan, D., Biver, N., Crovisier, J., & Rickman, H. 2008, A&A, 484, 537
Gutiérrez, P. J., Ortiz, J. L., Alexandrino, E., Roos-Serote, M., & Doressoundiram, A. 2001, A&A, 371, L1
Harrington Pinto, O., Kelley, M. S. P., Villanueva, G. L., et al. 2023, PSJ, 4, 208
Harrington Pinto, O., Womack, M., Fernandez, Y., & Bauer, J. 2022, PSJ, 3, 247
Harris, W., Woodney, L., & Villanueva, G. 2019, EPSC-DPS Joint Meeting 2019 (Washington, DC: AAS) 2019
Hoffmann, M., Fink, U., Grundy, W. M., & Hicks, M. 1993, JGR, 98, 7403
Horner, J., Evans, N. W., & Bailey, M. E. 2004, MNRAS, 354, 798
Ivanova, O., Afanasiev, V., Rosenbush, V., & Kiselev, N. 2016, 41st COSPAR Scientific Assembly, (Amsterdam: Elsevier) B0.4–25–16
Ivanova, O., Agapitov, O., Odstrcil, D., et al. 2019a, MNRAS, 486, 5614
Ivanova, O., Luk'yanyk, I., Kolokolova, L., et al. 2019b, A&A, 626, A26
Ivanova, O., Rosenbush, V., Luk'yanyk, I., et al. 2023, A&A, 672, A76
Ivanova, O. V., Skorov, Y. V., Korsun, P. P., Afanasiev, V. L., & Blum, J. 2011, Icar, 211, 559
Ivanova, O. V., Picazzio, E., Luk'yanyk, I. V., Cavichia, O., & Andrievsky, S. M. 2018, P&SS, 157, 34
Jaeger, M., Prosperi, E., Vollmann, W., et al. 2011, IAU Colloq., 9213, 2
Jewitt, D. C. 2002, AJ, 123, 1039
Jewitt, D. 2009, AJ, 137, 4296
Jewitt, D., Chizmadia, L., Grimm, R., & Prialnik, D. 2007, in Protostars and Planets V, ed. B. Reipurth, D. Jewitt, & K. Keil (Tucson, AZ: Univ. Arizona Press) 863
Juric, M., Heinze, A., Jones, R. L., et al. 2023, Asteroids, Comets, Meteors Conf. 2851, 2083
Kareta, T., Sharkey, B., Noonan, J., et al. 2019, AJ, 158, 255
Kareta, T., Woodney, L. M., Schambeau, C., et al. 2021, PSJ, 2, 48
Kiselev, N. N., & Chernova, G. P. 1979, SvAL, 5, 156
Kiselev, N., Shubina, E., Velichko, S., et al. 2017, NASA Planetary Data System, urn:nasa:pds:compil-comet:polarimetry: 1.0, 2017
Kleshchonok, V., & Sierks, H. 2025, Icar, 425, 116300
Kochergin, A., Zubko, E., Chornaya, E., et al. 2021, Icar, 366, 114536
Korsun, P. P., Ivanova, O. V., & Afanasiev, V. L. 2008, Icar, 198, 465
Kowal, C. T., Liller, W., & Chaisson, L. J. 1977, IAU Colloq., 3147, 1
Larson, S. M. 1980, ApJL, 238, L47
Lilly, E., Jevčák, P., Schambeau, C., et al. 2024, ApJL, 960, L8
Lin, Z.-Y. 2023, PASJ, 75, 462

Lin, H. W., Chen, Y. T., Lacerda, P., et al. 2014, AJ, 147, 114
Lisse, C. M., Steckloff, J. K., Prialnik, D., et al. 2022, PSJ, 3, 251
Luu, J. X., Jewitt, D. C., & Trujillo, C. 2000, ApJL, 531, L151
Mainzer, A. K., Masiero, J. R., Abell, P. A., et al. 2023, PSJ, 4, 224
Mazzotta Epifani, E., Perna, D., Dotto, E., et al. 2017, A&A, 597, A59
Mazzotta Epifani, E., Dotto, E., Ieva, S., et al. 2018, A&A, 620, A93
Mazzotta Epifani, E., Dall'Ora, M., Perna, D., Palumbo, P., & Colangeli, L. 2011, MNRAS, 415, 3097
McKay, A. J., Harrington-Pinto, O., Faggi, S., et al. 2023, Asteroids, Comets, Meteors Conf. 2851,; Houston, TX: LPI) 2513
Meech, K. J., & Svoren, J. 2004, in Comets II, ed. M. C. Festou, H. U. Keller, & H. A. Weaver (Tucson, AZ: Univ. Arizona Press) 317
Meech, K. J., Buie, M. W., Samarasinha, N. H., Mueller, B. E. A., & Belton, M. J. S. 1997, AJ, 113, 844
Miles, R., & Faillace, G. 2011, EPSC-DPS Joint Meeting 2011 (Washington, DC: AAS) 279
Miles, R., Faillace, G. A., Mottola, S., et al. 2016, Icar, 272, 327
Miles, R. 2016, Icar, 272, 387
Moreno, F. 2009, ApJS, 183, 33
National Academies of Sciences, Engineering, and Medicine (NASEM) 2023, Origins, Worlds, and Life: A Decadal Strategy for Planetary Science and Astrobiology 2023–2032 (Washington, DC: The National Academies Press)
Neslušan, L. 2014, P&SS, 101, 162
Paganini, L., Mumma, M. J., Boehnhardt, H., et al. 2013, ApJ, 766, 100
Peixinho, N., Lacerda, P., Ortiz, J. L., et al. 2001, A&A, 371, 753
Peixinho, N., Thirouin, A., Tegler, S. C., et al. 2020, in The Trans-Neptunian Solar System, ed. D. Prialnik, M. A. Barucci, & L. Young (Amsterdam: Elsevier) 307
Pereira, C. L., Braga-Ribas, F., Sicardy, B., et al. 2024, MNRAS, 527, 3624
Picazzio, E., Luk'yanyk, I. V., Ivanova, O. V., et al. 2019, Icar, 319, 58
Pinilla-Alonso, N., Licandro, J., Brunetto, R., et al. 2024, A&A, 692, L11
Prialnik, D., Brosch, N., & Ianovici, D. 1995, MNRAS, 276, 1148
Prialnik, D., Benkhoff, J., & Podolak, M. 2004, in Comets II, ed. M. C. Festou, H. U. Keller, & H. A. Weaver (Tucson, AZ: Univ. Arizona Press) 359
Prialnik, D. 2006, Asteroids, Comets, Meteors, 229, 153
Roemer, E. 1958, PASP, 70, 272
Rosenbush, V., Ivanova, A., Kiselev, N., et al. 2014, Asteroids, Comets, Meteors Conf., (Houston, TX: LPI) 450
Roth, N. X., Bonev, B. P., DiSanti, M. A., et al. 2021, PSJ, 2, 54
Rousselot, P., Korsun, P. P., Kulyk, I., Guilbert-Lepoutre, A., & Petit, J.-M. 2016, MNRAS, 462, S432
Rousselot, P., Levasseur-Regourd, A. C., Muinonen, K., & Petit, J.-M. 2005, EM&P, 97, 353
Rousselot, P. 2008, A&A, 480, 543
Rousselot, P., Kryszczyńska, A., Bartczak, P., et al. 2021, MNRAS, 507, 3444
Samarasinha, N.H., Schambeau, C.A., & Bauer, I 2023, Asteroids, Comets, Meteors Conference, 2851, 2192
Schambeau, C.A., Fernández, Y.R., Samarasinha, N.H., Woodney, L.M., & Kundu, A. 2019, AJ, 158, 259

Schambeau, C.A., Fernández, Y.R., Samarasinha, N.H., Womack, M., Bockelée-Morvan, D., Lisse, C.M., & Woodney, L.M. 2021, PSJ, 2, 126
Secull, T., Fraser, W. C., Puzia, T. H., Fitzsimmons, A., & Cupani, G. 2019, AJ, 157, 88
Sen, A. K., Botet, R., Vilaplana, R., Choudhury, N. R., & Gupta, R. 2017, JQSRT, 198, 164
Senay, M. C., & Jewitt, D. 1994, IAU Colloq., 5929, 1
Shubina, O., Kleshchonok, V., Ivanova, O., Luk'yanyk, I., & Baransky, A. 2023, Icar, 391, 115340
Sickafoose, A. A., Bosh, A. S., Emery, J. P., et al. 2020, MNRAS, 491, 3643
Sickafoose, A. A., Levine, S. E., Bosh, A. S., et al. 2023, PSJ, 4, 221
Sierks, H., Barbieri, C., Lamy, P. L., et al. 2015, Sci, 347, aaa1044
Singer, K. N., Stern, S. A., Stern, D., Verbiscer, A., & Olkin, C. 2019, EPSC-DPS Joint Meeting 2019 (Washington, DC: AAS) EPSC-DPS2019-2025
Stansberry, J., Grundy, W., Brown, M., et al. 2008, in The Solar System Beyond Neptune, ed. M. A. Barucci, et al. (Tucson, AZ: Univ. Arizona Press) 161
Stansberry, J. A., Van Cleve, J., Reach, W. T., et al. 2004, ApJS, 154, 463
Steckloff, J. K., Sarid, G., Volk, K., et al. 2020, ApJL, 904, L20
Szegö, K., Crifo, J.-F., Rodionov, A. V., & Fulle, M. 2002, EM&P, 90, 435
Tegler, S. C., Bauer, J. M., Romanishin, W., & Peixinho, N. 2008, in The Solar System Beyond Neptune, ed. M. A. Barucci, et al. (Tucson, AZ: Univ. Arizona Press) 105
Tholen, D. J., Hartmann, W. K., Cruikshank, D. P., et al. 1988, IAU Colloq., 4554, 2
Trigo-Rodríguez, J. M., García-Melendo, E., Davidsson, B. J. R., et al. 2008, A&A, 485, 599
Trigo-Rodríguez, J. M., García-Hernández, D. A., Sánchez, A., et al. 2010, MNRAS, 409, 1682
van Biesbroeck, G. 1927, PA, 35, 586
Vincent, J.-B., Böhnhardt, H., & Lara, L. M. 2010, A&A, 512, A60
Vincent, J.-B., Farnham, T., Kührt, E., et al. 2019, SSRv, 215, 30
Voitko, A., Zubko, E., Ivanova, O., et al. 2022, Icar, 388, 115236
Volk, K., & Malhotra, R. 2013, Icar, 224, 66
Wierzchos, K., Womack, M., & Sarid, G. 2017, AJ, 153, 230
Wierzchos, K., & Womack, M. 2020, AJ, 159, 136
Womack, M., & Stern, S. A. 1999, SoSyR, 33, 187
Wood, J., Horner, J., Hinse, T. C., & Marsden, S. C. 2017, AJ, 153, 245
Zubko, E., Videen, G., & Kulyk, I. 2020, RNAAS, 4, 75

Chapter 9

The Near-Centaur Environment: Satellites, Rings, and Debris

A A Sickafoose, S M Giuliatti Winter, R Leiva, C B Olkin, D Ragozzine and L M Woodney

The unexpected finding of a ring system around the Centaur (10199) Chariklo opened a new window for dynamical studies and posed many questions about the formation and evolutionary mechanisms of Centaurs as well as the relationship to satellites and outbursting activity. As minor planets that cross the orbits of the giant planets, Centaurs have short dynamical lifetimes: Centaurs are supplied from the trans-Neptunian region and some fraction migrates inward to become Jupiter-family comets. Given these dynamical pathways, a comparison of attributes across these classifications provides information to understand the source population(s) and the processes that have affected these minor planets throughout their lifetimes. In this chapter we review the current knowledge of satellites, rings, and debris around Centaur-like bodies, discuss the observational techniques involved, place the information into context with the trans-Neptunian Objects, and consider what the results tell us about the outer solar system. We also examine open questions and future prospects.

9.1 Introduction

This chapter focuses on the presence (or lack thereof) of satellites, rings, and/or surrounding debris found around bodies in the Centaur region. Table 9.1 contains all the Centaurs and closely-related objects for which these features have been detected.

9.1.1 Satellites, Rings, and Debris in the Centaur Region

There are no known Centaur binaries using the definition of Centaurs as objects with perihelion distances $q > 5.2$ au and heliocentric semimajor axes $a_{\odot} < 30$ au. However, there are two related objects that are interesting to consider: these are

doi:10.1088/2514-3433/ada267ch9

Table 9.1. Published Detections of Satellites, Rings, and Debris around Centaurs and Giant Planet Crossers (GPCs).

Object	q (au)	$a_\odot$ (au)	Satellites	Rings	Debris	Type[a]	References[b]
Centaurs:							
(2060) Chiron	8.6	13.7	–[c]	Maybe[d]	Yes	I:O:IR	Bus et al. (1996) Elliot et al. (1995) Ruprecht et al. (2015) Ortiz et al. (2015, 2023) Sickafoose et al. (2023)
(10199) Chariklo	13.1	15.8	–[c]	Yes	–[e]	O;IR	Braga-Ribas et al. (2014) Leiva et al. (2017) Bérard et al. (2017) Morgado et al. (2021) Santos-Sanz et al. (2023)
(60558) Echeclus	5.8	10.8	–[c]	No[f]	Yes	I;O;IR	Rousselot et al. (2016) Bauer et al. (2008) Pereira et al. (2024a)
29P/Schwassmann—Wachmann 1 (SW1)	5.8	6.0	–[c]	No[f]	Yes	I;O;IR	Jewitt (1990) Wierzchos & Womack (2020) Buie et al. (2023)
GPCs:							
(42355) Typhon	17.5	37.5	Binary	–[e]	–[e]	I	Noll et al. (2006) Grundy et al. (2008) Stansberry et al. (2012)
(65489) Ceto	17.7	99.2	Binary	–[e]	–[e]	I	Grundy et al. (2007)

Notes: Centaurs are defined as $q > 5.2$ au and $a_\odot < 30$ au and GPCs have 5.2 au $< q <$ 30 au. We report heliocentric semimajor axis ($a_\odot$) and perihelion distance (q). Most objects are detected near perihelion.

[a] Observation type: I—imaging (visible wavelength), O—stellar occultation, IR—infrared imaging or spectroscopy.

[b] See also references within the cited works; these targets have been extensively observed, as reported in Chapters 10 and 11.

[c] HST has observed these objects and would be sensitive to ~km sized satellites at separations of $\gtrsim$0.5 arcseconds (typically $\gtrsim$7000 km) and larger satellites at closer separations. For example, at ~700 km, an equal-brightness binary could be detected.

[d] For Chiron, secondary detections in stellar occultations have been interpreted as a jet (Bus et al. 1996; Elliot et al. 1995), a shell or arcs (Ruprecht et al. 2015), a two-ring system (Ortiz et al. 2015), and most recently as changing material (Ortiz et al. 2023; Sickafoose et al. 2023).

[e] Empty entries indicate rings and/or debris have neither been detected nor ruled out.

[f] Stellar occultations were used to place upper limits on ring material for Echeclus, and initial light curves did not show signs of ring-like dips for SW1 (see Section 9.3.1; Pereira et al. 2024a; Buie et al. 2023).

Giant Planet Crossers (GPCs), with perihelion distances $5.2 < q < 30$ au (see Chapter 3). The two known GPC binaries are (42355) Typhon-Echidna and (65489) Ceto-Phorcys.

The number of Centaurs that are known to have rings is likewise low, at one or possibly two objects (see Table 9.1): Chariklo (rings dubbed C1R and C2R) and (2060) Chiron. The ringed objects are two of the largest Centaurs, with effective diameters $\gtrsim 200$ km. The first discovery of a small-body ring system is relatively recent and was achieved via a stellar occultation in 2013 (Braga-Ribas et al. 2014). The success rate of this technique has benefited from advances in astrometric measurements as well as instrumentation and larger campaigns (see Chapter 11). Notably, the subsequent discovery of ring material around the two large trans-Neptunian objects (TNOs) (136108) Haumea and (50000) Quaoar (Ortiz et al. 2017; Morgado et al. 2023) prompts consideration of whether rings form in the Centaur region or survive the transition from the trans-Neptunian region (e.g., Araujo et al. 2016).

Finally, while there are tens of known active Centaurs (see, e.g., Chapter 8 and Chandler et al. 2020), only three Centaurs have had observable surrounding debris (Table 9.1 and references therein): Chiron, (60558) Echeclus, and 29P/ Schwassmann–Wachmann 1 (SW1). We consider debris to be material consisting of particles larger than the gas and dust typically released from comets and that could have sufficiently significant optical depth to be detected in stellar occultations. Fewer than a dozen Centaurs have been successfully observed using occultations, leaving ample space for more discoveries of rings and debris. Note that many comets could qualify as GPCs surrounded by debris. For the purposes of this chapter, we consider only the GPC binaries and the GPCs for which stellar occultations have been reported. More details are provided about the known satellites, rings, and debris of Centaur-like bodies in Sections 9.2-9.4.

9.1.2 Overview of Observing Techniques

Satellites, rings, and debris can be or have been detected around Centaur-like objects through both direct and indirect methods, including occultations, imaging, spectroscopy, and photometry-generated light curves. The applications of these techniques are discussed briefly in this section.

9.1.2.1 Direct-Detection Methods

Stellar occultations: Six Centaurs have published stellar occultation results: Chiron (e.g., Bus et al. 1996), (8405) Asbolus (Rommel et al. 2020), (54598) Bienor (Fernández-Valenzuela et al. 2023, 2017), Chariklo (e.g., Braga-Ribas et al. 2014), Echeclus (Pereira et al. 2024a), and (95626) 2002 GZ_{32} (Santos-Sanz et al. 2017); also see references in Table 9.1. Additionally, stellar occultations have been reported for Centaur 2008 YB_3, the GPC 2014 YY_{49}, and the resonant TNO (591376) 2013 NL_{24} (Strauss et al. 2021), the latter of which meets our criteria for being a GPC.

Stellar occultations are a well-established method for detecting and characterizing planetary rings (e.g., Elliot et al. 1978; Bosh et al. 2002). In a stellar occultation, the

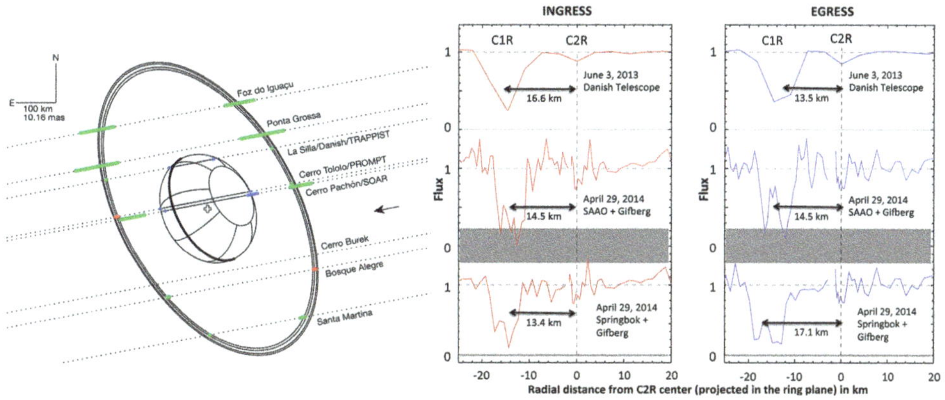

Figure 9.1. Example of ring detections from stellar occultation data. Left: Sky-plane view of Chariklo with occultation chords in 2013. (Reproduced from Braga-Ribas et al. 2014 with permission from Springer Nature.) The green segments show locations of ring detections, with the length of the segments representing the uncertainties (red represents ring locations during camera readouts). Blue segments detecting the central object allowed constraint of the geometry of the rings. Right: Radial profiles of Chariklo's rings during stellar occultations. (Reproduced from Bérard et al. 2017 © 2017. The American Astronomical Society. All rights reserved.) The flux of the star plus Chariklo is normalized to unity, and dips indicate ring material blocking the stellar flux. The ring profiles at the time of discovery are shown in the top panel for the inner and outer rings, named C1R and C2R, respectively, as a function of radial distance. The lower panels show more detailed structure a year later, indicating longitudinal variations in the distance between the rings.

planetary body is observed passing in front of a distant star. The starlight acts as a probe of the body and its vicinity. Extinction of the starlight due to rings or debris can provide astrometric information about the location and an indication of the amount of surrounding material (see Figure 9.1). In order to derive accurate relative positions, an occultation of the rings/debris and primary body are both needed. This can be accomplished with one observing site or with separate observations of each component by different observers if the separate sites have precise timing and well-known coordinates for the observing locations. Today, multiple-site scenarios are enabled by global navigation satellite systems that provide accurate position and timing services.

Stellar occultations are recorded with a series of images, preferably accurately timed. Each image in the sequence provides a unique probe of the small body system as the apparent location of the star moves relative to the body. High-speed imaging allows derivation of spatial information on a scale that is smaller than the diffraction limit of direct-imaging systems. The cadence of observations is driven by the required signal-to-noise ratio (SNR) of the data, which depends on the science objective of the observations. There is a trade between achieving high SNR and achieving high spatial sampling. Higher SNRs can be achieved without compromising the spatial sampling of the data for brighter stars, lower shadow velocities, and/or larger telescope apertures.

The observed extinction due to ring, dust, or atmospheric particles depends on the wavelength of light being observed, particularly for particles of the same size or smaller than this wavelength. For this reason, additional information may be retrieved from a stellar occultation if multiple wavelengths are used in the observations. From the reported occultation results, only two Centaurs are thought to possibly have rings (Chariklo and Chiron), and Chariklo-like rings have been ruled out for two other Centaurs (Echeclus and SW1). Observations with higher SNR and/or spatial sampling are still needed to detect or rule out rings like those around Chariklo at all other Centaurs and GPCs.

Stellar occultations by comets are generally more difficult to observe than stellar occultations by Centaurs and TNOs. Complications include (i) less accurate predictions of the shadow path, because non-gravitational forces and the difficulty in detecting the nucleus of an active comet both degrade the accuracy (Miles & Kretlow 2018), and (ii) the typically small size of the nucleus, making the shadow path on the Earth narrow (Combes et al. 1983; Miles & Kretlow 2018). The difficulty of using occultations to characterize debris around outbursting objects is demonstrated by reported stellar appulses for comets SW1, Hale-Bopp, and 17P/Holmes (Fernández et al. 1999; Lacerda & Jewitt 2012; Miles & Kretlow 2021). The stars passed near, but not behind, the nuclei during these events. An optically thick coma was only detected within 100 km of Hale-Bopp's nucleus and no other significant debris was observed in the occultation data.

Multichord occultations are useful for determining sizes and shapes of small bodies. Currently, only Chariklo, Chiron, and Echeclus have sufficient occultation data to determine well-constrained, triaxial shapes (see Section 9.3.1, Pereira et al. 2024a, and Chapter 4). For the other objects, upper limits on diameters and axis ratios have been placed. Stellar occultations can additionally be used to discover satellites, such as occurred for Neptune's moon Larissa (Reitsema et al. 1982). However, the odds of this are low due to the typically small sizes of any moons and the need for favorable geometry. Stellar occultations are more likely to be successful for a satellite once its orbit is well established, as has been done for the TNOs Quaoar (Weywot; Fernández-Valenzuela et al. 2023) and Orcus (Vanth; Sickafoose et al. 2019).

Imaging: Direct imaging is the most productive method to discover and characterize binary objects. The sensitivity to detecting companions depends on factors such as the components' angular separation, position angle, and brightness ratio, translating into significant variations in the satellite detection limit among the objects observed so far. When the two objects in a binary are separated by at least one Point Spread Function Full Width at Half Maximum (PSF FWHM), then they can typically be resolved as more than one object, especially if they are near-equal brightness. Observations with a stable and small PSF are thus desirable to find binaries as close as possible, which has led to the extensive use of the Hubble Space Telescope (HST) for direct imaging campaigns. Indeed, all the objects in Table 9.1 have been observed by HST at least once. HST can often resolve near-equal-brightness binary components that are separated by $\gtrsim$ 0.05 arcseconds, which translates to $\sim$700 km at 20 au. Faint binary components are harder to detect close-in; however, Centaur satellites a few kilometers in size could potentially be detected at wide separations of several thousand kilometers. Whether multiple components can be discovered depends on the separation at the time of the

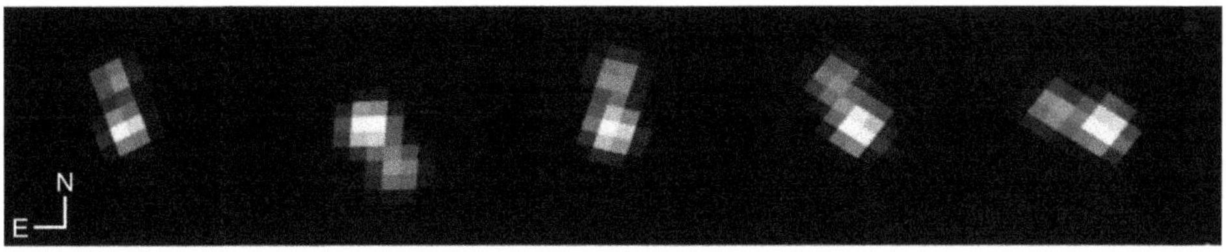

Figure 9.2. The binary GPC Ceto-Phorcys observed by HST. (Reprinted from Grundy et al. 2007, Copyright 2007, with permission from Elsevier.) The mosaic is 2.0 by 0.4 arcsec, with each frame centered on Ceto. From left to right, the images were taken between 2006 April (discovery) and May (followup).

observation: even orbits much larger than the resolution limit cannot be ruled out, because the observations could have occurred when the two objects happened to be aligned with respect to the observer. This is more common for edge-on orbits observed only once and might be avoided entirely for more face-on orbits.

Upon discovery of a binary, more HST time is typically requested and granted to make multiple observations in order to fully determine the parameters of the mutual orbit. For example, in Grundy et al. (2007), five HST observations of Ceto and its companion Phorcys allowed for the measurement of the period, semimajor axis, eccentricity, and total system mass (see Figure 9.2). In this case, a “mirror degeneracy” remains so that there are two possible orientations for the system in three-dimensional space. The standard has been to assume a Keplerian orbit, but new methods now allow for the exploration of non-Keplerian effects (e.g., Ragozzine et al. 2024; Proudfoot et al. 2024). These non-Keplerian effects are most typically caused by non-spherical shapes or unknown components. Modeling can thus provide additional information on shapes if there are sufficient observational constraints over a long enough time to detect the slow orbital changes due to non-Keplerian effects.

Planetary rings are not easily detectable in images because of the high contrast with the nearby parent body. Visible-wavelength imaging has been unsuccessful to date for observing Centaur rings, even with HST for the known system at Chariklo (Bérard et al. 2017). However, imaging of Centaur rings from space might be feasible at some point. Dusty rings have been observed by the *Spitzer* Space Telescope (albeit on a massive scale at Saturn; Verbiscer et al. 2009) and JWST has promising capabilities for studying ring systems (Tiscareno et al. 2016). Currently, direct detection of Chariklo’s rings is being investigated by subtracting out the signal from the nucleus for recent JWST observations (Santos-Sanz et al. 2023). Such measurements would provide a direct way to study the evolution of rings at higher temporal resolution than the currently-available method (stellar occultations).

Imaging at different wavelengths is currently the primary method to identify and characterize debris around Centaurs. Outbursting activity on Centaurs has been detected in visible-wavelength images (see Chapters 8 and 10). At these wavelengths, sunlight is scattered off of dust particles (as well as some ions and radicals) in the comae (Cochran & Cochran 1991). Dust and gas production rates can be determined by studying the brightness profiles of extended objects, where the PSFs do not match the field stars (e.g., A’Hearn et al. 1984; Fink et al. 2021). Changing color indices over time can further indicate varying dust-size distributions (e.g., Rousselot 2008). Infrared images and spectra can be used to study morphology, constrain particle sizes, and determine dust composition (e.g., Bauer et al. 2008; Stansberry et al. 2004). Furthermore, submillimeter data combined with visible data can be used to look for correlations between gas and dust outbursting (e.g., for CO; Wierzchos & Womack 2020). The detection of bigger debris in images would require searching for any separated components, like those seen in the break-up of active asteroid P/2013 R3 (Jewitt et al. 2014). An example analysis for ejected material is of a "secondary" source that was observed at Echeclus after outbursting, though this was determined to not be an ejected fragment but rather the result of localized cometary activity (Weissman et al. 2006; Bauer et al. 2008; Rousselot et al. 2016).

Spacecraft: The best way to directly detect small satellites and rings around bodies in the outer solar system would be to send a spacecraft (e.g., Singer et al. 2021). However, the cost and long timelines for spacecraft missions are often prohibitive (see Chapter 16 for a comprehensive discussion of Centaur missions). In addition, the debris environment must be sufficiently characterized to ensure a low risk of spacecraft damage (e.g., Fink et al. 2021).

9.1.2.2 Indirect Methods

Comparison of occultation and infrared data (binaries): An indirect way to detect the presence of satellites is comparing the sizes derived from infrared data and stellar occultations. Available thermal measurements for Centaurs are unresolved, providing only the total equivalent size. Among the infrared instruments used for thermal surveys of TNOs and Centaurs (e.g., Müller et al. 2009), *Spitzer*/MIPS had an angular resolution between 6 and 40 arcsec (Rieke et al. 2004), equivalent to a spatial resolution of 22,000–145,000 km at 5 au, while Herschel/PACS and SPIRE had 5 and 20–30 arcsec resolution, respectively (Poglitsch et al. 2010; Griffin et al. 2010). In comparison, stellar occultations have inherently higher spatial resolution, down to sub-km levels, allowing resolution of individual binary components. If the stellar occultation samples the main body of a multiple system (but not the companion), the multiplicity can still be evident as a smaller occultation size with respect to the thermal size. The presence of satellites due to such a discrepancy between thermal and occultation sizes has been proposed in a few cases for TNOs, although there is no known case so far for Centaurs (Rommel et al. 2023; Ortiz et al. 2020).

Light curves (binaries): The analysis of the shape, amplitude, and period of a rotational lightcurve is a complementary technique to search for satellites, by detecting multiple rotation periods, mutual events, or potential contact binaries. In the general case, a binary system can be revealed by the detection of the rotation period of a separate component, although this is rare in practice. Under favorable observing geometries, binaries can be detected by the occurrence of mutual events and contact binaries can often be revealed by their characteristic V-shape lightcurve if the geometry is near edge-on (see Figures 9.3 and 9.4). The statistical analysis of light curves has been used to deduce a population of equal-size synchronous binaries among Jupiter Trojans (Nesvorný et al. 2020) and to isolate a sample of contact-

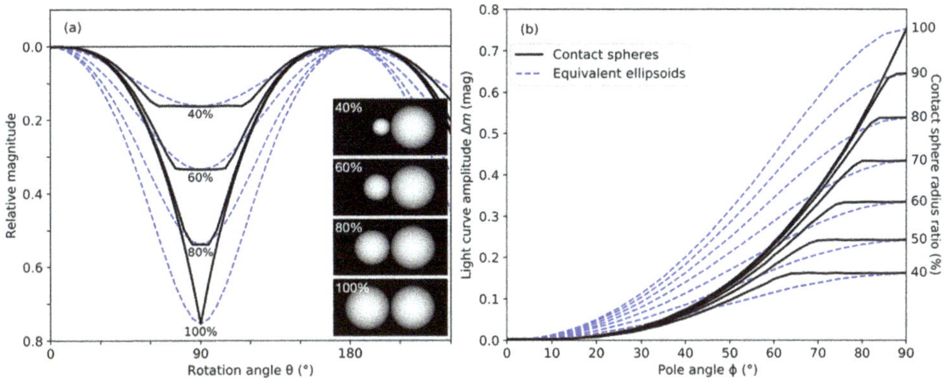

Figure 9.3. Left: Example model light curves for spherical (black) or ellipsoidal (blue) contact binaries of different relative sizes, exhibiting a characteristic V-shape. The pole angle is 90°. Right: Models of peak-to-peak lightcurve amplitude as a function of pole angle for contact binaries of different relative sizes. (Reprinted from Showalter et al. 2021, Copyright 2021, with permission from Elsevier.)

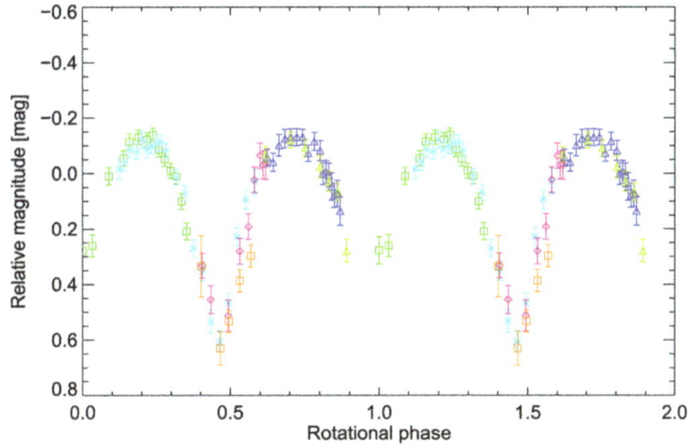

Figure 9.4. Example of a rotational lightcurve for the contact binary candidate and Plutino 2014 JQ_{80}, showing a characteristic V-shape. Data points of different colors were taken on different dates in May and June of 2017. (Reproduced from Thirouin & Sheppard 2018 © 2018. The American Astronomical Society. All rights reserved.)

binary candidates among the TNO population, revealing an unexpectedly high fraction among Plutinos (Thirouin & Sheppard 2018). An analysis of light curves to infer the shapes of bodies in the outer solar system further found that the contact-binary fraction is not well constrained and could be very high (Showalter et al. 2021).

So far, no contact binary has been reported among Centaurs or GPCs from lightcurve analyses (Thirouin et al. 2010). The main limitation for this technique is the need for dense photometry of a somewhat faint population, which in turn requires many hours of scarce, large-size telescope time. Based on the cumulative

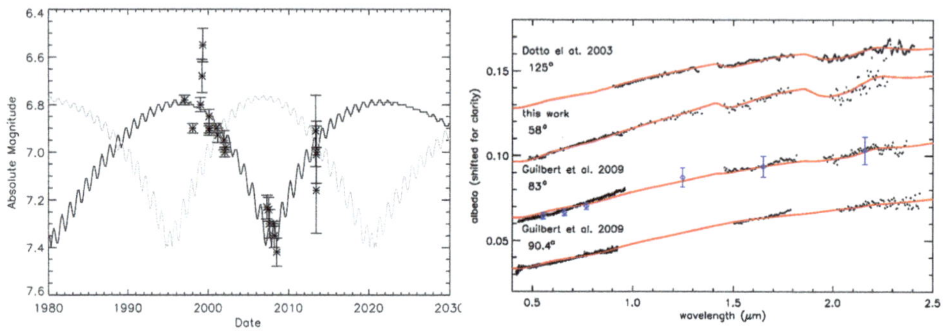

Figure 9.5. Variability over time in Chariklo's brightness (left) and reflectance spectra (right). Left: Model solutions for Chariklo's magnitude based on the two possible ring-plane poles are shown along with observational data. The darker line represents the preferred pole solution. Right: Model spectra are shown with data taken at different epochs, demonstrating changes in the typical water-ice absorption features at 1.5 and 2 μm. These results were attributed to the changing aspect angle of Chariklo's water-ice ring particles (Reproduced from Figures 1 and 5 in Duffard et al. 2014. CC BY 4.0).

distributions of lightcurve amplitudes, Centaur light curves (adopting the Minor Planet Center definition of Centaurs) have been found to be statistically different from the rest of the TNOs, presenting an abundance of low-amplitude objects (Showalter et al. 2021). The origin of this difference is currently unknown.

Magnitude and spectroscopic variations (rings or debris): As the viewing geometry of substantial surrounding material changes with time, there should be corresponding absolute magnitude and spectral changes. For example, moving from edge-on to fully open rings, the brightness should increase and there should be a deeper water-ice feature in the spectra (assuming icy ring particles, e.g., Hedman et al. 2013). The existence of rings around Chariklo was bolstered, and those of Chiron proposed, based on such data (see Figure 9.5 and Duffard et al. 2014; Ortiz et al. 2015).

9.2 Satellites

9.2.1 Characteristics of Known Centaur Binaries

There are two known Centaur-like binaries: the GPCs Ceto-Phoycys and Typhon-Echidna. Characteristics of these systems are listed in Table 9.2. After initial discovery, additional observations by HST were able to determine their mutual orbits. This process involves making multiple observations with precise relative astrometry and then fitting an orbital model, which includes the motion of the Earth (the HST-Earth relative position is unimportant), the heliocentric motion of the object, and the orbital parameters of the binary. Of particular interest is that the period and semimajor axis can be combined to measure the total mass of the system; measuring the masses of individual components requires absolute astrometry, which is generally not feasible. HST observations are the only practical way to measure astrometry for many binaries, so observations typically stop once a reasonably precise orbit can be inferred (e.g., Grundy et al. 2007). In some cases, there is a "mirror degeneracy" where the unknown direction of motion to/away from the

Table 9.2. Characteristics of Binary Giant Planet Crossers.

Parameter	Value	References
Binary: Ceto-Phorcys		
Mutual orbit period (days), P	9.544 ± 0.011	[1]
Semimajor axis (km), a	1840 ± 48	[1]
Eccentricity, e	0.008 ± 0.008	[1]
System Mass (10^{18} kg), M_{sys}	5.5 ± 0.3	[1]
Bulk density (g cm^{-3})^{a}, ρ	$1.37^{+0.66}_{-0.32}$; $0.64^{+0.16}_{-0.13}$	[1], [2]
System Albedo	$0.084^{+0.021}_{-0.014}$; 0.056 ± 0.006	[1], [2]
Radius of primary (km)^{b}	87^{+8}_{-9}; 223 ± 10	[1], [2]
Radius of secondary (km)^{b}	66^{+6}_{-7}; 171 ± 10	[1], [2]
Binary: Typhon-Echidna		
Mutual orbit period (days), P	18.971 ± 0.006	[3]
Semimajor axis (km), a	1628 ± 29	[3]
Eccentricity, e	0.526 ± 0.015	[3]
System Mass (10^{18} kg), M_{sys}	0.95 ± 0.05	[3]
Bulk density (g cm^{-3})^{a}, ρ	$0.60^{+0.72}_{-0.29}$; $0.36^{+0.08}_{-0.07}$	[4], [2]
System Albedo	$0.06^{+0.041}_{-0.021}$; 0.044 ± 0.003	[4], [2]
Radius of primary (km)^{b}	137 ± 30; 162 ± 7	[4], [2]
Radius of secondary (km)^{b}	77 ± 16; 89 ± 6	[4], [2]

[a] Under the assumption of equal densities for each body.
[b] Assuming equal-albedo spheres.
[1] Grundy et al. (2007)
[2] Santos-Sanz et al. (2012)
[3] Grundy et al. (2008)
[4] Stansberry et al. (2012)

observer leads to two precise orbits that cannot be distinguished in their orientation, even though key parameters such as mass and eccentricity can be determined. Recently, it has become possible to employ more advanced dynamical and statistical methods to look for hints of non-Keplerian motion, where the orbits are not assumed to be fixed ellipses, but can include precession most typically induced by the oblateness of the components, known as J_2 (Ragozzine et al. 2024; Proudfoot et al. 2024). These analyses are still quite limited by the small amount of data.

For Ceto-Phorcys, the binary period, semimajor axis, and eccentricity were determined even though the mirror degeneracy was not resolved because both the prograde and retrograde orbital solutions agreed (Grundy et al. 2007). The resulting total mass was combined with thermal modeling to return individual sizes for Ceto and Phorcys (Grundy et al. 2007). The system masses and sizes led to inferred densities, though these estimates did not include potential errors from systematic uncertainties in the thermal modeling or deviations from the assumption of equal-albedo, equal-density spheres. The implied system albedo is somewhat intermediate between TNOs and JFCs, and the bulk density from Grundy et al. (2007) requires that there must be a rock component. However, the bulk density derived by Santos-Sanz et al. (2012) is significantly lower and the sizes larger. The inferred compositions for this binary depend significantly on the poorly-understood porosity.

Originally discovered by Noll et al. (2006), Typhon-Echidna was then followed up for mutual-orbit determination by Grundy et al. (2008). The system mass in Table 9.2 assumes a nearly-Keplerian orbit. Although its size is similar to Ceto-Phorcys, the Typhon-Echidna binary has a larger eccentricity, which is very different from the effectively circular orbit of Ceto-Phorcys. Typhon's orientation was uniquely determined, and the system has a prograde orbit. The inferred system density is consistent with TNOs and JFCs of similar sizes.

Dynamical modeling of both binaries shows that there are not enough observations for a significant detection of non-Keplerian effects (Proudfoot et al. 2024). A single high-precision observation of Ceto would readily break the mirror degeneracy. Both binaries are good candidates for the discovery of slow non-Keplerian effects with only a small amount of new data, because any new observations would increase the baseline of observations by almost 20 years. The systems are also both on relatively compact orbits so that apsidal and nodal precessions due to non-spherical shapes could be significant and detectable.

9.2.2 Comparisons with Binary TNOs

Centaur-like objects typically start as "hot" TNOs (scattering, hot classical, resonant, and detached TNOs; see Chapter 3). Approximately 30% of Centaurs become JFCs (Levison & Duncan 1997) and the transition from Centaur to JFC is enabled by a low-eccentricity dynamical Gateway (Sarid et al. 2019). Comparisons of the binary frequency and characteristics between these related populations is thus interesting to consider. However, it is very challenging to make direct comparisons of the binary fractions because there are observational detection biases as well as uncertainties in the sizes of these populations.

Compared with the nearly 120 known binary TNOs, there are only two known binaries with Centaur-like orbits. Centaur binaries are not easy to detect. In 2006, direct imaging with HST detected Echidna around Typhon from a sample of eight objects, which included two Centaurs, (49036) Pelion and Bienor, and five other GPCs, 29981, 33128, 54520, 60608, and 87269 (Noll et al. 2006). A similar HST program targeting a sample of 12 Centaurs and six GPCs found Phorcys around Ceto (Grundy et al. 2007). Li et al. (2020) observed 23 Centaurs and 33 GPCs, finding no satellites and deriving an upper limit of 8% binaries in their sample. It's worth noting that Li et al. (2020) included the known binary Typhon-Echidna in the sample, but they did not detect any satellites in those observations. The HST archive indicates imaging observations for 58 Centaurs and 109 GPCs, but the associated programs were focused on more than just searching for satellites. Due to the different instrument configurations, filters, and observation strategies, it is difficult to accurately determine the rate of binary systems within the Centaur and GPC populations.

Ignoring biases and considering simply the frequency of binary detections versus numbers observed, the fractions of 0/58 Centaurs and 2/109 GPCs are interesting, especially given the estimate that ~10% of hot TNOs are known to be binaries (and ~30% of cold classical TNOs; Noll et al. 2020). There are also no known JFC binaries, although the majority of JFC nuclei are smaller than 10 km in effective diameter and have bilobate shapes (e.g., Thomas 2009; Safrit et al. 2021; Chapter 12). The decreasing binary fraction from the cold classical TNOs to the more dynamically excited populations in the trans-Neptunian region to the Centaur and to JFC populations could very easily be a real effect and not due entirely to observational biases.

Because the dynamical source of Centaur-like objects is known to have more observed binaries than the Centaurs themselves, we can speculate on the cause of the apparent differences in the binary fraction. One straightforward idea is that the binaries are affected by close encounters with giant planets, which are much more frequent for Centaur orbits compared to TNOs. Indeed, simulations have shown that widely-separated binaries are disrupted by scattering encounters with Neptune (Parker & Kavelaars 2010; Nesvorný & Vokrouhlický 2019; Stone & Kaib 2021). Most of these simulations have focused on the scattering that happens as hot TNOs are emplaced in the early solar system. That work is a helpful guide, but it is not as relevant as simulations that have looked at the present orbits of Centaurs specifically, such as Araujo et al. (2018).

In general, binary disruption will happen when external influences become more important than the binary's mutual gravitational attraction. One metric for this is the observed semimajor axis of the satellite, a_b, relative to the mutual Hill sphere of the binary, r_H, which is the distance at which the gravitational influence of the Sun would overwhelm the binary's self-gravity. Values of a_b/r_H of $\lesssim 1\%$ can be used to distinguish "tight" from "wide" binaries (although technically this metric is not strictly relevant for giant-planet encounters). Based on dynamical simulations and considering that the two known GPC binaries are both relatively tight systems, the general sense is that Ceto and Typhon have low probabilities of disruption after about 1 Myr, but their survival rate over tens of Myr (similar to their lifetimes) might be unlikely (e.g., Araujo et al. 2018). Perhaps they are the fortunate survivors of an initial binary fraction that is consistent with the source population.

Another possibility is that the direct comparison of binary fractions between Centaurs and TNOs is not appropriate because of observational biases. First, most discovery surveys find Centaur-like objects that are smaller than the corresponding TNOs simply due to the magnitude-limited nature of the observations. The binary frequency among TNOs as a function of size is not known, and perhaps at the ~100 km size range of Ceto and Typhon, TNOs actually have fewer binaries than the sample as a whole. Furthermore, binary discovery is often limited by angular separation, so that separations of Centaur-like binaries would be barely resolvable at TNO distances in images (see Section 9.1.2.1). For similar reasons, only binaries with larger a_b/r_H can currently be detected in the TNO region using direct imaging.

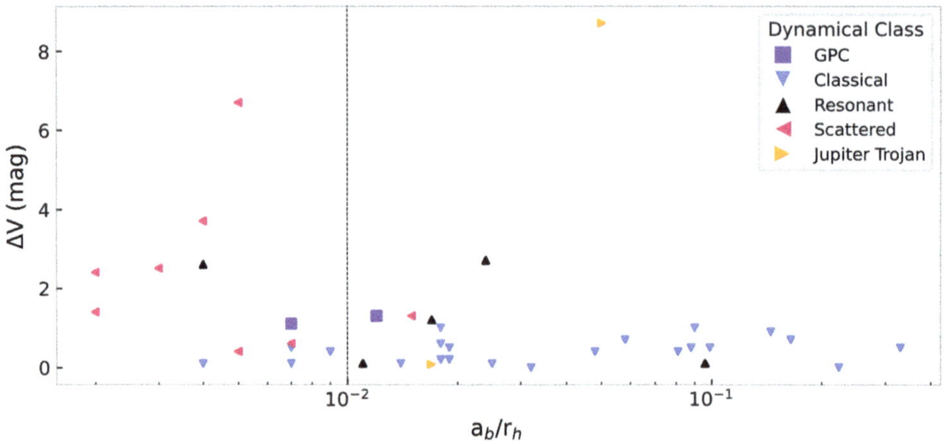

Figure 9.6. Comparison of binary characteristics for Centaur-like objects (GPCs) with other dynamical classes in the outer solar system. The horizontal axis is the ratio between the semimajor axis and the Hill radius and the vertical axis is the difference in apparent magnitudes of the binary components (as a proxy for their size difference). The vertical line indicates the $a_b/r_H = 1\%$ limit between "tight" and "wide" binaries. Data are from Grundy et al. (2019).

We can explore these possibilities by considering the properties of binaries from different dynamical classes. Gathering data on several binaries, we show in Figure 9.6 the size ratio of the binary (estimated by the difference in magnitudes, ΔV) versus the tightness of the binary measured by a_b/r_H, as a function of dynamical classification. The two GPC binaries have some of the smallest semimajor axes, appearing like an extension of tight TNO binaries. The GPC binary orbital characteristics are generally consistent with TNOs, including the moderate eccentricity of Typhon-Echidna. Though Centaur-like binaries are among the tightest known, there are a few TNO binaries that have been detected with similar characteristics, separations, and sizes, such as the Classical TNO (469514) 2003 QA_{91}. There are thus observational biases against finding TNOs with orbits like the known GPC binaries, but they are not overwhelming.

Nonetheless, relative observational bias implies that wide Centaur-like binaries must be more rare than tight Centaur-like binaries. This result seems consistent with estimates of dynamical destruction of these binaries. One potential observational signature that has not been explored is the discovery of Centaur or GPC "pairs," separate objects on very similar orbits, which could only plausibly be formed by a disrupted binary. We note that ultra-wide binaries that were the focus of Parker & Kavelaars (2010) are not particularly relevant to the Centaur-like population for two reasons. First, the widest binaries are generally cold classicals, which is likely a small source population for Centaurs. Second, there is new evidence that the observed ultra-wide binaries themselves could be modified from their original version by encounters with other TNOs (Campbell et al. 2023). The even more important influence of binary-binary encounters has not yet been studied in detail.

When objects reach a point in their orbit that activity becomes significant, it is also worth considering whether non-gravitational effects on one or both components would be enough to influence the binary orbit. This is an open area for research.

9.2.3 Binary Candidates

Tidal evolution can slow the spins of binary components until they are synchronous. This is especially likely for the tightest binaries that would be difficult to resolve. As a result, long-period ($\sim$ a few days) light curves may be indicative of an unknown binary. Two GPCs have been reported to have long-period light curves potentially indicative of binaries: (33128) 1998 BU_{48} (Sheppard & Jewitt 2002) and 2010 WG_9 (Rabinowitz et al. 2013). Both have been observed with HST with no companion reported. The existence of confined rings (or the presence of a clear gap in a ring system) may also imply shepherd or other small moons, but these cannot be directly inferred from present data.

9.3 Rings

9.3.1 Characteristics of Proposed Small-Body Ring Systems

Ring systems have been known around the giant planets for centuries: first Saturn's by Galileo Galilei in 1612 and later confirmed around Jupiter, Uranus, and Neptune with the advent of space missions and large telescopes (e.g., Charnoz et al. 2018). The data revealed different structures around the different planets, some of which is closely tied to satellites. Saturn has the most well-known system, surrounded by dense, tenuous, large, narrow, and arcing rings, with several moonlets in close connection with the ring particles. Notably, Saturn's narrow, dusty F-ring is located outside the Roche limit and its core is confined by the satellite Prometheus and precession (e.g., Cuzzi et al. 2014). Jupiter presents tenuous rings interior to the orbits of the Galilean satellites along with four small satellites, probably the source of these rings (e.g., Ockert-Bell et al. 1999). Uranus has several narrow rings, only a few kilometers wide, and the ϵ ring is confined by the shepherd satellites Cordelia and Ophelia (e.g., Elliot et al. 1977; Goldreich & Porco 1987; Porco & Goldreich 1987). Two faint rings (the μ and ν rings) were later discovered in HST images, lying outside the main ring system of Uranus (Showalter & Lissauer 2006). Ring arcs were discovered at Neptune via stellar occultations (e.g., Hubbard et al. 1986) and have continued to evolve (e.g., Souami et al. 2022). Spacecraft images have shown that the arcs are in fact the densest parts of the Adams ring, and they are likely radially confined by interactions with the moon Galetea (Namouni & Porco 2002) while small moonlets contribute to azimuthal confinement (Renner & Sicardy 2004; Giuliatti Winter et al. 2020). In all of these systems, small satellites (or moonlets) play an important role in defining the structures of the rings. The gravitational interactions between the ring particles and a satellite create waves and gaps, and they can confine the edges of the rings or the arcs against disruptive forces, which tend to scatter the ring particles.

Four centuries after Saturn's rings were first observed, the discovery of two narrow, dense rings around the largest Centaur Chariklo from occultation data in 2013 sparked a new field of study: rings around small bodies (Braga-Ribas et al. 2014). Basic characteristics for Chariklo's rings, and those of the other known small-body systems, are given in Table 9.3 and plotted in Figure 9.7. Chariklo's two-ring system has been well observed and characterized. Stellar occultations between 2014 and 2020: (i) confirmed the existence of C1R and C2R, along with the circular ring solution and pole position from 2013, (ii) indicated that the inner ring varied azimuthally in width by up to 4.3 km while the outer ring varied in width by up to 1 km, (iii) revealed W-shaped structure in the inner ring, and (iv) determined that Chariklo's shape is consistent with a triaxial ellipsoid with semi-axes $A = 143.8^{+1.4}_{-1.5}$, $B = 135.2^{+1.4}_{-2.8}$, $C = 99.1^{+5.4}_{-2.7}$ km (Bérard et al. 2017; Leiva et al. 2017; Morgado et al. 2021). The C1R ring-pole position has most recently been refined by Morgado et al. (2021). Additionally, the lack of differences in 2017 between the visible (0.45–0.65 μm) and red (0.7–1.0 μm) occultation data at the 1-σ level led to the conclusion that C1R contains particles mostly larger than a few microns in size (Morgado et al. 2021).

The situation at Chiron is not quite as straightforward. After its discovery in 1977, Chiron presented cometary-like activity (e.g., Meech & Belton 1989; Luu & Jewitt 1990; Elliot et al. 1995; see also discussion in Chapter 10). Data from a two-chord stellar occultation in 2011 were originally interpreted as showing a shell of material (Ruprecht et al. 2015). By combining these results with long-term photometry and spectroscopy, a two-ring system at ~324 km similar to that at Chariklo was proposed (Ortiz et al. 2015). Stellar-occultation data in 2018 and 2019 allowed better constraints on the size and shape of Chiron ($A = 126 \pm 22$, $B = 109 \pm 19$, $C = 68 \pm 13$ km, assuming a Jacobi equilibrium shape and considering the lightcurve amplitude; Braga-Ribas et al. 2023). Analysis of stellar occultations in 2018 and 2022 suggested that the proposed rings are instead surrounding material that is evolving in a very short period of time (Ortiz et al. 2023; Sickafoose et al. 2023). Based on the most recent observations, Ortiz et al. (2023) proposed that Chiron possesses a tenuous disk of material roughly 580 km wide, with concentrations of material located at 325 ± 16 km and 423 ± 11 km and pole ecliptic coordinates of $\lambda = 151° \pm 8°$ and $\beta = 18° \pm 11°$.

From a stellar occultation in 2017, a 70-km wide ring was also discovered around the TNO Haumea (Table 9.3; Ortiz et al. 2017). Compared to the Centaurs, Haumea is more distant ($q = 35$ au and $a_{\odot} = 43$ au), is an order of magnitude larger, and has a highly elongated triaxial shape with semi-axes $A = 1161 \pm 30$, $B = 852 \pm 4$, and $C = 513 \pm 16$ km (Ortiz et al. 2017). Haumea's ring is coplanar to the equator and to the orbit of the outer satellite Hi'iaka. The two known satellites are located tens of thousands of kilometers away from Haumea (e.g., Brown et al. 2007); therefore, their effects on the ring are minimal and the non-spherical shape of Haumea likely plays an important role in the dynamics of the particles (see Section 9.3.4).

Most recently, a ring system was reported around the TNO Quaoar (Table 9.3; Morgado et al. 2023). Quaoar's lightcurve has been interpreted to be single- or double-peaked, yielding a rotation rate of 8.84 or 17.69 hr (Ortiz et al. 2003). A non-

Table 9.3. Characteristics of Proposed Small-Body Ring Systems.

Object		Rotation (hr)	Location (km; radial distance)	Width (km)	Ring Eccentricity[a] e_r	Ring Pole α_p, δ_p (deg, J2000)[b]	References
Centaurs:							
(2060) Chiron[c]		5.92	325 ± 16 km	–	0	$160.2^{+13.2}_{-12.4}$, $27.8^{+13.9}_{-13.2}$	Marcialis & Buratti (1993)
			423 ± 11 km	–	0		Ortiz et al. (2023)
(10199) Chariklo	C1R	7.004	385.9 ± 0.4	~6.9[d]	$0.005 < e_r < 0.022$	151.03 ± 0.14, 41.81 ± 0.07	Fornasier et al. (2014)
							Morgado et al. (2021)
							Melita et al. (2017)
	C2R		399.8 ± 0.6	~3.5[e]	<0.017	150.91 ± 0.22, 41.60 ± 0.12	Braga-Ribas et al. (2014)
							Morgado et al. (2021)
TNOs:							
(50000) Quaoar							
(epoch 2018–2021)	Q1R	8.84; 17.69	4148.4 ± 7.4	8; 20–340[f]	0	259.82 ± 0.23, 53.45 ± 0.30	Ortiz et al. (2003)
							Morgado et al. (2023)
(epoch 2022)	Q1R		4,057.2 ± 5.8	5; 80–100[g]	0		Pereira et al. (2023, 2024b)
(epoch 2022)	Q2R		2,520 ± 20	~10[h]	0		Pereira et al. (2023, 2024b)
(136108) Haumea		3.9	2287^{+75}_{-45}	~70[i]	0	285.1 ± 0.5, –10.6 ± 1.2	Rabinowitz et al. (2006)
							Ortiz et al. (2017)

Notes: Opacities and optical depths are given in Table 9.4 and discussed in Section 9.3.2.

[a] Values of 0 are for rings assumed to be circular.

[b] Right ascension and declination of the ring-plane pole, α_p and δ_p, for the preferred pole solutions from Ortiz et al. (2023); Morgado et al. (2021); Pereira et al. (2023); Ortiz et al. (2017).

[c] Locations of concentrations of material, within an underlying ~580 km disk, from the most recent occultation observation in 2022. Data from previous occultations showed, among other detections, two distinct features at ~300 and ~310 km with widths ranging between 2.6 and 4.4 km (in 2011; Sickafoose et al. 2020) and between ~344 – 364 km with widths ranging from 2.2-4.5 km (in 2018; Sickafoose et al. 2023).

[d] The mean width for C1R was reported as 6.9 km in the text (not the abstract), ranging between 4.8 and 9.1 km (Morgado et al. 2021).

[e] Measured widths of C2R were $3.6^{+1.3}_{-2.0}$ km and $3.4^{+1.1}_{-1.4}$ km in Braga-Ribas et al. 2014. A typical radial width was not given for more recent observations, but measured values ranged from $0.095^{+0.015}_{-0.010}$km to $3.72^{+0.63}_{-0.53}$ km (Morgado et al. 2021).

[f] When detected, the adopted radial width was 8 km for the "densest part of the ring," while other measured widths ranged from 21.34 ± 3.13 to 336.34 ± 23.81 km (Morgado et al. 2023, 2024).

[g] When detected, the "dense part" of the ring was consistent with a Lorenztian shape extending 60 km with full-width-at-half-maximum of 5 km and, from the best light curves, the "tenuous part" was typically 80-100 km (Pereira et al. 2023, 2024b).

[h] Assuming coplanar with Q1R. The typical width was given as 10 km, with measured widths ranging from 6.8 ± 0.3 to 16.1 ± 3.3 km (Pereira et al. 2023, 2024b).

[i] Adopted width of 70 km, with measured ingress and egress values of ~74 & 44 km from the one site at which the ring profile was resolved (Ortiz et al. 2017).

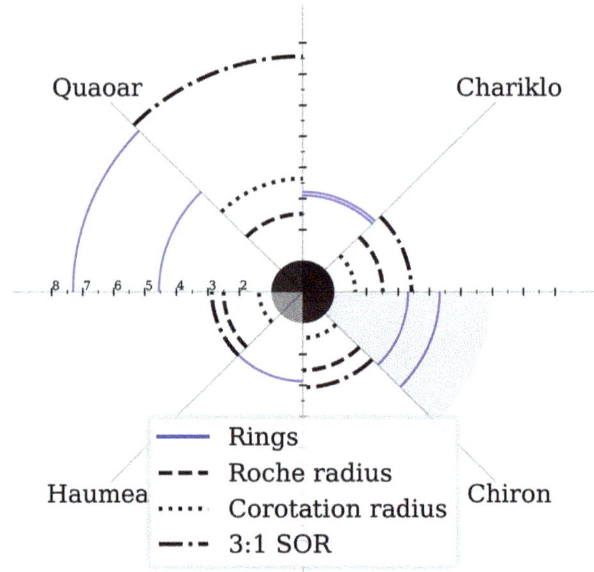

Figure 9.7. Comparison of proposed small-body ring systems for TNOs Quaoar and Haumea (left) and Centaurs Chariklo and Chiron (right). The radial distances are normalized by the volumetric-equivalent radii of the central bodies. The shading level of the central body represents the geometric albedo (from $\sim 0.1 - 0.7$). The locations of rings or concentrations of material are indicated by solid blue lines. The locations of the Roche radius, the 1:1 corotation radius, and the 3:1 spin–orbit resonance (SOR) are indicated with dashed, dotted, and dot-dashed lines, respectively (see Section 9.3.3 for the assumptions). For Chiron, the shaded area indicates the extension of a disk of material detected during a 2022 stellar occultation (Ortiz et al. 2023). Ring locations are from Table 9.3 and the Roche radii were calculated using the equation for a_{Roche} in Section 9.3.3 adopting $\rho_p/\rho_s = 2$ and $\gamma = 1.6$.

homogeneous ring, Q1R, with more non-detections in the occultation data than detections, presents an interesting scenario. It is located significantly beyond the Roche limit, regardless of the assumptions (see Section 9.3.3). A second proposed ring, Q2R interior to Q1R, is likewise beyond the Roche limit (Pereira et al. 2023). The variable nature of the occultation detections around Quaoar and the highly unusual locations of reported surrounding material make this a very compelling target for future study.

The locations of the small-body ring systems are intriguing because outside the Roche limit, particles can accrete into satellites in a very short period of time (see additional discussion in Section 9.3.3). Furthermore, the discovery of thin rings around small bodies is surprising because ring material will naturally disperse over relatively short timescales, (<1 Myr; e.g., Braga-Ribas et al. 2014; Pan & Wu 2016). An explanation is thus required for both how the rings can survive far from the Roche limit and how thin rings can be confined over long periods of time. Proposed mechanisms for formation and confinement are discussed in Section 9.5.2.

In terms of composition, small-body rings are thought to be made of water ice mixed with other materials (as are Saturn's rings; e.g., Nicholson et al. 2008).

For Chariklo and Chiron, correlations between proposed ring opening angles, visual brightness, and water-ice spectral features have implied that the rings contain water ice (see, e.g., Figure 9.5 and Duffard et al. 2014; Ortiz et al. 2015). A best-fit model for Chariklo's rings is 20% water ice, 40%–70% silicates, and 10%–30% tholins, with small amounts of amorphous carbon (Duffard et al. 2014). Mixtures of 20% or 30% water ice and 80% or 70% refractory material are compatible with near-infrared spectra for Chariklo and Chiron, respectively (Groussin et al. 2004), the water ice of which might all be in rings or surrounding material. Observations of Quaoar's rings are consistent with particles that have icy density (e.g., densities similar to that of the small inner saturnian satellites), and the rings have been modeled as frost-covered ice (Morgado et al. 2023).

There are four Centaurs for which upper limits have been placed on putative rings: Echeclus, SW1, Bienor, and 95626. Three stellar occultations by Echeclus have been published, with the best light curves at each epoch having 3σ upper limits on observed optical depth of 0.035 for features of width 2.6 km in 2019, 0.075 for features of width 2.9 km in 2020, and 0.08 for features of width 1.2 km in 2021 (Pereira et al. 2024a). For 100-km wide features, the upper limits for all observations ranged from 0.005 to 0.045 (Pereira et al. 2024a). Initial results from three successful occultation observations in 2022 and 2023 by SW1 did not show any obvious signs of extended structures (Buie et al. 2023). Stellar occultation data for Bienor in 2019 returned the strongest constraint for that body that a ring of width <14.1 km and opacity of 50% would not be detected (Fernández-Valenzuela et al. 2023). Finally, a stellar occultation by 95626 in 2017 returned the 3σ ring constraint (within the positive observations) that widths >3.7 and >1.8 km could have been detected for opacities of 50% and 100%, respectively (Santos-Sanz et al. 2017). For the best data near the limb, rings >1.1 or >0.5 km could have been detected for opacities of 50% and 100%, respectively (Santos-Sanz et al. 2017). These results rule out Chariklo-like rings at Echeclus and likely SW1, but the possibility of narrow and thin ring systems was not discarded for Bienor or 95626 (Pereira et al. 2024a; Fernández-Valenzuela et al. 2023; Santos-Sanz et al. 2017).

9.3.2 Published Values of Ring Optical Depths

It is worth noting that observed values of optical depths for Centaur rings have not been consistently reported in the literature. A stellar occultation lightcurve provides a measurement of the transmission, T, or the fractional amount of light observed when a star is blocked by ring material. Ring opacity along the line of sight, p, is related to transmission through $p = 1 - T$. Transmission is related to optical depth along the line of sight, τ_0, through $T = e^{-\tau_0}$ when assuming the ring is a gray screen many particles thick (e.g., Elliot et al. 1984). Converting to normal optical depth, τ_N, which provides the most physically-relevant information about the ring, requires knowing the ring opening angle, B, where $\tau_N = \tau_0 sin|B|$ for a polylayer ring (e.g., Elliot et al. 1984).

For Haumea, only line-of-sight opacities were reported (Ortiz et al. 2017). For Chiron, published values of optical depths for surrounding material were based on this standard calculation of transmitted flux (Sickafoose et al. 2020). In contrast, reported optical depths for Chariklo's rings and the proposed ring at Quaoar have included a factor of two. The relationship between transmission and optical depth for observations of Chariklo was defined as $T = e^{-2\tau_0}$ (Braga-Ribas et al. 2014; Bérard et al. 2017). For Quaoar, the normal optical depth was defined as $\tau_N = sin|B|\tau_0/2$ (Morgado et al. 2023; Pereira et al. 2023; noting that an additional factor of two was applied in the original manuscripts but revised in corrigenda; Morgado et al. 2024; Pereira et al. 2024b). According to Bérard et al. (2017) and Morgado et al. (2024), the factor of two stems from Cuzzi (1985), who found that the fill factor, τ_N/Q_e, had been underestimated for the Uranian rings and that the likely value of the Mie coefficient for those observations was $Q_e = 2$. Consistently, diffraction for a "zebra-striped" screen ($\tau = 0$ or ∞ for alternate stripes) returns an optical depth that is twice that of the optical depth produced by a gray screen of equivalent width (as described in section V.A of French et al. 1986).

The Mie coefficient comes into play when using optical depth to derive physical properties, such as particle sizes, densities, or reflectivities (following, e.g., Equation 1 in Cuzzi 1985). The ring characteristics of the small bodies are not well known; therefore, Table 9.4 contains ring opacities and optical depths calculated using the standard, gray-screen definition for all four systems so that nominal values can be compared. There is a wide range of optical depth measurements, especially when including error bars. We emphasize that the diffraction effects do need to be taken into consideration to infer characteristics of ring particles from the observed optical depths (French et al. 1986): if the rings are more "zebra-striped" than a polylayer gray screen, the factor of two described above comes into play and the *effective* optical depths can reach half of those in Table 9.4. Note that Table 9.4 does not contain the most recent stellar occultation by Chiron in 2022, as neither opacities nor optical depths were reported (Ortiz et al. 2023).

9.3.3 Consideration of the Roche Limit

By most classical calculations, the rings around small-bodies occur near or outside their Roche limits: the Quaoar system is exceptionally far, at more than twice this distance (see Figure 9.7 and Morgado et al. 2023). The Roche limit is the distance from the central body inside which tidal forces prevent the accretion of ring particles or can disrupt an orbiting body. It is expected that a ring system can exist inside this limit and a cluster of particles can accrete into a small moon beyond it. However, this limit depends on the mass of the body as well as the parameters of the object to be disrupted, such as density and internal material strength (Tiscareno et al. 2013).

For triaxial bodies, the Roche radius can be defined as $a_{Roche} = (4\pi ABC\frac{\rho_p}{\gamma\rho_s})^{1/3}$, where A, B, and C are the semimajor axes of the primary, ρ_p is the density of the primary, ρ_s is the density of the secondary, and γ is a dimensionless geometrical parameter describing the sphericity of the secondary (Tiscareno et al. 2013). Values of γ typically range from 0.85 to 1.6, the former as a limiting value for the equilibrium shape of an incompressible fluid (that may not be achievable for solid materials) and the latter representing a fully-filled Roche lobe with uniform density

Table 9.4. Consistently-Derived Opacities and Optical Depths for Proposed Small-Body Rings[a].

Body	Observation Date (UT)	Line-of-sight Opacity $(p)^b$	Line-of-sight Optical Depth $(\tau_0)^b$	Ring Opening Angle $(B;°)^c$	Range of Normal Optical Depths $(\tau_N)^b$	Nominal Normal Optical Depth $(\tau_N)^d$	Equivalent Width $(E_W;$ km$)^e$
Chariklo C1R[f]	2013 Jun 03	0.67–0.81	1.11–1.65	33.77	0.62–0.92	0.8	2.3–3.3
Chariklo C2R[f]	2013 Jun 03	0.13–0.35	0.14–0.43	33.77	0.08–0.24	0.12	0.1–1.0
Chiron F1[g]	1994 Mar 09	0.57–0.63	0.85–0.99	−57.9	0.72–0.84	0.78	2.7–4.9
Chiron inner[h]	2011 Nov 29	0.41–0.72	0.52–1.28	59.6	0.45–1.10	0.73	0.8–3.3
Chiron outer[h]	2011 Nov 29	0.33–0.62	0.40–0.97	59.6	0.35–0.84	0.63	0.8–2.2
Chiron[i]	2018 Nov 28	0.10–0.43	0.11–0.57	47.3	0.08–0.42	0.25	0.2–1.6
Haumea[j]	2017 Jan 21	0.55–0.56	0.80–0.82	−13.8	0.190–0.196	0.193	5.9–9.7
Quaoar Q1R[k]	2019 June 05[l]	0.04–0.11	0.04–0.12	−19.5	0.012–0.039	0.025	1.0–7.8[m]
	2021 Aug 27[n]	0.66–1.00	1.09–16.10	−19.3	0.36–5.32	2.84	1.1–3.9[p]
	2022 Aug 09[q]	0.03–0.85	0.03–1.87	−20.0	0.01–0.64	0.22	0.4–7.7[r]
Quaoar Q2R	2022 Aug 09[q]	0.02–0.06	0.02–0.06	−20.0	0.007–0.022	0.015	0.03–0.11[r]

[a] Assuming fractional transmission $T = e^{-\tau_0}$, $p = 1 - T$, and $\tau_N = \tau_0 \sin|B|$ for a polylayer ring, following Elliot et al. (1984).
[b] Full range of values when detected, including published error bars.
[c] Based the preferred ring-pole solutions from Braga-Ribas et al. (2014); Ortiz et al. (2015, 2017); Morgado et al. (2023) and published coordinates at the time of the occultations.
[d] Nominal value, either as cited or the average of the measured values.
[e] Defined as $E_W = W p_N$, where $p_N = p \sin|B|$ and the minimum and maximum radial ring widths, W, are from published values including errors.
[f] For each of the two rings, recalculated based on the values of τ_N in Braga-Ribas et al. (2014).
[g] Assuming minimum and maximum line-of-sight optical depths for the strongest feature (F1) resolved in the "KAO, optical" data in Elliot et al. (1995).
[h] Assuming minimum and maximum line-of-sight optical depths for each of the two features in the full-resolution "FTN" data from Sickafoose et al. (2020).
[i] Assuming minimum and maximum line-of-sight optical depths for all of the detected features from Sickafoose et al. (2023).
[j] Assuming line-of-sight ring opacities from the resolved ring profiles in Ortiz et al. (2017).
[k] Characteristics from the events with multiple detections are listed here: two additional events were reported in Morgado et al. (2023).
[l] Using the minimum and maximum apparent opacities on this date, specifically for "GTC i_s (bef.)" and "GTC r_s (aft.)" in the corrected version of Extended Data table 2 (Morgado et al. 2023, 2024).
[m] Using the minimum and maximum equivalent widths on this date, specifically for "GTC r_s (aft.)" and "GTC r_s (bef.)" in Morgado et al. (2023, 2024).
[n] Using the minimum apparent opacity on this date for "S. Valley (bef.)" and the maximum apparent optical depth from "Reedy Creek (bef.)" (we use the optical depth because the maximum apparent opacity for this date is given as $1.000 + 0.000$, which returns infinite optical depth[a]), in Morgado et al. (2023, 2024). No site observed ring material on egress.
[p] Using the minimum and maximum apparent opacities and ring widths for "Reedy Creek (bef.)" in Morgado et al. (2023, 2024).
[q] Using the minimum and maximum τ_N values for "Gemini (z")' for Q1R and "CFHT" ingress and "Gemini z"' egress for Q2R in Pereira et al. (2024b) (line-of-sight values were not provided). Note that Q2R was not detected during ingress for "Gemini (r")'.
[r] Using the minimum and maximum $W p_N$ from "Gemini (z")' on ingress and "TUHO" on egress for Q1R, and from "CFHT" on ingress and "Gemini (r")' on egress for Q2R. Note that fitted equivalent width values in Pereira et al. (2024b) differ from the simple equation used here.

(e.g., Tiscareno et al. 2013; Porco et al. 2007). From this equation, for example, in order for the Roche limit of Chariklo to be beyond the location of the rings, the density of the orbiting material needs to be exceptionally low with respect to the primary and/or γ needs to be low (Melita et al. 2017). The uncertainties in Chariklo's density and mass lead to a fairly wide range of possible Roche limits. Given the mass estimate for Chariklo of $6 - 8 \times 10^{18}$ kg (Leiva et al. 2017), assuming the ring material has density $\rho_s = 400$ kg m^{-3} (the value adopted for the small inner satellites of Saturn), and substituting the primary's mass $M_p = (4\pi/3)(ABC)^3\rho_p$, a_{Roche} at Chariklo ranges between 304 and 413 km.

As discussed in Tiscareno et al. (2013), for a ring system it is better to consider the critical density (ρ_{Roche}) instead of the critical distance, which is given by $\rho_{Roche} = 3M_p/\gamma a_b^3$, where a_b is the semimajor axis of the secondary orbit (Porco et al. 2007). For a given value of a_b, this is the critical density at which the object's size fills its region of gravitational dominance. As an example, for Quaoar Q1R, the value of $\rho_{Roche} \approx 30$ kg m^{-3}, assuming $\gamma = 1.6$ (Morgado et al. 2023). However, this value corresponds to very porous or fluffy material. Instead, the classical value of the Roche limit at Quaoar was found to be ~1780 km assuming ring-particle density of 400 kg m^{-3} (Morgado et al. 2023). Morgado et al. (2023) suggested that elastic collisions can maintain such a ring beyond the Roche limit. Material beyond the Roche limit can also be prevented from accreting if its radial velocity dispersion increases due to external perturbations. A spin–orbit resonance or shepherd satellite(s) might play roles in maintaining the observed rings (Sicardy et al. 2019; Sickafoose & Lewis 2024). At Quaoar, the 6:1 mean-motion resonance with Weywot may also be involved (Morgado et al. 2023; Pereira et al. 2023).

9.3.4 Ring Modeling

Thin rings around giant planets or small bodies do not survive throughout the age of our solar system (e.g., Tiscareno et al. 2013; Pan & Wu 2016). These rings require confinement mechanisms to prevent the spreading and/or a source to replenish the lost particles. Resonances between particles and the central body or nearby satellites can maintain ring confinement for a longer time. However, Neptune's arcs have been found to be transient (e.g., De Pater et al. 2018). There have been theoretical and numerical studies of the evolutionary dynamics in the small-body systems, more recently including the gravitational effects on ring particles around prolate, small, central bodies.

Since all the large-body ring systems are close to the giant planets (which are oblate bodies), the gravitational effect on the ring particles can be modeled using the J_n terms. It is well-known that the oblateness of the central body provokes large short-term variations on the osculating orbital elements of a particle around the primary, which can be corrected through the use of the geometrical elements (e.g. Renner & Sicardy 2006). However, for ellipsoid bodies, such as Chariklo and Haumea, in addition to the J_n coefficients, the C_{22} term (ellipticity of the equatorial region of the primary) has to be added in the gravitational potential. The C_{22} gravity coefficient can induce an increase in the eccentricity of the particle, which can be corrected using an appropriate choice of initial conditions (Ribeiro et al. 2021). Unfortunately, only adding terms in the geometrical-elements algorithm does not solve the problem. To deal with this, Ribeiro et al. (2021) developed a set of empirical equations as a function of C_{22}, which ensures that the particle will perform the nominal orbit around a prolate body.

Narrow, eccentric rings like those seen at Uranus have apse-alignment due to self-gravity (Goldreich & Tremaine 1979). A simple model that combined the ellipticity of the central body and the particles' self-gravity to maintain apse alignment was proposed for Chariklo (Pan & Wu 2016). This work determined a mass for the inner ring of 10^{16} g, a typical particle size of a few meters, and a spreading time of 10^5 yr. A theoretical model for apse-alignment was also developed to constrain Chariklo's ring surface density and eccentricity gradient, as well as the relative, minimum mass and location for a putative satellite (Melita & Papaloizou 2020). It turns out that the rings around Chariklo and Haumea are close to their 3:1 spin orbit resonances (SORs)[1]. These locations prompted studies of the effects of resonances between the spin of a non-axisymmetric body (elongated or with a topographic feature) and the orbital motion of the particles (Sicardy et al. 2019; Sicardy 2020). The shape of the central body pushed the ring material beyond the 2:1 SOR, clearing the region nearby: particles located inside the corotation radius migrated toward the central body, while those located outside were pushed outside the 2:1 resonance. Thus fast rotators, where the 2:1 SOR is within the Roche limit, could be most likely to host rings (Sicardy et al. 2019).

Several studies have analyzed the stability of the surrounding region around prolate bodies through the powerful technique of the Poincaré Surface of section (PSS); this technique allows identification of stable and chaotic regions as well as the locations of resonances. Figure 9.8 shows two PSSs for the Chariklo system covering the 3:1 SOR for different values of the Jacobi Constant. In these plots, resonant orbits are represented by fixed points, quasi-periodic orbits are represented by islands around the fixed points, and the points spread over the section are identified

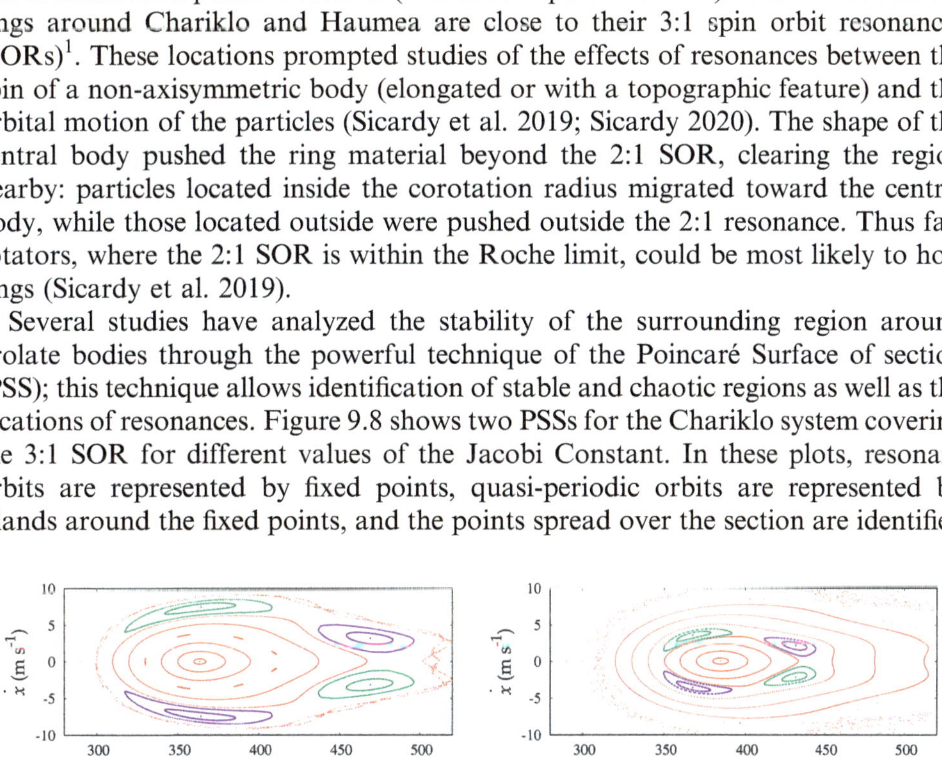

Figure 9.8. Examples of Poincaré Surface of sections (PSSs) within Chariklo's 3:1 SOR. A family of first kind periodic orbits is responsible for the quasi-periodic orbits shown as red islands. Purple and green represent two families of 3:1 resonant periodic orbits (periodic orbits of the second kind). The Jacobi constants for these plots were (left) $C_j = 7.374 \times 10^{-3}\,\text{km}^2\,\text{s}^{-2}$ and (right) $C_j = 7.483 \times 10^{-3}\,\text{km}^2\,\text{s}^{-2}$. Reproduced from Giuliatti Winter et al. (2023). CC BY 4.0.

[1] Here, we define SOR literally, as spin:orbit or *number of rotations:number of orbital revolutions*, while noting that some works reverse these values (effectively providing orbit:spin resonance numbers).

as chaotic trajectories. From a set of PSSs, a semimajor axis versus eccentricity map can be generated that identifies the locations and widths of the stable regions and resonances (e.g., Winter & Murray 1997; Ribeiro et al. 2021; Giuliatti Winter et al. 2023).

Application of the PSS technique suggested that Haumea's ring is located in a stable region associated with first-kind periodic orbits, and not with the 3:1 SOR (Winter et al. 2019). Similar results were found for Chariklo's C1R, which is located in a stable region due to first-kind period orbits, while C2R is located in an unstable region (Giuliatti Winter et al. 2023). Only large values of the eccentricity of C2R, and consequently a larger width, locate C2R in a stable region, prompting consideration of a three-moon system to confine the rings (Giuliatti Winter et al. 2023).

Additional analyses for a body with a topographical feature a few kilometers long (a mass anomaly like that proposed by Sicardy 2020 for Chariklo) used the adapted pendulum model and PSS to verify that the resonance locations are mainly affected by the mass and the spin-period of the primary, while the topographical feature can influence the width of the resonance (Madeira et al. 2022). For the Chariklo system, the results suggested that the stability of the C1R is associated with the periodic/quasi-periodic first kind orbits. An investigation of various spin rates and ellipticities of central bodies for Haumea, Chariklo, and five hypothetical systems found that periodic first-kind orbits are present in all systems with almost zero eccentricity of the particles and that resonant orbits have high eccentricities (Ribeiro et al. 2023).

Finally, global N-body simulations have been carried out to model the Chariklo system, with the conclusions that Chariklo should be more dense than the ring material and that the existence of narrow rings implied smaller than meter-sized particles or the existence of shepherd satellites (Michikoshi & Kokubo 2017). The most recently published N-body studies, developed from existing models for Saturn's rings, find that a single shepherd moon is capable of confining material into the thin widths and locations observed at Chariklo (Sickafoose & Lewis 2024). Such a moon would only need to be a few kilometers in diameter, below current imaging-detection limits.

9.4 Debris

Any active Centaur would be likely to have some level of debris environment from dust to boulders depending on the amount of activity and how recently an outburst may have occurred. Furthermore, given the mass of some of these objects, the larger Centaurs may have long-lived orbital debris left over from periods of activity (Fink et al. 2021). Debris can be detected and characterized through a variety of techniques, as also discussed in Chapters 8 and 10. The large-scale debris considered in this chapter could be detected through stellar occultations if it has sufficiently high optical density or large orbiting bodies. So far, observations of occultations by active Centaurs have only detected significant debris in Chiron's near environment.

As noted in Section 9.3.1, rings have been proposed for Chiron but consensus is lacking on whether what has been observed was rings, an active jet, or a shell of debris left over from past activity (Elliot et al. 1995; Bus et al. 1996; Ruprecht et al. 2015; Sickafoose et al. 2020). The most recent results suggest that there is actively-evolving material around Chiron, as opposed to a Chariklo-like ring system (Sickafoose et al. 2023; Ortiz et al. 2023). In terms of characterization, occultation data in 1994 were taken in visible wavelengths in ground-based data and in *K*-band from NASA's Kuiper Airborne Observatory (KAO); these data were used to place a lower limit of 0.25 μm on the radius of particles in the deepest occultation feature (Elliot et al. 1995). This result was noted to be different from comet Halley, for which analyses of spacecraft data found a significant population of particles smaller than this size (Elliot et al. 1995). In 2011, an occultation by Chiron was observed simultaneously in visible ($\sim$0.6 μm) and near-infrared wavelengths (low-resolution spectra from 1 to 2.5 μm); the SNR of the near-infrared data was insufficient to detect any surrounding material or color-wavelength trends (Sickafoose et al. 2020).

Echeclus and SW1 are known to have had significant surrounding material (see Chapters 8 and 10). Echeclus had a 7-magnitude increase in brightness in 2005 that included a detached coma and substantial dust production (Choi & Weissman 2006; Rousselot 2008; Bauer et al. 2008). From *Spitzer* and visible-wavelength data, the maximum size of the ejected particles was estimated to be 700 μm within an order of magnitude (Bauer et al. 2008). Maximum activity corresponded to a dust production rate of a few hundreds of kilograms per second, of the order of 30 times that seen in other Centaurs (Rousselot et al. 2016). Impacts and fragmentation were ruled out, but possibly large chunks of material were removed by erosion to expose a fresh surface (Bauer et al. 2008). After a 2017 outburst, visible images plus near-infrared spectra were used to detect a large cloud of ejected debris, leading to speculation that there may be several debris ejection or fragmentation events per year on other Centaurs that are going unnoticed (Kareta et al. 2019).

SW1 is extremely active, with a continuous dust coma and displaying at least one major outburst each year since 1927 (see Chapter 10). From 2002 to 2007, the average rate was 7.3 events per year, and visible-wavelength images showed that the coma was being continuously supplied with fine dust while grains larger than $\sim$1 μm were few in number (Trigo-Rodríguez et al. 2008). A debris trail was detected in *Spitzer* 24 μm data in 2003, with an upper limit placed on optical depth (Stansberry et al. 2004). More recently, a 1000-km debris cloud was proposed (Miles & Kretlow 2018), but substantial debris was ruled out within 500–1000 km of the nucleus from a stellar appulse in 2020 (Miles & Kretlow 2021).

More data are needed to detect and constrain the properties (and evolution) of large debris around Chiron, Echeclus, SW1, and other Centaurs.

9.5 Formation Mechanisms

In contemplating the formation mechanisms for satellites, rings, and debris, an initial question is whether the formation of these components is related. Possible interrelationships include simultaneous formation of debris and a satellite, debris as a source for ring particles, and a satellite as a source for ring particles. Within the Centaur-like population, there are no objects known with both satellites and rings;

on the other-hand, both TNOs known to have rings also have satellites. There is not yet enough data to tell whether ring formation is directly related to satellite formation, or whether debris can eventually coalesce into rings.

9.5.1 Satellite Formation

Satellite formation mechanisms can be inferred from the distribution of binary properties. As seen in Figure 9.6, many binaries are near-equal brightness and widely separated. Such objects have far too much angular momentum to form through collisions, and the collision rate is too low to easily produce such systems in any case. Instead, the dominant formation mechanism for TNO binaries is direct gravitational collapse of a large cloud of gas and small particles. This process naturally forms the types of binaries seen. The concentration mechanism causing the collapse is hypothesized to be the "streaming instability" where gas-dust-gas interactions concentrate material to the point where self-gravity causes gravitational collapse (see, e.g., Chapter 2). Indeed, the distribution of TNO binary angular momenta is a good match for simulations of clouds formed in the streaming instability, making TNO binaries a dominant observational constraint on this mode of planetesimal formation (Nesvorný et al. 2020).

The streaming instability is particularly effective in the cold classical region of the disk. The geomorphology of the TNO Arrokoth, a contact-binary in this region, was recently found to be consistent with a merger of similarly-sized planetesmials from a collapse cloud (Stern et al. 2023). The hot population of TNOs (the progenitors of most Centaurs and GPCs) is thought to originate in a primordial disk outside a compact configuration of the giant planets. The parameters for the streaming instability and binary formation in the primordial disk are poorly constrained. Still, the formation of Centaur-like binaries traces back to their formation as hot TNOs. Unlike for comet-like activity and debris, the orbital dynamics of Centaur-like objects are not likely to contribute to satellite formation, but rather serve as a destruction mechanism as discussed in Section 9.2.

9.5.2 Ring Formation

Soon after the discovery of Chariklo's rings, several studies addressed the question of their formation. The planet-crossing nature of Centaur orbits prompted Smoothed-Particle Hydrodynamics (SPH) simulations to investigate the possibility that the rings (and also small satellites) could be formed by tidal disruption of a differentiated Chariklo during the closest encounter with a giant planet (Hyodo et al. 2016). Such an encounter would remove material from the surface and create a surrounding disk, beyond the Roche limit at which particles could accrete into moons. This model is consistent with the presence of water ice in the rings, as suggested by Duffard et al. (2014). However, in order for the proposed disruption to occur, the distance between Chariklo and one of the giant planets at close-approach has to be within just a few planetary radii—much smaller than the likely approach distances (Araujo et al. 2016; Wood et al. 2017; see also Chapter 3). In fact, the majority of simulated close encounters (90%, Araujo et al. 2016, or 99%, Wood et al. 2017) do not even affect the stability of Chariklo-like rings, making tidal disruption an unlikely formation mechanism.

Rings could also be formed during a close encounter with a giant planet if the encounter perturbed an existing small moon inward. This scenario would require a single Neptune encounter and a close match between the satellite orbit and the encounter strength. This scenario weakly favors ring formation during the encounter, as opposed to within the trans-Neptunian region; to be feasible, it would require that most $\sim$100 km TNOs have ten-km sized moons, and it predicts that only a few percent of Centaurs would have rings (Pan & Wu 2016). If this scenario is prevalent, rings should be more common among large Centaurs but absent for small comets and large TNOs (Pan & Wu 2016). Along the same lines, rings might be formed from the disruption of a satellite after crossing the Roche limit: for Chariklo, a porous satellite with a radius of roughly 7 km would disaggregate at the ring location with sufficient mass to explain the rings as well as putative shepherd satellites (Melita et al. 2017). However, this model requires a mechanism to bring the satellite closer to Chariklo, as tidal effects are very weak.

Alternatively, the rings could be created by dusty outgassing or a collision. Chariklo's transition to the Centaur region would have increased its temperature, lofting dust particles off the surface. For particles that were not re-accreted, mutual collisions would settle them into an equatorial ring, and frequent collisions could eventually convert dust into meter-sized ring particles (Pan & Wu 2016). Collisional considerations include either cratering ejecta or the remnants of a destroyed satellite; however, assuming typical impact probabilities, the estimated timescales for these scenarios are longer than the dynamical lifetime of Centaurs (Melita et al. 2017).

If rings are common around TNOs, they would survive during the transition into the Centaur region (Wood et al. 2017), and formation mechanisms in the more distant solar system need to be considered. This is an open area of research, with rings only proposed around the TNOs Haumea and Quaoar. Haumea is a unique case, as it has a known collisional history (e.g., Brown et al. 2007) and is a highly-elongated, quickly-rotating body for which rotational fission might have stripped off material (Ortiz et al. 2017). N-body simulations of ring formation from fission at Haumea found that the ring likely formed between an unstable region of the orbit and the Roche radius, near the 3:1 SOR (Sumida et al. 2020). It has been suggested that ring material at Quaoar is the result of a primordial collisional system that settled into a disk, from which both rings and Weywot formed (Morgado et al. 2023). Due to the size difference between TNOs with rings and Centaurs with rings, it is possible that their ring formation mechanisms are not directly related. The discovery (or clear absence) of rings around smaller TNOs would be essential for understanding whether rings form in some (likely) primordial process and then survive the transition to Centaur-like orbits, or whether the transition itself generates new rings.

9.5.3 Debris Formation

If active jets are the source of large debris, then numerous cometary dust models are available to characterize the transient material (e.g., Tenishev et al. 2011). The larger

Centaurs, however, may have enough mass for some of the debris from activity and outbursts to end up on ballistic paths or become orbital (Fink et al. 2021). While a model such as that in Fink et al. (2021) cannot be used to understand morphology, it can be used to predict the size scale and density of orbiting particles. For example, SW1 could have orbiting particles ranging from 8 to 150 mm that were lifted during outburst events. To understand how these orbiting particles might accrete to form larger bodies, shells, arcs, or rings, additional dynamical modeling is needed.

Although collision rates are likely to be low in the Centaur region, debris could also be launched by collisions. Photometric observations of Chiron in 2014–2015 indicated microactivity that was thought to possibly be caused by existing debris impacting the surface, producing outbursts, and launching dust (Cikota et al. 2018).

9.6 Summary

Centaur-like bodies are located in a transitional dynamical region. There are many more TNO binaries than Centaur binaries, which could be due to the pathway from dynamically hot populations in the outer solar system to the giant planet region. However, there may be many close binary Centaurs that are as yet undetected. Ring systems have been proposed for two Centaurs and two TNOs, but observations are few and have mixed interpretations. Confirmation and characterization of stably-orbiting material over longer baselines is needed for all of these objects. Reliable, bulk statistics for satellites and rings for both Centaurs and TNOs will substantially aid understanding of formation mechanisms and locations. Debris has been detected around Centaurs but not at TNOs, suggesting that heliocentric distance plays a role. Importantly, observational biases mean that smaller and fainter features in counterpart TNOs may not yet have been detected. While we expect that there will be ties between satellites, rings, and debris for bodies in the outer solar system, we currently lack sufficient information to make concrete connections. The future prospects discussed below (and in Chapter 14) should help answer the many open questions in these fields.

9.6.1 Open Questions

Here we present a selection of open questions in our current understanding of satellites, rings, and debris around Centaur-like bodies.

- *What are the binary fractions of Centaurs and other related dynamical groupings?* Analyses of existing data are needed to determine the unbiased statistics of binary objects, the proportions of tight versus wide orbits, and the percentages of binaries as a function of object size.
- *What are the shapes and orientations of the known GPC binaries Ceto-Phorcys and Typhon-Echidna?* Even a small amount of new high-precision astrometry would likely detect non-Keplerian effects that would give better insight into the physical properties of these binaries.
- *Are there any Centaur "pairs" that indicate a previously-bound binary system?* Even with a lower limit of $\sim$10% for the binary fraction of hot TNOs, if the current, low frequency of Centaur-like binaries is attributable to the unbinding of TNO binaries in close encounters, then there would be a significant

number of Centaur "pairs": objects with orbits so similar that they must have been physically close in the past. These same close encounters cause the orbits of Centaurs to be chaotic, making it challenging to reliably trace orbits back far in time. Thus, Centaur pairs from binary formation would require a recent interaction. It is not clear whether the currently-known population of hundreds of Centaur-like objects is sufficient to detect any "pairs".

- *What are the effects of outbursting activity or binary-binary interactions on binary systems?* These processes could play roles in the apparent drop in binary fraction from TNOs to Centaurs and JFCs; however, binary-binary interactions are likely to be relevant only for very close encounters.
- *How can lightcurve observations be better employed to learn about Centaur and TNO satellites?* Light curves provide an indirect method of detecting contact binaries and there are indications that lightcurve amplitudes differ between dynamical populations. However, Centaur activity could be repressing lightcurve amplitudes and there remain observational biases.
- *How (and where) do Centaur rings form?* A major factor distinguishing Centaur ring-formation theories is whether (i) these rings are primordial and formed in the TNO region and were maintained throughout the dynamical transition to Centaurs, or (ii) the Centaur-precursors (small, hot TNOs) do not have rings, requiring either planetary encounters or heliocentric-distance-induced surface activity to generate them. It is unlikely that we happen to have detected rings during a short period of time when they exist; rather they are likely long-term features and may even be prevalent among Centaurs. The frequency of rings, especially with well-determined upper limits, is still poorly understood, particularly for small TNOs of similar sizes to ringed Centaurs.
- *What is the orbital evolution of material around active Centaurs?* More work is needed to understand the dynamics of larger-than-dust particles around active Centaurs, especially whether debris can coalesce into rings, and whether rings can be maintained or will disperse over relatively short timescales. A subset of this question is the following: *how is orbiting material evolving at Chiron?* Observations are specifically needed to estimate the particle sizes and locations of current debris at Chiron, as well as to continue to study its evolution.
- *Are there any promising Centaur targets for as-yet-undetected ring systems?* The spin–orbit coupling theory suggests that faster rotators may be promising targets to search for rings. Occultations and appulses by the Centaur Bienor (9.14 hr rotation) have accordingly been targeted, but Chariklo-like rings could not be ruled out (Fernández-Valenzuela et al. 2023). The limited number of known small-body ring systems makes it difficult to isolate ring-friendly characteristics.

9.6.2 Future Prospects

There are many possibilities for enhanced understanding of binaries, rings, and debris using existing data sets or observations expected in the near-term. Several of the open questions above can be addressed with current data, though anticipated discoveries in the next few years will significantly augment research opportunities.

With existing data, a clear step forward is better characterization of non-detections. For example, there are no detailed publications of non-detections or upper limits for binary components around the 109 Centaur-like objects observed with HST (or for the few hundred TNOs observed with HST). This makes it difficult to determine whether the binary frequency among similarly-sized objects is truly distinct.

Similarly, occultation observations of rings and debris suffer from a completeness problem. The occultation field is transitioning from isolated, anecdotal detections of individual objects to understanding statistics of multiple objects, which will require a new threshold of expectations. For example, published occultation observations do not always provide all the information necessary for homogeneous reanalysis of data, and non-detections or upper limits often remain unpublished. Although reporting null results and carefully determining upper limits is a challenging prospect for occultation observers, it is an important goal going forward.

Chapter 14 discusses new surveys and instruments expected to enhance Centaur science. Below, we briefly discuss how these would specifically be applied to Centaur satellites, rings, and debris.

9.6.2.1 JWST

JWST observations of Centaurs are already ongoing, but they are expected to be generally limited to only the most interesting objects. In its shortest wavelength filter, JWST's NIRCam can detect binaries closer and fainter than HST, but only by a factor of $\sim$2, so this is not expected to be a significant source of new binaries. Still, some high precision astrometry from JWST may complement continued observations with HST to learn more about binary fractions and orbits. JWST has been effectively used to detect Chariklo's rings during a stellar occultation (Santos-Sanz et al. 2023). It is also possible that JWST high-resolution imaging could resolve rings of Centaurs or TNOs. Direct imaging of the rings would provide new and unique information about their full orbital structure, composition, and temporal evolution.

9.6.2.2 Vera C. Rubin Observatory's Legacy Survey of Space and Time (LSST)

The upcoming LSST is expected to increase the number of known small bodies throughout the solar system by an order of magnitude. Furthermore, there are typically hundreds of detections processed to produce highly-precise astrometry and photometry in *ugriz* filters. Thus, the quality of small-body observations will also increase by more than an order of magnitude. The vast majority of small-body discoveries by LSST will occur in the first year or two of the survey, suggesting that the number of known Centaur-like objects will increase substantially in 2026–2027 (based on present estimates for the LSST schedule; see Chapter 14).

With both accurate (thanks to GAIA) and precise astrometry, the orbits of Centaurs will be improved significantly, allowing for increased success in occultation surveys. This improvement will be plausible within the first year of LSST for established objects, because new astrometry will significantly extend observational baselines and improve known positions. For new discoveries, such precise orbits are expected to take at least two years, but they may be possible for occultation predictions even earlier.

LSST's precise, multi-epoch, and multi-filter photometry will also enable detailed studies of many Centaur-like objects that are usually reserved for those with ground-based observing campaigns. A dramatically increased number of Centaur-like objects should have well-measured rotational amplitudes and periods, as well as colors. The rotational-period distribution then may give insights into binary frequency (as described in Section 9.2.3), while amplitudes will help constrain contact-binary fractions. Even low-level outbursting activity will be detectable at unprecedented levels with LSST, allowing for better characterization of surrounding material and understanding of possible debris-forming events.

9.6.2.3 Extremely Large Telescopes

Future extremely large telescopes open the possibility of significant advances in characterizing the near-Centaur environment. The Extremely Large Telescope in Chile will have direct imaging capabilities with 5 mas resolution and an intensity contrast of better than 10^{-8} at 30 mas (Kasper et al. 2010). This capability will enable a systematic search of Centaurs for satellites, rings and debris. As described in the TMT Detailed Science Case 2022[2], the Thirty Meter Telescope's IRIS instrument provides diffraction limited imaging and will be able to resolve ring structures similar to Chariklo's with its 7 mas spatial resolution at 1 μm.

9.6.2.4 Potential Spacecraft Missions

The Centaurs are arguably the main dynamical category of small bodies in the solar system that have not yet had the benefit of a close study enabled by a spacecraft mission. A mission to a Centaur is more practical than a visit to a distant TNO and could answer many similar questions. Further discussion is in Chapter 16, but we note here the special value of choosing a mission target that is a binary, has rings, or is known to have had debris. The potential insights into these phenomena from a close flyby are extensive and compelling.

9.7 Epilogue

Centaurs (including Giant Planet Crossers) are an interesting population themselves and further serve as a unique interface between the TNO and JFC populations. The properties of binaries, rings, and debris provide insights into the physical, dynamical, and evolutionary properties of these objects. Observations—both for individual objects and in terms of homogeneously-analyzed surveys—are quite limited and currently leave many questions open and unclear. There are ample prospects for future analyses of these features, especially with the anticipated improvements in quantity and quality of observations from JWST, LSST, and extremely large

[2] Available at https://www.tmt.org/documents.

telescopes. Further investigations will provide powerful insights into the formation and evolution of Centaurs as well as the outer solar system in general.

References

A'Hearn, M. F., Schleicher, D. G., Millis, R. L., Feldman, P. D., & Thompson, D. T. 1984, AJ, 89, 579

Araujo, R. A. N., Galiazzo, M. A., Winter, O. C., & Sfair, R. 2018, MNRAS, 476, 5323

Araujo, R. A. N., Sfair, R., & Winter, O. C. 2016, ApJ, 824, 80

Bauer, J. M., Choi, Y.-J., Weissman, P. R., et al. 2008, PASP, 120, 393

Bérard, D., Sicardy, B., Camargo, J. I. B., et al. 2017, AJ, 154, 144

Bosh, A. S., Olkin, C. B., French, R. G., & Nicholson, P. D. 2002, Icar, 157, 57

Braga-Ribas, F., Sicardy, B., Ortiz, J. L., et al. 2014, Natur, 508, 72

Braga-Ribas, F., Pereira, C. L., Sicardy, B., et al. 2023, A&A, 676, A72

Brown, M. E., Barkume, K. M., Ragozzine, D., & Schaller, E. L. 2007, Natur, 446, 294

Buie, M. W., Keller, J. M., Miles, R., et al. 2023, Asteroids, Comets, Meteors Conf., (Houston, TX: LPI) Abstract #2445

Bus, S. J., Buie, M. W., Schleicher, D. G., et al. 1996, Icar, 123, 478

Campbell, H. M., Stone, L. R., & Kaib, N. A. 2023, AJ, 165, 19

Chandler, C. O., Kueny, J. K., Trujillo, C. A., Trilling, D. E., & Oldroyd, W. J. 2020, ApJL, 892, L38

Charnoz, S., Crida, A., & Hyodo, R. 2018, in Handbook of Exoplanets, Vol 54, ed. H. J. Deeg, & J. A. Belmonte (New York: Springer International Publishing)

Choi, Y.-J., & Weissman, P. 2006, AAS/Division for Planetary Sciences Meeting Abstracts, (Vol. 38,; Washington, DC: AAS) 37.05

Cikota, S., Fernández-Valenzuela, E., Ortiz, J. L., et al. 2018, MNRAS, 475, 2512

Cochran, A. L., & Cochran, W. D. 1991, Icar, 90, 172

Combes, M., Lecacheux, J., Encrenaz, T., et al. 1983, Icar, 56, 229

Cuzzi, J. N. 1985, Icar, 63, 312

Cuzzi, J. N., Whizin, A. D., Hogan, R. C., et al. 2014, Icar, 232, 157

De Pater, I., Renner, S., Showalter, M. R., & Sicardy, B. 2018, in Planetary Ring Systems. Properties, Structure, and Evolution, ed. M. S. Tiscareno, & C. D. Murray (Cambridge: Cambridge Univ. Press) 112

Duffard, R., Pinilla-Alonso, N., Ortiz, J. L., et al. 2014, A&A, 568, A79

Elliot, J. L., Dunham, E., & Mink, D. 1977, Natur, 267, 328

Elliot, J. L., Dunham, E., Wasserman, L. H., Millis, R. L., & Churms, J. 1978, AJ, 83, 980

Elliot, J. L., French, R. G., Meech, K. J., & Elias, J. H. 1984, AJ, 89, 1587

Elliot, J. L., Olkin, C. B., Dunham, E. W., et al. 1995, Natur, 373, 46

Fernández, Y. R., Wellnitz, D. D., Buie, M. W., et al. 1999, Icar, 140, 205

Fernández-Valenzuela, E., Ortiz, J. L., Duffard, R., Morales, N., & Santos-Sanz, P. 2017, MNRAS, 466, 4147

Fernández-Valenzuela, E., Morales, N., Vara-Lubiano, M., et al. 2023, A&A, 669, A112

Fernández-Valenzuela, E., Holler, B. J., Ortiz, J. L., et al. 2023, EPSC-DPS Joint Meeting 2023, (Washington, DC: AAS) 202.04

Fink, U., Harris, W., Doose, L., et al. 2021, PSJ, 2, 154

Fornasier, S., Lazzaro, D., Alvarez-Candal, A., et al. 2014, A&A, 568, L11

French, R. G., Elliot, J. L., & Levine, S. 1986, Icar, 67, 134

Giuliatti Winter, S. M., Madeira, G., Ribeiro, T., et al. 2023, A&A, 679, A62
Giuliatti Winter, S. M., Madeira, G., & Sfair, R. 2020, MNRAS, 496, 590
Goldreich, P., & Porco, C. C. 1987, AJ, 93, 730
Goldreich, P., & Tremaine, S. 1979, Natur, 277, 97
Griffin, M. J., Abergel, A., Abreu, A., et al. 2010, A&A, 518, L3
Groussin, O., Lamy, P., & Jorda, L. 2004, A&A, 413, 1163
Grundy, W., Stansberry, J., Noll, K., et al. 2007, Icar, 191, 286
Grundy, W. M., Noll, K. S., Virtanen, J., et al. 2008, Icar, 197, 260
Grundy, W. M., Noll, K. S., Roe, H. G., et al. 2019, Icar, 334, 62
Hedman, M. M., Nicholson, P. D., Cuzzi, J. N., et al. 2013, Icar, 223, 105
Hubbard, W. B., Brahic, A., Sicardy, B., et al. 1986, Natur, 319, 636
Hyodo, R., Charnoz, S., Genda, H., & Ohtsuki, K. 2016, ApJ, 828, L8
Jewitt, D. 1990, ApJ, 351, 277
Jewitt, D., Agarwal, J., Li, J., et al. 2014, ApJL, 784, L8
Kareta, T., Sharkey, B., Noonan, J., et al. 2019, AJ, 158, 255
Kasper, M., Beuzit, J.-L., Verinaud, C., et al. 2010, Proc. SPIE, 7735, 77352E
Lacerda, P., & Jewitt, D. 2012, ApJL, 760, L2
Leiva, R., Sicardy, B., Camargo, J. I. B., et al. 2017, AJ, 154, 159
Levison, H. F., & Duncan, M. J. 1997, Icar, 127, 13
Li, J., Jewitt, D., Mutchler, M., Agarwal, J., & Weaver, H. 2020, AJ, 159, 209
Luu, J. X., & Jewitt, D. C. 1990, AJ, 100, 913
Madeira, G., Giuliatti Winter, S. M., Ribeiro, T., & Winter, O. C. 2022, MNRAS, 510, 1450
Marcialis, R. L., & Buratti, B. J. 1993, Icar, 104, 234
Meech, K. J., & Belton, M. J. S. 1989, IAU Circ., 4770, 1
Melita, M. D., Duffard, R., Ortiz, J., & Campo-Bagatin, A. 2017, A&A, 602, A27
Melita, M. D., & Papaloizou, J. C. B. 2020, Icar, 335, 113366
Michikoshi, S., & Kokubo, E. 2017, ApL, 837, L13
Miles, R., & Kretlow, M. 2018, JOA, 8, 11
Miles, R., & Kretlow, M. 2021, JOA, 11, 3
Morgado, B., Sicardy, B., Braga-Ribas, F., et al. 2023, Natur, 614, 239
Morgado, B. E., Sicardy, B., Braga-Ribas, F., et al. 2021, A&A, 652, A141
Morgado, B. E., Sicardy, B., Braga-Ribas, F., et al. 2024, Natur, 7997, E2
Müller, T. G., Lellouch, E., Böhnhardt, H., et al. 2009, EM&P, 105, 209
Namouni, F., & Porco, C. 2002, Natur, 417, 45
Nesvorný, D., & Vokrouhlický, D. 2019, Icar, 331, 49
Nesvorný, D., Vokrouhlický, D., Bottke, W. F., Levison, H. F., & Grundy, W. M. 2020, ApJL, 893, L16
Nicholson, P. D., Hedman, M. M., Clark, R. N., et al. 2008, Icar, 193, 182
Noll, K. S., Grundy, W. M., Nesvorný, D., & Thirouin, A. 2020, in The Trans-Neptunian Solar System, ed. D. Prialnik, M. A. Barucci, & L. Young (Amsterdam: Elsevier) 205
Noll, K. S., Levison, H. F., Grundy, W. M., & Stephens, D. C. 2006, Icar, 184, 611
Ockert-Bell, M. E., Burns, J. A., Daubar, I. J., et al. 1999, Icar, 138, 188
Ortiz, J., Santos-Sanz, P., Sicardy, B., et al. 2017, Natur, 550, 219
Ortiz, J. L., Gutiérrez, P. J., Sota, A., Casanova, V., & Teixeira, V. R. 2003, A&A, 409, L13
Ortiz, J. L., Duffard, R., Pinilla-Alonso, N., et al. 2015, A&A, 576, A18
Ortiz, J. L., Santos-Sanz, P., Sicardy, B., et al. 2020, A&A, 639, A134

Ortiz, J. L., Pereira, C. L., Sicardy, B., et al. 2023, A&A, 676, L12
Pan, M., & Wu, Y. 2016, ApJ, 821, 18
Parker, A. H., & Kavelaars, J. J. 2010, ApL, 722, L204
Pereira, C. L., Sicardy, B., Morgado, B. E., et al. 2023, A&A, 673, L4
Pereira, C. L., Braga-Ribas, F., Sicardy, B., et al. 2024a, MNRAS, 527, 3624
Pereira, C. L., Sicardy, B., Morgado, B. E., et al. 2024b, A&A, 683, C4
Poglitsch, A., Waelkens, C., Geis, N., et al. 2010, A&A, 518, L2
Porco, C. C., & Goldreich, P. 1987, AJ, 93, 724
Porco, C. C., Thomas, P. C., Weiss, J., & Richardson, D. C. 2007, Sci, 318, 1602
Proudfoot, B. C. N., Ragozzine, D. A., Thatcher, M. L., et al. 2024, AJ, 167, 144
Rabinowitz, D., Schwamb, M. E., Hadjiyska, E., Tourtellotte, S., & Rojo, P. 2013, AJ, 146, 17
Rabinowitz, D. L., Barkume, K., Brown, M. E., et al. 2006, ApJ, 639, 1238
Ragozzine, D., Pincock, S., Proudfoot, B. C. N., et al. 2024, arXiv e-prints, arXiv:2403.12785
Reitsema, H. J., Hubbard, W. B., Lebofsky, L. A., & Tholen, D. J. 1982, Sci, 215, 388
Renner, S., & Sicardy, B. 2004, CeMDA, 88, 397
Renner, S., & Sicardy, B. 2006, CeMDA, 94, 237
Ribeiro, T., Winter, O. C., Madeira, G., & Giuliatti Winter, S. M. 2023, MNRAS, 525, 44
Ribeiro, T., Winter, O. C., Mour ao, D., Boldrin, L. A. G., & Carvalho, J. P. S. 2021, MNRAS, 506, 3068
Rieke, G. H., Young, E. T., Engelbracht, C. W., et al. 2004, ApJS, 154, 25
Rommel, F. L., Braga-Ribas, F., Ortiz, J. L., et al. 2023, A&A, 678, A167
Rommel, F. L., Braga-Ribas, F., Desmars, J., et al. 2020, A&A, 644, A40
Rousselot, P. 2008, A&A, 480, 543
Rousselot, P., Korsun, P. P., Kulyk, I., Guilbert-Lepoutre, A., & Petit, J. M. 2016, MNRAS, 462, S432
Ruprecht, J. D., Bosh, A. S., Person, M. J., et al. 2015, Icar, 252, 271
Safrit, T. K., Steckloff, J. K., Bosh, A. S., et al. 2021, PSJ, 2, 14
Santos-Sanz, P., Lellouch, E., Fornasier, S., et al. 2012, A&A, 541, A92
Santos-Sanz, P., Ortiz, J. L., Sicardy, B., et al. 2017, MNRAS, 501, 6062
Santos-Sanz, P., Gomes Júnior, A., Morgado, B., et al. 2023, EPSC-DPS Joint Meeting 2023, (Washington, DC: AAS) 301.07
Sarid, G., Volk, K., Steckloff, J. K., et al. 2019, ApJL, 883, L25
Sheppard, S. S., & Jewitt, D. C. 2002, AJ, 124, 1757
Showalter, M. R., & Lissauer, J. J. 2006, Sci, 311, 973
Showalter, M. R., Benecchi, S. D., Buie, M. W., et al. 2021, Icar, 356, 114098
Sicardy, B. 2020, AJ, 159, 102
Sicardy, B., Leiva, R., Renner, S., et al. 2019, NatAs, 3, 146
Sickafoose, A. A., Bosh, A. S., Levine, S. E., et al. 2019, Icar, 319, 657
Sickafoose, A. A., Levine, S. E., Bosh, A. S., et al. 2023, PSJ, 4, 221
Sickafoose, A. A., & Lewis, M. C. 2024, PSJ, 5, 32
Sickafoose, A. A., Bosh, A. S., Emery, J. P., et al. 2020, MNRAS, 491, 3643
Singer, K. N., Stern, S. A., Elliott, J., et al. 2021, P&SS, 205, 105290
Souami, D., Renner, S., Sicardy, B., et al. 2022, A&A, 657, A134
Stansberry, J. A., Van Cleve, J., Reach, W. T., et al. 2004, ApJS, 154, 463
Stansberry, J. A., Grundy, W. M., Mueller, M., et al. 2012, Icar, 219, 676
Stern, S. A., White, O. L., Grundy, W. M., et al. 2023, PSJ, 4, 176

Stone, L. R., & Kaib, N. A. 2021, MNRAS, 505, L31
Strauss, R. H., Leiva, R., Keller, J. M., et al. 2021, PSJ, 2, 22
Sumida, I., Ishizawa, Y., Hosono, N., & Sasaki, T. 2020, ApJ, 897, 21
Tenishev, V., Combi, M. R., & Rubin, M. 2011, ApJ, 732, 104
Thirouin, A., Ortiz, J. L., Duffard, R., et al. 2010, A&A, 522, A93
Thirouin, A., & Sheppard, S. S. 2018, AJ, 155, 248
Thomas, N. 2009, P&SS, 57, 1106
Tiscareno, M. S., Hedman, M. M., Burns, J., & Castillo-Rogez, J. 2013, ApJL, 765, L28
Tiscareno, M. S., Showalter, M. R., French, R. G., et al. 2016, PASP, 128, 018008
Trigo-Rodríguez, J. M., García-Melendo, E., Davidsson, B. J. R., et al. 2008, A&A, 485, 599
Verbiscer, A. J., Skrutskie, M. F., & Hamilton, D. P. 2009, Natur, 461, 1098
Weissman, P. R., Chesley, S. R., Choi, Y. J., et al. 2006, AAS/Division for Planetary Sciences Meeting Vol 38, (Washington, DC: AAS) 37.06
Wierzchos, K., & Womack, M. 2020, AJ, 159, 136
Winter, O. C., Borderes-Motta, G., & Ribeiro, T. 2019, MNRAS, 484, 3765
Winter, O. C., & Murray, C. D. 1997, A&A, 319, 290
Wood, J., Horner, J., Hinse, T. C., & Marsden, S. C. 2017, AJ, 153, 245

Chapter 10

Notable and Well-Studied Centaurs

Theodore Kareta, Charles Schambeau and Kacper Wierzchos

The Centaurs are a varied population, spanning orders of magnitude in size and ranging from completely inactive to the most active objects in the solar system. This chapter aims to introduce the reader to five Centaurs that are particularly well studied and sample this populational diversity, and thus are useful comparison points for other objects that we know less about. 29P/Schwassmann-Wachmann 1 is the most outburst-prone object known, and is now retroactively recognized as the first Centaur discovered. 95P/Chiron is the namesake of the Centaurs and the largest object yet discovered in the solar system to show cometary activity. (10199) Chariklo is the largest Centaur and was the first small body to have a ring system found around it. 174P/Echeclus is an active Centaur, which is a similar size to 29P/Schwassmann-Wachmann 1 and has outbursts of similar strength, but appears to become inactive between its activity epochs. P/2019 LD2 (ATLAS) is currently a Jupiter co-orbital, but has previously been a Centaur and will make the jump to being a Jupiter-family comet (JFC) in 2063—the first time such a transition can be observed and studied. We compare these objects to each other, the broader population, and discuss what questions about them might be most fruitful to investigate with the next generation of telescopes and future spacecraft missions.

10.1 Introduction

While one might be interested in the bulk properties of a whole population—the distribution of compositions, activity states, sizes and surface properties—it is a natural intermediate step to compare newly discovered objects to those that have been known of for the longest and characterized in the most depth. The aim of this chapter is to introduce the reader to some of the most studied and most compelling objects among the Centaurs. These are naturally among the brightest of the Centaurs—and thus the most accessible to telescopic study—and with one exception they are all significantly larger than a typical Jupiter-family Comet.

doi:10.1088/2514-3433/ada267ch10

The structure of this chapter is to detail the properties of each of these compelling objects, including which properties might be common to the bulk population and which properties might be unique to them or fundamentally rare overall, followed by a discussion comparing these "notable oddballs." Which ones are in desperate need of more characterization, and what knowledge gaps might be filled by doing so? Which objects are being considered for spacecraft visits, and how would a visit to each object shed light on the bulk properties of the whole population? Among the active objects, do they even behave similarly? Particular attention will also be paid to discussing the inherent biases in characterizing individual members of a broad and diverse population. While the understanding of the Centaur population presented in this book has evolved and been informed by our knowledge of these objects and a small number of others, it will continue to evolve as we push to smaller sizes, to weaker activity, and to larger heliocentric distances.

The five objects which this chapter focuses on are 95P/Chiron, 29P/Schwassmann-Wachmann 1, 174P/Echeclus, (10199) Chariklo, and P/2019 LD2 (ATLAS), hereafter referred to as Chiron, SW1, Echeclus, Chariklo, and LD2, respectively. A brief summary of their properties is included as Table 10.1 for reference while reading. Chiron is the largest active object yet discovered in the solar system, while SW1 is the most outburst-prone object yet discovered. Chariklo is the largest known Centaur, famous for its ring system—the first discovered on a small body. Echeclus is also prone to outbursts, similar in size to SW1, and is best known for its large outbursts followed by apparent total inactivity (unlike SW1). Lastly, LD2 is a JFC-sized active Centaur which will become a JFC in 2063—the first time such a transition can be witnessed and planned for. These objects span from the most active to fully inactive and range over a factor of a hundred in diameter—they are a varied group even on their own. Readers will likely have encountered these objects in many other chapters throughout this book—all but Echeclus have already been observed with JWST (see Chapter 14); Chiron, Chariklo, and SW1 have been subjects of sustained occultation campaigns (see Chapter 9); SW1, Chiron, and LD2 have had or will have observational campaigns dedicated to their study and monitoring (see Chapter 11). A comparison of the average nuclear

Table 10.1. Overview of the Five Objects Discussed in Detail in this Chapter

Name	Discovery Year	q [au]	Q [au]	P_{orb} [yr]	D [km]	p_V [%]	Active?
29P/S-W 1	1927	5.77	6.31	14.9	62 ± 6	3.3 ± 1.5	Yes.
95P/Chiron	1977	8.54	18.88	50.7	$\approx$210	$\approx$10	Yes.
(10199) Chariklo	1997	13.11	18.42	62.6	$\approx$250	3.1-4.9	No.
174P/Echeclus	2000	5.85	15.66	35.2	60 ± 1	5.0 ± 0.3	Yes.
P/2019 LD2	2019	4.58	5.99	12.1	<2.4	Not known.	Yes.

Note: A brief overview of the names, discovery years, orbital properties, and physical properties of the five objects discussed in detail in this chapter. The references for each of these quantities are in the sections corresponding to each individual object. The approximate sizes of Chiron and Chariklo are due to different approaches on how to model the significant dust or ring material around these objects and is discussed at length in their respective sections. The size estimate for LD2 assumes a typical cometary albedo (5%), but the object's nuclear properties are not well known as of writing.

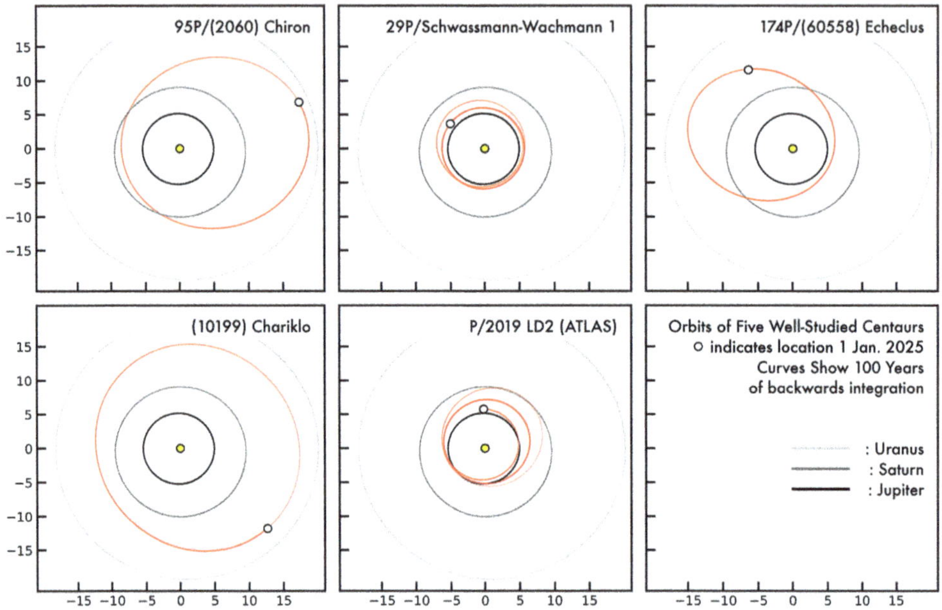

Figure 10.1. The orbits of each of the five well-studied Centaurs are shown as dark orange curves, superimposed on the orbits of Jupiter (black), Saturn (dark gray), and Uranus (light gray). The position of each of these objects on 1 January 2025 is shown as a white dot with a black outline, and the Sun is shown as a yellow dot with a black outline for context. More recent positions are shown in darker colors so as to distinguish the time evolution of each orbit and the directions of motion. The orbits of Chiron, Echeclus, and Chariklo show no notable evolution over the 1925–2025 period, but LD2 and SW1 have had their orbits evolve noticeably in the past century as discussed at some length in the text of this chapter and others. Given that all orbits are shown on the same scale, the significant changes in the smaller orbits of LD2 and SW1 are larger than they might appear at a glance.

magnitudes of these five objects against the known Centaurs is shown in Figure 10.2. Each object's orbit is described in detail in its subsection, and a comparison of the most recent hundred years of orbital evolution for all of the objects together is shown in Figure 10.1. The "Gateway" orbit that SW1 has and LD2 had (Sarid et al. 2019; Steckloff et al. 2020) is also specifically discussed in Chapter 3. Where overlap exists in content to be covered, we have tried to defer to other chapters and focus on activity patterns and surface properties here.

It is important to remind the reader that while these might be some of the best-studied objects, one should not expect them to be typical or representative. In fact, for many of these objects, we compare other objects to them *assuming* that they are functionally unique. We might not expect there to be a large population of Chiron-sized active Centaurs out beyond Saturn, but we can try to compare the activity profiles or surface properties of smaller objects to ascertain how size or orbit might play a role in their differences. If we take the observational biases in the problem seriously, we can try to make the most of the few objects that are bright enough for detailed study to understand the whole of the population.

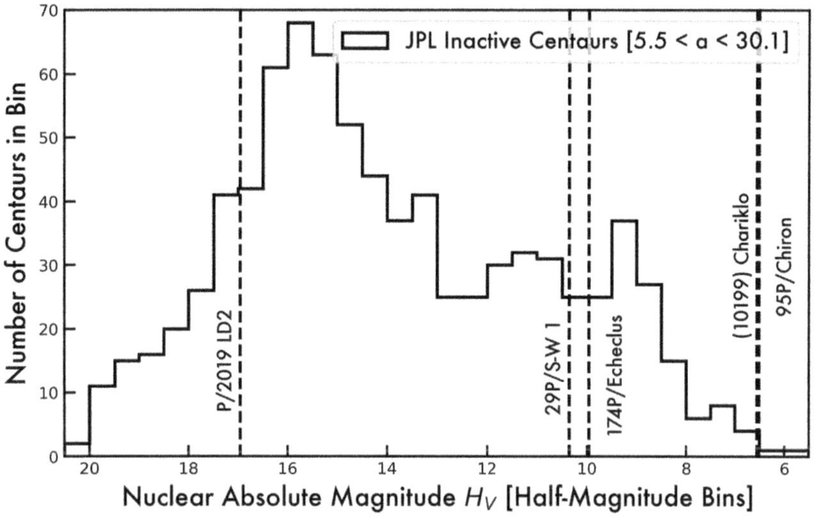

Figure 10.2. The distribution of absolute magnitudes, H_V, of the inactive Centaurs as shown as a histogram compared against the absolute magnitudes of the five objects discussed at length in this chapter plotted as vertical dashed lines. In order to compare apples to apples, the nuclear absolute magnitudes were specifically calculated for the active objects based on the information in Table 10.1. Two of these five objects required special treatment. First, the absolute magnitude of P/2019 LD2 is functionally a lower limit to its true nuclear magnitude as its nuclear size is only known as an upper limit. Second, the absolute magnitude for Chariklo was adjusted to be its long-term average brightness to incorporate some of the brightness from its rings and not just its dark surface. As can be seen from the plot, known Centaurs with absolute magnitudes greater than $H_V \sim 16$ decrease in abundance compared to brighter objects—this is almost certainly an observational bias against the detection of faint objects and not a feature of the true population. LD2's discovery must then be attributed primarily to its strong activity, which if included would move it several magnitudes brighter on this plot.

10.2 95P/Chiron

The first object recognized in, and thus namesake of, the population of Centaurs is Chiron, also cross-listed as asteroid 2060 and comet 95P. While a full review of the history of this object and its importance to the population is covered elsewhere in this book (Chapters 1 and 11), a discussion of how its nature was recognized and its properties were measured is warranted by how similar the story is to many of the other smaller active Centaurs discovered later.

After discovery as "Slow Moving Object Kowal" or 1977 UB by Charles Kowal (Kowal & Gehrels 1977), it took more than a decade for cometary activity to be discovered at Chiron. While the object clearly had a chaotic orbit (Oikawa & Everhart 1979)—the orbit must have been significantly different in the past and would be similarly different in the future—identifying an origin and a set of associated physical properties was challenging. The combined visible and thermal-infrared study of Lebofsky et al. (1984) showed that while Chiron's surface reflected visible light like a C-type asteroid would—that is, its reflectance spectrum was featureless and gently red-sloped—its bulk albedo was higher than a typical C-type asteroid. (They also suspected that the object was quite large, a topic we come back to below.) Fernandez (1980) predicted that Chiron may simply be among the brightest of the objects slowly migrating inwards from the yet-named-or-discovered trans-Neptunian belt—a prediction that seems to have more or less held up. Photometric measurements from the late 1980s (Tholen et al. 1988; Bus et al. 1989) showed Chiron brightening rapidly beyond what could be attributed to its rotational lightcurve or geometric effects. This culminated in the first unambiguous detections of cometary activity (Hartmann et al. 1990; Luu & Jewitt 1990) a year later. In order to actually detect the atmosphere of Chiron, it had to first grow to rival the nucleus in brightness (a challenge considering the considerable brightness of the solid body) and then become large enough to be spatially resolvable at such large geocentric distances ($\Delta > 11$ au; see a discussion in Meech & Belton 1990). Even during a period of activity in which the coma was easily detected from the ground, Meech et al. (1997) found that the nucleus contributed some $\sim$80 % of the light in smaller apertures.

Even though Chiron's comet-like nature had been suspected around the time of its discovery, it still took more than a dozen years to find unambiguous evidence for actual ongoing cometary activity. The nature of how new Centaurs are found means that this process has been repeated several times—it is easier to find larger (and thus brighter) objects, which are in turn challenging to search for comae around. This same process—discovery of a bright object with some suspicions of activity followed years later by a true confirmation of the ongoing mass loss—was repeated rather closely by another object in this chapter, Echeclus, throughout the early-to-mid-2000s.

10.2.1 Activity, History, and Coma Properties

Chiron is bright enough for many kinds of telescopic characterization throughout its entire orbit—a $\sim$4-meter class telescope today can obtain high-quality visible or near-infrared spectra of Chiron in a few hours or less even at aphelion—but there are only a few substances that have been detected on its surface or in its vicinity despite decades of searching. The first of these to be detected was the cyanogen radical (hereafter just "CN") in 1990 by Bus et al. (1991) at a heliocentric distance of about 11.3 au. A simple Haser model (which may or may not be applicable for a variety of reasons apparent throughout this section) inferred a CN production rate of $Q(CN) \sim 4 \times 10^{25}$ molecules s^{-1} at that distance. CN is typically the first species (e.g., atom, molecule, ion, or molecule) detected after a comet is discovered or becomes newly active, and thus the detection of CN toward Chiron was viewed as the final and most authoritative piece of evidence that Chiron was not some completely new kind of object, but instead the largest comet yet discovered.

The circumstances of the detection of CN toward Chiron are remarkable—a routine plan to observe the visible reflectivity of the object with a $\sim$2-meter telescope was completed several hours after the onset of a likely small outburst that had yet to be reported. The feature can even be discerned from a visual inspection of the 2D spectrum presented in Bus et al. (1991). However, despite the apparent ease by which those authors detected this radical, there have been no published detections of this feature since then despite increases in telescope collecting area and detector efficiency. Chiron and SW1 (Section 10.3) are the only two active Centaurs with unambiguous CN detections. While we leave a discussion of hunting for CN around these objects for the end of this chapter, we emphasize that the scenario of its detection toward Chiron does not provide unambiguous clues to how one might proceed with such a search—some luck may be required, or at the very least a very rapid Target-of-Opportunity follow-up program on several large telescopes.

The other material that has been found directly in Chiron's coma is CO (Womack & Stern 1999; Womack et al. 2017), with an apparent production rate of $Q(CO) = 1.3 \times 10^{28}$ molecules s^{-1} as measured at perihelion at $R_H = 8.5$au. Rauer et al. (1997) previously found no CO or CN in June through November of 1995 and January of 1996, respectively, and the longer-term study of Bockelée-Morvan et al. (2001) also found no CO over a similar time frame—developing an upper limit to Chiron's gas production a factor of 10× lower than the production rate inferred from the data of Womack & Stern (1999)—and concluded that if the sole CO detection was indeed real, then it *too* might have sampled a small outburst. From CN to CO, Chiron's gas production is clearly time-variable—it is unclear which phases of its activity facilitate their detection. Even if we accept that both CO and CN were both detected by chance shortly after the start of outbursts, do all outbursts produce these materials at these rates? Until more observations are obtained in the coming apparition, this is challenging to ascertain.

Part of why detecting the same substances at multiple epochs in Chiron's atmosphere has proven challenging is its relatively unpredictable activity pattern. Building upon the work of Bus et al. (2001) and Duffard et al. (2002), Belskaya et al. (2010) presented a secular lightcurve of Chiron from 1969 through 2008. Unlike a "typical" inner solar system comet, which becomes more active as it approaches the Sun and warms, Chiron's activity is apparently uncorrelated with heliocentric distance. Three distinct phases of activity were seen in that analysis, including activity at aphelion (Bus et al. 2001), the first detections of brightening and a coma described above, and a third phase that started in 1999 and had continued in some respect through the end of the study. A further persistent ~0.4–0.5-magnitude brightening started between February and June of 2021 (Dobson et al. 2021) which appears to still be ongoing as of writing. Even at Chiron's intrinsic dimmest in the late 1990s, Hubble Space Telescope imaging still clearly showed extended material around the object (Meech et al. 1997).

We note explicitly that Bus et al. (1991)'s detection of CN toward Chiron showed that that gas had a different and more-extended spatial profile than that of the reflecting material (a combination of dust and the nucleus). Even if one could explicitly tie a measure of Chiron's brightness to a sense of how active it has been recently, where the CN is coming from (e.g., a nuclear source versus from grains in the coma) is not fully clear. If CN is primarily released from the nucleus during outbursts, then one might expect it to last for very short periods of time. This is in agreement with there being a sole detection of the substance, but a single detection hardly rules in or out more complex scenarios. (If CN had even a partially distributed source, then the production rate of CN might be untethered from the bulk activity of the nucleus—smeared out or delayed over some extended period.)

Regardless of when, how, or how much Chiron shuts off between episodes of activity, it is clear that dust can and does stick around the object for prolonged periods. Occultation studies (see, e.g., Ortiz et al. 2015; Sickafoose et al. 2020, 2023) continue to see evidence for extended extincting material around Chiron with discrete dips—interpreted in several publications as rings, though a more full discussion of this topic is found in Chapter 9. Sickafoose et al. (2023) compared occultations of Chiron at different dates and while some structures remained, the optical depths and number of features was not constant. An occultation presented in Elliot et al. (1995) showed evidence for Chiron's coma being formed from the action of only a few small active areas spread across its surface. These kinds of analyses paint a picture of Chiron being surrounded by material for long periods of time, but the total amount of dust and the structures one can observe within it appear to be time-variable.

The existence of long-lived persistent material and Chiron's physically large size —and thus, a non-negligible surface gravity—have lead to suggestions that Chiron has a "ballistic" or "persistent" dust coma or atmosphere (Meech & Belton 1990; Meech et al. 1997). In essence, when a part of Chiron's surface becomes newly active and begins sublimating, much of the dust does not expand radially outwards as might be expected for a smaller comet but instead ends up on weeks-long orbits within a few thousand kilometers (still less than an arcsecond across in all parts of Chiron's orbit!). While radiation pressure acts on typically-sized comet dust grains ($D \sim 1 \mu m$) effectively, Chiron's great size and large heliocentric distance allow even small particles to stay bound to Chiron for some time within its "exopause." This kind of scenario can replicate a Chiron that slowly brightens over a period of weeks or months without immediately showing an extended appearance.

10.2.2 Nucleus and Colors

Chiron's apparent intermittent activity means that several studies of Chiron's surface features have been completed. It is unclear how much the variable-but-long-lasting dust atmosphere described in the previous paragraphs muddies, complicates, or even matters to each of the following conclusions and we comment on this when possible.

Chiron is the largest active body in the solar system (Fornasier et al. 2013; Ortiz et al. 2015; Lellouch et al. 2017), with an equivalent diameter of 210–220 km, or about the same size as Saturn's irregular moon Phoebe. Even when Chiron (apparently) had less dust around it and its lightcurve was the most apparent, the peak-to-trough amplitude of its lightcurve has never exceeded $\sim$0.09 magnitudes—it appears that Chiron is relatively close to round but not perfectly so. Analysis of observations of Chiron with ALMA (Lellouch et al. 2017) suggest the object is a tri-axial ellipsoid with semimajor axes of 114, 98, and 62 km respectively. Lellouch et al. (2017) speculate that Chiron's lightcurve must be diminished by dust around the object, either as a semi-bound coma or extended ring-like system. The shape derived by those workers also has the two axes in the plane ("a" and "b") as similarly sized, so the smaller third axis would be challenging to infer the existence of from lightcurve-only studies anyways. Even the largest Oort cloud comet yet known, C/2014 UN271 (Bernardinelli-Bernstein) is only about half as wide as Chiron (Hui et al. 2022; Lellouch et al. 2022). Perhaps it is to be expected then that so much about Chiron has remained mysterious and hard to pin down despite four decades of study—there really is not another object like it. Chiron's apparent uniqueness is a good part of its appeal as a future mission target, either as a fly-by or through a prolonged visit, and this topic is discussed briefly at the end of this chapter as well as in Chapter 16.

As mentioned above, Chiron's visible reflectivity as determined during a period of no or low activity in the early 1980s is relatively neutral (Lebofsky et al. 1984) and most similar to a C-type asteroid. This spectral neutrality continues until about $\sim$1.1–1.3 μm where the spectrum begins to become slightly blue sloped and show clear signs of water ice absorption (Foster et al. 1999; Luu et al. 2000), attributed to solid water ice on the surface as opposed to icy grains in its coma. Prior to JWST results (see Chapter 14), this is very nearly a summary of our knowledge of Chiron's surface properties. It is spectrally neutral, and near-infrared spectra obtained from 1996 to 1999 appear to be consistent with a few percent water ice. Near-infrared spectra taken in 1993 (Luu et al. 1994) do not show clear signatures of water ice absorption and are slightly redder, which Luu et al. (2000) interpreted as being covered or obscured by dust in a then-thicker dust coma. No part of Chiron's current orbit makes any part of the surface warm enough that water ice should be unstable, so unless Chiron has spent a significant amount of time closer to the Sun than it does now, one might not expect the surface to have changed its bulk water ice abundance too much since its escape from the trans-Neptunian belt.

Chiron's secular lightcurve as measured in Belskaya et al. (2010) and elsewhere shows orbit-long trends in its absolute magnitude not obviously attributable to activity changes. Specifically, Chiron was brighter during a period of inactivity in the early-to-mid 1980s compared to another period of inactivity in the late 1990s. Occultation studies (Ortiz et al. 2015; Sickafoose et al. 2020, 2023) have suggested links between the long-term brightness evolution of Chiron and the geometry with which ones views the rings—when the rings are viewed more face-on, the whole system appears brighter when unresolved. This is linked to the same idea put forward about the rings of Chariklo (see Section 10.5 and Chapter 9). While this is a fully plausible scenario that appears to be true at Chariklo, two points must be brought up before its application to Chiron. First, the dust structures around Chiron —be they rings, ring-arcs, or otherwise—do not appear to be constant in time (Sickafoose et al. 2023), so making long-term predictions about how a potential ring system should behave and reflect light given the apparent short-term variability of that system must have some caveats. Second, some of the variability seen in Chiron's absolute magnitude over time could be related to variations in surface properties. Maybe the body has a non-zero axial tilt and the part of Chiron best illuminated at aphelion is simply more reflective than that seen closer to perihelion? While such a scenario might explain aspects of the data at hand, new observations obtained more systematically will be needed to test these hypotheses. One particularly interesting project would be to combine the sparse long-term photometric coverage of Chiron (it is always bright enough for the major asteroid and comet surveys to detect it) with a long-term campaign to observe its reflectivity. One might expect that variations due to each of these kinds of processes—surface heterogeneity, evolution of a long-lived dust population, etc.—should vary on different timescales, and thus could provide a useful constraint on how to interpret the origin of its long-term and short-term brightness variations. Observations in the coming decade as Chiron begins to repeat the part of its orbit that was first observed in the early 1970s should clarify which parts of its secular lightcurve are reproduced and which need re-inspection.

10.3 29P/Schwassmann-Wachmann 1

Since its discovery almost a century ago, SW1 has been an enigmatic body of high observational priority whose flair for comae-manifestation theatrics has defied our understanding of active small icy bodies. We present in this section a historical perspective of SW1 and an overview for the current state of its understanding in relation to the early solar system's formation. A summary of key future observational priorities is reserved for the summary at the end of this chapter and is also discussed in Chapters 11 and 14.

10.3.1 Historical Perspective

SW1 (formerly comet 1925 II) was discovered on November 15, 1927 at an apparent magnitude of $m_V \sim 13.5$ while undergoing an outburst by Arnold Schwassmann and Arno Arthur Wachmann at the Hamburg Observatory, in Bergdorf, Germany (van Biesbroeck 1927). At the time, the discovery represented the most distantly discovered short-period comet that likely only was serendipitously detected due to the increased apparent magnitude and 2′ extended nebulosity while in outburst (Roemer 1958). However, once discovered, SW1 was routinely easily imaged on photographic plates even during periods of quiescent activity every year due to its nearly-circular orbit and relatively large nucleus size. (The dimmest that SW1 can get during a period of low activity is about $m_V \approx 18$.) The earliest known pre-discovery detections of SW1 were acquired in 1902 and were used to help refine its early orbit determination (Reinmuth 1931; Behrens 1932). The term "quiescent" has been used early on to describe SW1's periods of low-level activity where a conspicuous nebulosity surrounding the nucleus due to the outburst dust ejecta was not present. Typically, during the periods of quiescent activity SW1 appeared as a near stellar-like source by visual inspection and also on photographic plates (Roemer 1958), however it is likely (and will be discussed further below) that during these periods a dust coma was still present but below the detection thresholds of photographic plates. So while using the term is a misnomer due to the persistent and ongoing low-level activity present, the term for SW1's periods while outside of an outburst are still referred to as quiescent.

After SW1's discovery, the list of unique properties it possessed began to quickly grow. To start, after sufficient orbital coverage of positional astrometry, it was determined that its orbit was entirely beyond that of Jupiter's and that it had the smallest eccentricity for a known comet at the time, $e = 0.135$, with $q = 5.5$ au and $Q = 7.3$ au (Behrens 1932 and Herget 1947, with a large portion of the latter referenced orbit calculations being performed by Miss Antoinette Kettenacker at the Cincinnati Observatory). Furthermore, the minimum visual apparent magnitude while in quiescent activity ($\approx$ 18th magnitude) indicated that its nucleus was relatively large ($D = 70 \pm 30$ km; Whitney 1955). One of the more perplexing discoveries was of its outburst prone nature in terms of the frequency and severity (with Δm ranging from a few tenths of a magnitude to as large as ~6 magnitudes) of such events while possessing an orbit in a nearly constant thermal environment (Richter 1941; Whitney 1955), which still lacks a well-accepted explanation to this day.

Early studies of SW1 focused on monitoring its outburst nature by measuring its apparent brightness and also the ejecta comae morphology displayed during such events[1]. The first detailed work to analyze a large body of brightness measurements was by Nikolaus Richter (Richter 1941), who compiled 142 brightness measurements from the literature spanning from 1927 November 15, through 1939 June 21, from 13 observers, the bulk coming from George Van Biesbroeck, Richard Schorr, and Walter Baade. One of the first comprehensive studies of SW1's coma morphology and its evolution during different phases of activity and stages of outburst was presented in Roemer (1958). While these early observational studies were completed over 60 years ago, the observed behavior of SW1 is fundamentally identical to what we observe today indicating that the underlying outburst mechanism has not changed in almost 100 years. From the rate at which the object's brightness increases to the size of the largest outbursts to the apparent speed of the material released in them, the outbursts large enough to be detected and noticed decades ago seem broadly similar to those of similar strengths observed in the modern era. (Naturally, the more recent measurements are more precise—especially for the fainter events.) Figure 10.3 displays three examples of the typical coma morphology displayed by SW1 in the days to weeks after a major outburst event. The morphology at times appears as a nearly circular shell of material with the nucleus close to the center, while at other times presenting an asymmetric fan shape reminiscent of the arcade game character Pac-Man. The outer edge of the shapes expand radially from the nucleus' location consistently with skyplane linear speeds on the order of ~0.1 km s^{-1}. The expansion is in a manner which retains the overall initial bulk morphology, suggesting that the dust lofting event is impulsive with a duration much less than the nucleus' rotation period (Schambeau et al. 2015, 2017).

While the majority of early observations of SW1 were photographic imaging studies, a few early spectroscopic studies were undertaken. They are rare due to the faint nature (m_V= 15–18) of SW1 while not in outburst considering the telescopic

[1] Note that these early studies were almost entirely completed through broadband visible imaging that was sensitive to detections of the dust coma only as detections of gas comae species is difficult at SW1's distance.

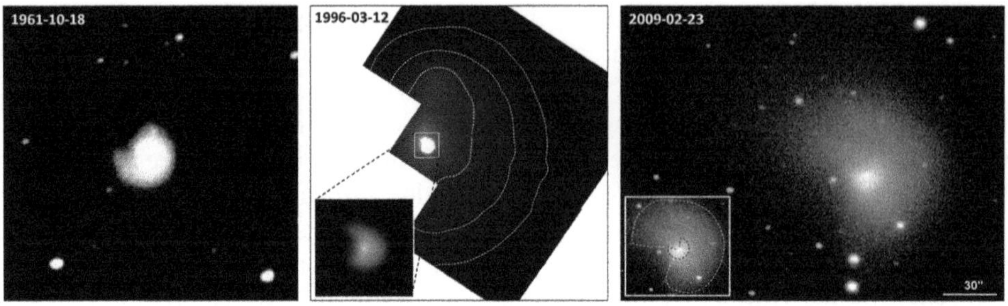

Figure 10.3. Shown are three examples of SW1's consistent asymmetric outburst coma morphology spanning nearly 50 years. Each individual panel has the same projected skyplane scale, as identified by the far-right panel, and is orientated with equatorial north up and east to the left. Left panel adapted from Roemer et al. (1963), with permission from Springer Nature. Middle panel adapted from Schambeau et al. (2019) © 2019. The American Astronomical Society. All rights reserved. Right panel adapted from Miles (2016). Copyright 2016, with permission from Elsevier.

resources at the time. One such early study was undertaken by Mayall (1941) using the 36-inch Crossley reflector at Lick Observatory, which detected a solar-like spectrum between 370 and 500 nm shortly after a reported outburst. This wavelength range includes the primary bandhead for CN emission, thus indicating little or no CN had been released during that outburst.

The source of SW1's outburst behavior have been investigated since its discovery. Early studies attempted to establish correlations between the solar activity production of outbursts, with Richter (1941) and Nicholson (1947) searching for connections to solar flares via measurements of ionospheric UV bursts detected through the Mögel–Dellinger effect and later works (Richter 1954) focused on extending the solar-driven connections to outbursts driven by coronal mass ejection events. While timing correlations were able to identify eight outbursts of SW1 with corresponding solar-induced terrestrial geomagnetic events, the later work of Whitney (1955) incorporating the at-the-time new icy-conglomerate nucleus model (Whipple 1951) identified that the energy required to produce the cometary outburst was not adequately provided by such solar driven events, and an internal source of energy was still required. Instead, Whitney (1955) mentions that solar activity could trigger such outbursts but first proposed a volatile driven mechanism as the primary energy source of SW1's outbursts. Prialnik & Bar-Nun (1987) were the first to suggest that amorphous water ice crystallization could drive cometary activity, and SW1 is well within the heliocentric distance where such a process could happen rapidly—but finding unambiguous evidence for this process in action has been challenging. While efforts have been made to link SW1's outburst morphology to the kinds of mechanisms that might help create them, the multiple confounding factors add significant challenge. In particular, even the complicated spiral-like feature called a "ringtailed snorter" by Roemer (1963) and Whipple (1980) is probably not diagnostic of a particular process or scenario as opposed to being an unexpected product of radiation pressure (Schambeau et al. 2021).

10.3.2 Orbital Properties

SW1 is currently in a nearly circular orbit just exterior to that of Jupiter, with eccentricity $e = 0.045$, semimajor axis $a = 6.047$ au, perihelion $q = 5.78$ au, and inclination $i = 9.364^{\circ}$[2]. This is one of the most circular orbits for a known comet, and was for many decades the lowest eccentricity of *any* active comet. When combined with SW1's wide range of activity levels described in the previous sections, the object's orbit was immediately perplexing. How does cometary activity vary so significantly with such a limited variation in heliocentric distance? While the nucleus' obliquity is not currently known, with such a low orbital eccentricity, even small to moderate obliquities may become increasingly important in the orbit-long modulation of an object's activity. At the time of writing, only three active small bodies have been found through the NASA, JPL Small-Body Database with smaller eccentricities: P/2020 MK4 (PANSTARRS) with $e = 0.021$, 158P/Kowal-LINEAR with $e = 0.030$, and 331P/Gibbs with $e = 0.042$[3].

Figure 10.1 shows the past one hundred years of orbital evolution for SW1. Over the course of the twentieth century, repeated moderate and close encounters with Jupiter have acted to slowly and progressively alter the orbit of SW1. When SW1 was discovered in 1927, its orbit (Berman & Whipple 1928) was more eccentric ($e \sim 0.14$) and its semimajor axis was slightly larger ($a \sim 6.43$ au) resulting in slightly lower perihelia ($q \sim 5.51$ au). In the subsequent hundred years, its semimajor axis and eccentricity have both decreased, acting to circularize and slightly shrink its orbit. While SW1's orbit has been recently repeatedly perturbed due to its close proximity to Jupiter, it has remained rather circular just exterior to Jupiter as long as it has been observed.

The orbital region of SW1 is chaotic (Carusi et al. 1995), where dynamical simulations constrain the half-life for orbits similar to SW1's at ~300 years (Horner et al. 2004) and a time before being expelled into interstellar space by Jupiter on the order $\sim 10^5$ years (Neslušan et al. 2017). SW1's orbit places it at the edge of several small body population categories that are defined by orbital properties alone. Some will call it a JFC, others a "Jupiter-Coupled Object" (Gladman et al. 2008), and most commonly SW1 is referred to as a Centaur using the definition of Jewitt (2009). It is noteworthy to mention that the population category schemes are more fluid in the astronomical community where often times the orbital boundaries between populations in published works differ between authors and have evolved through the decades. The challenge of accurately categorizing SW1's orbit highlights its unique boundary-case status and underscores the fact that the definitions of JFCs, Centaurs, and various TNO sub-populations lack universal consensus within the field.

Interest in the exact pathways between the JFCs and the Centaurs lead to recent investigations into SW1's orbit. Low eccentricity orbits just exterior to Jupiter, typified by SW1, seem to be particularly effective at moving objects from the

[2] NASA, JPL Small-Body Database, Reference JPL K192/80, Epoch 2459945.5.

[3] It is noted that 331P's activity has been attributed to rotational disruption and not sublimation similar to the others (Jewitt et al. 2021) so a heliocentric distance variation in activity is not necessarily expected.

Centaurs into the JFCs—this region has been given the nickname the (Orbital) “Gateway” (Sarid et al. 2019). That work suggests that a significant fraction (possibly even the *majority*) of JFCs made their way to the Jupiter-family by passing through a Gateway-like orbit first. However, Guilbert-Lepoutre et al. (2023) present a similar set of orbital integrations where a fraction of JFCs are found to transition between the Centaur to JFCs without experiencing a Gateway-like orbit. Furthermore, they point out that some of the Gateway objects should be those moving outwards, not inwards. The debate surrounding the Centaur-to-JFC Gateway hypothesis remains significant because it provides crucial insights into the thermal history of individual objects and the likelihood of their nuclei being more pristine. More discussion of this topic is included in Chapters 3 and 7.

10.3.3 Nucleus Properties

Limited knowledge exists regarding the properties of the nucleus of SW1 due to its persistent activity, which hinders direct observations of the nucleus without interference from additional contributions from the surrounding comae. To overcome this challenge, studies have employed various methods to separate the nucleus and coma flux contributions, allowing for the indirect constraint of the nucleus' physical and dynamical properties, including its size, shape, surface composition, and spin state. While many of these methods rely on the assumption that the coma is optically thin—a reasonable assumption for most comets—this assumption may be compromised shortly after one of SW1's major outburst events. Fortunately, the majority of observations used to constrain nucleus properties have been conducted during periods of quiescent activity or a sufficiently long duration (days to weeks following) after a recent major outburst. During these times, concerns related to optical thickness are likely alleviated[4], providing a more reliable basis for investigating its elusive nucleus.

The first property to be confidently constrained was its bulk nucleus size. SW1 has long been suspected of harboring a relatively large nucleus compared to the JFCs population (Bauer et al. 2015; Fernández et al. 2013) owing to its apparent visual magnitude threshold during periods of quiescent activity (as noted by Whitney 1955). Given the poorly constrained albedo of SW1, which impacts the credibility of nucleus size estimates based on visible magnitudes, the groundbreaking attempts to constrain the nucleus' size came from thermal emission observations. The seminal work by Cruikshank & Brown (1983) using the NASA Infrared Telescope Facility with a bolometer and Q filter (effective wavelength $\sim$21 μm and bandwidth FWHM $\sim$5.8 μm) marked a turning point by providing the first constraints on SW1's size and albedo through thermal observations at 20 μm ($R_N = 20 \pm 4$ km and $p_v = 0.13 \pm 0.04$), ushering in a new era of understanding cometary nuclei in the 1980s. However, a subsequent independent analysis of the photometry in the original

[4] It is noted that this statement may be true only outside of the near-IR emission bands of CO and CO_2. Analyses of JWST data of SW1 acquired in 2023 indicate optically thick CO and CO_2 comae based on observations acquired while in a state of quiescent activity (Faggi et al. 2024).

manuscript revealed inconsistencies in results, leading to the identification of an error. According to Lisse et al. (2022), the derived albedo and nucleus effective radius estimates are closer to 0.02 and 50 km, respectively, though uncertainties arise from the assumption that the original 1983 manuscript's photometry is based on a bare nucleus without contamination of additional flux from a dust coma. In the 2000s, with the emergence of new infrared space telescopes, a fresh wave of thermal emission-derived nucleus size estimates for SW1 surfaced. Given SW1's unique properties and high scientific importance, each new observatory (e.g., *Spitzer*, Wide-field Infrared Survey Explorer (WISE), *Herschel*) early on saw observations of SW1 approved. *Spitzer*, during its in-orbit checkout and science verification phase in 2003 November, made infrared observations of SW1 (continuum imaging at 5.730, 7.873, 15.80, 23.68, and 71.42 μm using the Infrared Array Camera (IRAC) and Multiband Imaging Photometry (MIPS)) where multiple analyses of these data have refined the nucleus radius estimate to $R_N = 32.3 \pm 3.1$ km (Stansberry et al. 2004, 2008; Schambeau et al. 2015, 2021). A radius measurement of 23 ± 7.5 km was estimated based on WISE 4-band thermal imaging data (3.368, 4.618, 12.082, and 22.194 μm) taken during the cryogenic mission in 2010 May (Bauer et al. 2013). Finally, analysis of three epochs of *Herschel* Photoconductor Array Camera and Spectrometer (PACS) 70 and 160 μm continuum images provided a $R_N = 31 \pm 3$ km (Bockelée-Morvan et al. 2022). Each of the three IR telescope-derived radius estimates required a coma modeling and removal process (Fernández et al. 2013), so the values possibly represent upper limits to the radius due to residual coma flux contributions. The consistency in the estimates provides evidence for a more spheroidal shape with little elongation.

Numerous endeavors have been made to constrain SW1's nucleus' spin state. Regrettably, a consistent set of constraints remains elusive. A spin period on the order of several hours, akin to measurements for spin states within a cohort of JFCs (Kolokolova et al. 2007), was derived from the analysis of photometric light curves while SW1 was in a period of quiescent activity (Meech et al. 1993). The photometry measurements of that work were measured using small apertures centered on the nucleus' position attempting to extract out modulations due to the underlying nucleus' rotation. However, the high and variable airmass coupled with variable seeing of the observations used in that effort (see Tables 1 and 2 in Meech et al. 1993) increase the uncertainties associated with the results of the investigation based on the small-aperture photometry. Furthermore, attempts to extract photometric light-curve modulations from a nucleus that appears more spheroidal in nature are inherently challenging. Using the current range of nucleus size estimates (radius values from 23 to 32 km) as the extremes for the projected cross section, the nucleus would present a $\Delta m \sim 0.5$ magnitudes at a typical observing geometry along SW1's orbit, which, while not an insurmountable task to measure, would undoubtedly be difficult given SW1's persistent and often variable dust coma activity. An independent estimate of a more typical rotation period was derived through the examination of gas coma morphology captured in Spitzer IRAC 4.5 μm images (Reach et al. 2013). Conversely, alternative methods for measurement propose a spin period on the order of several days to upwards of ~60 days. Some of these longer spin periods were derived from the analysis of observed trends in SW1's potential outburst periodicity (Trigo-Rodríguez et al. 2010; Miles et al. 2016b). The periods derived from this kind of technique are closer to long term averages of gaps between significant outbursts—the rotation periods inferred assume that the same spot on the surface or similar local lighting conditions produce outbursts repeatedly, which may or may not be true as the gaps between outbursts do seem to vary. Regardless, other groups have independently arrived at longer spin periods through analysis of outburst dust-comae morphology through applying 3D Monte Carlo coma modeling to reproduce features detected in dust continuum imaging (Schambeau et al. 2017, 2019). A complication of the latter approach arises from the uncertainty in the enhanced dust lofting duration during an outburst, where the modeling results actually place a constraint on the ratio of the spin period to outburst duration. Regardless of the inherent limitations of the modeling approach, assuming plausible outburst durations on the order of minutes to tens of hours requires a nucleus spin period on the order of several days or longer to replicate the evolution of outburst dust comae. The disparity in spin period estimates adds an additional layer of complexity to our understanding of SW1, emphasizing the need for further observational and analytical efforts to unravel its intriguing activity behaviors.

Information regarding whether SW1 is in a principal-axis or non-principal axis rotation state, as well as the angular momentum direction for any such state, remains scarce. Extreme spin pole obliquities for SW1—specifically, spin alignments within ± 10–20° of the orbital plane—seem unlikely. Such states would likely manifest in substantial seasonal variability in both quiescent and outburst activity behaviors, which, notably, is not observed (Miles et al. 2016b). Moreover, the substantial size of its nucleus suggests that a non-principal axis rotation state would likely dampen down to a principal-axis state within years to decades and the torques induced by the frequent outburst events are considered too small to induce excitation in the spin state (Schambeau et al. 2019). In summary, while the specifics of SW1's rotational dynamics remain unclear, certain possibilities, such as extreme spin pole obliquities and short rotation periods, on the order of hours to a few, can be reasonably ruled out based on the observed stability in its activity patterns and lack of comae morphology displaying signatures of rotation. The absence of definitive constraints on the spin state of this enigmatic Centaur poses a significant challenge, limiting our capacity to comprehensively explore the nature of its activity patterns through thermophysical modeling efforts. Chapter 7 includes further discussion on the topic of rotation state evolution.

No direct measures of the surface reflectance of SW1 have been made due to the persistent dust coma confusing such observations. The only surface material property to be constrained has been its infrared beaming parameter $\eta \sim 1$ (Bauer et al. 2013; Schambeau et al. 2021; Bockelée-Morvan et al. 2022), which is consistent with a surface of low thermal inertia similar to those of the JFCs (Fernández et al. 2013). These beaming parameter estimates do also suffer from the inherent drawbacks associated with being measured while a dust coma was present, however the consistency between similar values for this parameter resulting from three independent telescope and instrument combinations (*Spitzer*, WISE, *Herschel*) and that these independent values align well with measured values for other objects that were measured while inactive provides support for the $\eta \sim 1$ value. At the minimum of its brightness, SW1's surface may be directly detectable at longer wavelengths. As a result, one might imagine a target of opportunity (ToO) proposal to a telescope which does not activate when the target is bright but instead if it is dim.

10.3.4 Comae Properties

The comae of SW1 are often observed for attempts to understand its unique activity behaviors and their driving mechanisms. Observations while both in and out of outburst have been undertaken in order to compare the comae natures (both gas and dust comae) during each phase of activity strength to discern whether the underlying activity drivers are the same or different between phases. Historically, there is a larger set of quiescent comae characterizations in the literature due to the challenges associated with rapidly planning for and acquiring telescopic resources after an outburst event. This situation has been improving over the last few decades fortunately due to the almost nightly monitoring of SW1 through both dedicated campaigns and small body discovery surveys providing timely notifications of outburst events in which to trigger subsequent observations.

The gas comae of SW1 have had an increasing number of molecular species detected, both first-generation species and a number of photodissociation products. SW1 has the most molecular species detected out of the Centaurs population. The most abundant species detected is CO, whose outgassing is proposed to drive SW1's activity. The detection of CO is persistent, similar to that observed regarding its dust coma, with a typical production rate of $\sim$1–6 $\times$ 10^{28} molecules s^{-1} (Senay & Jewitt 1994; Crovisier et al. 1995; Festou et al. 2001; Gunnarsson et al. 2002, 2008; Paganini et al. 2013; Wierzchos & Womack 2020; Bockelée-Morvan et al. 2022). Results from Wierzchos & Womack (2020) suggest that the dust outbursts may not necessarily have a corresponding CO outburst, whereas results from Bockelée-Morvan et al. (2022) indicate that a strong correlation between the CO- and dust-production rates may exist. The difficulty in coordinating a targeted contemporaneous CO and dust outburst monitoring campaign makes determining whether a relation exists challenging. A detection of CO_2 has been reported to follow a spatial distribution differing from that of CO (Faggi et al. 2024). That work suggests that this heterogeneity of CO versus CO_2 coma, which appears to originate from the nucleus and not an extended icy grain coma, may reflect a similar heterogeneity of the nucleus. An H_2O coma has been detected on two separate occasions (Ootsubo et al. 2012; Bockelée-Morvan et al. 2022) where the source appears to be sublimation from an extended icy grain coma source after major outburst events. HCN has similarly been detected to be sourced from an extended icy grain coma (Bockelée-Morvan et al. 2022). There have also been a number of dissociation product species detected, CO^+, CN, and N_2^+ (Cochran & Cochran 1991; Korsun et al. 2008; Ivanova et al. 2016, 2018), where in particular the detection of N_2^+ is rare and notable among comets (Anderson et al. 2023).

Compositional properties of SW1's dust coma derived from infrared observations from *Spitzer*, *Herschel*, and the IRTF have indicated grains possessing water ice content that are actively sublimating, producing the subsequent H_2O gas coma (Schambeau et al. 2015, 2021; Bockelée-Morvan et al. 2022). A mid-infrared emissivity spectrum of SW1 based on *Spitzer* IRS data provided an opportunity to constrain plausible compositional and grain size properties of SW1's coma (see Figure 10.4; Schambeau et al. 2015). During those observations SW1 was in a state

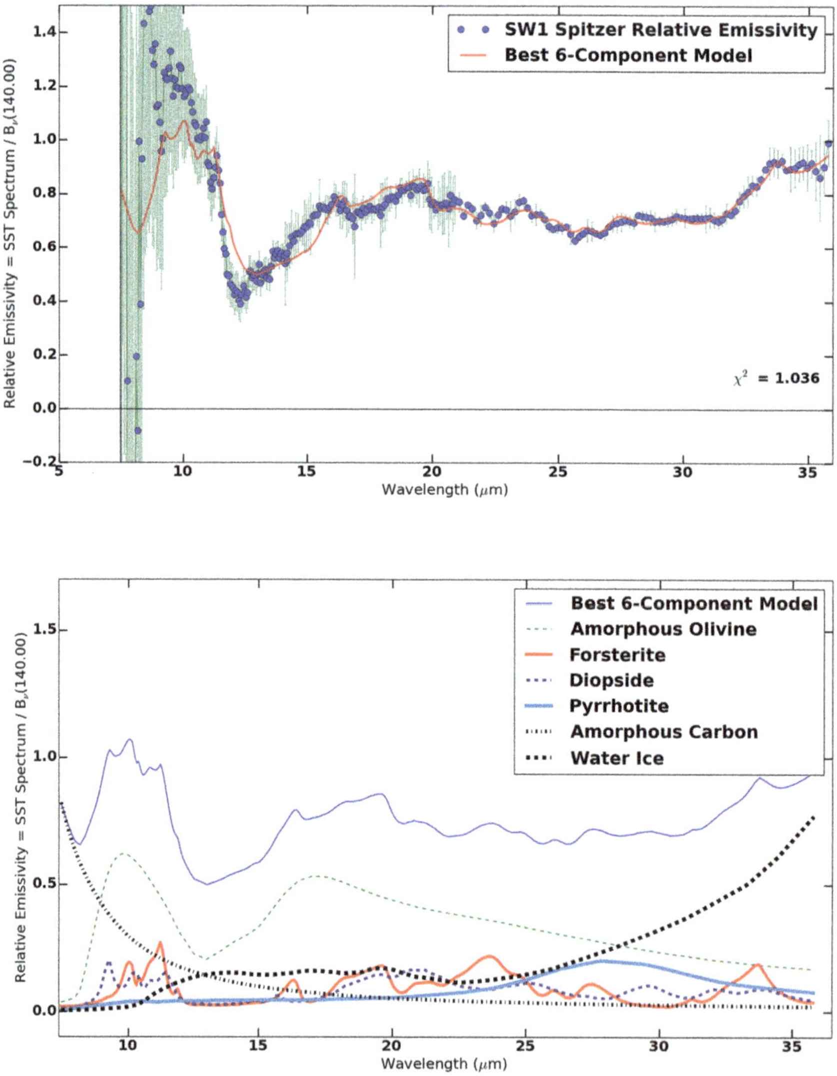

Figure 10.4. Top: the mid-infrared emissivity of 29P/Schwassmann-Wachmann 1 compared against the best-fit six-component model presented in Schambeau et al. (2015). Bottom: with the six components of the model plotted separately. SW1's coma appears similar to other comets observed at these wavelengths and discussed further in the text. Reprinted from Schambeau et al. (2015). Copyright 2015, with permission from Elsevier.

of quiescent activity. The emissivity spectrum clearly requires several components to be fit well, including multiple kinds of silicates, amorphous carbon, and water ice. The lack of high-quality optical constants for relevant materials taken under space-like conditions hampers efforts to identify all of these minerals as conclusively being a part of SW1's dust and ice coma, but SW1's relatively featureless and gently curving reflectance coma at visible and near-infrared wavelengths should not be taken to imply that we cannot ascertain the composition of SW1's dust at these longer wavelengths. This topic is discussed more in Chapter 14.

10.3.5 Discussion

As mentioned earlier, despite SW1's modest variation in energy input from the Sun, it frequently undergoes major outbursts of dust emission superimposed on its persistent quiescent dust production (Roemer 1958; Whipple 1980; Jewitt 1990; Trigo-Rodríguez et al. 2010; Kossacki & Szutowicz 2013; Hosek et al. 2013; Miles et al. 2016b; Schambeau et al. 2019). Additionally, the CO-production rate during periods of quiescent activity is more similar to long-period comets at similar heliocentric distances than JFCs (Bauer et al. 2015; Wierzchos et al. 2017; Womack et al. 2017; Bockelée-Morvan et al. 2022), and its dust outbursts may be uncorrelated with large fluctuations of its CO outgassing rate (Wierzchos & Womack 2020). Thus questions naturally arise as to what activity drivers explain its enigmatic activity, and do all JFCs experience a period of similar behaviors while they are in the Gateway region? Are SW1's activity behaviors reflective of outer solar system materials being thermally activated in the Gateway, after a long period of cryogenic storage? Or, are they an intrinsic property to it alone? While a description of various activity drivers and how they might relate to the physical structure of SW1 is left for other chapters (see, e.g., Chapter 12), we focus in the coming sections on one object of a similar size to SW1—Echeclus—and one object of a similar orbit to SW1—LD2.

10.4 174P/Echeclus

Along with SW1 and Chiron, Echeclus belongs to the class of outbursting active Centaurs. Echeclus does not present episodes of activity nearly as often as SW1, but it undergoes outbursts more frequently than Chiron. In the following section we present a historical overview of the activity history and the physical and orbital properties of Echeclus.

10.4.1 Activity History

Echeclus, cross-listed as asteroid 60558, was discovered in March of 2000 by Jim Scotti with the 0.9 m *SPACEWATCH* telescope on Kitt Peak as a magnitude $V = 20.8$ object at $R_H = 15.4$ au inbound. While the object did not receive a cometary designation until the end of 2005, precovery images obtained by *SPACEWATCH* suggest the presence of a dust coma (Choi & Weissman 2006), indicating that the object may have been discovered after a period of activity. Considering that Echeclus is best known for having just four moderate or large outbursts, finding further epochs of activity like this one in early 2000 or the potential for precovery activity detection (see the Orbital Properties subsection) is clearly of interest to understanding this object's activity. Searches in the early 2000s (Rousselot et al. 2005; Lorin & Rousselot 2007) found no evidence of ongoing activity at Echeclus until the outburst mentioned below (corresponding to a heliocentric distance range of 15.4–13.1 au inbound), even when observed with an 8 m telescope. We return to the possibility and nature of "quiescent" activity at Echeclus toward the end of this section.

The first recorded major outburst of Echeclus (and by far the best known) occurred in December of 2005 (Choi & Weissman 2006) when it was observed to brighten to a magnitude of $m_R \sim 15$ at a $R_H = 13.1$ au, as bright as a typical comet much closer to the Sun. This is a change of approximately $\sim$7 magnitudes, one of the largest cometary outbursts seen, though the object's pre-outburst brightness (m_V=21–22) obviously contributed. This outburst is notable not just because of how bright Echeclus became, but also for an apparent disconnection of the source of activity from the nucleus (Choi et al. 2006). After the initial detection of Echeclus's brightening, it became apparent that Echeclus's activity was not centered on Echeclus itself but instead on a separate secondary source moving relative to the primary nucleus. The source of activity presented a motion consistent with it being on a hyperbolic orbit relative to the nucleus (Weissman et al. 2006), but by March of 2006, a $15''$ coma was observed by Rousselot (2008) to be $8''$ (60,000 – 70,000 km) away from the nucleus and with a brightness distribution not consistent with emanating from a nuclear point source. Observations in the mid-IR (Bauer et al. 2008) before the dispersal of the secondary source showed a strong signal even at 70 μm, morphology similar to visible-wavelength imaging, and an apparent distribution of particle masses consistent with ongoing cometary activity. The question of what exactly happened in late 2005—did Echeclus release a fragment large enough to be active (Bauer et al. 2008; Fernández 2009), an agglomeration of dust which took weeks or months to disperse (Rousselot 2008), some combination thereof, or some other process yet to be considered—is still a matter of some debate, though the fragmentation scenario is probably the most popular explanation. Some of the images presented in Bauer et al. (2008) (the same data set as Choi & Weissman 2006; Choi et al. 2006; Weissman et al. 2006) really do seem challenging to explain qualitatively and quantitatively without a second discrete source in the coma. If the outburst was driven by the fragmentation of the primary nucleus and a large intact fragment getting ejected and experiencing a brief period of cometary activity, the fragment must have run out of volatiles or broken up by the time of Rousselot (2008)'s observations. The fragment could have been kilometers across and still not been directly detected even after the dust dispersed, so the lack of later detections of the putative fragment (should it even have survived) does not rule in or out these scenarios explicitly. The model presented in Rousselot et al. (2016) can explain much of the dust morphology of the 2005 outburst with two outbursts on the primary body with no fragmentation event at all—there are certainly other scenarios to be explored as well. Until Echeclus (or some other Centaur) experiences another similar outburst with similar coma morphology, it may be challenging to understand exactly what occurred in December 2005.

A key issue in this scenario, even if it is exciting and explains many aspects of the extant data sets, is the lack of modeling. If a fragment was ejected and its activity lasted only a couple of weeks, this timing constraint could be linked to the size of the fragment as well as its thermal conductivity and bulk volatile content. If the fragment disintegrated in those weeks after its ejection, this could place constraints on the object's mechanical properties. This is a clear path forward to understanding Echeclus's activity and to finding new insights that might help guide future observations. Furthermore, there are still observations of the 2005/2006 activity epoch that have yet to have been published or analyzed—we recommend strongly that those who observed Echeclus in that timeframe look back through their records and see what they can find.

Two smaller outbursts occurred closer to perihelion. An outburst occurred in May of 2011 (Jaeger et al. 2011) at $R_H = 8.5$ au inbound. During this outburst Echeclus had an increase of $\Delta m_V \sim 2$ magnitudes. Jaeger et al. (2011) reported the existence of a "jet-like" feature in the coma after outburst. A similar moderate outburst occurred once again in August of 2016 at $R_H = 6.2$ au outbound (Miles et al. 2016a), when Echeclus brightened $\Delta m_r = 2.6$ magnitudes in less than 24 hrs, though the coma appeared symmetric with no condensations or fans seen in other outbursts. Echeclus was observed six weeks after the 2016 outburst (Seccull et al. 2019) to search for evidence of surface changes induced by the outburst and none were detected. The dust that was still in the vicinity of Echeclus at that time was blue in color (Seccull et al. 2019), which would normally be assumed to be driven by small grains dominating the cross section. However, six weeks is long enough that radiation pressure would have driven many of them away, so the color of the dust was interpreted to be driven by composition instead. While Echeclus appeared point-source like in those observations, it is possible that undetected ongoing weak mass loss might complicate these interpretations of the origin of this spectral feature. In other words, the blue-means-large-grains explanation implicitly assumes that Echeclus had fully ceased activity by the time of the spectral observations as it had in other outbursts.

Echeclus's second-largest outburst happened in early December 2017, when the object brightened by $\sim$4.0–4.5 magnitudes (Kareta et al. 2019). Like with many outbursts of SW1, the outburst coma was clearly asymmetric and formed the classic "pac-man" or "two-fanned" shape previously discussed in the SW1 section and shown in Figure 10.5. The exact origin of these kinds of patterns is a topic of ongoing research, but they are frequently linked to the size of the nucleus and large changes in an object's activity state. For objects as big as Echeclus or SW1, gas and dust isotropically expanding from an outburst simply have a large amount of solid angle blocked by the nucleus, which when coupled with a thin or non-existent pre-outburst coma can produce a very asymmetric appearance. If Echeclus had been undergoing steady activity prior to the outburst, there simply would not be as much contrast in brightness between the "outburst side" of the object and the other.

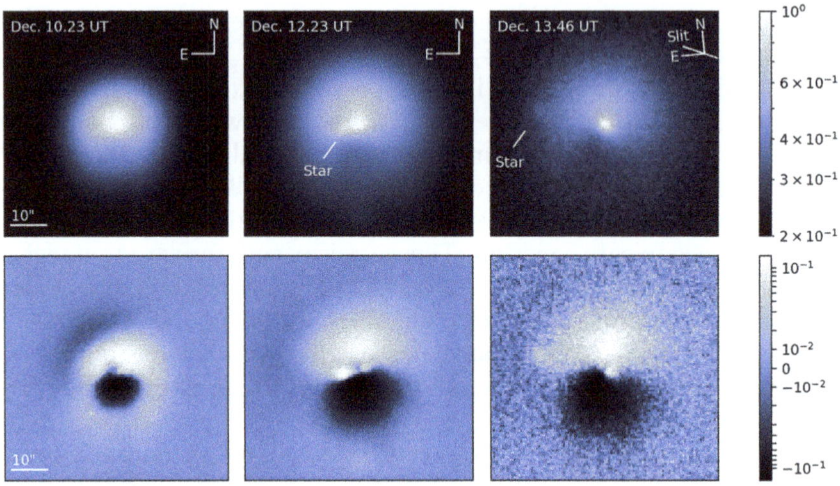

Figure 10.5. The evolution of Echeclus's 2017 outburst. Top row: three dates of images through the same filter normalized by their peak brightness (starting several days after the outburst was first reported). Bottom row: those same images with radial asymmetries enhanced. (An average radial profile was subtracted off, so what was previously the brightest pixel now has a value of zero.) Even though the "pac-man" like shape is less pronounced than in some images of similar outbursts at SW1, the radial asymmetries are very clear. The origin of this feature is discussed in the Echeclus and SW1 sections. Reproduced from Kareta et al. (2019) © 2019. The American Astronomical Society. All rights reserved.

Near-infrared observations of this outburst reported by Kareta et al. (2019) also showed evidence for a second "blob" in the coma, but only at longer wavelengths. Along-slit spatial profiles produced during the reduction and extraction of their data showed a second and fainter bump several arcseconds from the primary nucleus, the relative brightness of which increased toward longer wavelengths—by *K-band* ($\sim$2.2 μm) it was comparable to the primary body. No background stars or galaxies were in the background of the observations, and the relative positioning of the second peak remained constant over several frames with the slit in the same orientation meaning that it was moving at the same non-sidereal rates of the body. In other words, this secondary peak had to be related to Echeclus in some way. Those authors interpreted the second peak as likely consisting of a field of debris comprised of larger-than-dust particles as opposed to a fragment as many had proposed for the 2005 event. The increasing contrast at longer wavelengths is generally suggestive of the object having been obscured by dust at smaller wavelengths (thus explaining its non-detection in optical imaging), but a genuine fragment would be assumed to be smaller than Echeclus, so the two peaks having similar contrast at $K - band$ would be hard to imagine; so those authors suggested that this was the ejection of macroscopic debris—larger than dust and with a combined cross section roughly comparable to the nucleus, but not a discrete object. Those same near-infrared observations (Kareta et al. 2019) also supported a blue reflectance for the dust compared to the nucleus as inferred by Seccull et al. (2019), though follow-up work (Zubko et al. 2020) suggested this could be due to different scattering mechanisms at work in a thin coma compared to regolith.

The 2005 and 2017 events, a likely fragmentation and a "debris ejection event", certainly share many similarities but we do not know for certain that they were produced by the same process. Kareta et al. (2019) suggested these two events happening on the same body just twelve years apart—and in the only two of Echeclus's outbursts that were observed by larger aperture facilities close to the outburst—implies the likely existence of other similar events on other objects that are simply going undetected due to lack of detailed observations. While the real path toward understanding how Echeclus's activity works will come through more monitoring and characterization of its outbursts, characterizing the spread in outburst properties and features on other Centaurs with significant outbursts—like SW1 or Chiron—or other large outbursts like that seen at 108P/Ciffréo (Kim et al. 2023) clearly could provide some much-needed context as well.

While the origin of Echeclus's several outbursts is not perfectly clear, there are several properties which provide insights into what processes matter and which objects it can be compared to. Rousselot et al. (2021) presented a shape model for Echeclus based on many years worth of photometric observations that had a very significant obliquity of its rotational axis—fifty degrees or more—such that the object likely has very significant seasonal effects. Those authors infer that this seasonality must help initiate or at least play a significant role in Echeclus's activity. Perhaps other objects with significant outbursts like it might also have high obliquities? Pereira et al. (2024) recently presented analyses of stellar occultations taken in 2019–2021 toward Echeclus and found no conclusive evidence for dust or rings (explicitly ruling out a ring as optically thick as the main one around Chariklo), but their campaign did confirm an SW1-like size for Echeclus ($r_{eq} = 30.0 \pm 0.5$ km) and an albedo like a typical JFC ($\approx 5\%$). The lightcurve modeling of Rousselot et al. (2021) and the occultation analyses of Pereira et al. (2024) both support that Echeclus is moderately elongated as well. We comment on a more in-depth SW1/Echeclus comparison near the end of this chapter.

10.4.2 Orbital Properties

Echeclus's current orbit has a perihelion just external to Jupiter ($q = 5.84$ au), a moderate eccentricity ($e = 0.46$), and an aphelion between Saturn and Uranus ($Q = 15.66$ au). This significant temperature swing—some $\sim 65\%$ hotter at perihelion than aphelion—would normally be expected to drive significantly enhanced activity closer to the Sun than further out, but Echeclus's four well-documented major outbursts seem spread at random throughout its orbit. Two before perihelion, two after perihelion—both with one major outburst (2005/6, 2017) and one minor outburst (2011, 2016). Echeclus's orbit is sufficiently long that it has not even completed one revolution since discovery in 2000, so a clearer assessment of its activity state and how it varies over its orbit will clearly require continuing monitoring in the coming decades even as it becomes fainter. Unlike the brighter objects discussed elsewhere in this chapter like Chiron or Chariklo, there is not abundant precovery imaging on plates dating back decades. Echeclus's apparent magnitude at opposition and perihelion without any dust is $m_V \sim 16.5$, which for context is dimmer than Pluto and Charon even at their dimmest. All that said, Minor Planet Supplement (MPS) 139918[5] includes three archival observations of Echeclus from the Siding Spring Observatory in Australia that predate the object's discovery. Two from September 1979 ($R_H = 5.87$ au inbound, very close to perihelion) and one from September 1985 ($R_H = 10.19$ au outbound), and none of these three observations have an associated brightness estimate attached to their MPC submission. The apparent magnitude of Echeclus should it have *not* been in outburst on these two dates would have been $m_V \sim 16.4$ and $m_V \sim 19.3$, so the interpretation of whether or not the mere detection of Echeclus on those plates is enough to infer that it was caught undergoing an outburst is dependent on observational details that are not publicly available. That said, the latter dimmer magnitude would certainly have been too faint for many telescopes performing sky surveys at the time, so a renewed investigation of archival data and searches for precovery detections is certainly warranted. Considering that Echeclus's activity seems to be "on" or "off"—see a caveat above about carbon monoxide release—it is likely that any precovery detections of outbursts similar to what has been seen in the past two decades will be similarly unambiguous.

This long orbital arc from 1979 to present means that Echeclus's orbit is remarkably well constrained, allowing for a significantly longer period where Echeclus's orbit is well-determined than many other active Centaurs. Figure 10.6 shows the evolution of Echeclus's orbit forwards and backwards two-thousand years, with the orbit relatively well constrained from ~1100 to ~3000 CE (Kareta et al. 2019; Rousselot et al. 2021). In that time, Echeclus's perihelion has stayed relatively constant between 5.6 au $< q <$ 6.0 au, but its semimajor axis and eccentricity have evolved considerably. The detailed observations-and-dynamics study of Rousselot et al. (2021) integrated the orbits of 1000 clones sampled from the uncertainty matrix of the object's orbital fit using several different models (different integrators, different planetary ephemerides, etc.) and found that Echeclus was unlikely to have had a perihelion below ~4 au any time in the past ~12,000 years—and thus was unlikely to have undergone a period of water-dominated activity like that typical for the JFCs.

The dynamical portion of Kareta et al. (2019) focused on the perihelion evolution of Echeclus in part to compare to the suggestion of Fernández et al. 2018 that activity in the Centaurs might be driven by decreases in the perihelia of these objects. The hypothesis of Fernández et al. 2018 was that lowered perihelia drive higher peak surface temperatures, which then in turn drive higher sublimation rates and thus stronger (or at the very least more detectable) activity. Echeclus's perihelion distance hasn't changed in a significant way in a millennium or more (Kareta et al. 2019; Rousselot et al. 2021), but its activity is also not obviously centered at perihelion either—like many of the outburst-prone Centaurs, its activity appears well spread throughout its orbit. The recent dynamical study of Lilly et al. (2024) suggests that it

[5] Available from the Minor Planet Center's archive online at https://minorplanetcenter.net/iau/ECS/MPCArchive/2005/MPS_20050830.pdf.

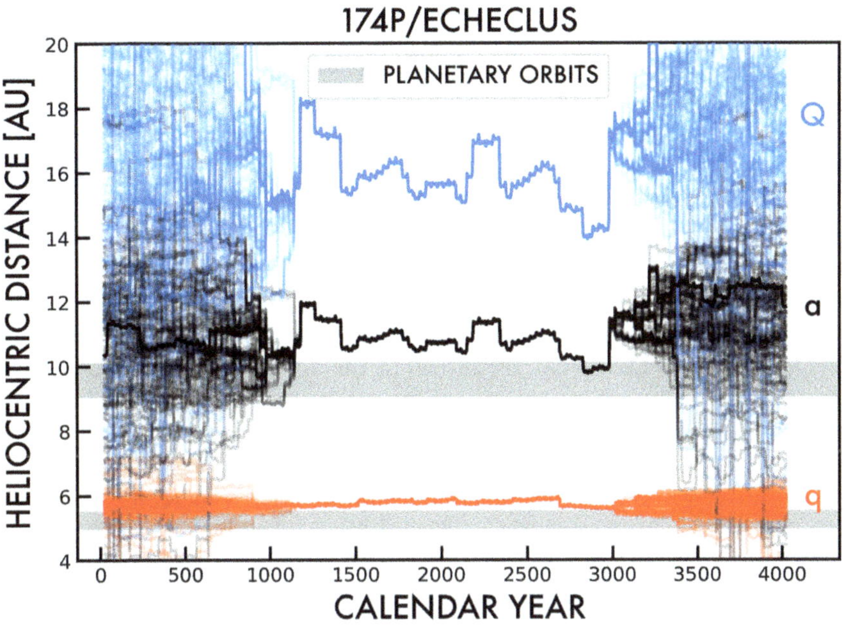

Figure 10.6. The evolution of Echeclus's orbit two-thousand years forwards and backwards from September 2023 (JPL Orbit Solution 112) is plotted, with perihelion (q) in orange, semimajor axis (a) in black, and aphelion (Q) in blue. This figure was updated from a similar one in Kareta et al. (2019) to make use of subsequent refinements to the objects orbit. The nominal orbit is shown as a solid line in each of these colors, and the evolution of 100 clones sampled appropriately from the uncertainty matrix of the orbit fit are shown in the same colors with significant opacity. While the object's perihelion distance has only changed slightly over the studied time period, its semimajor axis and aphelion distance varied considerably. The distinction between these two and how this changes our ability to interpret Echeclus's activity is discussed at length in the text.

is not perihelion decreases but instead semimajor axis decreases that drive the onset of Centaur activity. This is a preferable orbit-to-activity scenario to explain Echeclus's modern activity, both because Echeclus's semimajor axis *has* decreased between ~1100 CE and present and because it de-emphasizes the role that an object's perihelion plays. SW1, mentioned earlier, maintains vigorous activity despite a very low eccentricity orbit—it isn't a variation in solar insolation that's driving the majority changes in SW1's activity. That said, there is clearly more work to be done studying these various "orbital changes drive activity changes" hypotheses. For one, while Echeclus's semimajor axis has decreased since ~1100, there is a good chance that Echeclus's semimajor axis was *lower* prior to that date by inspection of the spread of black curves in Figure 10.6. If Echeclus previously had a semimajor axis of about ~10 au, how would a decrease from ~12 au to its current ~10.8 au affect the object's activity? How are the impacts of *recent* changes in an object's semimajor axis affected by the object's previous, longer-term history? At present, the answer is not clear—though Echeclus was and remains an interesting test-case for these different scenarios.

10.4.3 Quiescent Activity?

SW1 and Echeclus are approximately the same size and both show outburst dominated behavior, so it is natural to ask since SW1 appears to have a low-level of ongoing or "quiescent" activity—does the same occur at Echeclus? The likely detection of carbon monoxide emission toward Echeclus reported in Wierzchos et al. (2017) suggests the answer *might* be yes—but barely. Echeclus's CO production rate is some ~40 times less than SW1 at the same distance, was measured during a period in which no ongoing dust production was detected or noted, and took nearly a week of Arizona Radio Observatory Submillimeter Telescope time to detect at all. At its least active, SW1's coma and nucleus are about equal in brightness using very small apertures (comparing the typical brightness of the object during quiescent activity to an estimate of the apparent magnitude of the nucleus given its size and albedo), so if the coma-to-nucleus brightness ratio were scaled to Echeclus accounting for the differences in gas production rates, then the quiescent dust coma of Echeclus is likely well below most ground-based detection limits—just a percent or two of the brightness of the nucleus, and thus smaller than its lightcurve. The dimness of Echeclus during quiescent activity (it is currently $m_V \sim 21$ again on its way outbound as of writing, meaning that a scaled SW1-like quiescent coma would be magnitudes fainter than that) rules out many ground-based AO or high-resolution imaging options with extant facilities (future ELTs might very well be able to reach relevant contrast ratios), but a deeper stare with HST or JWST could be capable of setting useful upper limits on how Echeclus behaves between outbursts. JWST in particular could also serve double duty to verify the CO detection of Womack et al. (2017). However, the ground-based detection was taken near perihelion—a non-detection with JWST far from perihelion would not necessarily contradict the previous measurement. Another tactic to find evidence for quiescent activity would be to look for long-term variations in the object's absolute magnitude as has been discussed for other objects in this chapter. A preliminary inspection of MPC-submitted magnitudes suggests relatively little scatter away from the larger outbursts, and Dobson et al. (2023) was able to fit the phase curve of Echeclus quite reliably (although with a steep-by-comet-standards slope). This suggests that while we likely aren't missing any large outbursts and that short-term variability is often (typically?) minor, long-term variations in the total brightness would be very interesting to investigate in the same detail afforded to the larger and better studied objects. Comparing SW1 and Echeclus is clearly a comparison with limits—the two objects have very different orbits and thus thermal environments—but until we find more outburst-dominated objects more than a few kilometers in size, these two different objects will continue to be contrasted.

10.5 Chariklo

(10199) Chariklo is the largest Centaur yet discovered (see, e.g., Leiva et al. 2017; Morgado et al. 2021), with a shape approximate described as a tri-axial ellipsoid with dimensions of (144 km, 135 km, 99 km)—approximately the same size and similar in shape to Saturn's moon Epimetheus. Chariklo is undoubtedly best known for its complex ring system, the first detected on a small body (Braga-Ribas et al. 2014), revealed initially and then studied in depth through several occultation campaigns. As a result, we refer the reader to Chapter 9 for a much more in depth introduction to and discussion of those topics. In this chapter, we will instead focus on comparing Chariklo to other Centaurs and the results of non-occultation studies into its properties.

Shortly after Chariklo's discovery in 1997, water ice was clearly detected on its surface (Brown et al. 1998; Brown & Koresko 1998)—but disagreements about just how much water began nearly immediately. Brown et al. (1998) interpreted their near-infrared spectrum of Chariklo to be "dominated" by water ice, differentiating it from the few large TNOs which had been studied at the time, while Brown & Koresko (1998) could fit their spectrum with as little as $\sim$3% water ice by area. Spectra obtained in 2001 and 2002 were well fit by $\sim$2% water ice (Dotto et al. 2003), though those authors had started to speculate on possible surface heterogeneity at the object. This idea had also been raised by those studying the visible colors and lightcurves of Chariklo (Peixinho et al. 2001). Water ice was not seen in the spectra of Guilbert et al. (2009b, 2009a) taken in late 2007 and early 2008. Guilbert et al. (2009b) analyzed all of these observations using identical methodologies and showed conclusively that the depth of the 2.0-μm water ice band was decreasing in depth over time, from $\sim$20% dimmer than the continuum in 1997/1998 to $\sim$7% in 2002 to $\sim$0% in 2008. Luu et al. (2000) showed that water ice was more easily detected in the near-infrared spectrum of Chiron when its cometary activity was lower, adding another way to explain the changes in ice content over time. Was the surface variable, was Chariklo occasionally active, or was something else the cause?

Belskaya et al. (2010) compiled measurements of Chariklo's absolute magnitude over time (1997–2009) and found it had dimmed considerably and monotonically, from $H_V \sim 6.8$ around the time of its discovery to $H_V \sim 7.5$ by 2008. This corresponds to a comparatively small heliocentric distance range of 14.8–13.1 au thanks to Chariklo's moderate eccentricity, and thus a change in temperature of $\approx$ 5–10 Kelvin at most. Chariklo had come to perihelion in 2004, which, while its orbit is not particularly eccentric by comet standards ($e = 0.1678$), is when the most vigorous cometary activity might be expected due to higher surface temperatures. Chariklo's perihelion of $R_H \approx 13.1$ au is near where vigorous sublimation of CO_2 can be begin, so this small change in surface temperatures might result in a more substantial change in activity potential than might otherwise be expected. If cometary activity were to blame for the time-variability of the water ice features, then it should have been weaker in 2004 compared to 2008—and thus the surface should have "looked" icier—but this was not the case. Furthermore, even if cometary activity were to blame for the transience of the ice features, how would that be consistent with the system getting *dimmer* over time?

Braga-Ribas et al. (2014)'s detection of two rings around Chariklo provides a scenario that neatly explains both the dimming of the system over time and the disappearance of its water ice features. The orientation of the rings as solved for in that work suggest they were edge-on in 2008, meaning that the absorption features of solid water ice seen in the spectra of Chariklo were from light reflecting off the rings themselves—and that the system's dimming and the disappearance of the ice features were due to less and less of the rings being visible from the Earth (e.g., having a smaller visible cross-section over time). The surface of Chariklo could simply be red at visible wavelengths and slightly red at near-infrared ones, not very dissimilar to a traditional comet nucleus spectrum, and the variability could be attributed almost entirely to differences in ring visibility. Chariklo has brightened some since 2008, adding some credence to this scenario. A more detailed discussion of Chariklo's surface reflectivity and what can be said of the rings is in Chapter 14.

Two important points should be made here. First, the process by which individual measurements—in this case, near-infrared spectra and visible-wavelength photometry—from multiple groups coalesced to hint at some deeper, unresolved process ongoing at or property of Chariklo is deeply analogous to the discovery of activity at Chiron. Despite these objects being large, inherently bright, and with well-known orbits, it still took a decade or more of observing to start to assess their true natures. As with Chiron and unseen or unresolved activity, researchers finding variation in surface properties or bulk brightness will now occasionally suggest that changes in measured properties might be indicative of a ring system—or at the very least some long-lasting lofted material. This has yet to be born out—the only other small body with a verified ring system (see also the discussion of Chiron's lofted material above) is the TNO (136108) Haumea (Ortiz et al. 2017). Until a coma can be definitively spotted and resolved or extended material or rings can be detected with an occultation (or spacecraft visit), these kinds of inferences are likely too optimistic.

Second, one still needs to find a source for the material that makes up Chariklo's rings. Their long term stability is thought to require yet-detected small moons to "shepherd" them (Braga-Ribas et al. 2014), but what put the material currently confined to the rings into orbits that could be shepherded? Given that some Centaurs are active—like every other Centaur discussed in this chapter, for instance—the role of ongoing or transient cometary activity has been discussed as a way to loft significant material, especially at Chiron (see, e.g., Sickafoose et al. 2023; Chapters 8 and 9). However, the dust entrained in outflowing gas is typically assumed to also be moving primarily radially outwards—the vast majority of it lacking a significant transverse velocity that might help it enter into more long-term stable orbits around the body and thus a low efficiency of emplacement into the rings overall. If the outflowing dust were not primarily radial, such as from a jet with a non-radial orientation or from outbursts with a large opening area, the efficiency by which material could be implanted in the rings would be higher. However, given that no mass loss has been detected toward Chariklo at all, these would likely have to be sporadic or rare events. That said, even if the material in Chariklo's rings were to be from unnoticed or previous cometary activity, one might expect that the material in the rings should look much like the dust on the surface of Chariklo—dark and red. If Braga-Ribas et al. (2014)'s interpretation of their data are correct, the rings are bright and icy. Clearly some other scenario—the disintegration of a lost moon, a significant impact on Chariklo excavating icier material from below the surface—must be envisioned to explain the presence of the rings and the lack of known or suspected cometary activity. Given that Chariklo is the largest known Centaur, looking at smaller objects in the same orbital class might not necessarily provide the insights needed to solve the problem.

10.6 P/2019 LD2 (ATLAS)

P/2019 LD2 (ATLAS, hereafter just 'LD2') was both discovered more recently and is smaller than all the other objects in this chapter, yet it has drawn significant attention since discovery for its peculiar orbit and relatively strong activity for its size—a comparison we come back to later in this chapter. We will first discuss the nature of LD2's orbit and what kinds of orbits it might have had previously, then a discussion of the detailed campaigns to observe the object shortly after the announcement of its activity, and conclude with a discussion of how representative LD2 may be of the smaller Centaurs—and what future options exist to study it in more detail.

10.6.1 LD2's Orbit and History

LD2 is presently (2017–2029) a co-orbital of Jupiter, meaning that it orbits the Sun with the same period as that gas giant. This is a property of Jupiter's Trojan Asteroids as well, a group of inert and inactive objects which librate around Jupiter's L_4 and L_5 Lagrange points $\pm 60°$ away from the planet. Following discovery in 2019, LD2 was announced as active in early 2020 and initially categorized as a Jupiter Trojan—potentially the first active object in that population—but early dynamical studies (Kareta et al. 2020; Hsieh et al. 2021) showed that the true nature of LD2's current orbit was more complicated. While LD2 does share the orbital period of the Jupiter Trojans, it only became a co-orbital in 2017 following a close approach with Jupiter after previously having a Centaur orbit. A close encounter with Jupiter in 2029 will do the opposite, putting the object back into a classical Centaur-like orbit. (LD2 is typically grouped in with the other "Gateway" Centaurs, as many of the previous orbits it could have occupied are in or very close to the Gateway region; Steckloff et al. 2020.) Most excitingly, *another* close encounter in 2063 will push LD2 back inwards for good—it will no longer be transiently passing slightly interior to Jupiter, but instead will be more permanently stuck in a JFC orbit with a perihelion well inside the water ice line. While some JFCs have been found shortly after their insertion into the inner solar system (the *Stardust* target 81P/Wild 2 being the classic example; Wild & Marsden 1978; Królikowska & Szutowicz 2006), LD2 is the first object found for which this transition has been predicted—and can be observed when it happens.

This is at the heart of why LD2 is such a compelling object to characterize. While a reader perusing this book likely appreciates by now that the Centaurs are intermediate between the JFCs and the TNOs evolutionarily, it is another thing to watch an object move from one population into another and be able to observe it closely as it warms and the nature of its activity evolves. Furthermore, the detailed dynamical study of Steckloff et al. (2020) showed that LD2 has not been in the inner solar system in the recent past few thousand years. While knowing if LD2 has *ever* been in the inner solar system before is probably impossible (see, e.g., Chapter 3), it has at least not been inside of Jupiter in long enough that it cannot retain any heat from any previous passages closer to the Sun. (Even if one were to discover a Centaur that was going to get scattered onto a JFC-like orbit and then measure its orbit extremely precisely, it still would not be possible to "know" it had *never* done so previously.) One might expect some differences in how an object might respond to the Centaur-to-JFC transition depending on its past thermal history, such as the depth of various ice sublimation fronts, so we do not know if every Centaur would respond like LD2 will respond—but it is clearly still going to be a highly compelling observing opportunity (once everyone finishes observing Comet Halley's return two years earlier). Even more than just an observing opportunity, Seligman et al. (2021) showed that a spacecraft stationed near Jupiter could easily intercept LD2 during its 2063 close approach and follow the object inwards.

Both Steckloff et al. (2020) and Seligman et al. (2021) predict that while LD2's status as the first object discovered before transitioning might seem rare, it is really the tip of the iceberg—there ought to be a population of objects in near-Jupiter orbits experiencing close approaches to that gas giant and getting flung into JFC orbits—though how common exactly these encounters are is not yet clear. The models of Steckloff et al. (2020) suggested that objects move from these Gateway-like orbits into JFC-like orbits at a rate between three per decade and one-to-two per century depending on the size of LD2's nucleus. A second Jupiter co-orbital comet has recently been discovered, P/2023 V6 (PANSTARRS; see Kareta et al. 2024), about a third of a decade after LD2—in line with expectations if both are about the size of a typical JFC. P/2023 V6 is significantly fainter than LD2, so that might be an indication that future deeper surveys (e.g., the Vera Rubin Observatory, discussed in Chapter 14) might find this less active portion of the population. Regardless, these other objects would be equally accessible to spacecraft visitation as LD2 is.

10.6.2 Observations and Physical Properties

LD2's future entrance to the Jupiter-family comets and relative brightness for an active Centaur ($m_V \sim 18$ throughout 2020) made it a highly compelling target for telescopic characterization. It is worth contextualizing what has been learned about LD2 within the longer story of its orbital evolution. From discovery in 2019 through the time of writing, it has been interior to Jupiter's orbit or slightly outside of it—and thus significantly warmer than the Centaurs are typically thought to get. (Even calling LD2 a Centaur in its 2017–2029 orbit is hardly accurate!) Some differences ought to be expected between $R_H = 4.5$ au at LD2's perihelion and the traditional inner limit of Centaur orbits of $R_H = 5.2$ au, but how important this difference is in comparing LD2 to other active Centaurs is a matter of perspective.

During the 2020 observing season, LD2 appeared to be in steady-state activity (Kareta et al. 2021; Bolin et al. 2021; Licandro et al. 2021), only slowly evolving in brightness and showing coma features better explained through sustained dust ejection rather than episodic or variable activity. Kareta et al. (2021) also found no evidence for outbursts or sudden changes in activity rate in a search through the archive of the Catalina Sky Survey. Another indication that the object was in stable activity was that the visible-wavelength colors reported by these three teams agree within $\sim$1-σ despite being taken weeks apart. LD2's inner coma was slightly redder than the Sun at all epochs, which for a regular JFC would be unremarkable. While one assumes that LD2, given that it is an active pseudo-Centaur on the cusp of becoming a JFC, should be in the "less-red" group (Bauer et al. 2013), we do not have any constraints on the reflective properties of its nucleus.

As mentioned at the start of this section, LD2 is small. Comparing the spatial profile of LD2 against that of nearby stars in observations taken by the Hubble Space Telescope, Bolin et al. (2021) estimated a nucleus radius of $r_{nuc} = 1.8$ km assuming an albedo of $p_V = 0.08$. However, Schambeau et al. (2020) estimated a radius no larger than $r_{nuc} < 1.2$ km based on a search of archival ground-based images and assuming an albedo of $p_V = 0.05$, suggesting the former number includes some contamination by close-in dust. The JFCs in the sample of Bauer et al. (2017) have a mean radius of $r_{nuc} = 0.65$ km, so LD2 is either slightly larger than or about the same size as a typical JFC.

No searches for gas toward LD2 have thus far born fruit. Kareta et al. (2021) found no emission from CO toward LD2, placing an upper limit on the steady-state production rate of $Q(CO) < 4.4 \times 10^{27}$ molecules s^{-1}, and observations with the *Spitzer Space Telescope* (Bolin et al. 2021) found no extended emission attributable to CO or CO_2 in the filter sensitive to those species, resulting in upper limit estimates of $Q(CO) < \sim 10^{27}$ molecules s^{-1} and $Q(CO_2) < \sim 10^{26}$ molecules s^{-1} respectively. Neither Licandro et al. (2021) nor Bolin et al. (2021) found any visible-wavelength emission features in their spectra, setting upper limits for CN ($Q(CN) < (1.4 \pm 0.7) \times 10^{24}$ molecules s^{-1}; Licandro et al. 2021) and C_2 ($Q(C_2) < 7.5 \times 10^{24}$ molecules s^{-1}; Bolin et al. 2021). As noted in Licandro et al. (2021), extrapolating a typical JFC to LD2-like distances and assuming that production rates scale down accordingly would make one estimate that LD2's production rates should be a factor of a $\sim$few lower than most of these upper limits. In essence, these measurements rule out that LD2 is producing enough of any of these substances to be interestingly over-productive, but it does not rule out the object being in the realm of "typical" or even being gas-poor. On an $Af\rho$-only (a comet-specific measure of dust production rate, see A'Hearn et al. 1984) basis, LD2 is somewhat more active than a typical JFC at its perihelion distance (Kareta et al. 2021), which those authors linked to LD2's likely not having been in the inner solar system (Steckloff et al. 2020).

The sole substance detected thus far at LD2 is water ice in the near-infrared spectra of Kareta et al. (2021). If the dust and ice were intimately mixed, the volumetric ice fraction was best modeled as $f_{ice} = (9 \pm 2)\%$. If the ice and dust were physically separated (mixed "linearly"), a much smaller ice areal fraction of $f_{ice} = (2.0 \pm 0.6)\%$ was obtained. The intimately mixed models preferred larger dust grain sizes in the vicinity of $\sim 150\ \mu m$, similar to the dust sizes inferred from deep imaging (Bolin et al. 2021; Kareta et al. 2021), so the higher (volumetric) ice fraction might then be preferred. The challenges in obtaining high-quality near-infrared spectra of extended objects this faint using current ground-based facilities is one of the prime reasons LD2 was one of the first Centaurs observed with JWST (Chapter 14).

10.6.3 LD2 as a Proto-JFC?

In summary, LD2 is a JFC-sized (at most just slightly larger than a typical JFC) active object that is currently co-orbital with Jupiter. It was previously in a Centaur-like orbit and will return to one at the end of the 2020s, followed by a permanent entrance into the Jupiter-family following a close encounter with Jupiter in 2063. It is slightly to moderately more active than a JFC would be in its current orbit based on dust production rates, but no clear signs of gas emission have been detected as of yet. It seems to have an activity pattern unlike several of the other Centaurs described in this chapter—namely, it behaves somewhat like a typical inner solar system comet and becomes more active as it approaches the Sun. No obvious signatures of previous outbursts or discontinuous activity have been noted yet. LD2's dust coma has a typical reflective behavior for comets and active Centaurs, and its dust coma appears to have a small amount of solid ice mixed in. Outside of its odd orbit, LD2 *appears* very much like what one might expect a new JFC to look like extrapolated backwards in time by a few orbits.

All of this leads to the question: is LD2 a "typical" proto-JFC? Does it have the properties and behavior of what we would think a median soon-to-be-JFC would have just prior to being scattered inwards? The reason this question gets asked relates to how useful one thinks that monitoring its transition from Centaur inwards in 2063 will be—if LD2 is a strange outlier, then we might still learn a variety of interesting things, but not for how this transition affects the population at large. The answer is, of course, "maybe", but we again remind the reader of the caveats mentioned at the start of this chapter. If the models of Steckloff et al. (2020) and Seligman et al. (2021) are right, there ought to be several more LD2-sized objects in and around the Centaur Gateway waiting for their chance to enter the inner solar system—the recently discovered second Jupiter co-orbital comet P/2023 V6 (PANSTARRS) is probably one of many (Kareta et al. 2024). Discovering and characterizing more of them is the only way to know for sure.

10.7 Comparing Outliers and Future Research

In this section, we will try to address a few big-picture questions and topics that link multiple of our targets together. In particular, we will try to address how different the activity patterns of our four active objects are, what size trends can be discerned among all of the objects, other Centaurs not mentioned in this chapter which might be useful to study in comparison to an object discussed here, and what spacecraft visits to these targets might look like and entail.

10.7.1 Activity

While some active Centaurs display cometary activity like that seen in the inner solar system—mass loss rate and general activity state correlated with how warm an object is—this is only true for one of our four active objects, LD2. SW1 and Echeclus are both in outburst-dominated modes of activity, and even theirs are not identical. While Echeclus has had four outbursts observed in significant detail since its discovery more than two decades ago, SW1 can have that many strong outbursts in a single year. The exact nature of Chiron's activity is not even totally clear yet. While it was clearly active at perihelion, it might have been similarly active at its much colder aphelion. One could envision some scenario where the mystery of Chiron's activity comes down to the object having a significant obliquity and significant variation in ice content across its large surface—in essence, to the fact that the object is large enough to have appreciable "geology"—but even that is hard to assess. Even if one were able to produce evidence that Chiron's surface varied as it rotated (or over its orbit, more likely), how easily could that be disentangled from its already hard-to-pin-down activity variations? Spectroscopic variations were seen at Chariklo only for those to turn out to likely be driven by the geometry of the object's rings and not some evolution of the object's surface. In other words, one could look at these four objects and easily conclude that their activity patterns alone sorted them into four classes—or at least three, if one were to group SW1 and Echeclus together. In addition to the observational studies proposed throughout this chapter, further modeling efforts are clearly needed. Can models that explain SW1's activity be applied to Echeclus, or if not, what modifications are required? What role does size play in changing the character of an objects activity—should we expect that the outburst-prone active Centaurs are all large, or would the more common kilometer-scale objects also occasionally show this kind of activity? While these are primarily questions for thermophysical models discussed at length elsewhere in this book, the latter question about the frequency of large outbursts could likely be achieved, or at the very least addressed, through a statistical analysis of the orbital arcs of the known active Centaurs. If they were discovered during a period of enhanced activity or during an outburst and then dropped back down in brightness to below typical survey sensitivities, their collective abundance might be related to the fraction of active Centaurs with short orbital arcs. There may be small objects that are only discovered during periods of enhanced activity, and thus the limited data we have on these objects might be uniquely interesting.

10.7.2 Sizes

The sizes of these objects also break down into three groups, which we discuss from largest to smallest. Chariklo and Chiron are the largest objects not just in this chapter but among the Centaurs more broadly, and are closer in size to the smallest TNOs detected or some of the mid-sized moons of Saturn than they are to the other objects mentioned here. Chiron is the largest active object in the solar system, and Chariklo is the smallest object confirmed to have rings—we might very well expect both of their sets of properties to be filled with outliers in some respects. However, the question of the material that persistently surrounds these two objects makes them perhaps better analogues than their activity states might suggest. A sticky question raised at the end of the Chariklo section of this chapter was "how did the ring material even get up there?" While lofting material for a long time into what at Chiron may be described as a "ballistic atmosphere"

(Meech & Belton 1990) may be pretty feasible with traditional sublimation-driven cometary activity, finding enough transverse (as opposed to radial) motion to allow any of the particles to move to bound orbits is challenging theoretically. Continued occultation campaigns at both bodies have a chance to make some progress here, as the bigger-picture question of "how do bodies only 100–200 km across manage to hold on to fine-grained dust for long periods of time?" can be investigated at both targets for mutual benefit. While *the dust has yet to settle* on the exact structure of Chiron's semi-persistent dust shroud (Sickafoose et al. 2023; see also Chapter 9), any new constraints on its origin, sources, or sinks might very well prove useful to those investigating the origin of Chariklo's ring system. Future modeling of the ring/arc dynamics around these objects might be critical to identifying their sources and sinks. While the formation of large ring systems around the gas and ice giants or some large TNOs might require enough mass as to functionally require that a moon or small body was disrupted, there might be a wider range of phenomena that can produce them. Even the range of parameter space in which cometary activity could supply or affect a ring system could be useful in identifying future targets for occultation campaigns to search for rings. Another object that might prove an interesting comparison point to Chiron is C/2014 UN271 (Bernardinelli-Bernstein; Bernardinelli et al. 2021), an inbound Oort Cloud Comet with a diameter of ~130–150 km (Lellouch et al. 2022; Hui et al. 2022). While still somewhat smaller than Chiron, it is clearly the second-largest active object in the solar system as well as the largest Oort Cloud Comet. While Comet Bernardinelli-Bernstein is hardly in a similar thermal state to Chiron—an aphelion beyond the heliopause is certainly much colder than Chiron's $Q = 18.87$ au—it is a welcome addition to the list of objects which are experiencing ongoing ice sublimation but are also large enough that factors like self gravity or topography cannot be ignored. How frequently Bernardinelli-Bernstein outbursts and what the properties of those outbursts are is an area where comparisons might be more straightforward, but even the nature of Bernardinelli-Bernstein's long-term secular lightcurve would be key to interpreting Chiron's lightcurve as well. However, given that the object won't reach perihelion until the early 2030s, this kind of study won't really be possible for decades. Until then, the development of models which link the presence and duration of rings and ring arcs and other associated phenomena with the actual bodies the rings are around could be useful. Can Chiron's or Chariklo's topography, shape, or structure be inferred or commented upon based on the apparent long-lasting material around them? Is the dust that surrounds Chiron variable in abundance and distribution because of ongoing cometary activity, or is there some other reason that might explain why structures seen at one epoch appear to have changed or drifted significantly? Assumptions about other ring systems, or about smaller comets, may not apply as well as we expect them to when we look closer.

SW1 and Echeclus are both approximately ~60 km in diameter. While it might be tempting to link their similarities in size and activity pattern in some sort of causal relationship, objects far smaller than these two have been seen to display outbursts of similar magnitude—albeit less frequently. The Halley-Type Comet 12P/Pons-

Brooks has undergone several major outbursts on its way into an April 2024 perihelion as of writing, and other comets like 17P/Holmes in 2007 or P/2010 H2 (Vales; Jewitt & Kim 2020) in 2010 have undergone similarly larger outbursts closer to the Sun. However, there is preliminary evidence that not all of these large outbursts are the same. While 17P/Holmes giant outburst had a reflective spectrum clearly indicative of water ice (Yang et al. 2009), Echeclus's large outburst in 2017 showed no good evidence for ice at all (Kareta et al. 2019). The one tentative detection of ice in a SW1 outburst (Protopapa et al. 2021) remains to be properly published, and other near-infrared observations of SW1 have not reported significant water ice absorption. There is at least some evidence then for variation in the properties of the material lofted in these large outbursts beyond even the variation in strength of outbursts on single targets. We thus advocate for observational programs to assess the properties of these large outbursts regardless of the target in question so as to understand the basic mechanisms at play and to hopefully gain insight into objects like SW1 and Echeclus. This would primarily be ToO campaigns to watch the photometric and spectroscopic evolution of their comae, ideally to be activated within a day of the onset of outbursts. Modeling efforts to constrain when would be best to search for ice in these objects, and thus contextualize existing observations better, would also be critical. Do these large outbursts primarily throw material off the surface as opposed to significantly excavating colder material from depth? The question of what material is being excavated and from where would not just assist in interpreting telescopic observations, but might also assist in the interpretation of surface features should end of these objects be visited by spacecraft.

10.7.3 Future Spacecraft Visits and Larger Telescopes

While a detailed discussion of spacecraft visits to SW1 and Chiron in particular is left for Chapter 16, we can briefly discuss here what a visit to each of these objects might entail. SW1, Chiron, and LD2 are the three objects for which a visit to them would almost certainly encounter activity and ongoing mass loss. While SW1 is essentially always active and has frequent enough outbursts that a fly-by might still have a good chance of characterizing one in a quick visit, the answer is less clear for LD2 and Chiron. Chiron is clearly active in a large part of its orbit, but this activity is variable—and LD2 has simply not been observed long enough to have any real constraint on how active it might be at aphelion, let alone its new colder aphelion after its 2029 orbital change. Unlike these other objects, a visit to LD2 as a Centaur is on a timer—after 2063, it would "just" be a visit to an extremely young JFC. The modes by which each of these three active objects shed mass appear to be rather different as well, with SW1 being outburst dominated, LD2 experiencing ongoing stable sublimation, and Chiron seeming to show both behaviors. A visit to any of the three would be transformative, but naturally questions about the other kinds of activity would persist.

Size-wise, Chiron and Chariklo clearly provide the best opportunities to study the surface geologies of these two large objects. Furthermore, their sizes put them at the lower end of what has been found telescopically in the trans-Neptunian belt, so their evolutionary differences compared to your garden-variety TNO could very well be driven primarily by their orbital differences. A good reason for why the Centaurs are so compelling is their relative differences to the JFCs and the TNOs—their activity patterns, sizes, and surface properties are all different. (The case for a JFC-like size being a key comparison point elevates LD2's apparent science return as well.) Chiron and Chariklo also present the ideal opportunities—and the only ones inside of Neptune—to study the extended dust structures and rings that appear persistent around both objects. Chiron's time-variable dust structures (Sickafoose et al. 2023) could then present a compelling case for a more long-term visit to the object, and if Chariklo's rings are evolving in time (Giuliatti Winter et al. 2023; Santos-Sanz et al. 2023) then the same case could be made for it as well.

Echeclus, considering these above cases, might then seem like the least appealing target—no guarantee of high activity, and the ongoing quiescent activity that it might have (Wierzchos et al. 2017) might be very weak. Furthermore, even an orbiter might not last long enough to see an outburst happen—the gaps between outbursts appear to be years long. However, a sometimes-active ~60-km object is still a kind of solar system body that we have never seen before. Echeclus's giant outbursts may have left large scars on its surface, and thus the mystery of its 2005 outburst—fragment, debris ejection event, or some other process entirely—could be potentially solved through a careful investigation and mapping of its surface. The abundance and properties of surface features that might be linked to outbursts is still a critical aspect of diagnosing how these outbursts work as discussed above, so even a fly-by of Echeclus could be truly useful in investigating some of the most energetic cometary phenomena.

While we wait for a spacecraft mission to be planned, approved, and launched, the newest space-based telescopes and the next generation of ground-based extremely large telescopes presents a world of possibilities. While early JWST results are discussed in Chapter 14, many basic programs—such as capturing a significant outburst of SW1 or Echeclus at moderate to high resolution or mapping the geology through rotationally resolved spectroscopy of Chiron or Chariklo—have yet to be accepted or be implemented. Future deeper surveys, like all-sky efforts such as Rubin Observatory/LSST, will also greatly refine our understanding of the population of objects in near-Jupiter orbits—thus placing LD2 into a clearer context in terms of its size, activity state, and orbital history than is currently possible. Given that the five objects in this chapter are all bright enough for characterization with moderate-aperture ground-based facilities, the newest frontiers will be in pushing to new wavelengths and finer time resolutions in order to capture their nuances in more detail.

In summary, each of these objects appears to be the tip of an equally interesting iceberg—despite all being studied in detail that would make your average Centaur jealous, there are real questions that remain about their histories, modern properties, and activity states. The next Centaur book will have a different group of objects receiving special attention, but these are some of the bodies which have formed our modern understanding of the topic.

References

A'Hearn, M. F., Schleicher, D. G., Millis, R. L., Feldman, P. D., & Thompson, D. T. 1984, AJ, 89, 579
Anderson, S. E., Rousselot, P., Noyelles, B., Jehin, E., & Mousis, O. 2023, MNRAS, 524, 5182
Bauer, J. M., Choi, Y.-J., Weissman, P. R., et al. 2008, PASP, 120, 393
Bauer, J. M., Grav, T., Blauvelt, E., et al. 2013, ApJ, 773, 22
Bauer, J. M., Stevenson, R., Kramer, E., et al. 2015, ApJ, 814, 85
Bauer, J. M., Grav, T., Fernández, Y. R., et al. 2017, AJ, 154, 53
Behrens, J. G. 1932, AN, 245, 309
Belskaya, I. N., Bagnulo, S., Barucci, M. A., et al. 2010, Icar, 210, 472
Berman, L., & Whipple, F. L. 1928, LicOB, 394, 117
Bernardinelli, P. H., Bernstein, G. M., Montet, B. T., et al. 2021, ApJL, 921, L37
Bockelée-Morvan, D., Lellouch, E., Biver, N., et al. 2001, A&A, 377, 343
Bockelée Morvan, D., Biver, N., Schambeau, C. A., et al. 2022, A&A, 664, A95
Bolin, B. T., Fernandez, Y. R., Lisse, C. M., et al. 2021, AJ, 161, 116
Braga-Ribas, F., Sicardy, B., Ortiz, J. L., et al. 2014, Natur, 508, 72
Brown, M. E., & Koresko, C. D. 1998, ApJL, 505, L65
Brown, R. H., Cruikshank, D. P., Pendleton, Y., & Veeder, G. J. 1998, Sci, 280, 1430
Bus, S. J., A'Hearn, M. F., Bowell, E., & Stern, S. A. 2001, Icar, 150, 94
Bus, S. J., A'Hearn, M. F., Schleicher, D. G., & Bowell, E. 1991, Sci, 251, 774
Bus, S. J., Bowell, E., Harris, A. W., & Hewitt, A. V. 1989, Icar, 77, 223
Carusi, A., Kresák, Ľ, & Valsecchi, G. 1995, EM&P, 68, 71
Choi, Y.-J., & Weissman, P. 2006, DPS Meeting, Vol. 38, (Washington, DC: AAS) 37.05
Choi, Y. J., Weissman, P., Chesley, S., et al. 2006, CBET, 563, 1
Cochran, A. L., & Cochran, W. D. 1991, Icar, 90, 172
Crovisier, J., Biver, N., Bockelee-Morvan, D., et al. 1995, Icar, 115, 213
Cruikshank, D. P., & Brown, R. H. 1983, Icar, 56, 377
Dobson, M. M., Schwamb, M. E., Fitzsimmons, A., et al. 2021, RNAAS, 5, 211
Dobson, M. M., Schwamb, M. E., Benecchi, S. D., et al. 2023, PSJ, 4, 75
Dotto, E., Barucci, M. A., Leyrat, C., et al. 2003, Icar, 164, 122
Duffard, R., Lazzaro, D., Pinto, S., et al. 2002, Icar, 160, 44
Elliot, J. L., Olkin, C. B., Dunham, E. W., et al. 1995, Natur, 373, 46
Faggi, S., Villanueva, G. L., McKay, A., et al. 2024, NatAs, 8, 1237
Fernandez, J. A. 1980, MNRAS, 192, 481
Fernández, J. A., Helal, M., & Gallardo, T. 2018, P&SS, 158, 6
Fernández, Y. R. 2009, P&SS, 57, 1218
Fernández, Y. R., Kelley, M. S., Lamy, P. L., et al. 2013, Icar, 226, 1138
Festou, M. C., Gunnarsson, M., Rickman, H., Winnberg, A., & Tancredi, G. 2001, Icar, 150, 140
Fornasier, S., Lellouch, E., Müller, T., et al. 2013, A&A, 555, A15
Foster, M. J., Green, S. F., McBride, N., & Davies, J. K. 1999, Icar, 141, 408
Giuliatti Winter, S. M., Madeira, G., Ribeiro, T., et al. 2023, A&A, 679, A62
Gladman, B., Marsden, B. G., & Vanlaerhoven, C. 2008, in The Solar System Beyond Neptune, ed. M. A. Barucci, et al. (Tucson, AZ: Univ. Arizona Press) 43
Guilbert, A., Alvarez-Candal, A., Merlin, F., et al. 2009a, Icar, 201, 272
Guilbert, A., Barucci, M. A., Brunetto, R., et al. 2009b, A&A, 501, 777
Guilbert-Lepoutre, A., Gkotsinas, A., Raymond, S. N., & Nesvorny, D. 2023, ApJ, 942, 92

Gunnarsson, M., Bockelée-Morvan, D., Biver, N., Crovisier, J., & Rickman, H. 2008, A&A, 484, 537
Gunnarsson, M., Rickman, H., Festou, M. C., Winnberg, A., & Tancredi, G. 2002, Icar, 157, 309
Hartmann, W. K., Tholen, D. J., Meech, K. J., & Cruikshank, D. P. 1990, Icar, 83, 1
Herget, P. 1947, AJ, 53, 16
Horner, J., Evans, N. W., & Bailey, M. E. 2004, MNRAS, 354, 798
Hosek, , Matthew, W. J., Blaauw, R. C., Cooke, W. J., & Suggs, R. M. 2013, AJ, 145, 122
Hsieh, H. H., Fitzsimmons, A., Novaković, B., Denneau, L., & Heinze, A. N. 2021, Icar, 354, 114019
Hui, M.-T., Jewitt, D., Yu, L.-L., & Mutchler, M. J. 2022, ApJL, 929, L12
Ivanova, O. V., Luk`yanyk, I. V., Kiselev, N. N., et al. 2016, P&SS, 121, 10
Ivanova, O. V., Picazzio, E., Luk'yanyk, I. V., Cavichia, O., & Andrievsky, S. M. 2018, P&SS, 157, 34
Jaeger, M., Prosperi, E., Vollmann, W., et al. 2011, IAU Circ., 9213, 2
Jewitt, D. 1990, ApJ, 351, 277
Jewitt, D. 2009, AJ, 137, 4296
Jewitt, D., & Kim, Y. 2020, PSJ, 1, 77
Jewitt, D., Li, J., & Kim, Y. 2021, AJ, 162, 268
Kareta, T., Noonan, J. W., Volk, K., Strauss, R. H., & Trilling, D. 2024, ApL, 967, L5
Kareta, T., Sharkey, B., Noonan, J., et al. 2019, AJ, 158, 255
Kareta, T., Volk, K., Noonan, J. W., et al. 2020, RNAAS, 4, 74
Kareta, T., Woodney, L. M., Schambeau, C., et al. 2021, PSJ, 2, 48
Kim, Y., Jewitt, D., Luu, J., Li, J., & Mutchler, M. 2023, AJ, 165, 150
Kolokolova, L., Kimura, H., Kiselev, N., & Rosenbush, V. 2007, A&A, 463, 1189
Korsun, P. P., Ivanova, O. V., & Afanasiev, V. L. 2008, Icar, 198, 465
Kossacki, K. J., & Szutowicz, S. 2013, Icar, 225, 111
Kowal, C. T., & Gehrels, T. 1977, IAU Circ., 3129, 1
Królikowska, M., & Szutowicz, S. 2006, A&A, 448, 401
Lebofsky, L. A., Tholen, D. J., Rieke, G. H., & Lebofsky, M. J. 1984, Icar, 60, 532
Leiva, R., Sicardy, B., Camargo, J. I. B., et al. 2017, AJ, 154, 159
Lellouch, E., Moreno, R., Bockelée-Morvan, D., Biver, N., & Santos-Sanz, P. 2022, A&A, 659, L1
Lellouch, E., Moreno, R., Müller, T., et al. 2017, A&A, 608, A45
Licandro, J., de León, J., Moreno, F., et al. 2021, A&A, 650, A79
Lilly, E., Jevčák, P., Schambeau, C., et al. 2024, ApJL, 960, L8
Lisse, C. M., Steckloff, J. K., Prialnik, D., et al. 2022, PSJ, 3, 251
Lorin, O., & Rousselot, P. 2007, MNRAS, 376, 881
Luu, J., Jewitt, D., & Cloutis, E. 1994, Icar, 109, 133
Luu, J. X., & Jewitt, D. C. 1990, AJ, 100, 913
Luu, J. X., Jewitt, D. C., & Trujillo, C. 2000, ApJL, 531, L151
Mayall, N. U. 1941, PASP, 53, 340
Meech, K. J., & Belton, M. J. S. 1990, AJ, 100, 1323
Meech, K. J., Belton, M. J. S., Mueller, B. E. A., Dicksion, M. W., & Li, H. R. 1993, AJ, 106, 1222
Meech, K. J., Buie, M. W., Samarasinha, N. H., Mueller, B. E. A., & Belton, M. J. S. 1997, AJ, 113, 844

Miles, R. 2016, Icar, 272, 387
Miles, R., Camilleri, P., Birtwhistle, P., & Gonzalez, J. J. 2016a, CBET, 4313, 1
Miles, R., Faillace, G. A., Mottola, S., et al. 2016b, Icar, 272, 327
Morgado, B. E., Sicardy, B., Braga-Ribas, F., et al. 2021, A&A, 652, A141
Neslušan, L., Tomko, D., & Ivanova, O. 2017, CoSka, 47, 7
Nicholson, S. B. 1947, PASP, 59, 30
Oikawa, S., & Everhart, E. 1979, AJ, 84, 134
Ootsubo, T., Kawakita, H., Hamada, S., et al. 2012, ApJ, 752, 15
Ortiz, J. L., Duffard, R., Pinilla-Alonso, N., et al. 2015, A&A, 576, A18
Ortiz, J. L., Santos-Sanz, P., Sicardy, B., et al. 2017, Natur, 550, 219
Paganini, L., Mumma, M. J., Boehnhardt, H., et al. 2013, ApJ, 766, 100
Peixinho, N., Lacerda, P., Ortiz, J. L., et al. 2001, A&A, 371, 753
Pereira, C. L., Braga-Ribas, F., Sicardy, B., et al. 2024, MNRAS, 527, 3624
Prialnik, D., & Bar-Nun, A. 1987, ApJ, 313, 893
Protopapa, S., Kelley, M. S. P., & Yang, B. 2021, ATel, 14961, 1
Rauer, H., Biver, N., Crovisier, J., et al. 1997, P&SS, 45, 799
Reach, W. T., Kelley, M. S., & Vaubaillon, J. 2013, Icar, 226, 777
Reinmuth, K. 1931, AN, 241, 325
Richter, N. 1941, AN, 271, 207
Richter, N. 1954, AN, 281, 241
Roemer, E. 1958, PASP, 70, 272
Roemer, E. 1963, in The Moon Meteorites and Comets, ed. G. P. Kuiper, & B. M. Middlehurst (Chicago, IL: Univ. Chicago Press) 527
Rousselot, P. 2008, A&A, 480, 543
Rousselot, P., Korsun, P. P., Kulyk, I., Guilbert-Lepoutre, A., & Petit, J. M. 2016, MNRAS, 462, S432
Rousselot, P., Petit, J. M., Poulet, F., & Sergeev, A. 2005, Icar, 176, 478
Rousselot, P., Kryszczyńska, A., Bartczak, P., et al. 2021, MNRAS, 507, 3444
Santos-Sanz, P., Gomes Júnior, A., Morgado, B., et al. 2023, DPS Meeting, Vol. 55, (Washington, DC: AAS) 301.07
Sarid, G., Volk, K., Steckloff, J. K., et al. 2019, ApL, 883, L25
Schambeau, C., Fernandez, Y., Belton, R., et al. 2020, CBET, 4821, 1
Schambeau, C. A., Fernández, Y. R., Lisse, C. M., Samarasinha, N., & Woodney, L. M. 2015, Icar, 260, 60
Schambeau, C. A., Fernández, Y. R., Samarasinha, N. H., Mueller, B. E. A., & Woodney, L. M. 2017, Icar, 284, 359
Schambeau, C. A., Fernández, Y. R., Samarasinha, N. H., et al. 2021, PSJ, 2, 126
Schambeau, C. A., Fernández, Y. R., Samarasinha, N. H., Woodney, L. M., & Kundu, A. 2019, AJ, 158, 259
Seccull, T., Fraser, W. C., Puzia, T. H., Fitzsimmons, A., & Cupani, G. 2019, AJ, 157, 88
Seligman, D. Z., Kratter, K. M., Levine, W. G., & Jedicke, R. 2021, PSJ, 2, 234
Senay, M. C., & Jewitt, D. 1994, Natur, 371, 229
Sickafoose, A. A., Levine, S. E., Bosh, A. S., et al. 2023, PSJ, 4, 221
Sickafoose, A. A., Bosh, A. S., Emery, J. P., et al. 2020, MNRAS, 491, 3643
Stansberry, J., Grundy, W., Brown, M., et al. 2008, in The Solar System Beyond Neptune, ed. M. A. Barucci, et al. (Tucson, AZ: Univ. Arizona Press) 161

Stansberry, J. A., Van Cleve, J., Reach, W. T., et al. 2004, ApJS, 154, 463
Steckloff, J. K., Sarid, G., Volk, K., et al. 2020, ApJL, 904, L20
Tholen, D. J., Hartmann, W. K., Cruikshank, D. P., et al. 1988, IAU Circ., 4554, 2
Trigo-Rodríguez, J. M., García-Hernández, D. A., Sánchez, A., et al. 2010, MNRAS, 409, 1682
van Biesbroeck, G. 1927, PA, 35, 586
Weissman, P. R., Chesley, S. R., Choi, Y. J., et al. 2006, DPS Meetings, Vol. 38, (Washington, DC: AAS) 37.06
Whipple, F. L. 1951, ApJ, 113, 464
Whipple, F. L. 1980, AJ, 85, 305
Whitney, C. 1955, ApJ, 122, 190
Wierzchos, K., & Womack, M. 2020, AJ, 159, 136
Wierzchos, K., Womack, M., & Sarid, G. 2017, AJ, 153, 230
Wild, P., & Marsden, B. G. 1978, IAU Circ., 3167, 3
Womack, M., Sarid, G., & Wierzchos, K. 2017, PASP, 129, 031001
Womack, M., & Stern, S. A. 1999, SoSyR, 33, 187
Yang, B., Jewitt, D., & Bus, S. J. 2009, AJ, 137, 4538
Zubko, E., Videen, G., & Kulyk, I. 2020, RNAAS, 4, 75

Chapter 11

Observational Campaigns

L M Woodney, S Faggi, J Noonan and A A Sickafoose

Observational campaigns are defined as unique from surveys by being collaborations of observers working together across time, instruments, or objects. Their goal is not to discover new objects, but to target specific objects and scientific questions. This chapter reviews the history of how Centaurs were recognized as a distinct class of objects leading to the early campaigns to observe 95P/Chiron. Examples of both recent and ongoing campaigns and what we can learn from them are presented. This chapter concludes with a section looking to the future, with suggestions of types of campaigns observers will want to consider planning.

11.1 Introduction

This chapter explores how we can answer some of the big questions about Centaurs (as highlighted in other chapters) through observational campaigns. A campaign may recognize the importance of an event (e.g., 95P/Chiron perihelion, an occultation opportunity) or a property of a single object (e.g., 29P/Schwassmann-Wachmann 1's regular outbursts). Campaigns can be coordinated multiwavelength observations of individual objects, or of a group of objects done with one or two techniques. They may be differentiated from a survey by being reactionary, narrow in focus, and requiring a collaboration of multiple observers across time, instruments and/or targets, while a survey is based on characterizing an unknown element such as searching for new objects or properties (e.g., a binary companion search done by a single technique/instrument).

An example of one of the largest-scale campaigns to date in planetary science was conducted for the impact of D/Shoemaker-Levy 9 (SL9) into Jupiter[1]. In the months leading up to impact, hundreds of observers around the world joined an "email exploder" where they shared their observing plans and worked out how to hand off observations from one observatory to the next (Raugh 1994). Over the six days of

[1] The International Halley Watch for 1P/Halley's perihelion in 1986 may have been larger.

doi:10.1088/2514-3433/ada267ch11

impacts in July 1994, we shared our results live, allowing for quick pivots to address newly-discovered phenomena and maximize science return. Not only did we all get to thrill in each other's discoveries as they were made, we did better science by working together.

Not all campaigns need to be so large scale, but the SL9 Campaign makes it clear that collaborations on shorter timescales than conference presentations and publications can enhance our ability to address the biggest scientific questions. For Centaurs, campaigns can help us understand population statistics (size/shape), compositions, and outgassing mechanisms.

In this chapter, we begin by highlighting what has been learned from campaigns for the two most famous Centaurs: Chiron (also known as comet 95P or asteroid 2060) and 29P/Schwassmann-Wachmann 1 (hereafter 29P). This section includes an aside into the history of how Centaurs got their name. While not technically relevant to campaigns, it highlights a community coming together to recognize and classify a new type of object, in this case: Centaurs. We then summarize both what has been done in the past, and what is being learned in a number of ongoing campaigns for other Centaurs (Figure 11.1 summarizes campaign observations). These examples demonstrate the power of campaigns to advance our understanding. Finally, we conclude with looking to the future, discussing types of campaigns that should be organized and providing a table of observing opportunities over the next 50 years.

11.2 95P/Chiron: The First Centaur

Given the definition of a Centaur we adopt here: an object with both perihelion and aphelion within the semimajor axes of Jupiter ($\approx$5.2 au) and Neptune ($\approx$30 au; as discussed in Section 1.3 in Chapter 1), Chiron is technically not the first Centaur discovered[2]. However, the discovery of Chiron, and its curious behavior, played a central role in the early 1990s in the development of a new classification of objects that would come to be known as the Centaurs.

11.2.1 Origins of the Centaur Classification

Chiron, discovered as asteroid 2060 in 1977 (Kowal et al. 1977), was the most distant known asteroid for over a decade (Stern 1989). Surveys in the early 1990s found the distant object (5145) Pholus (Scotti et al. 1992)[3], and soon after the first Kuiper Belt object (15760) Albion (1992 QB_1; Jewitt & Luu 1993). The recognition that Chiron-like orbits were not dynamically stable and that the population was likely derived from the Kuiper Belt was quick to follow (Luu et al. 1994; Luu 1994), though the first Centaurs discovered tended to be classified as comets or asteroids depending on activity (Luu 1993; McFadden 1994). While it is commonly assumed the

[2] "What was the first object discovered in a Centaur orbit?" is not easy to answer given these object's orbits are frequently perturbed. Objects that meet the definition now may not have at discovery and vice versa. Our best guess is that first Centaur goes to 29P/Schwassman-Wachmann 1, discovered as a comet in 1927; see Kronk (2007).

[3] Pholus' aphelion of $q = 31.9$ au falls outside our Centaur definition, but it is commonly listed as a Centaur.

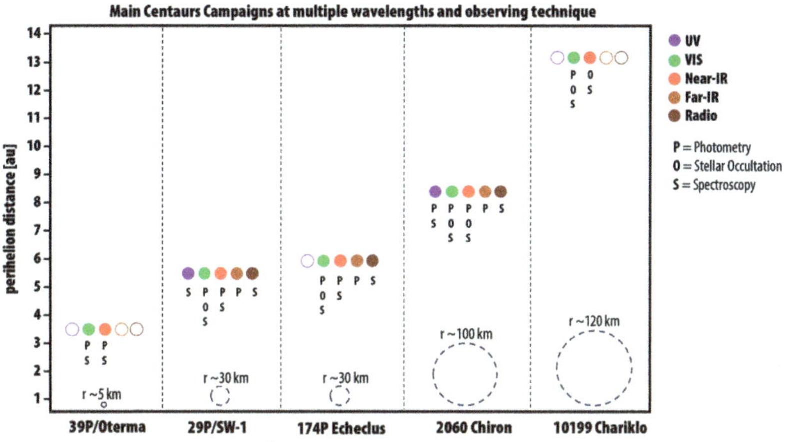

Figure 11.1. Observing campaigns help bring together observers using different techniques and exploring multiple wavelength ranges. This chart is a visual summary of the observing campaigns for the five most studied Centaurs (29P, 39P/Oterma, (60558) 174P/Echeclus, Chiron, and (10199) Chariklo) that are described in this chapter. The objects are ordered from smaller to larger in diameter [km] along the x-axis and from smaller to larger perihelion distance [au] in the y-axis. Colored circles are indicative of the wavelength used to study the objects (UV: purple, VIS:green, Near-IR:red, Far-IR:copper, Radio:brown), with filled circles representing existing data and open circles wavelengths that have not been observed. Letters indicate the technique employed for such investigations (P: Photometry, O: Stellar Occultation, S: Spectroscopy). For more details and associated campaign references, see the text. (Visible wavelength spectra of Oterma (Jewitt et al. 1982), Echeclus (Rousselot, 2008), Chiron (Romon-Martin et al. 2003), and Chariklo (Guilbert et al. 2009) have been published, and are indicated in the figure for completeness, but these data were not part of the campaigns we discuss).

classification "Centaur" is a reference to the dual asteroid/comet nature of these objects, the name seems to have actually been a serendipitous choice made by Chiron's discoverer Charles Kowal. Given it seemed likely that more objects in Chiron-like orbits would be discovered, Kowal thought it prudent to select a naming theme with a number of available options. The centaurs of Greek myth had not yet been used and Chiron, as the son of Saturn and grandson of Uranus, was a good fit for an object with Chiron's orbit (Kowal 1988). The use of "Centaur" as an object classification seems to have been adopted by the community by the early 1990s, with an early example of the term's use found in the title of a 1991 workshop on Chiron (Marcialis & Bus 1991). The term is ubiquitous in the literature by the middle of the decade (e.g., Luu et al. 1994; Luu 1994).

11.2.2 From Cometary Activity to the Chiron Perihelion Campaign

Chiron's cometary activity was discovered in 1988, and comparison to unpublished 1978 data made it clear it had been previously active and gone through a 10-year period of quiescence (Cruikshank et al. 1988; Bus et al. 1989). This discovery was soon followed by detection of a coma (Meech & Belton 1989). Observations of both long (months to years) and short (hours) timescale coma brightness changes were subsequently obtained (Hartmann et al. 1990; Luu & Jewitt 1990; West 1991). Perhaps most exciting was the discovery of extended neutral CN gas emission in the coma while Chiron was at over 11 au from the Sun. This was the most distant detection of gas sublimation at the time, and was demonstrated to be consistent with being driven by outbursts of CO_2 gas (Bus et al. 1991). These discoveries cemented the cometary nature of Chiron: it had a coma with secular brightness changes consistent with solar-driven activity, impulsive outbursts as are common in comets (the cause of which remains hotly debated), and that the coma contained a gas commonly found in comets (see Section 10.2 in Chapter 10 for a more detailed discussion of Chiron's properties).

These discoveries and the approaching perihelion of Chiron in February 1996 inspired a campaign for observations of this enigmatic object to attempt to discover its size, composition, and what made it active so far beyond the frost line[4]. On the heels of the highly successful Shoemaker-Levy 9 campaign in 1994, which had brought together observers with an email exploder and electronic "bulletin board" for sharing observing plans and data, a Chiron Perihelion Campaign (CPC) was organized with the same infrastructure run from the Planetary Data System Small Bodies Node at the University of Maryland. Approximately 50 astronomers participated in the CPC between 1994 and 1997 (Stern 1995). While a record of who participated in the CPC is not available, it is clear a wealth of new discoveries were made during this period. To list a few: new estimates of Chiron's size were made (Altenhoff & Stumpff 1995), Hubble Space Telescope (HST) imaging of coma structure was obtained (Meech et al. 1997), CO emission was detected by the NRAO 12-meter telescope (Womack & Stern 1999), and, as described in the next section, stellar occultation observations were obtained which revealed both nuclear size and coma properties.

11.2.3 Early Occultation Campaigns for Chiron

In 1992, efforts began to identify stellar occultations by Chiron (Bus et al. 1994). This work was motivated by Chiron's size not being well determined. Radiometric data indicated that it was relatively large, but those measurements were complicated by the unknown thermal contribution from surrounding dust. Understanding the dynamics of Chiron's coma depended on having a more accurate determination of its mass, specifically in terms of the influence of self-gravity. Two occultation campaigns were successful: 1993 November 07 and 1994 March 09.

In 1993, five sites in the western U.S.A. observed an occultation by Chiron of a 14th magnitude double star (Bus et al. 1996). The telescopes ranged in size from 0.35 to 0.9 m and portable instruments were deployed to three locations. Only one site recorded an occultation by Chiron's nucleus, with a second possibly grazing. From these data, Chiron's radius was constrained to be 89.6 ± 6.8 km if the chord were grazing or $>90.2 \pm 6.5$ km if not. Sharp, short drops in flux in two of the light curves (along with the grazing chord, if it were not a graze) were interpreted to be from a narrow, collimated jet. Longer, low-level dips in three of the light curves were

[4] At the time of publication of this chapter an associated website is still active: https://nssdc.gsfc.nasa.gov/planetary/chiron.html.

interpreted to be a larger region of asymmetric dust reaching out a few thousand kms from the nucleus.

In 1994, the occultation by Chiron of a 12th magnitude double star was successfully observed from the 0.9-m telescope aboard NASA's Kuiper Airborne Observatory (KAO) and a 0.5-m telescope in S. Africa (Elliot et al. 1995). There was no detection of the nucleus. Similar to the 1993 data, both sites recorded a sharp drop near closest approach and the KAO detected multiple, broader, shallower features, all within a few hundred kms of Chiron. The features were interpreted to be jets from a few active areas on the surface, along with a gravitationally-bound coma. Data were taken in visible and infrared wavelengths on the KAO, allowing Mie-scattering analysis to determine that the particle sizes from the most obvious feature were $>0.25\ \mu m$. Chiron's radius was only constrained to be between 83 and 156 km, based on the null detection of the nucleus.

Extensive astrometric observations and analyses were carried out leading up to the 1990s occultations in order to refine the predictions. For 1993, the shadow was first predicted to fall over northern California and Nevada. A few days before the event, it shifted to the southernmost part of California (Bus et al. 1996). In 1994 five sites in Brazil attempted the observations, but they were far from the final predicted path and also had bad weather (Elliot et al. 1995).

11.3 29P/Schwassmann-Wachmann 1: The Most Active Centaur

Comet 29P/Schwassmann-Wachmann 1 (29P) is perhaps one of the most fascinating Centaurs. Discovered in 1927 (van Biesbroeck 1927), 29P was the first small body known to have its entire orbit outside of Jupiter. Furthermore, not only is it the most active of all Centaurs with multiple outbursts per year, it was also the most distantly known active object for many years. In addition to periodic large outbursts that change its visual brightness 1–5 magnitudes on average 6–7 times a year, and smaller 0.5–1.0 magnitude outbursts occurring with great frequency in between, 29P exhibits permanent dust and gas activity. (e.g., Trigo-Rodríguez et al. 2008, 2010; Miles et al. 2016; Wierzchos & Womack 2020; Figure 11.2). Given this history, it is no surprise that 29P has been a frequent target for campaigns with goals of understanding its rotation and curious pattern of frequent outbursts. Being such an active transient object, 29P plays a key role in understanding the activity of distant comets and the evolutionary processes that affect their interiors and influence their surface compositions (see Section 10.3 in Chapter 10 for a more detailed discussion of 29P's properties).

11.3.1 Large-Scale Observing Campaigns for 29P

Figure 11.2 shows the frequent outbursting of 29P in a combined light curve, which is the result of multiple observing campaigns over the past two decades. As of 2023 there are two ongoing, public 29P campaigns. An international campaign was initiated by M. Womack and G. Sarid in 2018 (Womack et al. 2020; Womack & Sarid 2020). Here, astronomers are encouraged to submit their 29P observing plans to the searchable database to enable coordination of simultaneous observations

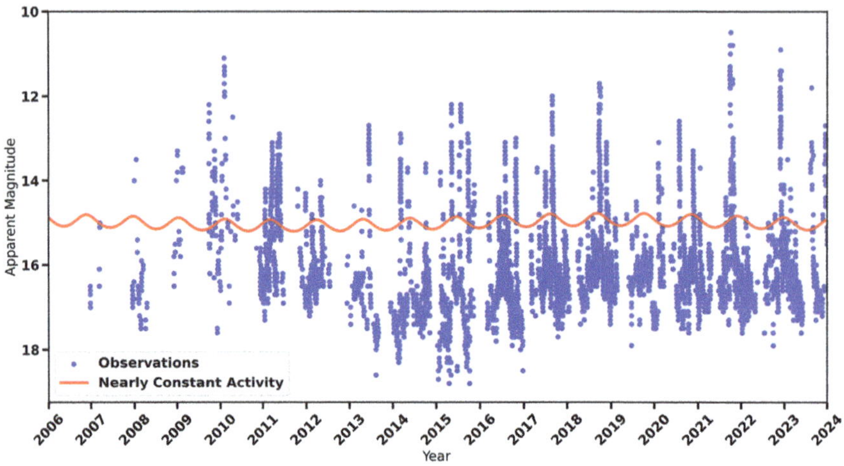

Figure 11.2. Photometry of 29P from 2006 to 2024 showing how frequently it outbursts. These data are nucleus photometry values as reported to the IAU Minor Planet Center. The red line represents normalized photometric behavior of 29P assuming constant, insolation-driven activity and corrections for changes in geocentric distance. Given the frequent outbursts of up to nearly eight magnitudes in brightness change, it is especially important for observers to coordinate into campaigns to compare results, and so that observers looking for difficult-to-detect phenomena will know when to trigger Target of Opportunity observations. Figure updated for this chapter by C. Schambeau following Schambeau (2018).

across multiple observing groups. While no data are publicly shared, the website serves as a gathering place to encourage collaboration across all wavelengths and observing techniques. While we are not aware of collaborations that have been sparked yet, the usefulness of more observers collaborating across instruments and wavelengths is highlighted by recent papers such as Wierzchos & Womack (2020); they combined millimeter-wavelength spectroscopy and visible magnitude observations and found that the periodic dust outbursts are not always correlated with CO outgassing[5].

Mission 29P, run by R. Miles and the British Astronomical Association, is an aggregate of observing photometry of 29P from a large group of regular observers, including many citizen scientist astronomers (Miles 2021). Their detailed light curve of 29P extends back to 2014. This database has enabled both studies of the rotation period and analysis of outburst properties (Miles 2016). Additionally, the timely and frequent updates here have allowed observers to plan and trigger Target of Opportunity (ToO) and Directors Discretionary Time (DDT) observations at major observatories.

In 2022, a ToO was triggered for spectroscopy of icy grain distribution at NASA's 3-m Infrared Telescope Facility (IRTF; Kareta et al. 2025). In 2021, imaging of coma color morphology and icy grain distribution was acquired at Gemini North

[5] We encourage observers to utilize the campaign website to coordinate observations: https://wirtanen.astro.umd.edu/29P/29P_obs.shtml

through DDT (Firgard et al., 2025, in prep). Two outburst events in early October 2019 and September 2021 were followed up with HST ToO and DDT proposals, respectively. These ToO observations were executed between 14 and 28 days after the outburst event, while the DDT suffered significant delays due to gyro issues on HST and was not executable until late November and early December 2021. Both campaigns used filter imaging strategies to search for color changes in the coma as well as broad filter imaging to search for fragments ejected by the major activity (Bodewits et al. 2019, 2021). Observations were also triggered on the Swift Ultraviolet and Optical Telescope (UVOT) with a quicker turnaround time, providing both UV-filter imaging and grism spectroscopy within days of 29P's outbursts (Moulane et al. 2024; Z. Xing et al., in preparation). The potent combination of high-resolution filter imaging and grism spectroscopy aimed at investigating changes to the dust properties following the outburst was tempered by the large delay in program execution due to HST's gyro. HST's aging infrastructure and status as the highest-spatial-resolution observatory at visible wavelengths highlights a key weakness in the current Centaur, as well as cometary and other astrophysical transient, response network. Given the value of understanding the mechanisms that cause outbursts beyond the frost line, it remains important to continue to trigger observations as near in time as possible to outbursts in order to characterize them.

11.3.2 Spectroscopic Campaigns for 29P

Spectroscopic observations of Centaurs can provide insights into the original composition of short period comets (SPCs), which are potentially their final dynamical end state. The volatile composition of SPCs is shaped by surface erosion and possibly thermal processing due to their close passages around the Sun, with Jupiter-family comets (JFCs) potentially being the most affected class of objects, due to their frequent close encounters with the Sun.

Simultaneous spectroscopic and imaging investigations at multi-wavelengths and with multiple instruments, from both ground- and space-based observatories, are a powerful synergistic tool to characterizing 29P gas and dust activity both during outbursts and quiescent states and investigating the drivers of cometary activity for distant objects.

The recent comprehensive observational campaign led by Bockelée-Morvan et al. (2022) showed the power of this approach. The team followed 29P with different instruments and observing techniques; from 2010, to 2013, by using multiple instruments on board of the Herschel space observatory, and across 2007, 2010, 2011 and 2021 with the IRAM 30 m antenna. Detections of water (first in the far-infrared, at 557 GHz) and searches for NH_3 (573 GHz) lines were performed with HIFI spectrometer (de Graauw et al. 2010); and simultaneously imaging of the dust coma were taken with the PACS (at 70 and 160 μm) and SPIRE (at 250, 350 and 500 μm) cameras on board of the Herschel space observatory (Pilbratt et al. 2010; Griffin et al. 2010), investigating the thermal flux from the nucleus and the dust coma. Ground-based radio observations with IRAM allowed first detection of HCN (89 GHz) and monitoring of well-studied CO outgassing activity (230 GHz). These long-term investigations confirmed a strong correlation between CO and dust activity and revealed that water is also correlated to such activity, possibly providing important constraints to studying the outburst-triggering mechanism that characterizes the surfaces and interiors of such distant objects. The study of the CO line shape also confirmed that outbursts occur in the subsolar region, where the CO outgassing predominantly and continuously operates, and this might point to surface evolutionary processes. The velocity shift offset observed for the water and HCN lines indicated that the nucleus did not contribute much, pointing to a release from sublimating icy grains. The H_2O and HCN line profiles suggested indeed that they are likely produced from dust particles that exceed a few micrometers in size.

Using a similar approach, Roth et al. (2023) coordinated a simultaneous multi-wavelength spectroscopic observation of 29P using iSHELL at the IRTF (Rayner et al. 2012, 2016) and nFLASH at the Atacama Pathfinder EXperiment (APEX; Güsten et al. 2006) to deeply study the September 2021 outburst and its evolution into the following months. This campaign reported strong CO detections, and stringent (3σ) upper limits on abundance ratios, relative to CO, for CH_4, C_2H_6, CH_3OH, H_2CO, CS, and OCS, demonstrating the synergistic power of coordinated radio and near-infrared spectroscopic measurements.

11.3.3 Current Limitations to Spectroscopic Campaigns

If our understanding of active Centaurs via telescopic imaging has been limited by their faint apparent magnitudes ($m_v \sim 20$), spectroscopic investigations have been even more challenging. Spectroscopy requires brighter targets than detection by imaging. A cometary atmosphere is characterized by multiple inter-related radiative processes (e.g., collisional excitation, fluorescence, radiative cooling, dissociative electron impact excitation, etc.) that lead to a complex exosphere. Gravity does not play a significant role in controlling the gas structure, so the expanding coma is mostly defined by the molecular outgassing velocity and photochemical/photodissociation decay (see Chapter 7).

The gas emissions detectable at near-UV through infrared wavelengths are primarily due to solar-pumped fluorescence, so they are particularly weak at large heliocentric distances. Molecular features are generally narrow and confined to at most twice the expansion velocity, requiring high-resolution infrared spectroscopy for detailed studies of rotational temperatures, and for identifying individual lines; however, the ro-vibrational bands can be quite large, providing an opportunity for low resolution spectroscopic investigations. The emitting fluxes are also heavily diluted by their large geocentric distances, ultimately requiring very sensitive instrumentation. Similarly ground-based mm-wave spectroscopy, often the sole source of information of CO emission from Centaurs, generally suffers from beam dilution due to the great distances of the Centaurs coupled with the large beam sizes. Moreover, the strongest molecular fundamental bands of H_2O and CO_2 are not accessible to ground-based observatories due to absent atmospheric transparency, confining molecular spectroscopy to key wavelength regions. Given these challenges, it is particularly important to continue to use target of opportunity observations to catch 29P when it is at its brightest and to look to new technology to continue to push the limits of what can be observed (see Section 11.6).

11.4 Using Campaigns to Understand Activity

Cometary *activity* is traditionally defined as the ensemble of physical processes leading to the release of gas and dust from the nucleus, forming features like tails and comae. Activity has a central role in controlling the evolution of a body surface, and it is generally the result of orbital changes that bring the object inward, where insolation is sufficient to trigger the outgassing of the relevant volatile species (e.g., Lilly et al. 2021, 2024).

Active Centaurs are objects that display comet-like features (e.g., tails, comae); they account only for a small percentage of the known Centaurs' population (see Chandler et al. 2020; Harrington Pinto et al. 2023, and references therein), and unlike comets, the activity does not peak at perihelion passage. Centaurs' activity can show a variety of behaviors, from a low-level but long-term outgassing, as seen in Chiron (Luu & Jewitt 1990; Foster et al. 1999), to sudden violent outbursts, as observed in Echeclus (Choi & Weissman 2006; Bauer et al. 2008; Rousselot 2008; Kareta et al. 2019; Seccull et al. 2019), to quasi-periodic outgassing as observed in 29P (Wierzchos & Womack 2020; Kareta et al. 2025; Bockelée-Morvan et al. 2022; Roth et al. 2023; Trigo-Rodríguez et al. 2010; Miles et al. 2016).

Although the basic principles of activity are well established, many details remain still elusive, especially regarding the mechanisms driving the activity of far objects, like active Centaurs. At such distances, it is too cold for water to readily sublimate. Water sublimation is generally the main driver of cometary activity within 2 au but it rapidly diminishes beyond 3 au (Mumma & Charnley 2011). Hyper-volatile species (e.g., CO, CO_2) have been considered possible, even likely, alternate drivers of Centaur outgassing, but they are not the only option; for example, energy from the crystalline to amorphous water ice state transition could be another possibility in generating activity (Jewitt 2009). At such cold surface temperatures (<150 K) and low pressures (below ~10–12 bar) many thermodynamical properties of volatile ices (e.g., sublimation) are nevertheless not well investigated from laboratory experiments (Fray & Schmitt 2009).

Studying comet-like activity, especially through observing campaigns with multi-instruments and multi-wavelengths, allows investigation into the intermixing of dust and ice into the nucleus, and it can reveal potential nuclei heterogeneities. As discussed in Section 11.3, discovering and studying activity on Centaurs is observationally very challenging because they are far, faint, and rare, so they require a lot of dedicated telescope time, very sensitive instrumentation and specific observing techniques. Broadband imaging at different spectral filters permits determination of the overall gas distribution and dust state and activity; when feasible, spectroscopic investigations provide complementary insights into the volatile composition and its outgassing rate activity. However, it is the combination of the two that allows further investigation of surface ice heterogeneities as well as their evolution as a result of the repeated perihelion passages.

Of the few active Centaurs discovered so far, Chiron and 29P are the most studied. In this section we will provide a summary of the observing campaigns focused on some of the other active bodies and the synergistic investigations of their activity. A more general review of the properties of Echeclus and P/2019 LD2 (Atlas) can be found in Chapters 8 and 10.

Centaur 2014 OG_{392}: This active Centaur was discovered as part of Citizen Science Active Asteroids. This project engages the public through a search for minor bodies that display comet-like activity. By mining the archival data from the DECam camera mounted at the Blanco 4-m telescope at the Cerro Tololo Inter-American Observatory in Chile, Chandler et al. (2020, 2022) detected faint activity emanating from Centaur 2014 OG_{392}. A two-year follow-up observational campaign from 2017 to 2019 was performed by using broadband *VR* filters at the Blanco, IMACS WB4800-7800 filter images at the *Magellan* 6.5-m Walter Baade Telescope in Chile, and by using LMI *g*, *r*, and *i* filters at the 4.3-m Lowell Discovery Telescope in Arizona, USA, to monitor the activity. By computing equilibrium temperatures and modeled mass-loss rates for seven ices (H_2O, NH_3, CH_3OH, CH_4, CO, CO_2, N_2), the team concluded that over the course of one orbit the sublimation rates for CO_2 and NH_3 could vary substantially, presumably producing significant variations in visible activity. These molecules would not effectively sublimate at Kuiper Belt distances, before 2014 OG_{392} would become a Centaur, providing an abundant surface reservoir. The highly volatile CO, N_2, and CH_4 ices would sublimate at lower temperatures, so their surface abundance might likely be depleted, but reservoirs could still be trapped below the surface. Following Harris & Harris (1997), the object has a radius of $\sim$20 km when assuming a slope parameter $G = 0.15$, as is typical for a dark surface, for a measured H magnitude of 11.03.

Centaur (60558) 174P/Echeclus: Discovered by the Spacewatch program (Larsen et al. 2001), Centaur Echeclus presents a quite unique and unstable orbit with a lifetime of the order of 10^5 years (a typical Centaur dynamical lifetime is $\sim 10^6 - 10^7$ years) and surprisingly strong sporadic outbursts (Kareta et al. 2019; Sarid et al. 2019). Since its discovery, four outbursts have been observed. A first major one at the end of 2005 (brightening by $\sim$7 magnitudes), two smaller ones in May 2011 and August 2016 August, and a $\sim$4 magnitude outburst in December 2017. The first outburst, which was the largest one ever detected in a Centaur, lasted a few months, and may have been a fragmentation event where the ejected piece remained active (Bauer et al. 2008; Fernández 2009; Rousselot 2008; Rousselot et al. 2016). It was observed with the 5-m Mount Palomar Observatory telescope (Choi & Weissman 2006), and it corresponded to a change in the visual magnitude from 21 to about 14, and occurred when the object was at a heliocentric distance of about 13 au. Due to the unique characteristics of this target, and its unpredictable outbursts, a large set of observational data were obtained before, during and after these two first outbursts allowing a detailed study of its light curve, and extracting information on the nucleus rotation axis. Comprehensive analyses covering from 2001 to 2015 and using multiple sets of observational data are presented in Rousselot et al. (2016), these data are mostly telescope imaging with broadband filter *BVR*, and only one observation is spectroscopy at optical wavelengths (345–590 nm). The main results of this long-term study are the absence of light curve in the 2013 data and the modeling of the main outburst with two short events and one long event, corresponding to three sources of dust. Subsequently, analysis on the lightcurve suggested a high obliquity of the rotation axis and elongated shape for Echeclus. A 3.6σ detection of the J = 2-1 CO line is reported in Wierzchos et al. (2017) using the Arizona Radio Observatory (ARO) Submillimeter 10-m telescope (SMT), to monitor the object from May to June 2016 when it was at smaller heliocentric distances. Follow-up observations with the IRAM 30-m telescope after the outburst that happened in August 2016 were also performed and led to a CO 3σ upper limit. The CO production rate obtained from SMT observations was $(7.7 \pm 3.3) \times 10^{26}$ mol s^{-1}, when the object was at $R_h = 6.1$ au and $\Delta = 6.6$ au, providing the lowest value ever measured for a Centaur. No line was detected 24 days after the outburst using the IRAM 30-m telescope but CO emission from the earlier would have dissipated by then.

Centaur P/2019 LD2 (ATLAS): Centaur P/2019 LD2 (hereafter LD2) is a uniquely interesting object as it will become a JFC in 2063, after an expected close encounter with Jupiter (Hsieh et al. 2020; Kareta et al. 2020). While this object pushes the boundaries of our perihelion/aphelion definition of a Centaur, it is widely accepted to be one (e.g., Steckloff et al. 2020; see also Chapter 10) so we will discuss it as one here. Studying LD2 will provide an opportunity to observe how primitive bodies respond to changes of thermal environment while they undergo dynamical evolution. Kareta et al. (2021) reported detailed, contemporaneous, multiepoch and multiwavelength observations of LD2 using Gemini North visible imaging, IRTF near-infrared spectroscopy, and ARO Submillimeter Telescope millimeter wavelength spectroscopy and added precovery DECam images as well as Catalina Sky Survey observations from 2016 to 2020. The study revealed a nucleus radius of ~1.2 km and no identification of large outbursts. The measured dust production rate was ~10–20 kg s^{-1} and dust outflow velocity was established to be $v \sim 0.6 - 3.3$ m s^{-1}. The reflectance spectra obtained with SpeX/IRTF show evidence for weak absorption features at 1.5 and 2.0 μm, which were interpreted as possibly a small amount of water ice. Submillimeter observations with the Arizona Radio Observatory SMT provided no detection of CO, and yielded a 3σ upper limit of about Q(CO) $<4.4 \times 10^{27}$ mol s^{-1}. Another multi-instrument observing campaign covering about one year of time, from September 2019 to August 2020, was presented by Bolin et al. (2021). The team targeted LD2 with both ground- and space-based observatories and using the HST/Wide Field Camera 3 (HST/WFC3), *Spitzer Space Telescope*/ Infrared Array Camera (Spitzer/IRAC), and the GROWTH telescope network, visible spectroscopy from Keck/Low Resolution Imaging Spectrometer (LRIS), and archival Zwicky Transient Facility. Observations revealed similar results as Kareta et al. (2021), a nucleus radius between 0.2 and 1.8 km, dust ejected at low speeds, and a total dust mass loss of ~6 kg s^{-1}. Spitzer observations provided a CO/CO_2 gas production of about 10^{27} mol s^{-1}/10^{26} mol s^{-1} while Keck/LRIS spectroscopy yielded a 3σ upper limit for C_2 of $<7.5 \times 10^{24}$ mol s^{-1}.

Centaur 39P/Oterma: 39P/Oterma (39P) is an active Centaur that experienced strong transitions of its orbit in a relatively short period of time. In the last 90 years, the orbit transferred from that of a Centaur to a closer-in JFC orbit and then outward again to Centaur status. 39P was observed with the JWST NIRSpec instrument on 2022 July 27, when it was at a heliocentric distance of 5.82 au. Thanks to the incredible sensitivity of the NIRSpec spectrometer and the performances of the JWST telescope, the first detection of CO_2 gas emission in a Centaur was reported (Harrington Pinto et al. 2023). The team presented the lowest detection of CO_2 in any comet so far. Neither CO and H_2O were detected, and stringent upper limits were reported. One of the strengths of the NIRSpec spectrometer is the Integrated Field Unit (IFU) mode, allowing extraction of 2D maps of the dust and specific gas emissions, investigating the spatial extension and distribution of the gas. For 39P, CO_2 column density (m^{-2}) and the continuum intensity (μJy) maps were reported and compared, and interestingly the CO_2 distribution showed a possible small asymmetry in the sunward direction when compared to the dust. As part of the observing campaign in support of the JWST data, simultaneous ground-based charge-coupled device (CCD) imaging observations were acquired in July 2022 with the Gemini North Telescope and followed up in September 2022 with the LDT. Photometric analysis of these data is consistent with an estimated effective nucleus radius for 39P of ~2.21–2.49 km (Harrington Pinto et al. 2023).

On the other hand, a non-detection of activity is also important to fully characterize the Centaur population and provide constrains on modeling of activity triggers. Lilly et al. (2021) recently reported no evidence of activity associated with any of the targeted 13 Centaurs orbiting beyond Jupiter that were observed over a 6-year period, from 2013 to 2019, by using new observations with the Gemini North telescope in 2017/18 and revisiting archival data from 2013 to 2019. The upper limits on dust production rates and fractional active surface areas, as revealed from their coma modeling, were significantly lower (1–2 orders of magnitude) than typical activity range in comets and were in agreement with values measured for other inactive Centaurs, indicating that either the dust and gas production are below their detection limits or that the objects were dormant. Most of the targets of this study had relatively stable orbits with no significant recent changes in perihelion distance, suggesting that the activity on Centaurs is most likely triggered by sudden drops in perihelion distance and/or semimajor axis, possibly induced by close encounters with giant planets, which change the thermal balance in the body. Being a transient population of active objects, studying Centaurs provides invaluable information on primitive ices incoming from the Kuiper Belt and not yet heavily altered by thermal evolution. Therefore coordinated multiwavelength and multi-observing technique campaigns are a powerful tool to investigate the trigger of activity.

11.5 Occultation Campaigns

Stellar occultations are one of the most accurate methods to study the physical characteristics of small bodies in the outer solar system, short of sending spacecraft. Object sizes and shapes can be measured at km-level accuracy, atmospheres can be detected down to the nanobar level, and surrounding material such as rings, debris, or moons can be discovered and characterized (see Chapter 9). Examples include (i) the Centaur Chariklo has been found to be a triaxial ellipsoid shape, with errors on

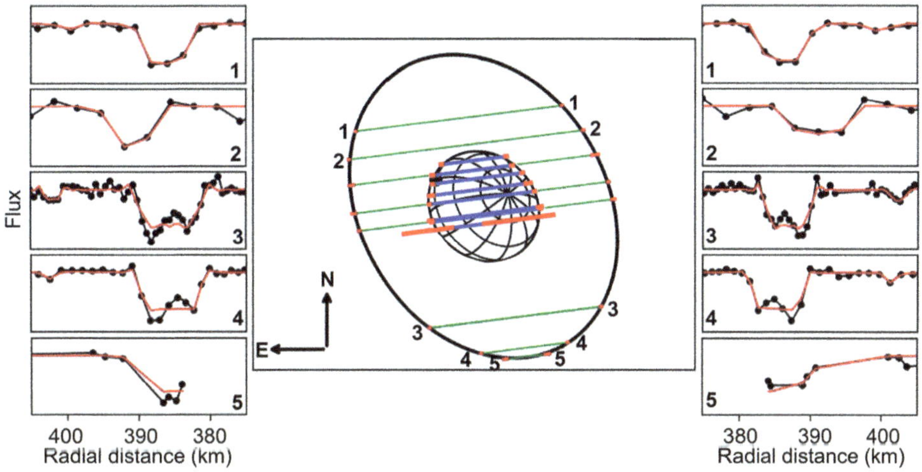

Figure 11.3. Example data from a Chariklo stellar occultation on 2017 July 23. In total, 29 different stations reported participating in this campaign and a subset of the results are shown here. The data set was used to constrain Chariklo's size and shape, as well as study the physical properties of the rings. (Center) A sky-plane view of the event, in which observed chords are represented by green lines, with blue representing occultations by the solid body and red indicating body or ring immersion and emersion with uncertainties. Chariklo is represented by the best ellipsoidal model fit to the solid-body data, and the black line is the best-fit ellipse for the thicker, interior ring. (Sides) Light curves of normalized flux versus radial distance in the ring plane, showing drops in the star signal caused by ring material. The panels correspond to the chord locations labeled 1 to 5 in the center, with immersion on the right. The varying data quality at different observing locations is apparent: a W-shaped ring feature is visible in the higher-quality data at sites 3 and 4. Reproduced from Morgado et al. (2021). CC BY 4.0.

each semiaxis of a few kms (see Figure 11.3 and Morgado et al. 2021), (ii) a global atmosphere of N_2, Ar, or CH_4 has been ruled out on the large trans-Neptunian object (TNO) Eris for surface pressures larger than ~1 nanobar (Sicardy et al. 2011), and (iii) a two-ring system was discovered at Chariklo, with ring widths of approximately 3 and 7 km and separated by 14 km (Braga-Ribas et al. 2014; example occultation data for Chariklo are shown in Figure 11.3). From occultation data, other properties can be inferred, such as albedo, density, atmospheric composition, and atmospheric pressure and temperature profiles. The most obvious example is Pluto, for which occultations have been employed for decades to study atmospheric composition (mostly N_2 with CH_4, CO, HCN, and other hydrocarbons as minor species), determine pressure profiles at the level of a few microbars, and even detect varying atmospheric haze (e.g., Elliot et al. 1989; Person et al. 2008; Gladstone et al. 2016; Meza et al. 2019). Table 11.1 contains a list of all the published successful stellar occultations by Centaurs: only ten Centaur-like bodies have been studied using this technique to date.

Stellar-occultation observations for the TNO Pluto were sporadically successful from the mid 1980s (e.g., Brosch & Mendelson 1985; Hubbard et al. 1988;

Table 11.1. Published Statistics for Stellar Occultations by Centaurs.

Body	No. of Successful Events	Epoch	Multichord?[a]	References[b]
(2060) 95P/Chiron	6	1993;1994;2011 2018;2019;2022	Y	1,2,3 4,5,6
(8405) Asbolus	1	2013	N	7
(10199) Chariklo	22	2013;2015;2016 2017;2019;2020;2023	Y	8,9 10,11,12
(54598) Bienor[c]	3	2017;2018;2019	Y	7,13
(591376) 2013 NL_{24}	1	2019	Y	14
(60558) 174P/Echeclus[c]	3	2012;2020;2021	Y	7,15
(95626) 2002 GZ_{32}[c]	1	2017	Y	16
2008 YB_3	1	2019	Y	14,17
2014 YY_{49}	1	2019	Y	14
29P/Schwassmann—Wachmann 1	3	2022; 2023	Y	18

Notes:
[a] Indicator of whether any of the observed occultations were successful from more than one telescope: Y—yes, N—no.
[b] Reference key: 1: Bus et al. (1994); 2: Elliot et al. (1995); 3: Ruprecht et al. (2015); 4: Braga-Ribas et al. (2023); 5: Sickafoose et al. (2023); 6: Ortiz et al. (2023); 7: Rommel et al. (2020); 8: Braga-Ribas et al. (2014); 9: Bérard et al. (2017); 10: Leiva et al. (2017); 11: Morgado et al. (2021); 12: Santos-Sanz et al. (2023); 13: Fernández-Valenzuela et al. (2023); 14: Strauss et al. (2021); 15: Pereira et al. (2024); 16: Santos-Sanz et al. (2021); 17: Strauss et al. (2020); and 18: Buie et al. (2023).
[c] Bodies that have additional, unpublished successful occultations listed at http://occultations.ct.utfpr.edu.br/results/ (Braga-Ribas et al. 2019). One additional Centaur, 2007 JK_{43}, has no published occultation data but is listed in this database as having an unpublished, single-chord observation in 2020.

Elliot et al. 1989), with campaigns focussed on this relatively large dwarf planet that has an evolving, microbar-pressure atmosphere (e.g., Elliot et al. 2007; Sicardy et al. 2003). As discussed in Section 11.2, the first successful stellar occultations by a Centaur were for Chiron in the late 1990s. The significance of these early occultation campaigns is highlighted by the fact that, despite many attempts, an observation of a stellar occultation by a TNO other than Pluto was not successful until 2010 (Elliot et al. 2010) and by any Centaur until 2011 (again by Chiron; Ruprecht et al. 2015). Small, distant bodies project narrow shadow paths onto the Earth. The ephemerides of both the target and the star require high levels of accuracy in order to predict the location of the path, within which observing locations can then be selected. Based on the diameters and perihelion distances of Centaurs in the JPL Small-Body Database, the maximum angular size is 35 mas; however, as of this publication, the normalized root-mean-squares of the orbital fits are typically on the order of 400 mas or larger.

Stellar occultations are predicted months to years in advance. More accurate stellar positions with the Gaia catalogs (Gaia Collaboration et al., 2021) have increased prediction accuracy, but refinement of Centaur orbits and the possibility of stellar multiplicity usually require supporting astrometric observations. Observing campaigns depend heavily on the availability of observers and equipment as well as the event locations. Data quality is primarily a function of telescope size, star brightness, and exposure cadence: smaller telescopes can return data of adequate quality for brighter stars or events for which longer exposure times are acceptable (having lower relative velocity between the target and the star and/or returning lower spatial resolution at the target). Instrumentation is typically visible or near-infrared imaging cameras with sub-second exposure capability and low deadtime, with accurate timing. The availability of off-the-shelf, modern, CCD and complementary metal-oxide semiconductor (CMOS) cameras with GPS-triggering has increased opportunities for obtaining good data sets.[6] The locations of the predicted occultation shadow paths are important, since they must contain acceptable fixed telescopes and/or feasible sites for hosting traveling equipment.

A variety of observing strategies has also contributed to the increased success rate for small-body occultations. Examples include (i) having an established network of specifically-located telescopes and observing as many events as possible when a shadow path is predicted to fall in that location (e.g., Strauss et al. 2021), (ii) selecting events that are predicted to have high signal-to-ratio data (with bright stars or slow relative velocities) and which occur over larger, fixed telescopes worldwide (e.g., Sickafoose et al. 2019), (iii) having a large collaboration over a number of densely-populated countries and observing many events (e.g., Santos-Sanz et al. 2021, 2022), and (iv) traveling with multiple sets of equipment to set up an observing network in the location where a particularly interesting event is predicted to occur (Buie et al. 2020). These different strategies have varying cost and resource-availability requirements. Also worth noting is the recently-decommissioned Stratospheric Observatory for Infrared Astronomy (SOFIA), which followed on the occultation successes from the Kuiper Airborne Observatory (KAO; e.g., Elliot et al. 1989) and was a particularly useful tool for occultation observations given its mobile platform (e.g., Person et al. 2021).

There are a few professional teams who are actively publishing occultation predictions and coordinating observations for stellar occultations by Centaurs: (i) the "Lucky Star" consortium led by Observatoire de Paris with members in Granada and Rio de Janeiro (https://lesia.obspm.fr/lucky-star/); (ii) the Research and Education Collaborative Occultation Network (RECON; http://tnorecon.net/), led by Southwest Research Institute (SwRI), and (iii) a team of collaborators from Massachusetts Institute of Technology (MIT), Lowell Observatory, and the Planetary Science Institute (PSI) (http://occult.mit.edu/). The Lucky

[6] Examples include the QHY174M-GPS CMOS (Buie et al. 2020) and Finger Lakes Instrumentation ML261E-25 CCD (Lockhart et al. 2010).

Star project greatly increased the number of campaigns, participants, and successful observations for small bodies (e.g., Ortiz et al. 2020), with nearly 300 reported occultations by more than 100 different objects to date (from the database described in Braga-Ribas et al. 2019). RECON is specifically a citizen-science network with 64 telescopes located in a picket-fence configuration along the western part of North America, from Arizona into Canada (e.g., Buie & Keller 2016). Notably, collaborations between professional scientists and citizen scientists have played an important role. The International Occultation Timing Association (IOTA, https://occultations.org/) also publish predictions, coordinate observing campaigns and reporting, and provide tools for data analysis and publication of results. Further, there are a number of regional and national occultation-observing organizations. Current observing campaigns typically involve dozens of professional and citizen scientists spanning multiple countries.

11.6 Designing Campaigns: Present and into the Future

The widespread availability of high sensitivity CCDs, global coverage of 1–2 m class telescopes, and vastly improved astrometry for small bodies has provided a notable foundation to build future Centaur observing campaigns upon. The ability to consistently monitor the sky for astrophysical transients across all energy scales, from comet outbursts to tidal disruption events around black holes, drastically improves our ability to trigger observational campaigns of rapidly-detected outbursts. As such, reviewing the current types of observational campaigns, both of Centaurs and of supernovae, their strategies, and their limitations is necessary to build future efforts. We then describe examples of rapid-response campaigns to illustrate the lessons learned and to provide a template for the community.

11.6.1 Current Campaigns

The major current and/or recent Centaur campaigns, notably the *Mission 29P* monitoring campaign, the TESS/29P fortuitous campaign, and the occultation campaigns by various groups described in Sections 11.2, 11.3, and 11.5 have all yielded vital information on the activity frequency, structure, and sizes of Centaurs. Here we divide the current campaigns into situational categories (astrometric, monitoring, and serendipitous) as well as the observable categories (physical versus compositional) to simplify the planning and execution of future efforts. For simplicity, we do not discuss discovery campaigns, which are more representative of surveys and do not involve multiple observatories.

11.6.1.1 Astrometric, Monitoring, and Serendipitous Event Campaigns

The observational parameters for the vast majority of Centaur campaigns fall into three groups: astrometric, monitoring, and serendipitous events.

Astrometric events would include optimal observing conditions of a Centaur or group of Centaurs for study. Stellar occultations, opposition events, and mutual occultations are all examples of astrometric events that have been seized on for observing campaigns (see Section 11.5). Most critically, astrometric events are predictable, and these campaigns take advantage of high-precision orbital elements and ephemerides to plan observations. The predictable nature of these events enables high-precision physical properties to be directly observed, most frequently in the form of target size, shape, and presence of rings or moons with occultation data. The success or failure of astrometric observations then leads to refined orbital elements and ephemerides, in turn improving the ability to plan for future astrometric events in a positive feedback loop.

Monitoring campaigns are resource intensive, as they require dedicated observations of designated targets on a timescale short enough to detect short term variations in the target brightness and provide the activity context around those changes. Perhaps the best example of a monitoring campaign is the *Mission 29P* campaign organized by Richard Miles and the British Astronomical Association to observe 29P, which is discussed in Section 11.3. The Centaur's relatively large brightness and activity variability means that the *Mission 29P* campaign can take advantage of smaller, and thus more numerous, telescopes of citizen scientist astronomers to obtain high-cadence coverage. Higher sensitivity is provided by observations with the 0.4-, 1.0-, and 2.0-m Las Cumbres Observatory (LCO) telescopes around the globe (Brown et al. 2013), but most frequently from the 2.0-m Faulkes Telescope at Haleakalā. Monitoring campaigns are essential for characterizing the base level activity of Centaurs, identifying the frequency and amplitude of outbursts, and providing triggers to serendipitous follow-up campaigns. Due to the high observing cadence required to execute an effective monitoring campaign, the participation of volunteer observers with additional telescopes can be considered a large benefit that is perhaps unique to the *Mission 29P* campaign; most other Centaurs are too faint for typical citizen scientist telescope/CCD setups. With first light of the Vera Rubin observatory approaching in 2025 it would be tempting to say that the monitoring of Centaurs will become a matter of data reduction rather than acquisition, but this would be misplaced. The average Centaur will only be observed every 3 to 4 nights, a cadence that would be unable to capture the rapid brightness increase and exponential decrease consistent with an outburst (Schwamb et al. 2023). As such, dedicated monitoring campaigns by smaller aperture observatories are necessary to establish activity baselines on timescales that can resolve outbursts.

Serendipitous follow-up campaigns require existing monitoring campaigns to be effective. These campaigns would activate upon the observation of a trigger, which we discuss in the following subsection. A comparison can be drawn to the campaigns that often follow the catastrophic breakup of a comet; the rapid brightening of a typically faint object invites more observations as more observatories with smaller aperture telescopes, and thus decreased sensitivity, are able to detect the object. However, for most comets these outburst campaigns are not coordinated and rely on mutual assurance that others will obtain observations that provide additional context. Centaur follow-up campaigns will face a similar problem, and it will require the coordination of the full community to ensure that a rapid response to a serendipitous event trigger is executed to obtain new information about the activity and composition of Centaurs.

11.6.2 Rapid-Response Triggers

The triggers for a rapid-response campaign are ubiquitously activity-related for Centaurs, as it is the short-term increases in both apparent magnitude and size that are most easily detected. Here we discuss the two common ways those activity changes are detected and, more critically, how the community is notified of the event. At the time of writing, there are several notification methods within the Centaur community, and a brief overview may be useful to the more casual reader.

11.6.2.1 Citizen Scientist Detected Outbursts

Citizen scientist observers obtaining targeted observations of active small bodies frequently discover outbursts events, especially in the case of 29P. The duty cycle required to catch the onset of an outburst is too high for professional planetary scientists to propose observing programs to larger telescopes, but the network of citizen scientist observers around the globe with sufficient technology to measure brightness and apparent size variation of comets and Centaurs has drastically increased the temporal resolution of changes to active object brightness. These activity variations are typically reported via email listservs and chat groups like the Comets Mailing List or websites like the *Mission 29P* webpage[7]. These distribution mechanisms require human monitoring of a datastream (i.e., checking email or refreshing a webpage) to minimize the time between detection and response as there is not a standardized reporting format.

11.6.2.2 Survey-Detected Outbursts

The era of ambitious wide-field all-sky surveys has also presented a new monitoring method for active objects like Centaurs. These surveys, like the Zwicky Transient Facility (ZTF; Bellm et al. 2018) observe by coordinates rather than by object, and detecting outbursts becomes a matter of automatically reducing and extracting the brightness of Centaurs in the data and monitoring this behavior with time. Other surveys, like the LCO Outbursting Object Key (LOOK) project (Lister et al. 2022), monitor specific targets for activity changes as well as respond to alerts from other surveys like ATLAS, PS1 and PS2, Catalina Sky Survey, and ZTF. Full stream brokers, like ANTARES (Matheson et al. 2021) and downstream brokers like the Solar System Notification Alert Processing System (SNAPS; Trilling et al. 2023), intake the alerts broadcast by the ZTF and, in the future, LSST surveys, filtering the moving-object alerts produced by default and perform additional inspection to determine if the brightness is beyond the normal for the object. If the brightness is

[7] https://britastro.org/section_information_/comet-section-overview/mission-29p-2

unusual an alert is broadcast via the SNAPS webpage and API. However, in its current form SNAPS is intended for point source science rather than extended objects like active Centaurs, and the ANTARES feed it receives has been filtered of extended objects marked as "bogus" by the ZTF feed due to their non-stellar PSF. At present, SNAPS does not have the ability to characterize activity itself, but this is anticipated in future versions. Combining the automation of SNAPS with the extended object sensitivity of LOOK is essential to a comprehensive Centaur monitoring network.

11.6.3 Rapid-Response Campaigns

The intake and processing of the alerts provided by citizen scientist observers, surveys, and intermediate brokers is only the first step; following the alert up with coordinated observations is what designates a rapid response as a campaign. The Centaur community can model the response on that of other astrophysical transient response campaigns, like black-hole mergers, tidal-disruption events, and supernovae. The ideal rapid-response campaign minimizes the time between detection and follow-up characterization, which in turn requires a designated flow of target of opportunity observations.

11.6.3.1 Learning from Supernovae and Transient Object Campaigns
An excellent example of a supernova response campaign can be found in Gal-Yam et al. (2011), regarding the Palomar Transient Factory detection and follow up of supernova PTF10vdl (SN2010id). Data from the wide-field PTF were reduced, processed, and the transient characterized within 40 minutes of data acquisition. All transients characterized as supernovae triggered observations by the robotic 60-inch (1.5-m) telescope on site at Palomar, independent of human decision. In addition to this automated follow up, a duty astronomer in Israel ten hours ahead of the PTF time zone was able to further screen and schedule follow-up observations via the Wise observatory telescope. This highlights another key component of the PTF follow up; the utilization of global citizen science to screen transient candidates and verify pipeline characterizations. While there have been more recent follow-up observations of transient events that make use of more advanced alert systems, the SN2010id case study is an excellent example of the breadth of observatories that were coordinated to make follow up observations via ToO observations.

The follow-up observations included Swift UV and X-ray observations, visible spectroscopy both Gemini North and the Telescopio Nazionale Galileo (TNG), and Extended Very Large Array (EVLA) radio observations. Between the supernova's detection just after midnight on September 15 and September 19, 2010 the PTF team was able to acquire data between UV and radio wavelengths from six individual observatories and then continued to monitor PTF10vdl for the next 100 days with ground- and space-based assets both with images and spectroscopy. Rapid ToO's for the larger observatories were executed as soon as possible after the event, with the longer-term monitoring done by smaller aperture telescopes like Wise, LCOGT, and the Palomar Observatory robotic 60- and 48-inch telescopes. The diversity in wavelength sensitivity and aperture size, which can be interpreted as a proxy for telescope availability, enables rapid characterization and long term monitoring.

While Centaur outbursts produce lightcurves that are quite short relative to this example supernova, the astrophysical community is also rapidly responding to kilonovae and Fast Blue Transient Objects (FBOT) such as GW170817 and AT2018cow (Cowperthwaite et al. 2017; Prentice et al. 2018), which decay on timescales similar to a Centaur outburst. The success of programs such as PTF, ZTF, and LOOK, as well as spectroscopy follow-up programs like ePESSTO (formerly PESSTO; Smartt et al. 2015) provides a substantial foundation for future Centaur follow-up programs to build from. A key aspect of a successful supernova campaign like that of our example for PTF10vdl is the effort required to prepare; numerous ToOs as well as standard scheduled nights which were proposed to a wide variety of both ground and space-based observatories to ensure follow-up observations. Such effort requires a large and cohesive team to write the necessary proposals, develop the requisite data reduction pipelines, and analyze the detected events. The first step for an effort of this magnitude should be dedicated to improving the detection frequency of Centaur Outbursts, followed by considering how Centaur outbursts are different from point source lightcurves.

By their very nature, the Centaur transients are extended and dynamic in structure. The motion of ejected dust represents the biggest constraint on the time to execute follow-up observations, as the dust will move radially away from the object and create an approximate r^{-1} brightness profile. As such, the brightness of the object will change depending on the extraction aperture used. This adds an additional consideration to the follow-up campaign that is not as essential for point source transients; spatial resolution. A similar constraint is imposed by the limited lifetimes of gases and icy grains. The destruction rate of the gas (via photo-dissociation/ionization) and of any icy grains (via sublimation) is governed most immediately by the distance from the Sun to the particles of interest. The timescale for acquiring spectroscopic observations of the gas and dust components of an outburst is then a function of the object's distance to the Sun as well as the velocity of the ejecta. A narrow field of view that may provide excellent point-source spectroscopy will essentially only capture nuclear emissions from a Centaur if enough time passes. This leads to an interesting paradox specific to the case of solar system transient campaigns. The highest spatial and spectral resolution observatories with the potential to make the most precise observations also have the longest times to execute disruptive observations. For example, although the JWST could obtain extremely high signal-to-noise detections of CO_2 and CO gas in Centaur outbursts, the time to execute an ultra-disruptive ToO for JWST is <3 days, compared to 2–5 days for HST, and $\sim$6–30 hr[8] for the Keck Observatories. Thus,

[8] This time depends on whether the ToO can be triggered before 15:00 HST, in which case observations could be scheduled for that night. See Keck Partnership ToO policy here: https://www2.keck.hawaii.edu/inst/common/too_policies.html.

there is an optimal size observatory and instrument suite for Centaurs that can be triggered within hours of the detected outburst and still has the collecting area, resolution, and instrumentation to acquire critical data on the physical properties of the outburst. These are observatories like NASA's IRTF, the Multi-mirror Telescope on Mt. Hopkins, Apache Point Observatory, the Large Binocular Telescope, and other queue-based and staffed mid-class telescopes equipped with visible and near-infrared imaging and spectroscopy capabilities. Additional key measurements of typical cometary gases like CO_2 and CO can be acquired with large single dish sub-mm antennae like the Arizona Sub-mm Telescope (SMT), IRAM 30-m telescope, and Nobeyama 45-m telescope via director's discretionary ToO requests. Effective Centaur outburst response campaigns should look to propose to these telescopes and others of similar size and capability to maximize odds of success.

Centaur follow-up campaigns thus need to have tight constraints on follow-up time in order to accurately capture the evolution of transient activity, or identify key science elements that can be accomplished within the likely response times of key observatories.

11.6.4 Example Response to Centaur Transient Event

To assist the community in planning future response campaigns to Centaur transient events like outbursts we have designed an example response campaign to an outburst detected from 174P/Echeclus during its excellent apparition during opposition near perihelion in 2050 (Table 11.2). We want to note that this is an idealized response that would be contingent on a community effort to coordinate ToO proposal triggers on multiple telescopes, which is not trivial and should not be interpreted as such. Rather than forecast what observatories would be operational, we model the response on the current capabilities of ground and space-based assets.

Example event: a four-magnitude outburst of 174P/Echeclus is observed by the Vera Rubin Observatory just after 02:00 UTC. The outburst is captured as part of the real-time image difference analysis and distributed to various alert brokers (e.g., Trilling et al. 2023). The alert is picked up by the ANTARES broker and is recognized as an outburst based on the expected magnitude of Echeclus by a separate Centaur-specific monitoring system. The Centaur-specific monitor system sends an alert to a global notification list for the first layer of the Centaur transient response team via email within minutes of receiving the alert from ANTARES. Within ten minutes of the initial detection of the outburst additional imaging observations are requested via the LCO "Rapid-Response mode", enabling immediate observations of the outburst's morphology with visible wavelength imaging. Upon successful acquisition of the outburst by LCO the outburst is verified further by an on-call volunteer, ideally at a location that is 120°–180° longitude (8–12 hr) ahead of the discovery observatory. Precise visible light curves of the outburst can then be obtained via a combination of LCO and telescopes 1–2 min size.

Following verification of the outburst by the volunteer the outburst alert is distributed to a second layer of the response team within thirty minutes of the detection. Spectroscopic observations to characterize the composition of ejected dust and gas, requiring triggers of ToOs on larger ground-based observatories like NASA's IRTF (iSHELL for gas emissions, SpeX for dust reflectance and ice grain search), and the SMT (gas emissions), are executed. Within four hours of initial detection, the composition of the outburst can be targeted, well before the escaping gas exits the optimal aperture for extended source extraction. The timing is critical for narrow slit spectroscopy; excessive delays in execution may see a significant fraction of the outbursting material move beyond the slit's boundaries. The same issue is less pressing for sub-mm spectroscopy, which uses a larger beam size on target. Extended follow-up of any dust is carried out via frequent small ($D \leqslant 2$ m) aperture telescopes like LCO, or larger surveys like LSST or the Large Array Survey Telescope (Ofek et al. 2023), and any identifiable debris will require tracking by either additional DDT proposals to large ($D \geqslant 6$ m) or space-based observatories.

This idealized scenario and response is attainable with improved communication between members of the Centaur community. More specifically, large-scale coordination between research teams is necessary to maximize observatory wavelength and geographical coverage. No one research group can (or should) be responsible for planning and scheduling the full response campaign[9]. Diversity among the proposing teams ensures success of a follow-up campaign by virtue of the range of experiences, observational strategies, instrumental familiarity, and modeling capability.

11.6.5 Planning Future Campaigns

To assist the Centaur community in identifying periods of interest for planning future campaigns we have assembled Table 11.2, a table of optimal observing dates for a sample of eight objects from 2025 until 2075. We have defined "optimal" as phase angles less than 15°, as queried from NASA's JPL Horizons online tool[10]. This table is not intended to serve as a complete guide, merely to assist in identifying periods of interest for the community to focus on. We encourage early career scientists interested in planning observing campaigns of their own to explore the options presented in the table below.

[9] To a similar extent, the ability of the authors of a book chapter to critically question what best serves the community's interests as put forth in a book also applies.

[10] https://ssd.jpl.nasa.gov/horizons/app.html#/

Table 11.2. Observation Windows for Future Campaigns.

Dates Available	Heliocentric Distance (au)	Δ (au)	Minimum V Mag	Phase Angle (°)	True Anomaly (°)
Chiron					
2042-Feb-01 to 2042-Mar-13	10.14–10.21	9.25–9.53	15.923	1.32–4.57	292.41–293.70
2043-Feb-06 to 2043-Mar-18	9.53–9.59	8.61–8.81	15.602	0.81–4.30	305.02–306.48
2044-Feb-21 to 2044-Apr-01	9.00–9.05	8.07–8.28	15.358	0.80–4.55	319.73–321.38
2045-Mar-07 to 2045-Apr-26	8.63–8.67	7.68–7.99	15.234	0.63–5.38	336.10–338.35
2046-Mar-22 to 2046-May-11	8.47–8.48	7.48–7.79	15.129	0.48–5.25	353.66–356.02
2047-Apr-06 to 2047-May-26	8.51–8.53	7.51–7.80	15.131	0.60–4.92	11.57–13.90
2048-Apr-20 to 2048-Jun-19	8.76–8.82	7.76–8.17	15.322	0.25–5.27	28.86–31.48
2049-May-05 to 2049-Jul-04	9.19–9.28	8.20–8.60	15.525	0.36–4.88	44.78–47.14
2050-May-20 to 2050-Jul-19	9.77–9.87	8.77–9.19	15.787	0.50–4.58	58.96–61.04
2051-Jun-04 to 2051-Jun-24	10.43–10.46	9.43–9.47	15.763	0.61–1.30	71.39–72.00
Chariklo					
2055-Jan-24 to 2055-Feb-13	15.04–15.05	14.17–14.35	18.512	1.71–2.77	274.69–275.03
2056-Jan-29 to 2056-Feb-18	14.77–14.78	13.88–14.06	18.427	1.63–2.75	281.13–281.49
2057-Feb-02 to 2057-Mar-04	14.49–14.51	13.60–13.90	18.388	1.54–3.19	288.07–288.65
2058-Feb-07 to 2058-Mar-09	14.24–14.26	13.34–13.62	18.305	1.45–3.18	295.42–296.04
2059-Feb-02 to 2059-Mar-14	14.0–14.02	13.05–13.35	18.225	0.63–3.16	302.87–303.70
2060-Feb-07 to 2060-Mar-18	13.78–13.80	12.83–13.11	18.148	0.59–3.12	310.64–311.48
2061-Feb-11 to 2061-Mar-23	13.58–13.60	12.63–12.88	18.076	0.65–3.07	318.38–319.22
2062-Feb-16 to 2062-Apr-07	13.41–13.43	12.46–12.81	18.061	0.82–3.49	326.10–327.14
2063-Feb-21 to 2063-Apr-12	13.27–13.29	12.33–12.63	18.004	1.05–3.41	333.82–334.87
2064-Feb-26 to 2064-Apr-16	13.17–13.18	12.23–12.50	17.958	1.30–3.32	341.61–342.67
2065-Feb-20 to 2065-Apr-21	13.10–13.11	12.17–12.40	17.922	1.46–3.20	349.31–350.59
2066-Feb-25 to 2066-May-06	13.08–13.08	12.15–12.44	17.944	1.62–3.51	357.35–358.88
2067-Mar-02 to 2067-May-11	13.09–13.09	12.17–12.41	17.932	1.75–3.35	5.53–7.09

(*Continued*)

Table 11.2. (*Continued*)

Dates Available	Heliocentric Distance (au)	Δ (au)	Minimum V Mag	Phase Angle (o)	True Anomaly (o)
2068-Mar-06 to 2068-May-25	13.13–13.15	12.22–12.53	17.976	1.81–3.58	13.86–15.66
2069-Mar-21 to 2069-May-30	13.22–13.24	12.32–12.58	17.987	1.88–3.38	22.50–24.09
2070-Mar-26 to 2070-Jun-04	13.35–13.37	12.44–12.66	18.010	1.84–3.17	30.88–32.45
2071-Mar-31 to 2071-Jun-19	13.50–13.54	12.60–12.88	18.086	1.81–3.35	39.06–40.76
2072-Apr-14 to 2072-Jun-23	13.69–13.73	12.78–13.03	18.127	1.71–3.12	47.04–48.46
2073-Apr-19 to 2073-Jul-08	13.91–13.96	12.99–13.31	18.221	1.59–3.30	54.35–55.88
2074-Apr-24 to 2074-Jul-13	14.14–14.20	13.21–13.51	18.276	1.48–3.10	61.25–62.68
2075-May-09 to 2075-Jul-28	14.40–14.46	13.46–13.85	18.382	1.31–3.29	67.95–69.33
Echeclus					
2047-Jun-05 to 2047-Aug-04	7.48–7.63	6.59–6.83	18.223	0.72–6.25	285.31–288.34
2048-Jun-29 to 2048-Aug-18	6.59–6.69	5.66–5.84	17.609	0.92–6.25	307.40–310.67
2049-Jul-24 to 2049-Sep-22	5.98–6.04	5.02–5.29	17.239	0.65–7.42	335.49–340.27
2050-Sep-07 to 2050-Oct-27	5.90–5.92	4.89–5.26	17.210	0.23–7.57	9.28–13.41
2051-Oct-12 to 2051-Nov-21	6.35–6.42	5.35–5.69	17.497	0.48–6.29	40.58–43.41
2052-Nov-05 to 2052-Dec-15	7.18–7.28	6.19–6.55	18.041	0.59–5.55	65.46–67.67
Okyrhoe					
2028-Dec-01 to 2029-Jan-10	7.15–7.23	6.28–6.45	19.699	2.36–5.77	277.84–279.91
2029-Dec-26 to 2030-Feb-04	6.46–6.52	5.58–5.77	19.276	2.72–6.64	300.10–302.65
2031-Jan-20 to 2031-Mar-01	5.94–5.98	5.02–5.20	18.879	2.46–6.81	327.22–330.26
2032-Feb-24 to 2032-Apr-04	5.76–5.76	4.77–5.03	18.765	1.39–7.35	0.63–359.81
2033-Mar-20 to 2033-May-09	5.94–6.00	4.95–5.31	18.974	0.64–7.49	30.21–34.02
2034-Apr-24 to 2034-Jun-03	6.48–6.54	5.49–5.82	19.327	1.89–6.63	58.61–61.17
2035-May-09 to 2035-Jun-28	7.16–7.25	6.18–6.56	19.783	2.24–6.14	80.91–83.53
2052-Dec-05 to 2053-Jan-14	7.07–7.15	6.20–6.38	19.66	2.42–5.96	280.04–282.17
2053-Dec-30 to 2054-Feb-08	6.39–6.46	5.51–5.71	19.240	2.75–6.77	302.86–305.47
2055-Jan-24 to 2055-Mar-05	5.91–5.95	4.99–5.17	18.859	2.36–6.87	330.47–333.54
2056-Feb-28 to 2056-Apr-08	5.77–5.77	4.78–5.05	18.779	1.25–7.36	2.37–5.63

2057-Mar-24 to 2057-May-13	5.99–6.05	5.00–5.37	19.018	0.71–7.50	33.27–37.01
2058-Apr-28 to 2058-Jun-07	6.55–6.62	5.57–5.91	19.384	2.04–6.68	61.12–63.62
2059-May-13 to 2059-Jul-02	7.23–7.33	6.26–6.65	19.839	2.22–6.21	82.94–85.50
2075-Nov-25 to 2075-Dec-25	7.66–7.72	6.77–6.91	19.956	2.18–4.93	265.03–266.39
29P/S-W 1					
2025-Feb-20 to 2025-Apr-01	6.26–6.27	5.28–5.56	17.412	0.82–6.8	143.20–145.70
2026-Mar-07 to 2026-Apr-26	6.31–6.31	5.33–5.61	17.445	1.39–6.9	166.94–170.07
2027-Mar-22 to 2027-May-21	6.31–6.31	5.33–5.61	17.443	1.70–7.01	190.86–194.63
2028-Apr-15 to 2028-Jun-14	6.26–6.27	5.28–5.55	17.403	1.90–7.06	215.37–219.14
2029-May-10 to 2029-Jul-09	6.16–6.18	5.18–5.44	17.330	1.91–7.04	239.66–243.39
2030-Jun-04 to 2030-Aug-03	6.04–6.06	5.05–5.29	17.233	1.41–6.90	263.66–267.29
2031-Jul-09 to 2031-Sep-07	5.91–5.93	4.92–5.25	17.175	0.57–7.80	287.78–291.31
2032-Aug-12 to 2032-Oct-01	5.81–5.82	4.81–5.11	17.080	0.72–7.54	310.95–313.78
2033-Sep-16 to 2033-Oct-26	5.74–5.74	4.74–5.00	17.010	1.32–7.17	332.87–334.99
2034-Oct-21 to 2034-Nov-30	5.71–5.71	4.74–5.05	17.023	2.29–7.87	353.83–355.87
2035-Nov-25 to 2035-Dec-25	5.72–5.72	4.77–5.02	17.009	3.05–7.38	13.39–14.62
2036-Dec-29 to 2037-Jan-28	5.74–5.74	4.82–5.12	17.061	3.82–8.03	21.61–21.87
2038-Jan-23 to 2038-Feb-22	5.77–5.77	4.83–5.10	17.063	3.06–7.63	17.04–17.17
2039-Feb-17 to 2039-Mar-19	5.82–5.83	4.86–5.11	17.085	2.46–7.21	26.33–27.53
2040-Mar-03 to 2040-Apr-22	5.92–5.94	4.94–5.30	17.203	1.17–7.90	44.07–46.66
2041-Mar-18 to 2041-May-17	6.08–6.10	5.10–5.43	17.308	1.73–7.46	64.24–67.51
2042-Apr-12 to 2042-Jun-11	6.27–6.30	5.28–5.59	17.434	1.90–7.06	85.60–88.91
2043-Apr-27 to 2043-Jul-06	6.45–6.49	5.47–5.77	17.561	1.70–6.78	106.80–110.76
2044-May-31 to 2044-Jul-30	6.62–6.64	5.62–5.94	17.671	1.22–6.68	129.49–132.92
2045-Jun-25 to 2045-Aug-24	6.74–6.75	5.73–6.08	17.753	0.60–6.76	151.64–155.02
2046-Jul-20 to 2046-Sep-08	6.80–6.80	5.78–6.05	17.754	0.46–6.02	173.41–176.17
2047-Aug-14 to 2047-Oct-03	6.78–6.78	5.77–6.08	17.758	0.83–6.40	194.70–197.41
2048-Sep-07 to 2048-Oct-27	6.69–6.70	5.70–6.04	17.720	1.12–6.82	215.77–218.5
2049-Oct-02 to 2049-Nov-11	6.55–6.56	5.58–5.82	17.597	1.55–6.25	237.19–239.43

(*Continued*)

Table 11.2. (*Continued*)

Dates Available	Heliocentric Distance (au)	Δ (au)	Minimum V Mag	Phase Angle (○)	True Anomaly (○)
2050-Oct-27 to 2050-Dec-06	6.36–6.38	5.41–5.67	17.483	1.81–6.62	259.57–261.93
2051-Dec-01 to 2051-Dec-31	6.17–6.19	5.24–5.48	17.350	2.82–6.85	283.96–285.86
2053-Jan-04 to 2053-Jan-24	6.00–6.01	5.11–5.29	17.222	4.11–6.92	310.09–311.44
2054-Jan-29 to 2054-Feb-18	5.89–5.90	4.98–5.15	17.127	3.74–6.78	337.10–338.51
2055-Feb-13 to 2055-Mar-25	5.87–5.87	4.89–5.21	17.142	1.52–7.72	4.34–7.22
2056-Feb-28 to 2056-Apr-18	5.92–5.94	4.95–5.23	17.175	1.54–7.34	31.73–35.32
2057-Mar-24 to 2057-May-13	6.06–6.08	5.08–5.34	17.263	1.78–6.90	59.20–62.64
2058-Apr-08 to 2058-Jun-17	6.23–6.26	5.25–5.61	17.430	1.82–7.52	84.72–89.22
2059-May-03 to 2059-Jul-12	6.41–6.45	5.42–5.78	17.552	1.58–7.22	109.19–113.41
2060-May-27 to 2060-Aug-05	6.58–6.60	5.57–5.95	17.659	1.13–7.09	132.00–135.97
2061-Jun-21 to 2061-Aug-20	6.69–6.70	5.68–5.96	17.693	0.77–6.18	153.57–156.81
2062-Jul-26 to 2062-Sep-14	6.74–6.74	5.73–6.04	17.735	0.58–6.43	175.01–177.67
2063-Aug-10 to 2063-Oct-09	6.72–6.73	5.72–6.06	17.736	0.68–6.77	195.32–198.56
2064-Sep-03 to 2064-Oct-23	6.63–6.64	5.64–5.90	17.650	1.25–6.19	216.65–219.46
2065-Oct-08 to 2065-Nov-17	6.48–6.50	5.52–5.79	17.567	1.73–6.59	239.57–241.94
2066-Nov-02 to 2066-Dec-12	6.31–6.33	5.35–5.63	17.453	1.97–6.92	263.35–265.86
2067-Dec-07 to 2068-Jan-06	6.12–6.14	5.20–5.45	17.323	3.08–7.11	289.39–291.41
2068-Dec-31 to 2069-Jan-30	5.97–5.98	5.03–5.28	17.205	2.74–7.14	316.28–318.4
2070-Jan-25 to 2070-Feb-24	5.88–5.89	4.93–5.16	17.126	2.20–6.97	344.15–346.3
2071-Feb-19 to 2071-Mar-31	5.88–5.89	4.91–5.24	17.161	1.80–7.84	12.09–14.91
2072-Mar-05 to 2072-Apr-24	5.96–5.98	4.98–5.29	17.212	1.48–7.46	38.54–41.94
2073-Mar-20 to 2073-May-19	6.11–6.14	5.13–5.41	17.313	1.90–7.02	63.70–67.55
2074-Apr-14 to 2074-Jun-13	6.30–6.33	5.32–5.58	17.439	1.85–6.64	88.010–91.59
2075-May-09 to 2075-Jul-08	6.49–6.52	5.50–5.76	17.566	1.58–6.40	110.88–114.3
P/2019 LD2					
2025-Jan-01 to 2025-Jan-31	5.78–5.79	4.83–5.05	20.430	2.83–6.90	156.26–158.42
2026-Jan-26 to 2026-Mar-07	5.84–5.84	4.89–5.20	20.543	2.48–7.87	183.77–186.55

2027-Feb-20 to 2027-Apr-01	5.70–5.72	4.75–5.03	20.330	1.91–7.96	211.65–214.76
2028-Mar-16 to 2028-Apr-25	5.36–5.4	4.41–4.64	19.813	0.92–8.07	283.43–326.73
2029-Apr-20 to 2029-May-30	5.60–5.65	4.61–4.92	20.235	1.88–7.72	45.28–48.83
2030-May-15 to 2030-Jul-04	6.13–6.2	5.14–5.53	21.018	1.97–7.49	75.65–78.98
2041-Nov-23 to 2042-Jan-02	6.59–6.65	5.68–5.86	21.488	1.80–6.04	261.95–264.22
2042-Dec-28 to 2043-Jan-27	6.03–6.07	5.12–5.30	20.765	2.6–6.71	286.68–288.73
2044-Jan-22 to 2044-Mar-02	5.55–5.59	4.62–4.87	20.110	1.91–8.00	315.6–318.83
2045-Feb-25 to 2045-Apr-06	5.31–5.32	4.34–4.63	19.756	1.64–8.53	349.68–353.24
2046-Apr-01 to 2046-May-11	5.39–5.42	4.41–4.73	19.916	1.88–8.35	25.09–28.53
2047-Apr-26 to 2047-Jun-15	5.76–5.83	4.77–5.16	20.512	1.87–7.98	56.81–60.54
2048-May-20 to 2048-Jul-09	6.3–6.37	5.31–5.66	21.216	2.10–6.91	83.78–86.89
2059-Dec-09 to 2060-Jan-18	6.55–6.6	5.65–5.91	21.47	2.18–6.92	267.75–269.99
2061-Jan-02 to 2061-Feb-11	6.02–6.07	5.11–5.38	20.787	2.27–7.59	291.04–293.64
2062-Jan-27 to 2062-Mar-08	5.57–5.61	4.63–4.87	20.135	1.43–7.71	318.55–321.62
2063-Mar-03 to 2063-Apr-12	5.37–5.39	4.39–4.72	19.884	1.61–8.48	169.79–172.15
2064-Mar-27 to 2064-May-06	5.30–5.34	4.36–4.63	19.748	2.34–8.68	194.94–197.72
2065-Apr-21 to 2065-May-31	4.73–4.81	3.83–4.00	18.788	2.87–9.19	224.29–227.71
2066-May-26 to 2066-Jul-15	3.73–3.86	2.84–2.98	16.929	3.71–11.81	265.05–271.63
2067-Aug-19 to 2067-Oct-08	2.89–2.91	1.9–2.14	14.637	2.44–15.27	344.73–356.1
2068-Dec-01 to 2069-Jan-10	3.56–3.66	2.6–2.98	16.706	4.34–12.34	79.35–85.2
2070-Jan-25 to 2070-Feb-24	4.61–4.67	3.65–3.96	18.688	3.34–9.18	127.72–130.38
2071-Feb-19 to 2071-Mar-31	5.22–5.25	4.24–4.59	19.677	1.65–8.71	158.18–160.96
2072-Mar-15 to 2072-Apr-24	5.36–5.37	4.39–4.68	19.829	1.96–8.44	184.4–187.06
2073-Apr-09 to 2073-May-19	5.0–5.06	4.08–4.28	19.246	2.55–8.76	211.53–214.56
2074-May-14 to 2074-Jun-23	4.19–4.29	3.29–3.43	17.808	3.43–10.28	246.44–250.72
2075-Jul-08 to 2075-Aug-27	3.12–3.22	2.19–2.35	15.335	2.41–13.86	305.01–314.54
Bienor					
2025-Feb-10 to 2025-Feb-20	13.38–13.38	12.59–12.69	18.914	2.63–3.14	336.69–336.91

(Continued)

Table 11.2. (*Continued*)

Dates Available	Heliocentric Distance (au)	Δ (au)	Minimum V Mag	Phase Angle (○)	True Anomaly (○)
2026-Feb-15 to 2026-Mar-07	13.27–13.27	12.44–12.65	18.909	2.41–3.45	344.62–345.05
2027-Feb-20 to 2027-Mar-12	13.20–13.21	12.34–12.54	18.869	2.15–3.29	352.60–353.03
2028-Feb-25 to 2028-Mar-16	13.19–13.19	12.29–12.48	18.843	1.88–3.10	0.67–1.1
2029-Mar-01 to 2029-Mar-31	13.21–13.22	12.29–12.58	18.887	1.59–3.42	8.82–9.49
2030-Mar-06 to 2030-Apr-05	13.29–13.29	12.34–12.61	18.893	1.31–3.24	17.04–17.7
2031-Mar-01 to 2031-Apr-10	13.40–13.41	12.41–12.69	18.915	0.28–3.05	25.07–25.95
2032-Mar-05 to 2032-Apr-24	13.56–13.58	12.57–12.95	19.004	0.19–3.37	33.27–34.38
2033-Mar-10 to 2033-Apr-29	13.75–13.78	12.76–13.11	19.053	0.41–3.20	41.34–42.42
2034-Mar-15 to 2034-May-04	13.98–14.01	13.00–13.31	19.112	0.66–3.04	49.16–50.18
2035-Mar-20 to 2035-May-09	14.23–14.27	13.26–13.54	19.180	0.78–2.89	56.55–57.52
2036-Mar-14 to 2036-May-23	14.50–14.56	13.54–13.93	19.304	0.89–3.19	63.26–64.51
2037-Mar-19 to 2037-May-28	14.80–14.86	13.85–14.21	19.382	1.00–3.05	69.69–70.86
2038-Mar-24 to 2038-Jun-02	15.11–15.17	14.17–14.50	19.464	1.11–2.93	75.73–76.85
39P/Oterma					
2025-Jan-01 to 2025-Jan-21	5.93–5.94	5.01–5.21	21.799	3.81–6.79	37.85–39.13
2026-Jan-26 to 2026-Feb-25	6.26–6.29	5.33–5.65	22.340	3.27–7.29	60.53–62.12
2027-Feb-10 to 2027-Mar-22	6.64–6.68	5.67–6.05	22.879	1.65–6.95	80.42–82.5
2028-Feb-25 to 2028-Apr-05	7.04–7.08	6.05–6.36	23.352	0.59–5.95	99.73–101.71
2029-Mar-11 to 2029-Apr-30	7.40–7.45	6.41–6.79	23.823	0.25–6.16	118.03–120.35
2030-Mar-26 to 2030-May-15	7.71–7.75	6.72–7.05	24.155	0.52–5.67	135.21–137.4
2035-Jun-08 to 2035-Aug-07	7.85–7.89	6.86–7.19	24.286	0.18–5.89	212.16–214.61
2036-Jun-22 to 2036-Aug-21	7.59–7.63	6.61–6.90	23.976	0.18–5.89	228.11–230.73
2037-Jul-17 to 2037-Sep-05	7.25–7.30	6.28–6.52	23.563	0.71–5.81	245.86–248.27
2038-Aug-01 to 2038-Sep-30	6.86–6.92	5.90–6.19	23.094	0.21–6.56	265.10–268.34
2039-Aug-26 to 2039-Oct-25	6.47–6.52	5.51–5.83	22.586	0.27–7.13	287.37–291.03
2040-Sep-29 to 2040-Nov-18	6.14–6.17	5.17–5.50	22.122	0.32–7.45	313.12–316.53
2041-Nov-03 to 2041-Dec-13	5.94–5.95	4.97–5.27	21.823	1.24–7.39	341.45–344.38

2042-Dec-08 to 2043-Jan-07	5.93–5.93	4.97–5.22	21.792	2.31–6.96	11.01–13.22
2044-Jan-02 to 2044-Feb-11	6.09–6.12	5.12–5.49	22.104	1.70–7.52	39.10–41.87
2045-Feb-05 to 2045-Mar-07	6.42–6.45	5.49–5.80	22.555	3.15–7.09	65.57–67.43
2046-Feb-20 to 2046-Apr-01	6.80–6.84	5.83–6.21	23.082	1.74–6.86	87.90–90.1
2047-Mar-07 to 2047-Apr-16	7.18–7.22	6.20–6.52	23.538	0.83–5.98	107.68–109.63
2048-Mar-21 to 2048-May-10	7.53–7.58	6.54–6.95	23.980	0.35–6.24	125.42–127.64
2049-Mar-26 to 2049-May-25	7.81–7.85	6.82–7.18	24.279	0.25–5.83	141.33–143.83
2054-Jun-08 to 2054-Aug-07	7.79–7.83	6.80–7.05	24.188	0.57–5.34	218.18–220.74
2055-Jul-03 to 2055-Sep-01	7.49–7.54	6.51–6.84	23.868	0.18–6.18	235.31–238.05
2056-Jul-17 to 2056-Sep-15	7.12–7.18	6.15–6.42	23.411	0.48–6.11	253.43–256.52
2057-Aug-11 to 2057-Oct-10	6.70–6.77	5.75–6.05	22.897	0.19–6.81	274.90–278.46
2058-Sep-15 to 2058-Nov-04	6.30–6.35	5.35–5.67	22.362	0.30–7.31	299.73–303.0
2059-Oct-10 to 2059-Nov-29	6.00–6.03	5.04–5.35	21.924	0.46–7.47	326.30–329.86
2060-Nov-13 to 2060-Dec-23	5.88–5.89	4.90–5.18	21.726	0.79–7.18	355.43–358.39
2061-Dec-18 to 2062-Jan-27	5.97–5.99	5.00–5.36	21.911	1.80–7.71	24.80–27.66
2063-Jan-22 to 2063-Feb-21	6.25–6.28	5.32–5.62	22.320	3.03–7.14	52.48–54.46
2064-Feb-06 to 2064-Mar-17	6.64–6.69	5.67–6.04	22.878	1.36–6.77	76.31–78.67
2065-Mar-02 to 2065-Apr-01	7.09–7.12	6.12–6.38	23.402	1.83–5.73	98.09–99.67
2066-Mar-07 to 2066-Apr-26	7.49–7.54	6.50–6.87	23.927	0.38–5.96	116.55–118.9
2067-Mar-22 to 2067-May-11	7.85–7.89	6.85–7.18	24.309	0.63–5.49	133.71–135.86
2073-Jul-18 to 2073-Aug-07	7.87–7.89	6.96–7.13	24.284	3.07–5.23	222.75–223.58
2074-Jul-03 to 2074-Sep-01	7.51–7.57	6.55–6.86	23.901	0.15–6.14	237.84–240.57
2075-Jul-28 to 2075-Sep-16	7.10–7.15	6.13–6.39	23.381	0.52–6.11	256.50–259.07

Notes:

Availability of optimal viewing geometries for Centaurs in the next fifty years based on phase angle, as queried from the JPL Horizons database. These Centaurs were chosen based on community interest as gauged from frequency of appearances in the literature.

11.7 Summary

Despite decades of investigations with different observational techniques, our understanding of the activity and composition of Centaurs remains elusive. There are still many unanswered questions, such as "What are the compositions of Centaurs?", "How are those compositions similar to or different from SPCs, and what does this tell us about the evolution of small icy bodies?", "What are the sizes and surface properties of Centaurs?", "What drives activity and outbursts well beyond the frost line?". In this chapter we collected the current knowledge and highlighted the future plans in place to address questions like these with coordinated observing campaigns. We showed how observing campaigns can provide invaluable hints regarding the properties of these far and difficult to access objects.

Campaigns find their strength in the collective efforts and the different expertise of the observers. Working together, in a coordinated fashion, and combining knowledge across the different observing techniques, the time domain, and/or wavelengths range, allows us to progress more quickly toward answering difficult questions. Importantly, in this chapter we have shown that campaigns can be triggered by different needs of the community. They might be long-term monitoring efforts, or be inspired by astrometric (such as perihelion), or serendipitous (such as an outburst) events. The large surveys coming online in the near future point to a need to learn from follow-up observing campaigns associated with such surveys in both the planetary and astrophysical communities. By developing communication networks, diverse teams using a wide range of expertise, strategies and facilities will be able to work together to maximize what we learn.

Acknowledgements

We would like to thank Charles Schambeau (a co-author of Chapter 10) for updating our Figure 11.2 from his PhD Thesis (Schambeau 2018) to include the 2019-2024 data from IAU Minor Planet Center data for 29P.

References

Altenhoff, W. J., & Stumpff, P. 1995, A&A, 293, L41
Bauer, J. M., Choi, Y.-J., Weissman, P. R., et al. 2008, PASP, 120, 393
Bellm, E. C., Kulkarni, S. R., Graham, M. J., et al. 2018, PASP, 131, 018002
Bérard, D., Sicardy, B., Camargo, J. I. B., et al. 2017, AJ, 154, 144
Bockelée-Morvan, D., Biver, N., Schambeau, C. A., et al. 2022, A&A, 664, A95
Bodewits, D., Jehin, E., Kelley, M. S., et al. 2019, HST Proposal. Cycle 27, ID. #15965
Bodewits, D., Stern, A., Farnham, T. L., et al. 2021, HST Proposal. Cycle 29, ID. #16852
Bolin, B. T., Fernandez, Y. R., Lisse, C. M., et al. 2021, AJ, 161, 116
Braga-Ribas, F., Sicardy, B., Ortiz, J. L., et al. 2014, Natur, 508, 72
Braga-Ribas, F., Crispim, A., Vieira-Martins, R., et al. 2019, JPhCS, 1365, 012024
Braga-Ribas, F., Pereira, C. L., Sicardy, B., et al. 2023, A&A, 676, A72
Brosch, N., & Mendelson, H. 1985, IAU Circ., 4097, 2
Brown, T. M., Baliber, N., Bianco, F. B., et al. 2013, PASP, 125, 1031
Buie, M. W., & Keller, J. M. 2016, AJ, 151, 73

Buie, M. W., Keller, J. M., Miles, R., et al. 2023, Asteroids, Comets, Meteors Conference, (Houston, TX: LPI) Abstract #2445
Buie, M. W., Porter, S. B., Tamblyn, P., et al. 2020, AJ, 159, 130
Bus, S. J., A'Hearn, M. F., Schleicher, D. G., & Bowell, E. 1991, Sci, 251, 774
Bus, S. J., Bowell, E., Harris, A. W., & Hewitt, A. V. 1989, Icar, 77, 223
Bus, S. J., Wasserman, L. H., & Elliot, J. L. 1994, AJ, 107, 1814
Bus, S. J., Buie, M. W., Schleicher, D. G., et al. 1996, Icar, 123, 478
Chandler, C., Trujillo, C., Kueny, J., et al. 2022, DPS Meetings, Vol. 54, (Washington, DC: AAS) 502.01
Chandler, C. O., Kueny, J. K., Trujillo, C. A., Trilling, D. E., & Oldroyd, W. J. 2020, ApL, 892, L38
Choi, Y.-J., & Weissman, P. 2006, AAS/Division for Planetary Sciences Meeting Abstracts, Vol. 38, AAS/Division for Planetary Sciences Meeting Abstracts #38, 37.05
Cowperthwaite, P. S., Berger, E., Villar, V. A., et al. 2017, ApL, 848, L17
Cruikshank, D. P., Hartmann, W. K., & Tholen, D. J. 1988, IAU Circ., 4653, 1
de Graauw, T., Helmich, F. P., Phillips, T. G., et al. 2010, A&A, 518, L6
Elliot, J. L., Dunham, E. W., Bosh, A. S., et al. 1989, Icar, 77, 148
Elliot, J. L., Olkin, C. B., Dunham, E. W., et al. 1995, Natur, 373, 46
Elliot, J. L., Person, M. J., Gulbis, A. A. S., et al. 2007, AJ, 134, 1
Elliot, J. L., Person, M. J., Zuluaga, C. A., et al. 2010, Natur, 465, 897
Fernández, Y. R. 2009, P&SS, 57, 1218
Fernández-Valenzuela, E., Morales, N., Vara-Lubiano, M., et al. 2023, A&A, 669, A112
Firgard et al. 2025, in preparation
Foster, M. J., Green, S. F., McBride, N., & Davies, J. K. 1999, Icar, 141, 408
Fray, N., & Schmitt, B. 2009, P&SS, 57, 2053
Gaia CollaborationBrown, A. G. A., Vallenari, A., et al. 2021, A&A, 649, A1
Gal-Yam, A., Kasliwal, M. M., Arcavi, I., et al. 2011, ApJ, 736, 159
Gladstone, G. R., Stern, S. A., Ennico, K., et al. 2016, Sci, 351, aad8866
Griffin, M. J., Abergel, A., Abreu, A., et al. 2010, A&A, 518, L3
Guilbert, A., Barucci, M.A., Brunetto, R., et al. 2009, A&A, 501, 777
Güsten, R., Nyman, L. Å., & Schilke, P. 2006, A&A, 454, L13
Harrington Pinto, O., Kelley, M. S. P., Villanueva, G. L., et al. 2023, PSJ, 4, 208
Harris, A. W., & Harris, A. W. 1997, Icar, 126, 450
Hartmann, W. K., Tholen, D. J., Meech, K. J., & Cruikshank, D. P. 1990, Icar, 83, 1
Hsieh, H. H., Novaković, B., Walsh, K. J., & Schörghofer, N. 2020, AJ, 159, 179
Hubbard, W. B., Hunten, D. M., Dieters, S. W., Hill, K. M., & Watson, R. D. 1988, Natur, 336, 452
Jewitt, D. 2009, AJ, 137, 4296
Jewitt, D., & Luu, J. 1993, Natur, 362, 730
Jewitt, D.C., Soifer, B.T., Neugebauer, G., et al. 1982, AJ, 87, 1854
Kareta, T., Schambeau, C. A., Firgard, M., & Fernández, Y. R. 2025, PSJ, in press
Kareta, T., Sharkey, B., Noonan, J., et al. 2019, AJ, 158, 255
Kareta, T., Volk, K., Noonan, J. W., et al. 2020, RNAAS, 4, 74
Kareta, T., Woodney, L. M., Schambeau, C., et al. 2021, PSJ, 2, 48
Kowal, C., Dressler, A., Adams, R., et al. 1977, IAU Circ., 3134, 6
Kowal, C. T. 1988, Asteroids: Their Nature and Utilization (New York NY: Halsted Press)

Kronk, G. W. 2007, Cometography: A Catalog of Comets, Volume 3, 1900-1932 Vol. 3, (Cambridge: Cambridge Univ. Press)
Larsen, J. A., Gleason, A. E., Danzl, N. M., et al. 2001, AJ, 121, 562
Leiva, R., Sicardy, B., Camargo, J. I. B., et al. 2017, AJ, 154, 159
Lilly, E., Hsieh, H., Bauer, J., et al. 2021, PSJ, 2, 155
Lilly, E., Jevčák, P., Schambeau, C., et al. 2024, ApL, 960, L8
Lister, T., Kelley, M. S. P., Holt, C. E., et al. 2022, PSJ, 3, 173
Lockhart, M., Person, M. J., Elliot, J. L., & Souza, S. P. 2010, PASP, 122, 1207
Luu, J. 1994, PASP, 106, 425
Luu, J., Jewitt, D., & Cloutis, E. 1994, Icar, 109, 133
Luu, J. X. 1993, Icar, 104, 138
Luu, J. X., & Jewitt, D. C. 1990, AJ, 100, 913
Marcialis, R. L., & Bus, S. J. 1991, Bull. Am. Astron. Soc., Vol. 23, 1157
Matheson, T., Stubens, C., Wolf, N., et al. 2021, AJ, 161, 107
McFadden, L. A. 1994, IAU Symp. 160, Asteroids, Comets, Meteors ed. A. Milani, & M. di Martino (Dordrecht: Kluwer Academic Publishers) 95
Meech, K. J., & Belton, M. J. S. 1989, IAU Circ., 4770, 1
Meech, K. J., Buie, M. W., Samarasinha, N. H., Mueller, B. E. A., & Belton, M. J. S. 1997, AJ, 113, 844
Meza, E., Sicardy, B., Assafin, M., et al. 2019, A&A, 625, A42
Miles, R. 2016, Icar, 272, 356
Miles, R. 2021, JBAA, 131, 70
Miles, R., Faillace, G. A., Mottola, S., et al. 2016, Icar, 272, 327
Morgado, B. E., Sicardy, B., Braga-Ribas, F., et al. 2021, A&A, 652, A141
Moulane, Y., Bodewits, D., & Bromley, S. 2024, IAU General Assembly, 329
Mumma, M. J., & Charnley, S. B. 2011, ARA&A, 49, 471
Ofek, E. O., Ben-Ami, S., Polishook, D., et al. 2023, PASP, 135, 065001
Ortiz, J. L., Sicardy, B., Camargo, J. I., Santos-Sanz, P., & Braga-Ribas, F. 2020, in The Trans-Neptunian Solar System, ed. D. Prialnik, M. A. Barucci, & L. Young (Amsterdam: Elsevier) 413
Ortiz, J. L., Pereira, C. L., Sicardy, B., et al. 2023, A&A, 676, L12
Pereira, C. L., Braga-Ribas, F., Sicardy, B., et al. 2024, MNRAS, 527, 3624
Person, M. J., Elliot, J. L., Gulbis, A. A. S., et al. 2008, AJ, 136, 1510
Person, M. J., Bosh, A. S., Zuluaga, C. A., et al. 2021, Icar, 356, 113572
Pilbratt, G. L., Riedinger, J. R., Passvogel, T., et al. 2010, A&A, 518, L1
Prentice, S. J., Maguire, K., Smartt, S. J., et al. 2018, ApJL, 865, L3
Raugh, A. C. 1994, AAS Meeting Abstracts, Vol. 185 (Washington, DC: AAS) 115.01
Rayner, J., Bond, T., Bonnet, M., et al. 2012, Proc. SPIE, 8446, 84462C
Rayner, J., Tokunaga, A., Jaffe, D., et al. 2016, Proc. SPIE, 9908, 990884
Romon-Martin, J., Delahodde, C., Barucci, M.A., et al. 2003, A&A, 400, 369
Rommel, F. L., Braga-Ribas, F., Desmars, J., et al. 2020, A&A, 644, A40
Roth, N. X., Milam, S. N., DiSanti, M. A., et al. 2023, PSJ, 4, 172
Rousselot, P. 2008, A&A, 480, 543
Rousselot, P., Korsun, P. P., Kulyk, I., Guilbert-Lepoutre, A., & Petit, J. M. 2016, MNRAS, 462, S432
Ruprecht, J. D., Bosh, A. S., Person, M. J., et al. 2015, Icar, 252, 271

Santos-Sanz, P., Ortiz, J. L., Sicardy, B., et al. 2021, MNRAS, 501, 6062
Santos-Sanz, P., Ortiz, J. L., Sicardy, B., et al. 2022, A&A, 664, A130
Santos-Sanz, P., Gomes Júnior, A., Morgado, B., et al. 2023, DPS Meetings, Vol. 55, (Washington, DC: AAS) 301.07
Sarid, G., Volk, K., Steckloff, J. K., et al. 2019, ApJL, 883, L25
Schambeau, C. A. 2018, PhD dissertation, Univ. Central Florida https://ui.adsabs.harvard.edu/abs/2018PhDT.......266S/abstract
Schwamb, M. E., Jones, R. L., Yoachim, P., et al. 2023, ApJS, 266, 22
Scotti, J. V., Rabinowitz, D. L., Shoemaker, C. S., et al. 1992, IAU Circ., 5434, 1
Seccull, T., Fraser, W. C., Puzia, T. H., Fitzsimmons, A., & Cupani, G. 2019, AJ, 157, 88
Sicardy, B., Widemann, T., Lellouch, E., et al. 2003, Natur, 424, 168
Sicardy, B., Bolt, G., Broughton, J., et al. 2011, AJ, 141, 67
Sickafoose, A. A., Bosh, A. S., Levine, S. E., et al. 2019, Icar, 319, 657
Sickafoose, A. A., Levine, S. E., Bosh, A. S., et al. 2023, PSJ, 4, 221
Smartt, S. J., Valenti, S., Fraser, M., et al. 2015, A&A, 579, A40
Steckloff, J. K., Sarid, G., Volk, K., et al. 2020, ApJL, 904, L20
Stern, S. 1995, S&T, 89, 32
Stern, S. A. 1989, PASP, 101, 126
Strauss, R., Leiva, R., Keller, J., et al. 2020, DPS Meetings, Vol. 52, (Washington, DC: AAS) 203.05
Strauss, R. H., Leiva, R., Keller, J. M., et al. 2021, PSJ, 2, 22
Trigo-Rodríguez, J. M., García-Hernández, D. A., Sánchez, A., et al. 2010, MNRAS, 409, 1682
Trigo-Rodríguez, J. M., García-Melendo, E., Davidsson, B. J. R., et al. 2008, A&A, 485, 599
Trilling, D. E., Gowanlock, M., Kramer, D., et al. 2023, AJ, 165, 111
van Biesbroeck, G. 1927, PA, 35, 586
West, R. M. 1991, A&A, 241, 635
Wierzchos, K., & Womack, M. 2020, AJ, 159, 136
Wierzchos, K., Womack, M., & Sarid, G. 2017, AJ, 153, 230
Womack, M., & Sarid, G. 2020, Comet 29P/S-W 1 - Observations, University of Maryland, https://wirtanen.astro.umd.edu/29P/29P_obs.shtml
Womack, M., Sarid, G., Harris, W., Wierzchos, K., & Woodney, L. 2020, MPBu, 47, 350
Womack, M., & Stern, S. A. 1999, SoSyR, 33, 187
Xing et al. in preparation, using data from Swift Cycle 16 proposal "Rapid follow up characterization of cometary outburst ejecta" by D. Bodewits

AAS | IOP Astronomy

Centaurs

Kathryn Volk, Maria Womack and Jordan Steckloff

Chapter 12

Centaurs' Deep Interiors

Masatoshi Hirabayashi, Paul Sánchez and Gal Sarid

Icy small objects are thought to preserve structures intact since their formation. Their deep interiors, unaffected by late thermal alteration excluding radiogenic heat, are a record of early solar system formation and evolution. Despite extensive numerical and theoretical efforts, the current state-of-the-art knowledge about deep interiors is extremely limited, given the lack of observational evidence. Mechanical behaviors are complex and consist of brittle and ductile deformation due to many factors, mechanically altering the deep interiors without thermal influences. The size, rotation, and shape are key contributors to controlling the mechanical behaviors of the deep interiors. This chapter summarizes the mechanical behaviors of small, irregularly shaped objects to infer the properties of Centaurs, which are the transition population between comets and icy objects beyond Neptune. Numerical analyses using the shape models of icy small bodies can characterize the complex stress fields in comets and trans-Neptunian objects to infer properties of the deep interiors of Centaurs.

12.1 Introduction

Icy small bodies are planetary bodies that formed beyond the snow line and have contained ice since their original formation in the protoplanetary disk (Weissman et al. 2020; Blum et al. 2022). Depending on different scenarios, they have either continuously evolved over the solar system's history or have never evolved and thus preserve their earlier records. These bodies include trans-Neptunian objects (TNOs) and Oort Cloud Objects with the TNOs being the dominant source of of the Centaur population (e.g., Luu & Jewitt 2002; Jewitt & Sheppard 2002; Schulz 2002). Depending on their orbits, TNOs are further classified into smaller groups such as Scattered Disk Objects and classical Kuiper Belt Objects (see, e.g., Fernández 2020 and Chapter 3). Many are generally considered remnants of icy planetesimals; some form by severe (but sometimes gentle) collisions and aggregation in turbulence in the protoplanetary disk and have perhaps kept their pristine conditions since then

doi:10.1088/2514-3433/ada267ch12

(Weissman et al. 2020; Blum & Wurm 2008; Blum et al. 2022). How rapidly they evolve depends on where and when they formed (see Chapter 2 on formation); some TNOs gradually migrate toward the inner solar system to eventually become comets, including Jupiter-family comets (JFCs); see, e.g., Dones et al. (2015), Nesvorný et al. (2017), Morbidelli et al. (2022), and Chapter 3. Icy small bodies in transition between these populations are called Centaurs.

This chapter's major focus is characterizing the deep interiors of icy bodies that have not been affected directly by solar heating and non-catastrophic impacts since the bodies' terminal events[1] and offering insights into those of Centaurs. These materials are generally considered pristine (but not necessarily primordial), meaning they have been preserved since their terminal event by either initial aggregations in the protoplanetary disk (Massironi et al. 2015; Davidsson et al. 2016; Davidsson 2023) or re-accumulation after violent collisions (Rickman et al. 2015; Jutzi & Benz 2017; Jutzi et al. 2017; Schwartz et al. 2018). Highly violent environments in the early Kuiper Belt likely caused collisional evolution of small TNOs (Morbidelli & Nesvorný 2020; Bottke et al. 2023b). Shapes may record these terminal events; for example, violent collisions have a low chance of forming a bilobate shape (Nesvorný et al. 2018) though recursive processes may help such a shape survive longer (Hirabayashi et al. 2016). Understanding their deep interiors may therefore directly connect icy bodies' formation and evolution mechanisms with the early stages of the solar system.

This chapter reviews the state-of-the-art knowledge of the deep interiors of TNOs and JFCs and uses these to offer insights into the deep interiors of Centaurs. The scope is the mechanical evolution of the deep interiors driven by three geophysical parameters: size, shape, and rotation. Potential factors influencing the mechanical properties include radiogenic heating-driven compaction, impacts, and rotational instability. Radiogenic heating-driven compaction is critical when objects are born big (Kataoka et al. 2013; Golabek & Jutzi 2021; Takir et al. 2023). Later thermal evolution can also alter the deep interiors, depending on existing materials, structures, and initial radiogenic heating (Sarid & Prialnik 2009). Violent collisions that cause catastrophic disruption may change their mechanical properties completely (Jutzi & Benz 2017; Jutzi et al. 2017; Schwartz et al. 2018). Because numerous studies have reviewed the deep interiors based on these factors (Davidsson et al. 2016; Davidsson 2023; Weissman et al. 2020; Blum et al. 2022), this chapter explores the mechanical behaviors at later stages when small, irregularly shaped objects have already settled from their terminal events. The details of thermal evolution are not the main topic here, but a summary quickly illustrates the influences of radiogenic heating on the deep interiors. The present summary also does not contain technical settings and methodologies, but rather compiles the current knowledge about the evolution of the deep interiors. Finally, the following discussions use *deep interior/interiors* as a general term based on the definition above unless specifying target objects.

[1] A terminal event specifies the most recent process that created the current structure of the deep interiors.

Detailing the formation scenarios and early-stage environments is beyond this chapter's scope; summaries of such topics may be found in the literature (e.g., Weissman et al. 2020; Blum et al. 2022) and in Chapter 2. A new geological and numerical study of (486958) Arrokoth (formally 2014 MU_{69}) proposes an alternative formation scenario based on the existence of the object's mounds (Stern et al. 2023). However, not discussing such processes obscures our central insights into deep interiors. While minimizing detailed explanations, referring to earlier studies helps the chapter interpret deep interiors. The following terminology applies throughout this chapter: *pristine* and *primordial* mean *unaltered* and *early-stage*, respectively. These two terms may be interchangeable in many cases. While *primordial* infers the time of icy planetesimal formation in the protoplanetary disk, *pristine* represents an unaltered condition after an object's major configuration change, i.e., its terminal event. For example, after catastrophic disruption, a newly born, icy rubble pile body is no longer primordial but can keep its pristine, deep interior after that event. This chapter also distinguishes TNOs, Centaurs, and JFCs. Earlier studies sometimes use *comet/comets* to describe icy small bodies in the outer solar system falling into the inner solar systems without clear classifications (Weissman et al. 2020; Blum et al. 2022). While this nomenclature may be useful in arguing for particular evolutionary pathways, such classification helps distinguish Centaurs from other icy small bodies.

We briefly mention the Jupiter Trojans, which reside in Jupiter's Lagrangian points (L4/L5), to offer possible similarities and contrasts between them and Centaurs. A giant planet instability, possibly 10s of Myr after the end of the solar nebula, scattered small bodies originally orbiting beyond Neptune at $\geqslant$25 au; these perturbed objects are so-called "destabilized" primordial TNOs (Bottke et al. 2023a, 2023b). Some of these objects went to regions further out, while others fell into inner regions. Thc former objects include TNOs, while a limited number of the latter objects were captured by Jupiter's L4/L5 points (Nesvorný et al. 2013) to become the Jupiter Trojans. The Jupiter Trojans further experienced intense collisions with each other, causing those smaller than 100 km in diameter to be significantly altered from their primordial conditions (Bottke et al. 2023a, 2023b). Centaurs and the Jupiter Trojans may share similar origins, with the contrast that the Jupiter Trojans fell into the inner solar system during the giant planet instability, while the Centaurs came only recently. How the orbit insertions at different timescales change these objects is not fully understood, but collisional and volatile sublimation processes may/may not differ between them. NASA's Lucy mission will detail the Jupiter Trojans (Levison et al. 2021), adding new data to characterize those objects originally coming from the outer solar system.

The chapter outline is as follows. First, the chapter discusses the classification of Centaurs based on their orbital states. Second, it defines how Centaurs are the bridge between two end member populations, TNOs and JFCs. The next step is to review the basic trends of the mechanical stresses within deep interiors using simplified conditions. The following section discusses the deep interiors of different classifications: TNOs, JFCs, and Centaurs. Finite Element Modeling (FEM) analyses apply detailed geophysical conditions to characterize the stress fields in TNOs and JFCs.

Such bodies include (486958) Arrokoth, 103P/Hartley 2, 67P/Churyumov–Gerasimenko, and 9P/Tempel 1. Arrokoth is a TNO, while the other three are JFCs. Finally, we provide interpretations of Centaurs' deep interiors from the findings for TNOs and JFCs. The key takeaway is that size, rotation, and shape are critical for how deep interiors evolve; in particular, the rotation and shape are newly identified as critical geological parameters for small icy bodies. Mechanical alterations can occur in structurally sensitive regions in deep interiors regardless of the lack of thermal influences. Throughout this paper, we introduce stress analyses using a typical engineering convention in which compression is negative, and tension is positive.

12.2 Centaur Orbital Classification

This section briefly discusses the definition of Centaurs based on their orbits (see also Chapters 1 and 3). While the detailed classification is unnecessary in this chapter, introducing it provides comprehensive comparisons among TNOs, JFCs, and Centaurs. In short, no clear definition exists for Centaurs. One study (Jewitt 2009) offers the following definition: icy small bodies may be called Centaurs if (1) the perihelion distance, q, and the semimajor axis, a, satisfy $a_J < q < a_N$ and $a_J < a < a_N$, where a_J and a_N are Jupiter's and Neptune's semimajor axes, which are 5.2 au and 30 au, respectively, and (2) the orbit is not in 1:1 mean-motion resonance with any planet. Later studies, however, provide a loose definition that only considers $a_J < a < a_N$ as a general agreement (Lisse et al. 2020). Another definition is that $a_J < q < a_N$ while $a < 100$ au (Brasser et al. 2012). In any case, all definitions are based on Jupiter's and Neptune's orbital locations. This ambiguity does not impact the discussions about deep interiors at all, even though their lifetime as Centaurs may change significantly, and the sources of Centaurs may or may not include Oort Cloud Comets.

A Centaur's life span, i.e., the timescale over which an icy small body enters the Centaur region and exits it to join the inner solar system or be ejected, can be as short as a few Myr, but may be up to ~100 Myr, depending on its origin (see Chapter 3 for a detailed discussion). The typical dynamical lifetime of Centaurs is 1-10 Myr; however, depending on their dynamical behavior (e.g., resonance sticking and dynamical diffusion due to chaotic motion), they may have various dynamical lifetimes (e.g., Tiscareno & Malhotra 2003). When an icy small body reaches a region with a semimajor axis ranging between 5.2 au and 30 au to become a Centaur, the dynamical evolution becomes quick, taking place possibly within a few Myr (e.g., Lisse et al. 2020), though some Centaurs experiencing resonance sticking may stay longer than 22 Myr (Bailey & Malhotra 2009). Scattered disk objects can become Centaurs with a dynamical lifetime of 72 Myr if there is no constraint on the semimajor axis or 7.6 Myr with the semimajor axis constraint, i.e., $a_J < a < a_N$ (Di Sisto & Brunini 2007). Some objects with $a_J < q < a_N$ and a high semimajor axis, $a > a_N$, may come from the Oort Clouds (Brasser et al. 2012).

Given Centaurs' unique orbits and origins, they are considered a gateway of transition from TNOs to JFCs (Sarid et al. 2019). Figure 12.1 shows a simple flow

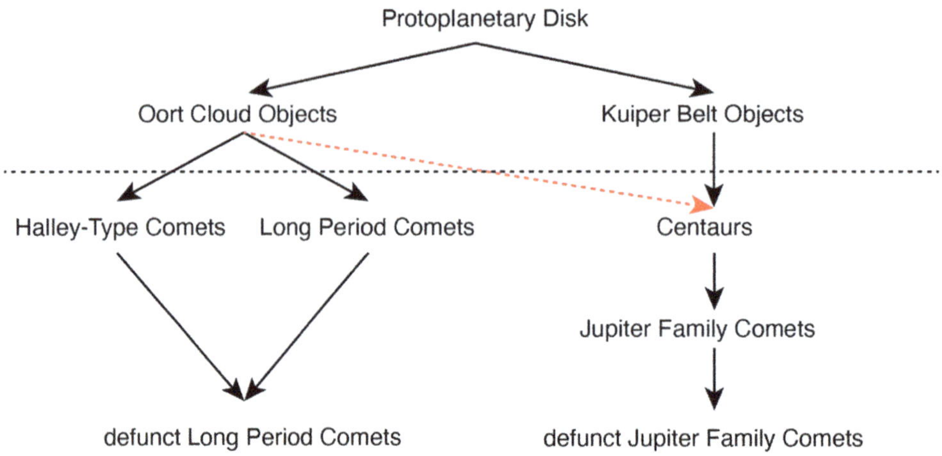

Figure 12.1. Evolutionary flow of icy small bodies following Jewitt (2015), with an addition made for some Oort cloud objects joining the Centaur population. In the flow here defined, Kuiper Belt Objects (a.k.a TNOs) includes Scattered Disk Objects.

diagram of icy small bodies (Jewitt 2015). TNOs are the major sources of Centaurs. With a looser definition, some Oort Cloud Objects may also become Centaurs, many of which may have higher eccentricities and inclinations (Brasser et al. 2012). At the early stage of the solar system, when Neptune and Uranus were gravitationally scattered outwards, many TNOs migrated into the outer main belt (Levison et al. 2009). Therefore, the flow argued in Figure 12.1 only applies to the current dynamical evolution.

12.3 Centaurs as Bridges between Endmembers

The series of comet observations by different spacecraft suggest that comets are generally highly processed, regardless of the potential implications of the exposure of primordial structures. For example, notable space missions to observe JFCs in detail include NASA's Deep Impact (A'Hearn et al. 2005), Stardust-NExt (Veverka et al. 2013), and Deep Impact eXtended Investigation (DIXI; A'Hearn et al. 2011). Their observations revealed diverse surface conditions, including heterogeneous surface morphologies. Although their spatial resolution is not high enough to detail local morphologies, images showed layering structures and other complex features on 9P/Tempel 1 (hereafter 9P; A'Hearn et al. 2005; Veverka et al. 2013), and variations in surface roughness on 103P/Hartley 2 (later denoted as 103P; A'Hearn et al. 2011). Active discussions offer new insights into how such complex surface morphologies evolved and thus infer deep interiors; however, no decisive conclusions were found to give insights into their formation and evolution processes.

ESA's Rosetta mission visited 67P/Churyumov–Gerasimenko (hereafter 67P), a ~4 km diameter JFC (Sierks et al. 2015; Thomas et al. 2015). This comet's nucleus consists of two lobes (a bilobate shape, similar to a rubber duck; El-Maarry et al. 2019) and has a bulk density of 538 kg m^{-3} (Pätzold et al. 2016; Jorda et al. 2016;

Pätzold et al. 2018), which implies that this body hosts a high volume of pore space. Rosetta's detailed observations revealed that primordial geological features might be exposed (Massironi et al. 2015), although the major part of the surface was likely to be severely altered due to its outgassing activities (Hu et al. 2017; Birch et al. 2017; El-Maarry et al. 2017, 2019; Leon-Dasi et al. 2021). The observations also captured possible structural features that record a soft merger of two lobes (Franceschi et al. 2020). Given the high outgassing activities of the nucleus, its spin rate changes drastically at every apparition, adding stress burdens to its deep interior (Mottola et al. 2014; Keller et al. 2015; Kramer et al. 2019).

NASA's New Horizons mission recently flew by Arrokoth, a cold classical TNO, and gave new insights into icy small bodies less evolved than JFCs (Stern et al. 2019). It does not exhibit remarkable erosion (Grundy et al. 2020; McKinnon et al. 2020; Marohnic et al. 2021; Spencer et al. 2020), though some geological features imply continuous/recent activities (Stern et al. 2021). From a geophysical perspective, Arrokoth's bilobate shape with a narrow neck suggests a soft merger between the lobes (McKinnon et al. 2020; Marohnic et al. 2021; Keane et al. 2022). This observation implies that this bilobate body has been kept relatively intact for a long time, but its surface layer has likely lost supervolatiles over its lifetime (Grundy et al. 2020; Steckloff et al. 2021). The supervolatile retention within the interior, however, may continue to the present day, as a tenuous atmosphere within the pore space of the body (Birch & Umurhan 2024). Arrokoth's relatively smooth surface (except for some geological variations; Schenk et al. 2021) and structurally sensitive shape (Hirabayashi et al. 2020a) are largely distinctive from comets' highly processed conditions.

The contrast between the observed JFCs and Arrokoth raised a new scientific question about how TNOs evolve to become highly processed JFCs (Figure 12.2). A processed object means a body significantly altered from its pristine condition. Thermal and collisional processes are among the main causes of such alteration. Depending on the scales, impacts may be violent enough to affect objects' deep interiors. Volatile ice sublimation, driven by solar heat, causes outgassing; a process unique to icy small bodies (Sarid & Prialnik 2009). This contrasts with small rocky bodies, which are influenced by solar radiation, e.g., Yarkovsky, YORP, and BYORP (if they are binaries), and the timescale for these is much longer than the thermal processes on icy small bodies (Ćuk & Burns 2005; Bottke et al. 2006). Centaurs are in the orbital transition between Neptune and Jupiter, representing those changing from icy planetesimals to comets. Studying Centaurs provides a comprehensive understanding of icy small bodies' evolution mechanisms to become processed icy bodies.

12.4 Basic Principles of Deep Interiors

Mechanical stresses influence small bodies' deep interiors, gradually (or sometimes rapidly) altering the structural conditions. The key physical contributors are size, rotation, and shape. This section applies a semi-analytical model to overview how these properties control the stress fields.

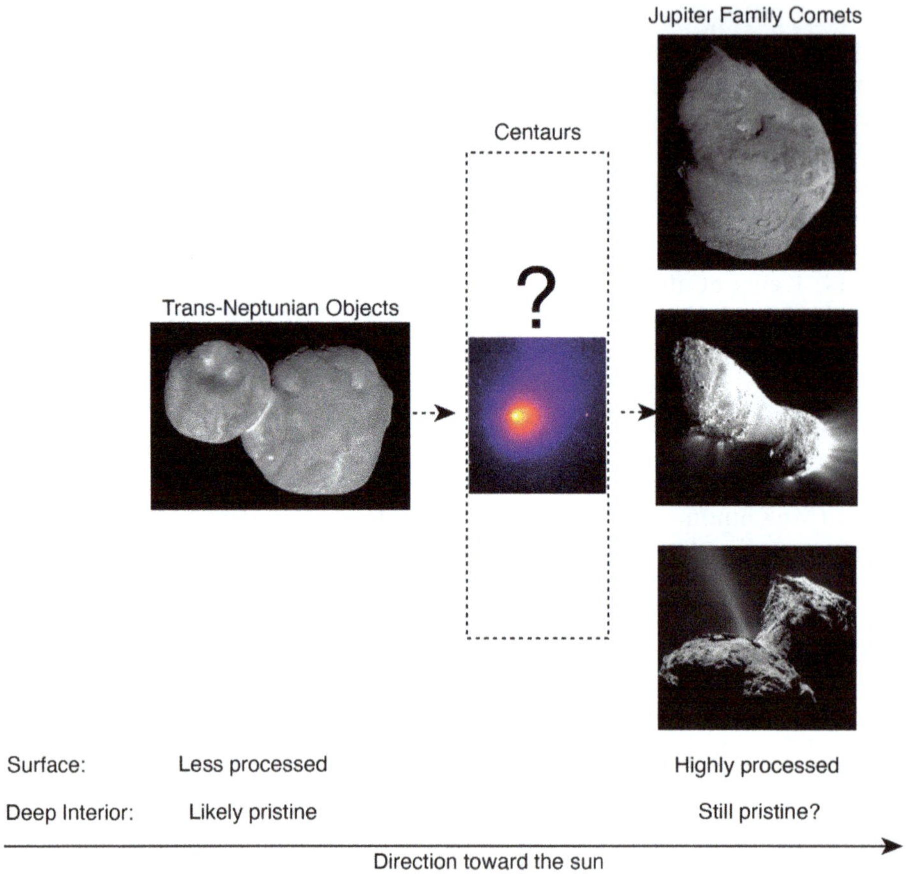

Figure 12.2. Schematic showing the transition from TNOs to JFCs. Centaurs are between these endmember icy bodies. JFCs are those typically targeted by planetary exploration missions, given their accessibility. The left image shows Arrokoth, the cold classical TNO, which may be a far end member given its relatively stable orbit over billions of years. The right images show 9P/Tempel 1, 103P/Hartley 2, and 67P/Churyumov–Gerasimenko from top to bottom. The Centaur section in the middle shows Schwassmann–Wachmann I. Image credits: NASA/JHUAPL/SwRI for Arrokoth, NASA/JPL-CalTech/UMD for 9P/Tempel 1 and 103P/Hartley 2, ESA/Rosetta/NAVCAM for 67P/Churyumov–Gerasimenko, and NASA/JPL/Caltech/Ames Research Center/University of Arizona for Schwassmann–Wachmann I.

12.4.1 Spherical Body with No Spin

Whether a deep interior keeps its pristine structure depends on what it experiences during and after its terminal process. Without radiogenic heating, which mainly contributes to the decrease of mechanical strengths, the average normal stress (the sum of all normal stress components divided by three) in a deep interior is a parameter that can characterize whether the structure is sensitive to compaction. A high magnitude of the average normal stress compacts pore space, altering its porous structure. This compaction process produces denser structures, perhaps enhancing thermal conduction in a deep interior.

We first overview how the stress field in a deep interior changes and affects its structural conditions without considering shape and rotation. A target is assumed to be perfectly spherical and have a zero spin state. This assumption gives an analytical solution of the average normal stress as a function of the radius (Love 2011). Figure 12.3 shows this as a function of the normalized radial distance (i.e., radial distance/radius) when the bulk density is $\rho = 500$ kg m^{-3}. The surface region, i.e., the unity normalized radius, experiences zero average normal stresses. The central region experiences the lowest average normal stress, which is proportional to R^2, where R is the radius. For example, it reaches about -0.1 kPa in a 5 km radius sphere and about -3.5 kPa in a 10 km radius sphere. Whether a deep interior is mechanically intact and can keep its structure depends on how mechanical strength works. Mechanical strength depends on a deep interior's evolution process, i.e., whether it contained icy or dry materials or experienced violent compression. Defining mechanical strength is a big problem that is beyond this study. Instead, we mainly introduce compressive, tensile, and cohesive strengths. This subsection discusses compressive and tensile strengths.

Earlier studies suggest that tensile and compressive strengths are highly dependent on icy structures at microscopic and macroscopic levels (Shimaki & Arakawa 2021; Blum et al. 2022). The tensile strength may reach MPa if the structure is non-porous (Blum et al. 2022). For 67P, the reported strength for a cometary body, which is highly porous, may range between ~100 Pa and ~kPa (Hirabayashi et al. 2016; Attree et al. 2018). Microscale (but fluffy) structures may have a strength of a few Pa (Skorov & Blum 2012). The surface compressive strength of 67P's nucleus at macro scales ranges from 10s to 100s of Pa (Heinisch et al. 2019; O'Rourke et al. 2020). This relatively low compressive strength comes from the measurements of

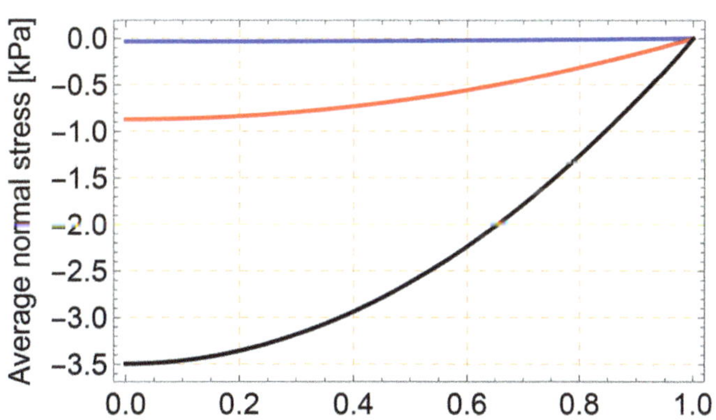

Figure 12.3. Average normal stress distribution as a function of the normalized radius, which is the distance from the center divided by the center-and-surface distance. The target body is spherical and does not have rotation. In the x-axis, the normalized radius is defined as the distance from the center to the surface. The y-axis is the average normal stress in kPa. The blue line shows the average normal stress in a spherical body with a radius of 1 km. The red line shows that in a 5-km radius spherical body. The black line shows a 10-km radius sphere with no rotation.

fluffy icy materials accumulated on the surface (O'Rourke et al. 2020). However, Philae's landing attempt, eventually causing it to bounce off the surface, identified a relatively high compressive strength of 1 kPa (Biele et al. 2015).

Another example may be a pile of snow. The compressive strength of snow at 250 K on Earth ranges between 0.1 MPa and 3.1 MPa (Jellinek 1959; Petrovic 2003), while the tensile strength under the same conditions is 0.7–3.1 MPa (Petrovic 2003). These strengths depend on their lifetime and density (Mellor 1974). If the bulk density is about 200 kg m^{-3}, cohesive and tensile strengths are about 5 kPa. When the density increases, these strengths increase. Temperature is another parameter that controls strength. Experimental research using icy grains suggests a large increase in the stickiness of mm-sized icy grains at temperatures ranging between 170 K and 200 K (Musiolik & Wurm 2019); see Chapter 13 for further discussion of laboratory studies.

Using the average normal stress in the 10 km radius sphere with no spin in Figure 12.3 and a 1 kPa compressive strength, for example, results in mechanical failures of any region below 80% of its radius. The affected region would lose its pore structure. Such mechanical alteration complicates the body's thermal evolution. Furthermore, if we assume that short-lived radiogenic nuclides are still active to continuously generate heat in this body, thermally driven mechanical weakening enhances structural compaction (Guilbert-Lepoutre et al. 2011; Takir et al. 2023). Compacted structures both trap sublimated gases and enhance thermal conductivity, helping heat to escape. A highly porous, sponge-like structure may possess low thermal conductivity but allow sublimated gases to escape from the deep interior and thus release heat efficiently. How deep interiors contribute to heat release is critical to constraining whether supervolatiles such as CO can stay for long periods of time or escape immediately (Davidsson et al. 2016; Davidsson 2023). However, the disclaimer of this scenario is that the discussion above does not account for the other two parameters: rotation and shape. We see different stories when considering them in Section 12.4.2.

12.4.2 Elongated Body with Rotation

Though considering a spherical body with no rotation may be simple to roughly characterize the structural conditions of a deep interior, this approach is problematic for detailed assessment. An approach considering a uniformly rotating triaxial ellipsoid (Love 2011) can give further insights into an icy small body's deep interior. This section briefly overviews this approach, which offers reasonable characterizations when only limited physical properties are available for a target body. While this chapter does not detail theoretical development, there exist many approaches in the literature, including averaging techniques (Sharma et al. 2009; Sharma 2013; Holsapple 2004; Holsapple & Michel 2006; Holsapple 2010) and extensions of limit analysis (Holsapple 2001). The basic principle assumes a homogeneous structure in a uniformly rotating triaxial ellipsoid and solves the structural equilibrium equation. The present model introduces the traction boundaries and small linear elastic deformation, where the stress field can be described in quadratic forms

(Dobrovolskis 1982; Hirabayashi 2015; Hirabayashi et al. 2020b; Nakano & Hirabayashi 2020).

A recent finding from this approach is that the stress field can change significantly, depending on spin and shape. When the spin is relatively slow, the surface needs small but non-zero strength to avoid failure. The surface tends to be more sensitive to failure under this condition. Mechanically sensitive areas concentrate on local surfaces so that structural modification may be limited. However, the interior is more susceptible to failure when rotation becomes fast. The failure mode may end up with a large-scale breakup, depending on the variations in strength, density, or both. The size and bulk density do not affect this behavior, though the stress level changes. Earlier studies detailed this structural behavior well by targeting rocky asteroids (Hirabayashi 2015; Hirabayashi et al. 2020b, 2022).

Figure 12.4 shows the average normal stress distribution in uniformly rotating biaxial ellipsoids with different shapes, sizes, and rotational states. A longer spin period provides negative stresses, meaning compression, while a shorter spin period tends to have positive stresses, meaning tension. When the size is larger, the stress magnitude becomes larger. When the body is spherical and the spin period is long,

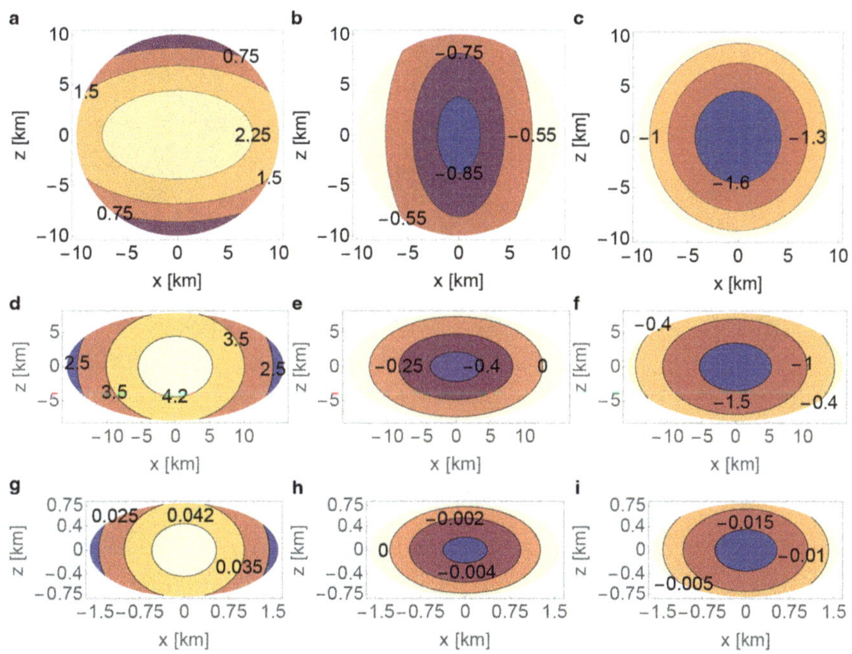

Figure 12.4. Average normal stress distribution in a uniformly rotating triaxial ellipsoid with a bulk density of 500 kg m^{-3} for different sizes, shapes, and spin periods. A negative value is a compression, while a positive value is a tension. The spin axis is along the vertical axis. Panels a, b, and c (top row): spherical object with a radius of 10 km. Panels d, e, and f (middle row): biaxial ellipsoid with $a/b = 2$ and an equivalent radius of 10 km. Panels g, h, and i (bottom row): biaxial ellipsoid with $a/b = 2$ and an equivalent radius of 1 km. Panels a, d, and g (left column): spin period of 2.5 hr. Panels b, e, and h (middle column): spin period of 5 hr. Panels c, f, and i (right column): spin period of 10 hr.

negative stresses remain in subsurface regions around the center and poles. If the spin period becomes shorter, the spherical body eventually experiences positive stresses at the center and lower (but still positive) stresses at the polar regions. An elongated body has similar trends; however, lower stresses appear at the tip along its longest axis. Figure 12.5 shows the spatial distribution of the minimum cohesive strength for a target body to keep its original structural configuration. The figure format is the same as that of Figure 12.4. In structural mechanics, cohesive strength defines shear strength at zero pressure; both normal and shear strengths control the failure of geological materials. Using the Drucker Prager yield criterion (Chen et al. 2007) offers a simple representation of the minimum cohesive strength (Hirabayashi et al. 2016). The figure describes the spatial distribution of the minimum strength required to keep local elements from structural failure, where such a mode causes irreversible deformation, yielding eventual reshaping processes (Hirabayashi 2015). If the minimum cohesive strength is zero, elements need no cohesive strength.

No cohesive strength is necessary for a deep interior if the spin period is long, although small but non-zero minimum cohesive strength appears around the surface

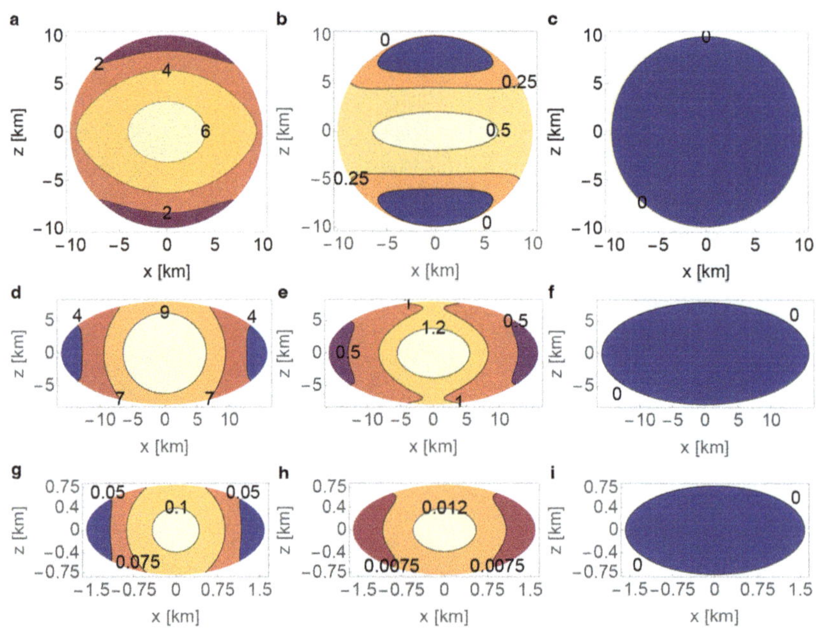

Figure 12.5. Minimum cohesive strength distribution in a uniformly rotating triaxial ellipsoid with a bulk density of 500 kg m^{-3} for different sizes, shapes, and spin periods. All negative values mean zero cohesive strength is necessary, though contour values are provided to show the equivalent stress level. The spin axis is along the vertical direction. Panels a, b, and c (top row): spherical object with a radius of 10 km. Panels d, e, and f (middle row): biaxial ellipsoid with $a/b = 2$ and an equivalent radius of 10 km. Panels g, h, and i (bottom row): biaxial ellipsoid with $a/b = 2$ and an equivalent radius of 1 km. Panels a, d, and g (left column): spin period of 2.5 hr. Panels b, e, and h (middle column): spin period of 5 hr. Panels c, f, and i (right column): spin period of 10 hr.

(Hirabayashi 2015; Hirabayashi et al. 2022). That means surface regions need some mechanical strength to keep the current configuration; otherwise, these locations may experience mass wasting. On rubble pile asteroids, such low cohesive strength correlates with mass wasting (Hirabayashi et al. 2020b, 2022). When the spin is fast, the minimum cohesive strength distribution changes, with the highest minimum cohesive strength appearing at the center. Even if the average normal stress level is negative at the center (Figure 12.4), the minimum cohesive strength can be positive, needing mechanical strength to keep the original configuration (Figure 12.5). This discrepancy comes from shear's critical role in the irreversible deformation of geological materials.

12.5 Deep Interiors of Trans-Neptunian Objects

This section explores TNOs' deep interiors. After their formation, violent or soft collisional processes are the most critical contributor to disturbing their deep interiors, if they happen (Morbidelli & Nesvorný 2020). While some objects in dynamically hot regions may be subject to violent collisions, others may be relatively intact in dynamically cold regions. Solar heat does not influence them much because of their distance from the Sun; however, radiogenic nuclides may contribute to the alteration of deep interiors (Tozer et al. 1965; Prialnik et al. 1987; Merk & Prialnik 2003; Sarid & Prialnik 2009; Lichtenberg et al. 2016; O'Neill et al. 2020). These processes can vary the magnitude of mechanical alterations of TNOs' deep interiors. TNOs are generally the remnants of icy planetesimals that have kept their pristine conditions after their terminal formation processes. After quickly reviewing the formation and early-stage evolution of icy planetesimals, this section discusses how the stress field contributes to deep interiors. As described in Section 12.4, the internal conditions highly depend on the size, shape, and rotation. The possible structural evolution of Arrokoth, a cold classical TNO, illustrates its structural response to a violent collisional process (if such a process happens after the soft merger of the two lobes).

12.5.1 TNOs as the Remnants of Icy Planetesimals

If TNOs have been pristine since their terminal processes, their deep interiors should preserve the conditions during such processes. Below is a quick review of the formation and evolution scenarios; there exist comprehensive reviews (e.g., Weissman et al. 2020, Blum et al. 2022, and Chapter 2).

During the accumulation in the protoplanetary disk, icy planetesimals continuously consume dusty and icy grains to become larger (or lose material depending on the violence of accumulation). Such processes contribute to the creation of deep interiors. There are two leading hypotheses for the formation of icy planetesimals, while a new one emerged from the observations of Arrokoth. First, hierarchical accretion is when small pebbles continuously collide with others in the protoplanetary disk so that planetesimals gradually grow. This process prefers to develop smaller icy planetesimals, possibly less than $\sim$1 km in diameter, depending on how effective the merging process is when the size becomes large (Weidenschilling 1977;

Weidenschilling et al. 1997; Weidenschilling 1997; Kenyon & Bromley 2004). The resulting structure may be a rubble pile with large voids (Weissman 1986; Donn 1989), while collisional compaction likely removes pore space from fluffy materials (Wada et al. 2008, 2009; Güttler et al. 2010; Zsom et al. 2010). Second, gravitational collapse following streaming instability creates large icy planetesimals (Youdin & Goodman 2005; Chapter 2). In a dense region of dusty and icy clouds in the protoplanetary disk, small grains, up to a few cm in diameter, gravitationally accrete to create a larger body (Youdin & Goodman 2005; Johansen et al. 2007). This mechanism produces larger bodies $\sim$1 km to $\sim$100 km in diameter. These processes may also occur when particle sizes differ; hierarchical accretion occurs first when particle sizes are small, while gravitational collapse after streaming instability happens later (Weidenschilling 1994; Davidsson et al. 2016). Recent observations of Arrokoth by New Horizons found mounds on its large lobe, suggesting that soft mergers of $\sim$km-sized clumps created its oblate shape. This formation scenario, i.e., the formation of $\sim$km-sized clumps followed by their soft mergers, does not belong to these leading formation hypotheses (Stern et al. 2023). Arrokoth's oblate lobes are also speculated to result from supervolatile sublimation (Zhao et al. 2021).

Post-accretion processes alter TNOs' deep interiors from their pristine conditions; two main contributors are impacts and thermal processes driven by radiogenic nuclides. If icy planetesimals exist in dynamically hot regions (Morbidelli & Nesvorný 2020) and experience non-catastrophic impacts, they may experience collisional compaction. If the stress level exceeds the local mechanical failure threshold due to impacts, compaction (and the resulting local heating) reduces pore space in the deep interiors (Davison et al. 2010). Depending on how often catastrophic disruption events occur to icy planetesimals, the deep interiors may or may not keep the primordial conditions of the location where their bodies were born in the protoplanetary disk (Jutzi & Benz 2017; Jutzi et al. 2017; Schwartz et al. 2018). If a catastrophic event is severe, it creates new smaller rubble piles whose deep interiors may likely differ from (or may occasionally be similar to) the deep interiors of their parent objects.

Radiogenic heat alters deep interiors, possibly melting and sublimating ice (Prialnik & Podolak 1995; Gregory et al. 2020). Short-lived radiogenic nuclides may play critical roles in this alteration. Such thermal evolution controls the deep interiors, depending on their size, structure, and the amount of radiogenic nuclides that existed at the time of formation. The half-life of major short-lived radiogenic nuclides, such as ^{26}Al and ^{60}Fe, is less than 1 Myr. The heating process becomes more significant for objects larger than $\sim$100 km in radius (Takir et al. 2023). These processes can cause deep interiors to lose volatiles or experience aqueous alteration, changing the deep interiors' mechanical structures. Some may form a few Myr after the Calcium-aluminium-rich inclusion (CAI) formation to avoid thermal alteration due to short-lived radiogenic nuclides (Mousis et al. 2017; Takir et al. 2023). For example, icy planetesimals with radii of $\sim$1.3 km to $\sim$35 km that were born $\sim$2.5 Myr to $\sim$5.5 Myr after CAI may host amorphous ice and (super) volatiles (Guilbert-Lepoutre et al. 2011). Either way, smaller icy planetesimals (up to 10s of km) can preserve their pristine conditions (or even their primordial conditions if no collisions alter their deep interiors).

12.5.2 Deep Interior of the observed TNO: (486958) Arrokoth

Section 12.4 focused on the basic principles of the stress distributions using a simplified model, which provides an overview of how the shape, size, and rotation can change a target's deep interior. This subsection discusses the deep interior of Arrokoth using the shape model and geomorphological information from New Horizons (Figure 12.6). The results show that the object's shape is critical to its structural condition. In addition, collisional processes at later stages may further impact the object's deep interior.

New Horizons's flyby observations of Arrokoth revealed the unique geological and geophysical features of this object (Stern et al. 2019). Its bizarre bilobate shape, also inferred by earlier occultation campaigns (Porter et al. 2018), gave critical insights into how such a shape formed (McKinnon et al. 2020; Marohnic et al. 2021; Keane et al. 2022). According to Keane et al. (2022), the extent of the full body of Arrokoth is $35.95 \times 19.9 \times 9.75$ km. The big lobe's dimensions are $21.2 \times 19.9 \times 9.05$ km, while the small lobe's are $15.75 \times 13.85 \times 9.75$ km; these dimensions correspond to equivalent radii of 15.63 km and 12.86 km, respectively. A preferred formation scenario suggests that the two lobes formed separately from gravitational collapse during stream instability, followed by a gentle merger (McKinnon et al. 2020; Marohnic et al. 2021; Keane et al. 2022; Brunini 2023) after mutual orbit evolution (Grishin et al. 2020; Lyra et al. 2021; Brunini 2023).

Arrokoth's bilobate shape features an extremely narrow neck with a cross-section of about 5 km by 5 km, compared to the two lobes. The object's bulk density may be low, 250 kg m^{-3} (Keane et al. 2022), almost half of the Jupiter-family comet 67P Churyumov–Gerasimenko, 538 kg m^{-3} (Pätzold et al. 2016; Jorda et al. 2016;

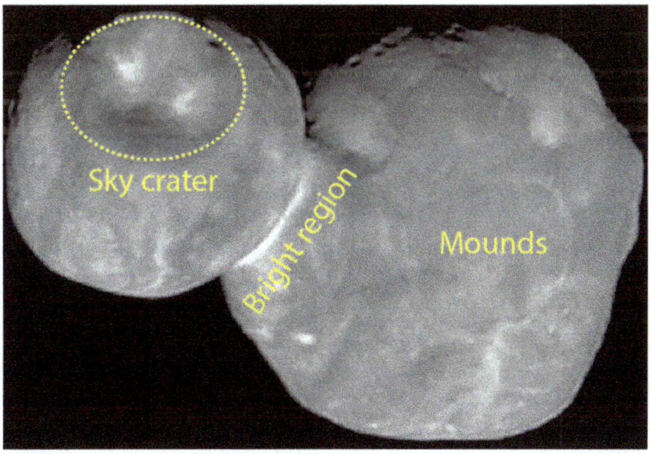

Figure 12.6. Geological features seen on Arrokoth's surface. The Sky crater is a circular depression on the small lobe. The narrow neck exhibits bright regions. The mound features are observable on the large lobe. Image credit: NASA/Johns Hopkins Applied Physics Laboratory/Southwest Research Institute.

Pätzold et al. 2018), suggesting an extremely high porosity for its deep interior. However, as Arrokoth's surface is observed to be relatively smooth (Spencer et al. 2020; Schenk et al. 2021; Keane et al. 2022), the deep interior has such high porosity, perhaps supported by cohesion and adhesion between ice and dust grains. Otherwise, such a highly porous structure may be vulnerable to structural collapse, causing surface depressions due to internal compaction.

Arrokoth's spin period is 15.93 h (Spencer et al. 2020; Keane et al. 2022), which seems slow; however, given its low bulk density, centrifugal forces under low self-gravity play a more important role in the internal stress, particularly around the neck (Hirabayashi & Scheeres 2013). Low compression of <1 kPa (i.e., negative average normal stress) in Arrokoth's major deep interior, regardless of its relatively large size, suggests that its highly porous structure may remain intact if its strength level is similar to the strengths of other rubble pile bodies (Figure 12.7(a)). Compaction becomes elevated around the neck region. Negative average normal stresses do not mean that strength is not needed. The cohesive strength level is high around the similar region, reaching 2.5 kPa (Figure 12.7(b)). Minor structural perturbations can lead to structural modification.

There are two scenarios for how the object's high porosity remains pristine under thermophysical processes driven by radiogenic heat. The first scenario is that this body forms within ~1 Myr after CAI and has enough radiogenic nuclides for thermal processes. However, because of its high porosity, sublimated gases would efficiently escape through pore space and reach the outside (Steckloff et al. 2021; Zhao et al. 2021). The dust-based structure could remain intact, keeping the original shape. If the sublimation process is severe, the body's shape may shrink (Zhao et al. 2021). This scenario generally suggests that Arrokoth does not possess significant volatile deposits at present. An issue of this scenario is that no outgassing features are identified on the present surface, so it lacks supporting evidence. However, the resulting morphologies due to outgassing are also not well-known for TNOs and are likely different from those on comets. The contrast between the predictions and observations does not rule out this scenario. The second scenario is that Arrokoth is

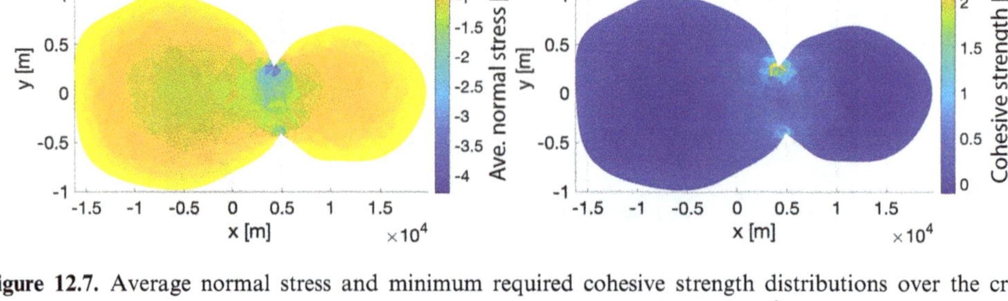

Figure 12.7. Average normal stress and minimum required cohesive strength distributions over the cross-section of Arrokoth at a spin period of 15.93 h with a bulk density of 250 $kg\,m^{-3}$. Courtesy of Yaeji Kim at UMD.

born after sufficient decay of short-lived radiogenic nuclides has already occurred. Without radiogenic nuclides, thermal processes do not happen. This scenario implies that Arrokoth does not experience thermal processing, so its deep interior contains volatiles in their pristine conditions. Currently, no sufficient evidence supports either of these scenarios, so future work is necessary.

Non-catastrophic impacts can still be critical to modifying Arrokoth's surface and interior. This object hosts a large circular crater-like depression feature on its small lobe; this crater-like feature is called Sky crater (Figure 12.6). If the feature is indeed the result of a violent impact event after the soft merger of the lobes, another problem is whether the formation of Sky crater modified Arrokoth's deep interior. Given its size, the Sky crater-forming impact likely shook the deep interior of the small lobe, neck, and large lobe (Hirabayashi et al. 2020a). Simulations using iSALE-2D Hydrocode (Amsden et al. 1980; Collins et al. 2004; Wünnemann et al. 2006; Collins et al. 2011) and assuming a bulk density of $\sim$500 kg m^{-3}, suggest that if a $\sim$730 m diameter spherical impactor hits a $\sim$13 km diameter spherical target (a simplified small lobe) at a vertical speed of 900 m s^{-1} (a typical impact speed in the cold classical Kuiper Belt; Greenstreet et al. 2015), the resulting crater diameter may be $\sim$6 km; this is comparable to the size of the observed circular depression feature. If the body has a cohesive strength of 10 kPa, the impact causes almost half of the small lobe to be damaged, inferring significant compaction within the lobe, regardless of material conditions. The temperature does not increase significantly. If the initial temperature is 55 K, the hottest region appears on the surface, and its temperature is about 60 K. When the bulk density is lower, damaged regions may become smaller because compaction can reduce shock propagation, but compaction increases.

Another issue is structural sensitivity during the Sky crater formation (Hirabayashi et al. 2020a). Earlier work suggested that the Sky crater formation may structurally disturb the neck region because of high shear stresses that result from the addition of torque to the body from the impact itself (Hirabayashi et al. 2020a; McKinnon et al. 2022). The neck fails if the mechanical strength is low, leading to shape reconfiguration. Whether the neck is broken and the lobes are detached is debatable given that this depends on the mechanical strength (Hirabayashi et al. 2020a; McKinnon et al. 2022; Kim & Hirabayashi 2022). Semi-analytical models suggest a shear stress of around 10s of kPa for the neck during the Sky crater formation (Hirabayashi et al. 2020a, McKinnon et al. 2022; Figure 12.8). This is higher than an upper limit on the cohesive strength of small bodies, $\sim$1 kPa. Preliminary work using FEM infers high variations in the necessary strengths around the object's neck (Kim & Hirabayashi 2022). Another study considered the coupling of dynamical-structural deformation to characterize how the normal stress along the neck's cross-section contributes to the neck's stress distribution. The study suggested that compression is comparable to shear during the impact, contributing to the mechanical disturbance (Hirabayashi 2023). It is still too early to conclude whether this process happened, and further studies are still necessary. For example, detailed analyses of the stress field evolution under Arrokoth's shape during the Sky crater formation event are still incomplete.

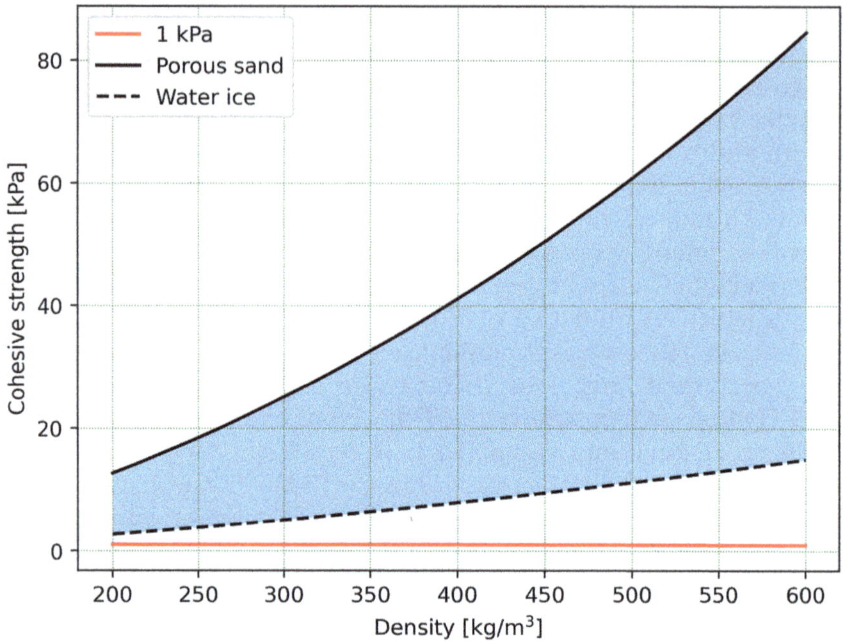

Figure 12.8. Cohesive strength level for Arrokoth's neck to endure the Sky crater-forming impact as a function of bulk density. The plot shows the cohesive strength necessary not to break the neck during the Sky crater-forming impact. The outcomes depend on Arrokoth's material conditions. The black line (upper boundary of the shaded region) results from the assumption that materials are mainly porous sands, usually representing weak materials. The black dashed line (lower boundary of the shaded region) gives the cohesive strength level when materials are water ice. The red line, the 1 kPa level, is an upper limit of cohesive strength for small icy bodies. The figure is reproduced from Hirabayashi et al. (2020b).

Shock wave propagation after the impact likely affects the neck and big lobe, which needs further investigation.

12.5.3 Late Stage Contributions to Deep Interiors

After completing their formation process in protoplanetary disks, some TNOs are exposed to violent collisional processes (Morbidelli & Rickman 2015; Morbidelli et al. 2022; Bottke et al. 2023b), while others keep their primordial conditions (Massironi et al. 2015; Davidsson et al. 2016). Such contrasts give them different evolutionary paths. Those having large-scale collisions change their structures perhaps after catastrophic disruption (Jutzi & Benz 2017; Jutzi et al. 2017; Schwartz et al. 2018). The dispersal of the planetesimal disk happens during the giant planet instability when comet-sized planetesimals have a low probability of survival (Morbidelli & Rickman 2015). In this model, planetesimals smaller than 10 km are likely reborn and not primordial; thus, they do not hold records of the original formation mechanisms in protoplanetary disks. This condition also applies to the cold classical Kuiper Belt as the crater size-frequency distribution at smaller

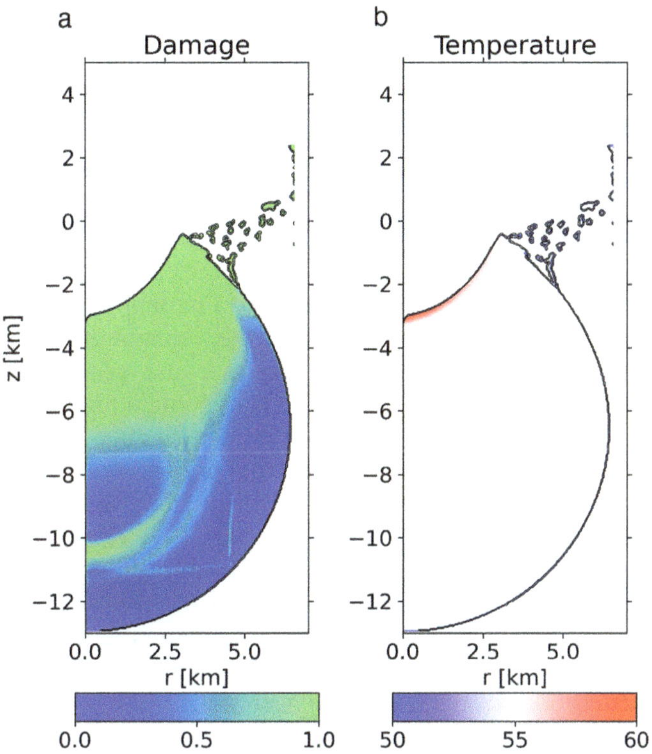

Figure 12.9. Snapshot of an axisymmetric iSALE-2D simulation for a Sky-crater-forming impact on a hypothetical spherical small lobe without connecting with the other lobe. Panel a (left): damage distribution (0= no damage, 1= complete damage). Panel b (right): temperature distribution (K). Both snapshots are from 10 min after the impact. The initial temperature is set to be 55 K.

sizes on Arrokoth infers collisional equilibrium (Morbidelli & Nesvorný 2020; Bottke et al. 2023b). Such intensive bombardments likely process the surfaces of larger bodies, completely altering their structures from the primordial conditions.

Collisional heating is another process that may alter deep interiors, depending on the target surface condition (Davison et al. 2010, 2012, 2014). Large-scale impacts sublimate volatiles significantly, leading to their depletion in deep interiors. While a smaller impactor may be less influential (Figure 12.9), a large number of small impactors may give non-negligible effects on the surface alteration (Davison et al. 2010). Recent numerical analysis, on the other hand, suggests that when a deep interior is kept at a low temperature, even catastrophic disruption may not induce enough heat to cause volatile sublimation (Schwartz et al. 2018), though it might be enough to crystallize amorphous water ice (Steckloff et al. 2023). Other models challenge such an analysis, arguing that there have been insufficient investigations for reasonable parameter ranges (Davidsson 2023).

If catastrophic disruption traps volatiles in a newly accreted body, while such an event destroys the precursor's primordial structure, the volatile composition may remain similar in the newly formed object. However, during the disruption, volatiles may widely spread under the influence of external forces and chemical reactions due to various energetic particles in cosmic rays and solar wind (Brown & Luu 1998; Stern & Trafton 2008; Marschall et al. 2020; Sunshine & Feaga 2021). Such conditions likely change the behavior of volatiles, resulting in a structure for the new body that differs from the original.

12.6 Deep Interiors of Comets

Once Centaurs fall inside Jupiter's orbit, they become comets. In contrast to TNOs, comets usually exhibit processed surfaces due to ongoing activity driven by solar heat. When solar heat warms the surface, subsurface ice sublimates and comes out of the surface as gases. Such outgassing processes cause geological alterations of cometary surfaces. A recent model accounted for geologic modifications that cause layering structures due to late events such as outgassing, surface heating, and impacts (Belton et al. 2007). Released gases carry momentum, which also changes the dynamics of comets (Mottola et al. 2014; Keller et al. 2015; Bodewits et al. 2018).

While deep interiors are unaffected by solar heat, they respond to the change in the rotational state driven by outgassing (e.g., Hirabayashi et al. 2016; Steckloff & Jacobson 2016; Safrit et al. 2021). Rapid rotation due to outgassing controls the centrifugal force acting on a cometary nucleus, modifying the stress field. If the stress field reaches the failure condition of a local element, mechanical failure causes irreversible deformation (Hirabayashi & Scheeres 2013). Depending on the magnitude of such failure, the outcome may be severe—some may experience breakups (Hirabayashi et al. 2016). For those experiencing continuous non-catastrophic events, deep interiors are likely to be under the influence of mechanical disturbance. The rotational change may occur due to violent impacts (Mao et al. 2021). This section focuses on the rotational change driven by outgassing.

The following subsections summarize recent studies of the internal structures of three cometary nuclei observed by spacecraft at high resolution: 103P, 67P, and 9P. Applying FEM approaches to their conditions, including irregular shapes, rotational states, and bulk densities, we infer the structural conditions of their deep interiors. This section focuses on the influence of shape and rotation on cometary deep interiors. Given that outgassing is critical in changing their spin states over a short period, the following discussions argue the internal conditions with various spin states.

12.6.1 103P/Hartley 2

103P/Hartley 2 (103P) is a JFC recently observed by the Deep Impact eXtended Investigation mission (DIXI, part of the EPOXI mission; A'Hearn et al. 2011). Figure 12.10 shows 103P's nucleus. DIXI's flyby was on November 4, 2010 (A'Hearn et al. 2011). The nucleus's spin period at the time of the flyby was 18.3 hr along its principal axis (Harmon et al. 2011; Drahus et al. 2011) and 27.79 hr along its long axis, suggesting that the object was tumbling (Belton et al. 2013). The spin state also changed by 1.3 min/day based on DIXI imagery and 1.0 min/day based on ground-based observations (Drahus et al. 2011). This spin variation implies

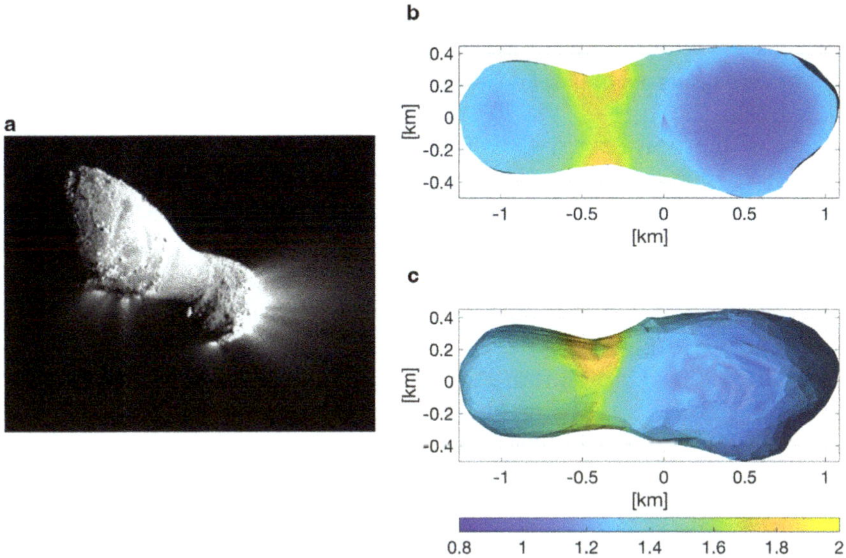

Figure 12.10. Comet 103P/Hartley's nucleus surface geology and outgassing activities (left) and the simulated average normal stress distribution at a spin period of 11 hr (right). Positive average normal stress means tension. Panel a: image of Hartley 2 taken by the DIXI spacecraft. Image credit: NASA/JPL-CalTech/UMD. Panel b: internal average normal stress [Pa]. Panel c: surface average normal stress [Pa].

that the nucleus's spin change was rapid, and the object likely experienced a rapid spin state within a short term (Steckloff et al. 2016). The nucleus was highly elongated; its size was 2.3 km along the principal axis and 800 m in the minimum and intermediate axes (Thomas et al. 2013).

FEM solutions of the stress field in the interior predict the stress field changes from compression to tension when the spin period is shorter than 11 hr (Steckloff et al. 2016). When the spin period is 11 hr, and the spin axis is along the maximum principal axis, the subsurface regions largely experience tension, and its narrowed region has the highest tension (Figure 12.10). The shape is highly elongated and has some local topography, including the object's neck; however, the neck's narrowness is milder than Arrokoth's and 67P's. This condition leads to widespread tension over the body. While the neck region experiences the highest tension, other regions are also exposed to milder tension. Such conditions cause the interior to be sensitive to structural failure.

The average normal stress distribution at the 11 hr spin period predicts tension becoming more significant everywhere when the spin period is shorter. If the stress level exceeds the yield conditions, local regions structurally fail and experience irreversible deformation. Unlike other highly bilobate bodies (Arrokoth and 67P), 103P experiences tension over a major area, and the failure mode may spread over its body. Structural alteration may differ depending on whether brittle or ductile deformation dominate such failure modes. If brittle deformation is dominant, cracks over its deep interior are the primary features expected to be seen. Ductile deformation may feature continuously deformed structures due to the centrifugal force. Either case alters the deep interior, differentiating it from its pristine state.

103P's nucleus may not have significant alterations after its terminal event in the past, at least in the tips of the lobes. Mechanical alterations may enhance its activity due to thermal energy driven by solar heat. Such mechanisms cause the deep interior to shrink rapidly. However, spacecraft and ground-based observations identified CO and CO_2 sublimation (Weaver et al. 2011; Meech et al. 2011), particularly CO_2 at the tips of the small lobe (A'Hearn et al. 2011). This finding implies that CO_2 must exist at a few thermal skin depths in a shadowed surface region if the diurnal temperature variation is a dominant contributor to its sublimation (Steckloff et al. 2016). CO_2 can be active even in the giant planet region (Jewitt 2009; Harrington Pinto et al. 2022). If mechanical alterations activate supervolatile outgassing, the body likely depletes such volatiles quickly. This scenario contradicts the observations, implying that the regions that exhibit CO_2 outgassing are unaffected by the widespread failure. The CO_2 outgassing event at the tip likely happens due to the exposure of pristine structures driven by landslides at a shorter spin period (Steckloff et al. 2016). Another possibility is that this comet experienced an unusually short migration from the outer solar system to the Jupiter family (Steckloff et al. 2016); this scenario is beyond the scope of our discussion here.

12.6.2 67P/Churyumov–Gerasimenko

Comet 67P/Churyumov–Gerasimenko (67P) was the target of ESA's Rosetta mission (Thomas et al. 2015; Sierks et al. 2015). 67P's nucleus is about 5 km in diameter and resembles a rubber duck, exhibiting a bilobate shape with a narrow neck (Preusker et al. 2015, 2017; Jorda et al. 2016; Chen et al. 2023). The reported spin period at the apparition in 2015 was 12.4 hr (Jorda et al. 2016). The spin period changed by 0.24 hr during the 2009 apparition, suggesting a rapid spin state evolution (Jorda et al. 2016). Numerical models also predicted a consistent spin period change, suggesting outgassing was the primary source of this change (Mottola et al. 2014; Keller et al. 2015). The nucleus exhibited complex and diverse geologic features.

Whether 67P's nucleus is primordial is under intense debate. 67P may have originated from hierarchical accretion, and the observed structures suggest they are primordial, not collisional in origin. The observed geological features, including the layered structures, and the presence of supervolatiles (Rubin et al. 2015; Combi et al. 2020) suggest the nucleus has a primordial condition (Massironi et al. 2015; Davidsson et al. 2016; Davidsson 2023). The observed volatile emission rate is relatively high, suggesting that formation via streaming instability followed by gravitational collapse may not be plausible because such objects tend to be born big and subject to radiogenic heat (Weissman et al. 2020). Alternatively, a collisional event rather than formation could be its terminal event (Morbidelli & Rickman 2015; Rickman et al. 2015); the suggested violent collisional processes in the trans-Neptunian region support this scenario (Morbidelli & Nesvorný 2020; Morbidelli et al. 2022; Bottke et al. 2023b). A collisional process can create a bilobate shape such as 67P (Schwartz et al. 2018), though it may be a rare process (Nesvorný & Vokrouhlický 2019; Nesvorný et al. 2022).

Regardless of the scenario that created 67P's present shape, another issue is whether the nucleus's deep interior has remained pristine since the terminal event. The nucleus's narrow neck makes the overall stress distribution complex. The following discussion summarizes recent work characterizing the stress distributions in deep interiors at different spin periods, accounting for the nucleus's rapid spin period changes. Numerical work employing FEM shows the stress variations in the nucleus at different spin periods (Hirabayashi et al. 2016). Panel (a) in Figure 12.11 shows the spin axis orientation, which is given as a green arrow (Jorda et al. 2016), while the other two arrows are orthogonal to it. Two spin period cases in Figure 12.11, 12.4 hr (panels (b) and (c)) and 8.0 h (panels (d) and (e)), show how the average normal stress level changes over the body. When the spin period is 12.4 hr, compression on the neck's surface increases to about -140 Pa. Other regions, except for surface regions, experience moderate compression of about -50 Pa. However, tension appears around the neck at a spin period of 8.0 hr. The appearance of such tension results from a bending moment; because of this, the surface region under tension likely undergoes opening-mode crack formation (Figure 12.12).

The variations in the stress field with rotation suggest a history where the nucleus's interior has been under the influence of various loadings over its lifetime. For example, the 2009 apparition caused the nucleus to change its spin period by 0.24 hr. Hypothetically, if the spin period becomes 0.24 hr shorter for every apparition, the nucleus can reach the 8-hr spin period that causes tensile stress within $\sim$100 yr and the 7-hr spin period that induces a complete breakup within $\sim$1000 yr.

This interpretation leads to the following scenario. 67P's deep interior, other than the neck, is relatively intact, given that the neck is the most sensitive to structural failure. For a spin period in which the neck mechanically fails, the deep interiors in the lobes are still under compression but below mechanical failures. These regions are thus relatively insensitive to structural failure. In contrast, the deep interior around the neck is most susceptible to experiencing mechanical alteration. The resulting mechanical failure is that a crack network (if brittle) or plastic deformation (if ductile) widely spreads over the neck region. Detailed observations of the fracture and crack distribution on the nucleus (Hirabayashi et al. 2016) and their possible growth (El-Maarry et al. 2017) support this view (Figure 12.12). Such altered structures can enhance thermal alteration as heat waves penetrate deeper regions. Not only does thermal conduction play a role in heat propagation, but thermal convection carried by sublimated gas through voids may contribute to it. Therefore, mechanically failed regions are further susceptible to thermal erosion, furthering mechanical alterations. A recent study suggests that subsurface regions below a few meters are still at temperature below 100 K (Hu et al. 2019), which is below the H_2O

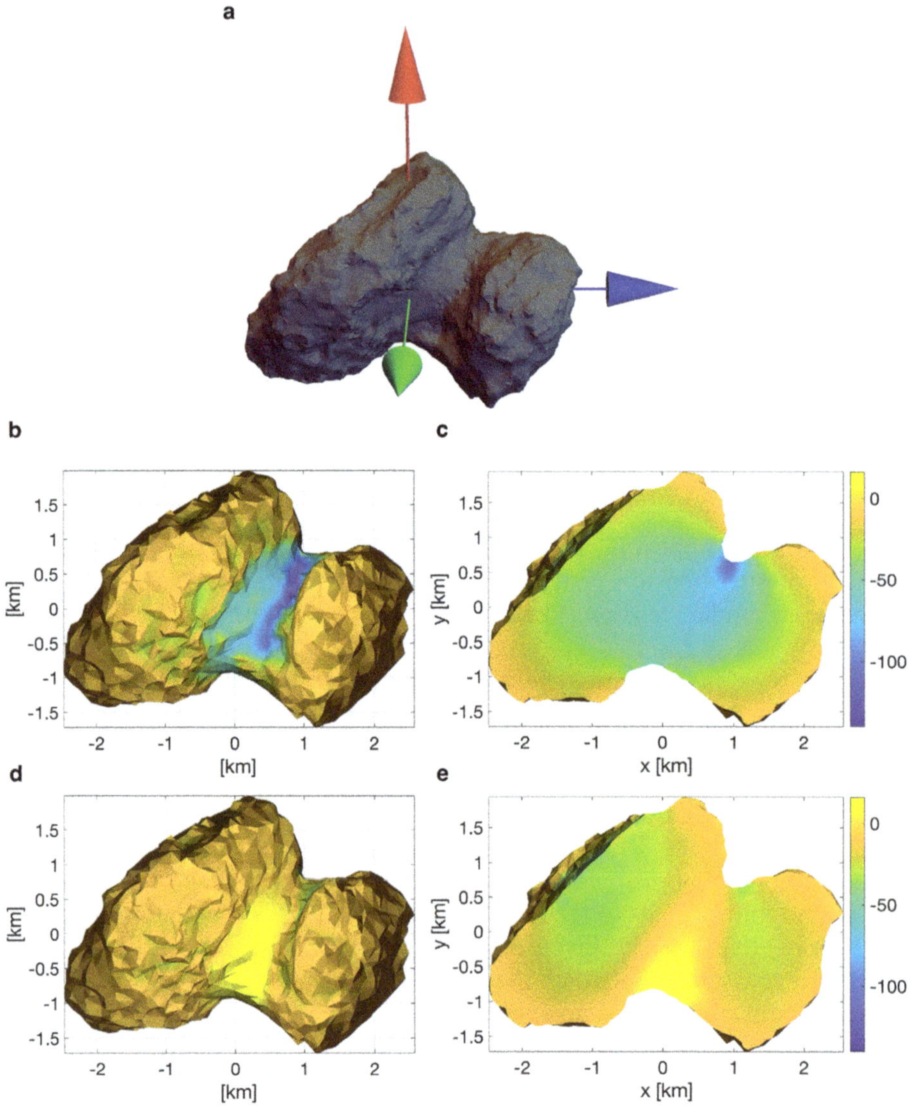

Figure 12.11. FEM simulations showing the average normal stress distribution in Pa on the surface and interior of 67P at spin periods of 12.4 hr and 8 hr. Panel a: schematic of 67P's nucleus. The green arrow shows the spin axis, which almost corresponds to the shortest axis. The red and blue arrows are orthogonal to the green arrow. Panel b: surface average normal stress for a 12.4-hr spin period. Panel c: internal average normal stress for a 12.4-hr spin period. Panel d: surface average normal stress for an 8-hr spin period. Panel e: internal average normal stress for an 8-hr spin period.

sublimation limit. However, other supervolatiles like CO and CO_2 should experience sublimation. If local regions have mechanically failed structures, heat propagation can be significant at a greater depth. If supervolatiles exist in the affected area, local structures may alter due to the resulting sublimation.

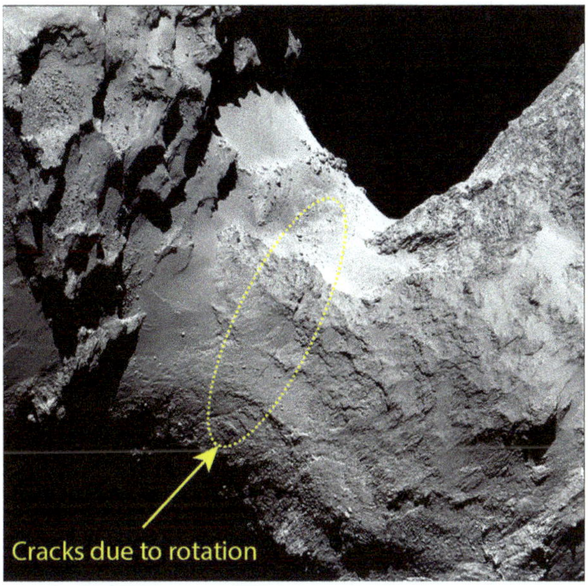

Figure 12.12. An open crack observed on the neck of the 67P nucleus. The crack extends over the narrow region of the neck, perpendicular to the tensile stress when the spin period is 8 hr. Image credit: ESA/Rosetta/NAVCAM for 67P/Churyumov–Gerasimenko.

12.6.3 9P/Tempel 1

The Jupiter-family comet 9P/Tempel 1 (9P) was the target of NASA's Deep Impact and Stardust-NExT missions. Deep Impact hit Comet Tempel 1 on July 4, 2005 (A'Hearn et al. 2005), while Startdust-NExT flew by Comet 9P/Tempel 1 on February 14, 2011 (Veverka et al. 2013). The observations covered two-thirds of the nucleus surface (Thomas et al. 2007, 2013). The average nucleus radius is 2.83 ± 0.1 km, and local elevation differences reach about 830 m (Thomas et al. 2013). The spin period at the encounter was 40.6 hr, and the nucleus produced 130 kg of dust per second (Veverka et al. 2013), causing its spin period to change (Belton et al. 2011; Chesley et al. 2013). Surface geologic features include pits, impact craters, and fluvial features (seen in panel (a) of Figure 12.13; Belton et al. 2007; Belton & Melosh 2009; Belton et al. 2013; Veverka et al. 2013). It is possible that some outbursts happened when the comet's perihelion decreased from 3.5 au to 1.5 au in 1609 AD (Veverka et al. 2013). The existence of impact craters implies the surface may be relatively old, suggesting it is less active than other active bodies such as 67P. The current orbit of 9P spans distances between 1.5 au and 4.5 au, while 67P is between 1.2 au and 5.7 au. The nucleus activity is less intense than observed on the nucleus of 67P. The smooth regions likely result from outgassing, perhaps causing material fluidization due to the release of CO and CO_2 (Belton & Melosh 2009). However, major fluidized materials do not exceed the escape speed, $\sim$1.3 m s^{-1} on this nucleus. These observations suggest that, rather than outgassing with dust mixtures, 9P's nucleus continues to have fluidized materials and smooth regions.

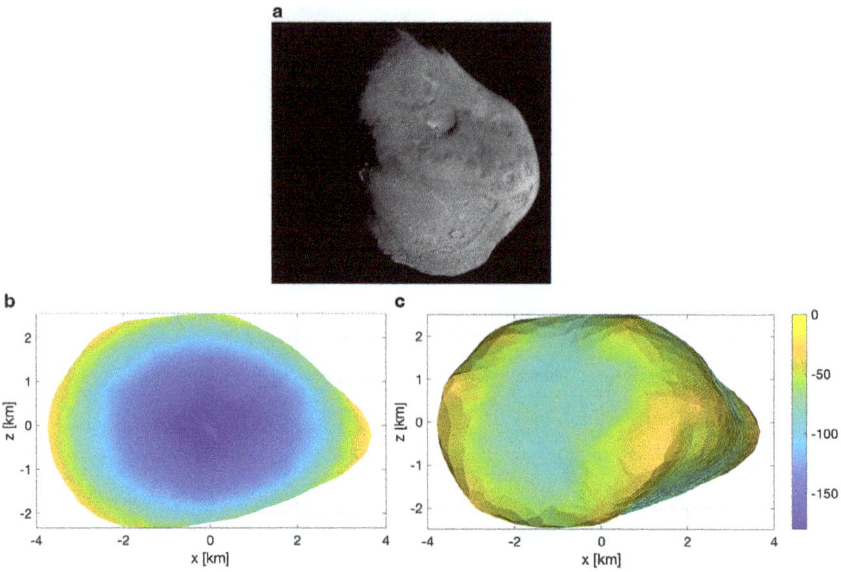

Figure 12.13. Comet 9P/Temple-1's nucleus surface geology (top) and average normal stress distribution at a spin period of 40.6 h (bottom). Panel a: image of 9P's nucleus taken by Startdust-NExT. Image credit: NASA/JPL/UMD. Panel b: average normal stress on the nucleus's cross-section. Panel c: average normal stress on the surface. The shape model refers to the NASA Planetary Data System (Farnham & Thomas 2013).

Such activities continue to create layered structures (Belton et al. 2007). The observed surface geology suggests that many features are less than 7000 orbits old (Belton & Melosh 2009).

Because the surface processes likely result from the phase change of H_2O, where CO and CO_2 are also released at 30–100 m deep (Belton & Melosh 2009), the majority of the interior of this 2.9 km diameter nucleus may remain unaffected by thermal effects. This implies that the deep interior beneath this surface layer is unaffected by heat propagation. Because 9P's nucleus is relatively round, the internal stress is similar to that in a spherical body. The nucleus's spin period of 40.6 h yields a stress field dominated by gravity and less affected by centrifugal forces. FEM simulations show that the average normal stress is compressive and is less than -150 Pa around the center (panels (b) and (c) in Figure 12.13). At this level of compression, no compaction occurs. Unless the spin changes significantly, which may not be critical for this nucleus, the stress field should remain similar.

The difference in the activity level between 9P and other JFCs discussed in this chapter is remarkable. 103P and 67P exhibit highly active outgassing processes, while 9P is less active than other nuclei. While there is no single accepted explanation for this variation, one possible explanation may be attributable to their shapes. While the nucleus shapes of 103P and 67P are bilobate, the nucleus shape of 9P is round. As seen above, bilobate shapes are susceptible to structural sensitivity, given their concave topography. 103P and 67P show stress concentrations in narrow areas when their spin periods are short. If a highly elongated shape like 103P's nucleus rotates at a critical spin period, rotation-driven landslides may occur at its tip (e.g., Steckloff et al. 2016). A round-shaped body like 9P usually does not have structural sensitivities at a longer spin period. At a short spin period, however, the internal structure becomes most structurally sensitive, perhaps inducing rotationally driven catastrophic disruption (Hirabayashi 2015; Hirabayashi et al. 2020b). This discussion leads to the following scenario. Since its terminal event, 9P's nucleus has not yet experienced a high spin state. The internal stress, where compression is dominant, does not reach a critical level (except for local regions) at a longer spin period. Thus, no essential processes of failure have occurred in the deep interior.

12.7 Deep Interiors of Centaurs

This section summarizes our interpretations of the deep interiors of Centaurs based on those of the TNO and JFCs discussed in the previous sections. The key takeaway is that structural sensitivity driven by size, shape, and rotation can influence deep interiors, even if heating does not play a role in altering them. If structural failure is significant, deep interiors likely host mechanical alterations such as crack formation and plastic deformation, which differ from their primordial structures. Unfortunately, none of the above discussions are yet supported by direct evidence in the Centaur population; they are just numerical predictions, regardless of recent progress in modeling techniques for possible geological and geophysical measurements. The biggest reason for that is the lack of measurements of Centaur deep interiors, so their mechanical response is not well-constrained. This challenge prevents us from filling knowledge gaps, so we have little solid knowledge about deep interiors. This caveat keeps the following discussions from being findings rather than simply speculations. With that in mind, this section concludes this chapter by discussing how the structural conditions in Centaurs likely change due to size, shape, and rotation.

12.7.1 Size, Shape, and Rotation as Key Parameters for Deep Interiors

Similar to other icy small bodies, the evolution pathways of Centaurs' deep interiors should highly depend on size, shape, and rotation. It is difficult to distinguish them as independent parameters because all these parameters cross-correlate. Nevertheless, it is still reasonable to attempt to describe how each parameter contributes to the structural conditions in icy small bodies.

The size controls the magnitude of the stress field. When a target body does not spin, the stress field is proportional to R^2, where R is the equivalent radius. This condition specifies that while the average normal stress does not change its sign, it changes its magnitude. For example, if the average normal stress is negative at a given location, the magnitude of compression becomes large when the size increases. The outcome is the same for tension. Because a larger body tends to have larger magnitudes of the stress field, a constant mechanical strength level can make the body more sensitive to structural failure. This is why a larger body loses a porous state in its deep interior if primordial structures are highly porous.

The spin controls the centrifugal force, which always contributes to outward loading. The centrifugal loading is proportional to $R\omega^2$, where ω is the angular velocity. The failure mode widely changes due to the spin state. A rapid rotation induces tensile stresses along the radial direction from the spin axis. If local elements cannot resist the tensile stress components, they fail structurally, causing either brittle or ductile deformation. Usually, internal structures are the most sensitive to failure when the spin is fast. The eventual failure mode may be catastrophic disruption. A slower rotation causes a lower stress field. Some locations become compressive, while others may still be tensile. This stress variation results from the stress field transitioning from tension to compression. A deep interior with a much longer spin eventually becomes compression-dominant.

The shape adds complexity to the stress field, which is simple when the shape is spherical or just elongated. However, local topography easily causes deviations in the stress field from the ideal one (Hirabayashi & Scheeres 2019). The stress field in a bilobate body is a great example of the stress distribution depending sensitively on its concavity, where stress concentrations appear (Hirabayashi et al. 2016). 29P/Schwassmann–Wachmann 1 is speculated to have a bilobate structure based on anisotropic outgassing features that suggest heterogeneous compositions (Faggi et al. 2024). This suggests some (many) Centaurs may have bilobate shapes. The shape can easily change the stress level and sign, i.e., compression or tension (Hirabayashi & Scheeres 2019), leading to difficulties in predicting the internal conditions without detailed observational measurements.

12.7.2 Chains between Activity and Structural Failure

Structurally failed regions can enhance thermal activities, causing outgassing. Deep interiors affected by compressive and tensile failure are susceptible to thermal processes if heat waves can reach them. If thermal processes happen, these regions are (by definition) no longer deep interiors. Because structurally failed regions tend to be weak, thermally driven processes disturb them more easily than mechanically strong regions. Thermal processes further weaken the failed and surrounding regions, accelerating structural failure. Figure 12.14 illustrates this possible pathway of an icy small body. Structural failure may occur in the TNO region, where the body may not have severe thermal processes. Collisional events also contribute to

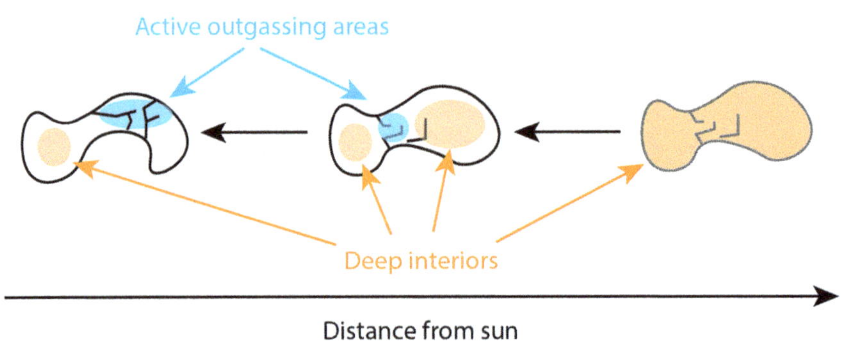

Figure 12.14. Schematic for the evolutionary pathway of an icy primitive body.

structural failure. Once a body enters the inner solar system and starts outgassing, regions that have already failed structurally are susceptible to thermal processes. These regions experience a higher rate of activity, causing intensive erosion. Such regions then further expose deep interiors to solar heat and alter them structurally. The interactions between structural failure and thermal processes continue to change the body's deep interior.

As Centaurs are actively transitioning from TNOs to JFCs, their deep interiors are possibly transitioning from their pristine states to more processed states. Because Centaurs inhabit the zones where supervolatiles such as CO and CO_2 can sublimate, structurally altered regions are likely susceptible to thermal processes. Thermal processes on these bodies are less severe than on JFCs. However, Centaurs' later phase significantly changes depending on how thermal processes control the size, shape, and rotation. There are multiple pathways based on the above discussions.

Larger bodies tend to have structurally altered deep interiors compared to smaller bodies. If they originally have a high amount of radiogenic nuclides, volatiles are likely depleted at early times. During this process, deep interiors are structurally altered as they warm up so that accumulated ice melts and/or sublimates. At later stages, they become relatively insensitive to thermal waves when they enter the inner solar system region. A large size can keep thermal waves from reaching deep interiors.

Bodies with shorter spins and spin variations tend to drive activity in their deep interiors much more easily than those with slower spins. A high spin creates centrifugal forces high enough to induce strong tension, leading to structure failure. Local elements fail structurally if the stress field reaches its yield condition. A high spin variation also causes a Coriolis force, adding additional stress sensitivity in a short time (Hirabayashi et al. 2021; Hirabayashi 2023). Spin variations occur due to various processes such as tidal encounters (without disruption; Souchay et al. 2018; Benson et al. 2023), outgassing (Mottola et al. 2014; Keller et al. 2015; Bodewits et al. 2018), and impacts (Mao et al. 2021).

Bilobate and elongated shapes tend to cause more activity. These bodies experience structural sensitivity due to the complexity of local topography. When the rotational period is constant but short, the centrifugal force can cause higher tension in concave regions than in other areas. If the rotation period is long, higher compression appears in such sensitive regions. Either way, the structural conditions in these regions are not ideal to keep them pristine. For example, an Arrokoth-like bilobate shape is not likely to keep its original shape when it enters the inner solar system. This scenario applies to 67P's nucleus, on which the neck was highly fractured.

There is a type of shape that keeps an icy small body relatively structurally intact compared to other bodies. Such a body is a round, slow rotator; 9P is possibly a good example. A spherical body usually tolerates a high spin state. However, there is one caveat. Once it reaches structural failure, the remaining process is catastrophic disruption. It cannot reach the critical spin without catastrophic disruption, creating numerous smaller fragments. A round shape with a low spin therefore evolves more slowly away from a pristine structure in its deep interior than other bodies.

Deep interiors are not necessarily intact but possibly mechanically altered without thermal effects. If they experience structural failure, they host large-scale cracks and structural deformation, which may be hidden under their surfaces and in deep interiors. Such structural alterations strongly depended on size, shape, and rotation. Because Centaurs are the transition between TNOs and JFCs, how JFCs mechanically evolve depends on what processes they have experienced. Some survive longer without critical activity, while others have severe activity and eventually lose their volatiles. Direct observations of these objects are essential to constrain the pathways of icy small bodies. Sending spacecraft to Centaurs is vital to answering the remaining questions about Centaurs' deep interiors (Chapter 16).

12.8 Conclusions

This chapter reviewed the mechanical aspects of the deep interiors of JFCs and TNOs, considering shape, size, and rotation, and offered a discussion of possible mechanical conditions of Centaurs' deep interiors (with as yet no supporting evidence). The major takeaway of this chapter is that we have no information about the deep interiors of any small icy Centaurs, given the lack of direct measurements. Still, we can give some quantitative interpretations using recent progress in numerical modeling. Arrokoth, representing a cold classical TNO, exhibits relatively smooth surface morphologies and an extremely mechanically sensitive neck; it was possibly affected by later structural disturbances such as meteoroid impacts. On the other hand, JFCs display highly processed surfaces. Their shapes, sizes, and spins have likely altered significantly since birth. Shape, size, and rotation play significant roles in these bodies' earlier and later evolution processes. These quantities can control the mechanical conditions that cause deep interiors to remain stable or fail structurally. Bilobate shapes, shorter spin states, and reasonably large sizes may make objects more sensitive to mechanical failure. Additional thermal input may add sensitivity to their mechanical conditions. In this view, TNOs (and other outer solar system objects) are less affected thermally, though mechanical failures may exist in their deep interiors. Comets are endmembers with high mechanical and thermal alteration; each comet may ultimately have its own unique history rather than holding the record of comets' formation mechanisms at large. Centaurs are the transition population bridging these two members. If Centaurs are indeed at the first stage (or at least the early stage) of solar heat-driven thermal alteration, we may observe mechanical damages related to it. Detailing and observing such processes can tightly connect primordial and processed objects to constrain their deep interiors and, thus, their earlier formation mechanisms.

Acknowledgements

M.H. is supported by NNH20ZDA001N-NFDAP/80NSSC24K0681. M.H. also thanks Bill Bottke at SwRI Boulder for thorough discussions about the formation and evolution of Trojans. M.S. gratefully acknowledges the developers of iSALE-

2D, including Gareth Collins, Kai Wünnemann, Dirk Elbeshausen, Tom Davison, Boris Ivanov, and Jay Melosh.

References

A'Hearn, M. F., Belton, M. J. S., Delamere, W. A., et al. 2005, Sci, 310, 258

Amsden, A. A., Ruppel, H. M., & Hirt, C. W. 1980, SALE: a simplified ALE computer program for fluid flow at all speeds, Technical Report Report No. LA-8095; TRN: 80-015132*, Los Alamos National Lab*

Attree, N., Groussin, O., Jorda, L., et al. 2018, A&A, 611, A33

A'Hearn, M. F., Belton, M. J. S., Delamere, W. A., et al. 2011, Sci, 332, 1396

Bailey, B. L., & Malhotra, R. 2009, Icar, 203, 155

Belton, M. J., & Melosh, J. 2009, Icar, 200, 280

Belton, M. J., Thomas, P., Veverka, J., et al. 2007, Icar, 187, 332

Belton, M. J., Meech, K. J., Chesley, S., et al. 2011, Icar, 213, 345

Belton, M. J., Thomas, P., Li, J.-Y., et al. 2013, Icar, 222, 595

Benson, C. J., Scheeres, D. J., Brozović, M., et al. 2023, Icar, 390, 115324

Biele, J., Ulamec, S., Maibaum, M., et al. 2015, Sci, 349, aaa9816

Birch, S. P. D., & Umurhan, O. M. 2024, Icar, 413, 116027

Birch, S. P. D., Tang, Y., Hayes, A. G., et al. 2017, MNRAS, 469, S50

Blum, J., Bischoff, D., & Gundlach, B. 2022, Univ, 8, 381

Blum, J., & Wurm, G. 2008, ARA&A, 46, 21

Bodewits, D., Farnham, T. L., Kelley, M. S. P., & Knight, M. M. 2018, Natur, 553, 186

Bottke, , William, F. J., Vokrouhlický, J., Rubincam, D. P., & Nesvorný, D. 2006, AREPS, 34, 157

Bottke, W. F., Marschall, R., Nesvorný, D., & Vokrouhlický, D. 2023a, SSRv, 219, 83

Bottke, W. F., Vokrouhlický, D., Marschall, R., et al. 2023b, PSJ, 4, 168

Brasser, R., Schwamb, M. E., Lykawka, P. S., & Gomes, R. S. 2012, MNRAS, 420, 3396

Brown, W. R., & Luu, J. X. 1998, Icar, 135, 415

Brunini, A. 2023, MNRAS, 524, L45

Chen, M., Huang, X., Yan, J., Lei, Z., & Barriot, J. P. 2023, Icar, 401, 115566

Chen, W., Han, D., & Han, D. 2007, Plasticity for Structural Engineers (Plantation, FL: J. Ross Publishing)

Chesley, S., Belton, M., Carcich, B., et al. 2013, Icar, 222, 516

Collins, G. S., Melosh, H. J., & Ivanov, B. A. 2004, M&PS, 39, 217

Collins, G. S., Melosh, H. J., & Wünnemann, K. 2011, IJIE, 38, 434

Combi, M., Shou, Y., Fougere, N., et al. 2020, Icar, 335, 113421

Davidsson, B. J. R. 2023, MNRAS, 521, 2484

Davidsson, B. J. R., Sierks, H., Güttler, C., et al. 2016, A&A, 592, A63

Davison, T., Collins, G., & Ciesla, F. 2010, Icar, 208, 468

Davison, T. M., Ciesla, F. J., & Collins, G. S. 2012, Geochim. Cosmochim. Acta, 95, 252

Davison, T. M., Ciesla, F. J., Collins, G. S., & Elbeshausen, D. 2014, MPS, 49, 2252

Di Sisto, R. P., & Brunini, A. 2007, Icar, 190, 224

Dobrovolskis, A. R. 1982, Icar, 52, 136

Dones, L., Brasser, R., Kaib, N., & Rickman, H. 2015, SSRv, 197, 191

Donn, B. 1989, IAU Colloq., 116, 335

Drahus, M., Jewitt, D., Guilbert-Lepoutre, A., et al. 2011, ApJL, 734, L4

El-Maarry, M. R., Groussin, O., Thomas, N., et al. 2017, Sci, 355, 1392
El-Maarry, M. R., Groussin, O., Keller, H. U., et al. 2019, SSRv, 215, 36
Faggi, S., Villanueva, G. L., McKay, A., et al. 2024, NatAs, 8, 1237
Farnham, T., & Thomas, P. 2013, Plate Shape Model of Comet 9P/Tempel 1 V2.0 DIF-C-HRIV/ITS/MRI-5-TEMPEL1-SHAPE-MODEL-V2.0, *NASA*
Fernández, J. A. 2020, in the Trans-Neptunian Solar System, ed. D. Prialnik, M. A. Barucci, & L. Young (Amsterdam: Elsevier) 1
Franceschi, M., Penasa, L., Massironi, M., et al. 2020, PNAS, 117, 10181
Golabek, G. J., & Jutzi, M. 2021, Icar, 363, 114437
Greenstreet, S., Gladman, B., & McKinnon, W. B. 2015, Icar, 258, 267
Gregory, T., Luu, T.-H., Coath, C. D., Russell, S. S., & Elliott, T. 2020, SciA, 6, 11
Grishin, E., Malamud, U., Perets, H. B., Wandel, O., & Schäfer, C. M. 2020, Natur, 580, 463
Grundy, W. M., Bird, M. K., Britt, D. T., et al. 2020, Sci, 367, eaay3705
Guilbert-Lepoutre, A., Lasue, J., Federico, C., et al. 2011, A&A, 529, A71
Güttler, C., Blum, J., Zsom, A., & Ormel, C. W. 2010, A&A, 513, A56
Harmon, J. K., Nolan, M. C., Howell, E. S., Giorgini, J. D., & Taylor, P. A. 2011, ApJL, 734, L2
Harrington Pinto, O., Womack, M., Fernandez, Y., & Bauer, J. 2022, PSJ, 3, 247
Heinisch, P., Auster, H. U., Gundlach, B., et al. 2019, A&A, 630, A2
Hirabayashi, M. 2015, MNRAS, 454, 2249
Hirabayashi, M. 2023, Icar, 389, 115258
Hirabayashi, M., Kim, Y., & Brozović, M. 2021, Icar, 365, 114493
Hirabayashi, M., & Scheeres, D. J. 2013, ApJ, 780, 160
Hirabayashi, M., & Scheeres, D. J. 2019, Icar, 317, 354
Hirabayashi, M., Trowbridge, A. J., & Bodewits, D. 2020a, ApJL, 891, L12
Hirabayashi, M., Scheeres, D. J., Chesley, S. R., et al. 2016, Natur, 534, 352
Hirabayashi, M., Nakano, R., Tatsumi, E., et al. 2020b, Icar, 352, 113946
Hirabayashi, M., Ferrari, F., Jutzi, M., et al. 2022, PSJ, 3, 140
Holsapple, K. 2001, Icar, 154, 432
Holsapple, K. A. 2004, Icar, 172, 272
Holsapple, K. A. 2010, Icar, 205, 430
Holsapple, K. A., & Michel, P. 2006, Icar, 183, 331
Hu, X., Gundlach, B., von Borstel, I., Blum, J., & Shi, X. 2019, A&A, 630, A5
Hu, X., Shi, X., Sierks, H., et al. 2017, A&A, 604, A114
Jellinek, H. H. G. 1959, JGlac, 3, 345
Jewitt, D. 2009, AJ, 137, 4296
Jewitt, D. 2015, AJ, 150, 201
Jewitt, D. C., & Sheppard, S. S. 2002, AJ, 123, 2110
Johansen, A., Oishi, J. S., Low, M.-M. M., et al. 2007, Natur, 448, 1022
Jorda, L., Gaskell, R., Capanna, C., et al. 2016, Icar, 277, 257
Jutzi, M., & Benz, W. 2017, A&A, 597, A62
Jutzi, M., Benz, W., Toliou, A., Morbidelli, A., & Brasser, R. 2017, A&A, 597, A61
Kataoka, A., Tanaka, H., Okuzumi, S., & Wada, K. 2013, A&A, 557, L4
Keane, J. T., Porter, S. B., Beyer, R. A., et al. 2022, JGRE, 127, e2021JE007068
Keller, H. U., Mottola, S., Skorov, Y., & Jorda, L. 2015, A&A, 579, L5
Kenyon, S. J., & Bromley, B. C. 2004, AJ, 127, 513
Kim, Y., & Hirabayashi, M. 2022, DPS Meetings, (Vol. 54,; Washington, DC: AAS) 410.01

Kramer, T., Läuter, M., Hviid, S., et al. 2019, A&A, 630, A3
Leon-Dasi, M., Besse, S., Grieger, B., & Küppers, M. 2021, A&A, 652, A52
Levison, H. F., Bottke, W. F., Gounelle, M., et al. 2009, Natur, 460, 364
Levison, H. F., Olkin, C. B., Noll, K. S., et al. 2021, PSJ, 2, 171
Lichtenberg, T., Golabek, G. J., Gerya, T. V., & Meyer, M. R. 2016, Icar, 274, 350
Lisse, C., Bauer, J., Cruikshank, D., et al. 2020, NatAs, 4, 930
Love, A. E. H. 2011, A Treatise on the Mathematical Theory of Elasticity (4th edn; New York: Dover)
Luu, J. X., & Jewitt, D. C. 2002, ARA&A, 40, 63
Lyra, W., Youdin, A. N., & Johansen, A. 2021, Icar, 356, 113831
Mao, X., McKinnon, W. B., Singer, K. N., et al. 2021, JGRE, 126, e2021JE006961
Marohnic, J. C., Richardson, D. C., McKinnon, W. B., et al. 2021, Icar, 356, 113824
Marschall, R., Skorov, Y., Zakharov, V., et al. 2020, SSRv, 216, 130
Massironi, M., Simioni, E., Marzari, F., et al. 2015, Natur, 526, 402
McKinnon, W. B., Richardson, D. C., Marohnic, J. C., et al. 2020, Sci, 367, eaay6620
McKinnon, W. B., Mao, X., Schenk, P. M., et al. 2022, GeoRL, 49, e2022GL098406
Meech, K. J., A'Hearn, M. F., Adams, J. A., et al. 2011, ApJL, 734, L1
Mellor, M. 1974, A review of basic snow mechanics (Hanover, NH: US Army Cold Regions Research and Engineering Laboratory)
Merk, R., & Prialnik, D. 2003, EM&P, 92, 359
Morbidelli, A., Baillié, K., Batygin, K., et al. 2022, NatAs, 6, 72
Morbidelli, A., & Nesvorný, D. 2020, in The Trans-Neptunian Solar System, ed. D. Prialnik, M. A. Barucci, & L. Young (Amsterdam: Elsevier) 25
Morbidelli, A., & Rickman, H. 2015, A&A, 583, A43
Mottola, S., Lowry, S., Snodgrass, C., et al. 2014, A&A, 569, L2
Mousis, O., Drouard, A., Vernazza, P., et al. 2017, ApJL, 839, L4
Musiolik, G., & Wurm, G. 2019, ApJ, 873, 58
Nakano, R., & Hirabayashi, M. 2020, ApJL, 892, L22
Nesvorný, D., Parker, J., & Vokrouhlický, D. 2018, AJ, 155, 246
Nesvorný, D., & Vokrouhlický, D. 2019, Icar, 331, 49
Nesvorný, D., Vokrouhlický, D., Dones, L., et al. 2017, ApJ, 845, 27
Nesvorný, D., Vokrouhlický, D., & Fraser, W. C. 2022, AJ, 163, 137
Nesvorný, D., Vokrouhlický, D., & Morbidelli, A. 2013, ApJ, 768, 45
O'Neill, C., O'Neill, H. S. C., & Jellinek, A. M. 2020, SSRv, 216, 37
O'Rourke, L., Heinisch, P., Blum, J., et al. 2020, Natur, 586, 697
Pätzold, M., Andert, T., Hahn, M., et al. 2016, Natur, 530, 63
Petrovic, J. J. 2003, JMatS, 38, 1
Porter, S. B., Buie, M. W., Parker, A. H., et al. 2018, AJ, 156, 20
Preusker, F., Scholten, F., Matz, K. D., et al. 2015, A&A, 583, A33
Preusker, F., Scholten, F., Matz, K. D., et al. 2017, A&A, 607, L1
Prialnik, D., Bar-Nun, A., & Podolak, M. 1987, ApJ, 319, 993
Prialnik, D., & Podolak, M. 1995, Icar, 117, 420
Pätzold, M., Andert, T. P., Hahn, M., et al. 2018, MNRAS, 483, 2337
Rickman, H., Marchi, S., A'Hearn, M. F., et al. 2015, A&A, 583, A44
Rubin, M., Altwegg, K., Balsiger, H., et al. 2015, Sci, 348, 232
Safrit, T. K., Steckloff, J. K., Bosh, A. S., et al. 2021, PSJ, 2, 14
Sarid, G., & Prialnik, D. 2009, Meteoritics & Planetary Sci, 44, 1905

Sarid, G., Volk, K., Steckloff, J. K., et al. 2019, ApJL, 883, L25
Schenk, P., Singer, K., Beyer, R., et al. 2021, Icar, 356, 113834
Schulz, R. 2002, A&ARv, 11, 1
Schwartz, S. R., Michel, P., Jutzi, M., et al. 2018, NatAs, 2, 379
Sharma, I. 2013, Icar, 223, 367
Sharma, I., Jenkins, J. T., & Burns, J. A. 2009, Icar, 200, 304
Shimaki, Y., & Arakawa, M. 2021, Icar, 369, 114646
Sierks, H., Barbieri, C., Lamy, P. L., et al. 2015, Sci, 347, aaa1044
Skorov, Y., & Blum, J. 2012, Icar, 221, 1
Souchay, J., Lhotka, C., Heron, G., et al. 2018, A&A, 617, A74
Spencer, J. R., Stern, S. A., Moore, J. M., et al. 2020, Sci, 367, eaay3999
Steckloff, J. K., & Jacobson, S. A. 2016, Icar, 264, 160
Steckloff, J. K., Graves, K., Hirabayashi, M., Melosh, H. J., & Richardson, J. E. 2016, Icar, 272, 60
Steckloff, J. K., Lisse, C. M., Safrit, T. K., et al. 2021, Icar, 356, 113998
Steckloff, J. K., Sarid, G., & Johnson, B. C. 2023, PSJ, 4, 4
Stern, S. A., Keeney, B., Singer, K. N., et al. 2021, PSJ, 2, 87
Stern, S. A., & Trafton, L. M. 2008, in The Solar System Beyond Neptune, ed. M. A. Barucci, et al. (Tucson, AZ: Univ. Arizona Press) 365
Stern, S. A., Weaver, H. A., Spencer, J. R., et al. 2019, Sci, 364, eaaw9771
Stern, S. A., White, O. L., Grundy, W. M., et al. 2023, PSJ, 4, 176
Sunshine, J. M., & Feaga, L. M. 2021, PSJ, 2, 92
Takir, D., Neumann, W., Raymond, S. N., Emery, J. P., & Trieloff, M. 2023, NatAs, 7, 524
Thomas, N., Sierks, H., Barbieri, C., et al. 2015, Sci, 347, aaa0440
Thomas, P., A'Hearn, M., Belton, M., et al. 2013, Icar, 222, 453
Thomas, P. C., Veverka, J., Belton, M. J., et al. 2007, Icar, 187, 4
Tiscareno, M. S., & Malhotra, R. 2003, AJ, 126, 3122
Tozer, D. C., Blackett, P. M. S., Bullard, E., & Runcorn, S. K. 1965, RSPTA, 258, 252
Veverka, J., Klaasen, K., A'Hearn, M., et al. 2013, Icar, 222, 424
Wada, K., Tanaka, H., Suyama, T., Kimura, H., & Yamamoto, T. 2008, ApJ, 677, 1296
Wada, K., Tanaka, H., Suyama, T., Kimura, H., & Yamamoto, T. 2009, ApJ, 702, 1490
Weaver, H. A., Feldman, P. D., A'Hearn, M. F., Russo, N. D., & Stern, S. A. 2011, ApJL, 734, L5
Weidenschilling, S. 1997, Icar, 127, 290
Weidenschilling, S., Spaute, D., Davis, D., Marzari, F., & Ohtsuki, K. 1997, Icar, 128, 429
Weidenschilling, S. J. 1977, MNRAS, 180, 57
Weidenschilling, S. J. 1994, Natur, 368, 721
Weissman, P., Morbidelli, A., Davidsson, B., & Blum, J. 2020, SSRv, 216, 6
Weissman, P. R. 1986, Natur, 320, 242
Wünnemann, K., Collins, G. S., & Melosh, H. J. 2006, Icar, 180, 514
Youdin, A. N., & Goodman, J. 2005, ApJ, 620, 459
Zhao, Y., Rezac, L., Skorov, Y., et al. 2021, NatAs, 5, 139
Zsom, A., Ormel, C. W., Güttler, C., & Blum, J. 2010, A&A, 513, A57
Ćuk, M., & Burns, J. A. 2005, Icarus, 176, 418

AAS | IOP Astronomy

Centaurs

Kathryn Volk, Maria Womack and Jordan Steckloff

Chapter 13

Laboratory Studies Applicable to Centaurs

Julie Brisset, Jürgen Blum, Elsa Hénault, Mark Burchell, Will Grundy, Murthy S Gudipati and Rosario Brunetto

In this chapter, we review current laboratory methods and outcomes for the study of ices and icy regoliths, which are applicable to our understanding of Centaur formation and evolution. We first present work on icy body formation, in particular how icy grains grow into larger aggregates and planetesimals. We show how the need for realistic simulant materials drove new techniques in the production of icy dust mixtures, which are used in the laboratory to study the early stages of icy body growth. We then dive into the thermal history of small icy bodies, such as Centaurs, and present the laboratory techniques associated with studying ice phase change dynamics and energies. We describe the current state of the art in our understanding of amorphous and crystalline states and transitions as well as the energy exchanges associated; phase transitions for several major ice types; and the behavior of ices composed of mixtures of more than one type of molecule, in particular looking at outgassing profiles. We also delve into laboratory spectroscopy of ices and ice mixtures, the various methodologies employed and how this work informs us on Centaur formation and evolution. Further, we describe work on the chemistry at the surface of ices, showing what instrumentation is used to study the influence of space weathering on the spectral signatures we observe on Centaurs among others. Another surface modification process of importance is cratering, and we then move on to describing how experimental work on impacts into ices has provided a wealth of insights into the cratering processes on icy bodies. Finally, we present work on the strength of icy granular materials and how surface strength alterations could be a mechanism for regular surface activity on rotating Centaurs.

13.1 Introduction

Now that our observational capabilities are ever increasingly enabled by new facilities, such as the NASA James Webb Space Telescope (JWST), the information we receive about distant small bodies of the Solar System is rapidly increasing.

doi:10.1088/2514-3433/ada267ch13

Centaurs, which allow us to link the trans-Neptunian Object (TNO) population to the Jupiter-family comets (JFCs; Sarid et al. 2019; Wood & Hinse 2022; Fraser et al. 2024), are being studied in more and more detail (Santos-Sanz et al. 2015; Lisse et al. 2020), and potentially even considered as mission targets (Howell et al. 2018; Singer et al. 2021) see Chapter 16. Lightcurves and occultation campaigns provide clues on size, rotational rate, and possible rings or satellite companions (Marton et al. 2020; Braga-Ribas et al. 2014). Spectra set constraints on composition and temperature (Barkume et al. 2008; Harrington Pinto et al. 2023; Faggi et al. 2024).

This increase in available information has resulted in a surging interest in laboratory data needed to support our efforts to understand Centaurs, their structure and composition, formation and evolution, and ultimately the history of our solar system. Indeed, the more details we are able to extract from telescope observation data the more we need to simulate materials and physical conditions experienced in and around Centaurs in the laboratory in order to decipher these observations. For example, new spectral information recently revealed the presence of icy silicate mixtures on TNOs and Centaurs (Cook et al. 2023; De Prá et al. 2025; Grundy et al. 2024a; Souza-Feliciano et al. 2024). Despite the great support offered by numerical simulations, only laboratory measurements can provide ground truth on the exact compounds and mixture compositions (Brunetto et al. 2006; de Bergh et al. 2008; Ruf et al. 2019). While laboratory efforts specifically focused on Centaurs (de Bergh et al. 2013) remain scarce, there are a number of studies aimed at other small bodies of the Solar System (active asteroids, comets, icy moons, etc.) that are relevant (e.g., Gundlach et al. 2011; Brisset et al. 2022a). Studies of water and other ices, the chemistry in and on them, and their thermophysical and structural behavior are of interest when investigating Centaurs.

Water ice is one of the most abundant materials in the outer Solar System. It is also the most conveniently, and therefore best, studied ice in the laboratory (Gundlach et al. 2011; Burchell 2013; Brisset et al. 2022a, 2022b; Kreuzig et al. 2023). However, with its phases of matter and other physical properties such as granularity, porosity, and heterogeneity (pure ice regions as well as ice and silicate mixtures), and any number of combinations of these states, a number of open questions remain to be answered before we can fully understand water ice in distant small bodies. Current areas of investigation include sublimation rates and thermal properties at low pressure and temperature, heat release during crystallization, and behavior of icy impactors and targets during impacts. Adding the presence of other ices (pure or mixed), such as CO, CO_2, and CH_4 (de Bergh et al. 2013), just to name a few, and the complexity of the chemistry happening on icy surfaces as Centaurs move further into the inner Solar System and get exposed to higher levels of radiation and temperature from the Sun (Hudson & Moore 2001; Baragiola 2003; Garrod 2019), the untangling of measured spectra still heavily relies on laboratory studies and contributions (Gudipati & Castillo-Rogez 2013). Laboratory studies towards understanding comets (closely related to Centaurs) have also been reviewed several years ago (Gudipati et al. 2015). In addition, unique Centaur features, such as distant surface activity (Jewitt 2009; Chandler et al. 2020) and the presence of rings (Braga-Ribas et al. 2014; Ortiz et al. 2015), also trigger questions on the mechanisms associated with bodies' overall structure, strength, and impacts, that can only be addressed in laboratory-calibrated (and scaled) simulations.

In this chapter, we are presenting the current state-of-the-art of laboratory studies that are relevant to the investigation of Centaurs. We start in Section 13.2 with works on the formation and structure of small icy bodies in the outer solar system. In Section 13.3, we take a look at ices in and on the surface of small bodies traveling closer to the Sun, followed by studies of chemistry on icy surfaces in Section 13.4. In Section 13.5, we present our current laboratory knowledge of impacts on icy surfaces. Finally, in Section 13.6, we address the question of body stability, surface activity, and rings around Centaurs and how these might be related. In our conclusion, Section 13.7, we summarize this overview and provide some insights on the laboratory work ahead to keep contributing to a better understanding of Centaurs.

13.2 Small Icy Body Formation and Structure

To fully understand present-day Centaurs, it is essential to take a holistic view on their formation in the early Solar System and their evolution over most of the ~4.6 Gyrs in the Kuiper Belt before they were implanted into their present orbits, as well as on their potential future as Jupiter-family comets. In this section, we will briefly review our current understanding of these phases and the contributions of laboratory works.

13.2.1 Collision and Growth: Formation of Planetesimals

Centaurs formed in the trans-Neptunian region of the outer Solar System (Di Sisto & Rossignoli 2020; see also Chapter 3). The current picture of the formation of planetesimals there has recently been reviewed many times (Birnstiel et al. 2016; Weissman et al. 2020; Blum et al. 2022; Simon et al. 2024; see also Chapter 2). It is based on the same processes as the formation of planetesimals in the inner protoplanetary disk (PPD), with the main difference being the presence of ices (in particular water ice) as constituent material. In summary, (sub-)μm-sized solid particles undergo sticking collisions at low speeds in the solar nebula, which leads to the formation of mm- to cm-sized agglomerates ("pebbles") that no longer stick to one another but bounce off ("bouncing barrier") or even break apart ("fragmentation barrier") upon mutual impacts. After spatial pre-concentration of the ensemble of pebbles in close orbits, due to, e.g., sedimentation or pressure bumps, the streaming instability (Youdin & Goodman 2005) concentrated the pebble cloud further until a gentle gravitational collapse formed up to ~1000 km-sized planetesimals (Johansen & Youdin 2007; Polak & Klahr 2023). This growth process forms planetesimals directly from mm to cm-sized aggregates and allows for the formation of large cores (>100 km) in the Kuiper Belt. Models have shown that such local aggregate concentration processes taking place in the outer solar system could leave 10%–25% of the original mm-sized aggregates not integrated in large bodies (Davidsson et al. 2016). The surface density of these remaining aggregates is too low for them to trigger streaming instabilities, so that binary collisions remain the main growth process for this population. At the same time, the disk conditions after the formation of large KBOs are such that remaining planetesimals experience a very different collisional environment than before large KBO formation, with average relative speeds below a few 10 m s^{-1} and a reduced collision rate. Alternatively, small KBOs could be formed in individual clouds (Lorek et al. 2016), in which numerous dissipative collisions between aggregates would reduce the relative velocities below 10 m s^{-1}. In both formation scenarios, icy aggregates are less likely to reach the bouncing and fragmentation barriers during collisions, and growth through gentle collisions becomes relevant again. The physical properties of small planetesimals formed by such processes matches those found in JFCs (Blum et al. 2017, 2022; Blum 2018), which are evolved Centaurs and still hold clues on their formation.

The contribution of laboratory work to the above planetesimal-formation scenario is manifold. A large body of work on (refractory) dust collisions and agglomeration went into the picture that pebbles are the largest possible collisionally grown solid objects (Güttler et al. 2010; Zsom et al. 2010). The extension to icy particles was initiated by Gundlach & Blum (2015) who showed that μm-sized crystalline water-ice particles stick at the low temperatures present in the solar nebula beyond the snowline at about tenfold higher collision velocities than their silica counterparts, which are used as the proxies for the silicates in the inner parts of the disk. Following this work, Lorek et al. (2018) showed that icy pebbles of the sizes relevant for the streaming instability to occur can indeed form beyond the snowline. Because Centaurs exist in a temperature range of $<$50 K to $\sim$110 K, water-ice phases – amorphous or crystalline – and trapping of other volatiles in amorphous water-ice play important roles in the formation and evolution of Centaurs over the past 4 billion yr.

Since 2011, laboratory studies have revealed the specific surface energy and the corresponding cohesion force (Gundlach et al. 2011), the inner structure of shock-frozen ice microspheres (Gärtner et al. 2017), the tensile strength of micro-granular dust-ice mixtures (Gundlach et al. 2018b; Haack et al. 2020), the mechanical properties of ice in cryogenic conditions (Henderson et al. 2019; Potter et al. 2020; Choukroun et al. 2020), the collision properties of ice agglomerates (Schräpler et al. 2022), optical properties of ice-dust samples (Jost et al. 2013, 2016; Pommerol et al. 2019), the sintering behavior of small ice grains (Gundlach et al. 2018a; Molaro et al. 2019) and the production and handling of large quantities of micro-granular water ice (Kreuzig et al. 2023).

In addition, Lowers et al. (2023) started investigating the behavior of icy-dust granular mixtures with irregular grain shapes in gentle collisions. To this purpose, spherical mm- and cm-sized aggregates are prepared in cryogenic conditions ($<$ 150 K) from $\sim$100 μm asteroid regolith simulant (Metzger et al. 2019) and water ice grains. Similarly to the investigation of dry dust aggregates (e.g., Beitz et al. 2011) and pure ice aggregates (Schräpler et al. 2022), a drop tower is used to produce gentle collisions at speeds ranging from a few 10 cm s^{-1} to a few m s^{-1} (Figure 13.1). By systematically measuring collision characteristics, such as the coefficient of restitution upon bouncing, or the onset of sticking, laboratory work provides input

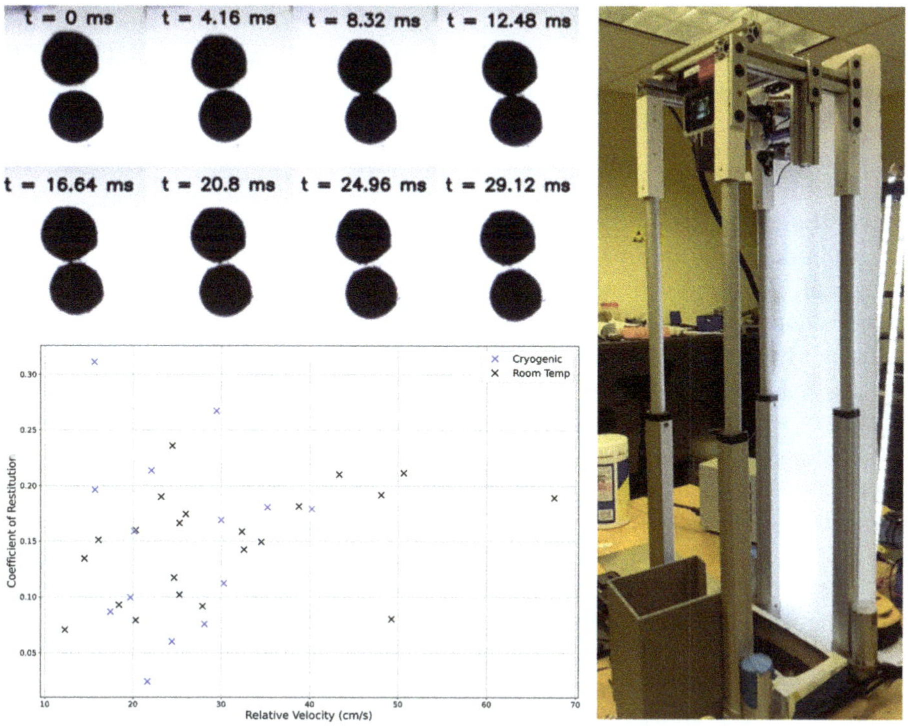

Figure 13.1. Icy aggregate collisions study in the laboratory. (Top left) Time sequence of a collision between 2 cm-sized aggregates composed of 0.1 mm grains; (Bottom left) Coefficient of restitution (COR) measured on bouncing collisions between 2 cm-sized aggregates at both room temperature (black) and cryogenic (blue) temperatures; (Right) Laboratory drop tower to perform icy dust collisions in the Florida Space Institute Laboratory at UCF. This data set shows that cryogenic aggregates have the same collision behavior as room-temperature dust aggregates. This indicates that we can transfer our knowledge on room-temperature aggregate behavior to outer Solar System environments. The COR graph also shows that, while the average coefficient of restitution of 2 cm-sized aggregates remains constant with collision velocity, the absolute deviation is higher at lower speeds. This behavior was also observed for mm-sized particles in Brisset et al. (2019).

parameters for numerical simulation work that can scale tens of orders of magnitude.

13.2.2 Laboratory Studies of Analogs of Primitive Icy Bodies in the Solar System

After having formed through a gentle gravitational collapse, icy planetesimals are internally heated through the decay of short-lived radionuclides, such as, e.g., ^{26}Al and ^{60}Fe, with half-lives on the order of a million years. For planetesimals with sizes below $\sim$100 km, the pebbles stay intact and form planetesimals with extremely low thermal conductivity, due to the small contact area per unit cross section (Bischoff et al. 2021). Thus, the heat generated by the radioactive decay cannot escape from the bulk of the planetesimals and may lead to an internal differentiation of the volatiles, depending on formation time, mineral abundance and planetesimal size, such that the ices diffuse outwards and condense in a shallow subsurface region with the most volatile ice (CO) enriched in the outermost and the least volatile ice (H_2O) in the innermost layer (Malamud et al. 2022). First indications that this model may be valid were recently published by Robinson et al. (2024) who confirmed by observational data the predicted correlation between the size of a comet and its CO/H_2O ratio in the coma. Therefore, the formation and early evolution of planetesimals leading to Centaurs and, at a later stage, to Jupiter-family comets, suggests that these bodies still mainly consist of the microscopic icy grains they formed from. Thus, it is of great importance to study, in the laboratory, the physics and chemistry of samples consisting of μm-sized ice particles.

Although the method for the generation of the micro-granular water-ice samples has developed since the first experiments in the early 2010s, a common feature is the formation of a mist of water micro-droplets, which then is guided into a cryo-liquid in which the droplets instantaneously freeze to micro-spheres. In the past, liquid nitrogen (lN_2) has been used as a cooling agent, which is a boiling cryo-liquid so that the Leidenfrost effect causes a rather bad thermal contact between droplet and cryo-liquid, leading to the formation of mostly *crystalline* water-ice spheres (Gärtner et al. 2017). The production of micro-granular *amorphous* water-ice particles requires a much higher cooling rate, which possibly can be provided by a non-boiling cryo-liquid, such as liquid methane or ethane, but an experimental proof of this concept for micro-particles is still missing. Kreuzig et al. (2023) have optimized the production, storage and handling process such that sample masses of up to ~1 kg can easily be treated. Mean particle radii are in the 2–3 μm range and the size distribution is relatively narrow (see Figure 13.2), with the central 80% of the particles being in the ~1.5–3.5 μm radius range and the central 80% of the sample mass in particles with ~2.2–4.5 μm radius (Kreuzig et al. 2023). Kreuzig et al. (2023) also showed how homogeneous and quantitative mixtures of water-ice grains with other micro-granular particles can be achieved.

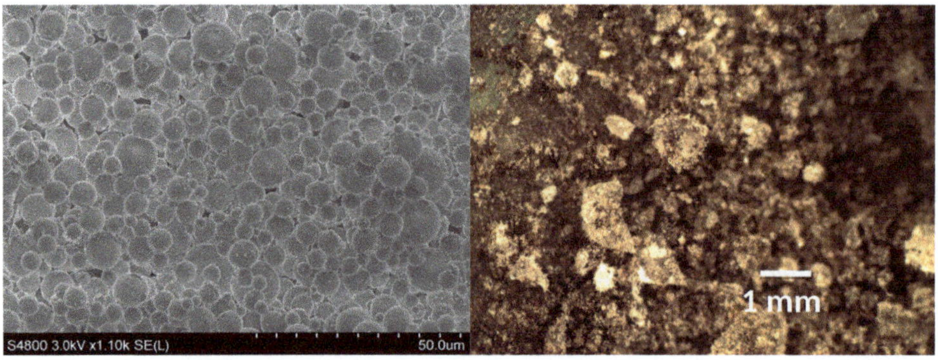

Figure 13.2. Icy laboratory samples. (Left) Cryo-SEM picture of water ice particles used for experimentation in the laboratory (image from the experiments reported in Kreuzig et al. 2023); (Right) Microscope picture of 1:3 dust to ice ratio sample with a grain size distribution centered around 0.5 mm. Reproduced from Brisset et al. (2022b). © 2022. The Author(s). Published by the American Astronomical Society.

For experiments relating to more gardened mixtures of ice and regolith, Brisset et al. (2022a) and Lowers et al. (2023) prepared irregular water ice grains by spraying liquid water onto the cold metal surface of a mortar (<150 K using liquid nitrogen) and crushing it into fine grains using a cooled pestle. Using a cold sieve, the ice grains can be size sorted. These ice grains are then mixed with irregular shaped asteroid simulant grains of similar sizes.

With these methods, laboratory experiments with a large variation of dust-to-ice mass ratios can be performed, even with very large sample quantities exceeding the kg limit. In the past few years, the first systematic studies with micro-granular ice and dust-ice samples were under way to understand the physics behind cometary activity. Haack et al. (2021a, 2021b) and Kreuzig et al. (2024) investigated the mechanical stability of outgassing and outgassed dust-ice mixtures with various dust-to-ice ratios. Several ongoing investigations are currently concentrating on understanding under which conditions (level of insolation, dust-to-ice ratio, packing density) micro-granular icy samples show comet-like activity, i.e. the emission of water molecules as well as of solid particles. Many of these experiments are being performed in the Comet Physics Laboratory (CoPhyLab) at TU Braunschweig (Kreuzig et al. 2021). One of the stunning results is that a pure micro-granular water-ice sample (as shown in Figure 13.2) shows signs of "dust" activity when illuminated (Molinski et al. 2022). Obviously, the absorbed irradiation produces sufficient gas flow to also eject small (typically 100 μm in size) clumps of water-ice particles.

13.3 Ices in and on Small Bodies

13.3.1 Thermal History of Small Bodies

Kuiper-Belt Objects (KBOs), Centaurs, and Jupiter-Family Comets (JFCs) have their surface and bulk nuclei composition evolve based on their size, orbit eccentricity, heliocentric distances of their perihelion and aphelion, and time spent in these orbits. These parameters determine how their surface evolves thermally in addition to the effect of solar wind and other radiation, as well as how their thermal equilibrium propagates into the nucleus over time. Owing to their large distances from the Sun, KBOs have low average equilibrium day-side surface temperatures around 30 K–60 K (e.g., De Sanctis et al. 2001; Stansberry et al. 2008; Moullet et al. 2011; Umurhan et al. 2022). Ambient day-side temperatures of Centaurs can get much warmer, with a typical range around 70 K to 110 K (e.g., Duffard et al. 2014). JFCs experience even higher day-side surface temperatures up to 350 K. Night-side temperatures can fall to as low as ~30 K on these bodies, and their winter hemispheres can get even colder. These extreme temperature variations cause Centaurs and comets to be highly active and thermally processed, with volatiles mobilized in warm regions diffusing through their interiors and eventually escaping to space or becoming trapped in colder regions (e.g., Prialnik et al. 2008; De Sanctis et al. 2010; Birch & Umurhan 2024).

Though water is known to be the major constituent of ice (based on comet outgassing data), detection of water ice on the surface of a comet happens only under a few special circumstances such as refrosting or cliff-collapse exposing fresh interior ice. Otherwise, cometary surfaces are composed of low-albedo refractory material (silicate and organic dust). Similarly, a few studies dedicated to detecting water-ice on the surface of Centaurs report mixed results – some show a direct detection of water ice and others only allow for the deduction of the presence of water ice through indirect spectral mixing with other, refractory materials (Brown et al. 1998; Romon-Martin et al. 2002; Dotto et al. 2003; Szabò et al. 2018).

If the origin of small icy bodies in our solar system were to be the same during the protoplanetary phase, all these bodies should have initially been made of significant amounts of other volatiles such as CO, O_2, CH_4, NH_3, CO_2, Ar, etc., with H_2O being the major volatile, and the non-volatile refractory materials being complex organics (potentially including molecules important for life) as well as silicate dust grains. Outgassing of long-period Oort-Cloud comets indeed shows significant amounts of CO and other highly volatile molecules, whereas short-period JFC comets are relatively depleted in these highly volatile molecules (Dello Russo et al. 2016). This is thought to be due to the loss of these volatiles as the higher equilibrium temperatures reach into the interior of cometary nucleus over their history. Similarly, Centaurs with an average lifetime of several million years in their orbit, irrespective of where they originated 4.6 billion years ago, would undergo thermal equilibration of the interior. However, modeling this equilibration is challenging both for comets and Centaurs due to their high-porosity ($\sim$70%) and grain-size distribution. Some models predict thermal equilibration and even conversion of amorphous water-ice to crystalline water-ice within a few million years (Guilbert-Lepoutre 2012), while others think the interior of a cometary nucleus could still be at $\sim$40–50 K trapping highly volatile molecules (Birch & Umurhan 2024).

Significant work is needed in understanding how Centaurs thermally evolve, their surface to interior thermal gradients, and how they correlate with comets. In particular, we need new laboratory data on thermal conductivity of porous amorphous and crystalline ice with different porosities and grain-sizes as well as mixed-ices such as $H_2O + CO + CO_2 + NH_3 + CH_4$ etc., with appropriate mixing ratios. We also need sublimation curves, binding energies of these molecules in homogeneous and heterogeneous ice forms. Finally, rigorous 3D models may need to be developed to use the new experimental data to accurately predict local temperatures, outgassing coefficients of highly volatile molecules with depth of a nucleus (Centaur or comet). Only then would it be possible to understand and predict how Centaurs and comets evolve over a period of time, why they undergo outbursts, and whether or not nuclei of Centaurs and comets are thermally processed.

13.3.2 Ice Phase Change Dynamics and Energies

Since Centaurs are objects in transition between the extremely low temperatures prevalent in the Kuiper belt and much warmer environs closer to the Sun, effects of heating on their materials are key to understanding their physical and chemical evolution. Phase changes are especially important as they enable migration and loss of material via sublimation from solid to vapor, and also expulsion of molecules trapped in amorphous ice when it crystallizes. Laboratory studies of these processes are essential, though they can be challenging at the low temperatures and pressures and long timescales relevant for Centaurs.

Equilibrium vapor pressure, the pressure where sublimation and condensation are balanced, is a key temperature-dependent property of materials that can be used to calculate their rate of sublimation loss into vacuum via the Hertz-Knudsen-Langmuir equation (e.g., Langmuir 1913). At temperatures and pressures sufficiently high for the gas to be collisional, vapor pressure can be measured using a pressure gauge on an isothermal chamber in which solid and vapor exist in equilibrium. At more relevant lower temperatures and pressures where the gas is non-collisional, pressure is not equalized throughout the apparatus by collisions, so other methods are used, such as measuring sublimation loss from a thin film of ice deposited onto a quartz crystal microbalance (e.g., Sack & Baragiola 1993; Allodi et al. 2013; Luna et al. 2014; Grundy et al. 2024b; Blakley et al. 2024). Such a system is illustrated in Figure 13.3. Vapor pressures are measured by freezing gas onto the quartz crystal at low temperature, and then, after warming to the temperature of interest, using the change in vibrational frequency of the quartz crystal to compute the rate of mass loss from sublimation into the surrounding vacuum.

A compendium of relevant vapor pressure measurements has been published by Fray & Schmitt (2009). For substances where measurements are not available at appropriate temperatures, it is possible to extrapolate using the Clausius-Clapeyron equation and the assumption that the latent heat of vaporization does not vary as a function of temperature, though for some materials, this extrapolation must be extended over many orders of magnitude in pressure to reach conditions relevant for Centaurs.

Another class of phase changes that are potentially important for Centaur evolution is between distinct solid states. For instance, crystallization of amorphous

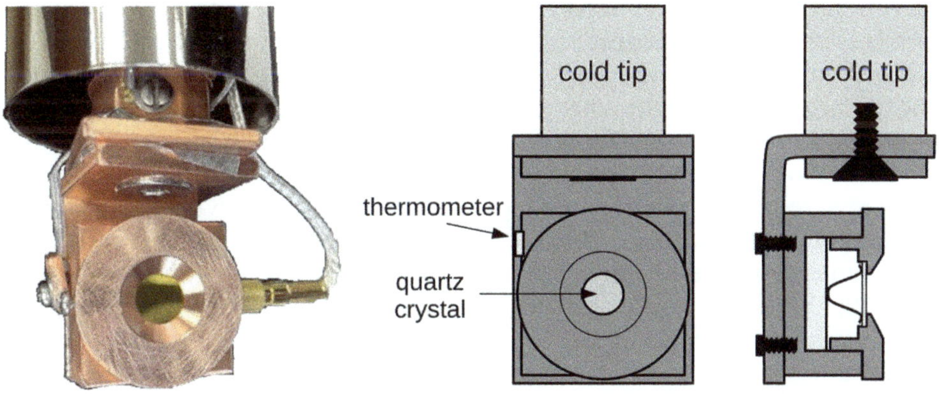

Figure 13.3. Laboratory system for measurement of vapor pressures at low temperatures and pressures: photographic view at left, schematic in center, and cross section at right. The quartz crystal microbalance is mounted on the cold tip of a closed cycle helium refrigerator using copper components to minimize temperature differences between the thermometer and the quartz crystal. Reprinted from Grundy et al. (2024b), Copyright 2024, with permission from Elsevier.

water ice can expel volatiles that had been trapped in the ice. The crystallization rate follows an Arrhenius type relation with an exponential dependence on the negative of an activation energy divided by temperature (e.g., Kouchi et al. 1994; Jenniskens & Blake 1996; Mastrapa et al. 2013). At warmer temperatures, crystallization proceeds much more rapidly. Again, substantial extrapolation is needed to get from laboratory timescales to Centaur orbital evolution timescales. On further warming, water ice recrystallizes from a cubic to hexagonal crystal structure (although the cubic phase of water ice has recently been challenged by Salzmann et al. 2024). As with the amorphous-crystalline transition, this is an irreversible phase change, in that cooling of hexagonal water ice does not cause it to revert to cubic or amorphous ice. Non-water ices also show irreversible amorphous to crystalline phase changes (Gerakines et al. 2023) but some potentially relevant solid-solid phase changes in these ices are reversible, such as occur in ices of CO, CH_4 and N_2 (e.g., Clayton & Giauque 1932; Bol'shutkin et al. 1971; Scott 1976). In some materials, such as CH_3OH, the phase transition on cooling may not occur immediately on laboratory timescales (Carlson & Westrum 1971).

In ices composed of mixtures of more than one type of molecule, phase changes become more complicated: phase transitions switch from occurring at a specific temperature to displaying bounded regions on a binary phase diagram where two or more phases coexist in equilibrium. Non-ideal mixing behavior occurs when the constituent molecules of the mixture have distinct affinities between like-like or between unlike pairings of molecules. Analogous to distillation, sublimation from a mixed ice tends to remove the more volatile constituent (Tan 2022), producing a change of composition at the surface of the ice. Molecular diffusion through the ice can eventually supply more of the volatile constituent to the surface, but at rates distinct from what would be predicted based on solid-vapor equilibrium thermodynamics for the bulk composition (e.g., Trafton 2015; Tan & Kargel 2018). The combined kinetics and thermodynamics of sublimation from mixed ices has received relatively little attention in laboratory studies, but is likely to be important for Centaur evolution. Ice mixtures also have the potential to form additional solid phases, beyond those of the pure species. These include various ammonia-water hydrate phases (e.g., Rupert 1909) and clathrate hydrates where volatile molecules are trapped in cage-like structures of water molecules, with important implications for retention of volatiles in warmer conditions than would be expected from their vapor pressures as pure substances (e.g., Blake et al. 1991; Devlin 2001; Marboeuf et al. 2012; Choukroun et al. 2013). Low temperature co-crystals between various volatile molecules have recently become an area of active study for Titan (Kirchner et al. 2010; Cable et al. 2014), and such materials could potentially also be important for volatile retention in Centaurs.

Laboratory studies of solid-state phase changes employ a variety of methods. Phase changes can often be detected in absorption or Raman spectra, typically appearing as small shifts or changes in the shapes of the vibrational bands (e.g., Vetter et al. 2007; Raposa et al. 2024). Determining the crystal structure of an ice is usually done using X-ray or neutron diffraction studies (e.g., Bertie & Shehata 1984; Maynard-Casely et al. 2013, 2020).

With Centaurs experiencing ambient temperatures varying significantly between perihelion and aphelion, a thermal wave propagates from the surface into the interior changing the equilibrium temperature of the interior, with a thermal gradient from the surface to the center. Recent laboratory work (Gudipati et al. 2023) provides some quantitative insight into how ice could evolve at various equilibrium temperatures (Figure 13.4).

Around 40 K, both CO and O_2 molecular ices outgas with a small fraction of CO_2 that is trapped in these ices. Unless the nucleus of the small body in consideration remains under equilibrium temperature to be below 40 K, it is likely that such an icy body loses highly volatile molecules such as CO, O_2, and CH_4, N_2, as well as Ar, all of which sublime at around 40 K. However, diffusion out of the interior of even a small planetesimal can be extremely slow when pressures remain low enough that the gas is in the non-collisional free molecular flow regime, as explored by Birch & Umurhan (2024). This could be the reason why Oort-Cloud long-term comets, that could be at ~30 K equilibrium temperatures before being scattered out of the Oort Cloud, outgas significant amounts of CO and other highly volatile molecules.

Centaurs operate in a temperature range that could trigger CO_2 outgassing or stop the same, based on whether their subsurface temperature reaches around 80 K. At aphelion (sometimes beyond 35 au) surface temperatures could go below 80 K,

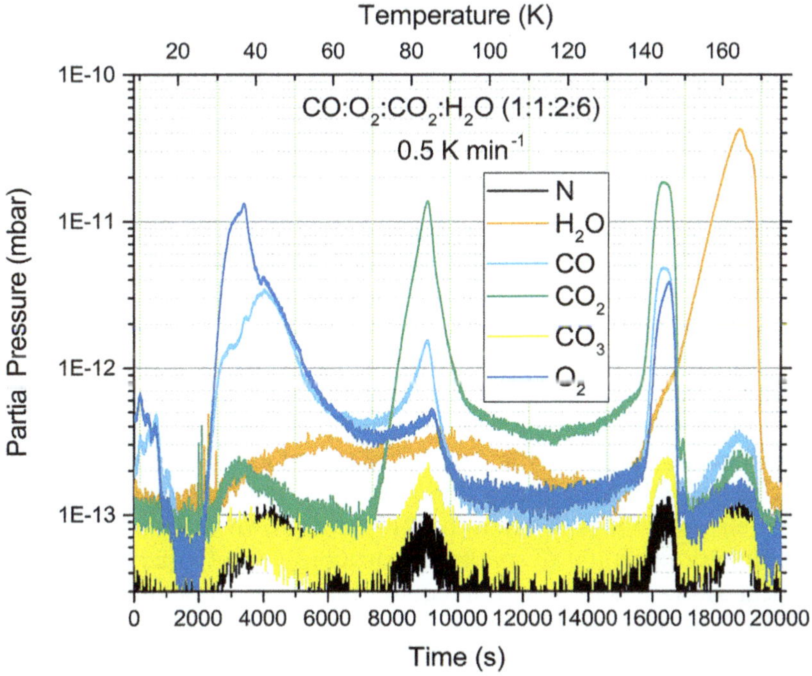

Figure 13.4. Laboratory experiment results demonstrating outgassing of amorphous water-ice with a deliberate excess of trapped highly volatile molecules CO, CO_2, and O_2 at a ratio of 6:2:1:1 (H_2O:CO_2:CO:O_2). Excess of volatiles form ice-domains of their own, in addition to being trapped in other ices. Reproduced from Gudipati et al. (2023). Copyright 2023, with permission from Elsevier.

where no CO_2 outgassing activity would occur. However, around perihelion (down to 5.5 au), ambient temperatures at ~110 K could lead to an increase in subsurface temperatures to > 80 K, resulting in outgassing of CO_2. Under unusual conditions if the Centaur's subsurface reaches temperatures of ~140 K for a short period of time, or remains around ~110–120 K for a long period of time, amorphous water-ice could undergo crystallization by expelling a significant amount (up to 30%) of trapped volatiles (Gudipati et al. 2023). Either of these two processes could lead to the exploding break-up of Centaurs that is often observed (e.g., Bauer et al. 2008; Rousselot 2008; Kareta et al. 2019; see also Chapter 8). It is likely that the surface of a Centaur is mostly covered with crystalline water ice mixed with silicate dust and non-volatile complex organics, somewhat similar to a JFC surface, except that Centaurs could retain higher quantities of water-ice than comets due to their respective thermal environments. Beyond that, likely molecules to be persistent on the surface are NH_3 and CH_3OH, which form strong bonding with H_2O and remain in ice to much higher temperatures. If larger complex organics, such as polycyclic aromatic hydrocarbons (PAHs), were to be present and trapped in amorphous water ice, they could get expelled when the ice crystallizes. This process would be similar to the release of volatiles seen above in Figure 13.4, but as these larger molecules are not volatile, they would remain on the ice and form aggregates (Lignell & Gudipati 2015). Such aggregates may lead to the formation of refractory organic surface layer, which could also change the color of the surface.

13.3.3 Optical and Spectral Properties

Observations of Centaurs' nuclei revealed their surface composition to be consistent with mixtures of ice, complex organics, and minerals. Some Centaurs are featureless in the near-infrared (NIR) while others have clear absorption bands, which allowed detection of water and methanol ice (e.g. Cruikshank et al. 1998; see Chapter 5). Laboratory spectroscopy of materials thought to be present at the surfaces of Centaurs is key for better understanding observations. Reference spectra of different materials allows for qualitative identifications. Optical constants are needed for light scattering models allowing access to quantitative estimates of surface composition. Yet, optical constants of appropriate materials at relevant temperatures (30–120 K) and in the proper wavelength range (0.5–2.5 or 5 μm) remain scarce. Most laboratory experiments for spectroscopy involve the formation of ice by condensation of a gas on a cooled substrate, which produces thin films particularly adapted to study intense absorptions (e.g. Gerakines & Hudson 2015). Ice grown in closed cells is typically thicker and allows the study of weaker absorptions (e.g., Grundy & Schmitt 1998). Proper optical constant determination requires the combination of both thin and thick ices to cover a broad wavelength range.

de Bergh et al. (2008) presents a thorough review of spectroscopic data for ices, organics (coals, tholins, and insoluble organic matter – IOM) and minerals. de Bergh et al. (2013) extended the ice review to newer measurements which add up to various data for: water (H_2O), methane (CH_4), carbon monoxide (CO), carbon dioxide (CO_2), methanol (CH_3OH), ammonia (NH_3), ethane (C_2H_6), and hydrogen cyanide (HCN) as well as some mixtures and irradiated ices.

Here, we highlight newer determinations of optical constants for amorphous CH_4 at 10K from 2 to 25 μm (Gerakines & Hudson 2015), for CH_3OH at temperatures ranging from 20 to 130 K in the 2–15.4 μm range (Luna et al. 2018), and for HCN at 10 and 120 K from 2 to 16.7 μm (Gerakines et al. 2022). Ammonia optical constants were obtained at 10 and 30 K by Zanchet et al. (2013) between 1.4 and 20 μm, and more recently by Hudson et al. (2022) at 10 and 100 K between 1.7 and 16.7 μm. Benzene might be also of interest for Centaur surfaces and optical constants are available for its amorphous and crystalline form (Hudson & Yarnall 2022).

Next, ice mixtures are necessary to investigate spectral changes arising from molecular mixtures which cannot be reproduced by light scattering models. Bernstein et al. (2005) studied the NIR bands of CO_2 in mixtures with H_2O and CH_3OH, focusing on changes in band intensity, position and shape as a function of mixture and temperature between 15 and 150 K. A similar study was performed for CH_4 and its mixture with H_2O (Bernstein et al. 2006). Some relevant mixtures have available optical constants like H_2O:CH_4 at 30 K from 2.5 to 10 μm (Bossa et al. 2015) and CH_4:C_2H_6 at 18 K and 30 K from 1.6 to 14.3 μm (Molpeceres et al. 2016). Band positions can also be used as a probe for temperature, particularly for the well characterized H_2O ice (Grundy & Schmitt 1998; Mastrapa et al. 2008, 2009). In the past years, progress on the determination of spectral properties on outgassed ice-dust mixtures has been ongoing (e.g., Pommerol et al. 2019).

Efforts have been made to make the extraction of optical constants more accessible to laboratory spectroscopists. Rocha & Pilling (2014) present their code and the indices obtained for many ices, pure and in mixtures, in the 2–16.6 μm range. Gerakines & Hudson (2020) provide an open-source code along with CH_3OH, CO_2, N_2O, and CH_4 optical constants each in an amorphous and crystalline phase in the 2.5–20 μm range. Tegler et al. (2024) have adapted the code from normal incidence transmission through an ice film deposited on a window to off-axis specular reflectance from ice deposited on a mirror, permitting optical constants to be measured for ices deposited on a quartz crystal microbalance.

Progress has also been made regarding the availability of spectroscopic data through the development of databases. A summary of current databases is presented in Table 13.1 along with a brief overview of their content. The LIDA database, hosted in the Netherlands, focuses on mid-infrared (MIR) spectra of a variety of ices, pure and in numerous mixtures, across broad temperature ranges along with optical constants for H_2O, CO_2, CO and two mixtures (Rocha et al. 2022). The NKABS database presents data obtained by LASA in Brazil, as well as associated laboratories (LNLS, GANIL and PUC-RIO) and focuses also on the MIR range but includes irradiated ices. They have a catalog of optical constants (2 to 16.6 μm, mostly at ~15 K but 2 experiments at 72 K) for various ice mixtures un-irradiated and then irradiated at two different fluence steps (Rocha et al. 2017). The OCdb (Optical Constants database, hosted by NASA) regroups ice and tholin optical constants. The Cosmic Ice Laboratory, part of the Astrochemistry Laboratory at NASA's Goddard Space Flight Center has available spectra of various ice mixtures

Table 13.1. Non-exhaustive List of Spectral Databases.

	Ices			Refractories		Data	
Database	Pure Species	Mixed (with Other Ices or Refractories)	Irradiated	Organics (Tholins, Residues, HAC…)	Minerals (Silicates, Meteorites…)	Spectra	Optical Constants
LIDA[a]	+	+				+	+
NKABS database[b]	+	+	+			+	+
Ocdb[c]	+	+		+		+	+
The Cosmic Ice Laboratory	+	+	+			+	
(spectra[d], OC[e])	+						+
SSHADE[f]	+	+		+	+	+	+
Database of Optical Constants for Cosmic Dust (DOCCD)[g]					+		+
Heidelberg – Jena – Database of Optical Constants (HEJDOC)[h]	+	+		+	+	+	+

A highlight of the content is marked by a cross in the column of materials relevant for Centaurs. Clicking on the database name will access the corresponding website.
[a] https://icedb.strw.leidenuniv.nl/spectrum_data
[b] https://www1.univap.br/gaa/nkabs-database/data.htm
[c] https://ocdb.smce.nasa.gov/dataset/36
[d] https://science.gsfc.nasa.gov/691/cosmicice/spectra.html
[e] https://science.gsfc.nasa.gov/691/cosmicice/constants.html
[f] https://www.sshade.eu/
[g] https://www.astro.uni-jena.de/Laboratory/OCDB/index.html
[h] https://www2.mpia-hd.mpg.de/HJPDOC/index.php

and optical constants of simple ices like CO and NH_3 to more complex ones like acetylene and benzene (Gerakines & Hudson 2020). Finally, SSHADE is a database infrastructure hosted in France to which various research groups contribute. It contains spectra and some optical constants of ices, and complex organics like tholins, minerals and rocks with varying parameters such as temperature, grain size and wavelength range (Schmitt et al. 2018). Additional complex organic optical constants include Titan tholins (N_2/CH_4 = 90/10) in the 2.5 to 25 μm range (Imanaka et al. 2012), Pluto tholins in the UV to NIR range (Jovanović et al. 2021) and organic residues from ice irradiation in the 0.125 to 20 μm range (Baratta et al. 2015).

In addition to ices and organics, minerals can be an important surface component of Centaurs. An intimate mixture of minerals and organics could be the main constituent of Centaurs' dust, similarly to what is observed in cometary dust. Remote mid-IR studies of cometary comae dust show spectral shapes in the 10-um region that are dominated by amorphous silicates, with variable contributions of crystalline peaks (Wooden et al. 2017, and references therein), similarly to laboratory spectra of particles originated from comets or primitive asteroids. Amorphous silicates seem to be an important surface component of Jupiter Trojans as well (Martin & Emery 2023). Thus, it would not be surprising to discover that Centaur dust contains a significant fraction of amorphous silicates.

Sulfides are believed to be an additional constituent of cometary dust, while high-temperature Ca-Al inclusions and ameboid-olivine aggregates are probably rare (Wooden et al. 2017). In analogy to what is observed in carbonaceous chondrites, carbonates and magnetites might be present if the parent bodies experienced aqueous alteration. In cometary grains and in primitive meteorites, the "opaque phases" (e.g., iron sulfides and oxides, carbon-rich materials, etc.) are intimately mixed with the silicates at scales that are often smaller than the wavelengths, suggesting that this may also be the case of Centaurs.

Optical properties of silicates and other minerals relevant to Centaurs dust have been widely studied, both from a solar system perspective and in the study of dust in the interstellar medium and in protoplanetary systems. Optical constants for many of these minerals are available from the Laboratory Astrophysics Group of the AIU Jena, Germany (DOCCD), from the Heidelberg – Jena database (HEJDOC; Henning et al. 1999) and from the SSHADE spectral databases (Schmitt et al. 2018). A lack of optical constants at low temperatures, relevant to the surface temperatures of Centaurs, should be addressed in future studies. Particular attention should also be given to overtone features of crystalline silicates that might be detected in the range of the NIRSpec instrument on JWST. Finally, the very intimate mixture of minerals and organics in primitive dust breaks the assumption of geometrical optics used in many scattering models (Hapke, Shkuratov). In addition, such models are limited to particle sizes that are larger than the wavelength of the incident light (often in the infrared where a number of relevant spectral features are found), which limits their applicability to fine powders and submicron grains. Under a few assumptions and by changing the way the scattering and absorption efficiency are computed (Hapke 1993), such models could also consider small particles and provide the possibility to take into account small inclusions within a given matrix (Hapke 2001). This kind of modification has shown to provide some benefits in fitting spectra of various objects, such as icy bodies for instance (Clark et al. 2012). Clark et al. (2012) also considers the role of nano-particles to explain the reddening effect seen in the optical part of the Iapetus spectrum.

Although progress has been made regarding the determination and the availability of optical constants, more laboratory work is still needed, specifically to address possible discrepancies between different sets of optical constants, target proper temperature ranges and investigate mixtures. Active communication between observers and experimentalists is necessary to best assess and address the needs of the community.

13.4 Chemistry on Icy Surfaces

Like many airless bodies in the Solar System, Centaurs are exposed to the so called "space weathering" processes, including irradiation by solar wind and cosmic ions, UV photons, and micrometeorite bombardment. These processes modify the composition and structure of the surfaces, and in turn their optical and spectral properties (Brunetto et al. 2015). In particular, a complex chemistry can be induced by ions and photons (energetic processing), similarly to what has been proposed on TNOs (Hudson et al. 2008). Here we review some laboratory research on the complex chemistry of icy surfaces in the outer solar system, when exposed to the space environment, and discuss how current results apply to Centaurs surface chemistry.

The times necessary for accumulating chemically significant dosages on icy surfaces of outer solar system objects depend on irradiation position within or outside the heliosphere (Cooper et al. 2003). Considering the relatively short lifetime of a Centaur ($< 10^7$ yr) with respect to TNOs or to primitive asteroids in the main belt, we can conjecture that energetic processing acts mildly at the surface of Centaurs. In addition, Centaurs orbit regions of the solar system where the cosmic ion flux is already lower than for many TNOs with large aphelion distances, and the solar wind ion flux is still not as high as for asteroids in the inner solar system. Given these premises, the Centaurs' surfaces probably recorded traces of the chemistry induced by energetic processing in their (pre-Centaur) TNO phase.

Some reviews of the effects of energetic processing on ices have been provided by Hudson et al. (2008), Bennett et al. (2013), Rothard et al. (2017), and Materese et al. (2021). Well documented irradiation effects reported on ices include: structural changes, such as crystalline to amorphous (Dartois et al. 2015), porous to compact (Mejía et al. 2015), sputtering (Johnson et al. 2013), chemical changes, such as the destruction and formation of new species (Rothard et al. 2017), and coloration (Brunetto et al. 2006).

Typical apparatuses used to obtain such results consist of a high vacuum chamber interfaced with a cryo-cooler. Ices are generated by condensation of a vapor on the cooled substrate, either in diffuse or direct deposition which influences physical parameters of the deposited ice (Dohnálek et al. 2003). Irradiation sources can vary from swift heavy ion facilities such as GANIL to lighter equipment like ion and electron guns and UV sources, typically Ly-α photons. The spectroscopic setup used to follow physico-chemical changes upon irradiation can also vary, typically from normal incidence transmittance spectroscopy to grazing incidence reflection-absorption infrared spectroscopy. Substrates can be simple IR-transparent windows or a quartz crystal microbalance to track the ice deposition rates and calculate density by coupling with laser interferometry (e.g., Tegler et al. 2024). Finally, adding a mass spectrometer allows to track molecules expelled in the gas phase and recover sputtering yields. A picture and a schematic representation of typical ice analog experimental equipment are given in Figure 13.5.

A significant amount of work has been carried out on the chemistry of water-ice containing other trapped molecules using UV-radiation (simulating solar UV photons) and electrons (simulating solar wind). When the trapped molecules are volatiles such as CO_2, CH_3OH, or NH_3, radiation (UV or electron) processing of these ices results in the formation of complex organic molecules including important prebiotic molecules such as HCN, formamide, and acetamide (Henderson &

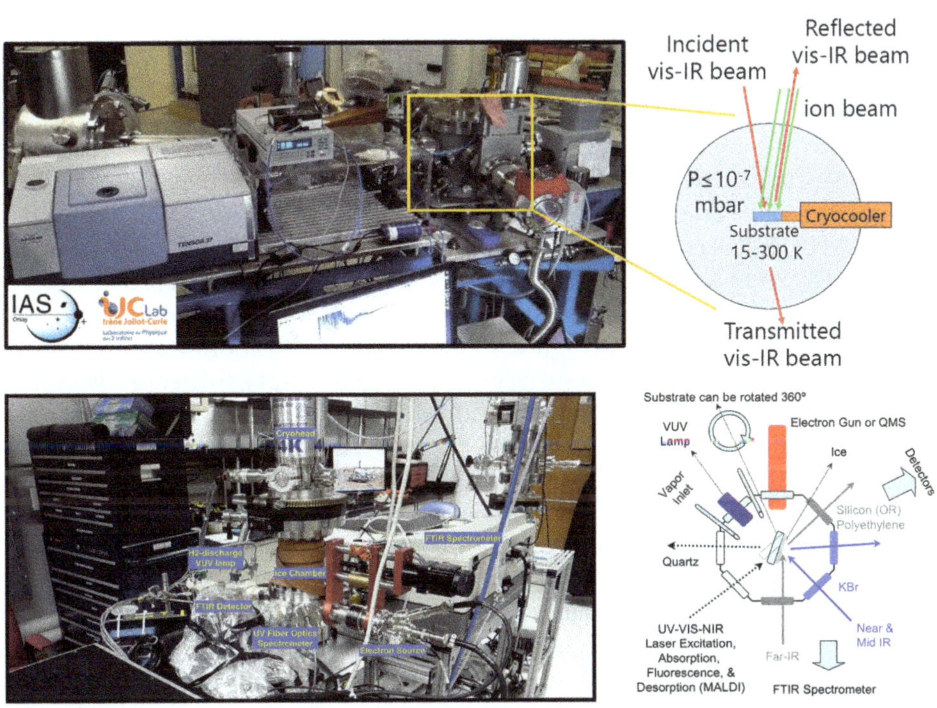

Figure 13.5. Top: Picture and schematic representation of the INGMAR experimental setup (IAS/IJCLab, Orsay, France; Figure adapted from Urso et al. 2020a, with permission from the Royal Society of Chemistry). Bottom: Typical modular astrophysical/planetary/cometary ice analog experimental equipment at the Ice Spectroscopy Lab (ISL) of the Jet Propulsion Laboratory (JPL). A turbomolecular pump and a dry scroll pump are used to achieve low 10^{-9} to high 10^{-10} mbar vacuum in the chamber, which is equipped with several ports that are modularly used for lasers, vacuum discharge lamps, electron-gun, ultraviolet and infrared spectrometers, mass spectrometry, and remote cryogenic microscopy, based on the experimental needs.

Gudipati 2015). Interestingly, most of these molecules found in the laboratory experiments have been detected by the ROSINA mass spectrometer on the Rosetta spacecraft from the outgassing of comet 67P/CG (Rubin et al. 2019). It is likely that Centaurs also store similar complex organics either from the protoplanetary phase or through surface chemistry. UV photons do not penetrate through the surface to greater depths where water ice may be still in amorphous form. However, cosmic rays as well as high-energy electrons could penetrate to meters depth and cause chemistry observed in the laboratory and confirmed by the Rosetta mission. As we have learned from various observational studies, CO_2 is the predominant minor component of ice in cometary nuclei (e.g., Ootsubo et al. 2012; Reach et al. 2013). JWST observations show that CO_2 is present at the surface of Centaurs (De Prá et al. 2025; see also Chapters 5 and 14) as well as in the gas phase from Centaur activity (Harrington Pinto et al. 2023; see also Chapter 6). This indicates that CO_2 is also strongly present on Centaurs, potentially to the same level as in comets. Radiation chemistry of CO_2/H_2O mixed ices has shown that with higher mixing ratio of CO_2 the formation of CO_3 as well as O_3 (ozone) increases, in addition to CO formation (Radhakrishnan et al. 2018). CO_3 as well as O_3 are less stable molecules in the gas phase due to their photodegradation. However, they may remain in equilibrium with CO_2 in the ice phase.

Polycyclic aromatic hydrocarbons (PAHs) are abundant in the interstellar medium (Tielens 2008) and are expected to be accreted on ice grains and transported to be a part of small icy bodies in the evolved solar system (e.g., Li 2009). Laboratory studies with PAHs trapped in ice demonstrated that these molecules are easily ionized (Gudipati & Allamandola 2006; Lignell et al. 2021), and the ionized PAHs undergo oxidation chemistry forming hydroxylated PAHs (Gudipati & Yang 2012; Yang & Gudipati 2014). It is observed that when trapped in water-ice, ionization energy of these PAHs drops by about 4 eV compared to their respective gas-phase ionization energies (Figure 13.6), making ionization accessible through solar UV photons reaching these surfaces (Lignell et al. 2021). Such chemistry makes organic molecules (e.g., PAHs) bond more strongly with water through OH-hydrogen bonding, which in turn could change the physical properties of water-ice based on the amount of these organic molecules trapped in ice. Further, ionization and oxidation could also change the color of the ice containing PAHs. So, coloration of Centaur surfaces could also be due to trapped PAH-like molecules and their chemical alteration.

Surface colors can be described as blue, neutral, or red, depending on the spectral slope in the visible to near-infrared range. Red surfaces like that of Pholus display an increase of reflectance with increasing wavelength (Cruikshank et al. 1998). This red coloring, a sign of organics and irradiation, has been studied in only a few reflectance experiments directly comparable to current observations of Centaurs. Following the finding that irradiation of methane, benzene, and methanol ice produces red slopes comparable to the diversity of those observed on TNOs and Centaurs (Brunetto et al. 2006), Zhang et al. (2023) irradiated methane and acetylene ice and performed simultaneous MIR and visible spectroscopy. They showed that as the red slope reaches an asymptote with increasing irradiation dose

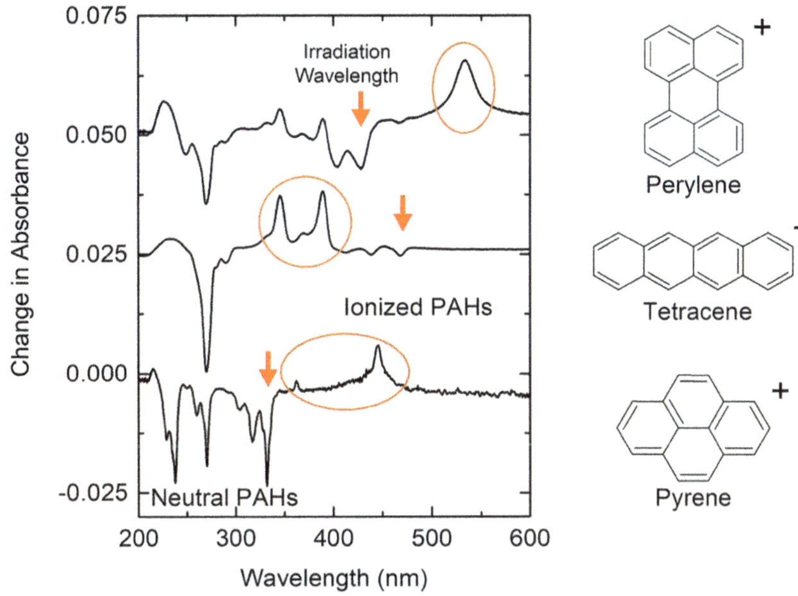

Figure 13.6. Ionization of various PAHs trapped in amorphous water-ice. Arrows indicate wavelengths at which ice was irradiated with a defocused laser that results in the formation of ionized PAH radical cations, whose absorption spectra are shifted to longer wavelengths (hence coloration). Negative bands are due to depleted PAH neutral molecules in the ice. Reproduced from Lignell et al. (2021). CC BY 4.0. Please note that tetracene radical cation (middle spectrum) has absorption bands at 750 nm and 865 nm that are not shown here (Szczepanski et al. 1995).

so does the quantity of aromatic compounds produced. They also noted no change of slope during the heating to 300 K, which is expected as pure hydrocarbon ices evolve under irradiation into refractory residues already at low temperatures (Ferini et al. 2004). This red material is not observed on active Centaurs, probably because it was quickly buried by ejecta material during activity (see discussions in Chapters 5 and 7).

Ice mixtures where oxygen, nitrogen, and even sulfur come into play might not follow the same pattern. Poston et al. (2018) irradiated H_2O:CH_3OH:NH_3 ice mixtures at 50 K and focused on the changes in optical properties with electron-irradiation, as well as adding H_2S as a starting component, building on a previous MIR characterization (Mahjoub et al. 2016). They showed the impact of sulfur addition and heating of the irradiated ices on their spectral properties. Hénault et al. (2025) ion-irradiated water and methanol ice mixtures in different proportions at 60 K to study the influence of water content in the reddening of irradiated ices.

Different numerical simulations of the energetic processing induced by ion bombardment find an irradiation gradient within the first meter of the surface (Loeffler et al. 2020, Quirico et al. 2023). This gradient stems from the different penetration ranges and sputtering efficiencies of solar wind, solar energetic particles, and cosmic ions. A representation of the chemical stratification induced by irradiation over 4.55 Gyrs is represented in panel A of Figure 13.7 and gives an idea of the chemical evolution experienced in the pre-Centaur phase. The top-most layers being the more irradiated could evolve to coloring complex organics while colorless precursor ices could survive in deeper and thus less irradiated layers (Quirico et al. 2023).

Before JWST observations, only water ice was detected across multiple Centaurs. The state of water ice, through its amorphous to crystalline fraction, holds information regarding the extent of irradiation the surface underwent (Loeffler et al. 2020). Concerning carbon-bearing species, only methanol was detected at the surface of Centaurs. Urso et al. (2020b) interestingly showed that irradiation effects can also be investigated by studying the band area ratio of the 2.27 and the 2.34 μm methanol bands. Panel B of Figure 13.7 shows the spectra of irradiated methanol ice, displaying absorption bands from irradiation products like CH_4 and CO, alongside the spectra of Pholus. The good match between the two spectra indicates that the surface of Pholus is likely irradiated. Yet, further characterization of Centaurs with higher signal to noise and spectral resolution in this wavelength range, as well as spectroscopy above 2.5 μm to access fundamental bands, will greatly improve our understanding of the ongoing chemistry by quantifying potential irradiation products.

Sputtering is expected to play a role in surface modification from ice-rich to crust-like as it preferentially removes ices over complex organics (Quirico et al. 2023). This effect could be more important on Centaurs than TNOs as sputtering efficiency increases drastically with increasing temperature (Raut & Baragiola 2013).

As temperature increases in the ice, many processes take place which will participate to the activity but also increase chemical complexity. Alongside

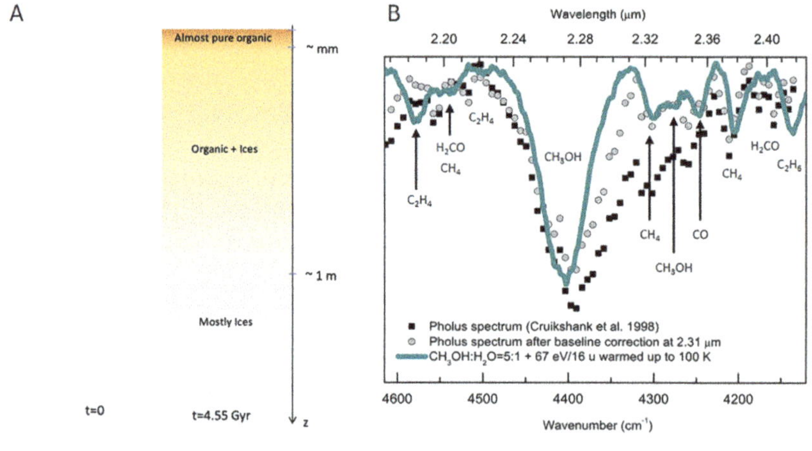

Figure 13.7. Panel A: Schematic representation of the distribution of irradiation products within the surface of an icy body exposed during 4.55 Gyr to energetic processing (reprinted from Quirico et al. 2023. Copyright 2023, with permission from Elsevier). Panel B: Spectra of an irradiated methanol ice (green line) compared to the spectra of Pholus (dots). Reproduced from Urso et al. (2020b). © 2020. The American Astronomical Society. All rights reserved. This highlights the detection of methanol ice at the surface of Pholus and the potential presence of its irradiation products, which would be evidence of ongoing chemistry.

segregation of volatiles from the water ice matrix (Öberg et al. 2009) and expulsion of trapped volatiles into the gas phase (Collings et al. 2004), structural evolution like pore collapse and crystallization will impact chemical evolution (Ghesquière et al. 2018). Finally, reactivity will be enhanced by the increased mobility of radicals (Butscher et al. 2017), which formed by irradiation and are limited by diffusion, and by the onset of thermal reactions of neutral molecules (Theulé et al. 2013).

Ion irradiation experiments on carbonaceous materials, analogs to the carbonaceous components of interstellar (Godard et al. 2011), protoplanetary (Brunetto et al. 2009) and planetary dust (Faure et al. 2021), have been conducted using swift ions, to simulate irradiation by cosmic ion or by high-energy solar ions. Overall, these experiments show that, at low dose, all carbonaceous precursors carbonize and transform into cross-linked polymeric disordered solids, with abundant olefinic and acetylenic bonds, but devoid of aromatic or polyaromatic species, while they evolve towards a sp^2-rich amorphous carbon above a certain critical nuclear dose (Faure et al. 2021). Given the dosages implied in these transformations (< 100 eV/atom), we expect these processes to be active at the surfaces of TNOs and possibly some Centaurs of longer lifetime.

Space weathering of silicates has been extensively studied in the context of asteroid surfaces (Brunetto et al. 2015). If exposed at the surfaces, silicates are expected to be somewhat altered also on TNOs and Centaurs (Brunetto et al. 2007). In particular, amorphization of crystalline anhydrous silicates and production of nanophase iron and/or iron sulfides are also possible in the outer solar system, although on longer timescales than for asteroids in the inner solar system. More recently, laboratory analysis of grains collected at the surface of *C*-type asteroid Ryugu by the Hayabusa2 mission, have provided direct evidence of the effects of space weathering on a carbon-rich rocky surface (Le Pivert-Jolivet et al. 2023). Weathered Ryugu grains show nm-sized vesicles and blisters, surface amorphization and partial melting of phyllosilicates, accompanied by dehydration (Noguchi et al. 2023). Some of these features might also be present in outer solar system objects, if some phyllosilicates are present due to potential aqueous alteration.

13.5 Impacts on Icy Surfaces

One of the most common long-term evolutionary drivers in the solar system are impact events. Impacts on atmosphere-less bodies occur on a microscopic level from dust, to cm and m scales from impacts by small objects, and even larger scales (see Figure 13.8 for examples of craters on icy solar system bodies). In most cases the result will be a crater, but if the impact energy density is too great, the target body can break apart and either reassemble under self-gravity, or, if the energy density is sufficient, the target fragments will achieve local escape velocity and disperse (catastrophic disruption). Given that Centaurs are considered relatively old objects, their surfaces and internal structures will record the consequences of past such impacts. Note that here we consider the Centaurs as the targets in the impacts, unlike the case in, e.g., Zahnle et al. (2003) and Nesvorný et al. (2023), where the Centaurs were considered as impactors on the outer planets and their satellites.

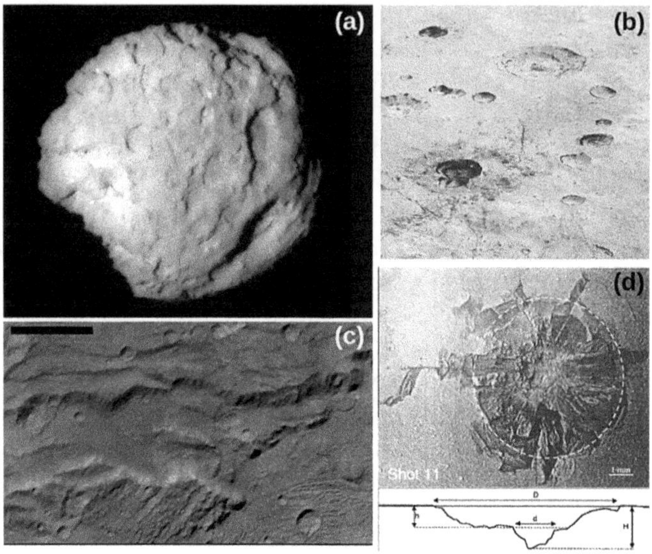

Figure 13.8. (a) Craters on comet Wild-2 seen by the NASA Stardust mission. Largest diameter of Wild-2 was about 2.75 km. Image credit: NASA/JPL-Caltech (b) Craters on Pluto imaged by the NASA New Horizons mission. Image credit: NASA/Johns Hopkins University Applied Physics Laboratory/Southwest Research Institute. (c) Surface features on Charon showing impact craters (scale bar is 50 km). Image credit: NASA/ Johns Hopkins University Applied Physics Laboratory/Southwest Research Institute. (d) Crater in ice from an impact in the laboratory of a 1 mm diameter aluminum projectile at 5.86 km s^{-1}. Reprinted from Shrine et al. (2002). Copyright 2002, with permission from Elsevier.

Traditionally, most impact studies concerned rock targets (e.g., see Melosh 1989). However, this has changed with craters on icy bodies now receiving more attention (e.g., Schenk 1989; Schenk et al. 2004; Kirchoff & Schenk 2010; Singer et al. 2019). A review of the general surface features on icy bodies is given in Burchell (2013), and Stephan et al. (2013) reviews cratering processes in ice in general. The critical parameters in impacts are the impact speed (v) and the masses of the impactor (m_{imp}) and target (m_t). In impact (rather than collision) events, it is usually assumed that $m_{imp} \ll m_t$, with the outcome a crater on the target (or its disruption). The fate of the projectile is often relatively ignored in both laboratory and computational studies, although this is changing. For example, see McDermott et al. (2016), who for impacts of copper projectiles impacting icy bodies with 53% porosity, found the projectile intact beneath the crater floor at 1 km s^{-1}, equivalent to 2–3 GPa peak pressure, with projectile fragmentation starting at 2 km s^{-1}, equivalent to some 10 GPa, and increasing thereafter.

A further complication in impact studies is whether the impact occurs in what is considered to be the strength- or gravity-dominated regime. This is dependent on the mass of the target. Then, dependent on impactor size (and hence crater size) in the gravity dominated regime, there is the same sequence of shapes that is observed on rocky bodies, i.e., simple (bowl) shaped craters at small sizes, then complex craters as crater size increases, followed by complex craters with a central peak and at larger sizes still, complex craters with a rimmed central pit (Stephan et al. 2013). By contrast, laboratory studies all take place in the strength dominated regime, i.e., small crater sizes. The result, in pure water ice, are craters with central pits surrounded by wider spall zones which form later in the impact process after the initial formation of the deeper impact crater in the center (Figure 13.8d). Laboratory experiments in a variety of icy targets have been carried out extensively (see Table 13.2 for examples) and the contrast between laboratory scale craters and craters imaged by space missions on a variety of icy objects can be seen in Figure 13.8.

Table 13.2. Laboratory Experiments Showing Sensitivity of Impacts to Varying Conditions.

Impact Speed (km s^{-1})	Target Composition and State	Comment	References
0.04–0.85	255 K	Observed cratering and fragmentation.	Kato et al. (1995)
0.16–0.41	261 K–263 K	Observed crater size and ejecta size distribution	Cintala et al. (1985)
1–7	259 K	Observed that crater depth, diameter and volume increased with impact speed	Shrine et al. (2002)
0.1–0.64	81 K–257 K	Observed that crater depth, diameter and volume increased slightly with increased temperature	Lange & Ahrens (1987)
5	100 K–253 K	Observed that crater diameter fell as temperature increased, but crater depth and volume increased with falling temperature	Grey & Burchell (2004)
5	Oblique Incidence 253 K	Observed that crater depth fell as angle decreased from normal incidence, but that crater width and length were unchanged until the impact angle exceeded 45° from the normal.	Grey et al. (2002)
0.9–3.8	Porous (50%) 213 K–253 K	Observed that excavated crater mass did not change with porosity.	Koschny et al. (2001)
1.1–7	Porous (50%) 255 K–265 K	Observed that excavated crater mass did not change with porosity, but crater depth increased as porosity increased.	Burchell et al. (2002); Burchell & Johnson (2005)
2.9–5.5	Porous (0%–70%) 255 K–265 K	Reported both cratering expt.'s and also modelled porosity effects on crater size	Burchell & Johnson (2005)
5	Ejecta angular distribution 255K	For aluminium projectiles, the peak angle of ejecta was found to be 73°	Burchell et al. (2003)

(*Continued*)

Table 13.2. (*Continued*)

Impact Speed (km s^{-1})	Target Composition and State	Comment	References
4.8	Ammonia-rich H_2O ice 142 K–162K	Used 0–50% ammonia by wt. Observed that crater depth fell as ammonia content increased from 0–10% ammonia, but was then independent of increasing ammonia content. Crater depth and volume were found to be independent of ammonia content.	Grey & Burchell (2003)
5	230 K CO_2	Crater depth, diameter and volume are smaller than in H_2O ice	Burchell et al. (1998, 2005)
2.3–6.3	Ice saturated sand 203 K, 273 K	Crater diameter fell with increased silicate content.	Croft (1981)
0.9–11.8	H_2O – silicate 245 K–255 K	5–20% silicate content by wt. Crater size decreases with increasing silicate content	Koschny & Grün (2001a, 2001b)
0.299–3.684	H_2O – silicate 263 K	0–50% silicate by wt. Crater size decreases with increasing silicate content	Hiraoka et al. (2007)
0.13–0.63	H_2O-CO_2-silicate Sample preparation temperatures 170 K–270 K	Porosity was ~48%. Crater depth, diameter and volume are smaller than in H_2O ice, but larger than in CO_2 ice	Arakawa et al. (2000)

Impacts are at normal incidence on solid (i.e. non-porous) H_2O targets unless stated otherwise.

As well as crater formation, impacts can also result in the distribution of material via ejection processes. Some of the ejecta is re-distributed over the local surface around the crater during the impact process. However, some ejecta travels into local space around the target body, with a fraction returning under self-gravity and which can thus be distributed over the surface of the whole body, with the remainder reaching local escape velocity. Given that Centaurs span a size range of some 1–250 km, the surface gravity will be low on many (particularly the smaller ones) and so any impact ejecta that is retained will not be concentrated at the impact site, but will be distributed across the entire surface of the body. The mean angle of ejection after an impact is often taken as 45°, but in laboratory experiments for impacts on ice at normal incidence this was not found to be the case. For example, Koschny & Grün (2001b) reported a peak ejection angle on water ice targets of 69.5° from the horizontal, with peak ejecta speeds occurring at a slightly lower angle of 60–65°, and fastest ejecta speeds of some 300–800 m s^{-1}. Burchell et al. (2003) reported a similar mean angle of ejection on water ice of 73° (impacts by aluminium projectiles at 5 km

s^{-1}). This was in contrast to impacts on CO_2 ice (impacts by stainless steel projectiles at 5 km s^{-1}) where there was little ejecta below 40° (from the target surface) but above 40° there was a near constant amount of ejecta vs. angle of ejection (Burchell et al. 1998).

As stated, if the impact energy density (taken as $\frac{1}{2} m_{imp} v^2/m_t$) exceeds a critical value (Q*), the result is that the target breaks apart. Studies of Q* in the laboratory have been carried out and find that Q* is in the range 1–2 J kg^{-1} at impact speeds around 1 km s^{-1}, rising to some 18 J kg^{-1} at higher speeds of 7 km s^{-1} (Dawe et al. 2005; Leliwa-Kopystyński et al. 2008). Q* on porous icy targets was found to be substantially greater in impacts on 25% and 50% porous ice targets at speeds of 1–7 km s^{-1} (Dawe et al. 2005).

Typical impact speeds for impacts on Centaurs can be estimated from those assumed for impacts on Pluto and Charon (albeit that these bodies are larger than Centaurs, with Pluto having a diameter of 2302 km and Charon of 1214 km), where it is assumed the velocity of the impactor at infinity is 1.9 km s^{-1}, and that this rises to 2.0 and 2.2 km s^{-1} for impacts on Charon and Pluto, respectively (Singer et al. 2019). By contrast, Dell'Oro et al. (2001), predicted that the mean impact speed on trans-Neptunian-objects (TNOs) was in the range 1.08–1.44 km s^{-1}. Taken together, these values suggest that a speed range of 1–2 km s^{-1} would be typical for impacts on Centaurs. This is well within the range of laboratory experiments. However, the crater size scale is still an issue, i.e., does it form in the strength or gravity dominated regime?

The boundary between strength and gravity dominated impact regimes can be estimated by considering the ratio R between gravitational and strength stresses. This is given by Equation (7.5.4) in Melosh (1989) as

$$R = \rho g l / Y, \tag{13.1}$$

where ρ is the target density, g is local gravity, l is a characteristic length, and Y is the tensile strength. This is a dimensionless quantity, and if, for example, SI units are used, no constant is required to adjust scales. If $R > 1$, the event is in the gravitational regime, but if $R < 1$, strength dominates the process. Melosh (1989) suggests taking the crater diameter D as the characteristic length involved in cratering. Then if we substitute for the local gravity by $g = Gm_t/r^2$, where r is the radius of the target and G is the gravitational constant, and replace m_t by $4/3\pi r^3 \rho$, and setting $R = 1$ (i.e., at the boundary between strength and dominated cratering regimes), then, again using SI units, we can obtain:

$$D = 3Y/(4\pi G \rho^2 r). \tag{13.2}$$

At small target sizes, the limiting crater size exceeds the target size, i.e., all impact craters are in the strength dominated regime (see Figure 13.9). The value of Y is usually determined at low strain rates (e.g., Hiraoka et al. 2008; Leliwa-Kopystyński et al. 2008) and is typically taken as 1.5 MPa. However, in dynamic processes such as impacts, i.e., at higher strain rates, Y increases by as much as a factor of 10 as the process which breaks the ice up changes from fracture propagation at low strain

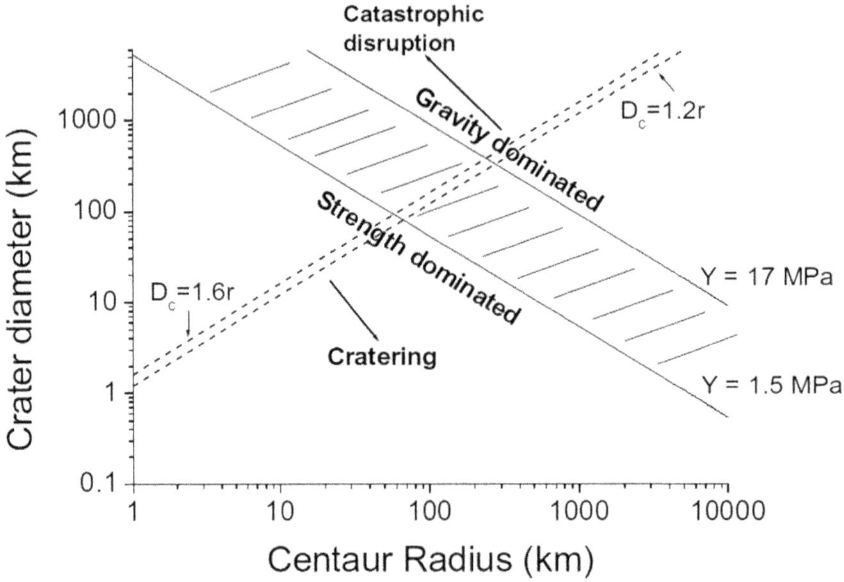

Figure 13.9. Crater diameter is shown vs. Centaur radius for a variety of conditions. The division between gravity and dominated regimes is shown by diagonal lines (predicted by Equation 13.2) where two choices are made for tensile strength Y (the value of 1.5 MPa corresponds to static strength, whereas 15 MPa is more typical of dynamic strength test results for pure water ice). The boundary between cratering and catastrophic disruption is shown by dashed lines. This is derived from the limits reported by Leliwa-Kopystyński et al. (2008), of $D_c = 1.6r$ in the strength dominated regime (based on experiments) and $D_c = 1.2r$ in the gravity-dominated regime (based on observations of icy bodies in the outer solar system) where D_c is the largest impact crater than can form on a body of radius r.

rates to fracture nucleation at high strain rates (e.g., Lange & Ahrens 1983 find $Y = 17$ MPa at high strain rates). The influence of other factors on the tensile strength of water ice, such as porosity, silicate content, and temperature, are discussed by various authors including Hiraoka et al. (2008), Yasui & Arakawa (2008), Litwin et al. (2012), Haack et al. (2020), and Brisset et al (2022a). The effect of silicate content on ice tensile strength is particularly important, and the strength is found to increase with increasing silicate content up to 50% (Hiraoka et al. 2008), which causes a corresponding decrease in crater size (Koschny & Grün 2001b; Hiraoka et al. 2007). Whichever value is taken for Y, the boundary between the gravitation and strength dominated regimes falls as target size increases (see Figure 13.9), and above around 100 km radius, larger craters on the targets form in the gravitational regime. This relationship is an approximation, so only sets a scale for this behavior, but it does explain why craters observed on icy satellites and large icy bodies, such as Pluto, have the traditional raised rim and bowl shape (which at larger scales have central peaks) of craters formed in the gravitational regime.

The size of craters found in experiments can be fit and used to extrapolate to other scales (often using power law scaling relationships). For example, a simple fit is vs. kinetic energy (E) and Fendyke et al. (2013), reported that (based on fits to data in Shrine et al. 2002; Burchell & Johnson 2005)

$$\text{Crater depth} = (2.25 \pm 0.03)\text{E}^{(0.477\pm0.002)}, \tag{13.3}$$

$$\text{Crater diameter} = (23.6 \pm 0.6)\text{E}^{(0.303\pm0.006)}, \tag{13.4}$$

where depth and diameter were in mm and kinetic energy in J (over the range 1–250 J).

See Melosh (1989) for a detailed discussion of crater scaling. The main projectile and target parameters which an impact event involves are taken, i.e., crater diameter D may depend on ρ (density of both projectile and target), m (mass, projectile), Y (target tensile strength), g (local gravity), v (impact speed), with subscripts t and imp denoting target and impactor respectively. Then dimensionless ratios are formed (the π groups) and the coefficients C and β of the power law relationships between the various π groups are found experimentally.

A more sophisticated form of fitting of data is pi-scaling, which permits scaling to a wide range of scales. This is dimensionless scaling, using ratios of variables (Table 13.3). The underlying idea is that if all terms on which an experimental outcome depends are considered, they can be formed into groups of dimensionless ratios. Data taken under one set of experimental conditions can then be readily compared to that obtained under other conditions, and the combined results used to predict impact outcomes at any scale (provided the underlying physics does not change). Despite the standard pi-scaling relationships given in Table 3, when fitting laboratory data, Shrine et al. (2002) actually fitted π_D vs. π_2 and not π_3 despite being in the strength dominated regime. This was to permit comparison with earlier work (Lange & Ahrens 1987), who had noted that, unless gravity is varied, π_2 and π_3 are simply related. However, here for convenience we have refit the data of Shrine et al. (2002) in the standard form of π_D vs. π_3 and, taking $Y = 17$ MPa, we obtain that $C = 3.66\pm1.30$ and $\beta = -0.35\pm0.04$, with goodness of fit criteria $r^2 = 0.8938$. This result can be compared to fits to a larger data set of impacts by Burchell & Johnson (2005), who reported $C = 0.108\pm0.001$ and $\beta = -0.305\pm0.004$. In principle, using these values of C and β then permits extrapolation of crater size in impacts over a wide range of scales.

By contrast to experiments, computer (hydrocode) simulations can be run at any size and speed scales, providing outcomes with no need for extrapolation. For example, at laboratory scales, Fendyke et al. (2013) used a standard hydrocode (iSALE2) for the central crater pit, combined with a simplified two-step Johnson-Holmquist model for the later forming surrounding spall zone. In this way they were able to well reproduce the experimental data of Shrine et al. (2002) and Burchell &

Table 13.3. Description of Pi (Dimensionless Scaling) as Applied to Crater Size

π Groups	Relations Between Groups
$\pi_D = D\,(\rho_t/m_{imp})^{1/3}$	Strength dominated regime: $\pi_D = C\pi_3{}^{-\beta}$
$\pi_2 = 2g(m_{imp}/\rho_{imp})^{1/3}/v^2$	Gravity dominated regime: $\pi_D = C\pi_2{}^{-\beta}$
$\pi_3 = Y/(\rho_{imp}v^2)$	

Johnson (2005). Thus it is possible to model cratering in brittle ice in the strength dominated regime.

At larger scales, in the gravity dominated regime, hydrocode simulations are run without the need for later-stage brittle spallation. Typical results are shown in Kraus et al. (2011) for prediction of crater sizes in impacts on H_2O ice. By fitting the results of their simulations, they found $\pi_D = C\pi_2^{-\beta}$ where $C = 2.5 \pm 0.4$, and $\beta = 0.16 \pm 0.02$. Catastrophic disruption can also be simulated in the gravity dominated regime using hydrocodes and Q* found, e.g., Benz & Asphaug (1999) and Leinhardt & Stewart (2009). However, when extrapolated to laboratory scales the hydrocode results tend to overestimate the Q* values reported in laboratory experiments.

Taking Jupiter-family comets as being similar to Centaurs, the surface of 81P/Wild-2 imaged by the NASA Stardust mission (Brownlee et al. 2004) shows a variety of surface features (Figure 13.8a), many of which were interpreted as the result of impact cratering into highly porous material of moderate cohesion, under a low surface gravity. Given the small crater size (m) and the small size of the comet nucleus (the comet nucleus was irregular with an average radius of some 2 km), this is compatible with the pi scaling shown in Figure 13.9 which suggests all the craters on the comet should form in the strength dominated regime. Interestingly, rather than using laboratory experiments featuring ice targets to simulate their observations, Brownlee et al. (2004), performed laboratory experiments using resin coated sand grains with bulk target porosities of 30%–40% impacted by 3.2 mm diameter projectiles at 2 km s^{-1}. Impact experiments on highly porous ice have been reported. For example, both Koschny et al. (2001) and Burchell & Johnson (2005) reported that at 50% porosity compared to solid ice, the excavated crater mass remained roughly unaltered. Burchell & Johnson (2005) used modeling to show that as porosity was introduced to targets, crater depth grows and scales with the porosity, suggesting that an equal amount of ice is being excavated independent of porosity. However, when porosity exceeds around 80% this breaks down and deeper craters form. They also suggested that crater diameter was unchanged as porosity was introduced, although there was some subsurface widening of the crater as porosity grew.

The Deep Impact mission impact on comet Tempel-1 at 10.3 km s^{-1} (see A'Hearn et al. 2005; Schultz et al. 2013) triggered a whole range of studies of laboratory impacts on ices. For example, Burchell & Johnson (2005) made a pre-impact prediction of the crater size based on impacts on porous ice, Schultz et al. (2007) reported experiments on crater growth in granular media to explain the observed crater, and Ernst & Schultz (2007) described analysis of the Deep Impact light flash compared to laboratory experiments.

The energy added to a crater during an impact can also drive chemistry on an icy body. It has long been known, for example, that shock processes can drive increasing chemical complexity in water ice mixed with various other ices (Nna-Mvondo et al. 2008; Bowden et al. 2009; Singh et al. 2022). The altered materials can be found in both the crater and ejected material. It has also been shown that amino acids such as glycine can be produced in laboratory impact experiments (Martins et al. 2013), providing a possible mechanism to explain the presence of glycine on comet Wild-2 for example (Elsila et al. 2009). Such impact driven chemistry is also likely to be found on the surfaces of Centaurs.

Overall, laboratory experiments of impact processes, and similar computer simulations, offer a wealth of insights into the cratering processes on icy bodies such as Centaurs. Both experiments and simulations can probe a wide range of phenomena and this section can only indicate some of the many different questions that can be asked.

13.6 Surface Activity and Rings around Small Bodies

Surface activity is observed on a variety of small bodies in the Solar System, ranging from comets (Vincent et al. 2019), where it was first discovered, to active asteroids in the Main Belt and Near-Earth space (Jewitt et al. 2015). From these closer bodies, a variety of activity mechanisms have been identified, including volatile sublimation, impact ejecta, and rotational disruption (Jewitt et al. 2015, see Table 13.4).

One of the challenges we face understanding the activity on Centaurs is their great distance to the Sun when triggered (Jewitt 2009; Chandler et al. 2020). Residing between 5 and 30 au, water sublimation, which is one of the main mechanisms for cometary activity, is not an obvious mechanism for the observed activity on Centaurs. For this reason, activity due to volatile sublimation would either require a heat source that is not the Sun, such as micro-impacts or thermodynamic processes of water crystallization (as is also suspected on pre-perihelion comets; Meech et al. 2009), or involve large reservoirs of volatile species that sublimate at much lower temperatures than water ice (Jewitt 2009). For example, Section 13.3.2 of this chapter describes a number of processes that could result in outgassing at these cold temperatures.

While CO_2 has recently been detected in two close ($\sim$6 au) Centaurs (Harrington Pinto et al. 2023; Faggi et al. 2024), the solar distances at which Centaur activity is

Table 13.4. Examples of Surface Activity Mechanisms Deduced from the Observation of Solar System Small Bodies in the Asteroid Belt and Near-Earth Space (Jewitt et al. 2015)

Activity Mechanism	Potential Evidence	Small Body Example
Volatile sublimation	Spectroscopic detection of volatile species Regular timing on the orbit, ideally around perihelion	Comets close to perihelion
Impact ejecta	Shape of the dust cloud	Active asteroid (596) Scheila
Rotational disruption	Fast rotation Impulsive ejection patterns not associated with perihelion Visible body disruption	Active asteroid 311P/PANSTARRS (P/2013 P5)

observed are such that surface CO_2 reservoirs would have been sublimated on much shorter timelines (Jewitt 2009). The low computed rate of collisions of pebbles within the Centaur region (Durda & Stern 2000) leaves rotational disruption as a likely mechanism for activity. The size-rotation rate relation of a number of Centaurs places them in a structurally unstable region (Toth & Lisse 2006), further supporting the idea of the possibility of strength-related weaknesses leading to some of the activity patterns observed.

As described in Hirabayashi & Scheeres (2014), disruption of a rotating small body can come in two forms: either total structural failure, with the destruction of the parent body and the creation of a number of new, smaller ones (such as P/2013 R3; Jewitt et al. 2014), or surface material shedding, in which the parent body survives, only undergoing localized material loss. No Centaur has been observed disintegrating at the time of writing of this chapter. Surface mass loss, on the other hand, can occur due to strength inhomogeneities between the top and sublayers of small, rotating bodies. Simulations by Hirabayashi et al. (2015) show that a less cohesive top layer covering a more cohesive core preferentially leads to surface mass shedding when rotational instability is reached.

Cox et al. (2024) and Brisset et al. (2022a) investigated the role of grain size and water ice grains on material strength in irregularly-shaped granular samples. In particular, they measured the angle of repose as well as shear resistance, deducing the samples' bulk cohesion and angle of internal friction. They found that, in confined environments, such as the sublayers of a planetary surface, larger grains contribute to higher strength. However, in unconfined conditions, such as the very top layer of a surface, larger grains make for weaker materials (Figure 13.10). This means that a surface depletion of finer grains, as observed on asteroids Ryugu and Bennu (Lauretta et al. 2019; Michikami et al. 2019) can lead to strength differences in the top layers of a small body, and thus surface material shedding if rotationally unstable.

Figure 13.10. Strength measurements on simulated asteroid granular material. (left) Image of the surface of Bennu (Credit: NASA/Goddard/University of Arizona); (middle) Cm-sized simulated asteroid grains (reproduced from Brisset et al. 2020); (right) Laboratory measurements of the bulk cohesion, angle of internal friction (AIF), and angle of repose (AOR) of simulated asteroid regolith at various grain sizes (data from Brisset et al. 2022a).

When including water ice grains into the samples, Brisset (2023) does not see significant changes in strength for coarse-grained samples (mm- and cm-sized grains). For fine grain samples, confined (sublayer) materials show lower strength in shear, while unconfined (top layer) show higher strength in angle of repose with increasing percentages of water ice grains included (Figure 13.11). This indicates that the loss of surface ice would not in itself contribute to strength differences facilitating surface mass loss.

However, the sublimation of surface ice would lead to the formation of a frost layer at some depth under the surface. Brisset et al. (2022a) showed that frosted grains displayed higher strengths, both in compression and shear, compared to room-temperature and cryogenic non-frosted grains. Since a porous and water-

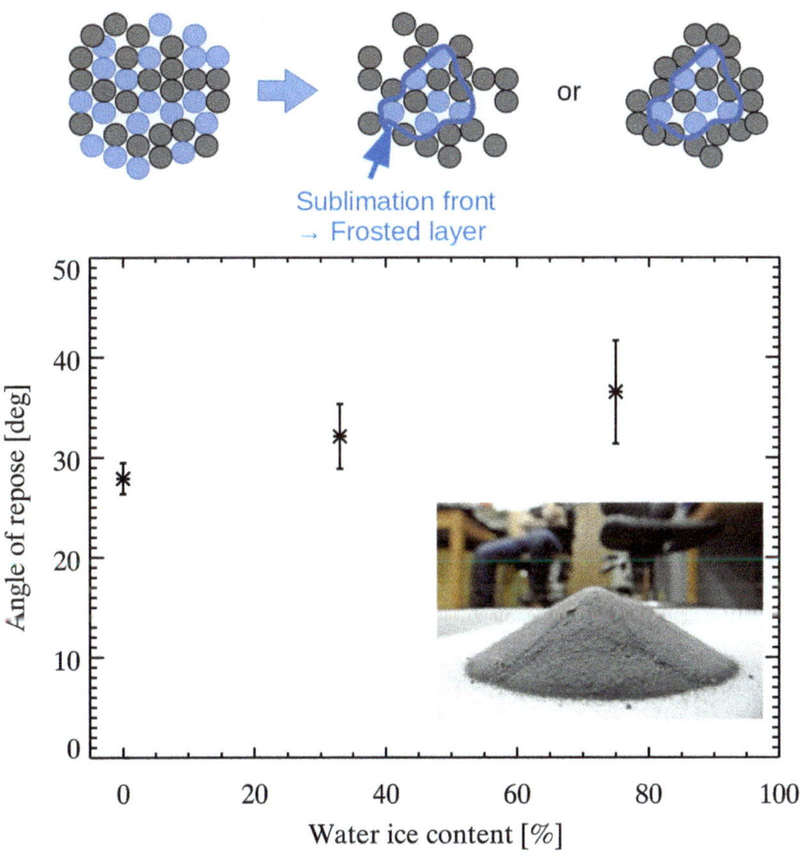

Figure 13.11. Strength measurements on icy dust mixtures of fine grains. (top) Schematic of how a frosted sublayer of regolith can be generated in a granular aggregate from volatile depletion in the top layer. Depending on the size of the aggregate, the desiccated top layer remains porous or collapses into a more compacted layer. In any case, the frosted layer and internal volatiles are thermally protected by the top layer insulating properties; (bottom) angle of repose dependence on the icy fraction of the granular sample. The inset shows a pile of 1:1 volumetric dust to ice ratio granular pile composed of 0.1 mm-sized grains.

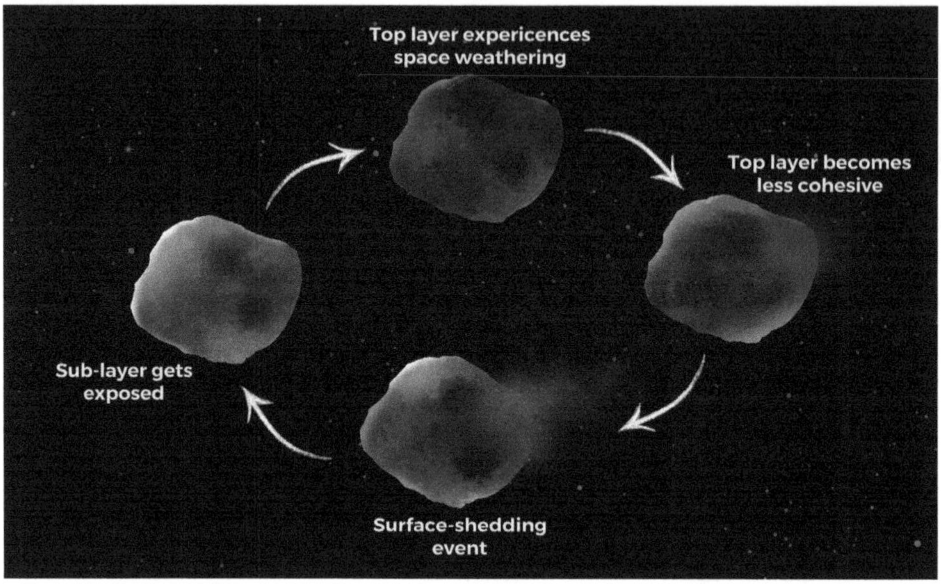

Figure 13.12. Diagram illustrating how changes in surface material strength can lead to recurring activity on Centaurs. Icy material gets weathered and modified by long-term exposure to the space environment of Centaur orbit heliocentric distances (top). This leads to a change in material strength at the surface (right), which becomes mechanically unstable due to the rotation of the body (bottom). Surface shedding events can be observed as activity and potentially feed ring systems. After a shedding event, icier sublayers are exposed (left), which, in turn, get weathered.

depleted top layer serves as an efficient thermal insulator (e.g., Gulkis et al. 2012), the loss of surface water ice due to the exposure to the space environment (Figure 13.12) would then generate soil strength differences inductive of surface material shedding upon mechanical instability. While a water ice depletion is not expected to be occurring at Centaur solar distances, we expect that other ices would lead to a similar layered structure.

Based on these findings, the exposure of an icy granular surface to space, either by fines or volatile depletion, seems to be leading to strength differences between its top and sublayers. As seen above, such a configuration is prone to surface mass shedding (Hirabayashi et al. 2015), which would be observed as dust activity from Earth. In a similar fashion to how dust jets and subsurface volatile sublimation were observed upon localized mechanical failure at the surface of 67P/Churyumov–Gerasimenko (Vincent et al. 2016), such mass shedding events can also lead to intense volatile sublimation events from exposed patches of the subsurface. As noted in Vincent et al. (2016), the dust component of the activity was observed first, soon followed by the volatile detection.

Once the exposed, volatile-rich subsurface material surfaces and sublimation depletes it of its ices, it becomes in its turn an insulating top layer, albeit now weaker in structural strength than its sublayers. This sets it up for failure as surface material shedding at the next mechanical instability, leading to a cycle of material exposure, weathering, and shedding (Figure 13.4). For Centaurs lying in the mechanical instability range (Toth & Lisse 2006), this mechanism can support our understanding of regular, as well as irregular, distant Centaur activity patterns that are currently difficult to explain with the sublimation of standard ices known to be composing them.

Regular surface shedding due to mechanical instabilities also has the potential to feed rings around the parent body. Indeed, the origin of the rings observed around some Centaurs (Braga-Ribas et al. 2014; Ortiz et al. 2015) is still debated. Melita et al. (2017) considered several scenarios involving tidal interactions or impacts in the presence of a satellite. However, the magnitude of tidal forces at the location of the observed rings, as well as the low probability of impacts in the Centaur region, do not lend strong conviction to such scenarios.

While the capture of material from the surface into a ring is not an obvious process, Laipert et al. (2014) and more recently Anand et al. (2023) showed computationally that oblate, rotating parent bodies can reliably achieve the formation of a captured disk, precursor to a ring. In particular, in comparison to capturing ejecta from impacts on the surface of the parent body, the lower velocity and equatorial escape of surface granular material displays the dynamical properties allowing for orbital capture rather than escaping or falling back on the surface.

The estimated time scale analysis of the dispersion of rings around Chariklo performed by Braga-Ribas et al. (2014) shows that they are either very young or confined by shepherding satellites. While satellites are not an uncommon feature in the outer solar system (e.g., Kiss et al. 2017; Fernández-Valenzuela et al. 2019), the regular mass shedding from the surface could explain young rings, as the parent body spin triggers both the escape of particles from its surface and their capture in a disk in its orbit. See Chapter 9 for further discussion of rings and debris around Centaurs.

13.7 Conclusion

In this chapter, we have presented various aspects of laboratory work that produce data supporting our efforts to better understand Centaurs, their surfaces, interiors, history, and evolution. As they are transitioning from the TNO to the JFC population, we can draw significantly from laboratory work targeting cometary or TNO studies to contribute to our knowledge on Centaurs. Cometary studies in the laboratory, which are exceptionally relevant now that the Rosetta mission has collected large amounts of data at 67P/Churyumov–Gerasimenko, are supporting our ever increasing understanding of Centaurs, and thus ultimately TNOs.

Yet, as Centaurs move from the outer Solar System closer to the Sun, the pristine materials composing them, in particular ices, are being exposed and transformed, leading to unique space weathering processes and resulting in new surface chemistry as well as potential surface structural failure and activity. For this reason, it will be important that future studies target Centaurs specifically, rather than simply benefiting from cometary investigations. In particular, we have yet to understand the formation and evolution of amorphous water ice, and its importance in driving surface chemistry and activity. The production of amorphous ice at scale in the laboratory is still a challenge that needs to be addressed. The preparation and study of granular icy samples is another laboratory technique that needs to be developed more systematically, for studying a wide range of phenomena from spectral properties to mechanical behavior. Recent work is showing clearly that the form and crystallinity of the ices studied is of critical relevance to our potential for understanding Centaur surfaces using data from our ground observations.

We note here too that, during their journey towards the inner Solar System, Centaurs can also be captured as moons by the giant planets, so that their study is relevant to understanding small icy moons. The possibility that missions to the gas and ice giant planets may be able to identify a satellite as a captured Centaur offers another motivation for continued and more advanced laboratory studies.

Acknowledgements

J. Brisset acknowledges support in part from the National Science Foundation (NSF) AST program, grant # 1830609 and the National Administration for Space and Aeronautics (NASA) Solar System Workings program, grant # 80NSSC20K0857.

J. Blum acknowledges the continuous support of various funding agencies, foremost the Deutsche Forschungsgemeinschaft (DFG) and the Deutsches Zentrum für Luft- und Raumfahrt (DLR Space Agency).

W. Grundy acknowledges support in part from NASA Solar System Workings Program grant 80NSSC19K0556.

M. S. Gudipati acknowledges funding from NASA DDAP and NFDAP programs for the part of the work that was carried out at the Jet Propulsion Laboratory, California Institute of Technology, under a contract with the National Aeronautics and Space Administration.

References

A'Hearn, M. F., Belton, M. J. S., Delamere, W. A., et al. 2005, Sci, 310, 258

Allodi, M. A., Baragiola, R. A., Baratta, G. A., et al. 2013, SSRv, 180, 101

Anand, K., Minton, D., & Brisset, J. 2023, DPS Meetings Vol. 55, (Washington, DC: AAS) 110.02

Arakawa, M., Higa, M., Leliwa-Kopystyński, J., & Maeno, N. 2000, P&SS, 48, 1437

Baragiola, R. A. 2003, P&SS, 51, 953

Baratta, G. A., Chaput, D., Cottin, H., et al. 2015, P&SS, 118, 211

Barkume, K. M., Brown, M. E., & Schaller, E. L. 2008, AJ, 135, 55

Bauer, J. M., Choi, Y.-J., Weissman, P. R., et al. 2008, PASP, 120, 393

Beitz, E., Güttler, C., Blum, J., et al. 2011, ApJ, 736, 34

Bennett, C. J., Pirim, C., & Orlando, T. M. 2013, ChRv, 113, 9086

Benz, W., & Asphaug, E. 1999, Icar, 142, 5

Bernstein, M. P., Cruikshank, D. P., & Sandford, S. A. 2005, Icar, 179, 527

Bernstein, M. P., Cruikshank, D. P., & Sandford, S. A. 2006, Icar, 181, 302

Bertie, J. E., & Shehata, M. R. 1984, JChPh, 81, 27

Birch, S. P. D., & Umurhan, O. M. 2024, Icar, 413, 116027
Birnstiel, T., Fang, M., & Johansen, A. 2016, SSRv, 205, 41
Bischoff, D., Gundlach, B., & Blum, J. 2021, MNRAS, 508, 4705
Blake, D., Allamandola, L., Sandford, S., Hudgins, D., & Freund, F. 1991, Sci, 254, 548
Blakley, B. P., Grundy, W. M., Steckloff, J. K., et al. 2024, P&SS, 244, 105863
Blum, J., Bischoff, D., & Gundlach, B. 2022, Univ, 8, 381
Blum, J. 2018, SSRv, 214, 52
Blum, J., Gundlach, B., Krause, M., et al. 2017, MNRAS, 469, S755
Bol'shutkin, D. N., Gasan, V. M., & Prokhvatilov, A. I. 1971, JStCh, 12, 670
Bossa, J.-B., Maté, B., Fransen, C., et al. 2015, ApJ, 814, 47
Bowden, S. A., Parnell, J., & Burchell, M. J. 2009, IJAsB, 8, 19
Braga-Ribas, F., Sicardy, B., Ortiz, J. L., et al. 2014, Natur, 508, 72
Brisset, J., Cox, C., Anderson, S., et al. 2020, A&A, 642, A198
Brisset, J. 2023, Asteroids, Comets, Meteors Conf. (Houston, TX: LPI) 2510
Brisset, J., Miletich, T., Metzger, J., et al. 2019, A&A, 631, A35
Brisset, J., Cox, C., Metzger, J., et al. 2022b, PSJ, 3, 176
Brisset, J., Sánchez, P., Cox, C., et al. 2022a, P&SS, 220, 105533
Brown, R. H., Cruikshank, D. P., Pendleton, Y., & Veeder, G. J. 1998, Sci, 280, 1430
Brownlee, D. E., Horz, F., Newburn, R. L., et al. 2004, Sci, 304, 1764
Brunetto, R., Orofino, V., & Strazzulla, G. 2007, MSAIS, 11, 159
Brunetto, R., Pino, T., Dartois, E., et al. 2009, Icar, 200, 323
Brunetto, R., Barucci, M. A., Dotto, E., & Strazzulla, G. 2006, ApJ, 644, 646
Brunetto, R., Loeffler, M. J., Nesvorný, D., Sasaki, S., & Strazzulla, G. 2015, in Asteroids IV, ed. P. Michel, F. E. DeMeo, & W. F. Bottke (Tucson, AZ: Univ. of Arizona Press) 597
Burchell, M. J., Johnson, E., & Grey, I. D. S. 2002, Asteroids, Comets, and Meteors: ACM 2002 500, (Houston, TX: LPI) 859
Burchell, M. J., & Johnson, E. 2005, MNRAS, 360, 769
Burchell, M. J. 2013, in The Science of Solar System Ices: Astrophysics and Space Science Library, ed. M. Gudipati, & J. Castillo-Rogez (Berlin: Springer) 356, 253
Burchell, M. J., Brooke-Thomas, W., Leliwa-Kopystynski, J., & Zarnecki, J. C. 1998, Icar, 131, 210
Burchell, M. J., Leliwa-Kopystyński, J., & Arakawa, M. 2005, Icar, 179, 274
Burchell, M. J., Galloway, J. A., Bunch, A. W., & Brandão, P. F. B. 2003, OLEB, 33, 53
Butscher, T., Duvernay, F., Rimola, A., Segado-Centellas, M., & Chiavassa, T. 2017, PCCP, 19, 2857
Cable, M. L., Vu, T. H., Hodyss, R., et al. 2014, GeoRL, 41, 5396
Carlson, H. G., & Westrum, E. F. 1971, JChPh, 54, 1464
Chandler, C. O., Kueny, J. K., Trujillo, C. A., Trilling, D. E., & Oldroyd, W. J. 2020, ApJL, 892, L38
Choukroun, M., Molaro, J. L., Hodyss, R., et al. 2020, GeoRL, 47, e88953
Choukroun, M., Kieffer, S. W., Lu, X., & Tobie, G. 2013, in The Science of Solar System Ices: Astrophysics and Space Science Library, ed. M. Gudipati, & J. Castillo-Rogez (Berlin: Springer) 356, 409
Cintala, M. J., Smrekar, S., Horz, F., & Cardenas, F. 1985, Lunar and Planetary Science Conf. Vol. 16, (Houston, TX: LPI) 131
Clark, R. N., Cruikshank, D. P., Jaumann, R., et al. 2012, Icar, 218, 831

Clayton, J. O., & Giauque, W. F. 1932, JAChS, 54, 2610
Collings, M. P., Anderson, M. A., Chen, R., et al. 2004, MNRAS, 354, 1133
Cook, J. C., Brunetto, R., De Souza Feliciano, A. C., et al. 2023, Asteroids, Comets, Meteors Conf. Vol. 2851, (Houston, TX: LPI) 2526
Cooper, J. F., Christian, E. R., Richardson, J. D., & Wang, C. 2003, EM&P, 92, 261
Cox, C., Brisset, J., Partida, A., Madison, A., & Bitcon, O. 2024, P&SS, 240, 105829
Croft, S. K. 1981, LPSC, 190
Cruikshank, D. P., Roush, T. L., Bartholomew, M. J., et al. 1998, Icar, 135, 389
Dartois, E., Augé, B., Rothard, H., et al. 2015, NIMPB, 365, 472
Davidsson, B. J. R., Sierks, H., Güttler, C., et al. 2016, A&A, 592, A63
Dawe, W., Murray, M., & Burchell, M. J. 2005, 36th Annual Lunar and Planetary Science Conf. (Houston, TX: LPI) 1096
de Bergh, C., Schmitt, B., Moroz, L. V., Quirico, E., & Cruikshank, D. P. 2008, in The Solar System Beyond Neptune, ed. M. A. Barucci, et al. (Tucson, AZ: Univ. Arizona Press) 483
de Bergh, C., Schaller, E. L., Brown, M. E., et al. 2013, ApSSL, 356, 107
De Prá, M. N., Hénault, E., Pinilla-Alonso, N., et al. 2025, NatAs
De Sanctis, M. C., Capria, M. T., & Coradini, A. 2001, AJ, 121, 2792
De Sanctis, M. C., Lasue, J., & Capria, M. T. 2010, AJ, 140, 1
Dell'Oro, A., Marzari, F., Paolicchi, P., & Vanzani, V. 2001, A&A, 366, 1053
Dello Russo, N., Kawakita, H., Vervack, R. J., & Weaver, H. A. 2016, Icar, 278, 301
Devlin, J. P. 2001, JGR, 106, 33333
Di Sisto, R. P., & Rossignoli, N. L. 2020, CeMDA, 132, 36
Dohnálek, Z., Kimmel, G. A., Ayotte, P., Smith, R. S., & Kay, B. D. 2003, JChPh, 118, 364
Dotto, E., Barucci, M. A., Boehnhardt, H., et al. 2003, Icar, 162, 408
Duffard, R., Pinilla-Alonso, N., Santos-Sanz, P., et al. 2014, A&A, 564, A92
Durda, D. D., & Stern, S. A. 2000, Icar, 145, 220
Elsila, J. E., Glavin, D. P., & Dworkin, J. P. 2009, M&PS, 44, 1323
Ernst, C. M., & Schultz, P. H. 2007, Icar, 190, 334
Faggi, S., Villanueva, G.L., McKay, A., et al. 2024, NatAs, 8, 1237
Faure, M., Quirico, E., Faure, A., et al. 2021, Icar, 364, 114462
Fendyke, S., Price, M. C., & Burchell, M. J. 2013, AdSpR, 52, 705
Ferini, G., Baratta, G. A., & Palumbo, M. E. 2004, A&A, 414, 757
Fernández-Valenzuela, E., Ortiz, J. L., Morales, N., et al. 2019, ApJL, 883, L21
Fraser, W. C., Dones, L., Volk, K., Womack, M., & Nesvorný, D. 2024, in Comets III, ed. K. J. Meech, et al. (Tucson, AZ: Univ. of Arizona. Press) 121
Fray, N., & Schmitt, B. 2009, P&SS, 57, 2053
Garrod, R. T. 2019, ApJ, 884, 69
Gerakines, P. A., & Hudson, R. L. 2015, ApJL, 805, L20
Gerakines, P. A., Yarnall, Y. Y., & Hudson, R. L. 2022, MNRAS, 509, 3515
Gerakines, P. A., Materese, C. K., & Hudson, R. L. 2023, MNRAS, 522, 3145
Gerakines, P. A., & Hudson, R. L. 2020, ApJ, 901, 52
Gärtner, S., Gundlach, B., Headen, T. F., et al. 2017, ApJ, 848, 96
Ghesquière, P., Ivlev, A., Noble, J. A., & Theulé, P. 2018, A&A, 614, A107
Godard, M., Féraud, G., Chabot, M., et al. 2011, A&A, 529, A146
Grey, I. D. S., & Burchell, M. J. 2003, JGR (Planets), 108, 5019
Grey, I. D. S., & Burchell, M. J. 2004, Icar, 168, 467

Grey, I. D. S., Burchell, M. J., & Shrine, N. R. G. 2002, JGR (Planets), 107, 5076
Grundy, W. M., & Schmitt, B. 1998, JGR, 103, 25809
Grundy, W. M., Wong, I., Glein, C. R., et al. 2024a, Icar, 411, 115923
Grundy, W. M., Tegler, S. C., Steckloff, J. K., et al. 2024b, Icar, 410, 115767
Gudipati, M. S., Fleury, B., Wagner, R., et al. 2023, FaDi, 245, 467
Gudipati, M. S.Castillo-Rogez, J. (ed) 2013, The Science of Solar System Ices: Astrophysics and Space Science Library (Berlin: Springer) 356
Gudipati, M. S., & Yang, R. 2012, ApJL, 756, L24
Gudipati, M. S., & Allamandola, L. J. 2006, ApJ, 638, 286
Gudipati, M. S., Abou Mrad, N., Blum, J., et al. 2015, SSRv, 197, 101
Guilbert-Lepoutre, A. 2012, AJ, 144, 97
Gulkis, S., Keihm, S., Kamp, L., et al. 2012, P&SS, 66, 31
Gundlach, B., Schmidt, K. P., Kreuzig, C., et al. 2018b, MNRAS, 479, 1273
Gundlach, B., Kilias, S., Beitz, E., & Blum, J. 2011, Icar, 214, 717
Gundlach, B., & Blum, J. 2015, ApJ, 798, 34
Gundlach, B., Ratte, J., Blum, J., Oesert, J., & Gorb, S. N. 2018a, MNRAS, 479, 5272
Güttler, C., Blum, J., Zsom, A., Ormel, C. W., & Dullemond, C. P. 2010, A&A, 513, A56
Haack, D., Kreuzig, C., Gundlach, B., Blum, J., & Otto, K. 2021b, A&A, 653, A153
Haack, D., Otto, K., Gundlach, B., et al. 2020, A&A, 642, A218
Haack, D., Lethuillier, A., Kreuzig, C., et al. 2021a, A&A, 649, A35
Hapke, B. 1993, Theory of Reflectance and Emittance Spectroscopy, Topics in Remote Sensing (Cambridge: Cambridge Univ. Press) 1
Hapke, B. 2001, JGR, 106, 10039
Harrington Pinto, O., Kelley, M. S. P., Villanueva, G. L., et al. 2023, PSJ, 4, 208
Henderson, B. L., Gudipati, M. S., & Bateman, F. B. 2019, Icar, 322, 114
Henderson, B. L., & Gudipati, M. S. 2015, ApJ, 800, 66
Henning, T., Il'In, V. B., Krivova, N. A., Michel, B., & Voshchinnikov, N. V. 1999, A&AS, 136, 405
Hénault, E., Brunetto, R., Pinilla-Alonso, N., et al. 2025, A&A, 694, A126
Hirabayashi, M., & Scheeres, D. J. 2014, ApJ, 780, 160
Hirabayashi, M., Sánchez, D. P., & Scheeres, D. J. 2015, ApJ, 808, 63
Hiraoka, K., Arakawa, M., Yoshikawa, K., & Nakamura, A. M. 2007, AdSpR, 39, 392
Hiraoka, K., Arakawa, M., Setoh, M., & Nakamura, A. M. 2008, JGR (Planets), 113, E02013
Howell, S. M., Chou, L., Thompson, M., et al. 2018, P&SS, 164, 184
Hudson, R. L., Palumbo, M. E., Strazzulla, G., et al. 2008, in The Solar System Beyond Neptune, ed. M. A. Barucci, et al. (Tucson, AZ: Univ. Arizona Press) 507
Hudson, R. L., & Moore, M. H. 2001, JGR, 106, 33275
Hudson, R. L., Gerakines, P. A., & Yarnall, Y. Y. 2022, ApJ, 925, 156
Hudson, R. L., & Yarnall, Y. Y. 2022, Icar, 377, 114899
Imanaka, H., Cruikshank, D. P., Khare, B. N., & McKay, C. P. 2012, Icar, 218, 247
Jenniskens, P., & Blake, D. F. 1996, ApJ, 473, 1104
Jewitt, D., Hsieh, H., & Agarwal, J. 2015, in Asteroids IV, ed. P. Michel, F. E. DeMeo, & W. F. Bottke (Tucson, AZ: Univ. Arizona Press) 221
Jewitt, D. 2009, AJ, 137, 4296
Jewitt, D., Agarwal, J., Li, J., et al. 2014, ApJL, 784, L8
Johansen, A., & Youdin, A. 2007, ApJ, 662, 627

Johnson, R. E., Carlson, R. W., Cassidy, T. A., & Fama, M. 2013, ApSSL, 356, 551
Jost, B., Pommerol, A., Poch, O., et al. 2016, Icar, 264, 109
Jost, B., Gundlach, B., Pommerol, A., et al. 2013, Icar, 225, 352
Jovanović, L., Gautier, T., Broch, L., et al. 2021, Icar, 362, 114398
Kareta, T., Sharkey, B., Noonan, J., et al. 2019, AJ, 158, 255
Kato, M., Iijima, Y.-I., Arakawa, M., et al. 1995, Icar, 113, 423
Kirchner, M., Bläser, D., & Boese, R. 2010, ChEuJ, 16, 2131
Kirchoff, M. R., & Schenk, P. 2010, Icar, 206, 485
Kiss, C., Marton, G., Farkas-Takács, A., et al. 2017, ApJL, 838, L1
Koschny, D., Kargl, G., & Rott, M. 2001, AdSpR, 28, 1533
Koschny, D., & Grün, E. 2001b, Icar, 154, 391
Koschny, D., & Grün, E. 2001a, Icar, 154, 402
Kouchi, A., Yamamoto, T., Kozasa, T., Kuroda, T., & Greenberg, J. M. 1994, A&A, 290, 1009
Kraus, R. G., Senft, L. E., & Stewart, S. T. 2011, Icar, 214, 724
Kreuzig, C., Kargl, G., Pommerol, A., et al. 2021, RScI, 92, 115102
Kreuzig, C., Bischoff, D., Molinski, N. S., et al. 2023, RASTI, 2, 686
Kreuzig, C., Bischoff, D., Meier, G., et al. 2024, A&A, 688, A177
Laipert, F. E., Minton, D. A., & Longuski, J. M. 2014, 45th Annual Lunar and Planetary Science Conf. (Houston, TX: LPI) 2319
Lange, M. A., & Ahrens, T. J. 1983, JGR, 88, 1197
Lange, M. A., & Ahrens, T. J. 1987, Icarus, 69, 506
Langmuir, I. 1913, PhRv, 2, 329
Lauretta, D. S., Dellagiustina, D. N., Bennett, C. A., et al. 2019, Natur, 568, 55
Le Pivert-Jolivet, T., Brunetto, R., Pilorget, C., et al. 2023, NatAs, 7, 1445
Leinhardt, Z. M., & Stewart, S. T. 2009, Icar, 199, 542
Leliwa-Kopystyński, J., Burchell, M. J., & Lowen, D. 2008, Icar, 195, 817
Li, A. 2009, Deep Impact as a World Observatory Event: Synergies in Space, Time, and Wavelength, (Berlin: Springer) 161
Lignell, A., & Gudipati, M. S. 2015, JPCA, 119, 2607
Lignell, A., Tenelanda-Osorio, L. I., & Gudipati, M. S. 2021, CPL, 778, 138814
Lisse, C., Bauer, J., Cruikshank, D., et al. 2020, NatAs, 4, 930
Litwin, K. L., Zygielbaum, B. R., Polito, P. J., Sklar, L. S., & Collins, G. C. 2012, JGR (Planets), 117, E08013
Loeffler, M. J., Tribbett, P. D., Cooper, J. F., & Sturner, S. J. 2020, Icar, 351, 113943
Lorek, S., Gundlach, B., Lacerda, P., & Blum, J. 2016, A&A, 587, A128
Lorek, S., Lacerda, P., & Blum, J. 2018, A&A, 611, A18
Lowers, R., McNair, D., & Brisset, J. 2023, DPS Meeting Abstracts 55, (Washington, DC: AAS) 219.05
Luna, R., Satorre, M. Á., Santonja, C., & Domingo, M. 2014, A&A, 566, A27
Luna, R., Molpeceres, G., Ortigoso, J., et al. 2018, A&A, 617, A116
Mahjoub, A., Poston, M. J., Hand, K. P., et al. 2016, ApJ, 820, 141
Malamud, U., Landeck, W. A., Bischoff, D., et al. 2022, MNRAS, 514, 3366
Marboeuf, U., Schmitt, B., Petit, J.-M., Mousis, O., & Fray, N. 2012, A&A, 542, A82
Martin, A. C., & Emery, J. P. 2023, PSJ, 4, 153
Martins, Z., Price, M. C., Goldman, N., Sephton, M. A., & Burchell, M. J. 2013, NatGe, 6, 1045
Marton, G., Kiss, C., Molnár, L., et al. 2020, Icar, 345, 113721

Mastrapa, R. M., Bernstein, M. P., Sandford, S. A., et al. 2008, Icar, 197, 307
Mastrapa, R. M., Sandford, S. A., Roush, T. L., Cruikshank, D. P., & Dalle Ore, C. M. 2009, ApJ, 701, 1347
Mastrapa, R. M. E., Grundy, W. M., & Gudipati, M. S. 2013, in The Science of Solar System Ices: Astrophysics and Space Science Library, Volume 356, ed. M. Gudipati, & J. Castillo-Rogez (Berlin: Springer) 371
Materese, C. K., Gerakines, P. A., & Hudson, R. L. 2021, AChRv, 54, 280
Maynard-Casely, H. E., Wallwork, K. S., & Avdeev, M. 2013, JGRE, 118, 1895
Maynard-Casely, H. E., Hester, J. R., & Brand, H. E. A. 2020, IUCrJ, 7, 844
McDermott, K. H., Price, M. C., Cole, M., & Burchell, M. J. 2016, Icar, 268, 102
Meech, K. J., Pittichová, J., Bar-Nun, A., et al. 2009, Icar, 201, 719
Mejía, C., de Barros, A. L. F., Seperuelo Duarte, E., et al. 2015, Icar, 250, 222
Melita, M. D., Duffard, R., Ortiz, J. L., & Campo-Bagatin, A. 2017, A&A, 602, A27
Melosh, H. J. 1989, Impact Cratering: A Geologic Process (Oxford: Clarendon) 1
Metzger, P. T., Britt, D. T., Covey, S., et al. 2019, Icar, 321, 632
Michikami, T., Honda, C., Miyamoto, H., et al. 2019, Icar, 331, 179
Molaro, J. L., Choukroun, M., Phillips, C. B., et al. 2019, JGR (Planets), 124, 243
Molinski, N., Kreuzig, C., Blum, J., et al. 2022, EPSC, EPSC2022–383
Molpeceres, G., Satorre, M. A., Ortigoso, J., et al. 2016, ApJ, 825, 156
Moullet, A., Lellouch, E., Moreno, R., & Gurwell, M. 2011, Icar, 213, 382
Nesvorný, D., Dones, L., De Prá, M., Womack, M., & Zahnle, K. J. 2023, PSJ, 4, 139
Nna-Mvondo, D., Khare, B., Ishihara, T., & McKay, C. P. 2008, Icar, 194, 822
Noguchi, T., Matsumoto, T., Miyake, A., et al. 2023, NatAS, 7, 170
Öberg, K. I., Fayolle, E. C., Cuppen, H. M., van Dishoeck, E. F., & Linnartz, H. 2009, A&A, 505, 183
Ootsubo, T., Kawakita, H., Hamada, S., et al. 2012, ApJ, 752, 15
Ortiz, J. L., Duffard, R., Pinilla-Alonso, N., et al. 2015, A&A, 576, A18
Polak, B., & Klahr, H. 2023, ApJ, 943, 125
Pommerol, A., Jost, B., Poch, O., et al. 2019, SSRv, 215, 37
Poston, M. J., Mahjoub, A., Ehlmann, B. L., et al. 2018, ApJ, 856, 124
Potter, R. S., Cammack, J. M., Braithwaite, C. H., Church, P. D., & Walley, S. M. 2020, Icar, 351, 113940
Prialnik, D., Sarid, G., Rosenberg, E. D., & Merk, R. 2008, SSRv, 138, 147
Quirico, E., Bacmann, A., Wolters, C., et al. 2023, Icar, 394, 115396
Radhakrishnan, S., Gudipati, M. S., Sander, W., & Lignell, A. 2018, ApJ, 864, 151
Raposa, S.M., Sugata, T.P., Grundy, W.M., et al. 2024, PSJ, 5, 275
Raut, U., & Baragiola, R. A. 2013, ApJ, 772, 53
Reach, W. T., Kelley, M. S., & Vaubaillon, J. 2013, Icar, 226, 777
Robinson, J. E., Malamud, U., Opitom, C., Perets, H., & Blum, J. 2024, MNRAS, 531, 859
Rocha, W. R. M., Rachid, M. G., Olsthoorn, B., et al. 2022, A&A, 668, A63
Rocha, W. R. M., & Pilling, S. 2014, AcSpA, 123, 436
Rocha, W. R. M., Pilling, S., de Barros, A. L. F., et al. 2017, MNRAS, 464, 754
Romon-Martin, J., Barucci, M. A., de Bergh, C., et al. 2002, Icar, 160, 59
Rothard, H., Domaracka, A., Boduch, P., et al. 2017, JPhB, 50, 062001
Rousselot, P. 2008, A&A, 480, 543
Rubin, M., Altwegg, K., Balsiger, H., et al. 2019, MNRAS, 489, 594

Ruf, A., Bouquet, A., Boduch, P., et al. 2019, ApJL, 885, L40
Rupert, F. F. 1909, JAChS, 31, 866
Sack, N. J., & Baragiola, R. A. 1993, PhRvB, 48, 9973
Salzmann, C. G., Murray, B. J., Fox-Powell, M. G., et al. 2024, Icar, 410, 115897
Santos-Sanz, P., French, R. G., Lin, Z.-Y., et al. 2015, EPSC, (Washington, DC: AAS) EPSC2015–309
Sarid, G., Volk, K., Steckloff, J. K., et al. 2019, ApJL, 883, L25
Schenk, P. M. 1989, JGR, 94, 3813
Schenk, P. M., Chapman, C. R., Zahnle, K., & Moore, J. M. 2004, in Jupiter. The Planet, Satellites and Magnetosphere, ed. F. Bagenal, T. E. Dowling, & W. B. McKinnon (Cambridge: Cambridge Univ. Press) 427
Schmitt, B., Bollard, P., Albert, D.the SSHADE Partner's Consortium, et al. 2018, SSHADE: Solid Spectroscopy Hosting Architecture of Databases and Expertise (OSUG Data Center. Service/Database Infrastructure), 10.26302/SSHADE
Schräpler, R. R., Landeck, W. A., & Blum, J. 2022, MNRAS, 509, 5641
Schultz, P. H., Hermalyn, B., & Veverka, J. 2013, Icar, 222, 502
Schultz, P. H., Eberhardy, C. A., Ernst, C. M., et al. 2007, Icar, 190, 295
Scott, T. A. 1976, PhR, 27, 89
Shrine, N. R. G., Burchell, M. J., & Grey, I. D. S. 2002, Icarus, 155, 475
Simon, J. B., Blum, J., Birnstiel, T., & Nesvorný, D. 2024, in Comets III, ed. K. J. Meech, M. R. Combi, D. Bockelee-Morvan, S. N. Raymond, M. E. Zolensky, & R. Dotson (Tucson, AZ: Univ. of Arizona. Press) 63
Singer, K. N., McKinnon, W. B., Gladman, B., et al. 2019, Sci, 363, 955
Singer, K. N., Stern, S. A., Elliott, J., et al. 2021, P&SS, 205, 105290
Singh, S. V., Dilip, H., Meka, J. K., et al. 2022, eLife, 12, 508
Souza-Feliciano, A. C., Holler, B. J., Pinilla-Alonso, N., et al. 2024, A&A, 681, L17
Stansberry, J., Grundy, W., Brown, M., et al. 2008, in The Solar System Beyond Neptune, ed. M. A. Barucci, et al. (Tucson, AZ: Univ. Arizona Press) 161,
Stephan, K., Jaumann, R., & Wagner, R. 2013, in The Science of Solar System Ices: Astrophysics and Space Science Library, ed. M. Gudipati, & J. Castillo-Rogez (Berlin: Springer) 356, 279
Szabó, G. M., Kiss, C., Pinilla-Alonso, N., et al. 2018, AJ, 155, 170
Szczepanski, J., Wehlburg, C., & Vala, M. 1995, CPL, 232, 221
Tan, S. P. 2022, MNRAS, 515, 1690
Tan, S. P., & Kargel, J. S. 2018, MNRAS, 474, 4254
Tegler, S. C., Grundy, W. M., Loeffler, M. J., et al. 2024, PSJ, 5, 31
Theulé, P., Duvernay, F., Danger, G., et al. 2013, AdSpR, 52, 1567
Tielens, A. G. G. M. 2008, ARA&A, 46, 289
Toth, I., & Lisse, C. M. 2006, Icar, 181, 162
Trafton, L. M. 2015, Icar, 246, 197
Umurhan, O. M., Grundy, W. M., Bird, M. K., et al. 2022, PSJ, 3, 110
Urso, R. G., Alemanno, G., Baklouti, D., et al. 2020a, Proc. IAU, 350, 399
Urso, R. G., Baklouti, D., Djouadi, Z., Pinilla-Alonso, N., & Brunetto, R. 2020b, ApJL, 894, L3
Vetter, M., Jodl, H.-J., & Brodyanski, A. 2007, JLTP, 33, 1052
Vincent, J.-B., Oklay, N., Pajola, M., et al. 2016, A&A, 587, A14
Vincent, J.-B., Farnham, T., Kührt, E., et al. 2019, SSRv, 215, 30
Weissman, P., Morbidelli, A., Davidsson, B., & Blum, J. 2020, SSRv, 216, 6

Wood, J., & Hinse, T. C. 2022, ApJ, 929, 157
Wooden, D. H., Ishii, H. A., & Zolensky, M. E. 2017, RSPTA, 375, 20160260
Yang, R., & Gudipati, M. S. 2014, JChPh, 140, 104202
Yasui, M., & Arakawa, M. 2008, GeoRL, 35, L12206
Youdin, A. N., & Goodman, J. 2005, ApJ, 620, 459
Zahnle, K., Schenk, P., Levison, H., & Dones, L. 2003, Icar, 163, 263
Zanchet, A., Rodríguez-Lazcano, Y., Gálvez, Ó., et al. 2013, ApJ, 777, 26
Zhang, C., Zhu, C., Turner, A. M., et al. 2023, SciA, 9, eadg6936
Zsom, A., Ormel, C. W., Güttler, C., Blum, J., & Dullemond, C. P. 2010, A&A, 513, A57

Chapter 14

New Observational Studies: Early Results from JWST and Future Prospects with JWST and the Vera C. Rubin Observatory

Estela Fernández-Valenzuela, Aurélie Guilbert-Lepoutre, Megan E Schwamb, Bryan J Holler, Flavia L Rommel and Charles Schambeau

The Centaurs remain one of the least explored small body populations within the solar system. This is partly due to the lack of dedicated discovery surveys and the relatively small number of currently known objects (a few hundred at the time of writing). However, the Vera Rubin Observatory with the Legacy Survey of Space and Time (LSST) is expected to double the known population during its first year of observations. Additionally, the James Webb Space Telescope (JWST) is yielding fascinating results, including the first detection of new molecules and their isotopologues. We anticipate that the combination of these two observatories will significantly enhance our understanding of this population, offering new insights into the physical properties and processes in the outer solar system, particularly the behavior of ices as they approach the Sun. In this chapter, we summarize early results and discuss future prospects and opportunities for Centaur science with JWST and the LSST.

14.1 Introduction

As we stand on the cusp of a new era in observational astronomy, propelled by groundbreaking observatories such as the *James Webb Space Telescope* (JWST) and the Vera C. Rubin Observatory, the opportunity to advance in our knowledge about Centaurs has never been more promising. Centaurs, with their diverse compositions, complex surface features, and unstable orbits (see, e.g., Barucci et al. 2011, Horner et al. 2004, and Chapters 1, 3, 5, and 6) offer a unique window into the processes of planetary migration, volatile transport, and surface evolution within our solar system. The JWST, with its unprecedented sensitivity and spectral range, is proving its potential to transform our understanding of Centaur objects by providing detailed spectroscopic and imaging observations across the infrared spectrum (e.g., Licandro

doi:10.1088/2514-3433/ada267ch14

et al. 2023; Harrington Pinto et al. 2023; Pinilla-Alonso et al. 2024). These observations have already provided valuable insights into surface compositions by detecting isotopologues of CO_2 and CO never observed before in Centaurs, and new molecules such as CH_4 (methane) and C_2H_6 (ethane) (e.g., Wong et al. 2023; Pinilla-Alonso et al. 2024), and characterizing their physical properties as never before.

The Rubin Observatory's Legacy Survey of Space and Time (LSST) will play a pivotal role in advancing our understanding of Centaur objects by conducting comprehensive surveys of the night sky. The wide-field imaging capabilities of the Rubin Observatory will enable large-scale surveys of Centaur populations. There is no specific estimate for the number of new Centaurs that will be discovered by the LSST. However, considering the number of currently known Centaurs ($\sim$400, using the JPL Horizons definition of a semimajor axis between 5.5 and 30.1 au and excluding objects they also define as Jupiter-family or Halley-type comets) and the expected numbers of new trans-Neptunian objects (TNOs) and asteroids, LSST should increase the number of known Centaurs by an order of magnitude. Although the brighter objects experiencing short-term variability phenomena will be followed-up using ground-based, relatively small facilities, the high-resolution imaging and spectroscopic capabilities of JWST will allow for detailed studies of interesting individual objects, especially for wavelengths beyond 2 μm, where the materials found on Centaurs (and TNOs) have fundamental absorption bands and/or emission lines (see Fernández-Valenzuela et al. 2021, and references therein).

In this chapter, we discuss the future prospects for Centaur research enabled by JWST in synergy with the capabilities of the Rubin Observatory. By leveraging the combined power of these cutting-edge facilities, astronomers are poised to unlock new insights into the origins, evolution, and dynamics of Centaur objects, further shaping our understanding of the outer solar system and its rich diversity of icy worlds.

14.2 Early JWST Results and Future Observations

14.2.1 Observations with NIRSpec

The largest contribution from JWST will be made with NIRSpec, a near-infrared spectrograph that enables 0.6–5.3 μm spectroscopy at resolving powers ($\lambda/\Delta\lambda$) of $\sim$100, $\sim$1000, and $\sim$2700 (Böker et al. 2023). Of interest to the study of Centaurs, two observing modes can be used: fixed-slit spectroscopy and integral-field spectroscopy. At the time this chapter was written, several Centaurs had been observed in the framework of both GTO (Guaranteed Time Observations) and GO (General Observer) programs: 29P/Schwassmann-Wachmann 1, 39P/Oterma, 423P/Lemmon, 450P/LONEOS, 2008 FC_{76}, C/2014 OG_{392}, 2013 XZ_8, P/2019 LD_2, Amycus, Chariklo, and Chiron.

Centaurs are typically studied alongside TNOs, their progenitors in the outer solar system (Fernandez 1980). TNOs are solar system bodies whose heliocentric orbits have semimajor axes (a) greater than that of Neptune but less than the region where the Oort Cloud begins, i.e., $30.07 < a < 2000$ au (Gladman et al. 2008). Centaurs are believed to originate from TNOs due to gravitational perturbations by Neptune and other giant planets (e.g., Fernandez 1980; Levison & Duncan 1997; see Chapter 3 for a detailed discussion of Centaur dynamics). Consequently, both populations are studied together, with an understanding that Centaur surfaces are more likely to have undergone recent modifications as they migrate from the trans-Neptunian region toward the Sun (see Chapter 7).

Indeed, the trans-Neptunian region is an essential probe for models of the formation and evolution of the solar system, insofar as this remnant of the Sun's protoplanetary disk was shaped by complex dynamical processes. Constraining the surface composition of TNOs can thus lead to a deeper understanding of the chemical and physical conditions that prevailed in the early solar system in this critical region of the protoplanetary disk where various ice lines were located (see also discussion in Chapter 5). In addition to probing their molecular composition, JWST has the capability to provide constraints for the isotopic composition of each species detected on the surfaces of these bodies. For instance, the D/H and $^{13}C/^{12}C$ ratio in methane were constrained for the dwarf planets Eris and Makemake (Glein et al. 2024; Grundy et al. 2024). Provided that the corresponding molecular abundance is large enough, such isotopic ratios could be constrained for other species (water, for instance), and atoms (like oxygen or nitrogen), leading to a deeper understanding of the link between the interstellar medium and icy objects in the solar system (see, e.g., Chapter 15).

However, Centaurs and TNOs are subject to a number of evolutionary processes that can modify their initial composition (e.g., Métayer et al. 2019). Interpreting colors and spectra therefore must involve a detailed understanding of irradiation (e.g., Brunetto et al. 2006; Bennett et al. 2013; Poston et al. 2018), volatile loss and retention (e.g., Schaller & Brown, 2007; Brown et al. 2011; Johnson et al. 2015; Wong & Brown, 2017), differentiation and surface renewal (e.g., McKinnon et al. 2008; Guilbert-Lepoutre et al. 2011; Shchuko et al. 2014; Malamud & Prialnik, 2015; Guilbert-Lepoutre et al. 2020), and collisions (see, e.g., Barr & Schwamb, 2016; Abedin et al. 2021; Bottke et al. 2023; Morbidelli & Nesvorný, 2020, for reviews). In the pre-JWST era, surface compositions were constrained for only a limited sample of TNOs and Centaurs, biased toward the largest objects in the population (e.g., Barkume et al. 2008; Fornasier et al. 2009; Guilbert et al. 2009; Barucci et al. 2011; Brown et al. 2012). The JWST has proven to be a game-changer as one can target for the first time intermediate-sized objects ($\sim$200 – 800 km in diameter) in the outer solar system (e.g., Harrington Pinto et al. 2023; Souza-Feliciano et al. 2024; Emery et al. 2024). Three spectral types have been identified, each dominated by specific chemical species: water ice and silicates for one group, CO_2 and CO ices for the second group, and carbon ices and complex organics for the third group. These three groups likely provide indications of the original formation distances of Centaurs and TNOs in the primordial disk.

As part of the dynamical cascade that links TNOs, Centaurs in the giant planet region, and Jupiter-family comets in the inner solar system, Centaurs are expected to inherit their surface characteristics from objects observed in the outer solar system. However, because Centaurs experience increased surface temperatures (being closer to the Sun), additional modifications can occur; in particular, when superficial and near sub-surface ices start undergoing phase transitions which may trigger cometary activity. The sources of such activity have not been definitively identified yet (see Chapters 7 and 8 for further discussion), in part because pre-JWST observations had failed to identify volatile signatures for most active Centaurs. Additionally, CO_2

detections are only accessible from space observations and prior space observatories that could detect CO_2 were not able to separate the CO versus CO_2 flux received (e.g., Spitzer, WISE) or did not have the instrumental sensitivity to detect CO_2 in the Centaur region (e.g., Akari). This status is rapidly changing with recent JWST spectroscopic observations. The Cycle 1 GO program 2416, "Measuring Volatile Production in Active Centaurs with JWST NIRSpec" (PI: A. McKay; McKay et al. 2021) acquired snapshot observations of six Centaurs with known periods of previous coma activity. Indeed, volatile species have been detected in the comae of every active Centaur observed in this JWST program.[1] The coma of 29P/Schwassmann-Wachmann 1 presents CO, CO_2, and their isotopologues (Faggi et al. 2023; McKay et al. 2023; Faggi et al. 2024); however, only CO_2 was detected in the coma of 39P/Oterma (see Figure 14.1; Harrington Pinto et al. 2023) and 450P/

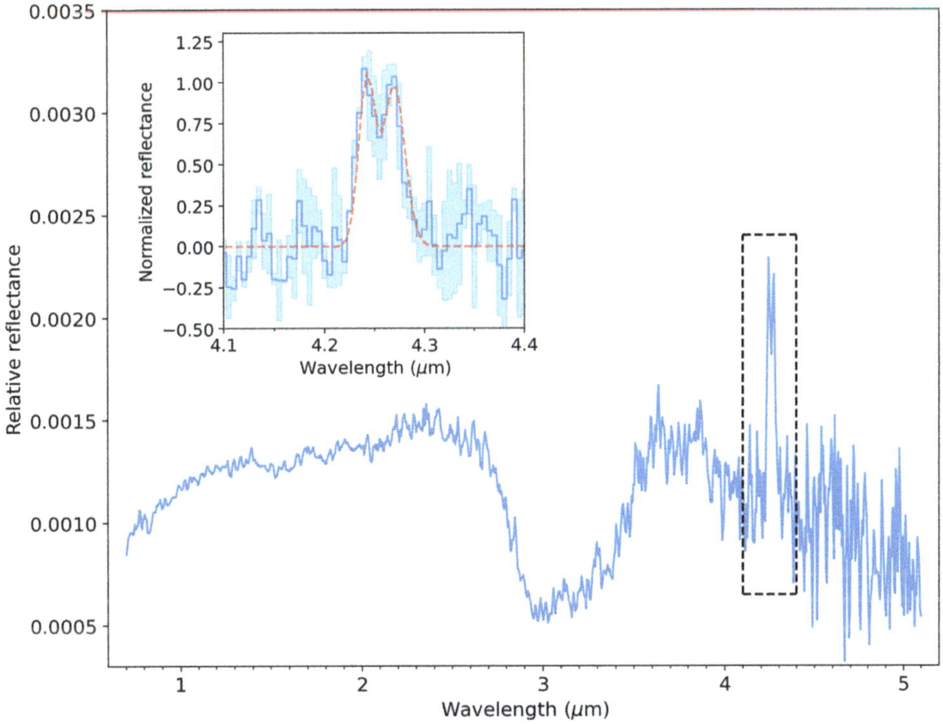

Figure 14.1. NIRSpec PRISM mode spectrum of Centaur 39P/Oterma, the first Centaur to be observed with JWST on 2022 July 27 UTC. This spectrum exhibits a conspicuous 3 μm absorption feature attributed to water ice and a CO_2 emission feature, indicating the presence of a gas coma. The panel in the upper left corner depicts the continuum-removed portion of the 39P/Oterma spectrum centered on the ν_3 CO_2 emission band at 4.26 μm, representing the first detection of CO_2 emission in a Centaur. Reproduced from Harrington Pinto et al. (2023). © 2023. The Author(s). Published by the American Astronomical Society. CC BY 4.0.

[1] The NIRSpec IFU pointing was offset from the ephemeris position for the observations of P/2019 LD_2 (ATLAS), resulting in the IFU Field of View (FoV) only containing a portion of the tail. Detection of volatile emission from this data set is still inconclusive.

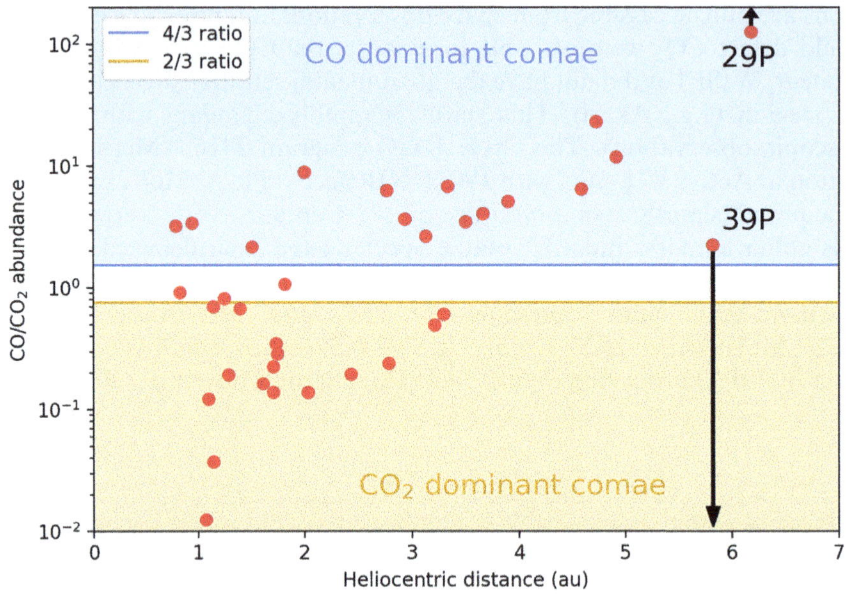

Figure 14.2. There is a lack of CO/CO_2 comae abundance ratios for Centaurs, with only 29P and 39P having limits presented in the literature as of August 2024. Interestingly, the diversity seen in abundance ratios for comets (including both Jupiter-family comets and long period comets) is already suggested by those two Centaurs, consistent with loss of CO relative to CO_2. More measurements are needed for individual objects as well as long baseline monitoring over large portions of an individual object's orbit to fully understand the current status of these volatiles in the interiors of Centaurs. Figure reproduced from Harrington Pinto et al. (2023). © 2023. The Author(s). Published by the American Astronomical Society. CC BY 4.1. See also references therein.

LONEOS (Schambeau et al. 2023), and only CO emission was detected in 423P/Lemmon and C/2014 OG_{392} (Schambeau et al. 2023). These early snapshot observations of active Centaur comae have identified a diversity in the range for the CO to CO_2 abundance ratio (see Figure 14.2). What is unclear is whether these measured abundance ratios are reflective of primordial nuclei bulk compositions or are the result of thermal/orbital evolutionary history differences between individual objects (see Chapter 7). Another possibility is that the abundance ratio changes for an individual object over the course of its orbit, representative of potential seasonal effects due to a heterogeneous nucleus or by the activation of different volatile activity drivers at depth due to a thermal wave penetrating deeper into a nucleus' interior.

For inactive Centaurs, spectra allow direct access to their surface properties, in particular how much of their surface composition they have retained from their time in the trans-Neptunian region, and how any previously sustained activity might have modified it. JWST has targeted a number of inactive Centaurs thus far, both as part of GTO programs and as part of the DiSCo-TNOs large program (see Figure 14.3; Licandro, 2025; Licandro et al. 2023; see also discussion in Chapter 5). Two of the aforementioned TNO spectral types are clearly detected in the Centaur population:

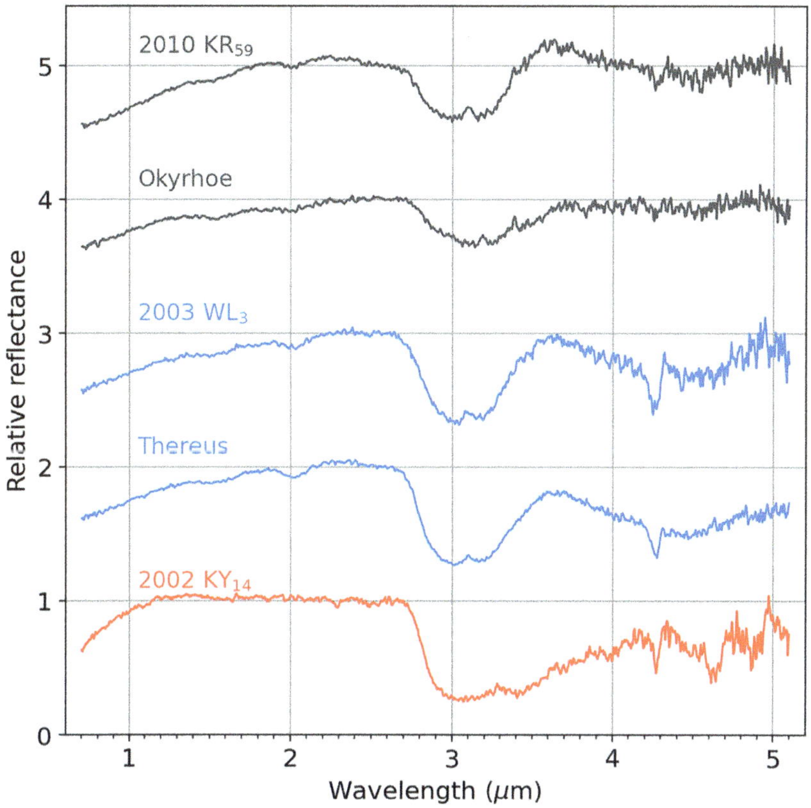

Figure 14.3. NIRSpec spectra of 5 inactive Centaurs presented in Licandro et al. (2023); Licandro, (2025). The spectrum in red corresponds to a carbon-dominated surface, and the blue spectra correspond to water-bearing surfaces. The two spectra in grey do not match other surface types observed in the trans-Neptunian population. Reproduced from Licandro et al. (2023). CC BY 4.0.

water ice and silicates and carbon ices and complex organics. Indeed, about a third of the current Centaur sample resembles water-bearing objects found in the trans-Neptunian region, however, the water ice absorption bands detected in their spectra appear shallower than for their TNO counterparts. In addition, the TNO spectral type dominated by organics and carbon-bearing species is also observed in the Centaur population. For those, deviations from the spectral average of the class are also observed, which might reflect some degree of thermal processing since lower amounts of CO, CO_2, and CH_3OH (methanol) are detected. Overall, we can argue that for those two spectral types, Centaur surfaces appear thermally processed, being less icy and more refractory than their TNO cousins of the same size. The reason behind the lack of objects within the third spectral group (those with CO_2 and CO) is not clear. Indeed, size does not seem to be the reason behind the different spectral types, as discussed in De Prá et al. (2025).

Of note, the TNO spectral type dominated by CO_2 and CO ices is absent from the current (admittedly small) Centaur sample, although it represents 43% of the TNO population (De Prá et al. 2025; Pinilla-Alonso, 2024; Pinilla-Alonso et al. 2025). These two volatile species are thermodynamically unstable in the giant planet region (see Chapters 6 and 7), so the absence of this spectral type might testify to the surface modification undergone by icy objects as they orbit closer to the Sun. Indeed, a number of Centaurs targeted by JWST do not match any spectral type seen in the trans-Neptunian region (i.e., "misfits" in the classification; Licandro, 2025). Whether these misfits represent a processed version of the CO and CO_2 dominated objects remains unresolved at this stage.

A special mention should be given to Chiron, which is challenging every aspect of the current knowledge of activity in the Centaur region. Chiron's spectra taken with NIRSpec present emission lines of CH_4, which are observed for the first time at such distances together with CO_2 emission lines (see Figure 14.4). Surprisingly, no CO emission lines have been found although solid-state absorption bands of this compound are clear, and there are previously published detections of CO emission lines (Womack & Stern 1999). This is puzzling due to the fact that the volatility of CO is higher than that of CH_4 and CO_2 (e.g., Fray & Schmitt, 2009). The nature of Chiron's activity could reveal important clues about the ice evolution of outer small solar system bodies. It is interesting to note that its cometary activity is not related to its heliocentric distance but with its position within the orbit, meaning, the two detected outbursts have occurred at the same orbital (seasonal) location with a time separation of approximately one orbital period (Ortiz et al. 2023). A localized heated region due to the geometry of its orbit could release internal material from more primitive times, explaining the absence of CO in Chiron's spectrum. Moreover, Chiron has been found to be surrounded by a tenuous disk ~580 km in radius, with dense ring or ring-like concentrations at 325 ± 16 km and 423 ± 11 km (Ortiz et al. 2015; Braga-Ribas et al. 2023; Sickafoose et al. 2023; Ortiz et al. 2023). Although the last activity period of Chiron was in August 2021 (Dobson et al. 2021) and it has since decreased in brightness by more than 50%, it was still active during the JWST

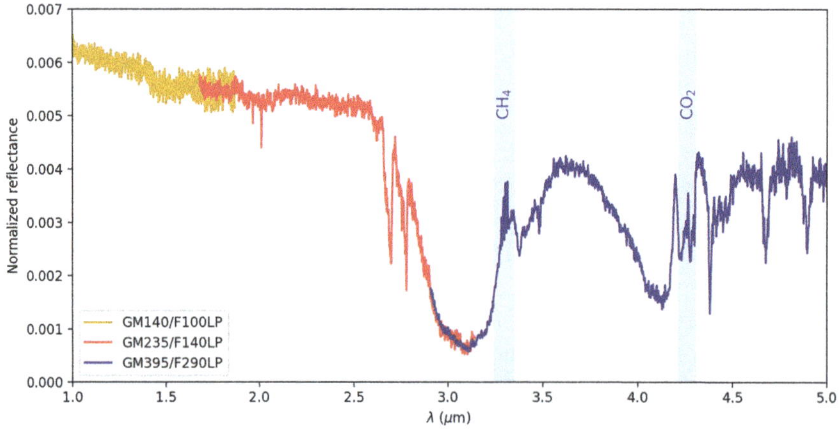

Figure 14.4. Spectrum of Chiron taken with NIRSpec at JWST. Emission lines are highlighted by the blue shadowed region. Data from Pinilla-Alonso et al. (2024).

observations in July 2023, and therefore, the spectral data clearly show cometary activity with CH_4 and CO_2 emission lines (Wong et al. 2023; Pinilla-Alonso et al. 2024). Is CH_4 responsible for Chiron's activity? Are the rings fed by these outbursts? Future NIRSpec observations once Chiron returns to quiescence may shed light on these questions. For a review on notable Centaurs such as Chiron, see Chapter 10.

14.2.2 NIRCam Observations

NIRCam's sensitivity and observing modes (Rieke et al. 2023) provide a great opportunity to measure events that occur on sub-second timescales, and therefore it is perfectly suited to observe stellar occultations of very faint stars with an exquisite time resolution. Although detecting stellar occultations with JWST is challenging due to its constantly changing orbit about the Earth-Sun L2 point (Santos-Sanz et al. 2016), on 2022 October 18, an occultation was observed of a star by Chariklo's rings with impressive results.

While two rings were previously detected using ground-based observations (Braga-Ribas et al. 2014; see also Chapter 9), only the interior, more dense ring was seen in the NIRCam images (Figure 14.5; Santos-Sanz et al. 2023). The time resolution of the observations was 0.3 s (with negligible deadtime), which is equivalent to a width as narrow as 750 m at Chariklo's distance (compared to the 3 km from previous studies of the exterior ring). Therefore, the non-detection of this ring is not due to a lack of sensitivity or time resolution, but to the physical properties of the ring. While ground-based observations occur at visible wavelengths ($\lambda < 1$ μm), NIRCam can observe at longer wavelengths ($\lambda > 1$ μm) through two different filters simultaneously. For the occultation by Chariklo, observations were carried out at 1.5 and 3.2 μm, using the F150W2 and F322W2 filters, respectively. The external ring seems not to be seen at 1.5 μm or 3.2 μm. The combination of NIRCam with ground-based observations and radiative transfer models may provide important information about the physical properties of the particles that form the rings. This will provide important advances in understanding the formation

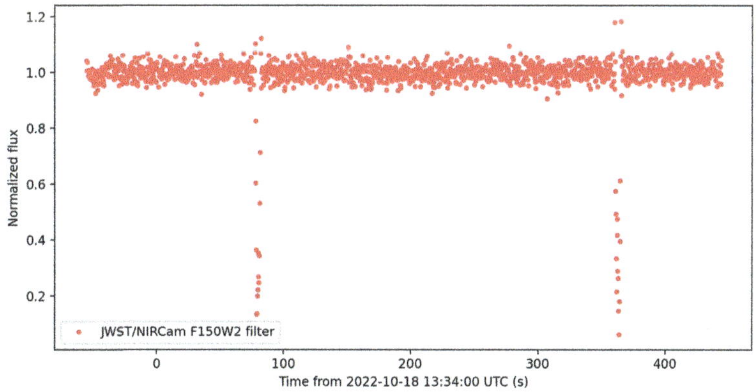

Figure 14.5. JWST NIRCam observations of the stellar occultation by the rings of Chariklo. Points represent observational data from PID 1272 (Hines et al. 2017).

and stability of rings around Centaurs and other small, irregularly shaped bodies, which is an open question since their discovery in 2013. For instance, are the rings formed from particles with different physical properties and/or compositions? How does the density of the ring particles affect the detection from stellar occultations?

Another important parameter to measure is the existence of binary objects in the Centaur population (see, e.g., Chapter 9 for a review). To date, there have been no firm confirmations of binary Centaurs or satellites orbiting Centaurs, at least for those objects that can be unambiguously classified as Centaurs. For instance, the binary systems Typhon-Echidna and Ceto-Phorcys have perihelia in the giant planet region but semimajor axes beyond Neptune, resulting in divergent classifications as Centaurs and scattered disk TNOs. However, based on the binary fractions of the parent population of scattered disk TNOs, binary Centaurs should exist. In fact, the existence of sharp-edge rings around Chariklo could indicate a shepherd satellite between the rings (Braga-Ribas et al. 2014). Also, binary systems are common in the trans-Neptunian region (e.g., Noll et al. 2008; Grundy et al. 2019), the progenitors of Centaurs, and are also found in the Trojan population (e.g., Marchis et al. 2006; Buie et al. 2022), which may ultimately be derived from the same primordial reservoir as Centaurs (e.g., Fernandez, 1980). The study of binary objects beyond Jupiter is a relatively new area of study, and due to their far distances, not many facilities are able to provide the spatial resolution needed for this detection. The combination of NIRCam's spatial resolution (31 mas in the short-wavelength channel), which exceeds that of the UVIS channel of the Wide Field Camera 3 (40 mas) on the Hubble Space Telescope (HST), and NIRCam's sharp point-spread function (PSF) could provide important constraints on the binary fraction in the Centaur population.

14.2.3 MIRI Observations

Spectroscopic observations with the Mid-InfraRed Instrument (MIRI, Wright et al. 2023) over the 5–28 μm range can enhance our understanding of the overall compositional diversity among primitive outer solar system small bodies, complementing observations of these bodies with NIRSpec. Specifically, this wavelength range includes emission lines from silicates (e.g., at 10 μm for the strongest features) typically observed on the surfaces of primitive asteroids and comets. The spectral structure in this region is indicative of the composition of the specific silicate mixture, which in turn provides valuable information about conditions in the primordial solar nebula, as well as the formation environment and early thermal evolution of Centaurs (e.g., Hanner, 1999). At the time this chapter was written, only a few Centaurs had been or were planned to be targeted with MIRI, including 2013 XZ_8 in PID 1272 (PI: Hines; Hines et al. 2017) and Bienor, Chariklo, 1999 OX_3, 2002 GG_{166}, 2013 LU_{28}, and 2020 VF_1 in PID 2820 (PI: Vernazza; Vernazza et al. 2023), with no results yet available to report.

While not part of an accepted JWST proposal in the first 3 cycles, the combination of NIRCam and MIRI imaging presents the prospect of obtaining photometry of the reflected and thermal components, respectively, of a large sample

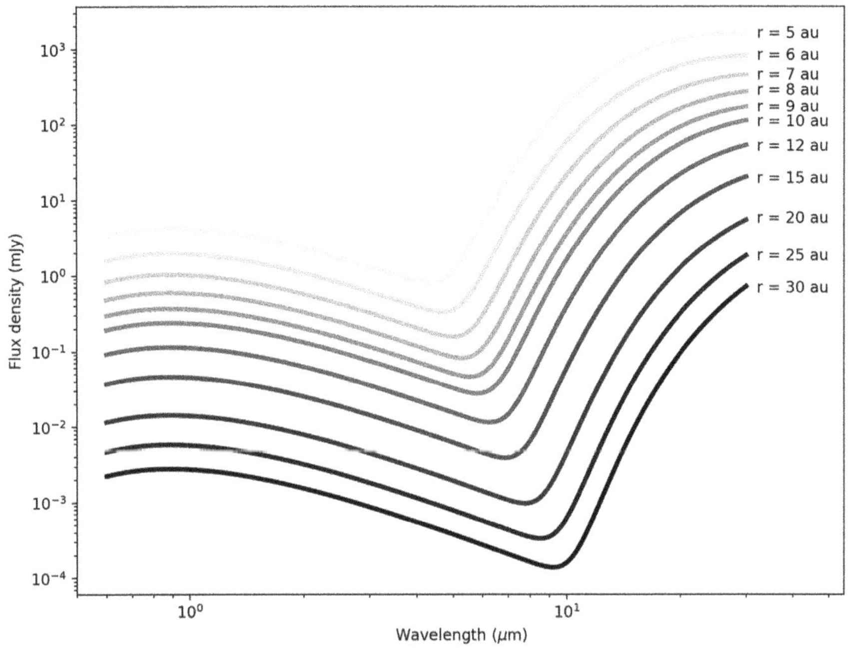

Figure 14.6. Thermophysical models of a 100-km diameter Centaur with a visible geometric albedo of 8% at the heliocentric distances listed and a solar elongation angle of 110°. The thermal blackbody peak shifts to longer wavelengths with increasing heliocentric distance and is largely within the MIRI wavelength range (5–28 μm) out to the semimajor axis of Saturn ($\sim$9.5 au).

of Centaurs. The heliocentric distances of Centaurs lead to thermal blackbody peaks in the MIRI wavelength range (Figure 14.6), which enables accurate thermophysical modeling for the purpose of constraining visible geometric albedos and diameters of a larger number of Centaurs. Such an effort could expand on previous work with the Herschel Space Observatory (Duffard et al. 2014) by increasing the sample size of Centaurs with measured physical properties and providing additional information to test previously observed trends, including the lack of correlation between orbital parameters, diameters, and albedos.

14.3 The Vera C. Rubin Observatory and the Legacy Survey of Space and Time (LSST)

The Vera C. Rubin Observatory will usher in a new era in Centaur discovery and follow-up when it comes online. Based on the most recent construction schedule at the time of writing, Rubin Observatory is expected to begin science operations in late 2025/early 2026. The observatory, currently being built on Cerro Pachón in Chile, comprises the 8.36-m Simonyi Telescope and the 9.6 square degree LSST Camera (LSSTCam). For its first ten years, Rubin Observatory will execute the Legacy Survey of Space and Time (LSST; Ivezić et al. 2019; Bianco et al. 2022), a multi-filter, wide-field survey that will uniformly cover $\sim$18,000 square degrees of the night

Table 14.1. Expected Median LSST Single Visit Image Depths

Filter	u	g	r	i	z	y
AB mag†	23.18	24.42	24.05	23.56	22.96	22.02

Based on the `baseline_v3.3` cadence simulation (Yoachim & Becker 2023).
† 5−σ Limiting Magnitude.

sky. The LSST will also include a full survey of the ecliptic plane. The science requirements for the survey are summarized in Ivezić & the LSST Science Collaboration (2013), Ivezić et al. (2019), Bianco et al. (2022), and references within. Effectively, every three nights the LSST will cover the visible sky from Chile in 6 broad band filters (*ugrizy*). The expected median limiting magnitude per filter is reported in Table 14.1. LSST will be the widest and deepest discovery survey to date for solar system objects. Indeed, inventorying the solar system is one of the four main LSST science drivers (Ivezić et al. 2019; Bianco et al. 2022). LSST is expected to discover close to 5 million new small solar system bodies, and provide our best ever view of the Centaur region (Ivezić et al. 2019).

14.3.1 LSST Survey

The LSST survey observing strategy was not yet set at the time of writing. The majority of decisions surrounding the LSST cadence have been made, however there are a few remaining questions that need to be answered before the baseline observing strategy can be finalized (Ivezić & the SCOC 2021; Bianco & the SCOC 2022, 2024). Figure 14.7 shows the total number of visits skymap in the LSST footprint for the latest 10-year baseline observing strategy (`baseline_v4.0`) simulation (Naghib et al. 2019; Jones et al. 2021; Yoachim et al. 2022; Yoachim & Becker 2023; Bianco & the SCOC 2024). The LSST is comprised of multiple surveys, with the majority of the observing time dedicated to the Wide-Fast-Deep survey (WFD; Bianco et al. 2022). The WFD survey includes the Southern hemispheric extent of the ecliptic, but a separate survey within the LSST will observe the Northern parts of the ecliptic. This mini-survey is known as the Northern Ecliptic Spur (NES) and is expected to survey the sky northward of the WFD $+10^\circ$ ecliptic latitude. The NES will receive fewer visits compared to portions of the sky in the WFD part of the footprint, with the observations taken in the g, r, i, and z filters only. Analysis of the various options for the LSST cadence show that most observing strategies perform well for the metrics associated with the discovery and characterization of solar system objects (Schwamb et al. 2023).

Although previous surveys dedicated to TNOs have served as pathways for the discovery of Centaurs, to date, there has been no dedicated Centaur discovery wide-field survey specifically designed with one set of detection biases and efficiencies, which affect the population statistics. The Rubin Observatory Solar System Processing (SSP) pipeline will be sensitive to the entire Centaur region, making Rubin Observatory an unprecedented Centaur discovery machine (Myers et al. 2013; Jurić et al. 2020). There are no detailed Centaur discovery rates published in

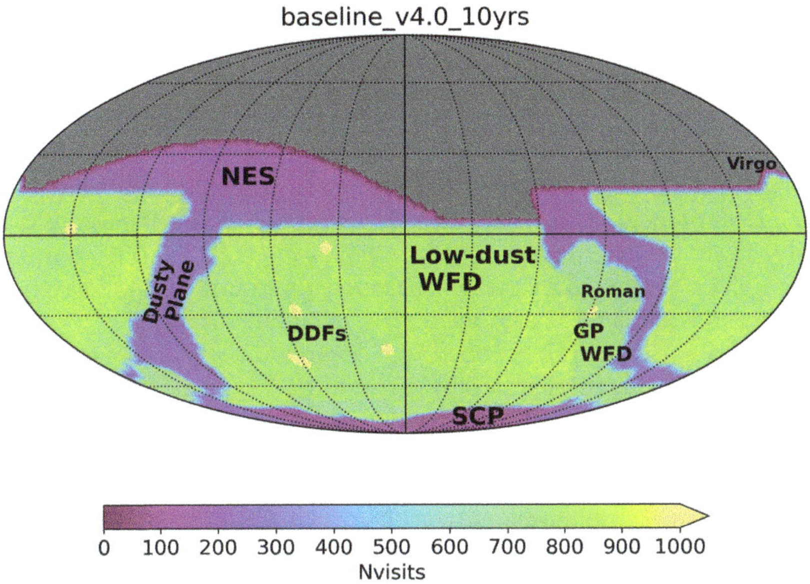

Figure 14.7. Number of visits after 10 years across the LSST footprint from the LSST `baseline_v4.0` cadence simulation. The LSST is a collection of surveys and the various regions of the footprint are labeled as Northern Ecliptic Spur (NES); Wide-Fast-Deep (WFD); Deep Drilling Fields (DDFs), which are pointings with more dedicated and concentrated visits; Galactic Plane (GP); Dusty Plane of the Galaxy (Dusty Plane); and South Celestial Pole (SCP). The WFD survey includes the Southern hemispheric extent of the ecliptic, but a separate survey within the LSST will observe the Northern parts of the ecliptic; most of the Centaur population is expected to be found relatively close to the ecliptic. Reproduced from Bianco & the SCOC (2024). CC BY 4.0.

the literature to date, but based on the expected main-belt asteroid and TNO discovery rates, LSST should discover an order of magnitude more Centaurs than are known today (LSST Science Collaboration et al. 2009; Ivezić et al. 2019; Jones et al. 2018). The majority of Centaurs will be discovered in the first two years of the survey, and all Rubin astrometry and photometry for moving solar system objects will be reported to the Minor Planet Center (MPC) on a nightly basis (Myers et al. 2013; Jurić et al. 2020; Schwamb et al. 2023).

Beyond just finding new Centaur objects, the LSST will provide some monitoring and characterization over a 10-year period as well. Each object observed in the LSST will receive hundreds of photometric and astrometric observations distributed over a decade and across multiple filters (Ivezić et al. 2019). This includes opportunities to study the optical surface colors, phase curves, and incomplete rotational light curves (see Chapter 11 for a detailed discussion) in bulk for a large number of Centaurs, significantly more than have been characterized to date (e.g., Rabinowitz et al. 2007; Tegler et al. 2008; Sheppard et al. 2008; Peixinho et al. 2015; Schwamb et al. 2019; Tegler et al. 2016). Although we do not expect a significant change regarding the bimodal color distribution (e.g., Peixinho et al. 2015; Chapter 5), we will be able to detect interesting outliers and do follow-up observations. We will also be able to explore if there is any change of colors due to increasing activity, which could potentially indicate what are the best facilities and instrumentation for follow-up observations.

With monitoring over ten years, LSST will be the best placed facility to detect the onset and subsequent evolution of cometary-like outbursts in the Centaur population, both through the analysis of the measured photometry and phase curves (e.g., Duffard et al. 2002; Dobson et al. 2021, 2023) and the search for visible coma in LSST images (e.g., Luu & Jewitt 1990; Kareta et al. 2019; Chandler et al. 2020). The most interesting and outlier LSST Centaur discoveries will become prime targets for follow-up initiatives with other ground- and space-based telescopes, including JWST.

Regardless of not being fully defined, the observation strategy should be two 30-second exposures per night separated by 33 minutes. The same field will be observed every three nights using different filters, but the gap between the same filter observations of the same FoV depends on the observation strategy that considers the lunar transit (Bianco et al. 2022). Therefore, despite the expected photometric accuracy of ten milli-magnitudes, Centaurs' absolute magnitudes, colors, and rotational parameters will need years of observations to be precisely determined. Once enough sparse data are collected to determine an object's colors, a combined analysis could reveal rotational light curve amplitudes as low as $\sim$0.10 mag, based on the expected photometric accuracy given above. Such small amplitudes may indicate spheroidal shapes with variability due to albedo spots or tri-axial ellipsoid shapes (or contact binaries, such as Arrokoth) with aspect angles close to pole-on orientations. The aspect angle of Centaurs change relatively fast (compared to TNOs), so changes in their light curve amplitudes could be detected over the duration of the survey, thus obtaining information about their pole orientations and shapes (e.g., Magnusson 1986; Tegler et al. 2005; Fernández-Valenzuela 2022). These sparse rotational light curves will also help to increase the precision on the objects' rotational periods, which allows the determination of the rotational phase during a given stellar occultation. This is important for most Centaurs since their shapes are expected to be irregular (see Chapters 4 and 11) and from stellar occultations only a snapshot at a very specific rotational phase of the three-dimensional shape is obtained. A combined analysis of stellar occultation plus rotational light curves reveals an object's three-dimensional shape, bulk density, etc. (e.g., Vara-Lubiano et al. 2022; Fernández-Valenzuela et al. 2023).

14.3.2 Stellar Occultations in the LSST Era

Currently, stellar occultations have been detected for approximately 20 Centaurs. However, to our knowledge, multi-chord events that effectively constrain the shape and size of these objects have been achieved for only six Centaurs: Bienor, Chariklo, Chiron, 2002 GZ_{32}, 2002 TC_{302}, and 2007 JK_{43} (see, e.g., Braga-Ribas et al. 2014; Ortiz et al. 2015; Santos-Sanz et al. 2021; Fernández-Valenzuela et al. 2023; Leiva et al. 2023; see also Chapter 11). The success of these multi-chord detections is largely due to the prior detection of single-chord events, which significantly improve the precision of the object's astrometric position. Gaia photometric and astrometric data for a subset of stars will be used to calibrate LSST data (Ivezić et al. 2019). As a

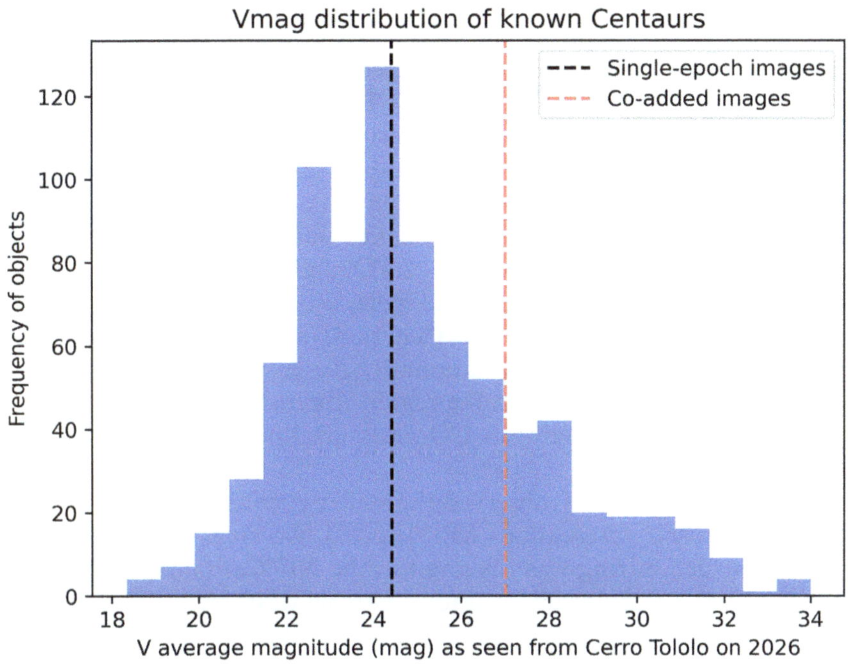

Figure 14.8. Average V magnitude distribution of the ~800 known objects classified as Centaurs by the JPL Small-Body database as of 2024 August 21. We note that JPL has a very relaxed definition of Centaur, only requiring semimajor axes between Jupiter and Neptune with no constraints on perihelion distance, so this list includes ~400 objects that are also classified as Jupiter family and Halley-type comets. The LSST survey can provide astrometry for at least 72% of these known objects ($V \lesssim 24$ mag).

result, we expect to have accurate photometric and astrometric data from LSST for an even fainter sample of stars than Gaia (up to ≈27 mag in co-added images in the r filter). The astrometric precision is extremely important to predict and detect stellar occultations of small bodies; in this context, the main source of uncertainty to detect stellar occultations will be the objects' orbits.

LSST will provide astrometric accuracy ranging from 11 to 74 mas for $m_r \sim 21$ to $m_r \sim 24$ mag (Schwamb et al. 2018). The largest Centaur known to date, (10199) Chariklo, has an angular diameter seen from Earth at the sky plane of ≈ 20 mas with $m_r \sim 18.5$ mag. Therefore, we expect astrometric positions for the brightest Centaurs having sufficient precision to directly obtain predictions of stellar occultations by Centaurs. However, uncertainties in astrometric positions increase as a function of time after the most recent observations used in the determination of their orbit from tens to hundreds of mas in only a few years (Ortiz et al. 2020). Smaller and fainter Centaurs, like (8405) Asbolus, which has an angular diameter of 5 mas at its current location and $m_r \sim 23$ mag, require follow-up observations to decrease the uncertainty in the orbit below their angular diameter. Nonetheless, considering the single-epoch upper limits on g magnitude, LSST can provide astrometry for at least 72% of Centaurs and comets with semimajor axes in the giant planet region (Figure 14.8).

14.4 Conclusions

JWST observations have detected volatile species in the comae of active Centaurs, including CO, CO_2, and their isotopologues. The abundance ratios of these species vary among different Centaurs, raising questions about their primordial compositions versus evolutionary histories. These observations provide valuable insights into the volatile activity and evolution of Centaurs, shedding light on their role in the broader context of solar system dynamics and how they refer to the three spectral types detected by the JWST's GTO and DiSCo-TNOs programs for intermediate-sized TNOs ($\sim$500 km). The spectral type dominated by CO_2 and CO ices, common in TNOs, is absent from Centaurs, indicating surface modifications as they orbit closer to the Sun. Additionally, the case of Chiron challenges our current understanding of activity in the region. Its spectra reveal emission lines of CH_4 and CO_2, but surprisingly no CO emission lines, despite clear solid-state absorption bands.

On 2022 October 18, JWST observed the stellar occultation by Chariklo's rings. Previous ground-based observations had detected two rings around Chariklo, but only the interior, denser ring was observed in the NIRCam images. The high time resolution of the observations allowed for a narrow width equivalent to 750 meters at Chariklo's distance to be achieved, revealing details about the physical properties of the rings. By observing at longer wavelengths than ground-based observations, NIRCam may provide insights into the rings' particle size and composition. Thus, combining ground-based and NIRCam observations offers significant advancements in understanding the formation and stability of rings around Centaurs.

The Vera C. Rubin Observatory, scheduled to begin operations in late 2025/ early 2026, will have a significant role in the discovery and study of Centaurs thanks to the designed wide-field survey that will include the ecliptic plane. The LSST will increase the number of known Centaurs by an order of magnitude. LSST observations will be promptly reported to the Minor Planet Center, facilitating follow-up studies with other telescopes such as JWST and ground-based telescopes. The LSST is anticipated to offer astrometric accuracy ranging from 11 to 74 mas for stars with magnitudes ranging from $m_r \sim 21$ to $m_r \sim 24$, which will increase the number of stellar occultation predictions considerably.

The opportunities to expand the study of Centaurs in the near future with JWST and the Rubin Observatory are extensive. Combination with current tools like PanSTARRS and Las Cumbres Observatory could also help address key questions on activity (see Chapter 8) and evolution (see Chapter 7). We also invite the reader to see Chapter 11 for a deeper understanding on current observational campaigns that could provide important synergies in our understanding of Centaurs. Indeed, JWST and Rubin observatory will build a strong foundation in the areas of discovery and characterization via imaging, spectroscopy, and stellar occultations that will set the stage for future facilities like the Nancy Grace Roman Space Telescope and the Extremely Large Telescope class observatories, in order to help answer outstanding questions and those that will be posed in the coming decade.

Acknowledgements

EFV acknowledges financial support from the Space Research Initiative operated by the Florida Space Institute. AGL received funding from the European Research Council (ERC) under the European Union's Horizon 2020 research and innovation programme (Grant Agreement No 802 699). MES was supported by the UK Science Technology Facilities Council (STFC) grant ST/V000691/1, and she acknowledges travel support provided by STFC for UK participation in LSST through grant ST/N002512/1. FLR acknowledges financial support from the Preeminent Postdoctoral Program (P^3) at the University of Central Florida.

References

Abedin, A. Y., Kavelaars, J. J., Greenstreet, S., et al. 2021, AJ, 161, 195
Barkume, K. M., Brown, M. E., & Schaller, E. L. 2008, AJ, 135, 55
Barr, A. C., & Schwamb, M. E. 2016, MNRAS, 460, 1542
Barucci, M. A., Alvarez-Candal, A., Merlin, F., et al. 2011, Icar, 214, 297
Bennett, C. J., Pirim, C., & Orlando, T. M. 2013, ChRv, 113, 9086
Bianco, F. & the SCOC 2022, Survey Cadence Optimization Committee's Phase 2 Recommendations LSST *PSTN-055*
Bianco, F. & the SCOC 2024, Survey Cadence Optimization Committee's Phase 3 Recommendations, LSST *PSTN-056*
Bianco, F. B., Ivezić, Ž., Jones, R. L., et al. 2022, ApJS, 258, 1
Böker, T., Beck, T. L., Birkmann, S. M., et al. 2023, PASP, 135, 038001
Bottke, W. F., Vokrouhlický, D., Marschall, R., et al. 2023, PSJ, 4, 168
Braga-Ribas, F., Sicardy, B., Ortiz, J. L., et al. 2014, Natur, 508, 72
Braga-Ribas, F., Pereira, C. L., Sicardy, B., et al. 2023, A&A, 676, A72
Brown, M. E., Schaller, E. L., & Fraser, W. C. 2011, ApJL, 739, L60
Brown, M. E., Schaller, E. L., & Fraser, W. C. 2012, AJ, 143, 146
Brunetto, R., Barucci, M. A., Dotto, E., & Strazzulla, G. 2006, ApJ, 644, 646
Buie, M., Keeney, B., Levison, H., & Olkin, C., & the Lucy Occultations Team 2022, DPS Meeting Abstracts Vol. 54, (Washington, DC: AAS) 512.03
Chandler, C. O., Kueny, J. K., Trujillo, C. A., Trilling, D. E., & Oldroyd, W. J. 2020, ApL, 892, L38
De Prá, M. N., Hénault, E., Pinilla-Alonso, N., et al. 2025, NatAs, 9, 252
Dobson, M., Fitzsimmons, A., Schwamb, M. E., et al. 2021, ATel, 14903, 1
Dobson, M. M., Schwamb, M. E., Benecchi, S. D., et al. 2023, PSJ, 4, 75
Duffard, R., Lazzaro, D., Pinto, S., et al. 2002, Icar, 160, 44
Duffard, R., Pinilla-Alonso, N., Santos-Sanz, P., et al. 2014, A&A, 564, A92
Emery, J. P., Wong, I., Brunetto, R., et al. 2024, Icar, 414, 116017
Faggi, S., Villanueva, G. L., McKay, A., et al. 2023, DPS Meeting Abstracts Vol. 55, (Washington, DC: AAS) 400.05
Faggi, S., Villanueva, G. L., McKay, A., et al. 2024, NatAs, 8, 1237
Fernandez, J. A. 1980, Icar, 42, 406
Fernández-Valenzuela, E. 2022, FrASS, 9, 796004
Fernández-Valenzuela, E., Pinilla-Alonso, N., Stansberry, J., et al. 2021, PSJ, 2, 10
Fernández-Valenzuela, E., Morales, N., & Vara-Lubiano, M. 2023, A&A, 669, A112
Fornasier, S., Barucci, M. A., de Bergh, C., et al. 2009, A&A, 508, 457

Fray, N., & Schmitt, B. 2009, P&SS, 57, 2053
Gladman, B., Marsden, B. G., & Vanlaerhoven, C. 2008, in The Solar System Beyond Neptune, ed. M. A. Barucci, et al. (Tucson, AZ: Univ. Arizona Press) 43
Glein, C. R., Grundy, W. M., Lunine, J. I., et al. 2024, Icar, 412, 115999
Grundy, W. M., Noll, K. S., Roe, H. G., et al. 2019, Icar, 334, 62
Grundy, W. M., Wong, I., Glein, C. R., et al. 2024, Icar, 411, 115923
Guilbert, A., Alvarez-Candal, A., Merlin, F., et al. 2009, Icar, 201, 272
Guilbert-Lepoutre, A., Lasue, J., Federico, C., et al. 2011, A&A, 529, A71
Guilbert-Lepoutre, A., Prialnik, D., & Métayer, R. 2020, in The Trans-Neptunian Solar System, ed. D. Prialnik, M. A. Barucci, & L. Young (Amsterdam: Elsevier) 183
Hanner, M. S. 1999, SSRv, 90, 99
Hines, D. C., Holler, B. J., Mueller, M. M., & Wright, G. 2017, JWST Proposal, 1, 1272
Harrington Pinto, O., Kelley, M. S. P., Villanueva, G. L., et al. 2023, PSJ, 4, 208
Horner, J., Evans, N. W., & Bailey, M. E. 2004, MNRAS, 355, 321
Ivezić, Ž. & the LSST Science Collaboration 2013, LSST Science Requirements Document, *LSST LPM-17*
Ivezić, Ž. & the SCOC 2021, Survey Cadence Optimization Committee's Phase 1 Recommendation, LSST PSTN-053
Ivezić, Ž., Kahn, S. M., Tyson, J. A., et al. 2019, ApJ, 873, 111
Johnson, R. E., Oza, A., Young, L. A., Volkov, A. N., & Schmidt, C. 2015, ApJ, 809, 43
Jones, R. L., Yoachim, P., Ivezic, Z., Neilsen, E. H., & Ribeiro, T. 2021, Survey Strategy and Cadence Choices for the Vera C. Rubin Observatory Legacy Survey of Space and Time (LSST), LSST pstn-051
Jones, R. L., Slater, C. T., Moeyens, J., et al. 2018, Icar, 303, 181
Jurić, M., Eggl, S., Moeyens, J., & Jones, L. 2020, Proposed Modifications to Solar System Processing and Data Products, *LSST DMTN-087*
Kareta, T., Sharkey, B., Noonan, J., et al. 2019, AJ, 158, 255
Leiva, R., Ortiz, J. L., Gómez-Limón, J. M., et al. 2023, Asteroids, Comets, Meteors Conf. Vol. 2851, (Houston, TX: LPI) 2527
Levison, H. F., & Duncan, M. J. 1997, Icar, 127, 13
Licandro, J., Pinilla-Alonso, N., Stansberry, J., et al. 2023, Asteroids, Comets, Meteors Conf. Vol. 2851, (Houston, TX: LPI), 2851, 2255
Licandro, J., Pinilla-Alonso, N., Holler, & Bryan, J. 2025, NatAs, 9, 245
LSST Science Collaborations 2009, arXiv e-prints, arXiv:0912.0201
Luu, J. X., & Jewitt, D. C. 1990, AJ, 100, 913
Magnusson, P. 1986, Icar, 68, 1
Malamud, U., & Prialnik, D. 2015, Icar, 246, 21
Marchis, F., Berthier, J., Hestroffer, D., Descamps, P., & Merline, W. J. 2006, IAU Circ., 8666, 1
McKay, A., Bauer, J. M., DiSanti, M. A., et al. 2021, JWST Proposal, Cycle 1, ID. # 2416
McKay, A., Faggi, S., Villanueva, G. L., et al. 2023, DPS Meeting Abstracts Vol. 55, (Washington, DC: AAS) 400.04
McKinnon, W. B., Prialnik, D., Stern, S. A., & Coradini, A. 2008, in The Solar System Beyond Neptune, ed. M. A. Barucci, et al. (Tucson, AZ: Univ. Arizona Press) 213
Métayer, R., Guilbert-Lepoutre, A., Ferruit, P., et al. 2019, FrASS, 6, 8
Morbidelli, A., & Nesvorný, D. 2020, in The Trans-Neptunian Solar System, ed. D. Prialnik, & M. A. Barucci (Amsterdam: Elsevier) 25

Myers, J., Jones, L., & Axelrod, T. 2013, Moving Object Pipeline System Design *LSST LDM-156*
Naghib, E., Yoachim, P., Vanderbei, R. J., Connolly, A. J., & Jones, R. L. 2019, AJ, 157, 151
Noll, K. S., Grundy, W. M., Chiang, E. I., Margot, J. L., & Kern, S. D. 2008, in The Solar System Beyond Neptune, ed. M. A. Barucci, et al. (Tucson, AZ: Univ. Arizona Press) 345
Ortiz, J. L., Sicardy, B., Camargo, J. I. B., Santos-Sanz, P., & Braga-Ribas, F. 2020, in The Trans-Neptunian Solar System, ed. D. Prialnik, & M. A. Barucci (Amsterdam: Elsevier) 413
Ortiz, J. L., Duffard, R., Pinilla-Alonso, N., et al. 2015, A&A, 576, A18
Ortiz, J. L., Pereira, C. L., Sicardy, B., et al. 2023, A&A, 676, L12
Peixinho, N., Delsanti, A., & Doressoundiram, A. 2015, A&A, 577, A35
Pinilla-Alonso, N. 2024, AAS Meeting Abstracts, Vol. 244, (Washington, DC: AAS) 116.01
Pinilla-Alonso, N., Brunetto, R., De Pra, M. N., et al. 2025, NatAs, 9, 230
Pinilla-Alonso, N., Licandro, J., Brunetto, R., et al. 2024, A&A, 692, L11
Poston, M. J., Mahjoub, A., Ehlmann, B. L., et al. 2018, ApJ, 856, 124
Rabinowitz, D. L., Schaefer, B. E., & Tourtellotte, S. W. 2007, AJ, 133, 26
Rieke, M. J., Kelly, D. M., Misselt, K., et al. 2023, PASP, 135, 028001
Santos-Sanz, P., French, R. G., Pinilla-Alonso, N., et al. 2016, PASP, 128, 018011
Santos-Sanz, P., Ortiz, J. L., Sicardy, B., et al. 2021, MNRAS, 501, 6062
Santos-Sanz, P., Gomes Júnior, A., Morgado, B., et al. 2023, DPS Meeting Abstracts Vol. 55, (Washington, DC: AAS) 301.07
Schaller, E. L., & Brown, M. E. 2007, ApJ, 659, L61
Schambeau, C., McKay, A., Faggi, S., et al. 2023, AGUFM, 55, P44B–06
Schwamb, M. E., Jones, R. L., Chesley, S. R., et al. 2018, arXiv e-prints, arXiv:1802.01783
Schwamb, M. E., Fraser, W. C., Bannister, M. T., et al. 2019, ApJS, 243, 12
Schwamb, M. E., Jones, R. L., Yoachim, P., et al. 2023, ApJS, 266, 22
Shchuko, O. B., Shchuko, S. D., Kartashov, D. V., & Orosei, R. 2014, P&SS, 104, 147
Sheppard, S. S., Lacerda, P., & Ortiz, J. L. 2008, in The Solar System Beyond Neptune, ed. M. A. Barucci, H. Boehnhardt, D. P. Cruikshank, A. Morbidelli, & R. Dotson (Tucson, AZ: Univ. Arizona Press) 129
Sickafoose, A. A., Levine, S. E., Bosh, A. S., et al. 2023, PSJ, 4, 221
Souza-Feliciano, A. C., Holler, B. J., Pinilla-Alonso, N., et al. 2024, A&A, 681, L17
Tegler, S. C., Bauer, J. M., Romanishin, W., & Peixinho, N. 2008, in The Solar System Beyond Neptune, ed. M. A. Barucci, et al. (Tucson, AZ: Univ. Arizona Press) 105
Tegler, S. C., Romanishin, W., Consolmagno, G., & J, S. 2016, AJ, 152, 210
Tegler, S. C., Romanishin, W., Consolmagno, G. J., et al. 2005, Icar, 175, 390
Vara-Lubiano, M., Benedetti-Rossi, G., Santos-Sanz, P., et al. 2022, A&A, 663, A121
Vernazza,, et al. 2023, JWST Proposal, 2, 2820
Womack, M., & Stern, S. A. 1999, SoSyR, 33, 187
Wong, I., & Brown, M. E. 2017, AJ, 153, 145
Wong, I., Protopapa, S., Brunetto, R., et al. 2023, JWST Proposal, Cycle 2, ID. #4621
Wright, G. S., Rieke, G. H., Glasse, A., et al. 2023, PASP, 135, 048003
Yoachim, P., & Becker, M. R. 2023, lsst-sims/sims_featureScheduler_runs3.3, Zenodo, doi:10.5281/zenodo.10126869
Yoachim, P., Jones, L., Eric, H., Neilsen, J., et al. 2022, lsst/rubin_sim: Version 1.0, Zenodo, doi:10.5281/zenodo.10028614

Chapter 15

Cosmochemistry and Astrobiology of Centaurs as Remnants of Icy Planetesimals

M Telus and Z Martins

Nebular volatiles such as H_2O and CO_2 can be preserved as ice in trans-Neptunian objects (TNOs), but we have limited access to these volatiles for detailed chemical and isotopic analyses. Centaurs are remnants of icy planetesimals, formerly TNOs; they now reside in dynamically unstable orbits between Jupiter and Neptune. Some Centaurs migrate inward, becoming Jupiter-family comets (JFCs). Centaurs may hold a more pristine and accessible record of nebular ices compared to comets, which should be more processed by solar radiation. Chemical analyses of Centaurs could provide important constraints on the composition of nebular ice, a chemical reservoir that is not well-understood but has important implications for the dynamical and chemical evolution of the solar nebula and for the delivery of volatiles and prebiotic organic matter to the inner solar system and terrestrial planets. Here, we evaluate the potential for geochemical exploration of Centaurs to address outstanding questions in cosmochemistry (i.e., the study of the chemical composition of and the processes that led to those compositions in the universe) and astrobiology (i.e., the study of the origin, evolution, distribution, and future of life in the universe).

15.1 Introduction

The protoplanetary disk formed from the collapse of a molecular cloud core. Conditions in the protoplanetary disk varied with radial distance and time (e.g., Williams & Cieza 2011). Immediately, nebular gas, ice and dust accreted, resulting in km-sized objects called planetesimals (see, e.g., Chapter 2) with compositions that were rocky or ice-rich depending on their formation locations (e.g., Genda 2016). Today, remnants of these planetary building blocks predominantly reside in the Asteroid Belt (2–3 au) and trans-Neptunian populations (including the classical Kuiper Belt at 30–50 au and more distant populations), reservoirs of primordial rocky and icy material, respectively (see Figure 15.1). Spectral designations of small

doi:10.1088/2514-3433/ada267ch15

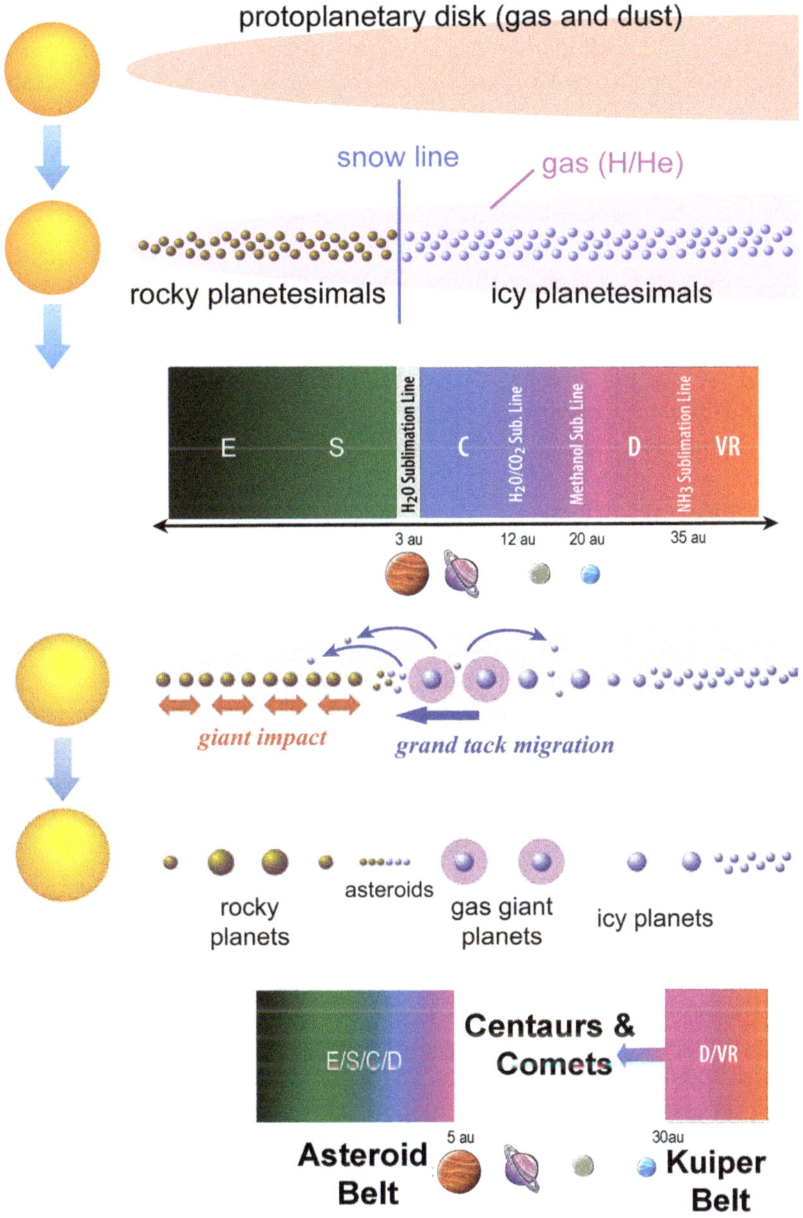

Figure 15.1. Summary of planet formation and giant planet migration (modified from Genda 2019, with permission from Springer Nature, and the CORAL mission concept study). Spectral types of small bodies are linked to their formation locations. Centaurs are primitive small bodies that were originally trans-Neptunian objects (which includes the classical Kuiper belt population just beyond Neptune), but currently reside between the orbits of Jupiter and Neptune. Centaurs are remnants of icy planetesimals that potentially provide a pristine and accessible record of nebular ices.

bodies in these regions (E, S, C, D, etc.) are linked to their formation locations. Giant planet migration and radial mixing played an important role in redistributing icy planetesimals and their volatiles (Levison et al. 2009; Brown 2012; O'Brien et al. 2014; Wong & Brown 2017). Spectroscopic observations of trans-Neptunian objects (TNOs) and Centaurs vary significantly, many are featureless, some have water ice absorption bands, while others do not (Cruikshank 2005; Barucci et al. 2008; Barkume et al. 2008). Advances in ground-based (e.g., NSF-DOE Vera Rubin Observatory) and space-based telescopes (e.g., the NASA James Webb Space Telescope) will certainly revolutionize our understanding of the composition of these distant faint objects (Pinilla-Alonso et al. 2024; see also Chapter 14). The possibility of landed *in situ* compositional exploration of a Centaur (as described in the CORAL[1] mission concept study) would facilitate interpretation of spectral features and provide geochemical constraints on the ice, organics, and refractory dust that accreted to form icy planetesimals and the fluids that altered them (see Chapter 15 on Centaur missions).

Meteorites, extraterrestrial rocks that fall to the Earth's surface, provide a detailed chemical record of the accretion, composition, timing, and thermal evolution of planetesimals (e.g., Hutchison 2004), but this is almost exclusively rocky material from the Asteroid Belt. Nebular volatiles such as H_2O and CO_2 are preserved as ice in TNOs (e.g., Schaller & Brown 2007; Brown 2012), but we have limited access to these objects for detailed chemical analyses in the form of interplanetary dust particles (IDPs) and comet 81P/Wild samples returned via the Stardust mission (Brownlee et al. 2012; Keller & Flynn 2022). The lack of detail regarding ice-rich planetesimals is an outstanding issue in cosmochemistry (i.e., the study of the chemical composition and the processes that led to those compositions in the universe; McSween & Huss 2022) and astrobiology (i.e., the study of the origin, evolution, distribution, and future of life in the universe; Kolb 2019) but is key to understanding the composition of nebular ice and the delivery of volatiles and prebiotic organic matter to the inner solar system and terrestrial planets.

Centaurs are an enigmatic class of primitive small bodies that were originally TNOs, but currently reside between the orbits of Jupiter and Neptune. Centaurs can migrate further inward and transition to Jupiter family comets (JFCs), or they can scatter back outward (see Chapter 3). Therefore, Centaurs represent a potentially more pristine and accessible record of icy planetesimals compared to comets that have been processed by solar heating (e.g., Gkotsinas et al. 2022). The lack of differences in colors, spectra and water ice fraction between Centaurs and TNOs of similar sizes is a strong indicator of the pristine nature of these objects (Brown 2012; Wong & Brown 2017; see also Chapters 5 and 6). Given their closer proximity to Earth compared to the distant TNOs and their relatively pristine chemical record compared to comets, Centaurs offer a unique opportunity to study the composition of ice, dust and organics in ice-rich planetesimals and to constrain the isotopic composition of nebular ice in unprecedented detail. In this chapter, we evaluate the

[1] CORAL Decadal mission concept study https://smd-cms.nasa.gov/wp-content/uploads/2023/10/coral-Centaur-orbiter-and-lander.pdf

potential for geochemical exploration of Centaurs to address outstanding questions in cosmochemistry and astrobiology.

15.2 Composition of Centaurs Compared to Chondritic Material

Chondrites are very similar in composition to the Sun's photosphere except for highly volatile elements like H, C, O, N, and noble gasses (Hutchison 2004). The bulk composition of chondrites can vary significantly, which is partly related to the formation location of their parent bodies in the protoplanetary disk (Genda 2016, 2019). Ordinary chondrites are the most abundant meteorites (>90%). They are rocky in composition, mostly containing olivine, pyroxene, and Fe-metal. Based on compositional similarities with S-type asteroids (Nakamura et al. 2011), ordinary chondrites likely formed within 3 au with water ice abundances of <1 wt.% (Alexander et al. 2010) and with varying amounts of metal (Hutchison 2004). Enstatite chondrites are volatile-poor with highly reduced compositions consisting of Fe-metal, pure enstatite ($MgSiO_3$), and Fe-sulfide (Keil 1989; Righter et al. 2006).

Carbonaceous chondrites vary widely in their composition (Hutchison 2004). CIs and CMs, the most volatile-rich carbonaceous chondrites, are composed of over 60% phyllosilicates, hydrated silicates (Howard et al. 2009; King et al. 2015). Their mineralogies are consistent with an environment with significant amounts of water ice that subsequently melted, leading to extensive aqueous alteration, including the formation of phyllosilicates, the oxidation of metal, and the formation of carbonates (McSween et al. 2002). CI chondrites are considered compositionally primitive meteorites, those with compositions most similar to the Sun's photosphere (McSween & Huss 2022). Pristine carbonaceous material returned from asteroids Bennu and Ryugu share many chemical and mineralogical similarities with CI and CM chondrites but are enriched in fluid-mobile elements (Lauretta et al. 2024).

Chondritic IDPs and ultracarbonaceous antarctic micrometeorites are small (< 50 μm) and highly primitive material based on their abundances of carbon, presolar grains, amorphous silicates and isotopically anomalous organic matter (Koschny et al. 2019; Busemann et al. 2009; Dobrica et al. 2009). Chondritic porous IDPs have been linked to TNOs based on their relatively long exposure ages of > 1 Myr, observed for both anhydrous and hydrous IDPs (Keller & Flynn 2022), supporting spectral observations that show signatures of aqueously altered minerals on some TNOs (de Bergh et al. 2004; Seccull et al. 2024). Geochemical analyses of Centaurs can provide further constraints on the link between chondritic material and TNOs.

15.3 Potential Occurrence of Rocky Chondritic Components in Centaurs

Detailed exploration of Centaurs could reveal whether high-temperature chondritic components occur in TNOs as observed for Wild 2 cometary particles from the NASA Stardust mission and chondritic IDPs. Chondritic meteorites are cosmic sedimentary material that formed in the nebula. They did not experience melting, preserving the rocky components that went into building planetesimals. Chondrites contain the oldest rocky material, calcium aluminum rich inclusions (CAIs), which constrain the age of the solar system (4.568 Ga; Bouvier & Wadhwa 2010; Connelly et al. 2012; Desch et al. 2023). CAIs are high-temperature condensates formed from a gas of Solar composition at >2000 K, likely very close to the protosun followed by rapid transport to planetesimal forming regions. Chondrules are the most abundant igneous inclusions in chondrites. They also formed at high temperatures (~1200 K–1400 K) from molten dust and particles that cooled rapidly in the nebula only a few million years after CAIs (Connolly & Jones 2016). Hydrated silicates, called phyllosilicates, permeate the matrix of chondrites along with submicron circumstellar condensates called presolar grains (Scott & Krot 2014; Zinner 2014). These heat-sensitive materials require peak temperatures <600°C (Huss et al. 2006). Thus, chondrites are a mixture of high- and low-temperature nebular components that formed under a wide range of conditions.

CAI- and chondrule-like material have been reported for IDPs and Stardust samples returned from the coma of comet Wild 2 (Zolensky 1987; Brownlee et al. 2012). The occurrence of refractory inclusions in Wild 2 was unexpected, since CAIs are thought to have formed within 0.1 au and Wild 2 likely originated in the TNO populations. The occurrence of high-temperature condensates in cometary material points to extensive radial mixing of high-temperature material from the inner solar system with low-temperature material in the outermost regions of the protoplanetary disk (Brownlee et al. 2012). Thus far, olivine has been inferred from the spectra of Centaur (5145) Pholus (Cruikshank et al. 1998). The abundance of refractory material (e.g., CAI- and chondrule-like material) in Centaurs is unknown, but could answer questions about the extent of radial mixing between the inner and outer regions of the solar nebula.

15.4 Accretion Times and Thermal History of Icy Planetesimals

Petrological and geochemical studies of the thermal metamorphism of ordinary and enstatite chondrites indicate that their bodies formed within 2 Myr after CAIs and were 100–200 km in radius, with the heat source for thermal metamorphism being primarily from the decay of ^{26}Al (Blackburn et al. 2017; Trieloff et al. 2022; Siron et al. 2022). Enstatite chondrite parent bodies formed within 1 au, consistent with their reduced mineralogy and compositional similarities with the Earth (Shukolyukov & Lugmair 2004; Piani et al. 2020). Thermal evolution of carbonaceous chondrite parent bodies is controlled by numerous factors, especially accretion time, size, ice fraction, and permeability (Castillo-Rogez & Young 2017; Bland & Travis 2017). Models indicate that aqueously altered carbonaceous chondrite parent bodies must have accreted within ~2.5 Myr after CAIs, otherwise heating from the decay of ^{26}Al would have been insufficient to melt ice (Wakita & Sekiya 2011). These chondrites may be derived from ice-rich planetesimals that formed beyond the Asteroid Belt, but interior to the orbit of Saturn (Alexander et al. 2018). The accretion times for Centaurs can potentially be inferred through the detection of phyllosilicates such as those observed in primitive IDPs (Keller & Flynn 2022). Although the surfaces of TNOs do not reach temperatures necessary to achieve and maintain liquid water (>273 K), impact cratering at the surface of Centaurs may reveal signatures of aqueous alteration signatures at depth, which would be important for constraining their accretion times (Golabek & Jutzi 2021; Seccull et al. 2024).

15.5 Centaurs as a Record of the Composition of Nebular Ice

In the outer solar system, nebular gas was locked into different ices (H_2O, CO_2, CO, etc.) that are thought to have condensed at various distances from the Sun and then incorporated into growing planetesimals. Some carbonaceous chondrites (CI, CM, and CR) show evidence for accretion of up to 80% water ice (Clayton & Mayeda 1999). Studies have attempted to constrain the composition of this accreted ice from analyses of secondary minerals, those that formed during fluid-rock alteration, also referred to as aqueous alteration (Alexander et al. 2015; Telus et al. 2019). However, the uncertainties are large for these indirect approaches. *In situ* geochemical analyses of Centaurs could provide a more direct approach for studying the composition of nebular gas and ice reservoirs. In addition to water ice, absorption bands for methanol and organic solids have been detected on Centaurs (Barucci et al. 2008). Spectral similarities for Centaurs and TNOs indicate that their compositions have been relatively undisturbed by inward scatter from the trans-Neptunian region (Brown 2012). Analyses of such ices from a Centaur could provide insight into nebular gas chemistry and the chemical evolution of the protoplanetary disk. H, C, O, and N isotope analyses of these ices could be compared to chondrites and comets as a first order approach. Furthermore, the isotopic composition of water ice and organics from a Centaur would provide constraints on the origin of hydrated asteroids, which may have been scattered into the Asteroid Belt during giant planet migration (Figure 15.1; Walsh et al. 2011). Expectations for H, N, and C isotope ratios for scenarios where TNOs are either unfractionated from solar, similar to comets, or isotopically heavier than comets are shown in Figures 15.2 and 15.4.

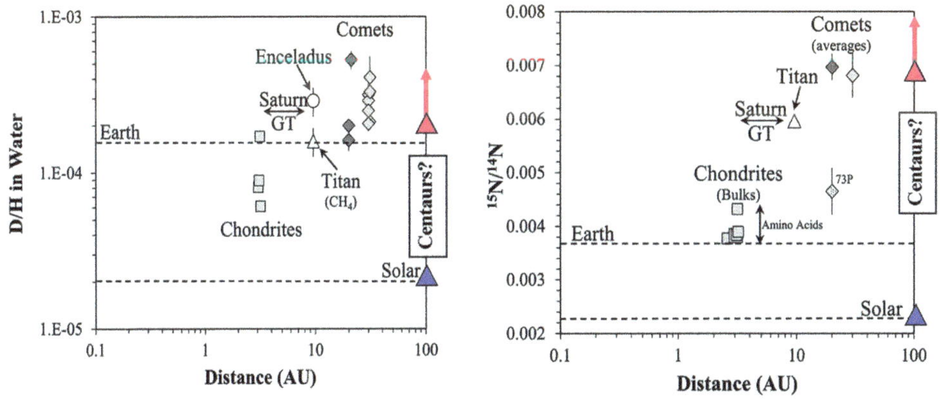

Figure 15.2. Hydrogen-isotope (left) and nitrogen-isotope (right) compositions for Jupiter-family comets (dark gray symbols), Oort cloud comets (light gray symbols), and various other solar system objects. The potential range for Centaurs is shown, ranging from unfractionated solar (blue triangle), similar to comets (pink triangle), and/or isotopically heavier than comets (pink arrow). Reproduced from Alexander et al. (2018), with permission from Springer Nature.

Models that simulate the isotope effects of volatile loss during the migration of TNOs are necessary for understanding how devolatilization could affect the isotopic composition of icy planetesimals.

15.5.1 H and N Isotopes

The hydrogen isotope ratio (D/H) of water in the nebula is expected to increase systematically with distance from the Sun because H is preferentially lost relative to deuterium (D) at low temperatures (Roberts et al. 2004). However, there are debates about whether D/H increased monotonically with distance from the Sun. The data for chondrites and comets show a general increase in D/H; however, there is a large spread in the D/H ratio of comets that may be related to solar heating (Alexander et al. 2018; also see the left panel of Figure 15.2). The nitrogen isotopic composition of the sun, chondrites, and comets follow a similar pattern as those for D/H ratios (right panel of Figure 15.2). Centaurs have not crossed inward past Jupiter's orbit and have not experienced the extensive devolatilization of water ice that comets have been subjected to (Brown 2012; Gkotsinas et al. 2022). Therefore, Centaurs likely preserve a more pristine record of nebular water ice. Analysis of primordial ice from Centaurs could provide insight into the factors influencing the observed variations in D/H and N isotopes (Figure 15.2), whether this is related to secondary processing on chondrite parent bodies or is a nebular signature associated with planetesimal formation regions.

15.5.2 O Isotopes

The bulk O-isotope composition of chondrites ($\delta^{17}O$ and $\delta^{18}O$, relative to ^{16}O) vary by ~20‰ (‰ = per thousand) and are the predominant method for classifying chondrites into groups (left panel of Figure 15.3). The source of these variations is unclear, but they may be associated with distinct nebular reservoirs that were influenced by varying degrees of CO self-shielding and/or mixing between ^{16}O-rich anhydrous dust and ^{16}O-poor nebular water (Lyons & Young 2005). CO self-shielding in the nebula would result from photodissociation of CO in the inner solar system that preferentially affects $C^{16}O$ due to the abundance of ^{16}O (99.8%) compared to ^{17}O (0.04%) and ^{18}O (0.2%), resulting in a reservoir of dust that is depleted in ^{17}O and ^{18}O and a reservoir of nebular H_2O vapor that is enriched in these isotopes (Lyons & Young 2005). This is consistent with mass-independent O-isotope variations observed for chondrites, chondrules, and CAIs in chondrites (Tartèse et al. 2018). Although variations in the O-isotope signatures are associated with nebular reservoirs, there is no obvious correlation with formation location. The O-isotope compositions of cometary dust grains are similar to terrestrial and CAI composition, while that for cometary water vapor exhibit a wider range (Paquette et al. 2018; right panel of Figure 15.3). Nebular water is expected to be ~200‰ (for both $\delta^{17}O$ and $\delta^{18}O$) or heavier if estimates from primitive chondrites are accurate (Sakamoto et al. 2007; also see inset in the left panel of Figure 15.3). Primordial water ice is expected to be preserved near the surface of Centaurs since they have not been subjected to high temperatures (Brown 2012). However, due to the relatively

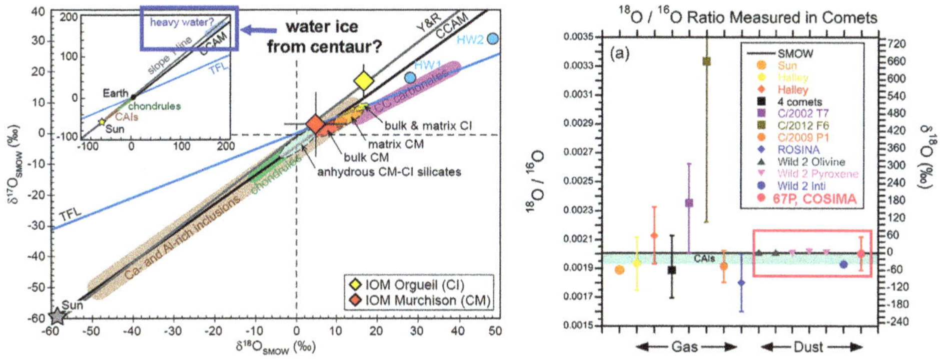

Figure 15.3. Oxygen-isotope composition of chondrite components (left, reproduced with permission from Tartèse et al. 2018) and comets (right, reproduced with permission from Paquette et al. 2018). Nebular ice preserved near the surface of a Centaur is expected to be enriched in the heavy isotopes of oxygen. Delta notation (δ^{17}O and δ^{18}O) is the deviation of ^{18}O/^{16}O or ^{17}O/^{16}O from a reference value (SMOW) multiplied by 1000. IOM refers to inorganic matter. HW1 and HW2 are estimates for the composition of water on the CM chondrite parent body. The composition of the Sun is determined from the Genesis mission. The CCAM and Y&R are slope ∼1 lines that describe the O-isotopic composition of many solar system objects (e.g., Sun, CAIs, chondrules, and terrestrial planets). The cometary gas measurements are from spectroscopic analyses of H_2O from various comets, while cometary dust analyses are from returned samples from the Stardust mission and from *in situ* analyses of 67P from Rosetta. See Tartèse et al. (2018) and Paquette et al. (2018) for details.

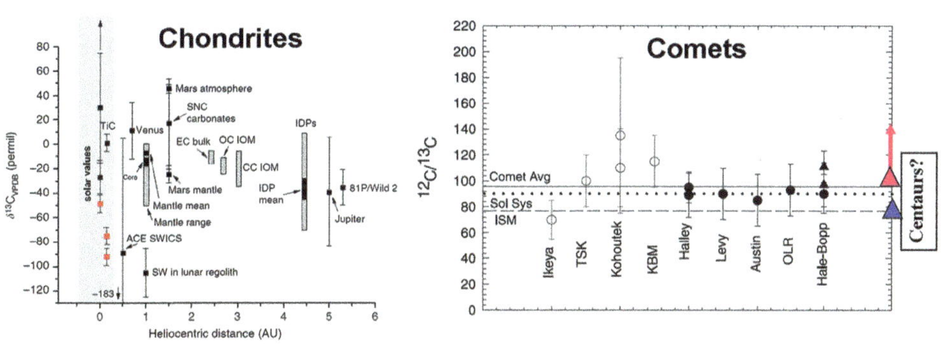

Figure 15.4. Carbon isotopes for various planetary materials and the Sun (left; reproduced from Lyons et al. 2018. CC BY 4.0) and comets (right; from Wyckoff et al. 2000; © 2000. The American Astronomical Society. All rights reserved. Printed in U.S.A.). Solar values are from measurements of solar wind from various sources, including the Genesis mission (red squares) and CO absorption in the sun's photosphere (black squares). The comet isotopes are based on analyses of C_2 (open circles), CN (filled circles) and HCN (filled triangles) from Wyckoff et al. (2000). Delta notation, δ^{13}C, is the deviation of ^{13}C/^{12}C from a reference value (VPDB) multiplied by 1000. Note the differences in notation for the y-axes for these two plots. The potential range in C isotope ratios for Centaurs is indicated; e.g., similar to ISM and/or solar (blue triangle), similar to average comets (pink triangle) or enriched in ^{13}C relative to comets (pink arrow).

small magnitude of the O-isotopic variations (i.e., ∼100‰–200‰ variation for δ^{17}O vs ∼1000‰ for δD and δ^{15}N), analysis of the O-isotopic composition of nebular ice from a Centaur may be challenging without a high-resolution mass spectrometer or a cryogenic sample-return mission.

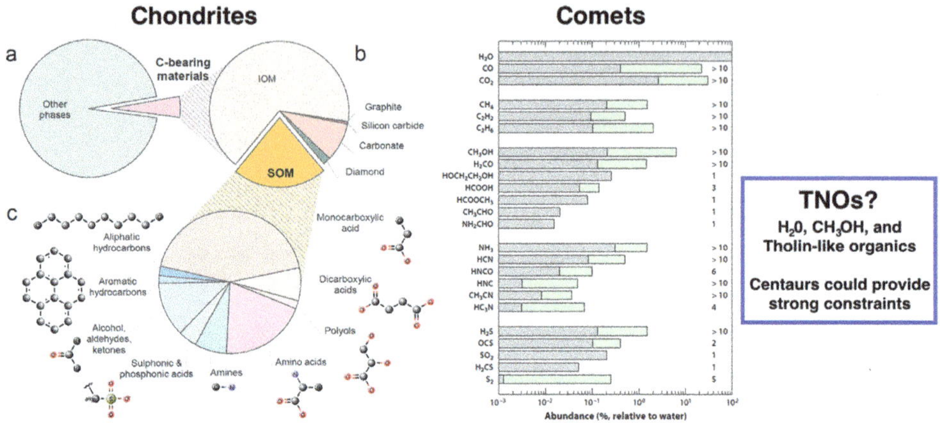

Figure 15.5. Left: The carbon phases in carbonaceous chondrites include the insoluble organic matter (IOM), soluble organic matter (SOM), graphite, silicon carbide, carbonates and diamonds. The SOM fraction is based on the distribution and molecular structures of organic molecules present in the Murchison meteorite (from Chan & Zolensky 2022. Copyright 2022, with permission from Elsevier). Center: Average molecular composition of comets with comet-to-comet range in variation shown in green (Reproduced with permission from Mumma & Charnley 2011, © Annual Reviews). Knowledge of the organic content of TNOs is relatively limited (Dalle Ore et al. 2015).

15.5.3 C Isotopes

Similar to oxygen, carbon isotopes do not vary systematically with radial distance (left panel of Figure 15.4). The C-isotopic composition $\delta^{13}C$ for insoluble organic matter (IOM) in meteorites (see Figure 15.5), planets, and comets do not vary by more than ~100‰. The C-isotopic composition of the Sun varies between spectroscopic analyses and those determined from the Genesis mission. If solar wind values from the Genesis mission are taken to be the most accurate (red squares in the left panel of Figure 15.4), then values for most rocky material and refractory organics show ~20‰–40‰ enrichment in ^{13}C (right panel of Figure 15.4). The $\delta^{13}C$ composition of nebular ice (e.g., CO_2 and CO) is unclear, but Centaurs could provide more constraints. They may be similar to solar if they retained an unfractionated composition. They may be similar to comets if devolatilization of CO and other carbon-bearing ices has significantly influenced the $\delta^{13}C$ composition. Another possibility is that Centaurs are enriched in ^{13}C compared to solar because they retain more interstellar organic components (Hashizume et al. 2004). C isotope analyses of Centaurs are beginning to be investigated with the ground-breaking capabilities of JWST (Faggi et al. 2024). Detailed geochemical analyses of a Centaur would provide valuable insight into the factors controlling the C-isotopic variations of planetary materials.

15.6 The Origin of Organic Matter in the Solar Nebula

15.6.1 Organic Matter in Chondrites

The fine-grained matrix of pristine chondrites contains organic compounds in the form of solvent-insoluble and solvent-soluble organic matter (referred to as IOM and SOM, respectively). Organic matter in chondrites mainly consists of polycyclic aromatic hydrocarbons (up to 5 wt% C). IOM contains two fractions (Figure 15.6): one labile, which is enriched in D, ^{13}C and ^{15}N, and containing an aliphatic-rich and heteroatom component, and one refractory, depleted in ^{13}C and ^{15}N, and containing an aromatic-rich and heteroatom-poor component (Sephton et al. 2003, 2004; Remusat et al. 2007). The D and ^{15}N enrichments measured in bulk carbonaceous chondrites and hotspots of the IOM (i.e., regions that are extremely isotopically enriched, with δD values of up to ∼20,000‰ and $\delta^{15}N$ values of up to 3200‰; Busemann et al. 2006; Remusat et al. 2007; Floss et al. 2014; Figure 15.7), suggest

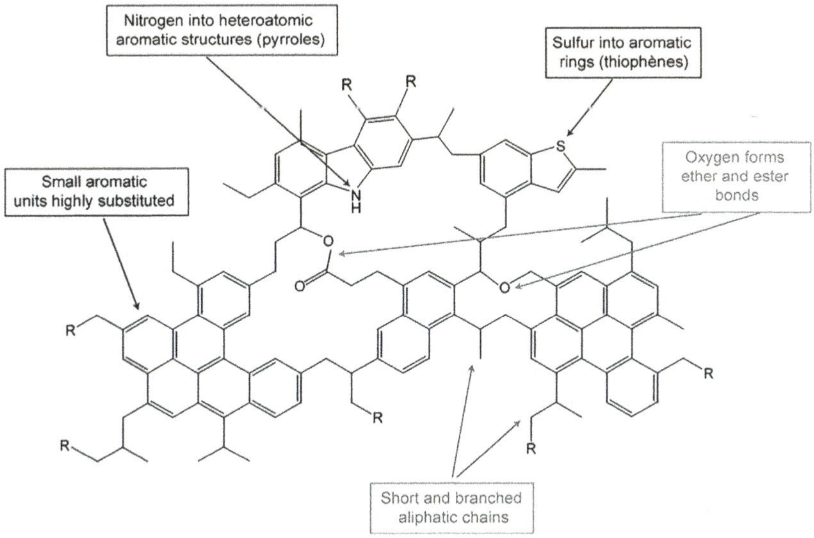

Figure 15.6. Molecular information on the chemical structure of insoluble organic matter (IOM) of carbonaceous chondrites. R refers to an organic moiety. Reprinted with permission from Remusat et al. (2007). © 2007 Académie des sciences.

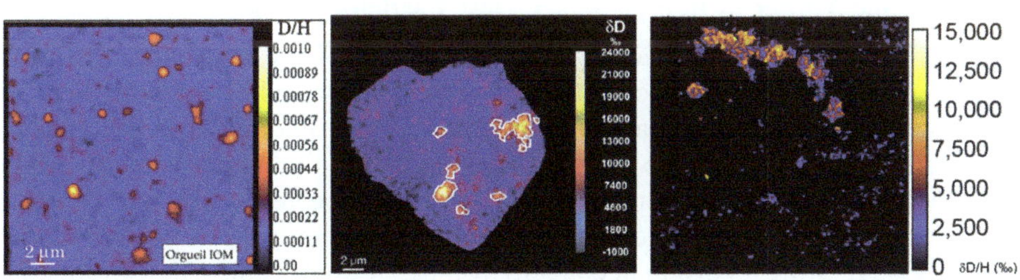

Figure 15.7. D/H isotopic maps showing hotspots in the IOM fraction of carbonaceous chondrites: Orgueil (left), EET 92042 (center), and QUE 99177 (right). The isotopic maps were measured by Nanoscale secondary ion mass spectrometry (NanoSIMS). Left panel reprinted with permission from Remusat et al. (2007) © 2007 Académie des sciences. Middle panel from Busemann et al. (2006). Reprinted with permissions from AAAS. Right panel reprinted from Alexander et al. (2017) with permission from Elsevier. © 2017 Elsevier GmbH. All rights reserved.

that the IOM of carbonaceous chondrites was formed in the ISM, then later modified by nebular and parent body processes, such as aqueous alteration and thermal metamorphism (Robert & Epstein 1982; Yang & Epstein 1983; Remusat et al. 2006; Alexander et al. 2007).

The abundance of free polycyclic aromatic hydrocarbons (PAHs) in CM chondrites may not be correlated with the degree of alteration (Lecasble et al. 2022). Relatively large PAHs are depleted in ^{13}C (for the pyrene series, the $\delta^{13}C$ range from −0.7 to −14.7‰ for the Koland meteorite, 2.3 to −14‰ to the Mukundpura meteorite, and 0.5 to −14.8‰ to the Aguas Zarcas), indicating an ISM origin (Lecasble et al. 2022). On the other hand, free PAHs were depleted in D, which points to secondary processes involving hydrogen exchange with water during aqueous alteration. This is consistent with analyses of the Paris meteorite showing a lack of alkylated PAHs, which may be related to the low degree of aqueous alteration on its parent body (Martins et al. 2015).

Prebiotic soluble organic compounds such as amino acids, proteins, and carbohydrates also occur in pristine chondrites in relatively trace concentrations (Cooper et al. 2001; Martins & Sephton 2009). Stable isotope analyses of the total amino acid fractions of the Murchison meteorite showed $\delta D = +1370‰$, $\delta^{15}N = +90‰$ and $\delta^{13}C = +23.1‰$ (Epstein et al. 1987), while the stable isotope results of individual amino acids in different carbonaceous meteorites were highly enriched in D, ^{15}N, and ^{13}C (for a review see Martins & Sephton 2009). Elsila et al. (2007) determined whether Strecker synthesis or radical–radical reactions were responsible for amino acid formation in interstellar ices by using isotopic labeling techniques in interstellar ice analogues. They found that the amino acid formation in interstellar ices by UV photolysis is unlikely because an unreasonably high UV flux would be necessary, which is in clear contrast with the low resistance to UV photolysis of amino acids. It is more likely that the precursors of the soluble meteoritic amino acids were synthesized in the ISM, while amino acids were synthesized during secondary processes in the planetesimal.

15.6.2 Organic Matter in Interplanetary Dust Particles (IDPs)

The organic matter in carbonaceous meteorites and IDPs have differences (e.g., carbonaceous chondrites have only a few wt% of carbon, while the bulk elemental abundance of carbon in IDPs can be as high as 45 wt%; Thomas et al. 1993) as well as similarities (for a review, see Martins et al. 2020). While it is possible to analyze the soluble and insoluble matter in meteorites, due to the very small mass available for IDPs, bulk analyses are used instead. The organic matter in IDPs is enriched in D and ^{15}N, suggesting an origin in cold environments, such as the ISM or outer solar nebula (Alexander et al. 2007, 2010). Furthermore, Busemann et al. (2009) found hydrogen and nitrogen isotopic anomalies of the bulk organic matter in IDPs, similar to the most primitive meteorites, suggesting a similar original mixture of molecular cloud material, followed by significant inward transport (Ciesla 2009). In parallel, the analysis by Busemann et al. (2009) supported a common origin for the most primitive anhydrous IDPs and comets, as previously suggested (Messenger 2002; Bradley 2003).

15.6.3 Organic Matter in Comets and Centaurs

The refractory carbon in comet 67P/Churyumov-Gerasimenko is similar to chondritic IOM (Fray et al. 2016). The elemental composition of $C_{100}H_{80}N_4O_{20}S_2$ was determined for particles from comet Halley (Kissel & Krueger 1987). Hydrocarbon and methanol ices and organics with tholin-like spectral signatures have been identified on the surfaces of Centaurs (Dalle Ore et al. 2015). Given that some Centaurs evolve into JFCs, it is not surprising that comets and Centaurs share some spectral and physical similarities. Spectral models indicate that Centaurs can be nearly identical to the mean composition of a comet nucleus, i.e., water ice, silicate dust, complex organic matter, and light hydrocarbons.

As an example, the Centaur 2012 DR_{30} contains surface ice and tholins, with 60% of complex organics (30% of Titan and 30% of Triton tholins), 30% of ice (pure water and water with inclusions of complex organics), 10% of silicates, and a low limit of amorphous carbons (Szabó et al. 2018). Also, the spectrum of sunlight reflected from the Centaur (5145) Pholus was observed and modeled by Cruikshank et al. (1998), indicating the presence a mixture of Titan tholins, water ice, silicate dust (olivine), amorphous carbon, and methanol ice. Approximately 15% of all known Centaurs display comet-like activity (e.g., Jewitt 2009; Galiazzo et al. 2019; see also Chapter 8). The coma of Centaur (60558) 174P/Echeclus sporadically becomes active (Wierzchos et al. 2017), and other Centaurs have presented some activity throughout their history, e.g., 29P/Schwassmann–Wachmann 1, (2060) Chiron, (10199) Chariklo, (5154) Pholus, 165P/Linear, 39P/Oterma, 165P/Linear, (8405) Asbolus, and C/2001 M10 (NEAT) (Bockelée-Morvan et al. 2001; Jewitt 2009; Duffard et al. 2014; see also Chapter 10).

The abundances, distribution, and isotopic composition of primitive organic compounds from Centaurs would provide crucial information on whether they originate from the interstellar medium (e.g., Namouni & Morais 2018, 2020), from the outer solar system (Morbidelli et al. 2020), and/or from secondary processes (aqueous alteration and/or thermal metamorphism) in planetesimals. If organics in Centaurs are interstellar in origin, then their composition should be enriched in D and ^{15}N due to low-temperature gas-phase ion-molecule reactions and solid-phase reactions on interstellar ice grains by high-energy particles (e.g., Tielens 1983; Yang & Epstein 1983; Millar et al. 1989; Terzieva & Herbst 2000; Sandford et al. 2001; Robert 2003; Aléon & Robert 2004). Conclusive evidence of the abundance, distribution, and isotopic composition of organic matter in Centaurs remains to be established by future space missions. In the meantime, analysis of carbonaceous meteorites, IDPs, and comets provide fundamental clues to rebuild the synthesis and evolution of organic matter in Centaurs.

15.7 Centaurs as Storage and Transport of Prebiotic Material

Ice is ubiquitously present in the outer solar system, with bodies composed or covered in ice, such as comets, "primitive" asteroids, TNOs, and Centaurs

(Bockelée-Morvan et al. 2000, 2004; Crovisier et al. 2004; Ehrenfreund & Charnley 2000; Greenberg & Hage 1990; Mumma et al. 1993; Barucci et al. 1996; Tegler 2007). Ice samples would preserve volatiles and organic compounds delivered to the early Earth-Moon system by comets and/or asteroids, providing a window into the past of the inner solar system. Analyses of these ices would also provide clues into the chemical reactions that happen on the surface of icy outer solar system bodies. Chemical reactions occurring on the surface of icy bodies in the outer solar system are driven by UV photons (to depths of up to one centimeter), ionizing radiation (to as deep as a meter or two below the surface; Barnett et al. 2012), and thermal processes. Radiation driven processes that occur on the surface of outer solar system bodies have been extensively studied by using photochemical laboratory experiments on icy materials (Baratta et al. 2002; Cooper et al. 2003; Hand & Carlson 2011; Hudson & Moore 2001; Johnson et al. 2012; Kaiser 2002; see also Chapter 13). However, low-temperature thermochemistry of the icy bodies has not been studied in detail. This gap in our knowledge is crucially relevant because the thermochemistry studies of icy bodies are particularly important to determine whether organic compounds present at different depths (and therefore different temperatures) have been thermally altered over time, and if so, which are the degradation products. The study of Centaur samples would therefore provide a fundamental opportunity to analyze prebiotic material available in the early solar system, as Centaurs are intermediate between TNOs and active comets, as discussed in the previous section. The existing gaps in the organic content of primitive celestial bodies can be filled through a future space mission to Centaurs.

15.8 Geochemical Exploration of Centaurs

15.8.1 Mission Concept Studies

Cosmochemistry often focuses on objects that provide the most pristine record of nebular material that inform us about the starting compositions and reservoirs of the protoplanetary disk. Thus, for a mission to have the most cosmochemical impact, it will be important to target Centaurs that preserve the most pristine record of nebular gas, ice and dust near the surface. Spectral analyses of Centaurs and similarly sized TNOs do not show any systematic differences in color or spectral signatures, which indicates most Centaurs should be viable pristine records of icy planetesimals (Wong & Brown 2017; see also Chapter 5). Chemical analyses of Centaurs could be compared to the composition of chondrites, comets, and TNOs to address questions about the isotopic composition of nebular reservoirs and the implications for nebular and/or parent body conditions that led to the observed chemical variations.

Multiple mission concept studies have been evaluated to explore Centaurs, with most focusing on understanding comet-like activity (see Chapter 16). Centaurs are up to hundreds of kilometers in size, relatively large compared to comets. Centaurs have varying features (e.g., comet-like activity, rings, color, size) therefore fly-by missions such as that proposed for Centaurus (Singer et al. 2019) would allow for a broad characterization of this population of small bodies. The mission concept study

Chiron Orbiter[2] (National Research Council 2011) and the proposed Chimera mission (Harris et al. 2019) evaluated remote orbital observations of a single Centaur. Centaur characterization in orbit would allow for detailed spectral studies of the surface along with potential for analyses of dust and gas during outgassing events. From a cosmochemistry and astrobiology perspective, studies of outgassing of Centaurs could shed light on the composition of the interior, especially regarding the composition of gases being released and potentially the size and composition of dust particles. Mission objectives that focus on analyzing gas and dust from comet-like activity could be applied to understanding ice and refractory components of icy planetesimals.

Centaur ORbiter and Lander (CORAL), a mission concept study to carry out geochemical analyses of the surface of a Centaur was evaluated for the 2020 Decadal Survey and found to be viable within a New Frontiers 6 budget (NASEM 2023). CORAL was designed for landed analyses at two different locations on the surface. This mission would use a mass spectrometer on a lander to carry out H, C, O, and N isotope analyses, and characterize molecular abundances and distribution. A mission such as CORAL could provide revolutionary insight into our understanding of the composition and evolution of icy planetesimals, the isotopic composition of nebular ice, and the concentration of prebiotic organics. Given the current challenges with cryogenic sample return, such a mission has not yet been tested for Centaurs. If samples could be well-preserved for return to the Earth, then this would provide opportunities for detailed and high precision laboratory analyses of nebular ice.

15.8.2 Analyzing H, C, O, and N Abundances and Isotopic Composition

The CORAL lander would drill the surface and use PlanetVac (i.e., a flexible, low-cost sample collection technology) to bring material into a sample carousel for analyses. To analyze the composition precisely, care must be taken to standardize each analysis. One idea would be to have a position on the carousel for solid reference materials, similar to the standard mounts used for laboratory electron microprobe analyses. The frequency of standard analyses should also be considered (e.g., at the beginning of each session). The capabilities of mass spectrometers that have operated on recent missions (e.g., Rosetta) could readily distinguish differences of $\sim$500–1000‰ expected for H and N isotopes. However, isotopic standards of known compositions will be important for C and O isotope analyses since they are not expected to vary by more than 100‰.

15.8.3 Analyzing Organic Compounds

To fully understand the chemical and physical conditions of the early solar system and determine what was the prebiotic inventory delivered to the early Earth, it is necessary to analyze the organic content of primitive samples, ideally, samples

[2] Chiron Orbiter Decadal mission concept study https://nap.nationalacademies.org/reports/13117/App%20G%2022_Chiron_Orbiter.pdf

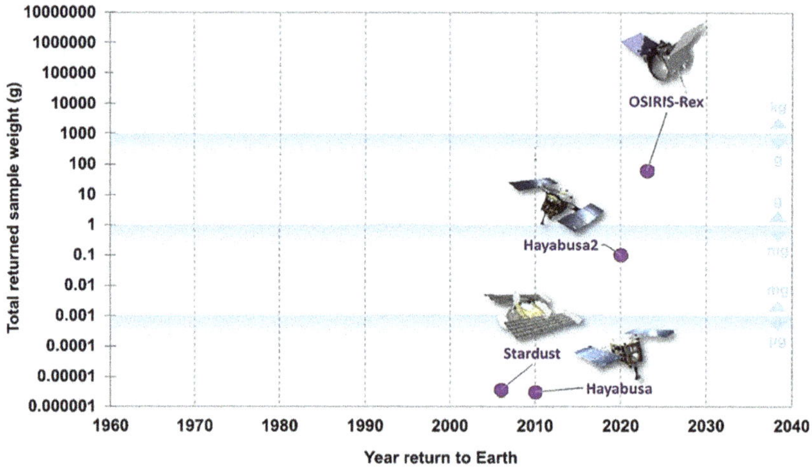

Figure 15.8. Summary of sample return missions, in which the samples are analyzed for organic matter in terrestrial laboratories, and the corresponding total targeted mass of the returned samples. Minimum target total mass of returned samples is shown for Hayabusa 2 (target: 0.1–10 g) and OSIRIS-Rex (target: 0.06–2 kg). Adapted from Chan & Zolensky et al. (2022), Copyright 2022, with permission from Elsevier. NASA, JAXA, CNSA.

returned from an asteroid or comet still not analyzed, or a meteorite missing from museum collections. In that sense, an in-situ or a sample return mission from a Centaur would fill in a gap in knowledge, informing the scientific community about the comet-asteroid continuum (e.g., Hsieh 2017) and Centaurs as the transition state between TNOs and JFCs (e.g., Levison & Duncan 1997). A cryogenic sample return of at least a few dozen grams of material from a Centaur would enable a full set of laboratory analyses, including measurement of a wide range of organic compounds as well as isotopic analysis (e.g., Chan et al. 2020; Yada et al. 2021; Figure 15.8). Given that cryogenic sample return is challenging, the CORAL mission concept study proposed using a combination of GCMS and UV/Raman spectrometry for landed analyses of organic composition and isotopic analyses at the surface of a Centaur. Again, repeated standard analyses will be key for ensuring reliable results.

15.8.4 Curation for Possible Sample Return

A sample return mission from a Centaur would have to pay special attention to preserving the samples, while at the same time minimizing terrestrial organic contamination, by using several mitigation actions (Martins et al. 2020; Chan et al. 2020). These include controlling the conditions of the sample return capsule (e.g., temperature, and gas composition), the handling procedure from landing site up to curation facility, and of course the need for a curation facility in which the specimens would be curated under low temperature and neutral atmosphere. Despite all these precautions, a minimal level of terrestrial contamination will always be present, but the use of state-of-the-art equipment in the laboratory will allow distinguishing between extra-terrestrial organic matter and terrestrial contamination (Chan et al. 2020).

15.9 Future Perspectives and Conclusions

Centaurs represent a more pristine record of icy planetesimals compared to comets and more accessible targets compared to TNOs. Detailed geochemical analyses of a single Centaur could revolutionize our understanding of the composition of nebular ice and the accretion and evolution of icy planetesimals. Below we list a series of questions that could be addressed with either landed geochemical analyses and/or cryogenic sample return from a Centaur:

- What are the compositions of nebular ice, dust, and organics in icy planetesimals?
- How does the composition relate to the observed variations in surface colors of TNOs?
- What was the extent of radial mixing between the innermost and outermost regions of the nebula?
- How did H, C, O, and N isotopes vary across the nebula?
- What is the origin of carbonaceous chondrites, especially CI and CM, and are they related to TNOs?
- Does impact cratering at the surface reveal differences between the interior and surface of icy planetesimals?
- Which volatiles, if any, control activity on Centaurs and how does this affect the composition of the surface?
- How does irradiation affect the composition of the surface, especially prebiotic organics?
- Do TNOs have an interstellar origin or interstellar components?
- What are the geochemical links between JFCs and TNOs?

Geochemical exploration of Centaurs should target those that preserve the most pristine record of nebular ice near the surface but are also accessible. It will be important to take advantage of state-of-the-art tools such as JWST and the Rubin Observatory to identify the best candidates. Cryogenic sample return would be ideal, but issues with costs and challenging curation need to be addressed. A 2020 Planetary Science Decadal mission concept study to carry out orbital and landed science of a Centaur (CORAL) is feasible within the New Frontiers 6 budget and would be of broad interest to the planetary science community and the public.

Acknowledgements

We thank the editors and two anonymous reviewers for their thoughtful comments, which resulted in significant improvements to this chapter. M. Telus is supported by NSF CAREER GLOW (AST-2239225) and NASA's Interdisciplinary Consortia for Astrobiology Research (NNH19ZDA001N-ICAR) under grant number 80NSSC21K0597. Z. Martins, the Institute of Molecular Science, and Centro de Química Estrutural acknowledges financial support from Fundação para a Ciência e

Tecnologia (FCT) through projects UIDB/00100/2020, UIDP/00100/2020, and LA/P/0056/2020.

References

Aléon, J., & Robert, F. 2004, Icar, 167, 424
Alexander, C. M. O. D., Newsome, S. D., Fogel, M. L., et al. 2010, GeCoA, 74, 4417
Alexander, C. M. O., Bowden, R., Fogel, M. L., & Howard, K. T. 2015, M&PS, 50, 810
Alexander, C. M. O., Fogel, D., Yabuta, M., & Cody, G. D., H. 2007, GeCoA, 71, 4380
Alexander, C. M. O., Cody, D., De Gregorio, G. D., Nittler, B. T., & Stroud, R. M., L. R. 2017, ChEG, 77, 227
Alexander, C. M. O., McKeegan, K. D., & Altwegg, K. 2018, SSRv, 214, 36
Baratta, G. A., Leto, G., & Palumbo, M. E. 2002, A&A, 384, 343
Barkume, K. M., Brown, M. E., & Schaller, E. L. 2008, AJ, 135, 55
Barnett, I. L., Lignell, A., & Gudipati, M. S. 2012, ApJ, 747, 13
Barucci, M. A., Brown, M. E., Emery, J. P., & Merlin, F. 2008, in The Solar System Beyond Neptune, ed. M. A. Barucci, H. Boehnhardt, D. P. Cruikshank, A. Morbidelli, & R. Dotson (Tucson, AZ: Univ. Arizona Press) 143
Barucci, M. A., Fulchignoni, M., & Lazzarin, M. 1996, P&SS, 44, 1047
Blackburn, T., Alexander, C. M. O., Carlson, R., & Elkins-Tanton, L. T. 2017, GeCoA, 200, 201
Bland, P. A., & Travis, B. J. 2017, SciA, 3, e1602514
Bockelée-Morvan, D., Crovisier, J., Mumma, M. J., & Weaver, H. A. 2004, in Comets II, ed. M. C. Festou, H. U. Keller, & H. A. Weaver (Tucson, AZ: Univ. Arizona Press) 391
Bockelée-Morvan, D., Lis, D. C., Wink, J. E., et al. 2000, A&A, 353, 1101
Bockelée-Morvan, D., Lellouch, E., Biver, N., et al. 2001, A&A, 377, 343
Bouvier, A., & Wadhwa, M. 2010, NatGe, 3, 637
Bradley, J. P. 2003, in Treatise on Geochemistry–Vol. 1, ed. A. M. Davies (Amsterdam: Elsevier) 711
Brown, M. E. 2012, AREPS, 40, 467
Brownlee, D., Joswiak, D., & Matrajt, G. 2012, M&PS, 47, 453
Busemann, H., Young, A. F., Alexander, O. D., C. M., et al. 2006, Sci, 312, 727
Busemann, H., Nguyen, A. N., Cody, G. D., et al. 2009, E&PSL, 288, 44
Castillo-Rogez, J., & Young, E. D. 2017, in Planetesimals: Early Differentiation and Consequences for Planets, ed. L. T. Elkins-Tanton, & B. P. Weiss (Cambridge: Cambridge Univ. Press) 92
Chan, Q. H. S., & Zolensky, M. E. 2022, in New Frontiers in Astrobiology, ed. R. Thombre, & P. Vaishampayan (Amsterdam: Elsevier) 67
Chan, Q. H. S., Stroud, R., Martins, Z., & Yabuta, H. 2020, SSRv, 216, 56
Ciesla, F. J. 2009, Icar, 200, 655
Clayton, R. N., & Mayeda, T. K. 1999, GeCoA, 63, 2089
Connelly, J. N., Bizzarro, M., Krot, A. N., et al. 2012, Sci, 338, 651
Connolly, H. C., & Jones, R. H. 2016, JGRE, 121, 1885
Cooper, G., Kimmich, N., Belisle, W., et al. 2001, Natur, 414, 879
Cooper, J. F., Christian, E. R., Richardson, J. D., & Wang, C. 2003, EM&P, 92, 261
Crovisier, J., Bockelée-Morvan, D., Biver, N., et al. 2004, A&A, 418, L35
Cruikshank, D. P. 2005, in The Outer Planets and their Moons, ed. T. Encrenaz, et al. (Berlin: Springer) 421

Cruikshank, D. P., Roush, T. L., Bartholomew, M. J., et al. 1998, Icar, 135, 389
Dalle Ore, C. M., Barucci, M. A., Emery, J. P., et al. 2015, Icar, 252, 311
de Bergh, C., Boehnhardt, H., Barucci, M. A., et al. 2004, A&A, 416, 791
Desch, S. J., Dunlap, D. R., Williams, C. D., Mane, P., & Dunham, E. T. 2023, Icar, 402, 115611
Dobrica, E., Engrand, C., Duprat, J., et al. 2009, M&PS, 44, 1643
Duffard, R., Pinilla-Alonso, N., Santos-Sanz, P., et al. 2014, A&A, 564, A92
Ehrenfreund, P., & Charnley, S. B. 2000, ARA&A, 38, 427
Elsila, J. E., Dworkin, J. P., Bernstein, M. P., Martin, M. P., & Sandford, S. A. 2007, ApJ, 660, 911
Epstein, S., Krishnamurthy, R. V., Cronin, J. R., Pizzarello, S., & Yuen, G. U. 1987, Natur, 326, 477
Faggi, S., Villanueva, G. L., McKay, A., et al. 2024, NatAs, 8, 1237
Floss, C., Le Guillou, C., & Brearley, A. 2014, GeCoA, 139, 1
Fray, N., Bardyn, A., Cottin, H., et al. 2016, Natur, 538, 72
Galiazzo, M. A., Silber, E. A., & Dvorak, R. 2019, MNRAS, 482, 771
Genda, H. 2016, GeocJ, 50, 27
Genda, H. 2019, AsBio, 197
Gkotsinas, A., Guilbert-Lepoutre, A., Raymond, S. N., & Nesvorny, D. 2022, ApJ, 928, 43
Golabek, G. J., & Jutzi, M. 2021, Icar, 363, 114437
Greenberg, J. M., & Hage, J. I. 1990, ApJ, 361, 260
Hand, K. P., & Carlson, R. W. 2011, Icar, 215, 226
Harris, W., Woodney, L., & Villanueva, G. 2019, EPSC-DPS Joint Meeting 2019 (Washington, DC: AAS) EPSC–DPS2019–1094
Hashizume, K., Chaussidon, M., Marty, B., & Terada, K. 2004, ApJ, 600, 480
Howard, K. T., Benedix, G. K., Bland, P. A., & Cressey, G. 2009, GeCoA, 73, 4576
Hsieh, H. H. 2017, RSPTA, 375, 20160259
Hudson, R. L., & Moore, M. H. 2001, JGR, 106, 33275
Hutchison, R. 2004, Meteorites: A Petrologic, Chemical and Isotopic Synthesis (Cambridge: Cambridge Univ. Press) 520
Huss, G. R., Rubin, A. E., & Grossman, J. N. 2006, in Meteorites and the Early Solar System II, ed. D. S. Lauretta, & H. Y. McSween Jr. (Tucson, AZ: Univ. Arizona Press) 567
Jewitt, D. 2009, AJ, 137, 4296
Johnson, P. V., Hodyss, R., Chernow, V. F., Lipscomb, D. M., & Goguen, J. D. 2012, Icar, 221, 800
Kaiser, R. I. 2002, ChRv, 102, 1309
Keil, K. 1989, Metic, 24, 195
Keller, L. P., & Flynn, G. J. 2022, NatAs, 6, 731
King, A. J., Schofield, P. F., Howard, K. T., & Russell, S. S. 2015, GeCoA, 165, 148
Kissel, J., & Krueger, F. R. 1987, Natur, 326, 755
Kolb, V. M. (ed) 2019, Handbook of Astrobiology (1st ed.; Boca Raton, FL: CRC Press)
Koschny, D., Soja, R. H., Engrand, C., et al. 2019, SSRv, 215, 34
Lauretta, D. S., Connolly, H. C., Aebersold, J. E., et al. 2024, M&PS, 59, 2453
Lecasble, M., Remusat, L., Viennet, J.-C., Laurent, B., & Bernard, S. 2022, GeCoA, 335, 243
Levison, H. F., Bottke, W. F., Gounelle, M., et al. 2009, Natur, 460, 364
Levison, H. F., & Duncan, M. J. 1997, Icar, 127, 13

Lyons, J. R., & Young, E. D. 2005, Natur, 435, 317
Lyons, J. R., Gharib-Nezhad, E., & Ayres, T. R. 2018, 49th Annual Lunar and Planetary Science Conf. (Houston, TX: LPI) 2907
Martins, Z., & Sephton, M. A. 2009, in Amino Acids, Peptides and Proteins in Organic Chemistry, ed. A. B. Hughes (New York: Wiley-VCH)
Martins, Z., Modica, P., Zanda, B., & D'Hendecourt, L. L. S. 2015, M&PS, 50, 926
Martins, Z., Chan, Q. H. S., Bonal, L., King, A., & Yabuta, H. 2020, SSRv, 216, 54
McSween, H. Y., & Huss, G. R. 2022, Cosmochemistry (Cambridge: Cambridge Univ. Press)
McSween, H. Y., Ghosh, A., Grimm, R. E., Wilson, L., & Young, E. D. 2002, in Asteroids III, ed. W. F. Bottke, Jr., et al. (Tucson, AZ: Univ. Arizona Press) 559
Messenger, S. 2002, M&PS, 37, 1491
Millar, T. J., Bennett, A., & Herbst, E. 1989, ApJ, 340, 906
Morbidelli, A., Batygin, K., Brasser, R., & Raymond, S. N. 2020, MNRAS, 497, L46
Mumma, M. J., Weissman, P. R., & Stern, S. A. 1993, in Protostars and Planets III, ed. E. H. Levy, & J. I. Lunine. (Tucson, AZ: Univ. Arizona Press) 1177
Mumma, M. J., & Charnley, S. B. 2011, ARA&A, 49, 471
Nakamura, T., Noguchi, T., Tanaka, M., et al. 2011, Sci, 333, 1113
Namouni, F., & Morais, M. H. M. 2020, MNRAS, 494, 2191
Namouni, F., & Morais, M. H. M. 2018, MNRAS, 477, L117
National Academies of Sciences, Engineering, and Medicine (NASEM) 2023, Origins, Worlds, and Life: A Decadal Strategy for Planetary Science and Astrobiology 2023–2032 (Washington, DC: The National Academies Press)
National Research Council 2011, Vision and Voyages for Planetary Science in the Decade 2013–2022 (Washington, DC: The National Academies Press)
O'Brien, D. P., Walsh, K. J., Morbidelli, A., Raymond, S. N., & Mandell, A. M. 2014, Icar, 239, 74
Paquette, J. A., Engrand, C., Hilchenbach, M., et al. 2018, MNRAS, 477, 3836
Piani, L., Marrocchi, Y., Rigaudier, T., et al. 2020, Sci, 369, 1110
Pinilla-Alonso, N., Licandro, J., Brunetto, R., et al. 2024, A&A, 692, L11
Remusat, L., Palhol, F., Robert, F., Derenne, S., & France-Lanord, C. 2006, E&PSL, 243, 15
Remusat, L., Robert, F., & Derenne, S. 2007, CRGeo, 339, 895
Righter, K., Drake, M. J., & Scott, E. R. D. 2006, in Meteorites and the Early Solar System II, ed. D. S. Lauretta, & H. Y. McSween Jr. (Tucson, AZ: Univ. Arizona Press) 803
Robert, F., & Epstein, S. 1982, GeCoA, 46, 81
Robert, F. 2003, SSRv, 106, 87
Roberts, H., Herbst, E., & Millar, T. J. 2004, A&A, 424, 905
Sakamoto, N., Seto, Y., Itoh, S., et al. 2007, Sci, 317, 231
Sandford, S. A., Bernstein, M. P., & Dworkin, J. P. 2001, M&PS, 36, 1117
Schaller, E. L., & Brown, M. E. 2007, ApJ, 659, L61
Scott, E. R. D., & Krot, A. N. 2014, in Meteorites and Cosmochemical Processes, Volume 1 of Treatise on Geochemistry, (2nd ed.) ed. A. M. Davis (Amsterdam: Elsevier) 65
Seccull, T., Fraser, W. C., Kiersz, D. A., & Puzia, T. H. 2024, PSJ, 5, 42
Sephton, M. A., Love, G. D., Watson, J. S., et al. 2004, GeCoA, 68, 1385
Sephton, M. A., Verchovsky, A. B., Bland, P. A., et al. 2003, GeCoA, 67, 2093
Shukolyukov, A., & Lugmair, G. W. 2004, GeCoA, 68, 2875

Singer, K. N., Stern, S. A., Stern, D., Verbiscer, A., & Olkin, C. 2019, EPSC-DPS Joint Meeting 2019 (Washington, DC: AAS) EPSC-DPS2019-2025
Siron, G., Fukuda, K., Kimura, M., & Kita, N. T. 2022, GeCoA, 324, 312
Szabó, G. M., Kiss, C., Pinilla-Alonso, N., et al. 2018, AJ, 155, 170
Tartèse, R., Chaussidon, M., Gurenko, A., Delarue, F., & Robert, F. 2018, PNAS, 115, 8535
Tegler, S. C. 2007, in Encyclopedia of the Solar System, ed. L. McFadden (Amsterdam: Elsevier) 605
Telus, M., Alexander, C. M. O., Hauri, E. H., & Wang, J. 2019, GeCoA, 260, 275
Terzieva, R., & Herbst, E. 2000, MNRAS, 317, 563
Thomas, K. L., Blanford, G. E., Keller, L. P., Klock, W., & McKay, D. S. 1993, GeCoA, 57, 1551
Tielens, A. G. G. M. 1983, A&A, 119, 177
Trieloff, M., Hopp, J., & Gail, H.-P. 2022, Icar, 373, 114762
Wakita, S., & Sekiya, M. 2011, EP&S, 63, 1193
Walsh, K. J., Morbidelli, A., Raymond, S. N., O'Brien, D. P., & Mandell, A. M. 2011, Natur, 475, 206
Wierzchos, K., Womack, M., & Sarid, G. 2017, AJ, 153, 230
Williams, J. P., & Cieza, L. A. 2011, ARA&A, 49, 67
Wong, I., & Brown, M. E. 2017, AJ, 153, 145
Wyckoff, S., Kleine, M., Peterson, B. A., Wehinger, P. A., & Ziurys, L. M. 2000, ApJ, 535, 991
Yada, T., Abe, M., Okada, T., et al. 2021, NatAs, 6, 214
Yang, J., & Epstein, S. 1983, GeCoA, 47, 2199
Zinner, E. 2014, in Meteorites and Cosmochemical Processes, Volume 1 of Treatise on Geochemistry, (2nd ed.) ed. A. M. Davis (Amsterdam: Elsevier) 181
Zolensky, M. E. 1987, Sci, 237, 1466

Chapter 16

Centaur Missions

Walter Harris, S Alan Stern and Geronimo L Villanueva

The combination of their intermediate evolutionary state and relative accessibility of the Centaurs compared to their trans-Neptunian progenitor population makes them compelling targets for spacecraft exploration. Over the past two decades, multiple concept studies and proposals have investigated potential Centaur missions in the Discovery and New Frontiers class. These studies demonstrate that multiple mission architectures are possible, ranging from flybys to landers. This chapter reviews the history of Centaur mission design within the context of our improving understanding of their diverse properties, the lessons learned from spacecraft encounters with the Jupiter-family comets, and the various trajectory, power, and propulsion options available. We then describe a strategy for Centaur exploration that addresses scientific priorities, includes several near-term targets of opportunity, and identifies areas where technical advances and observational resources are necessary.

16.1 Introduction

Ice-rich planetesimals are understood to be among the earliest objects of significant size in the early solar system and are collectively the building blocks of differentiated protoplanets and the cores of the giant planets. As such, the modern remnants of this ancient population are the most pristine recordings of the composition and conditions within the protoplanetary disk and of the processes governing how planetary systems form and evolve.

The size, shape, and interior structure of ice-rich planetesimals, combined with the physical state and relative abundances of various kinds of dust and volatiles of which they are composed, are our primary windows through which we can probe the epoch of formation. For small icy bodies in the deep cold of the extreme outer solar system, this physical and chemical record is believed to be retained in near-primordial form, subject only to modification by rare large impacts and slow surface evolution from micro-impact gardening, sublimation of highly volatile ices (i.e., CO and CO_2), and irradiation by cosmic UV and energetic particles. However,

doi:10.1088/2514-3433/ada267ch16

most of these objects are too small to be imaged directly and too faint for detailed characterization of their composition from even the most powerful modern telescopes. Instead, the vast majority of what we know about the composition of icy planetesimals has come from remote sensing and a series of spacecraft encounters with the tiny fraction of the residual population that are perturbed into orbits that penetrate deep into the inner solar system as long and short period comets. A common characteristic of these objects is that they represent an evolutionary endgame, where solar driven activity drives physical evolution that accumulates with repetitive encounters to gradually erase their surface features and modify the state, composition, and abundance of interior volatiles. This process is particularly acute for short period, Jupiter-family comets (JFCs), which have an active lifespan on the order of $\sim 10^4$ years (e.g., Levison & Duncan 1997). Hence, the comets that are most frequently observed and easiest to reach with spacecraft are also the most physically altered, complicating our ability to invert their properties back to a primordial state.

Within the above context, the significance of the Centaurs as a focus for the next stage of icy planetesimal exploration is clear. The outer solar system between Jupiter and Neptune is well beyond the water sublimation line, such that the Centaurs have not experienced the primary alteration processes that define the JFCs. In the nearly five decades since their presence was first confirmed, the number of known Centaurs has increased dramatically, along with our understanding of their physical characteristics, which represent a combination of the properties of their parent population and the range of intermediate evolutionary processes they experience as they migrate inward. This diversity of properties showcases their potentially transformative scientific value as targets for spacecraft encounter missions that are enabled by their relative proximity compared with trans-Neptunian objects (TNOs). This chapter reviews the history of Centaur discovery, provides a retrospective of their prioritization as targets for *in situ* exploration, and describes the various mission concepts developed to explore them. We then propose a framework for Centaur exploration spanning the next several decades that incorporates this earlier work, identifies key science questions and near-term targets of opportunity, and leverages the lessons learned from multiple JFC encounters. We also describe the technical challenges that will be faced by future mission designers and propose proactive actions the scientific community can pursue to mitigate them.

16.1.1 Centaur Discovery

Prior to the 20th century, the source regions of comets were a mystery. The first suggestion that 'non-periodic' comets may in fact be solar system bodies in closed elliptical orbits was made by A. O. Leuschner (Leuschner 1907). In 1932, E. Öpik showed that the gravitationally bound solar system potentially extended to heliocentric distances exceeding a parsec, and that objects within this spherical region could be the source of the distant-elliptical comets (Öpik 1932). This earlier work was subsequently carried forward by Oort (1950) into the now accepted theory that the long period comets originated from a remote cloud of icy planetesimals that had been scattered outward by interactions with the giant planets. Attempts to further link the Oort cloud as the source of the JFCs identified theoretically viable pathways (e.g., Everhart 1972), but failed to produce the number observed by several orders of magnitude (e.g., Joss 1973; Kresak & Pittich 1978). Alternative sources considered include migration from the Jupiter Trojan clouds (e.g., Rabe 1972) and a proposed reservoir of planetesimals located beyond Neptune (Edgeworth 1943, 1949; Kuiper 1951, 1974). The latter of these implied the existence of a minimally evolved transitional population of objects occupying the region between Jupiter and Neptune. As it turns out, the first member of this group had already been found some decades earlier.

In 1927, Arnold Schwassmann and Arno Wachmann discovered a comet with several unusual characteristics. 29P/Schwassmann-Wachmann 1 (SW1) was the first small body found with an orbit located entirely beyond Jupiter. The first observations captured it in outburst, and it was subsequently shown to have persistent activity throughout its low eccentricity orbit, with multiple brightening events per year (e.g., van Biesbroeck 1928; Whipple 1930; Roemer 1958, Jewitt 1990). Even though its size is about an order of magnitude larger than most comet nuclei, SW1's initial designation was as a JFC. However, its high level of activity at a heliocentric distance beyond the water sublimation region was clearly anomalous compared to other comets that faded to near-inactivity at large distances.

SW1 would retain its status as a unique presence in the outer solar system for 50 years until the discovery of another small body at large heliocentric distance in the giant planet region, (2060) Chiron (Kowal & Gehrels 1977; Kowal et al. 1979); this was the first object to bear the designation of Centaur. Chiron's unstable Saturn-crossing orbit (e.g., Hahn & Bailey 1990) implied that it must have been perturbed into the giant planet region from an external reservoir. Fernandez (1980) then explored the efficiency of producing the JFCs via perturbations of a trans-Neptunian population in the region from 35–50 au and concluded that Chiron was possibly a particularly bright member of a vast undetected collection of objects filtering through the giant planet region toward the inner solar system. The subsequent detection of photometric variability (Hartmann et al. 1990) and a coma (Meech & Belton 1990) from Chiron (now also designated 95P) demonstrated that it was cometary in nature, consistent with the trans-Neptunian origin hypothesis. This dynamical pathway would receive further confirmation from the near simultaneous discoveries of the first TNO, 1992 QB_1 (Jewitt & Luu 1993) and the third Centaur, (5145) Pholus (Scotti et al. 1992).

During the three decades following the discovery of 1992 QB_1 (now 15760 Albion) the pace at which new objects were identified, both in the giant planet and the trans-Neptunian regions, increased steadily as the quality and depth of surveys improved. By the end of 2000, 25 Centaurs had been found meeting the dynamical definition of Gladman et al. (2008; $q > 5.2$ au, $a < 30.1$ au, and not strongly dynamically coupled to Jupiter), and the number of known TNOs ($q > 30$ au) was 386.[1] Between then and 2024, the inventories in both regions have expanded by well

[1] Based on queries to JPL's small body database at ssd.jpl.nasa.gov/sbdq_query.html

over another order of magnitude. With even more capable facilities (e.g., the *Vera C. Rubin Observatory*) beginning to contribute over the next decade, the discovery rate is expected to increase further (see Chapter 14). This history is highly relevant to how missions to the Centaurs would be prioritized by the Planetary Decadal Surveys produced over this period.

As the number of Centaurs and TNOs has increased, so has our understanding of their physical and dynamical characteristics as they relate to and drive theoretical predictions (see e.g., Chapters 4, 5, and 7). Among TNOs, the orbital distribution, relative numbers, and photometric properties of the cold classical, resonant, and scattered disk populations have been revealed, providing physical context to predictions of their relationship to late-stage planetary migration and as the source of the Centaurs and JFCs (e.g., Malhotra 2019; Morbidelli & Levison 2007; Duncan & Levison 1997; see also Chapters 2 and 3). For Centaurs, the greater number of objects has facilitated characterization of the extent, type, and sources of activity, their photometric properties including color distributions, and the dynamical pathways leading toward the JFCs (e.g., Jewitt 2009; Tegler et al. 2008; Tiscareno & Malhotra 2003; Seligman et al. 2021; see also Chapters 3, 4, and 5 and Chapter 8). Moreover, a large inventory provides more potential mission designs with a closed trajectory for what, as the sections below show, is a challenging target with existing technology.

16.1.2 Centaur Characteristics

All the above has led to our current understanding of the distribution and evolutionary state of icy bodies shown in Figure 16.1. The detailed properties of the Centaurs are discussed in the previous chapters of this volume and will only be summarized here. As of this writing, there are 393 objects that meet the dynamical definition of Gladman et al. (2008) and several more that exceed its boundaries but have Centaur-like properties. The known population is observationally biased toward the largest objects (e.g., Bauer et al. 2013), especially at the greatest heliocentric distances (see Figure 1.4 in Chapter 1). Various models that follow perturbations from the Scattered Disk into the giant planet region imply a total population of 1 to 100 million Centaurs larger than 1 km in diameter (e.g., Di Sisto & Brunini 2007; Volk & Malhotra 2008). Their size distribution has been estimated using a variety of methods (e.g., Sheppard et al. 2000; Adams et al. 2014; see also Chapters 2 and 4) to follow a power-law distribution in the range of $0.4 \leqslant \alpha \leqslant 0.6$, with a possible break at smaller diameters (e.g., Lawler et al. 2018). Once they enter the region inside Neptune's orbit, the Centaurs random walk their way through the outer solar system via close encounters with the individual giant planets for a typical dynamical lifetime of 1–10 Myr, ending with $\sim$30% becoming JFCs and the rest either colliding with a planet or being ejected (Tiscareno & Malhotra 2003). A significant fraction of Centaur-JFC exchanges appears to occur in a 'gateway' region near Jupiter (e.g., Sarid et al. 2019; Steckloff et al. 2020), while others are be diverted inward from a series of Jupiter-Saturn mean motion resonances (Seligman et al. 2021).

As the inventory of Centaurs has increased, significant physical diversity has been revealed. Of the known objects, 39 ($\sim$10%) bear a periodic comet designation due to

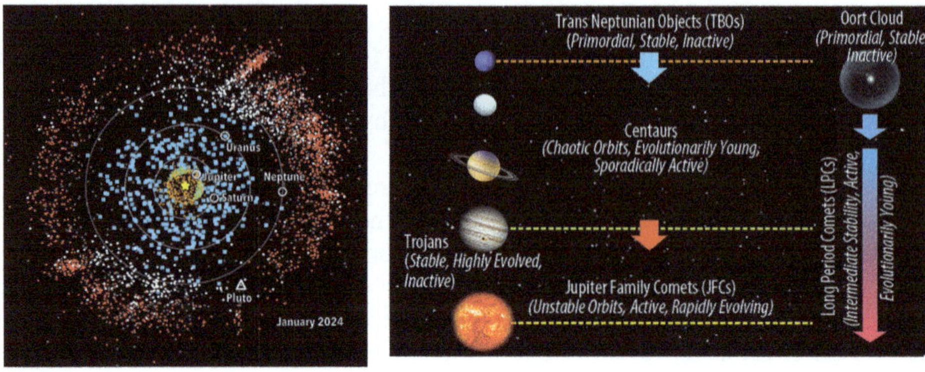

Figure 16.1. Left: A map showing the distribution of JFCs (orange triangles), Centaurs (blue squares), Jupiter Trojans (green), and TNOs (classical TNOs in red, 3:2 resonant TNOs in solid white, and other TNOs in open white circles) relative to the giant planets and Pluto in January 2024 (credit: K. Volk). Right: The distribution and state of icy planetesimals from the inner solar system to the Oort cloud is shown, along with the ranges over which major volatiles dominate sublimation and the expected level of physical evolution (Adapted from Harris et al. 2021. CC BY 4.0). The most significant evolutionary demarcation is the crystalline water-ice sublimation region inside of 3 au. Beyond this, CO and CO_2 are volatile out to the edge of the Kuiper Belt at 50 au.

evidence of activity. It isn't clear why some Centaurs are active when others aren't, though there is an identified correlation with perihelion distance (Jewitt 2009). Centaur activity occurs at all heliocentric distances along their orbits, with the extent varying from the century-long sustained quiescent and outburst behavior of 29P (e.g., Jewitt 1990; Trigo-Rodríguez et al. 2008), to the more sporadic activity of Chiron (e.g., Womack et al. 2017), to the outburst dominated 174P/Echeclus (e.g., Kareta et al. 2019); see Chapter 8 for further discussions of activity and Chapter 10 for details about specific notable active Centaurs. The volatile composition has been measured spectroscopically for only a few active objects (e.g., 29P, Chiron, 174P/Echeclus, 39P/Oterma; see also discussion in Chapter 6), though the higher sensitivity of JWST promises to improve this significantly (e.g., Faggi et al. 2024; Harrington Pinto et al. 2023; see Chapter 14). The sources of Centaur activity are not well understood. CO and CO_2 are the most common species detected, despite both being volatile well into the trans-Neptunian region (e.g., Bauer et al. 2015; Womack et al. 2017). Annealing and the transformation of amorphous water ice is frequently invoked as a potential source for both quiescent and outburst activity (e.g., Prialnik 1992), with recent observations potentially capturing this process for Chiron (Pinilla-Alonso et al. 2024). In addition to activity, the Centaurs display a diversity of color ranging from solar-like (gray) to extremely reddened. The Centaur color distribution is thought to be bimodal and may be distinct from their TNO progenitor population (e.g., Peixinho et al. 2012; see also Chapter 5). The origin of the color distribution remains a subject of ongoing study, with several processes (e.g., radiation, chemistry, impacts, activity) thought to play a role. A small number of giant-planet-crossing bodies have been observed to be binary objects (e.g., Grundy et al. 2007; whether they are classified as Centaurs depends on the choice of definition), while other Centaurs have been shown from stellar occultations to possess ring systems or debris shrouds (e.g., Braga-Ribas et al. 2014; Fernández-Valenzuela et al. 2017; Sickafoose et al. 2020); see Chapter 9 for a full discussion of Centaur rings, debris, and satellites.

16.1.3 A Case for Centaur Missions

It was understood at the time of Chiron's discovery that a transitional Centaur population would be an evolutionary middle ground between two bookend populations (i.e., the TNOs and JFCs) at the extremes of physical evolution (Figure 16.1). The period since has revealed a large collection of volatile-rich bodies in the Centaur region with a variety of physical characteristics that reflect both the properties of their trans-Neptunian source region and an evolutionary continuum that extends from the near-pristine to just up to the onset of the water sublimation that defines the JFCs. The palette of potential targets in the Centaur region is already substantial and will continue to grow with new and better observing tools at our disposal, as will our understanding of their activity patterns and surface composition. However, as we have learned from multiple robotic missions to the JFCs, *in situ* exploration provides a level of detail on composition, surface properties, activity patterns, and interior structure that Earth-based remote sensing cannot. This is especially true for Centaurs, which are far more distant and therefore more challenging for remote observation. The epoch of previous observations has provided a roadmap for developing mission traceability around several key questions, including, but not limited to

- What are the mechanisms driving Centaur quiescent and outburst activity?
- How are active regions distributed on Centaur surfaces?
- What is the composition and heterogeneity of outgassed volatiles?
- Why are some Centaurs active while others are not?
- What is the chemical composition and origin of color-defining surface properties?
- How does surface evolution vary among Centaurs relative to TNOs and the JFCs?
- What is the internal structure of the Centaurs?
- How are Centaurs physically similar or dissimilar to the Jupiter Trojans and TNOs?
- How do Centaur rings originate?
- How do icy planetesimals form?

The answers to these questions will reside in the mission design and target selections for a future series of missions to these unique objects. Here we provide a retrospective summary of the prioritization of Centaur exploration over the periods covered by the last three decadal surveys and the wide range of concepts that have been developed at various levels to address the questions stated above. The evolution of these concepts as the infrastructure of mission design has improved and the range of potential targets has expanded, demonstrates that a new era of outer solar system small body exploration is both feasible and necessary.

16.2 A Retrospective on Centaur Mission Prioritization and Concept Development

Serious discussion of the Centaurs as a community priority for robotic exploration dates to the first formal Planetary Science decadal survey carried out by the U.S. National Research Council (National Research Council (NRC) 2003). The subsequent 20+ years have been transformative for our understanding of the physical characteristics of primordial bodies and their numbers and size distributions, evolution, and activity patterns of the outer solar system population. To date, no mission involving a Centaur as a primary or secondary target has been advanced for flight development. However, a diverse collection of Centaur concepts has been explored, and in a few cases proposed formally. Encounter scenarios have run the gamut from flybys to orbiters to landed spacecraft employing a variety of instrument suites, trajectories, power systems, and propulsion designs at mission levels from Discovery-class to flagship. The assortment of the options explored reflects the large parameter space of environments and the science potential that Centaurs represent. As existing and future surveys discover more and smaller Centaurs and our ability to determine the extent and duration of activity improves, the portfolio of concepts will continue to evolve. In what follows the history of these efforts is broken down by the decadal survey periods that set the priorities for Centaur exploration. These earlier efforts serve as a template upon which they will be based.

16.2.1 Decadal Period from 2003 to 2013

The 2003 decadal survey exercise occurred during a still-early phase of small body robotic exploration, characterized by the 1986 Halley 'armada' of spacecraft encounters (e.g., Reinhard 1986), the flyby encounters of (951) Gaspra and (243) Ida-Dactyl system by *Galileo* (Belton et al. 1996), and subsequent Discovery-Class and smaller missions such as the Near-Earth Asteroid Rendezvous (*NEAR*) mission to orbit (433) Eros (Cheng 1997) and the flyby of 19P/Borrelly (Boice et al. 2002; Soderblom et al. 2002) by *Deep Space 1* (Figure 16.2). At the time the report was assembled, roughly 30 Centaurs had been identified, including most of the largest known objects and many of those have subsequently been studied as targets for exploration. Also, more ambitious JFC exploration was in process in the form of both small and large missions. The Discovery-Class *Stardust* dust collection mission (Brownlee et al. 2003) was en-route to a 2004 encounter with 81P/Wild, while ESA's Cornerstone-Class *Rosetta* (Glassmeier et al. 2007) awaited a 2004 launch after problems with the Ariane V vehicle forced a target switch from hyperactive comet 46P/Wirtanen to comet 67P/Churyumov-Gerasimenko. The *Deep Impact* mission (A'Hearn et al. 2005) to 9P/Tempel and the *Dawn* mission (Rayman et al. 2006; Russell & Raymond 2011) to the large asteroid Vesta and dwarf planet Ceres were in active development. Unfortunately, another JFC mission, the Comet Nucleus Tour (*CONTOUR*), was lost during a solid-rocket motor burn shortly after its mid-2002 launch (NASA 2023) and just prior to the release of the Decadal Survey. Acknowledging the ongoing role of smaller missions for achieving many science

Figure 16.2. Highlights of small body exploration prior to 2003. Upper left: 1P/Halley as imaged by *Giotto* (credit ESA/MPS). Upper right: (243) Ida and Dactyl, the first observed asteroid satellite (credit NASA/JPL/SSI). Lower left: (433) Eros from *NEAR* (credit NASA/JPL/JHUAPL). Lower right: *1*9P/Borrelly from *Deep Space 1* (Credit NASA/JPL).

objectives, the survey's primitive bodies panel put forward five 'medium class' objectives emphasizing surface sample return and the outer solar system. This list included a Trojan Asteroid/Centaur Reconnaissance (*TA/CR*) flyby mission using a notional solar powered spacecraft with a science suite consisting of imaging, spectroscopic, and radio science experiments that would target a Centaur inside of 6.5 au. *TA/CR* was ranked as a 'deferred priority', and it was not included in the portfolio for the second New Frontiers (NF2; NASA, AO_03_OSS_03) announcement of opportunity (AO) in 2003.

As an exercise leading up to the third New Frontiers (NF3) AO (NASA, NNH09ZDA007O), NASA requested an NRC mid-decadal review to provide guidance for the determining the next portfolio. The subsequent New Opportunities in Solar System Exploration (NOSSE) report (National Research Council (NRC) 2008) revisited the *TA/CR* concept in the light of advances in our understanding of small bodies from the now completed *Stardust* and *Deep Impact* primary missions, as well as the orbital and surface exploration of (25143) Itokawa by JAXA's *Hayabusa*. The report recommended *TA/CR* as one of eight candidate missions, and it was included in the NF3 portfolio as a result. While a *TA/CR* concept was not among the step-1 selections, the NOSSE report recommendation led to several design exercises that explored mission options in the NF class. Successive JPL Planetary Summer School Team X studies in 2008 and 2009 found viable Trojan-Centaur flyby solutions that included Centaur targets inside of 7 au. The 2008 mission summary (Klesh 2009; *SHOTPUT*) was based on a 2015 trajectory opportunity that enabled flybys of a main belt asteroid (2001 HM_{10}), the largest Trojan (624 Hektor) and its satellite (Skamandrios), and the active Centaur 39P/Oterma. The *SHOTPUT* spacecraft would have been solar powered and used chemical propulsion and a single Earth–Moon gravity assist to reach its targets over 7 years. It included a capable instrument suite consisting of a multi-spectral imager, a wide-angle camera, a thermal-infrared spectrometer, a mass spectrometer, and a UV spectrograph. A white paper based on the 2009 effort (*TRACER*; Ryan et al. 2009a, 2009b) detailed an 8 year mission that included two Earth gravity assists followed by flyby encounters of the Trojan (2207) Antenor and the Centaur 2001 BL_{41} with a spacecraft using solar electric propulsion (SEP) powered by a four panel 19.8 kW MultiFlex solar array and carrying science package consisting of a visible camera, visible/NIR spectrometer, a mass spectrometer, and a dust counter. In 2009, the Collaborative Modeling and Parametric Assessment of Space Systems (COMPASS) team published the results of a feasibility study for a more ambitious Trojan flyby/Centaur Rendezvous mission for New Frontiers (Oleson et al. 2009; Oleson & McGuire 2011). The COMPASS design effort focused on the use of Radioisotope Electric Propulsion (REP) for an 11 year mission launching in late 2024 that included a flyby of the Trojan (15651) Tlepolemus followed by orbit insertion around the Centaur (32532) Thereus near its perihelion at 8.5 au. The spacecraft concept used six Advanced Stirling Radioisotope Generators (ASRGs) containing a total of 5.4 kg of Pu-238 to provide 900 W of beginning of life (BOL) power. It carried a 57 kg science suite including wide and narrow field cameras, a near infrared spectrometer, a thermal emission imager, LIDAR, and a mass spectrometer. ASRGs, which were in active development at the time of this report, were the enabling technology for keeping the COMPASS mission design within the New Frontiers cost cap due to their much higher power generation efficiency and reduced plutonium requirement compared with existing Multi-Mission Radioisotope Thermal Generators (MMRTGs; NASA 2020). NASA awarded multiple concept studies for missions using an ASRG power system (Colozza & Cataldo 2015), and the technology showed promise for lower-cost exploration in the Discovery and New Frontiers classes. However, while ASRG development reached a late phase (Dudzinski et al. 2013), NASA ended the program in 2013 for budgetary reasons without ever constructing a flight version (Reich 2013).

16.2.2 Decadal Period from 2013 to 2023

The 2013–2023 decadal report (National Research Council (NRC) 2011) was written as *Dawn* (Russell & Raymond 2011) and *New Horizons* (Stern 2008) were en-route to encounters with Vesta, Ceres, and the Pluto-Charon system and shortly after the repurposed *Deep-Impact (EPOXI)* flyby of 102P/Hartley in 2010 (Figure 16.3).

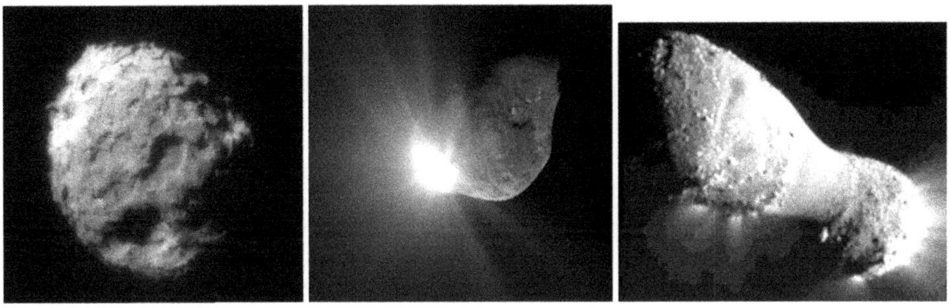

Figure 16.3. Some key advances in small body exploration from 2003 to 2013. Left: a nucleus image from *Stardust's* 2004 encounter with 81P/Wild (credit NASA/NSSDCA). Center: the impact event in 2005 triggered by *Deep Impact*'s impactor striking 9P/Tempel (credit NASA/JPL-Caltech/UMD). Right: the flyby of the hyperactive nucleus of 103P/Hartley in 2010 by the repurposed *Deep Impact* spacecraft (as *EPOXI*, credit NASA/JPL-Caltech/UMD).

The population of known Centaurs had quadrupled to more than 120 objects and it was now thought that ~10%–15% of them were active in a manner more consistent with crystallization of amorphous ice than sublimation as its source (e.g., Jewitt 2009). The survey treated the Trojan asteroids and Centaurs as independent exploration priorities, with the primitive bodies sub-panel commissioning concept studies for a *Trojan Tour and Rendezvous* survey mission (Brown et al. 2010) and a *Chiron Orbiter* mission to the very large and first-named Centaur to investigate the sources of Centaur activity (Buie & Veverka 2010). The Chiron study focused primarily on the deliverable mass to orbit for different trajectory, power system, and propulsion scenarios for launches between 2015 and 2025; they modeled an 11–13 year cruise and a 14 year total mission that conformed to the expected lifetime of an ASRG Radioisotope Power System (RPS). A notional low mass instrument suite of high-heritage instruments was included, consisting of an IR mapping spectrometer, a thermal mapper, an ion/neutral mass spectrometer, and a UV spectrometer. As a backup to Chiron, the study team also performed lower-level analyses of possible missions to 174P/Echeclus and (52872) Okyrhoe.

The initial phase of that Decadal Study evaluated five propulsion/power options for an 11 year cruise phase, including fully chemical, chemical propulsion with a Star 48 solid rocket, chemical/SEP/Earth gravity assist with a 15 kW solar array, a SEP/REP combination including a detachable solar array and two 134 W ASRGs, and the fully REP design with six ASRGs from the earlier COMPASS report (Oleson et al. 2009). None of the scenarios produced a closed solution for an 11 year cruise, either due to an unacceptably high encounter velocity differential (ΔV) at Chiron or too small a delivered mass. The low-thrust two ASRG REP version failed to reach Chiron at all. A study of Echeclus trajectories also resulted in similarly high ΔV's. However, multiple configurations did close for Okyrohe, with cruise times as short as 9 years. To find a viable mission scenario for Chiron, the study team relaxed the cruise requirement to 13 years and added REP configurations that either varied the number of ASRGs or used a high-power ASRG (HP-ASRG) concept that was in early development. The expanded parameter space resulted in multiple closed trajectories, with the HP-ASRG version even meeting the original 11 year cruise requirement. Cost modeling of the chemical/Star 48 and chemical/SEP combinations rated them as achievable within the resources of New Frontiers, but their delivered masses (342–576 kg) were considered marginal for achieving the scientific objectives. The REP scenarios delivered far greater masses, but the hypothetical nature of ASRG development and the anticipated availability and cost of Pu-238 led the study team to conclude they would likely require Flagship Class resources.

In retrospect, *Chiron Orbiter* could be viewed as an instructive case study for the challenges of outer solar system small body mission design, where distances and orbital periods are large and favorable opportunities for giant planet gravity assists are widely spaced in time. Centaurs are particularly challenging targets compared to the giant planets because they lack a significant gravity well and their orbital eccentricity and inclination are larger. Timing is a major factor for mission feasibility, and, despite their scientific significance as targets, it was particularly poor timing for both Chiron and Echeclus. The studied range of launch dates from 2015 to 2025 bracket the 2021 aphelion of Chiron's 50.8 year orbit, resulting in encounters at large heliocentric distances from 13.5 to 16.15 au that drove the mission parameters toward longer cruise times, larger arrival ΔV and lower delivered mass. The circumstances for Echeclus were nearly identical to Chiron for the 2015–2025 window, with all possible encounters occurring near aphelion at heliocentric distances greater than 15 au. However, a subsequent study (Telus & Amato 2021; Section 16.2.3) investigating launch dates starting in the mid 2030's found solutions for much more massive spacecraft that encountered Chiron well inside 10 au, where it is more likely to be active (e.g., Womack et al. 2017). Within this context, the study team's success in finding more viable solutions for Okyrhoe was likely due in no small part to the fact that the encounters would have all occurred near perihelion and inside of 7.5 au.

Irrespective of the reasons why, the failure of the *Chiron Orbiter* study to find a viable trajectory/mass combination led the Decadal Study primitive bodies subpanel to conclude that such a mission was not feasible within the New Frontiers cost cap using existing technology. The more favorable encounter scenarios from Okyrhoe were also rejected without further study of its activity and physical properties. Instead, the decadal report advocated for a program of technology development and Centaur exploration as a secondary flyby target as part of other outer solar system missions. The subsequent NF4 AO (NASA, NNH16ZDA011O) dropped the *TA/CR* concept from its portfolio in favor of a *Trojan Tour and Rendezvous mission*.

Following the release of NF4, another Planetary Science Summer Seminar (PSSS) conducted a Team X investigation of potential Centaur missions conforming to the resources identified in the AO (Howell et al. 2018). Their study developed trajectories for a variety of mission types including single object flybys, orbital encounters, and tour/orbital combinations, all for launch dates between 2024 and 2030. The PSSS team ultimately settled on a flyby encounter with (10199) Chariklo, the largest known Centaur and the first small body observed to have a ring system (see Chapter 9). The final concept, *Camilla*, would have launched in 2026 and used chemical propulsion combined with multiple inner solar system gravity assists and a plane changing Jupiter flyby to reach Chariklo's 23.4° orbital inclination. The 9.3 km s^{-1} relative velocity encounter with Chariklo would occur near aphelion in nearly 2039 at a heliocentric distance of 18.3 au and 7 au above the ecliptic plane. The spacecraft power system used three MMRTGs (the maximum available in the NF4 AO) with a combined end of life (EOL) output of $\leqslant$210 W. *Camilla's* science suite consisted of high-heritage instruments including a camera-Vis/IR spectrometer, sub-mm and UV spectrometers, and a radio science investigation. In addition, the spacecraft included an unguided 100 kg tungsten impactor that would be released 5 days before the encounter. *Camilla* conformed to the cost, schedule, and launch vehicle limits of the NF4 AO, thereby demonstrating viability for future consideration. The study team also investigated whether a modified version of *Camilla* that descoped the UV and sub-mm spectroscopic instruments could fit under the projected cost-cap of the next Discovery AO. Their preliminary modeling suggested that the mission may be possible, but only if NASA provided the required three MMRTGs as government furnished equipment (GFE).

The 2019 Discovery competition (NASA, NNH19ZDA010O) marked a sea change in the prospects for robotic exploration of the Centaurs and TNOs. The previous competition in 2014 had led to the selection of *Lucy*, a groundbreaking multi-target tour of the Trojan asteroids (Levison et al. 2021). *Lucy*, which achieves multiple science objectives of the New Frontiers *Trojan Tour and Rendezvous* mission, was the first outer solar system mission selected for flight in Discovery, and it demonstrated that small body exploration at and beyond the orbit of Jupiter is possible with the relatively modest resources of the program.

The 2019 proposal round expanded on this success. Submitted concepts featured a nuclear-powered mission to Triton (*Trident*, Sharma et al. 2022) and *two* missions that targeted Centaurs, including a 29P/Schwassmann-Wachmann orbiter (*Chimera*, Harris et al. 2019) and a flyby tour of 29P and Chiron (*Centaurus*, Singer et al. 2019). *Trident* would have launched in 2025 or 2026 for a 13 year cruise using multiple gravity assists to reach Neptune following an almost fully ballistic trajectory. The concept carried a science suite consisting of an infrared spectrometer/camera, a wide-angle imager, a magnetometer and plasma science package, and a radio science experiment. *Centaurus* and *Chimera* were both solar powered missions with available launch windows between 2026 and 2029, using chemical propulsion and gravity assists to reach their respective targets at heliocentric distances between 5 and 10 au. *Centaurus* carried a science suite consisting of spectrometers and imagers optimized for flyby encounters (Singer et al. 2019), while *Chimera's* suite consisted of wide and narrow field imagers, a thermal camera, a mass-spectrometer, an IR spectrometer, a radar, and a radio science experiment (Harris et al. 2019).

It is significant that all three of the outer solar system missions were found to be technically viable within the Discovery class. *Chimera* and *Centaurus* were scored in the selectable categories I and II, respectively, although neither were advanced for Phase A study for the reason of maintaining scientific balance in the NASA mission portfolio (NASA, Discovery Step-1 selection announcement 2019). *Trident* was scored in category II and *was* selected for a concept study. Multiple factors contributed to the high ranking of these concepts. Both *Chimera* and *Trident* relied on infrequent gravity assist configurations to reach their targets with minimal carried ΔV (Sharma et al. 2022; Harris et al. 2019). Fortuitous orbital phasing likely played a key role in *Centaurus* being able to reach two high science-value Centaurs in a single mission. In addition to trajectory, two structural elements of the 2019 AO were important for increasing the competitiveness of outer solar system exploration. First, the time period between the 2010 and 2019 Discovery AOs coincided with dramatic changes in the capacity and efficiency of the launch vehicles available to the program (Table 16.1). This resulted in substantial reductions in the charge against the Discovery cost cap for all vehicle options, with the savings increasing with performance class and accelerating substantially after 2014. Vehicle performance curves of characteristic energy (C3) vs. launch mass also changed during this period, so the comparisons are not exact. However, the trend toward higher capacity launches at lower cost is nevertheless clear. As an example, the largest delivered payload for a solar powered scenario in the *Chiron Orbiter* study (Buie & Veverka 2010; Appendix A, Option C) was obtained by launching a 3000 kg wet mass to an initial $C3 = 22\ km^2\ s^{-2}$. As a 2010 Discovery AO submission, this mission would have required a high-performance class vehicle with a $68M charge against the $425M cost cap. In the 2019 competition, there would have been *no charge* for the same vehicle class against a $500M cap.

The second important change in the 2019 AO was the availability of the MMRTGs used to power *Trident* (Sharma et al. 2022). While not provided as GFE, their cost ($54M for one and $69M for two) was significantly lower than in the 2014 New Frontiers AO ($77M for one and $94M for two). Prior to this, RPS, except for radioisotope heater units, had been intentionally excluded as an option in Discovery AOs dating back to 2004 (NASA, NNH04ZSS002O) due to the cost of the units and the limited availability of plutonium. This restriction sharply curtailed the options for outer solar system and small body landed missions, which put their targets at the mercy of their inclusion in the NF portfolio. However, history does show that scientifically compelling RPS missions are possible within Discovery. The lone exception to the Discovery RPS prohibition was the 2010 competition (NASA, NNH10ZDA007O) when NASA offered an intermediate TRL ASRG as GFE. This opportunity generated multiple proposals, including two concepts that were selected for Phase A study (*Comet Hopper*, Clark et al. 2008; *Titan Mare Explorer*, Stofan et al. 2009). The Phase A selection of *Trident* further demonstrates the potential for

Table 16.1. A Comparison of Reductions (Additions) to the Discovery Phase A–D Cost Cap vs. the Fairing Size and Performance Class (Payload Mass vs. C3) of the Launch Vehicle During the Last three Competitions.

	Discovery 2010		Discovery 2014		Discovery 2019	
Performance Class	4 m	5 m	4 m	5 m	4 m	5 m
Low	No Change	$14M	($16M)	$13	($15M)	($15M)
Medium	$22M	$25M	No Change	$28M	N/A	($10M)
High	$32M	$68M	$14M	$43M	N/A	No Change

RPS in the Discovery class, and, more significantly, that all the Centaurs and potentially both classical and scattered disk TNOs are accessible within the program resources.

16.2.3 Decadal Period from 2023 to 2033

The most recent decadal survey exercise (NASEM 2023) was conducted against the backdrop of another decade of transformative advances in our understanding of small bodies and the evolution of planetary systems, with more expected from upcoming robotic missions and observing facilities. *Rosetta* had completed its 17-month orbital exploration of the physical properties, evolution, and volatile production from comet 67P (Figure 16.4). After its primary encounter with Pluto-Charon in 2015, *New Horizons* then flew by the cold-classical TNO, (486958) Arrokoth, providing an evolutionary bookend to the more heavily processed JFCs visited in the inner solar system. The *Lucy* mission was launched in 2021 and was en-route to an encounter with the first of a suite of Trojan asteroids in 2027. The *Hayabusa 2* mission (Watanabe, et al. 2017) had returned 5 g of material from the near-Earth asteroid Ryugu, while *OSIRIS-REx* (Lauretta et al. 2017; Figure 16.4) was en-route to the successful return of 121.5 g of material from Bennu in October 2023. Additionally, existing surveys (e.g., Weryk et al. 2016; Bannister et al. 2018) had expanded the known population of small bodies orbiting between 5 and 30 au to more than 250, while the total number of known TNOs had nearly tripled. Soon to be commissioned facilities, including JWST and the *Vera C. Rubin Observatory* were expected to greatly improve our understanding of Centaur activity (e.g., Faggi et al. 2024; Harrington Pinto et al. 2023; see Chapter 14) and increase the rate at which new objects are discovered. Occultation studies had identified ring systems around two Centaurs (Chariklo; Braga-Ribas et al. 2014; Chiron; Sickafoose et al. 2020) and two large TNOs (Quaoar, Morgado et al. 2023; Haumea, Ortiz et al. 2017). Finally, at the more basic level of solar system formation, the Atacama Large

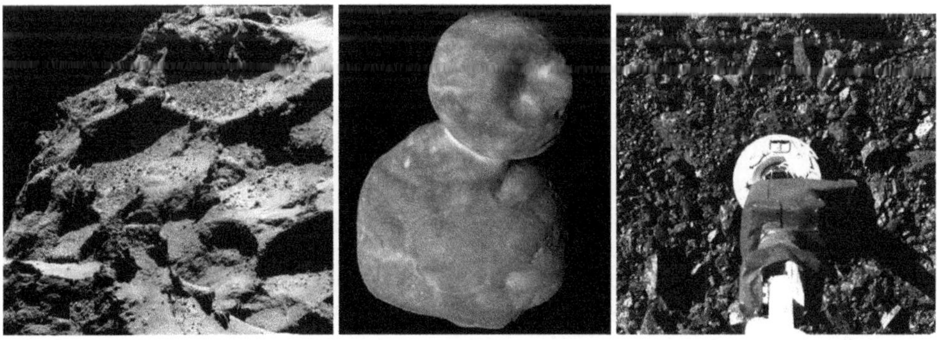

Figure 16.4. Images from a transformative decade of discovery with missions of increasing complexity and reach. Left: the first orbital exploration of a comet nucleus by *Rosetta* in 2014 (Credit Rosetta/ESA/NAVCAM). Center: the first encounter with a Trans-Neptunian small body in 2019 by *New Horizons* (Credit NASA/JHUAPL/SWRI/NOAO). Right: the retrieval of surface samples from the near-Earth asteroid Bennu by *OSIRIS-REx* in 2020 (NASA/GSFC/UA).

Millimeter Array (*ALMA*) had been providing detailed studies of the structure, dynamics, and composition of planet-forming disks since 2013 (e.g., Cieza et al. 2019, Öberg et al. 2021).

Working from the advances of the previous decade and the expected findings from new facilities and missions, the small solar system bodies panel of the Decadal Survey adopted a broad set of key questions focused on initial conditions and radial mixing in the protoplanetary disk, the mechanisms of planetesimal formation and subsequent evolution, the origins of small body compositional and physical diversity, and how giant planet migration contributed to the modern orbital distributions of the trans-Neptunian population. The Centaurs were identified as a compelling population for study given their compositionally primitive nature, dynamic evolution from the TNOs, and their relative accessibility. Primitive bodies were not recommended at the flagship level, but the survey did commission a detailed study of a *Centaur Orbiter and Lander* (*CORAL*) concept for New Frontiers (Telus & Amato 2021; see also discussion in Chapter 15).

The traceability of the *CORAL* concept flowed from four science goals to better understand (1) primordial composition (mineralogical, isotopic, and volatile), (2) accretion (cratering and physical properties of the nucleus), (3) geology & evolutionary processes, and (4) biological potential (mineral alteration, organic materials). As with the earlier *Chiron Orbiter* study, the primary design objective for *CORAL* was to explore propulsion and trajectory options, but in this case for a range of Centaurs as opposed to a pre-selected target. The study emphasized maximum delivered mass for landed RPS powered spacecraft launched during a window from 2036 to 2040 with a Centaur encounter inside 10 au at an inclination less than 60°. The cruise duration was limited to 13 years or less by the MMRTG lifetime, and a Falcon Heavy Expendable launch vehicle was assumed. A group of 550 candidate targets meeting JPL definitions for Centaur and Chiron-type comets were filtered according to the distance, inclination, and duration requirements. This group was evaluated for ballistic trajectories using either chemical or REP + chemical propulsion. The top performing targets, along with several high-value objects (e.g., Chiron, SW1, Chariklo) were then studied for additional propulsion designs (e.g., REP or SEP only, SEP + chemical). From a final list of the 15 best performers, three targets, including Okyrohe (1998 SG_{35}), 2008 SJ_{236}, and 2015 BQ_{311}, were selected based on their science potential and delivered mass (>2000 kg) for more detailed characterization of gravity assist options, cruise duration, and encounter configuration.

The design studies for 2008 SJ_{236} and Okyrohe (Figure 16.5, top and middle) both produced optimal trajectories with launch dates near the end of the required window. Their common characteristics were cruise durations at the 13 year maximum, deep space chemical thrust corrections $\Delta V > 0.5$ km s^{-1}, and an arrival $\Delta V > 1.7$ km s^{-1}. The large in-flight ΔVs required the addition of an additional propulsion stage that increased the cost and complexity of each mission and reduced the delivered dry mass. By contrast, 2015 BQ_{311} produced closed trajectories with much more propulsion requirements across the entire launch window and delivered >2000 kg mass for cruise durations as short as 7 years. While 2015 BQ_{311}'s scientific

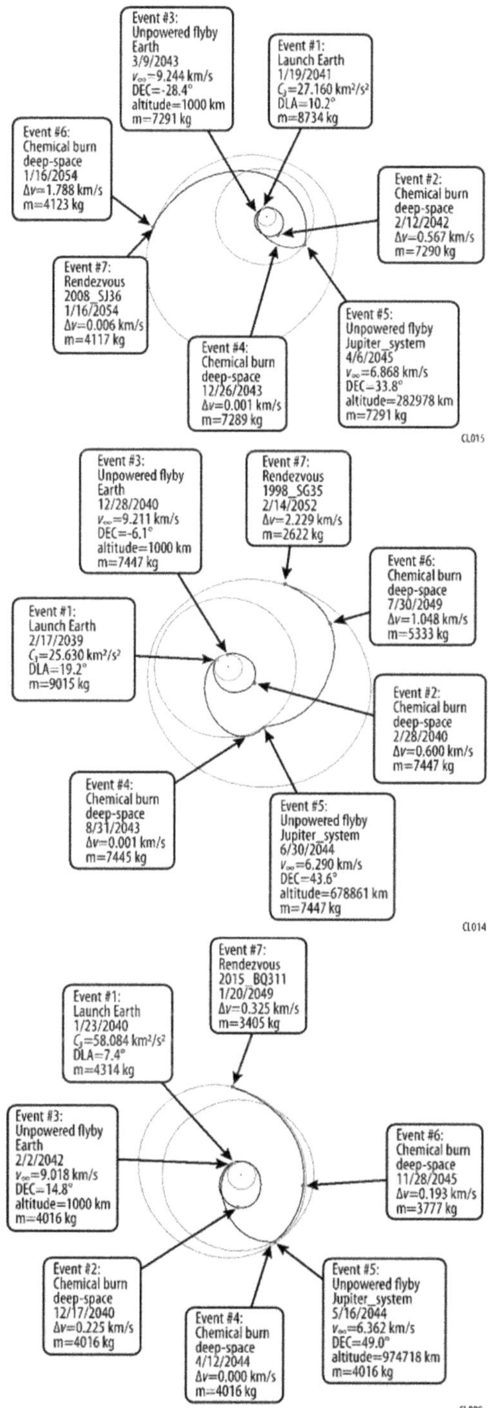

Figure 16.5. Optimal trajectories for *CORAL* candidate objects. Top: the result for 2008 SJ_{236}. Middle: the optimal trajectory for Okyrohe. Bottom: the best solution for 2015 BQ_{311}, the selected target for the design reference mission (reproduced from Telus & Amato 2021).

value was ranked well below the other targets, the variety and quality of the trajectory options, combined with a much larger delivered mass, reduced complexity, and the opportunity for longer proximity operations led the study team to select it for the *CORAL* design reference mission (DRM).

The final *CORAL* trajectory (Figure 16.5, bottom) began with a launch in early 2040 using the Falcon Heavy to reach a $C_3 = 58\ km^2\ s^{-2}$, followed by a 9 year cruise with Earth and Jupiter gravity assists, and ending with a low velocity encounter ($\Delta V = 325\ m\ s^{-1}$) with 2015 BQ_{311}. The lander was powered by two 16-GPHS STEM-RTGs (e.g., Colozza & Cataldo 2015) that provided 580 W (EOL), nearly triple the combined output ($\leqslant$200 W, EOL; NASA 2023) of the three MMRTGs made available in NF4. *CORAL* had an extensive science suite including a gas chromatograph-isotope ratio mass spectrometer, an X-ray lithochemistry instrument, a UV Raman spectrometer, a visible/IR spectrometer, a UV imaging spectrometer, wide and narrow field cameras, a magnetometer, a radio science experiment, and a sample acquisition and handling system, all of which were based on previous heritage. Once it reached 2015 BQ_{311}, the spacecraft would begin an extended 3.6 years mapping phase to measure the gravity field and map surface topography and composition at progressively higher detail. After selection of a series of landing sites, the spacecraft would then descend to the surface for 8 weeks of investigation at one or more locations.

CORAL was modeled against an estimated $1.1 B (FY 25) Phase A–D cost cap extrapolated from NF 4. The optimized DRM came in at $1.3B for phases A–D, with 30% reserves. The study team also modeled a descoped version that removed the magnetometer and UV imaging spectrometer to obtain a more compliant $1.16B cost, again with 30% reserve. Pricing of the high-power GPHS was based on NF4 2 MMRTGs ($95M for two units), but neither the availability nor the actual cost of the next generation STEM-RTGs were known at the time of the study.

As a final coda to the *CORAL* study, the team discussed options for a possible Chiron landed mission. Preliminary models produced a best-case trajectory using chemical propulsion that delivered 1832 kg to Chiron, just missing the 2000 kg study requirement, within the launch window and cruise limits. To obtain more closed trajectories, the team then relaxed the launch window, cruise duration, and encounter distance restrictions of the study. This resulted in a variety of solutions with maximum delivered masses exceeding 2600 kg for launches in the 2042–2043 timeframe and cruise durations up to 14 years (Figure 16.6). The number and quality of possible trajectories increased significantly throughout and beyond the study launch window, with delivered mass increasing by 400%–500% between 2036 and 2042 (see Figure 16.6). This reinforces the significance and complexity of gravity assist and orbit phasing for Centaur exploration.

Based on the successful identification of potential targets and responsiveness of the mission traceability to the priority science questions of the decadal survey, *CORAL* was recommended as one of four portfolio additions for the NF6 call expected in the latter half of the 2020s. Following a subsequent delay in the release of NF5 AO to no earlier than 2026, NASA approached the NAS Committee on Astrobiology and Planetary Science (CAPS) to consider modifications to the New

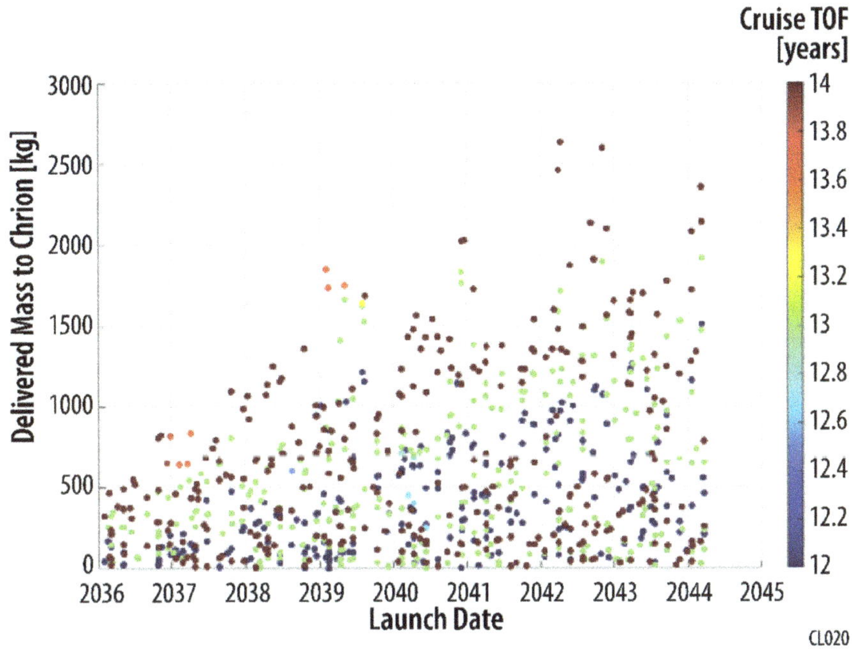

Figure 16.6. A comparison of launch date vs. delivered mass for a potential Chiron lander mission shows the importance of proper phasing with respect to gravity assist configurations and the location of the target in its orbit. Cruise durations up to the 14 year MMRTG lifetime limit were investigated (from Telus & Amato 2021).

Frontiers portfolio reflecting the newer decadal recommendations. The completed report was released on February 25, 2025 with a recommended portfolio that included Centaur Orbiter and Lander. As of this writing, NASA had not announced whether they would follow the CAPS recommendations as they did in 2008. Should Centaur Orbiter and Lander be included in the delayed NF5 announcement, the window of launch opportunities will be close to that of the original *CORAL* study.

16.3 A Blueprint for Future Exploration

The 30 years since (15760) Albion (1992 QB_1) established the physical connection between the Centaurs, TNOs, and JFCs has seen an explosion of small body discovery in the outer solar system. These discoveries were mirrored by the transformative exploration of the JFCs with a series of inner solar system robotic probes and our first foray into the TNO region with *New Horizons*. When the *Lucy* mission encounters the Eurybates-Queta system in 2027, the era of missions *dedicated* to small body exploration in the region beyond the modern ice line will formally begin. As the most broadly transitional population of icy planetesimals in the solar system, the Centaurs are an important key to understanding how they form and evolve, and they will assume a central role in this next phase of exploration (Figure 16.7). Our experience with the JFCs provides a strategic roadmap of mission designs and investigations that can be used to develop a comparable program for the Centaurs. The concepts and proposals described in the sections above have built on

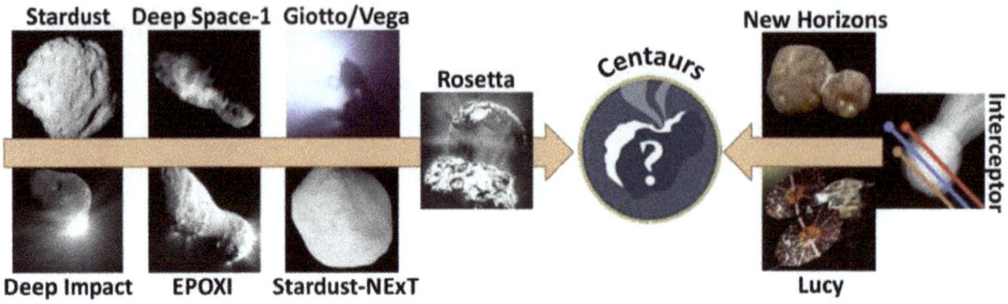

Figure 16.7. The targets of previous, ongoing, and in-development missions will have visited all the populations shown in Figure 16.1. The Centaurs are the evolutionary connection between the near-pristine TNOs and the highly evolved Trojans and short period comets, and they are the last major group that can be explored by modern spacecraft. These earlier missions provide a template for exploration on which a Centaur exploration program can be based.

this experience within the context of the unique opportunities and challenges posed by expanding our exploration into the giant planet region (Table 16.2). They also identify areas both for near-term emphasis and where technical advances can increase the reach and return of new missions.

16.3.1 Mission Options and Goals

Since 2000 there have been five spacecraft encounters with JFCs and four unsuccessful attempts (2P/Encke and JFCs 73P/Schwassmann-Wachmann and 6P/d'Arrest with *CONTOUR*; 107P/Wilson-Harrington with *Deep Space 1*). One object (9P/Temple) was visited *twice* on successive apparitions. This cohort was assembled using a variety of mission designs including targets of opportunity during a technology demonstration (*Deep Space 1*), a flyby-sample return (*Stardust*), an active experiment encounter (*Deep Impact*), a multi-object tour (*CONTOUR*), as extended mission targets (*EPOXI* and *Stardust-NExT*), and an orbiter-lander (*Rosetta*). By leveraging the opportunities available, these missions have sampled the broad diversity of JFCs, including target masses covering two orders of magnitude, activity levels from hyperactive to near-asteroidal, different physical shapes (bilobate and quasi-spheroidal) and surface properties, and one object that had undergone a recent fragmentation event (73P/Schwassmann-Wachmann). Ongoing (*Lucy*) and in-development (*Comet Interceptor*; Jones et al. 2024) missions are on track to encounter additional populations (the Jupiter Trojans and an Oort cloud comet), which, combined with the Arrokoth flyby (*New Horizons*), leaves the Centaurs as the last major dynamical group that are accessible with existing spacecraft technology. Except for *Rosetta* and *New Horizons*, all of these encounters were or will be achieved with low-cost missions at or below the Discovery class.

The current inventory of known Centaurs is as or more diverse than the JFCs. Observational bias emphasizes the largest members of the population, with more than half of known objects having a diameter larger than every known JFC; only $\sim$10% of the Centaurs discovered to date have diameters less than 10 km. The largest and smallest known Centaurs span a range of mass of order 10^6. Their color

Table 16.2. A Comparison of the Previous Concepts and Proposals Developed for Centaur Exploration.

Name	Launch Date/Range	Mission Type	Mission Class	Target	Instruments[a]	Propulsion	Mission Duration	Power
SHOTPUT	03/2025	Flyby Tour	New Frontiers	624 Hektor, 39P/Oterma	MSI, TIS, INMS, UVS, RS, WFI	Chemical Gravity Asst.	7.6 yr	Solar
TRACER	03/2019	Flyby Tour	New Frontiers	Antenor, 2001 BL41	IRS, NAC, WAC, INMS, DC, RS	SEP, Gravity Asst.	9 yr	Solar
COMPASS	11/2024	Flyby Orbiter	New Frontiers	Tlepolemus, Thereus	NAC, WAC, INMS, IRS, LIDAR, TI	REP, STAR 48	11 yr	6 x ASRG
Chiron Orbiter	05/2024, 05/2025	Orbiter	New Frontiers, Flagship	Chiron	MSI/IRS, INMS, UVS, TI, RS	REP, STAR 48	14, 16 yr	6 x ASRG 2 x HP-ASRG
Camilla	10/2026	Flyby	New Frontiers	Chariklo	NAC/VIRS, SMM, UVS, IMP, RS	Chemical Gravity Asst.	12.5 yr	3 x ASRG
Chimera	2025–2026	Orbiter	Discovery	29P/Schwassmann-Wachmann	NAC, WAC, IRS, INMS, TI, RADAR, RS	Chemical Gravity Asst.	14–15 yr	Solar
Centaurus	2026–2029	Flyby Tour	Discovery	29P/Schwassmann-Wachmann, Chiron	Imagers, Spectrometers	Unknown	Unknown	Solar
CORAL	01/2040	Orbiter & Lander	New Frontiers	2015 BQ311	GC/INMS, WAC, NAC, UVS, XRL, PAC, MAG, SAS, LIDAR	Chemical, Gravity Asst.	13.3 yr	2 × 16-GPHS STEM-RTG

Notes [a] MSI: Multispectral Imager, TIS: Thermal IR Spectrometer, INMS: Ion/Neutral Mass Spectrometer, UVS: UV Spectrometer, RS: Radio Science, WAC: Wide Angle Camera, NAC: Narrow Angle Camera, DC: Dust Counter, IRS: IR Spectrometer, TI: Thermal Imager, SMM: Sub-mm Spectrometer, VIRS: Vis/IR Spectrometer, GC: Gas Chromatograph, SAS: Sample Acquisition System, MAG: Magnetometer, PAC: Panoramic Camera, UVRS: UV Raman Spectrometer, XRL: X-Ray Lithography.

distribution is distinct from the JFCs and TNOs. The range of Centaur activity is also vast, with 29P's continuous production rate contrasted against the ~80% of Centaurs that are either inactive or are below our current ability to detect activity with remote sensing. Overall, Centaur nuclei occupy a continuum of poorly understood evolutionary states that stretch from TNO-like to hyperactive (e.g., SW1), to objects that have spent significant time in the inner solar system as JFCs before scattering back into the giant planet region (e.g., 39P/Oterma), to mini-planets like Chiron and Chariklo with rings and extensive dust structures surrounding them. Investigating this variability must be a major objective of a Centaur exploration strategy. However, as previous concept studies have shown, the challenge will be to match encounter opportunities with a target list of distinct objects representing the diverse set of Centaur physical properties.

Thanks to the substantial reduction in launch costs over the previous decade, it is now possible to conceive a Centaur mission series comparable in scale and diversity to that of the JFCs with a combination of low (Discovery) and medium (New Frontiers) sized missions, coupled with other serendipitous encounters. With its flexible scientific focus, the Discovery program can facilitate the development of innovative Centaur concepts to conduct single flybys, tours, and orbital missions. These opportunities would complement the results of a New Frontiers *Centaur Orbiter and Lander* mission. Beyond dedicated missions, the scope of Centaur exploration could be leveraged further by incentivizing en-route and extended mission encounters as part of the wider outer solar system exploration program. In addition to potential Discovery-class missions (e.g., *Trident* or *TiME*), the 2023 Planetary Decadal survey adds three new outer solar system missions (*Enceladus Multiple Flyby*, *Titan Orbiter*, *Triton Ocean World Surveyor*), along with *Centaur Orbiter and Lander*, to New Frontiers 6 and 7, adding to the existing portfolio of *Saturn Probe* and *Ocean Worlds*. The survey also prioritizes two Flagship class missions to outer solar system targets, including *Uranus Orbiter* and *Enceladus Orbilander*. While only a subset of these missions will advance to flight over the next several decades, those that do can add to the diversity of Centaur exploration provided appropriate cruise phase or extended mission encounter opportunities are identified.

16.3.2 Challenges and Enabling Development

While the above investigation strategy is possible within the limits imposed by existing technology and using only the current Centaur inventory, there are two areas where new advances can significantly enhance both the scope and scientific return of future missions.

16.3.2.1 Centaur Discovery and Characterization

The current number of known Centaurs is roughly half that of the JFCs, with the Centaurs occupying a far larger volume of space in orbits with a typical period ten times as long. Their relative isolation, combined with the slow rate of change in their configuration with respect to orbital phase and gravity assist opportunities means that the spacing between optimal encounter scenarios can be generationally long.

Because of this, only a fraction of the Centaur population is well-positioned for exploration at any given time and that fraction is relatively static from the perspective of near-term mission planning. The number of available objects depends on the mission type, with single-target flyby encounters being the least affected, particularly if a giant-planet gravity assist or a specific orbital phase is not required. However, it poses a significant constraint on multi-object tours, cruise-phase encounters during outer solar system missions, the identification of extended mission targets, and for any concept that requires orbital insertion. Multiple Centaur concept studies over the past 20 years have been challenged by the availability of targets well matched to their scientific objectives. Both *Chiron Orbiter*, and *CORAL* started with a specific target or a set of objects of interest that were ultimately found to be poorly placed for meeting the mission objectives and timelines. Other proposed missions (e.g., *Chimera, Centaurus*) were able to exploit enabling configurations to reach high-value Centaur targets, but their windows of opportunity will not repeat for several decades. These constraints on object availability may ultimately doom efforts to build the diverse mission portfolio described above.

The most obvious way to improve target availability is to increase both the total number of known Centaurs and the number for which their spectroscopic properties and activity levels have been measured. Fortunately, the next decade should provide such opportunities if the scientific community advocates for the resources required to take advantage of them. In the last ~20 years, outer solar system surveys (e.g., Elliot et al. 2005; Weryk et al. 2016; Alexandersen et al. 2016; Bannister et al. 2018; Bernardinelli et al. 2022), have discovered Centaurs at a rate of roughly 10 yr^{-1}. This rate is poised to increase dramatically when the *Vera C. Rubin Observatory* begins its 10 year Legacy Survey of Space and Time (LSST) in 2025 (see discussion Chapters 14 and 17). To take full advantage of the depth and coverage it provides, a large-scale coordinated effort must be assembled to first identify new Centaurs in the database, and then to conduct follow up observations to determine their orbit, spectral type, and activity levels (see, e.g., Chapter 11 for a discussion of Centaur observational campaigns). The other area of near-term emphasis is to increase the number of Centaurs for which the production rate and coma composition are known. The last two decadal surveys have emphasized increased understanding of the origins of activity as a major goal for future missions, yet, while ~10% of Centaurs are believed to be active, volatile species have been detected in only a few objects. Recent observations of 29P and 39P/Oterma with *JWST-NIRSpec* demonstrate its exceptional sensitivity to the most abundant parent molecules in comet comae (CO, CO_2, and H_2O; Faggi et al. 2024; Harrington Pinto et al. 2023). Advocating for a dedicated JWST campaign to obtain deep spectra from the other known active Centaurs would increase the number of potential mission targets for activity-focused study by up to an order of magnitude, while placing much stricter limits on production rate for others.

16.3.2.2 Radioisotope Power Systems

While RPS systems are now being made available to NASA Discovery-class missions, the current generation of MMRTGs is not well matched to Centaur

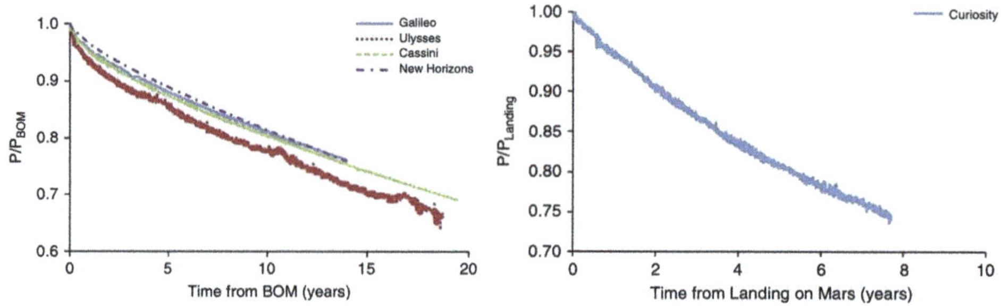

Figure 16.8. The degradation rate for GHPS-RTGs (left) compared with that of the MMRTGs currently being supplied for NASA missions (right). As can be seen, the PbTe-TAGS based designs decay to 75% of their BOM levels more than twice as fast as the earlier Si-Ge units. The measured power degradation rate for the Curiosity MMRTG has been ~3.3% yr^{-1}, compared with 1.5% yr^{-1} from Galileo's GPHS-RTG. At the end of their 14 year mission life, MMRTG power will drop by ~50%, which limits the duration of any mission using them (adapted from Whiting & Woerner 2023. © 2023 John Wiley & Sons, Inc).

exploration or to the outer solar system in general. Hermetically sealed MMRTGs were designed for operation in the Martian atmosphere (Whiting & Woerner 2023) and are based on the PbTe-TAGS thermoelectric module that was used for the *Viking* landers. Their power production (W kg^{-1}) is only 50%–60% that of the Si-Ge based General Purpose Heat Source RTGs (GHPS-RTG) used on *Cassini*, *Galileo*, and *New Horizons* (Zillmer 2023), and they degrade *twice* as fast (Figure 16.8; Whiting & Woerner 2023). These compromises make sense for the *Curiosity* and *Perseverance* rovers and for the *Dragonfly* Titan mission (Lorenz et al. 2009), because Si-Ge modules need to be open to vacuum (Zillmer 2023). However, for space based robotic probes, MMRTGs add unnecessary mass for the power they produce, and their short performance life limits the range of missions that arc possible.

Under the current NASA agreement with the Department of Energy (DOE), the agency is expected to increase production of Pu-238 to 1.5 kg yr^{-1} by 2026. This amount can be used to build three MMRTG units with ~115 W of Beginning of Mission (BOM) power and a 14 year design life. These units must meet the RPS needs of the entire NASA portfolio of missions from Flagship to Discovery, in addition to the Lunar and Mars exploration programs. To date, all the MMRTG missions have used a single unit. With a BOM power <25% of the RTGs on *Galileo, Cassini*, and *Voyager* and half that of *New Horizons*, they have the lowest output of any RPS powered spacecraft since *Viking*, and the more rapid performance degradation of the MMRTGs places much stricter limits on both their operational lifetime and any potential extended mission. Indeed, within the orbit of Saturn they may only be justifiable for landed missions. For example, the *Lucy* solar array would generate approximately the same power at 10 au as three MMRTGs after a 14 year mission.

The Discovery-class *Trident* concept provides an illustrative example for potential MMRTG powered Centaur mission design. *Trident* relied on a combination of thrusters and an Earth–Venus–Earth–Earth–Jupiter gravity assist trajectory to arrive at Triton after a 13.2 year cruise, followed by 1.5 years of data transmission. The spacecraft used two MMRTGs with a BOM of 228 W that dropped to 138 W at the end of mission (EOM) assuming an optimistic 2.7% yr^{-1} degradation rate (Sharma et al. 2022). The *Trident* trajectory is most directly transferable to a single flyby encounter like the *Camilla* Chariklo flyby-impactor concept or possibly a multi-Centaur tour where the mission could be completed either within the 14 year lifetime or with the second target as an extended mission. Orbital or landed Centaur missions are a poorer fit to MMRTGs unless the 14 year lifetime limit is violated significantly or a closer-in Centaur is targeted where the cruise duration is in the range from 9–11 years. Indeed, *none* of the Centaur concept studies commissioned by the decadal surveys (*Chiron Orbiter, Centaur Orbiter and Lander*) or for New Frontiers (*Trojan Flyby-Centaur Rendezvous*) used an MMRTG for their DRM. They all relied on power systems under development for future use (e.g., variations of the ASRG and GPHS-STEM RTG), and, with EOM power-levels significantly exceeding that produced by three MMRTGs, could not be implemented as designed.

The NASA RPS program has been actively engaged in the development of new technologies since it was formed in 2010. To date no new flight-rated units have been completed. The earlier ASRG development program ended in cancellation in 2013 (Reich 2013). A second effort to incorporate upgraded thermocouple materials in the existing MMRTG design (eMMRTG) to achieve a ~20–30% increase in BOM/EOM power (Matthes et al. 2019; Table 16.3) was downgraded from an active project to a technology maturation effort in 2019. A recent NASA Office of Inspector General report (IG-20-10; NASA OIG 2023) was sharply critical of Technology Readiness Assessment (TRA) within the RPS program, which it blamed for cost overruns and development delays in both projects. At present, NASA is pursuing development of two RPS technologies (Table 16.2), including the Next-Gen RTG, which transitions back to a Si-Ge thermoelectric module with improved power efficiency (W kg^{-1} of Pu-238) and slower degradation (Barklay & Woerner 2023), and the Dynamic Radioisotope Power System (DRPS), a high-efficiency piston driven RPS (Sandifer II et al. 2023). A single 293 W (BOM) Next-Gen RTG

Table 16.3. A Comparison of the Properties of the MMRTG with Potential Successor RPS Systems (adapted from NASA IG-20-10; NASA OIG 2023).

Name	Pu-238 Req. (kg)	Fueled Clads	BOL Watts	EOL Watts	System Mass (kg)
Multi-Mission RTG (MMRTG)	4.8	32	110	63	44
Enhanced MMRTG (eMMRTG)	4.8	32	145	77	44
Next-Gen RTG Mod-0	10.8	72	293	208	57
Next-Gen RTG Mod-1	9.6	64	245	177	56
Next-Gen RTG Mod-2	9.6	64	400	290	56
Dynamic Radioisotope Power System (DRPS)	3.6	24	300–400	241–321	150–200

(Mod 0), to be constructed from a heritage defueled GPHS-RTG, is anticipated to be qualified no earlier than 2028; with the timing, cost, and technical maturity of later versions (Mods 1 & 2) and the DPRS uncertain, the timing will likely be much later and dependent on future funding levels.

Even assuming these new technologies are fully developed and the 1.5 kg yr^{-1} target delivery rate is met by the DOE, the supply of Pu-238 will still severely limit the availability of RPS. The available fuel would only be enough to build a single Next-Gen RTG or three DRPSs each decade, which is unlikely to meet the demands of the solar system exploration and human spaceflight programs. Moreover, the total power generation potential based on the current Pu-238 supply is well below that required for the earlier Decadal Survey and COMPASS Centaur mission concepts (*TA/CR*, *Chiron Orbiter*, and *CORAL*) and marginal for the use of REP as a propulsion source, regardless of the technology used. Until the next generation RTGs become available with an adequate Pu-238 supply, Centaur mission designs requiring radioisotope power will need to operate within limitations of existing MMRTGs or use a non-US alternative such as the long-lived ^{241}Am based RPS under late-stage development at the University of Leicester (Ambrosi et al. 2019).

The 2023 Decadal Survey included two recommendations for NASA, to (1) revisit the DOE arrangement and increase the amount of Pu-238 available if necessary and (2) invest in higher efficiency RPS technologies like the DRPS to stretch the existing supply. The Centaur exploration community should collectively advocate for both recommendations and for the consideration of RPS alternative technologies produced outside the United States. While the future of RPS unfolds over the next 1–2 decades, a program emphasizing Centaur encounter opportunities inside Saturn's orbit will allow greater flexibility in mission design. Solar power is now a widely accepted option for exploration at and beyond Jupiter's orbit, with modern high kW systems (e.g., *JUNO, JUICE, Lucy, and Europa Clipper*) capable of delivering power comparable to or exceeding the output of MMRTGs. Moreover, their higher power production inside of 5 au enables the use of SEP systems to supplement chemical propulsion. Even in cases where an RTG is required (e.g., landers), SEP-chemical hybrid trajectories, where the solar panels detach (e.g., *Chiron Orbiter*) can expand the number of viable targets and mission options.

16.3.3 Near-term Centaur Exploration Priority

While broader exploration of the Centaur population can be expected to proceed over the next several decades, near-term concept development may yield ground-breaking results by prioritizing an upcoming confluence of available targets of high significance. Three of the best characterized active Centaurs, 29P/Schwassmann-Wachmann 1, 95P/Chiron, and 174P/Echeclus, are or will soon enter periods of greater accessibility and/or activity where they can be studied by missions with a launch date in the next 15–20 years. Chiron and Echeclus will reach perihelion in 2046 (8.5 au) and 2050 (5.86 au) respectively, and one or both will be within the orbit of Saturn from 2042 to 2055. SW1 occupies a low-eccentricity orbit inside of 6.5 au and continues to be the most active icy body in the outer solar system. It is currently in proper phasing for periodic encounters making use of a Jupiter gravity assist. These opportunities bracket its 2035 Jupiter-opposition, when transfer cruise times are shortest. Following opposition, SW1 will then fall behind Jupiter in its orbit at a rate of $\sim 6^{\circ}$ yr^{-1}, which slowly increases the post-assist cruise period until it exceeds 14 years. All three of these objects are targetable at different times for launch dates through the early 2040s. Moreover, they will occupy heliocentric distances where solar powered orbital missions and MMRTG powered lander missions of 14 years or less are possible. Exploration of any of these objects would provide key insight into the broader evolution of icy bodies unmatched by any other known Centaur. However, if these opportunities are missed, they will not repeat until nearly the end of this century.

16.4 Summary

The Centaurs will be the last known accessible dynamical group in the solar system to be visited by a spacecraft, and they have the potential to transform our understanding of planetesimal formation and evolution. After a long period of concept and technological development, a robust program of Centaur exploration is poised to begin. As the number of well-characterized Centaurs increases in the coming decade, opportunities for missions and secondary encounters with a diverse set of objects will be more common. However, until the next generation of RTGs appear, the properties of MMRTGs will restrict the range of possible RPS missions in the outer solar system to short duration flyby and tour encounters, with orbital and lander missions limited to closer objects reachable with cruise periods of 11 years or less. Emphasizing mission development targeting one or more of the active trio of SW1, Chiron, and Echeclus will provide the best near-term opportunity for significant discoveries. The other key priority must be to sample the diversity of the Centaur population as a whole.

References

Adams, E. R., Gulbis, A. A. S., Elliot, J. L., et al. 2014, AJ, 148, 55

A'Hearn, M. F., Belton, M. J. S., Delamere, A., & Blume, W. H. 2005, SSRv, 117, 1

Alexandersen, M., Gladman, B., Kavelaars, J. J., et al. 2016, AJ, 152, 111

Ambrosi, R. M., Williams, H., Watkinson, E. J., et al. 2019, SSRv, 215, 55

Bannister, M. T., Gladman, B. J., Kavelaars, J. J., et al. 2018, ApJS, 236, 18

Barklay, C. D. 2023, in The Technology of Discovery, ed. D. F. Woerner (New York: Wiley) 245

Bauer, J. M., Stevenson, R., Kramer, E., et al. 2015, ApJ, 814, 85

Bauer, J. M., Grav, T., Blauvelt, E., et al. 2013, ApJ, 773, 22

Belton, M. J. S., Chapman, C. R., Klaasen, K. P., et al. 1996, Icar, 120, 1

Bernardinelli, P. H., Bernstein, G. M., Sako, M., et al. 2022, ApJS, 258, 41

Boice, D. C., Soderblom, L. A., Britt, D. T., et al. 2002, EM&P, 89, 301

Braga-Ribas, F., Sicardy, B., Ortiz, J. L., et al. 2014, Natur, 508, 72

Brown, M. & the Trojan Tour Decadal Study Team 2010, Trojan Tour Decadal Mission Study Final Report SDO-12346, *NASA*

Brownlee, D. E., Tsou, P., Anderson, J. D., et al. 2003, JGRE, 108, 8111

Buie, M., & Veverka, J. 2010, Chiron Orbiter Mission Study, Mission Concept Study Report to the NRC Planetary Decadal Survey, Primitive Bodies Panel
Cheng, A. F. 1997, SSRv, 82, 3
Cieza, L. A., Ruíz-Rodríguez, D., Hales, A., et al. 2019, MNRAS, 482, 698
Clark, B. C., Sunshine, J. M., A'Hearn, M. F., et al. 2008, Asteroids, Comets, Meteors 2008, (Houston, TX: LPI) 8131
Colozza, A. J., & Cataldo, R. L. 2015, Applicability of STEM-RTG and High-Power SRG Power Systems to the Discovery and Scout Mission Capabilities Expansion (DSMCE) Study of ASRG-Based Missions NASA/TM-2015-21, *NASA*
Di Sisto, R. P., & Brunini, A. 2007, Icar, 190, 224
Dudzinski, L. A., Hamley, J. A., McCallum, P. W., et al. 2013, 11th Int. Energy Conversion Engineering Conf. (Reston, VA: AIAA) IAIAA 2013–3924
Duncan, M. J., & Levison, H. F. 1997, Sci, 276, 1670
Edgeworth, K. E. 1943, JBAA, 53, 181
Edgeworth, K. E. 1949, MNRAS, 109, 600
Elliot, J. L., Kern, S. D., Clancy, K. B., et al. 2005, AJ, 129, 1117
Everhart, E. 1972, ApL, 10, 131
Faggi, S., Villanueva, G. L., McKay, A., et al. 2024, NatAs, 8, 1237
Fernandez, J. A. 1980, MNRAS, 192, 481
Fernández-Valenzuela, E., Ortiz, J. L., Duffard, R., Morales, N., & Santos-Sanz, P. 2017, MNRAS, 466, 4147
Gladman, B., Marsden, B. G., & Vanlaerhoven, C. 2008, in The Solar System Beyond Neptune, ed. M. A. Barucci, et al. (Tucson, AZ: Univ. Arizona Press) 43
Glassmeier, K.-H., Boehnhardt, H., Koschny, D., Kührt, E., & Richter, I. 2007, SSRv, 128, 1
Grundy, W. M., Stansberry, J. A., Noll, K. S., et al. 2007, Icar, 191, 286
Hahn, G., & Bailey, M. E. 1990, Natur, 348, 132
Harrington Pinto, O., Kelley, M. S. P., Villanueva, G. L., et al. 2023, PSJ, 4, 208
Harris, W., Woodney, L., & Villanueva, G. 2019, EPSC-DPS Joint Meeting 2019 (Washington, DC: AAS) EPSC–DPS2019–EPSC–DPS1094
Harris, W., Fernandez, Y. R., Sarid, G., et al. 2021, BAAS, 53, 296
Hartmann, W. K., Tholen, D. J., Meech, K. J., & Cruikshank, D. P. 1990, Icar, 83, 1
Howell, S. M., Chou, L., Thompson, M., et al. 2018, P&SS, 164, 184
Jewitt, D. 1990, ApJ, 351, 277
Jewitt, D., & Luu, J. 1993, Natur, 362, 730
Jewitt, D. 2009, AJ, 137, 4296
Jones, G. H., Snodgrass, C., Tubiana, C., et al. 2024, SSRv, 220, 9
Joss, P. C. 1973, A&A, 25, 271
Kareta, T., Sharkey, B., Noonan, J., et al. 2019, AJ, 158, 255
Klesh, A. T. 2009, 40th Annual Lunar and Planetary Science Conf. (Houston, TX: LPI) 1223
Kowal, C. T., Liller, W., & Marsden, B. G. 1979, DSS, 81, 245
Kowal, C. T., & Gehrels, T. 1977, IAU Circ., 3129, 1
Kresak, L., & Pittich, E. M. 1978, BAICz, 29, 299
Kuiper, G. P. 1951, PNAS, 37, 1
Kuiper, G. P. 1974, CeMec, 9, 321
Lauretta, D. S., Balram-Knutson, S. S., Beshore, E., et al. 2017, SSRv, 212, 925
Lawler, S. M., Shankman, C., Kavelaars, J. J., et al. 2018, AJ, 155, 197

Leuschner, A. O. 1907, PASP, 19, 67
Levison, H. F., & Duncan, M. J. 1997, Icar, 127, 13
Levison, H. F., Olkin, C. B., Noll, K. S., et al. 2021, PSJ, 2, 171
Lorenz, R. D., Stofan, E. R., Lunine, J. I., et al. 2009, AGU Fall Meeting Abstracts 2009, (New York: Wiley) P51G–1199
Malhotra, R. 2019, GSL, 6, 12
Matthes, C. S. R., Woerner, D. F., Caillat, T., & Pinkowski, S. 2019, 2019 IEEE Aerospace Conf. (Piscataway, NJ: IEEE) 1
Meech, K. J., & Belton, M. J. S. 1990, AJ, 100, 1323
Morbidelli, A., & Levison, H. F. 2007, in Encyclopedia of the Solar System (Amsterdam: Elsevier) 589
Morgado, B. E., Sicardy, B., Braga-Ribas, F., et al. 2023, Natur, 614, 239
NASA 2003, Comet Nucleus Tour (CONTOUR) Mishap Investigation Board Report, NASA
NASA 2020, Multi-Mission Radioisotope Thermoelectric Generator, NASA NF-2020-05-619-HQ
NASA Office of Inspector General 2023, NASA's Management of Its Radioisotope Power Systems Program NASA IG-20-010 Assignment No. A-22-02-00-SAR, *NASA*
National Academies of Sciences, Engineering, and Medicine (NASEM) 2023, Origins, Worlds, and Life: A Decadal Strategy for Planetary Science and Astrobiology 2023–2032 (Washington, DC: The National Academies Press)
National Research Council (NRC) 2003, New Frontiers in the Solar System: An Integrated Exploration Strategy (Washington, DC: The National Academies Press)
National Research Council (NRC) 2008, Opening New Frontiers in Space: Choices for the Next New Frontiers Announcement of Opportunity (Washington, DC: The National Academies Press)
National Research Council (NRC) 2011, Vision and Voyages for Planetary Science in the Decade 2013–2022 (Washington, DC: The National Academies Press)
Öberg, K. I., Guzmán, V. V., Walsh, C., et al. 2021, ApJS, 257, 1
Oleson, S., McGuire, M., Sarver-Verhey, T., et al. 2009, 44th AIAA/ASME/SAE/ASEE Joint Propulsion Conf. & Exhibit (Reston, VA: AIAA) https://arc.aiaa.org/doi/abs/10.2514/6.2008-5179
Oleson, S., & McGuire, M. 2011, COMPASS Final Report: Radioisotope Electric Propulsion (REP) Centaur Orbiter New Frontiers Mission NASA/TM-2011-216971, CD-2007-16, *NASA*
Oort, J. H. 1950, BAN, 11, 91
Öpik, E. 1932, PAAAS, 67, 169
Ortiz, J. L., Santos-Sanz, P., Sicardy, B., et al. 2017, Natur, 550, 219
Peixinho, N., Delsanti, A., Guilbert-Lepoutre, A., Gafeira, R., & Lacerda, P. 2012, A&A, 546, A86
Pinilla-Alonso, N., Licandro, J., Brunetto, R., et al. 2024, A&A, 692, L11
Prialnik, D. 1992, ApJ, 388, 196
Rabe, E. 1972, The Motion, Evolution of Orbits, and Origin of Comets, 45, 55
Rayman, M. D., Fraschetti, T. C., Raymond, C. A., & Russell, C. T. 2006, AcAau, 58, 605
Reich, E. S. 2013, Natur
Reinhard, R. 1986, Natur, 321, 313
Roemer, E. 1958, PASP, 70, 272
Russell, C. T., & Raymond, C. A. 2011, SSRv, 163, 3

Ryan, E. L., Hörst, S. M., Benfield, M. P., et al. 2009a, White Paper Submitted to the 2013 Planetary Science Decadal Survey, NASA
Ryan, E. L., Horst, S. M., Benfield, M. P. J., et al. 2009b, DPS Meeting Abstracts, (41,; Washington, DC: AAS) 16.26
Sandifer II, C. E., Van Lear, C., Newman, J. M., et al. 2023, American Nuclear Society Meeting: Nuclear and Emerging Technologies for Space (NETS) (Red Hook, NY: Curran Associates) 53110
Sarid, G., Volk, K., Steckloff, J. K., et al. 2019, ApJ, 883, L25
Scotti, J. V., Rabinowitz, D. L., Shoemaker, C. S., et al. 1992, IAU Circ., 5434, 1
Seligman, D. Z., Kratter, K. M., Levine, W. G., & Jedicke, R. 2021, PSJ, 2, 234
Sheppard, S. S., Jewitt, D. C., Trujillo, C. A., Brown, M. J. I., & Ashley, M. C. B. 2000, AJ, 120, 2687
Sharma, P., Lawler, C. R., Mitchell, K. L., et al. 2022, 2022 IEEE Aerospace Conf. (AERO) (Piscataway, NJ: IEEE) https://doi.org/10.1109/AERO53065.2022.9843368
Sickafoose, A. A., Bosh, A. S., Emery, J. P., et al. 2020, MNRAS, 491, 3643
Singer, K. N., Stern, S. A., Stern, D., Verbiscer, A., & Olkin, C. 2019, EPSC-DPS Joint Meeting 2019 (Washington, DC: AAS) EPSC–DPS2019–EPSC–DPS2025
Soderblom, L. A., Becker, T. L., Bennett, G., et al. 2002, Sci, 296, 1087
Steckloff, J. K., Sarid, G., Volk, K., et al. 2020, ApJ, 904, L20
Stern, S. A. 2008, SSRv, 140, 3
Stofan, E. R., Lunine, J., Lorenz, R., et al. 2009, AAS/Division for Planetary Sciences Meeting Abstracts #41 (Washington, DC: AAS) 45.04
Tegler, S. C., Bauer, J. M., Romanishin, W., & Peixinho, N. 2008, in The Solar System Beyond Neptune, ed. M. A. Barucci, et al. (Tucson, AZ: Univ. Arizona Press) 105
Telus, M., & Amato, M. 2021, CORAL: Centaur Orbiter and Lander, Mission concept study report for the Planetary Science and Astrobiology Decadal Survey 2023-2032 (Greenbelt, MD: NASA GSFC) https://smd-cms.nasa.gov/wp-content/uploads/2023/10/coral-centaur-orbiter-and-lander.pdf
Tiscareno, M. S., & Malhotra, R. 2003, AJ, 126, 3122
Trigo-Rodríguez, J. M., García-Melendo, E., Davidsson, B. J. R., et al. 2008, A&A, 485, 599
van Biesbroeck, G. 1928, PA, 36, 69
Volk, K., & Malhotra, R. 2008, ApJ, 687, 714
Watanabe, S., Tsuda, Y., Yoshikawa, M., et al. 2017, SSRv, 208, 3
Watters, T. R., Thomas, P. C., & Robinson, M. S. 2011, GeoRL, 38, L02202
Weryk, R. J., Lilly, E., Chastel, S., et al. 2016, arXiv e-prints, arXiv:1607.04895
Whipple, F. L. 1930, LicOB, 15, 21
Whiting, C. E., & Woerner, D. F. 2023, in The Technology of Discovery, ed. D. F. Woerner (New York: Wiley) 183
Womack, M., Sarid, G., & Wierzchos, K. 2017, PASP, 129, 031001
Zillmer, A. 2023, in The Technology of Discovery, ed. D. F. Woerner (New York: Wiley) 7

Chapter 17

Highlights and the Next Ten Years for Centaur Research

Maria Womack, Kathryn Volk and Jordan Steckloff

Centaurs are an unstable, transitional population of small, icy bodies connecting the Jupiter-family comets (JFCs) to their reservoir population in the trans-Neptunian region. Due to the more extreme thermal environment of the giant-planet region relative to that of the trans-Neptunian objects (TNOs), Centaurs experience rapid dynamical and physical changes, thermophysically evolving these otherwise well-preserved objects. Many of these physical changes may be seen in the surface topography, ices, and refractory composition and volatiles emitted from active Centaurs. The dawn of several new observational assets and laboratory techniques whose data will be augmented with dramatically increased computing power indicates that Centaur research is at the start of a transformational era. Here we briefly summarize some of the foundational discoveries covered in earlier chapters and review what is needed to make significant progress on outstanding problems. We also present and discuss eight Centaur Priority Goals for the next ten years that we put in context of the 2023 U.S. National Academies of Sciences, Engineering, and Medicine's Planetary Science and Astrobiology Decadal Survey.

17.1 Identifying Priority Goals for the Next 10-Years of Centaur Research

In addition to providing an overview of the research to date in a Centaur sub-field, each of the chapters in this book identify critical "next steps" and recommend research goals that are relevant to the chapter's main topic. Many recommendations were brought up in multiple chapters and we consider them to be overarching for the field, and thus we selected eight of them as Centaur Priority Goals for the next ten years (Table 17.1).

These goals overlap significantly with most of the Priority Science Questions from the Planetary Science and Astrobiology 2023-2032 report carried out by the

doi:10.1088/2514-3433/ada267ch17

Table 17.1. Centaur Priority Goals (2025–2034)

Goal	Chapters	NASEM "OWL" Decadal Questions
1. Significantly constrain the small-end size distribution of Centaurs and TNOs	2, 4, 7, 13, 14	Q1, Q2, Q4, Q5
2. Comprehensive census of Centaurs	3, 4, 5, 7, 9, 11, 14	Q1, Q2, Q9
3. Constrain nucleus bulk properties	2, 4, 7, 9, 10, 11, 12	Q1, Q2, Q4, Q5, Q8
4. Colors, spectra, activity constraints, and monitoring for 50–100 Centaurs	2, 5, 6, 7, 8, 13, 14, 15	Q1, Q2, Q4, Q5
5. Establish how physical properties relate to dynamical properties and evolution	3, 5, 6, 7, 8, 10, 12, 13	Q1, Q2, Q3, Q4, Q5, Q7, Q9
6. Case study of 29P/SW1's composition, activity, and behavior	6, 8, 10, 11, 14, 16	Q1, Q2, Q5, Q6, Q9
7. Characterize LD2 before its JFC transition	3, 10, 11, 14	Q1, Q2, Q5
8. Centaur Spacecraft Mission	9, 12, 15, 16	Q1, Q2, Q3, Q4, Q5, Q6, Q8, Q9, Q10, Q12

Many chapters discuss ideas relevant to each Centaur Priority Goal and NASEM OWL Decadal Question; we have identified the chapters with the strongest relevance to each goal.

National Academies of Sciences, Engineering, and Medicine (NASEM) titled *Origins, Worlds, and Life: Planetary Science and Astrobiology in the Next Decade* (OWL; National Academies of Sciences & Medicine 2023). For the reader's convenience, the NASEM OWL goals are provided in Figure 17.1, and are referred to in column 3 of Table 17.1 by their number value in the OWL (Q1, Q2, etc.) when we estimate that achieving one of our identified Centaur Priority Goals will make a substantial or greater advance toward answering that decadal question. In the rest of this chapter, we briefly summarize some of the foundational discoveries covered in earlier chapters, review what is needed to make significant progress on outstanding problems, and discuss the selected Centaur Priority Goals for the next ten years. Each goal is covered briefly in a separate subsection.

17.1.1 Significantly Constrain the Small-End Size Distribution of Centaurs and TNOs

As discussed in Chapter 2, the past ~20 years have brought important new insights into small body formation models. The streaming instability (SI) is a phenomenon caused by the sunward radial drift of many dust particles and their collisions with slower-moving and generally less abundant volatiles and is linked to planetesimal formation. The models predict a transition from a power-law to an exponential taper at sizes larger than the characteristic formation size, which is consistent with observations of the cold TNOs. The Centaurs originate from the hot TNO populations, which formed closer to the Sun but are thought to share a similar size distribution with the cold TNOs in the 100–400 km diameter range; however,

National Academy of Sciences, Engineering, & Medicine Decadal Planetary Science and Astrobiology Priority Questions (OWL 2023-2032)

Q1. Evolution of the protoplanetary disk What were the initial conditions in the Solar System? What processes led to the production of planetary building blocks, and what was the nature and evolution of these materials?

Q2. Accretion in the outer solar system How and when did the giant planets and their satellite systems originate, and did their orbits migrate early in their history? How and when did dwarf planets and cometary bodies orbiting beyond the giant planets form, and how were they affected by the early evolution of the solar system?

Q3. Origin of Earth and inner solar system bodies How and when did the terrestrial planets, their moons and the asteroids accrete? What processes determined initial properties? To what extent were outer Solar System materials incorporated?

Q4. Impacts and dynamics How has the population of Solar System bodies changed through time, and how has bombardment varied across the Solar System? How have collisions affected the evolution of planetary bodies?

Q5. Solid body interiors and surfaces How do the interiors of solid bodies evolve, and how is this recorded in a body's physical and chemical properties? How are solid surfaces shaped by subsurface, surface, and external processes?

Q6. Solid body atmospheres, exospheres, magnetospheres, and climate evolution What establishes the properties and dynamics of solid body atmospheres and exospheres, and what governs material loss to space and exchange between the atmosphere and the surface/interior? Why did planetary climates evolve to their current varied states?

Q7. Giant planet structure and evolution What processes influence the structure, evolution, and dynamics of giant planet interiors, atmospheres, and magnetospheres?

Q8. Circumplanetary systems What processes and interactions establish the diverse properties of satellite and ring systems, and how do these systems interact with the host planet and the external environment?

Q9. Insights from Terrestrial Life What conditions and processes led to the emergence and evolution of life on Earth, what is the range of possible metabolisms in the surface, subsurface and/or atmosphere, and how can this inform our understanding of the likelihood of life elsewhere?

Q10. Dynamic Habitability Where in the solar system do potentially habitable environments exist, what processes led to their formation, and how do planetary environments and habitable conditions co-evolve over time?

Q11. Search for life elsewhere Is there evidence of past or present life in the solar system beyond Earth; how to detect it?

Q12. Exoplanets What does our planetary system and its circumplanetary systems of satellites and rings reveal about exoplanetary systems, and what can circumstellar disks and exoplanetary systems teach us about the solar system?

Figure 17.1. National Academy of Science, Engineering, and Medicine's Planetary Science and Astrobiology (OWL) Priority Science Questions. All have substantial overlap with the Centaur Priority Goals in Table 17.1 with the exception of Q7 and Q11.

we have few direct constraints on the Centaur size distribution, especially down to small sizes. The size (and mass) distribution remains a large knowledge gap.

The known Centaur population is observationally biased toward the largest and brightest objects. We have some constraints on the size distributions of smaller Centaurs and scattering TNOs from both direct observations (see Section 4.2 in Chapter 4) and from outer solar system cratering records (discussed in Section 2.4.1 in Chapter 2), but these data sets are limited due to a lack of surveys dedicated to Centaurs and by the difficulty of detecting and characterizing the faint, small Centaurs. To improve our knowledge of the Centaur size distribution to very small sizes, we need significantly more crater distribution measurements and cratering rate models down to smaller crater sizes for both Centaurs and TNOs. Improved models of crater formation from impacts onto icy bodies (see discussion in Chapter 13) will also help more confidently connect crater sizes to impactor sizes. Also needed are observational surveys (see Chapter 14; also discussed further in Section 17.1.2) that increase the overlap between size distribution constraints based on craters and those based on direct observations of the Centaur and TNO populations. As discussed in Chapter 2, crater measurements for Pluto, Charon, and Arrokoth have yielded important new constraints on the TNO size distribution down to sizes much smaller than observable by optical surveys. Ultimately, it will be important to link Centaur size distributions to those of TNOs and JFCs to develop a model of their formation and subsequent collisional and dynamical histories (see, e.g., discussion in Section 4.5.2 of Chapter 4 and in Chapter 7). Collisional evolution can produce fragments, often rubble piles, whose deep interiors may differ significantly from those of their parent object; such differences may lead to significant variations in a Centaur's bulk density, porosity, strength, and ultimately shape (see Chapter 12). Future crater measurements from bodies in other TNO subpopulations or from a Centaur mission (see Chapter 16) combined with dynamical models will help link the size distributions inferred from cratering records to specific formation regions. Laboratory experiments are also needed to provide additional constraints on how icy grains can grow into larger aggregates and planetesimals. This is discussed in Chapter 13, which addresses the importance of realistic simulant materials when studying how icy dust mixtures are produced in the early stages of icy body growth. As discussed in Section 2.5 of Chapter 2, improvements in computational approaches and methods are needed to increase the resolution of formation simulations and extend streaming instability models to the formation of smaller (diameter <10 km) bodies to compare to the small-end of the TNO and Centaur size distributions probed by cratering records. Coupled with upgraded observations of the physical properties of the relatively accessible Centaur population, we expect the next decade to yield new constraints on small body formation in the outer solar system.

17.1.2 Comprehensive Census of Centaurs

Our ability to test models depends critically on how many Centaurs have been observed, how completely the population is sampled, and how well-characterized the observed objects are. Thus far, only a small fraction of all Centaurs have been detected. Significantly increasing the number of Centaurs that are observed and characterized by surveys with well-understood biases is a compelling goal for solar system research.

As with all small body populations, we preferentially detect the larger/brighter objects, and then typically only the biggest and brightest among those are subject to detailed characterization studies to measure lightcurve properties, activity, photometric colors, or spectra (see discussions in Chapters 4 and 5). Because most of our Centaur detections have come from a wide variety of surveys, many of which do not have well-determined biases for Centaur detections, it has been difficult to understand how representative and complete the observational sample is (see discussion in Chapter 4). Searches for Centaurs over long time periods in wide-area surveys with well-understood biases will provide a more complete census of Centaurs, firmly establish their orbits, and record brightness variations related to their shape and rotation state and activity levels.

The NSF-DOE Vera C. Rubin Observatory's Legacy Survey of Space and Time (LSST) will provide us with such an improved census of the Centaur population (see Section 14.3 in Chapter 14). LSST will observe the entire southern sky and up to 10° north of the northern hemisphere ecliptic (see Figure 14.7 in Chapter 14) down to magnitudes of 23-24 in typical solar system filters (g, r, i, z). This will provide a roughly one order of magnitude improvement in the number of known Centaurs, with enough detections per object to yield very precise orbits. We will get colors, sparse lightcurves, and activity constraints for many of the brighter objects (those that are bright enough to be detectable in most filters will have the most observations and thus the most opportunities for characterization). The Centaur detections and characterization from LSST will be critical in identifying objects worth following up with other ground- and space-based facilities for additional characterization and monitoring, including via organized observing campaigns (see Chapter 11). Section 9.6.2 in Chapter 9 discusses how the LSST Centaur observations will also enable more occultation studies both by providing new Centaur detections and improving the orbits (and thus occultation predictions) of already-known objects. Importantly, the fact that this large LSST Centaur data set will come from a single well-understood survey will allow us to infer intrinsic population properties from the biased observational sample in a much more robust way than ever before (see Section 14.3.1 in Chapter 14 and Section 7.5.1.2 in Chapter 7). This will also enable even more detailed dynamical studies of the Centaur population. Research linking physical properties to dynamical pathways and source regions in the TNO populations, particularly resonant source regions, will benefit from LSST observations of TNOs and Centaurs (see discussion in Chapter 3 and Section 17.1.5).

17.1.3 Constrain Nucleus Bulk Properties: Sizes, Shapes, Rotation, Rings, and Binarity

Nucleus bulk properties are tied to the porosity, density, and strength of material in the deep interior of the nucleus. These bulk properties are difficult to determine, and there are only a handful of measurements of the sizes, shapes, and the near-nucleus regions of Centaurs, but some interesting trends are emerging. Lightcurve data are largely consistent with Centaurs having spherical nucleus shapes for sizes $D \gtrsim 100$ km; this cutoff size for elongated bodies is about four times smaller than what is found for the giant planets' icy moons (see discussion in Section 4.3.5 of Chapter 4). If this holds with a much larger Centaur data set, it could indicate a significant difference in material strength between Centaurs and icy moons. The cutoff size for round Centaurs could also be affected by collisions in the early solar system, in which case the same trend should be observed for TNOs. If stresses from such collisions exceeded material strength, this would reduce the porosity of the deep interiors. If the trend holds for Centaurs, but not TNOs, then perhaps the physical processing that Centaurs experience as they enter the giant planet region due to gravitational interactions with the giant planets, thermal alteration and activity, changes in their rotation rates, etc., alters the population's shape distribution (see, e.g., discussions in Chapters 7 and 12). In particular, the prevalence of bilobate shapes in the Centaur population compared to the JFC population (where such shapes are common) and the TNO population (where, e.g., Arrokoth's bilobate shape has been linked to formation scenarios; see Section 12.5.2 in Chapter 12) has important implications for the formation and evolution of icy bodies (see Figure 17.2).

An improved range of Centaur shape models and measurements is needed to test planetesimal formation and physical evolution scenarios. For example, violent collisions between planetesimals have been shown to have a low chance of forming bilobate shapes (Nesvorný et al. 2018), so the percentage of Centaur nuclei that are

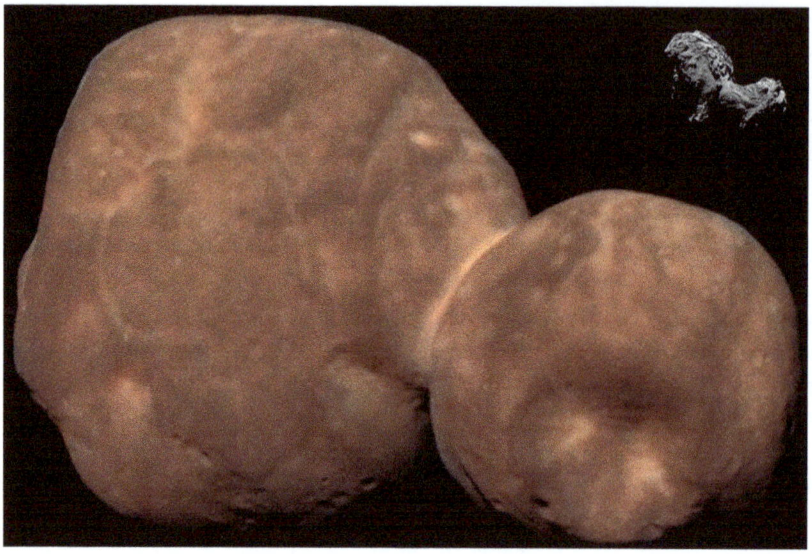

Figure 17.2. This composite image shows a smoothed-faced TNO Arrokoth (left) and the jagged and pitted JFC 67P/Churyumov–Gerasimenko (upper right) approximately to relative scale. Both objects have bilobate shapes. Although it is not known what percentage of TNOs, Centaurs, and comets have bilobate nuclei, this value is expected to put tight observational constraints on solar system formation models. Another outstanding question is whether Centaur activity creates the pitted and layered topography of JFCs. This image has been obtained by the author(s) from the Wikimedia website where it was made available by Hyakutake (1996) under a CC BY-SA 4.0 licence. It is included within this article on that basis. It is attributed to NASA/JHUAPL and ESA/Rosetta/NAVCAM.

bilobate is highly relevant to constraining formation and collisional evolution models (see Chapters 2 and 12). It will also be helpful to look for any changes in the shape distribution with orbital parameters to see if bilobate shapes become more prevalent with evolution in the Centaur region. Ideally, we need measured shapes for Centaurs down to about 1 km in diameter for comparison with the JFCs that we have visited with spacecraft; additional shape information for TNOs will also be very useful. Shapes can have a significant influence on activity (see discussion in Section 12.7.2 of Chapter 12). So far, one Centaur, which happens to be the most active Centaur, outgasses CO and CO_2 in patterns which are consistent with a bilobate shape (29P/Schwassmann-Wachmann, aka 29P/SW1; see Figure 17.3 and Faggi et al. 2024). Overall, very few Centaur shapes are well-constrained. Future measurements of many other Centaurs are needed to ascertain how common bilobate and spherical shapes are in the population.

Well-sampled lightcurves and other time-series observations are powerful tools for measuring a Centaur nucleus' rotation state, activity level, shape and the location of active areas (see Chapters 4, 7, and 8). Thus far, Centaur lightcurve properties display no clear relationship with perihelion distance, orbital eccentricity, or inclination, but observational constraints are weak (less than ∼5% of the known Centaurs have had their lightcurves recorded; see Chapters 10 and 11, and Peixinho

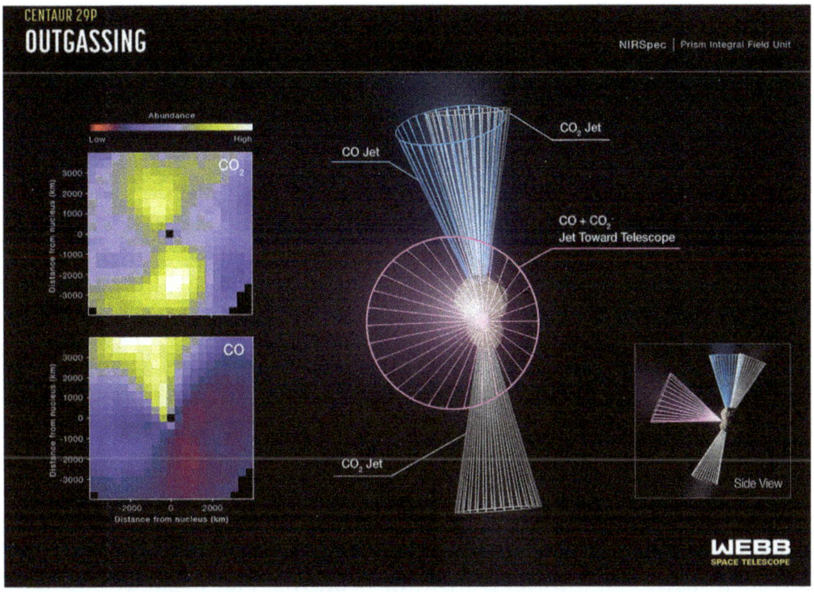

Figure 17.3. JWST observations of the very productive Centaur 29P/SW1 show outgassing patterns of CO and CO_2 that indicate heterogeneous outgassing and the first evidence for a Centaur with a bilobate nucleus (Faggi et al. 2024). The image contains JWST data (left) and an artist's illustration (middle, right), credit NASA, ESA, CSA, Leah Hustak (STScI), Sara Faggi (NASA-GSFC, American University).

et al. 2020). We anticipate that the number of lightcurves will increase significantly with LSST observations (see Chapter 14) in coordination with other observational assets. For example, observations using JWST and other new large telescopes will provide critical constraints on activity as it relates to lightcurve properties to help disentangle shape and activity. Helpful observational tips can be found in Chapter 11, including a list (Table 11.2) of observing opportunities for eight Centaurs to help identify time periods when objects will be favorable for observing campaigns.

Once substantial lightcurves are in hand, leveraging those data with surface compositions inferred from colors and spectra will significantly expand our knowledge. Interestingly, for the 16 Centaurs that have size and rotation period measurements, models predict that it is unlikely that any of these Centaurs will break up due to their rotation rate (Chapter 4). It will also be critical to intercompare lightcurve results for TNOs, Centaurs, and JFCs. For example, thus far, the correlation that smaller TNOs are more elongated than larger TNOs is not seen in the Centaur population, though the Centaur sample size is admittedly small (see Peixinho et al. 2020). Whether this holds with a much larger pool of Centaur data will be useful for comparing the collisional histories of both populations (see Section 4.5 in Chapter 4 and Chapter 7 for more discussion on this topic). As reviewed in Chapter 12, shape and rotation are also key to understanding the interior states of Centaur nuclei. Given that only a few Centaurs have their rotation periods and axial measurements established, achieving rotational lightcurves for even 10-20 Centaurs would significantly advance constraints on the population.

Inspecting the near-nucleus environment of Centaurs has the potential to tightly constrain solar system formation models. There is evidence for a ring, ring system, and fragments found around some Centaurs (Chariklo, Chiron, 29P/SW1, Echeclus) and rings around two TNOs (Haumea and Quaoar). No Centaurs are known to be binaries (although see discussion of Typhon and Ceto, a binary on a Centaur-like giant-planet-crossing orbit, in Chapter 9). Binaries are common, however, among the cold classical TNOs (binary fraction of $\sim$30%) and less common, but still present in the hot TNO population that feeds the Centaurs (see, e.g., Noll et al. 2020). The streaming instability has been very successful in explaining the high rate of binary objects and the maximum sizes of the cold classical TNOs (see, e.g., Figure 2.2 in Chapter 2) under the widespread assumption that they formed *in situ* at $\sim$45 au. As discussed in Chapter 2, the size distribution of the hot TNO population is also broadly consistent with formation through streaming instability, which could indicate that the differences in the binarity rate of the hot and cold TNO populations are due to collisional and dynamical evolution post-formation. Similarly, any differences in the rate of binaries, satellites, and debris/rings between hot TNOs and Centaurs that remain after greatly expanding our data sets would indicate further physical processing in the transition from the TNO region into the Centaur region; see Section 9.2.2 in Chapter 9 for a more detailed discussion of TNO binaries and their potential relationship to Centaur binaries. Evolutionary processes in the giant planet region could also lead to significant physical and chemical changes to the nucleus surface layers; this is explored in more detail in Chapter 7 (see also Figure 17.2). New observational tools will be critical in determining whether there are as-yet undetected Centaur binaries, small moons, or additional debris or rings systems.

More and improved searches to detect new ring systems and satellites will allow us to test how likely they are to survive the various dynamical and collisional pathways Centaurs follow from their formation, implantation into the TNO populations, and transition into the giant planet region. In particular, further theoretical, numerical, and observational work is needed to understand how ring systems form and evolve, including the interactions between a rotating Centaur and particles orbiting the nucleus in a surrounding disk. Stellar occultations of known ring systems may be able to provide the width and optical depth of material (see Chapter 9). Extremely large telescopes will probably have the angular resolution needed to discern rings, moons, and other near-nucleus material of some of the closer Centaurs.

17.1.4 Colors, Spectra, Activity Constraints, and Monitoring for 50–100 Centaurs

The surface colors of Centaurs and small TNOs range from neutral/grey ($B - R \sim 1$ mag) to very red ($B - R \sim 2$ mag). The color distribution of $\sim$60 Centaurs is consistent with bimodal peaks in this range (the effect is strongest for Centaurs with diameters under 120 km), but not with high confidence. In addition to proposed ties between color and activity (see, e.g., Chapter 7), there is a possible correlation between Centaur color and inclination which may indicate some ties between color and formation region (see Figure 5.2 and Chapter 5). Red Centaurs tend to be observed at lower orbital inclinations than gray ones, which is also a trend found throughout the TNO populations (e.g., Marsset et al. 2019). The surface properties of TNOs can be divided into two broad categories that somewhat overlap with typical red and gray colors, with dynamically hot TNOs (which have a larger range of inclinations), typically falling into one category and cold TNOs into the other (see, e.g., Fraser et al. 2023). Combined with constraints on the dynamical history of the hot and cold TNOs, their surface properties likely reflect that the dynamically hot TNOs formed closer to the Sun than the dynamically cold TNOs (see, e.g., Gladman & Volk 2021). It is possible that the trend between color and inclination in the Centaurs is related to these two groups of TNOs that formed in different regions of the planetesimal disk. But this interpretation is complicated by several factors, including that the cold TNOs should be a relatively small contributor to the Centaur population dynamically (see Chapter 3), that activity in the giant planet region likely alters surface properties (the active Centaurs almost all fall into the gray category; see Figure 5.2 in Chapter 5), and that the Centaur data set is still small and has unknown observational biases.

The biases in the current Centaur color measurements are particularly important for interpreting the apparent bimodality. Smaller objects from the low-perihelion scattering and resonant TNO populations are also bimodal in color, and the sizes of the observed Centaurs overlap with these smaller-sized TNOs (see Section 5.2.1 in Chapter 5). Thus, nucleus size may play a role in the observed color bimodality for both TNOs and Centaurs. If bimodality persists with a much larger data set, it will be important to test the extent to which the bimodality is related to size and/or inclination, both of which have implications for formation.

The fact that active Centaurs fall into one part of the bimodal color distribution further complicates things by raising the question of how TNO, Centaur, and JFC surfaces change as they become more processed over time; i.e., are color differences better explained by original composition ('nature'; e.g. formation location, see Chapter 2) or by physical processing in the giant planet region ('nurture')? Laboratory studies indicate that the range of colors observed for TNOs and Centaurs may be explained with an appropriate combination of initial albedo, meteoritic bombardment, and space weathering (see Chapter 13). Activity can change the surfaces of Centaurs and JFCs (see Section 7.4.3 in Chapter 7), which is likely reflected in their color properties.

In the pre-JWST era, surface compositions were measured for only a limited sample of TNOs and Centaurs and were biased toward the very largest (brightest) objects. Some of the first results from JWST reveal three spectral types of TNOs: one dominated by water ice and silicates, another by CO_2 and CO ices, and the third by carbon ices and complex organics (see Chapter 14). Early JWST observations of Centaurs reveal that some appear relatively unprocessed with only moderate volatile depletion compared to TNO JWST spectra, while others are not TNO-like, hinting at more extensive alterations (see discussions in Chapters 5, 7, and 14). It will be critical to have more data to see whether the TNO spectral groups are also represented in the overall Centaur population. Estimates are that measurements of approximately 100-200 more Centaurs are needed to more accurately capture the color distribution for comparison to TNO observations, better determine the correlations between color and size or orbital properties, and to test models of physical processing. Over its ten years of operation, LSST will greatly expand the color data set for Centaurs, TNOs, and JFCs (see Section 17.1.2 and Section 14.3 of Chapter 14). Additional characterization of a wide range of these objects with JWST will further help clarify how surface colors are tied to spectra and composition, and thus to the formation region, size, and subsequent physical evolution of Centaurs.

In addition to dozens of new Centaurs, observations should be continued for established Centaurs, such as Chariklo, Chiron, and 29P/SW1. In many ways, Chiron can be considered the mascot of the population. It kick-started the field shortly after its orbit was determined to be chaotic soon after its discovery. It is one of the largest known Centaurs and displays most of the characteristics observed in other Centaurs. Because of its activity variations, we have had opportunities to characterize its dust and gaseous emission when active and surface ice features, nucleus size, and nucleus rotation properties when inactive; we also have details about the material in its near-nucleus environment, which has been attributed to a ring/debris system (see Section 9.3 in Chapter 9). As described in more detail in Chapter 10, Chiron has a very long history of documented changes in brightness and outbursts that are not well-correlated with heliocentric distance. Continuing the long timeline of observations for well-known Centaurs will be needed for comparisons with the newly detected Centaurs.

Chiron recently showed us that Centaurs need not generate a large coma for activity to be detected. A multi-year analysis of Chiron observations obtained from many instruments revealed that it was much brighter than in the previous 5 years; this has become known as the "2021 Brightening Event." Chiron's brightness increased by about 1 magnitude without changing its color index or generating a noticeable visible dust coma (Dobson et al. 2024). The team ruled out Chiron's rotational lightcurve, solar phase angle effects, possible high-albedo features, and the ring system as significant contributors to the brightness changes. Instead, they proposed that new or increased distant activity, perhaps caused either by the amorphous-to-crystalline water-ice transition or by newly exposed ice pockets, may have generated a coma that is too faint or small to be detected by large telescopes, but is able to change the albedo of the surface while the Centaur is near aphelion ($\sim$18–19 au). The LSST will provide a decade-long high-cadence data set of hundreds of Centaurs that will document their activity levels, even perhaps to very low levels.

The improved constraints on Centaur activity from LSST will lead to improved characterization of active Centaurs and their comae. In contrast to dozens of surface color measurements, comae colors have been measured for only a few Centaurs, including 29P/SW1, Chiron, and 174P/Echeclus. For these Centaurs, the visible coma was bluer than the solar radiation, which may be consistent with small dust grains (see, e.g., Chapter 10). See Chapter 8 for coma color values for these and other Centaurs.

Studies of the volatile compositions of Centaurs have only really just begun, largely because of the difficulty in obtaining spectroscopic information of the comae surrounding such faint objects at large distances. In addition to there being relatively few known active Centaurs (only ~13% are documented to have activity), most of those are only sporadically active, which complicates attempts to characterize their dust and gas comae with telescopes (see discussion of active Centaurs in Chapter 8). We also stress that the Centaurs that are the easiest to observe (i.e. are closest to the Sun and therefore the brightest) are those that are mostly likely to be known to be active. There is a strong perihelion trend with activity, such that, for example, more than 40% of the observed Centaurs with perihelia within 7 au are known to be active. This is discussed in more detail in Chapter 8. For decades, the only volatiles detected in Centaurs were CN, CO^+, CO, H_2O, and HCN, and most of them were seen only in 29P/SW1 (see Chapter 6), partly because it is always active and so makes a reliable target (see Chapter 10). For 29P/SW1, CO is produced in large enough quantities to drive the nearly continuous dust coma, and a few CO outbursts have been observed (see Chapter 8; Wierzchos & Womack 2020; Bockelée-Morvan et al. 2022). CO rotational spectral emission was also detected in smaller amounts than 29P/SW1 for Echeclus at ~6 au and Chiron at ~8.5 au (see Chapter 10) but in no other Centaurs to date.

JWST has already provided transformative results for Centaur chemical compositions with the first detection of CO_2 emission in a Centaur (39P/Oterma; Harrington Pinto et al. 2023), detailed morphologies that indicate the presence of heterogeneous jets of CO and CO_2 in 29P/SW1 from a bilobate nucleus (Faggi et al. 2024; Figure 17.3), and the widespread detection of CO_2, CO, CH_4, and H_2O ices, silicates, and complex organics on many Centaur (and TNO) surfaces (Pinilla-Alonso et al. 2024; De Prá et al. 2025). Other recent observations of Chiron with JWST detected CH_4 and CO_2 emission in the coma, frozen CO_2 on the surface, and provided dust particle size constraints; the space telescope has the potential to continue providing spectra of many more Centaurs (see Pinilla-Alonso et al. 2024, Wong et al. 2024 and Chapter 14). A key characteristic for H_2O ice measurements is to identify the phase of that water ice when possible. The relative abundances of the noble gases Xe and Ar can aid in distinguishing between pure condensates, full clathration, and amorphous water ice (Chapter 6). More important results from JWST are expected as additional observations and further analyses are completed (see Chapter 14).

Another promising resource for Centaur spectra is the upcoming Spectro-Photometer for the History of the Universe, Epoch of Reionization, and Ices Explorer (SPHEREx) mission. SPHEREx is a NASA Astrophysics Medium-Class Explorers (MIDEX) mission that will provide the first unbiased all-sky survey at near-infrared wavelengths with a spectral range from 0.75–5.0 μm and 6 arcsecond resolution. This instrument is capable of recording abundances of gaseous CO, CO_2, H_2O, PAHs, and simple organics like CH_4, H_2CO, and CH_3OH. Solid ices of CH_4, N_2, H_2O, and CO, as well as Fe-bearing olivine/pyroxines will also be detected (see Figure 17.4 and Lisse et al. 2024). The SPHEREx science return is estimated to yield reflectance and comae spectra of ~100 Centaurs every 6 months (for a total of 4

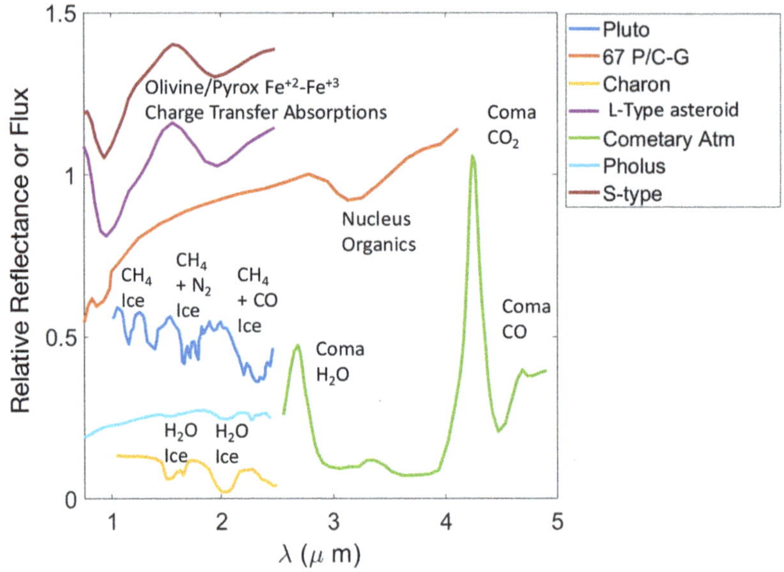

Figure 17.4. Representative expected infrared spectra for a range of small bodies including asteroids (upper left), comets (right), and TNOs and Centaurs (lower left) from the upcoming SPHEREx all-sky survey which may record ~100 Centaurs. In principle, any of these features may be detected on Centaur surfaces or in active Centaurs' comae, and some of these spectral signatures have been identified for several Centaurs with other telescopes. Note that these are all reflectance spectra with the exception of the green line on the lower right that shows emission flux features. Figure modified from Lisse et al. (2024).

times during the prime 2-year mission). Also captured in the SPHEREx survey will be 100s of comets and TNOs at the same cadence, which will be invaluable for comparison with the Centaurs (Lisse et al. 2024). Between SPHEREx, JWST, and other facilities, the next decade should see a dramatic increase in the number of spectra we have for Centaur surfaces and comae. This will undoubtedly yield new insights into how Centaurs evolve as they transit from TNOs to JFCs. Also see Chapter 7 for more discussion of SPHEREx and other upcoming observatories. It will be important to combine this with information about total Centaur productivity, including patterns of activity or outbursting behavior, from lightcurves or any other measurements repeated over time, to put surface and comae properties in context to see the impact of activity on Centaur surface colors and ice compositions.

Next generation ground-based telescopes (such as the ELT, TMT, GMT), will also be of great use for characterizing more Centaurs. These telescopes will give us faster reaction times to outbursts and access to wavelengths shorter than those of the JWST. Additionally, high resolution broadband imaging of the inner coma will be useful for mapping dust jets, and IFU and/or narrowband filters similar to Hale-Bopp filters (for OH, CN, etc.). The Near Earth Object Surveyor mission (*NEO Surveyor*) will also provide a rich survey of thermal emission from large-grain dust of active Centaurs and will be sensitive to gaseous emission from Centaurs emitting CO_2 and/or CO (see Chapter 8 for more discussion). Other ground-based campaigns with smaller telescopes (1–2 m diameter aperture) will be valuable for ongoing monitoring of Centaurs. These facilities will be particularly important for following up on sparse LSST observations that hint at activity and for monitoring predicted occultation events (see Chapter 11).

Because Centaurs are subjected to changing energetic environments, capturing information about their activity and comae is critical. CO and CO_2 are the two most abundant carbon-bearing molecules in comae and are considered to have major roles in driving activity of Centaurs (see Chapter 8). CH_4 is also positioned as a significant player in thermophysical models of Centaur nuclei due to methane ice's sublimation temperature being well-matched to the Centaur region; its presence is also noted in cometary spectra (see Chapter 6, including Figure 6.2). The relative abundances of these and other key species (including isotopes of C-, H-, O-, and N-bearing molecules) will provide strong observational constraints on models of solar system formation and evolution. These carbon-bearing species have never been studied together in a large sample of Centaurs, and their measurements in dozens of objects would be transformative for the field of planetesimal formation and evolution and also for cosmochemistry and astrobiology studies related to delivery of organic materials to Earth (see Chapter 15).

While activity in Centaurs offers opportunities for observing detailed chemical compositions, interpreting comae measurements is not straightforward. A Centaur nucleus' chemical composition is likely determined by a combination of conditions where it formed and subsequent processing that may deplete some species. Moreover, what is observed in the coma is also controlled to some extent by how much heating it receives from the Sun. This well-known effect is clearly shown in a comprehensive study of CO, CO_2, and H_2O production rates in more than 20 cometary objects (Harrington Pinto et al. 2022). As the study shows, the relative amounts of two species in the same comet can be very different at 3 au or 6 au due to differences in activity rates for different species. Inside 3-4 au, this heliocentric dependence is mostly due to a comet approaching close enough for water-ice to sublimate significantly. As comets and Centaurs move farther away, they cross different ice-lines (CO, CH_4, CO_2, etc.) which may affect what is released into the coma. If not taken properly into account, this heliocentric dependence for gas production rates can obscure knowledge of a Centaur's true driving processes and chemical composition, thus complicating our efforts to distinguish between 'nature' or 'nurture' based on observations. Chemical reactions on the surface and in the coma may also play significant roles in the amount and relative abundances of CO and CO_2 (and other volatiles; see Chapter 6) detected in comae. We urgently need new models that can be used to more accurately connect gaseous comae production rates to nucleus compositions (at least in the surface layers) that also address heliocentric variability of a Centaur nucleus to insolation. This will require a better understanding of the drivers of Centaur activity.

In addition, given the heliocentric distance overlap of some Centaurs with other well known distantly active comets (e.g., Hale-Bopp, C/2017 K2, UN271) comparison of activity mechanisms and chemical composition will likely be fruitful even though most of these are Oort Cloud comets. For example, similarities have been noted between CO emission line structure in 29P/SW1 and Hale-Bopp (see detailed line modeling and jets from Gunnarsson et al. 2008 and Bockelée-Morvan et al. 2022 and review by Womack et al. 2017). In addition to outgassing from the nucleus, detailed comparisons of gaseous contributions from the sublimation of icy grains within the coma is needed to constrain models of activity, which has been investigated for H_2O and HCN in 29P/SW1 (Bockelée-Morvan et al. 2022).

17.1.5 Establish How Physical Properties Relate to Dynamical Properties and Evolution

The scattered disk/scattering TNO population is the dominant source of Centaurs, with smaller contributions from other TNO populations and small body reservoirs (see discussion in Chapter 3 and Table 3.1). Once Centaur orbits are redirected into the giant planet region, their trajectories are chaotic and dominated by close encounters with the giant planets and some Centaurs end up under Jupiter's control as JFCs. The myriad dynamical lifetimes and pathways from the TNO populations through the giant planet region and the JFC population inevitably lead to significant differences in the thermal and physical evolution of Centaurs (see, e.g., Figure 7.1 and Chapter 7). Improved models of how orbital evolution is connected to physical evolution will be key to fully understanding trends between orbital and physical properties in the Centaur region.

Solar radiation and radioactive decay are the dominant sources of energy for Centaurs (for nuclei over $\sim$10-15 km radius; Womack et al. 2017; Steckloff et al. 2021). At large heliocentric distances in the trans-Neptunian region, the energy flux from Galactic Cosmic Radiation can also be important for the top layers of the nucleus and deplete CO in the outermost layers (Gronoff et al. 2020; Maggiolo et al. 2020; Harrington Pinto et al. 2022). Other energy contributions to the near-surface of Centaur nuclei may include the latent heat of sublimation, heat released during the crystallization of amorphous water ice, other relevant chemical reactions, and even impacts. For some objects, tidal deformation during particularly close encounters with the giant planets could also contribute, however such events are rare (see discussion in Section 7.3.3 of Chapter 7). The physical evolution of deep interiors reflects the nature of their respective terminal event, the most recent major process that altered them. Deep interiors are otherwise isolated from the changing thermal environment above, and therefore retain information about their terminal event; such information includes the size, shape, and the rotation state of the nucleus (although there can be feedback between surface-driven activity and *some* of these bulk-nucleus properties). For example, non-spherical nuclei are more likely to experience changes to their internal stress state, which can lead to deformation, mass loss and/or rotational instability, and even disruption (see discussions in Chapters 7 and 12).

Given that solar radiation is the dominant energy source, the rapid orbital evolution Centaurs experience in the giant planet region is likely the driving force behind the predominant physical changes that they experience. When a Centaur evolves inward to a much closer-in orbit, this significantly affects both the peak amount and secular average amount of sunlight its nucleus receives and thus affects its total energy budget.

Interestingly, Centaurs with perihelia beyond Saturn have very little, if any, detected activity; all known active Centaurs have $q < 12$ au (e.g., Jewitt 2009; Chapter 8), providing clues to this being a critical threshold for solar-radiation-driven activity (either directly or indirectly). However, it is important to consider how all potential energy sources might affect a Centaur nucleus as it evolves from the TNO region to the JFC population. With enough energy, any ice near the surface can sublimate, although some may do so very weakly at Centaur distances. One major knowledge gap is that we do not know the surface compositions for many Centaurs (though as discussed in Section 17.1.4, JWST is improving constraints here; see Chapters 5 and 14). Due to the low temperatures of their formation environments, Centaurs likely formed with a great deal of amorphous water ice with trapped gases, such as CO and CO_2, along with many other frozen volatiles (see Chapters 6 and 7). However, it is presently unknown how significantly subsequent processing during emplacement into their TNO reservoirs may remove this primordial amorphous water ice.

Centaur orbital transitions are also not always sunward; i.e., the Centaurs we observe are not always on the journey from TNO to JFC, some Centaurs are objects evolving back outward from JFC orbits (e.g., 39P/Oterma; see Figure 3.13 in Chapter 3). This diversifies the thermal histories of all three dynamically linked populations. Thermophysical models predict that nuclei that have undergone inward and then outward transitions (after significant time spent in the JFC region) will likely be much dimmer and less active than those entering the Centaur region for the first time. Additionally, while they are in the Centaur region from either direction, nuclei likely preferentially lose ices that are volatile at very low temperatures (such as CO, CO_2, and CH_4) and retain H_2O, which drives JFC activity.

How volatiles become depleted is critical to constraining formation and evolution models for icy bodies. A possible useful example of evolutionary depletion is seen in the relative amounts of CO and CO_2 produced by two Centaurs thought to have very different orbital histories: 39P/Oterma and 29P/SW1. 39P/Oterma's gas coma was dominated by CO_2 with no evidence for CO, while 29P/SW1's coma produced significantly more CO than CO_2 (Harrington Pinto et al. 2023; Faggi et al. 2024). The CO/CO_2 production rate ratio difference between 29P/SW1 and 39P/Oterma, measured when both were at approximately the same heliocentric distance, is striking and may arise from their different orbital and thus thermal histories. 39P/Oterma recently transitioned from a JFC orbit back onto a Centaur orbit (see discussion in Chapter 3), whereas orbital calculations imply that 29P/SW1 has probably not yet been closer to the Sun than ~5–6 au (see, e.g., Sarid et al. 2019). One scenario under consideration is that CO may have been preferentially depleted in the top layers of 39P/Oterma from its recent past as a JFC. Studies in the lab also indicate that strength differences between pristine and weathered icy materials could be a source of regular surface activity at large heliocentric distances (see Chapter 13).

While water ice does not sublimate vigorously beyond ~3–4 au, it may still sublimate in much smaller amounts at ~6–8 au and even further. Amorphous water ice can undergo a phase change to the crystalline state and release trapped gases if enough energy is added, and this has been proposed as an efficient source of activity in cometary objects, including Centaurs, that are 5–12 au from the Sun (see discussion in Chapter 7). Some models predict that Centaurs undergoing this crystallization process must have moved inward within the last $\sim 10^4$ years due to the short timescale for this process (Guilbert-Lepoutre 2012), while others calculate longer timescales for amorphous water ice driven activity (e.g., Lisse et al. 2022). There are still open questions about how the crystallization of amorphous water ice drives distant Centaur activity (see Chapter 8). Additional observations and laboratory studies (see Chapter 13) are needed.

A region of near-circular "holding pattern" orbits just beyond Jupiter's orbit was found for many Centaurs moving into JFC orbits (Sarid et al. 2019; Guilbert-Lepoutre et al. 2023; see also Chapter 3). In addition to advancing knowledge of the Centaur–JFC transition, this helps explain the behavior of 29P/SW1, which has long been considered an outlier among active icy bodies. Despite its low-eccentricity orbit being too far from the Sun ($\sim$6 au) for water-ice to sublimate efficiently, it is seemingly always active with multiple strong, regular outbursts each year (see Chapter 10). This implies some other driving mechanism is at work to explain the activity, such as hypervolatile sublimation or amorphous water ice crystallization. The Gateway orbital parameter space that 29P/SW1 occupies has important thermophysical consequences, potentially initiating significant activity for Centaurs that pass through it on the way to becoming a JFC (see, e.g., discussion of 2019 LD2, in Chapter 10).

Centaurs provide rich observational opportunities to constrain models of their coupled dynamical and thermal evolution, and further comparative analysis with other Centaurs is critical. Long-term tracking of the activity behavior and orbits of many more individual Centaurs is needed to test some of the hypotheses about distant activity drivers. Activity constraints combined with a larger data set of colors and spectra (see Section 17.1.4) for Centaurs across a wide range or orbital evolution states will enable a much deeper understanding of how Centaurs evolve physically from their primordial states.

17.1.6 Case Study of 29P/SW1's Composition, Activity, and Behavior

As discussed above (and in many of the chapters), one specific Centaur of great interest is 29P/SW1. This object resides in the relatively nearby so-called Gateway orbital region just beyond Jupiter, has exhibited a dust coma for more than 100 years, and undergoes explosive outbursts several times a year (see Chapter 10). Models indicate that 29P/SW1 has never been in the inner solar system and is therefore anticipated to be relatively pristine, which may explain its current prolific activity levels despite being well beyond the water sublimation distance. Its activity must be driven primarily by non-water based volatile outgassing.

29P/SW1 is a strong candidate for a future space mission (having already been a target for two such proposals, see Chapter 16), and it is the subject of observational campaigns from both professional and amateur astronomers (see Chapter 11). Such campaigns are needed to document its long-term baseline activity and outbursting behavior, including monitoring any substantial chunks of material ejected. Characterizing and monitoring 29P/SW1's near-nucleus region at multiple wavelengths and with different modalities (e.g., lightcurves, photometry, broadband filter imaging, astrometry, spectroscopy, and interferometry), including tracking of its many heterogeneous jets of CO and CO_2 gas, will be especially critical to understanding why this object is so anomalously active. Its rotation rate has been particularly difficult to pin down, largely due to its near-constant activity, which obscures the nucleus' surface; this also limits the accuracy of nucleus size estimates (see discussion in Section 10.3.3 of Chapter 10). Through long-term monitoring we may succeed in constraining the nucleus' rotation period, spin pole orientation, and size. Additional data, combined with better measurements on its nucleus properties, will be important for understanding why 29P/SW1 is so active and how its activity and evolution can inform us about the Centaur to JFC transition.

17.1.7 Characterize P/2019 LD2 before Its JFC Transition

Another object deserving a critical and prolonged gaze is P/2019 LD2 (ATLAS), which is the first active Centaur discovered during its transition into the JFC population. It occupies a particularly unstable orbit as an extremely temporary Jupiter co-orbital, which it entered after a close encounter with Jupiter in 2017. It will experience another close approach to Jupiter in 2028, which will temporarily boost its orbit back to the gateway region just beyond Jupiter and then another close encounter in late 2062/early 2063 will place it onto a JFC orbit (see Figure 17.5). Models indicate that LD2 is very likely on its first trip into the JFC region and that any material deeper than ~20–100 m in its nucleus has not yet been affected by heating greater than what it experienced in the TNO region (see Section 10.6 in Chapter 10 for a detailed discussion of LD2's characteristics and orbital history; see also discussion in Chapter 3). Pre-discovery images indicate that LD2 had sustained a dust coma continuously for at least a year at levels consistent with its activity being

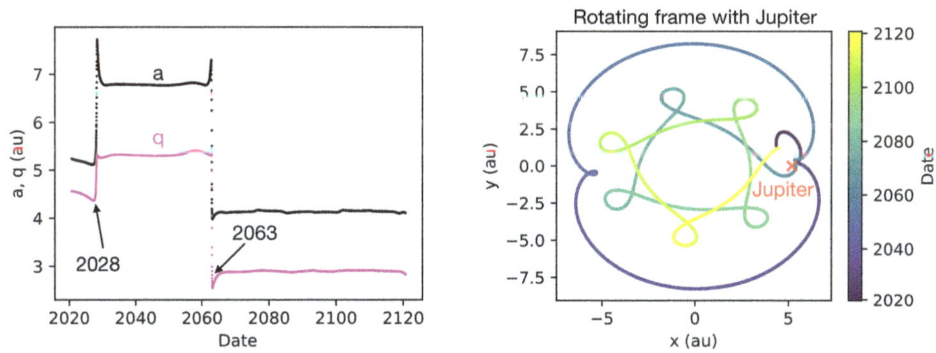

Figure 17.5. Near-term evolution of P/2019 LD2 in semimajor axis and perihelion distance (left) and in the rotating frame with Jupiter (right) based on the best-fit orbit in JPL's small body database at the time of writing (orbit solution date 2024-Apr-13). As seen in the rotating frame with Jupiter (where Jupiter's position is marked with a red 'X'), LD2 will have a close encounter in 2028 that temporarily places it back onto a more distant orbit for two orbital periods. Then at the end of 2062/beginning of 2063, it will have another close encounter with Jupiter that transfers it to a JFC orbit with $a \approx 4$ au and $q \approx 2.9$ au. This figure was made using the Small Body Dynamics Tool (https://github.com/small-body-dynamics/SBDynT).

thermally driven (as opposed to being driven by impacts or mass-wasting, which if they occurred would have been more intermittent; Steckloff et al. 2021). Until it gets much closer to the Sun, its outgassing is predicted to be driven by ices which sublimate at lower temperatures, such as CO, CO_2, and CH_4. LD2 is predicted to transition into the JFC population in 2063, giving us the first known opportunity to observe the evolution of an active Centaur nucleus as it transitions from distant activity to the more typical water-ice-sublimation-driven activity of JFCs.

Therefore, LD2 is a compelling target for monitoring over the next decade using multiple wavelengths and techniques (e.g., imaging and spectroscopy); see Table 11.2 in Chapter 11 for a list of favorable observation windows for LD2. Observations of its activity, including long-term behavior (lightcurves) and possible outbursts (dust and gaseous), will be invaluable for documenting its response over time to the dramatically increased thermal environment of the JFC region. Such repeated measurements can made at optical wavelengths as part of LSST or other similar programs with telescopes of at least moderate aperture. Measured production rates (or significant upper limits on production rates) of volatiles such as CO, CO_2, and H_2O made with JWST (see Chapter 14) in the next year or two will help either challenge or support the assertion that LD2's nucleus is probably pristine. Measurements of these species and overall activity, absolute production rates, and relative production rates over the next ten years (including from other new assets such as SPHEREx, Roman, LUVOIR, and ELTs) will document the changing thermal response of the nucleus' top layers. These data will address foundational knowledge gaps for Centaur nucleus structure and hence would also test models of solar system formation and evolution.

LD2 provides a rare opportunity to document how this dynamical transition from Centaur—JFC affects the activity, thermal evolution, and behavior of an icy body as it enters the JFC thermal environment. As LSST and other surveys continue to discover new Centaurs, we expect additional compelling targets to be identified. A second object in active transition has already been discovered: P/2023 V6 (PANSTARRS). This object is considerably less active than LD2, consistent with it being more thermally processed and less pristine (Kareta et al. 2024). Additional discoveries like LD2 and P/2023 V6 and improved characterization of their activity states will offer important clues to understanding how many transition objects are pristine versus processed and how that affects their physical properties. While LD2 is certainly a priority for sustained observational follow-up so future comet researchers can compare its behavior as a JFC to its behavior in the gateway transition region, we expect the list of such objects to expand rapidly in the coming years.

17.1.8 Spacecraft Mission to a Centaur

Longterm Earth-based observations of Centaurs that utilize photometry, spectroscopy, and polarization techniques are of great value and can address many of the questions and goals discussed above. We are excited to find out what will be revealed about the Centaurs by JWST, LSST, LUVOIR, HWO, SPHEREx, ELTs, and other planned observatories. However, throughout this book, one common point made is that while Earth-based observational assets will significantly move the field forward, a spacecraft mission with *in situ* analyses is critical for providing breakthrough science and ground-truth. Some parts of Centaurs, such as properties of their deep interiors (density, porosity, strength; see Chapter 12), are notoriously difficult and/or limited in terms of what can be learned from the ground. *In situ* analysis is particularly critical for fully understanding the cosmochemical and astrobiological context of Centaurs; detailed comparisons between Centaur compositions and cosmochemical constraints from, e.g., meteorites, asteroid samples, interplanetary dust particles, can only be done with up-close spacecraft measurements or returned Centaur samples (see Chapter 15 for more details).

TNOs and JFCs have been explored with spacecraft, but not the Centaurs. The *New Horizons* spacecraft flew past the trans-Neptunian objects Pluto and its moons in 2015 and Arrokoth in 2019. These flybys pinned down these objects' sizes and shapes (including Arrokoth's remarkably flat bilobate structure; see Figure 17.2), revealed complex terrain, water and nitrogen ices, and other complex chemical compositions, and hints of geological activity. With the Arrokoth flyby, *New Horizons* has given us the first look at a truly pristine icy body that formed and is still at a very large distance (> 40 au) from the Sun and is not in a stage of physical or dynamical transition. In contrast, all comet missions outside of the "Halley Armada" have been to JFCs whose small perihelia ($\sim$1–1.5 au) and relatively short orbital periods have caused them to be among the most physically altered objects on the TNO-Centaur–JFC continuum (see Figure 17.2). These comet missions have been a boon for studying small bodies actively undergoing significant solar processing, including documenting landslides and other surface changes. However, we have never seen such objects close-up *before* they make the JFC transition and start undergoing significant heating and vigorous processing, but after they have exited the trans-Neptunian region. Thus, the processes driving their activity and evolution have never been directly observed; how Centaurs evolve between the TNO and JFC stages is poorly understood.

Chapter 16 provides an enlightening historical retrospective on Centaur mission prioritization and concept development over the last three planetary decadal study periods. It also outlines several mission design requirements including trajectory, power, and propulsion possibilities as well as strategy that addresses scientific priorities and current targets of opportunity. The chapter also highlights how *in situ* exploration concepts of Centaur missions have steadily progressed as this knowledge gap was identified. Centaur spacecraft mission design fully matured with the submission of two competing mission proposals to the 2019 Discovery AO: *Chimera* and *Centaurus*. The *Chimera* mission was led by Walter Harris and developed by the Lunar and Planetary Laboratory at the University of Arizona, Goddard Space Flight Center, and Lockheed Martin. Its main goal was to orbit and explore the Centaur 29P/SW1 while it undergoes many outbursts. *Centaurus* was led by Alan Stern and was a joint proposal of the Southwest Research Institute and Jet Propulsion Lab with the goal of flying by multiple Centaurs, including Chiron and 29P/SW1. Both proposed missions were favorably reviewed, proving that Centaur exploration is feasible, even within the relatively low-cost Discovery mission portfolio, as described in Chapter 16.

Chapters 15 and 16 both discuss the mission feasibility concept study referred to as "Centaur ORbiter And Lander" (CORAL). This concept study described a dedicated NASA New Frontiers level spacecraft mission to a Centaur and was included in the NASEM OWL 2023-2032 decadal report (National Academies of Sciences & Medicine 2023). Its goals were to investigate the physical and chemical properties of a Centaur from orbit and on the surface. It was identified as one of the highest priority goals from the decadal report, reflecting the consensus view of the that a Centaur mission would provide an impressive science return.

17.2 Community Resources

Over the past 50 years, the field of Centaur research has made remarkable strides, evolving from studying an interesting oddball named Chiron into a multifaceted scientific domain. The discovery of these enigmatic transitional objects between the trans-Neptunian Objects and Jupiter-family comets has expanded our understanding of the solar system's formation and subsequent evolution, including into the modern era.

The steadily growing catalog of known Centaurs is set to experience a significant surge once the Vera C. Rubin Observatory's Legacy Survey of Space and Time (LSST) begins. These groundbreaking observations will dramatically enhance our ability to detect and monitor these elusive objects. Complementing this, the James Webb Space Telescope (JWST) is now delivering unparalleled high-resolution spectra, providing critical insights into the compositional makeup of Centaurs, including volatiles, ices, and dust. These advancements are bolstered by a suite of other emerging observational technologies, which will also capture cutting-edge measurements of these distant, faint objects. Laboratory experiments, theoretical models, and advanced computational tools are well-poised to work with this influx of data, setting the stage for a transformative decade in Centaur research. The insights gleaned from these efforts will deepen our understanding of these transitional bodies and their role in the solar system's history. The motivation for this book is deeply rooted in this exciting juncture—a moment of extraordinary potential for discovery and innovation.

The journey of Centaur research began almost half a century ago within the asteroid and cometary science communities. The discovery of the first Centaurs, their orbital characterization, and their classification as a distinct dynamical population laid the groundwork for this field. Subsequent early milestones, such as the discovery of more Centaurs and the TNO 1992 QB_1 (now 15760 Albion), sparked a broader recognition of Centaurs as a critical link between Jupiter-family comets and their distant sources. Chiron's 1995 perihelion passage galvanized the scientific community, culminating in the Chiron Perihelion Campaign, which was coordinated by Michael A'Hearn and Anne Raugh of the University of Maryland and drew global participation and cemented the importance of collaborative efforts in this emerging discipline. For a detailed account of these formative years, see Chapter 11, which offers a rich historical perspective on the origins and naming of the Centaur population. Observations and modeling continued through the 1990s,

2000s, and 2010s, which constrained surface compositions, activity levels and triggers, orbital dynamics, and uncovered the first Centaur rings systems. In the last ten years, enough Centaurs have been measured to commence with statistical studies of the population.

The momentum continued with a series of workshops from 2016 to 2019, organized by Gal Sarid and Maria Womack, which brought together researchers initially from within the state of Florida and later from around the world. The most recent, the "Centaur Exploration Workshop: The Roots of Activity," attracted over 60 participants (Figure 17.6) and produced a High Priority Questions list that has guided subsequent research initiatives. Fortunately for those who are starting out in the field, the presentations and discussions for 2019 workshop were recorded by the NASA CLASS SSERVI group and are shared at this link: https://tinyurl.com/bddkdpum.[1]

Future conferences and workshops devoted to Centaurs are needed at least every 3–5 years to advance new research, explore new directions, promote collaboration, and strengthen interdisciplinary ties with TNO and JFC research. Brainstorming at these workshops will also be needed to coordinate community standards for data and other computational resources.

Notably, the 2019 workshop pioneered hybrid participation a year before the COVID-19 pandemic, demonstrating the field's commitment to inclusivity and innovation. The outcomes of this workshop, including a multi-wavelength coordinated observing campaign for 29P/Schwassmann-Wachmann 1, exemplify the collaborative spirit driving Centaur research: https://wirtanen.astro.umd.edu/29P/29P_obs.shtml. We are pleased to share that many of the participants from that

Figure 17.6. The Centaur Exploration Workshop: The Roots of Activity held March 6-8, 2019 at the University of Central Florida's Florida Space Institute. Online participants' images can be seen on the background screen. The meeting was recorded by NASA's SSERVI and is publicly available (see the text for address). Photo credit: M. Womack.

[1] Original URL: https://www.youtube.com/playlist?list=PLKg3EyXg9SjqYyAQM_fN5tUvfMxCP7Q-e

workshop are chapter co-authors (and reviewers!) for this book, demonstrating the strong dedication and commitment from this research community.

As we look to the future, the field's potential for transformative breakthroughs is immense. Upcoming conferences and workshops will be vital for advancing research, fostering interdisciplinary collaborations, and establishing community standards for data and computational resources. The integration of state-of-the-art observational data with sophisticated computational capabilities will enable researchers to answer critical questions about the solar system's formation and evolution. Observing how Centaurs respond to solar radiation and heating as they migrate inward provides valuable clues about their chemical and physical properties, offering another window into the primordial conditions of the early solar system.

The development of open, accessible data sets and computational ecosystems adhering to the FAIR/CARE principles will democratize access to the tools and data necessary for advancing Centaur research. This includes comprehensive databases cataloging the properties of a growing number of Centaurs, such as photometric data, phase functions, lightcurves, albedos, and sizes. Such resources will be instrumental in accommodating the anticipated explosion of data from initiatives like LSST and ensuring the broader scientific community can fully participate in this exciting field.

Funding agencies will play pivotal roles in shaping the future of Centaur research. For instance, Vera Rubin Observatory's data pipelines (funded by the U.S. National Science Foundation) will deliver an unprecedented volume of data products and analytical tools, including the Rubin Science Platform with web-based computational resource, while NASA's Planetary Data System will continue to archive mission-related data sets and provide access to critical resources. These two, and other, agencies also provide valuable funding for researchers to carry out the work. The recent move by the American Astronomical Society (AAS) to make its journals fully open access further underscores the importance of accessibility in accelerating scientific progress. The AAS Journals recently published a special focus issue on Centaurs which is housed with the relatively new Planetary Science Journal: https://iopscience.iop.org/journal/2632-3338/special/2632-3338_Centaurs.

The next generation of researchers is poised to take Centaur science to extraordinary heights. With the tools, data, and collaborative frameworks now in place, the opportunities for discovery are immense. The promise of Centaur research lies not only in what we have already achieved but in the tantalizing possibilities that lie ahead. The eight Priority Goals in Table 17.1 are our recommendations intended to guide us over the next decade.

There are also many other publicly available relevant resources for Centaurs. A non-exhaustive list of other possibly useful websites includes the following:

- Observing Circumstances of Active Centaurs, Y. Fernandez: https://physics.ucf.edu/yfernandez/actcentobs.html
- Frequently updated database on comets sorted by category, Y. Fernandez: https://www.physics.ucf.edu/¢yfernandez/cometlist.html
- Seppcon database, Y. Fernandez: https://arxiv.org/pdf/1307.6191.pdf
- TNOs are Cool database: http://public-tnosarecool.lesia.obspm.fr

- Centaur characteristics stored at NASA PDS SBN: https://sbnarchive.psi.edu/pds4/non_mission/neowise_diameters_albedos_V1_0/data/neowise_centaurs.tab
- Centaur physical properties from Spitzer: https://www.lpi.usra.edu/books/ssbn2008/7017.pdf
- British Astronomical Association long-term lightcurve archive for 29P/SW1, R. Miles: https://britastro.org/section_information_/comet-section-overview/mission-29p-2/mission-29p-centaur-comet-observing-campaign
- Amateur Observers Campaign http://aop.astro.umd.edu

17.3 Closing Thoughts

As we have reviewed here and as discussed in all the chapters of this book, Centaurs are a key subset of minor bodies that remain significantly understudied. These small, dark objects composed of ice, dust, and rock are on short-lived transitional orbits between the ancient, minimally processed TNOs and the highly evolved JFCs. A significant fraction of Centaurs are active now, and many are on the cusp of physical transformation by solar-driven activity as they experience significant changes in their orbits. A wide range of activity/inactivity levels, rings and other debris have been observed and their orbits, sizes, shapes, and physical and chemical compositions are key observational constraints for formation and evolution models of our solar system and exoplanetary systems. New and highly anticipated observational assets will undoubtedly provide an immense amount of data and transformative observations of Centaurs over the next decade. Laboratory experiments, theoretical models, and computational tools are well-poised to make substantial progress with this plethora of new data, and thus the field of Centaurs itself is also on the cusp of a transformative period. Finally, despite their ability to greatly increase our knowledge of solar system formation and evolution, Centaurs have never been visited by a spacecraft and warrant high prioritization as targets for exploration. Our hope is that the chapters and recommendations will prove to be a useful reference and provide guidance to the community in this crucial expansionary time of Centaur research.

Acknowledgments

The authors thank Theodore Kareta and Yanga Fernández for valuable comments which improved this chapter. This material is based on work supported by the U.S. National Science Foundation under grant AST-1945950. This material is based in part on work done by M.W. while serving at the U.S. National Science Foundation. K.V acknowledges support from NASA (grants 80NSSC21K0376, 80NSSC23K0680, 80NSSC23K1169, and 80NSSC23K0886).

References

Bockelée-Morvan, D., Biver, N., Schambeau, C. A., et al. 2022, A&A, 664, A95

De Prá, M. N., Hénault, E., Pinilla-Alonso, N., et al. 2025, NatAs, 9, 252

Dobson, M. M., Schwamb, M. E., Fitzsimmons, A., et al. 2024, PSJ, 5, 165

Faggi, S., Villanueva, G. L., McKay, A., et al. 2024, NatAs, 8, 1237
Fraser, W. C., Pike, R. E., Marsset, M., et al. 2023, PSJ, 4, 80
Gladman, B., & Volk, K. 2021, ARA&A, 59, 203
Gronoff, G., Maggiolo, R., Cessateur, G., et al. 2020, ApJ, 890, 89
Guilbert-Lepoutre, A. 2012, AJ, 144, 97
Guilbert-Lepoutre, A., Gkotsinas, A., Raymond, S. N., & Nesvorny, D. 2023, ApJ, 942, 92
Gunnarsson, M., Bockelée-Morvan, D., Biver, N., Crovisier, J., & Rickman, H. 2008, A&A, 484, 537
Harrington Pinto, O., Womack, M., Fernandez, Y., & Bauer, J. 2022, PSJ, 3, 247
Harrington Pinto, O., Kelley, M. S. P., Villanueva, G. L., et al. 2023, PSJ, 4, 208
Jewitt, D. 2009, AJ, 137, 4296
Kareta, T., Noonan, J. W., Volk, K., Strauss, R. H., & Trilling, D. 2024, ApJL, 967, L5
Lisse, C. M., Bauer, J. M., & Kim, Y. 2024, 55th Lunar and Planetary Science Conf. (Houston, TX: LPI) 2039
Lisse, C. M., Steckloff, J. K., Prialnik, D., et al. 2022, PSJ, 3, 251
Maggiolo, R., Gronoff, G., Cessateur, G., et al. 2020, ApJ, 901, 136
Marsset, M., Fraser, W. C., Pike, R. E., et al. 2019, AJ, 157, 94
National Academies of Sciences & Medicine 2023, Origins, Worlds, and Life: A Decadal Strategy for Planetary Science and Astrobiology 2023-2032 (Washington, DC: The National Academies Press)
Nesvorný, D., Parker, J., & Vokrouhlický, D. 2018, AJ, 155, 246
Noll, K., Grundy, W. M., Nesvorný, D., & Thirouin, A. 2020, in The Trans-Neptunian Solar System, ed. D. Prialnik, M. A. Barucci, & L. Young (Amsterdam: Elsevier) 201
Peixinho, N., Thirouin, A., Tegler, S. C., et al. 2020, in The Trans-Neptunian Solar System, ed. D. Prialnik, M. A. Barucci, & L. Young (Amsterdam: Elsevier) 307
Pinilla-Alonso, N., Licandro, J., Brunetto, R., et al. 2024, A&A, 692, L11
Sarid, G., Volk, K., Steckloff, J. K., et al. 2019, ApJL, 883, L25
Steckloff, J. K., Lisse, C. M., Safrit, T. K., et al. 2021, Icar, 356, 113998
Wierzchos, K., & Womack, M. 2020, AJ, 159, 136
Womack, M., Sarid, G., & Wierzchos, K. 2017, PASP, 129, 031001
Wong, I., Protopapa, S., Holler, B., et al. 2024, DPS Meeting Abstracts Vol. 56, (Washington, DC: AAS) 405.07

www.ingramcontent.com/pod-product-compliance
Ingram Content Group UK Ltd.
Pitfield, Milton Keynes, MK11 3LW, UK
UKHW051910100926
8556IPUK00005B/828

9 780750 355865